Physical Metallurgy and Advanced Materials

Physical Metallurgy and Advanced Materials

Seventh edition

R. E. Smallman, CBE, DSc, FRS, FREng, FIMMM

A. H. W. Ngan, PhD, FIMMM, CSci, CEng

ELSEVIER

AMSTERDAM • BOSTON • HEIDELBERG • LONDON • NEW YORK • OXFORD
PARIS • SAN DIEGO • SAN FRANCISCO • SINGAPORE • SYDNEY • TOKYO
Butterworth-Heinemann is an imprint of Elsevier

BH

Butterworth-Heinemann is an imprint of Elsevier
Linacre House, Jordan Hill, Oxford OX2 8DP
30 Corporate Drive, Suite 400, Burlington, MA 01803

Seventh edition 2007

British Library Cataloguing in Publication Data
A catalogue record for this book is available from the British Library

Library of Congress Cataloging-in-Publication Data
A catalog record for this book is available from the Library of Congress

ISBN: 978 0 7506 6906 1

For information on all Butterworth-Heinemann publications visit our website at http://books.elsevier.com

Typeset by Charon Tec Ltd (A Macmillan Company), Chennai, India
www.charontec.com

Contents

Preface

Physical Metallurgy and Advanced Materials has evolved from the earlier editions of *Modern Physical Metallurgy* (1962, 1970, 1985) and later editions of *Modern Physical Metallurgy and Materials Engineering* (1995, 1999). The present treatment contains much of the previous editions and follows the same overall philosophy and aims. It has, however, been updated again in both presentation and content. Additions have been made to almost every chapter, which now include a number of worked examples in the text to illustrate and emphasize a particular aspect of the subject. At the end of each chapter there is a set of questions, most of which are numerical. These are included to give the reader an opportunity to apply the scientific background presented in the chapter, but also to emphasize important material properties, e.g. elastic moduli, atomic dimensions, etc. The solutions to these problems are worked out in a Solutions Manual, which may be obtained from Elsevier by teachers and lecturers who use the book.

To keep the book a manageable size some text from the previous edition has been omitted together with associated diagrams, and some of the text has been totally recast in a different format. The early chapters are predominantly directed towards metals (physical metallurgy) but the principles are equally relevant to non-metals, which are specifically dealt with in the later chapters. Characterization using X-rays, electron microscopy, etc. is important to all areas of materials and several new techniques such as scanning tunneling microscopy (STM), atomic force microscopy (AFM), nanoindentation and so on have been described. The book ends with a focus on some newer areas which are developing rapidly and are being incorporated to a greater or lesser extent in a number of university courses. The presentation of biomaterials, sports materials and nanomaterials is very much illustrative of the essential and significant application of a wide variety of materials and associated materials science to the successful development of these new fields.

R. E. Smallman
A. H. W. Ngan

January 2007

Solutions Manual

This provides a set of fully worked solutions, available for lecturers only, to the Problems found at the end of chapters.

To access the Solutions Manual go to: http://www.textbooks.elsevier.com and search for the book and click on the 'manual' link. If you do not have an account on textbooks.elsevier.com already, you will need to register and request access to the book's subject area. If you already have an account on textbooks, but do not have access to the right subject area, please follow the 'request access' link at the top of the subject area homepage.

About the authors

Professor R. E. Smallman

After gaining his PhD in 1953, Professor Smallman spent five years at the Atomic Energy Research Establishment at Harwell before returning to the University of Birmingham, where he became Professor of Physical Metallurgy in 1964 and Feeney Professor and Head of the Department of Physical Metallurgy and Science of Materials in 1969. He subsequently became Head of the amalgamated Department of Metallurgy and Materials (1981), Dean of the Faculty of Science and Engineering, and the first Dean of the newly created Engineering Faculty in 1985. For five years he was Vice-Principal of the University (1987–92).

He has held visiting professorship appointments at the University of Stanford, Berkeley, Pennsylvania (USA), New South Wales (Australia), Hong Kong and Cape Town, and has received Honorary Doctorates from the University of Novi Sad (Yugoslavia), University of Wales and Cranfield University. His research work has been recognized by the award of the Sir George Beilby Gold Medal of the Royal Institute of Chemistry and Institute of Metals (1969), the Rosenhain Medal of the Institute of Metals for contributions to Physical Metallurgy (1972), the Platinum Medal, the premier medal of the Institute of Materials (1989), and the Acta Materialia Gold Medal (2004).

He was elected a Fellow of the Royal Society (1986), a Fellow of the Royal Academy of Engineering (1990), a Foreign Associate of the United States National Academy of Engineering (2005), and appointed a Commander of the British Empire (CBE) in 1992. A former Council Member of the Science and Engineering Research Council, he has been Vice-President of the Institute of Materials and President of the Federated European Materials Societies. Since retirement he has been academic consultant for a number of institutions both in the UK and overseas.

Professor A. H. W. Ngan

Professor Ngan obtained his PhD on electron microscopy of intermetallics in 1992 at the University of Birmingham, under the supervision of Professor Ray Smallman and Professor Ian Jones. He then carried out postdoctoral research at Oxford University on materials simulations under the supervision of Professor David Pettifor. In 1993, he returned to the University of Hong Kong as a Lecturer in Materials Science and Solid Mechanics, at the Department of Mechanical Engineering. In 2003, he became Senior Lecturer and in 2006 Professor. His research interests include dislocation theory, electron microscopy of materials and, more recently, nanomechanics. He has published over 120 refereed papers, mostly in international journals. He received a number of awards, including the Williamson Prize (for being the top Engineering student in his undergraduate studies at the University of Hong Kong), Thomas Turner Research Prize (for the quality of his PhD thesis at the University of Birmingham), Outstanding Young Researcher Award at the University of Hong Kong, and in 2007 was awarded the Rosenhain Medal of the Institute of Materials, Minerals and Mining. He also held visiting professorship appointments at Nanjing University and the Central Iron and Steel Research Institute in Beijing, and in 2003, he was also awarded the Universitas 21 Fellowship to visit the University of Auckland. He is active in conference organization and journal editorial work.

Acknowledgments

The contribution made by Dr R. Bishop to two previous editions of the book has helped significantly in the development of the present treatment. The authors wish to acknowledge this with thanks. Acknowledgment is also made to a number of publishers and researchers for kind permission to reproduce a number of diagrams from other works; these are duly noted in the captions.

Illustration credits

Figure 4.1	Askeland, D. R. (1990). *The Science and Engineering of Materials*, 2nd edn. p. 723. Chapman and Hall, London.
Figures 4.3, 4.5	Zeiss, C. (Dec 1967). *Optical Systems for the Microscope*, **15**. Carl Zeiss, Germany.
Figure 4.12	Vale, R. and Smallman, R. E. (1977). *Phil. Mag.*, **36**, p. 209.
Figure 4.13	Barrett, C. S. and Massalski, T. B. (1980). *Structure of Metals and Alloys*, McGraw-Hill.
Figure 4.35a	Gilman, J. (Aug. 1956). *Metals*, 1000.
Figure 4.35b	Dash, J. (1957). *Dislocations and Mechanical Properties of Crystals*, John Wiley and Sons.
Figure 4.38	Hirsch, P. B., Howie, A. and Whelan, M. (1960). *Phil. Trans.*, **A252**, 499, Royal Society.
Figure 4.40	Hirsch, P. B., Howie, A., Whelan, M., Nicholson, R. B. and Pashley, D. (1965). *Electron Microscopy of Thin Crystals*, Butterworths, London.
Figure 4.41	Howie, A. and Valdre, R. (1963). *Phil. Mag.*, **8**, 1981, Taylor and Francis.
Figure 4.43	*A Practical Guide to Scanning Probe Microscopy*, Park Scientific Instruments 1997.
Figure 4.44b	Courtesy J. B. Pethica.
Figure 4.46	Courtesy J. B. Pethica.
Figures 4.47, 4.48	*A Practical Guide to Scanning Probe Microscopy*, Park Sc. Inst. 1997.
Figure 4.50	Hoffmann, P. M., Oral, A., Grimble, R. A., Ozer, H.O., Jeffrey, S. and Pethica, J. B. (2001). *Proc. Roy. Soc.*, London **A457**, 1161.
Figure 4.51	Courtesy C. S. Lee.
Figure 4.52	From the website of Hysitron Inc (http://www.hysitron.com).
Figure 4.56	Feng, G. and Ngan, A. H. W. (2002). *J. of Materials Research*, **17**, 660–668.
Figure 4.58	Wo, P. C. and Ngan, A. H. W. (2004). *Phil. Mag.* **84**, 314–315.
Figure 4.59	Barnes, P. (1990). *Metals and Materials.* Nov. 708–715, Institute of Materials.
Figure 4.61a,b	Hill, M. and Nicholas, P. (1989). *Thermal Analysis on Materials Development, Metals and Materials*, Nov, 639–642.
Figure 4.61c	Hay, J. N. (1982). *Thermal Methods of Analysis of Polymers. Analysis of Polymer Systems*, Ed. Bark, G. S. and Allen, N. S. Chapter 6, Applied Science, London.
Figure 5.1	Ashby, M. (2005). *Materials Selection in Design*, 3rd Edn, p. 54. Elsevier.
Figure 5.10	Wert, C. and Zener, C. (1949). *Phys. Rev.*, **76**, 1169, American Institute of Physics.
Figure 5.14	Morris, D., Besag, F. and Smallman, R. E. (1974). *Phil. Mag.* **29**, 43. Taylor and Francis, London.
Figures 5.15, 5.16	Pashley, D. and Presland, D. (1958–9). *J. Inst. Metals.* **87**, 419. Institute of Metals.
Figure 5.17	Barrett, C. S. (1952). *Structure of Metals*, 2nd edn. McGraw-Hill.
Figures 5.26, 5.28	Rose, R. M., Shepard, L. A. and Wulff, J. (1966). *Structure and Properties of Materials*. John Wiley and Sons.
Figure 5.31	Raynor, G. V. (1958). *Structure of Metals,* Inst. of Metallurgists, 21, Iliffe and Sons, London.
Figure 5.35	Shull, C. G. and Smart, R. (1949). *Phys. Rev.*, **76**, 1256.

Figure 6.1	Churchman, T., Mogford, I. and Cottrell, A. H. (1957). *Phil. Mag.*, **2**, 1273.
Figure 6.13	Lücke, K. and Lange, H. (1950). *Z. Metallk*, **41**, 65.
Figure 6.15	Johnston, W. G. and Gilman, J. J. (1959). *J. Appl. Phys.*, **30**, 129, American Institute of Physics.
Figure 6.16	Stein, J. and Low, J. R. (1960). *J. Appl. Physics*, **30**, 392, American Institute of Physics.
Figure 6.25	Hahn (1962). *Acta Met.*, **10**, 727, Pergamon Press, Oxford.
Figure 6.27	Morris, D. and Smallman, R. E. (1975). *Acta Met.*, **23**, 573.
Figure 6.28	Cottrell, A. H. (1957). Conference on Properties of Materials at High Rates of Strain. Institution of Mechanical Engineers.
Figure 6.29	Adams, M. A., Roberts, A. C. and Smallman, R. E. (1960). *Acta Metall.*, **8**, 328. Hull, D. and Mogford, I. (1958). *Phil. Mag.*, **3**, 1213.
Figure 6.33	Adams, M. A., Roberts, A. C. and Smallman, R. E. (1960). *Acta Metall.*, **8**, 328.
Figure 6.34	Hull, D. (1960). *Acta Metall.*, **8**, 11.
Figure 6.36	Adams, M. A. and Higgins, P. (1959). *Phil. Mag.*, **4**, 777.
Figure 6.38	Hirsch, P. B. and Mitchell, T. (1967). *Can. J. Phys.*, **45**, 663, National Research Council of Canada.
Figure 6.40	Steeds, J. (1963). *Conference on Relation between Structure and Strength in Metals and Alloys*, HMSO.
Figure 6.44	Dillamore, I. L., Smallman, R. E. and Wilson, D. (1969). *Commonwealth Mining and Metallurgy Congress*, London, Institute of Mining and Metallurgy.
Figure 6.45	Wilson, D. (1966). *J. Inst. Metals*, **94**, 84, Institute of Metals.
Figure 6.48	Clareborough, L. M., Hargreaves, M. and West (1955). *Proc. Roy. Soc.*, **A232**, 252.
Figure 6.49	Cahn, R. (1949). *Inst. Metals*, **77**, 121.
Figure 6.56	Buergers, *Handbuch der Metallphysik*. Akademie-Verlags-gesellschaft.
Figure 6.57	Burke and Turnbull (1952). *Progress in Metal Physics*, 3, Pergamon Press.
Figure 6.60	Hancock, J., Dillamore, I. L. and Smallman, R. E. (1972). *Metal Sci. J.*, **6**, 152.
Figure 6.64	Puttick, K. E. and King, R. (1952). *J. Inst. Metals*, **81**, 537.
Figure 6.67	Ashby, M. F. (1972). *Acta Metal.*, **20**, 887.
Figure 6.70	Broom, T. and Ham, R. (1959). *Proc. Roy. Soc.*, **A251**, 186.
Figure 6.73	Cottrell, A. H. (1959). *Fracture*. John Wiley & Sons.
Figure 7.2	Silcock, J., Heal, T. J. and Hardy, H. K. (1953–4). *J. Inst. Metals*, **82**, 239.
Figures 7.3, 7.4, 7.6	Nicholson, R. B., Thomas, G. and Nutting, J. (1958–9). *J. Inst. Metals*, **87**, 431.
Figure 7.5	Guinier, A. and Walker, R. (1953). *Acta Metall.*, **1**, 570.
Figure 7.10	Fine, M., Bryne, J. G. and Kelly, A. (1961). *Phil. Mag.*, **6**, 1119.
Figure 7.14	Greenwood, G. W. (1968). *Institute of Metals Conference on Phase Transformation*, Institute of Metals.
Figure 7.18	Metals Handbook, American Society for Metals.
Figures 7.19, 7.20	Mehl, R. F. and Hagel, K. (1956). *Progress in Metal Physics*, 6, Pergamon Press.
Figure 7.22	Kelly, P. and Nutting, J. (1960). *Proc. Roy. Soc.*, **A259**, 45, Royal Society.
Figure 7.23	Kurdjumov, G. (1948). *J. Tech. Phys. SSSR*, **18**, 999.
Figures 7.25, 7.26, 7.27	Brookes, J. W., Loretto, M. H. and Smallman, R. E. (1979). *Acta Met.*, **27**, 1829.

Figure 7.33	Cottrell, A. H. (1958). Brittle Fracture in Steel and Other Materials. *Trans. Amer. Inst. Mech. Engrs.*, **April**, p. 192.
Figure 7.35, 7.36, 7.37, 7.38	Ashby, M. F. *et al.* (1979). *Acta Met.*, **27**, 669.
Figure 7.39	Gandhi, C. and Ashby, M. F. (1979). *Acta Mater.*, **27**, 1565.
Figure 7.41a	Li, J.
Figure 7.41b	Abraham, F.
Figure 7.42a	Wen, M. and Ngan, A. H. W. (2000). *Acta Mater.*, **48**, 4255–4265.
Figure 7.42b	Li, J., Ngan, A. H. W. and Gumbsch, P. (2003). *Acta Mater.*, **51**, 5711–5742.
Figure 7.43a	Li, J., Ngan, A. H. W. and Gumbsch, P. (2003). *Acta Mater.*, **51**, 5711–5742.
Figure 7.43b	Wen, M. and Ngan, A. H. W. (2000). *Acta Mater.*, **48**, 4255–4265.
Figure 8.2	Smithells, C. J., *Smithells Metals Reference Book*, 7th edn. Butterworth-Heinemann.
Figure 8.4	Balliger, N. K. and Gladman, T. (1981). *Metal Science*, March, 95.
Figure 8.5	Balliger, N. K. and Gladman, T. (1981). *Metal Science*, March, 95.
Figure 8.8	Sidjanin, L. and Smallman, R. E. (1992). *Mat. Science and Technology*, **8**, 105.
Figure 8.9	Driver, D. (1985). *Metals and Materials*, June, 345–54, Institute of Materials, London.
Figure 8.12	Brandes, E. A. and Brook, G. B. (Ed) *Smithells Metal Reference Book*. Butterworth-Heinemann, Oxford.
Figure 8.13	Woodfield, A. P., Postans, P. J., Loretto, M. H. and Smallman, R. E. (1988). *Acta Metall.*, **36**, 507.
Figure 8.15	Noguchi, O., Oya, Y. and Suzuki, T. (1981). *Metall. Trans.* **12A**, 1647.
Figure 8.16	Kim, Y-W. and Froes, F. H. (1990). *High-Temperature Aluminides and Intermetallics*, TMS Symposium, ed. by Whang, S. H., Lin, C. T. and Pope, D.
Figure 8.18	Gilman, P. (1990). *Metals and Materials*, Aug, 505, Institute of Materials, London.
Figure 9.3	Ashby, M. F. and Jones, D. R. H. (2005). *Engineering Materials, Elsevier.*
Figure 9.14	Banshah, R. F. (1984). Industrial Materials Science and Engineering, L. E. Murr (Ed) Chapter 12, Marcel Dekker, N.Y.
Figure 9.15	Barrell, R. and Rickerby, D. S. (1989). Engineering coatings by physical vapour deposition. *Metals and materials*, August, 468–473, Institute of Materials.
Figure 9.16	Kelly, P. J., Arnell, R. D. and Ahmed, W. *Materials World.* (March 1993), pp. 161–5. Institute of Materials.
Figures 9.17, 9.18	Weatherill, A. E. and Gill, B. J. (1988). Surface engineering for high-temperature environments (thermal spray methods). *Metals and Materials*, September, 551–555, Institute of Materials.
Figure 10.2	Hume-Rothery, W., Smallman, R. E. and Haworth, C. W. (1988). *Inst. of Metals*, London.
Figure 10.3	Kingery, W. D., Bowen, H. K. and Uhlmann, D. R. (1976). *Introduction to Ceramics*, 2nd edn. Wiley-Interscience, New York.
Figures 10.5, 10.6, 10.7, 10.8	Jack, K. H. (1987). Silicon nitride, sialons and related ceramics. In *ceramics and Civilisation*, **3**. American Ceramic Society Inc., New York.

Figure 10.11	Kingery, W. D., Bowen, H. K. and Uhlmann, D. R. (1976). *Introduction to Ceramics*, 2nd edn. Wiley-Interscience, New York.
Figure 10.16	Bovenkerk, H. P. *et al.* (1959). *Nature*, **184**, 1094–1098.
Figure 10.17	Cahn, R. W. and Harris, B. (1969). Newer forms of carbon and their uses. *Nature*, 11 January, **221**, 132–141.
Figure 10.20	Hexveg, S., Carlos, H., Schested, J. (2004). *Nature*, 427–429.
Figure 10.23	Creyhe, W. E. C., Sainsbury, I. E. J. and Morrell, R. (1982). *Design with Non-Ductile Materials*, Elsevier/Chapman and Hall, London.
Figure 10.24	Davidge, R. W. (1986). *Mechanical Behaviour of Ceramics.* Cambridge University Press, Cambridge.
Figure 10.25	Richetson (1992). *Modern Ceramic Engineering: Properties, Processing and Use in Design*, Marcel Dekker, New York.
Figure 10.26	Korb, H. J., Morant, C. A., Calland, R. M. and Thatcher, C. S. (1981). Ceramic Bulletin No 11. American Ceramic Society.
Figure 11.1	Ashby, M. F. and Jones, D. R. H. (2005). *Engineering Materials*, Elsevier.
Figure 11.2	Mills, N. J. (1986). *Plastics: Microstructure, Properties and Applications*, Edward Arnold, London.
Figures 11.3, 11.4	Ashby, M. F. and Jones, D. R. H. (2005). *Engineering Materials*, Elsevier.
Figure 11.6	Askeland, D. R. (1990). *The Science and Engineering of Materials*, p. 534, Chapman Hall, London.
Figure 11.11	Hertzberg, R. W. (1989). *Deformation and Fracture Mechanics of Engineering Materials*, 3rd Edn. John Wiley & Sons.
Figure 11.14	Ngan, A. H. W. and Tang, B. (2002). *J. of Material Research*, **17**, 2604–2610.
Figure 11.17	Ashby, M. F. and Jones, D. R. H. (2005). *Engineering Materials*, Elsevier.
Figure 11.18	Mills, N. J. (1986). *Plastics: Microstructure Properties Application*, Edward Arnold, London.
Figure 11.20	Ashby, M. F. and Jones, D. R. H. (2005). *Engineering Materials*, Elsevier.
Figure 11.24	Hughes, J. D. H. (1986). *Metals and Materials*, p. 365–368, Inst. of Materials, London.
Figure 11.25	Feet, E. A. (1988). *Metals and Materials*, p. 273–278, Inst. of Materials, London.
Figure 11.26	King, J. E. (1989). *Metals and Materials*, p. 720–726, Inst. of Materials.
Figure 12.1	Vincent, J. (1990). *Metals and Materials*, June, 395, Institute of Materials.
Figure 12.6a	Walker, P. S. and Sathasiwan, S. (1999). *J. Biomat.* **32**, 28.
Figure 12.10	Horwood, G. P. (1994). Flexes, bend points and torques. In *Golf: the Scientific Way* (ed. A. Cochran) Aston Publ. Group, Hemel Hempstead, Herts, UK. pp. 103–108.
Figure 12.11	Jenkins, M. (Ed) (2003). Materials in Sports equipment, Woodhead Publishing Ltd., UK
Figure 12.13	Easterling, K. E. (1990). *Tomorrow's Materials.* Institute of Metals, London.
Figures 12.16, 12.17	Ashby, M. F. (2005). *Materials Selection in Mechanical Design*, Elsevier.
Figure 12.19	K. Y. Ng, A. Muley, A. H. W. (2006). Ngan and co-workers. *Materials Letters*, **60**, 2423.
Figure 12.20	Ng, H. P. and Ngan, A. H. W. (2002). *Journal of Materials Research*, **17**, 2085.
Figure 12.21	Poole, C. P. Jr. and Owens, F. J. (2003). *Introduction to Nanotechnology*, Wiley-Interscience.
Figure 12.23	Webpage of Centre for Quantum Devices, Northwestern University: http://cqd.ece.northwestern.edu/
Figure 12.24b	M. L. Povinelli and co-workers (2001). *Physical Review B*, **64**, 075313.

Figure 12.25	Ngan, A. H. W. unpublished.
Figure 12.26	Uchic, M. U., Dimiduk, D. M., Florando, J. N. and Nix, W. D. (2004). *Science*, **305**, 986–989.
Figure 12.27a	Feng, G. and Ngan, A. H. W. (2001). *Scripta Mater.*, **45**, 971–976.
Figure 12.27b	Chiu, Y. L. and Ngan, A. H. W. (2002). *Acta Mater.*, **50**, 1599–1611.
Figure 12.28a	Feng, G. and Nix, W. D. (2004). *Scripta Mater.*, **51**, 599–603.
Figure 12.28b	Nix, W. D. and Gao, H. (1998). *Journal of the Mechanics and Physics of Solids*, **46**, 411–425.
Figure 12.30	Erb, U. (1995). *NanoStructured Materials*, **6**, 533.
Figure 12.31	Cai, B., Lu, K. and co-workers (1999). *Scripta Mater.*, **41**, 755.

Chapter 1
Atoms and atomic arrangements

1.1 The realm of materials science

In modern-day activities we encounter a remarkable range of different engineering materials, metals and alloys, plastics and ceramics. Metals and alloys are still predominant by both the tonnage and the variety of applications in which they are used, but increasingly plastics and ceramics are being used in applications previously considered the domain of the metals, often within the same engineered structure, e.g. cars, aeroplanes, etc. The science of plastics and ceramics is more recently developed than that of metals, but has its origin in the study of structure and structure–property relationships. In this it has developed from the science of metals, physical metallurgy and a structure–property–processing approach. In such an approach, a convenient starting point is a discussion on the smallest structural entity in materials, namely the atom, and its associated electronic states.

1.2 The free atom

1.2.1 The four electron quantum numbers

Rutherford conceived the atom to be a positively charged nucleus, which carried the greater part of the mass of the atom, with electrons clustering around it. He suggested that the electrons were revolving round the nucleus in circular orbits so that the centrifugal force of the revolving electrons was just equal to the electrostatic attraction between the positively charged nucleus and the negatively charged electrons. In order to avoid the difficulty that revolving electrons should, according to the classical laws of electrodynamics, emit energy continuously in the form of electromagnetic radiation, Bohr, in 1913, was forced to conclude that, of all the possible orbits, only certain orbits were in fact permissible. These discrete orbits were assumed to have the remarkable property that when an electron was in one of these orbits, no radiation could take place. The set of stable orbits was characterized by the criterion that the angular momenta of the electrons in the orbits were given by the expression $n\mathbf{h}/2\pi$, where $\mathbf{h}$ is Planck's constant and n could only have integral values ($n \equiv 1, 2, 3$, etc.). In this way, Bohr was able to give a satisfactory explanation of the line spectrum of the hydrogen atom and to lay the foundation of modern atomic theory.

In later developments of the atomic theory, by de Broglie, Schrödinger and Heisenberg, it was realized that the classical laws of particle dynamics could not be applied to fundamental particles. In classical dynamics it is a prerequisite that the position and momentum of a particle are known exactly: in atomic dynamics, if either the position or the momentum of a fundamental particle is known exactly, then the other quantity cannot be determined. In fact, an uncertainty must exist in our knowledge of the position and momentum of a small particle, and the product of the degree of uncertainty for each quantity is related to the value of Planck's constant ($\mathbf{h} \equiv 6.6256 \times 10^{-34}$ J s). In the macroscopic world, this fundamental uncertainty is too small to be measurable, but when treating the motion of electrons revolving round an atomic nucleus, application of Heisenberg's Uncertainty Principle is essential.

The consequence of the Uncertainty Principle is that we can no longer think of an electron as moving in a fixed orbit around the nucleus, but must consider the motion of the electron in terms of a wave function. This function specifies only the probability of finding one electron having a particular energy in the space surrounding the nucleus. The situation is further complicated by the fact that the electron behaves not only as if it were revolving round the nucleus, but also as if it were spinning about its own axis. Consequently, instead of specifying the motion of an electron in an atom by a single integer n, as required by the Bohr theory, it is now necessary to specify the electron state using four numbers. These numbers, known as electron quantum numbers, are n, l, m and s, where n is the principal quantum number, l is the orbital (azimuthal) quantum number, m is the magnetic quantum number and s is the spin quantum number. Another basic premise of the modern quantum theory of the atom is the Pauli Exclusion Principle. This states that no two electrons in the same atom can have the same numerical values for their set of four quantum numbers.

If we are to understand the way in which the Periodic Table of the chemical elements is built up in terms of the electronic structure of the atoms, we must now consider the significance of the four quantum numbers and the limitations placed upon the numerical values that they can assume. The most important quantum number is the principal quantum number, since it is mainly responsible for determining the energy of the electron. The principal quantum number can have integral values beginning with $n = 1$, which is the state of lowest energy, and electrons having this value are the most stable, the stability decreasing as n increases. Electrons having a principal quantum number n can take up integral values of the orbital quantum number l between 0 and $(n - 1)$. Thus, if $n = 1$, l can only have the value 0, while for $n = 2$, $l = 0$ or 1, and for $n = 3$, $l = 0$, 1 or 2. The orbital quantum number is associated with the angular momentum of the revolving electron, and determines what would be regarded in non-quantum mechanical terms as the shape of the orbit. For a given value of n, the electron having the lowest value of l will have the lowest energy, and the higher the value of l, the greater will be the energy.

The remaining two quantum numbers m and s are concerned, respectively, with the orientation of the electron's orbit round the nucleus, and with the orientation of the direction of spin of the electron. For a given value of l, an electron may have integral values of the inner quantum number m from $+l$ through 0 to $-l$. Thus, for $l = 2$, m can take on the values +2, +1, 0, −1 and −2. The energies of electrons having the same values of n and l but different values of m are the same, provided there is no magnetic field present. When a magnetic field is applied, the energies of electrons having different m values will be altered slightly, as is shown by the splitting of spectral lines in the Zeeman effect. The spin quantum number s may, for an electron having the same values of n, l and m, take one of two values, that is, $+\frac{1}{2}$ or $-\frac{1}{2}$. The fact that these are non-integral values need not concern us for the present purpose. We need only remember that two electrons in an atom can have the same values for the three quantum numbers n, l and m, and that these two electrons will have their spins oriented in opposite directions. Only in a magnetic field will the energies of the two electrons of opposite spin be different.

1.2.2 Nomenclature for the electronic states

Before discussing the way in which the periodic classification of the elements can be built up in terms of the electronic structure of the atoms, it is necessary to outline the system of nomenclature which enables us to describe the states of the electrons in an atom. Since the energy of an electron is mainly determined by the values of the principal and orbital quantum numbers, it is only necessary to consider these in our nomenclature. The principal quantum number is simply expressed by giving that number, but the orbital quantum number is denoted by a letter. These letters, which derive from the early days of spectroscopy, are s, p, d and f, which signify that the orbital quantum numbers l are 0, 1, 2 and 3, respectively.[1]

[1] The letters, s, p, d and f arose from a classification of spectral lines into four groups, termed sharp, principal, diffuse and fundamental, in the days before the present quantum theory was developed.

Table 1.1 *Allocation of states in the first three quantum shells*

Shell	*n*	*l*	*m*	*s*	*Number of states*	*Maximum number of electrons in shell*
1st *K*	1	0	0	±1/2	Two 1*s*-states	2
2nd *L*	2	0	0	±1/2	Two 2*s*-states	8
		1	+1	±1/2	Six 2*p*-states	
			0	±1/2		
			−1	±1/2		
3rd *M*	3	0	0	±1/2	Two 3*s*-states	18
		1	+1	±1/2	Six 3*p*-states	
			0	±1/2		
			−1	±1/2		
		2	+2	±1/2	Ten 3*d*-states	
			+1	±1/2		
			0	±1/2		
			−1	±1/2		
			−2	±1/2		

When the principal quantum number $n=1$, l must be equal to zero, and an electron in this state would be designated by the symbol 1*s*. Such a state can only have a single value of the inner quantum number $m=0$, but can have values of $+\frac{1}{2}$ or $-\frac{1}{2}$ for the spin quantum number *s*. It follows, therefore, that there are only two electrons in any one atom which can be in a 1*s*-state, and that these electrons will spin in opposite directions. Thus, when $n=1$, only *s*-states can exist and these can be occupied by only two electrons. Once the two 1*s*-states have been filled, the next lowest energy state must have $n=2$. Here l may take the value 0 or 1, and therefore electrons can be in either a 2*s*- or a 2*p*-state. The energy of an electron in the 2*s*-state is lower than in a 2*p*-state, and hence the 2*s*-states will be filled first. Once more there are only two electrons in the 2*s*-state, and indeed this is always true of *s*-states, irrespective of the value of the principal quantum number. The electrons in the *p*-state can have values of $m=+1, 0, -1$, and electrons having each of these values for *m* can have two values of the spin quantum number, leading therefore to the possibility of six electrons being in any one *p*-state. These relationships are shown more clearly in Table 1.1.

No further electrons can be added to the state for $n=2$ after two 2*s*- and six 2*p*-states are filled, and the next electron must go into the state for which $n=3$, which is at a higher energy. Here the possibility arises for l to have the values 0, 1 and 2 and hence, besides *s*- and *p*-states, *d*-states for which $l=2$ can now occur. When $l=2$, *m* may have the values +2, +1, 0, −1, −2 and each may be occupied by two electrons of opposite spin, leading to a total of ten *d*-states. Finally, when $n=4$, l will have the possible values from 0 to 3, and when $l=3$ the reader may verify that there are 14 4*f*-states.

Table 1.1 shows that the maximum number of electrons in a given shell is $2n^2$. It is accepted practice to retain an earlier spectroscopic notation and to label the states for which $n=1, 2, 3, 4, 5, 6$ as *K*-, *L*-, *M*-, *N*-, *O*- and *P*-shells, respectively.

Table 1.2 *The Periodic Table of the elements (from Puddephatt and Monaghan, 1986; by permission of Oxford University Press).*

1	2	3	4	5	6	7	8	9	10	11	12	13	14	15	16	17	18	←New IUPAC notation
IA	**IIA**	**IIIA**	**IVA**	**VA**	**VIA**	**VIIA**	**VIII**			**IB**	**IIB**	**IIIB**	**IVB**	**VB**	**VIB**	**VIIB**	**O**	**← Previous IUPAC form**
$_{1}$H 1.008																	$_{2}$He 4.003	
$_{3}$Li 6.941	$_{4}$Be 9.012											$_{5}$B 10.81	$_{6}$C 12.01	$_{7}$N 14.01	$_{8}$O 16.00	$_{9}$F 19.00	$_{10}$Ne 20.18	
$_{11}$Na 22.99	$_{12}$Mg 24.31											$_{13}$Al 26.98	$_{14}$Si 28.09	$_{15}$P 30.97	$_{16}$S 32.45	$_{17}$Cl 35.45	$_{18}$Ar 39.95	
$_{19}$K 39.10	$_{20}$Ca 40.08	$_{21}$Sc 44.96	$_{22}$Ti 47.90	$_{23}$V 50.94	$_{24}$Cr 52.00	$_{25}$Mn 54.94	$_{26}$Fe 55.85	$_{27}$Co 58.93	$_{28}$Ni 58.71	$_{29}$Cu 63.55	$_{30}$Zn 65.37	$_{31}$Ga 69.72	$_{32}$Ge 72.92	$_{33}$As 74.92	$_{34}$Se 78.96	$_{35}$Br 79.90	$_{36}$Kr 83.80	
$_{37}$Rb 85.47	$_{38}$Sr 87.62	$_{39}$Y 88.91	$_{40}$Zr 91.22	$_{41}$Nb 92.91	$_{42}$Mo 95.94	$_{43}$Tc 98.91	$_{44}$Ru 101.1	$_{45}$Rh 102.9	$_{46}$Pd 106.4	$_{47}$Ag 107.9	$_{48}$Cd 112.4	$_{49}$In 114.8	$_{50}$Sn 118.7	$_{51}$Sb 121.8	$_{52}$Te 127.6	$_{53}$I 126.9	$_{54}$Xe 131.3	
$_{55}$Cs 132.9	$_{56}$Ba 137.3	$_{57}$La 138.9	$_{72}$Hf 178.5	$_{73}$Ta 180.9	$_{74}$W 183.9	$_{75}$Re 186.2	$_{76}$Os 190.2	$_{77}$Ir 192.2	$_{78}$Pt 195.1	$_{79}$Au 197.0	$_{80}$Hg 200.6	$_{81}$Tl 204.4	$_{82}$Pb 207.2	$_{83}$Bi 209.0	$_{84}$Po (210)	$_{85}$At (210)	$_{86}$Rn (222)	
$_{87}$Fr (223)	$_{88}$Ra (226.0)	$_{89}$Ac (227)	$_{104}$Unq	$_{105}$Unp	$_{106}$Unh	$_{107}$Uns												
←*s*-block→		←				*d*-block					→	←		*p*-block			→	
Lanthanides			$_{57}$La 138.9	$_{58}$Ce 140.1	$_{59}$Pr 140.9	$_{60}$Nd 144.2	$_{61}$Pm (147)	$_{62}$Sm 150.4	$_{63}$Eu 152.0	$_{64}$Gd 157.3	$_{65}$Tb 158.9	$_{66}$Dy 162.5	$_{67}$Ho 164.9	$_{68}$Er 167.3	$_{69}$Tm 168.9	$_{70}$Yb 173.0	$_{71}$Lu 175.0	
Actinides			$_{89}$Ac (227)	$_{90}$Th 232.0	$_{91}$Pa 231.0	$_{92}$U 238.0	$_{93}$Np 237.0	$_{94}$Pu (242)	$_{95}$Am (243)	$_{96}$Cm (248)	$_{97}$Bk (247)	$_{98}$Cf (251)	$_{99}$Es (254)	$_{100}$Fm (253)	$_{101}$Md (256)	$_{102}$No (254)	$_{103}$Lr (257)	
			←					*f*-block									→	

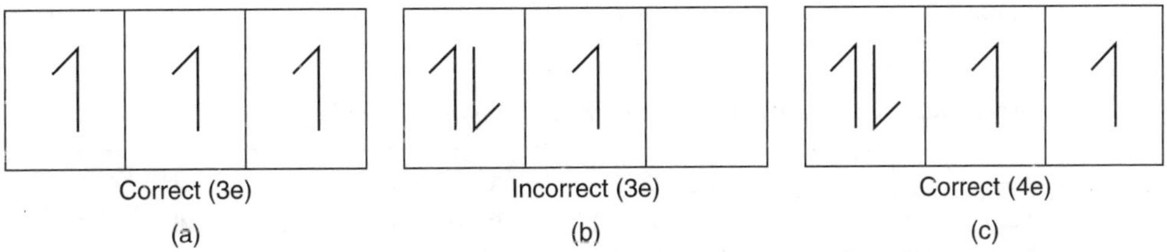

Figure 1.1 *Application of Hund's multiplicity rule to the electron filling of energy states.*

1.3 The Periodic Table

The Periodic Table provides an invaluable classification of all chemical elements, an element being a collection of atoms of one type. A typical version is shown in Table 1.2. Of the 107 elements which appear, about 90 occur in nature; the remainder are produced in nuclear reactors or particle accelerators. The atomic number (Z) of each element is stated, together with its chemical symbol, and can be regarded as either the number of protons in the nucleus or the number of orbiting electrons in the atom. The elements are naturally classified into periods (horizontal rows), depending upon which electron shell is being filled, and groups (vertical columns). Elements in any one group have the electrons in their outermost shell in the same configuration and, as a direct result, have similar chemical properties.

The building principle (*Aufbauprinzip*) for the table is based essentially upon two rules. First, the Pauli Exclusion Principle (Section 1.2.1) must be obeyed. Second, in compliance with Hund's rule of maximum multiplicity, the ground state should always develop maximum spin. This effect is demonstrated diagrammatically in Figure 1.1. Suppose that we supply three electrons to the three 'empty' $2p$-orbitals. They will build up a pattern of parallel spins (a) rather than paired spins (b). A fourth electron will cause pairing (c). Occasionally, irregularities occur in the 'filling' sequence for energy states because electrons always enter the lowest available energy state. Thus, $4s$-states, being at a lower energy level, fill before the $3d$-states.

We will now examine the general process by which the Periodic Table is built up, electron by electron, in closer detail. The progressive filling of energy states can be followed in Table 1.3. The first period commences with the simple hydrogen atom, which has a single proton in the nucleus and a single orbiting electron ($Z = 1$). The atom is therefore electrically neutral and, for the lowest energy condition, the electron will be in the $1s$-state. In helium, the next element, the nucleus charge is increased by one proton and an additional electron maintains neutrality ($Z = 2$). These two electrons fill the $1s$-state and will necessarily have opposite spins. The nucleus of helium contains two neutrons as well as two protons, hence its mass is four times greater than that of hydrogen. The next atom, lithium, has a nuclear charge of three ($Z = 3$) and, because the first shell is full, an electron must enter the $2s$-state, which has a somewhat higher energy. The electron in the $2s$-state, usually referred to as the valency electron, is 'shielded' by the inner electrons from the attracting nucleus and is therefore less strongly bonded. As a result, it is relatively easy to separate this valency electron. The 'electron core' which remains contains two tightly-bound electrons and, because it carries a single net positive charge, is referred to as a monovalent cation. The overall process by which electron(s) are lost or gained is known as ionization.

The development of the first short period from lithium ($Z = 3$) to neon ($Z = 10$) can be conveniently followed by referring to Table 1.3. So far, the sets of states corresponding to two principal quantum numbers ($n = 1$, $n = 2$) have been filled and the electrons in these states are said to have formed closed shells. It is a consequence of quantum mechanics that, once a shell is filled, the energy of that shell falls to a very low value and the resulting electronic configuration is very stable. Thus, helium, neon, argon and krypton are associated with closed shells and, being inherently stable and chemically unreactive, are known collectively as the inert gases.

Table 1.3 *Electron quantum numbers (Hume-Rothery, Smallman and Haworth, 1988).*

Element and atomic number	*Principal and secondary quantum numbers*									
$n=$	**1**	**2**		**3**			**4**			
$l=$	**0**	**0**	**1**	**0**	**1**	**2**	**0**	**1**	**2**	**3**
1 H	1									
2 He	2									
3 Li	2	1								
4 Be	2	2								
5 B	2	2	1							
6 C	2	2	2							
7 N	2	2	3							
8 O	2	2	4							
9 F	2	2	5							
10 Ne	2	2	6							
11 Na	2	2	6	1						
12 Mg	2	2	6	2						
13 Al	2	2	6	2	1					
14 Si	2	2	6	2	2					
15 P	2	2	6	2	3					
16 S	2	2	6	2	4					
17 Cl	2	2	6	2	5					
18 Ar	2	2	6	2	6					
19 K	2	2	6	2	6		1			
20 Ca	2	2	6	2	6		2			
21 Sc	2	2	6	2	6	1	2			
22 Ti	2	2	6	2	6	2	2			
23 V	2	2	6	2	6	3	2			
24 Cr	2	2	6	2	6	5	1			
25 Mn	2	2	6	2	6	5	2			
26 Fe	2	2	6	2	6	6	2			
27 Co	2	2	6	2	6	7	2			
28 Ni	2	2	6	2	6	8	2			
29 Cu	2	2	6	2	6	10	1			
30 Zn	2	2	6	2	6	10	2			
31 Ga	2	2	6	2	6	10	2	1		
32 Ge	2	2	6	2	6	10	2	2		
33 As	2	2	6	2	6	10	2	3		
34 Se	2	2	6	2	6	10	2	4		
35 Br	2	2	6	2	6	10	2	5		
36 Kr	2	2	6	2	6	10	2	6		

Continued

Table 1.3 *(Continued)*

Element and atomic number	*Principal and secondary quantum numbers*										
n =	**1**	**2**	**3**	**4**				**5**			**6**
l =	–	–	–	**0**	**1**	**2**	**3**	**0**	**1**	**2**	**0**
37 Rb	2	8	18	2	6			1			
38 Sr	2	8	18	2	6			2			
39 Y	2	8	18	2	6	1		2			
40 Zr	2	8	18	2	6	2		2			
41 Nb	2	8	18	2	6	4		1			
42 Mo	2	8	18	2	6	5		1			
43 Tc	2	8	18	2	6	5		2			
44 Ru	2	8	18	2	6	7		1			
45 Rh	2	8	18	2	6	8		1			
46 Pd	2	8	18	2	6	10		–			
47 Ag	2	8	18	2	6	10		1			
48 Cd	2	8	18	2	6	10		2			
49 In	2	8	18	2	6	10		2	1		
50 Sn	2	8	18	2	6	10		2	2		
51 Sb	2	8	18	2	6	10		2	3		
52 Te	2	8	18	2	6	10		2	4		
53 I	2	8	18	2	6	10		2	5		
54 Xe	2	8	18	2	6	10		2	6		
55 Cs	2	8	18	2	6	10		2	6		1
56 Ba	2	8	18	2	6	10		2	6		2
57 La	2	8	18	2	6	10		2	6	1	2
58 Ce	2	8	18	2	6	10	2	2	6		2
59 Pr	2	8	18	2	6	10	3	2	6		2
60 Nd	2	8	18	2	6	10	4	2	6		2
61 Pm	2	8	18	2	6	10	5	2	6		2
62 Sm	2	8	18	2	6	10	6	2	6		2
63 Eu	2	8	18	2	6	10	7	2	6		2
64 Gd	2	8	18	2	6	10	7	2	6	1	2
65 Tb	2	8	18	2	6	10	9	2	6		2
66 Dy	2	8	18	2	6	10	10	2	6		2
67 Ho	2	8	18	2	6	10	11	2	6		2
68 Er	2	8	18	2	6	10	12	2	6		2
69 Tm	2	8	18	2	6	10	13	2	6		2
70 Yb	2	8	18	2	6	10	14	2	6		2
71 Lu	2	8	18	2	6	10	14	2	6	1	2
72 Hf	2	8	18	2	6	10	14	2	6	2	2

Continued

Table 1.3 *(Continued)*

Element and atomic number	*Principal and secondary quantum numbers*											
n =	**1**	**2**	**3**	**4**	**5**				**6**			**7**
l =	–	–	–	–	**0**	**1**	**2**	**3**	**0**	**1**	**2**	**0**
73 Ta	2	8	18	32	2	6	3		2			
74 W	2	8	18	32	2	6	4		2			
75 Re	2	8	18	32	2	6	5		2			
76 Os	2	8	18	32	2	6	6		2			
77 Ir	2	8	18	32	2	6	7		2			
78 Pt	2	8	18	32	2	6	9		1			
79 Au	2	8	18	32	2	6	10		1			
80 Hg	2	8	18	32	2	6	10		2			
81 Tl	2	8	18	32	2	6	10		2	1		
82 Pb	2	8	18	32	2	6	10		2	2		
83 Bi	2	8	18	32	2	6	10		2	3		
84 Po	2	8	18	32	2	6	10		2	4		
85 At	2	8	18	32	2	6	10		2	5		
86 Rn	2	8	18	32	2	6	10		2	6		
87 Fr	2	8	18	32	2	6	10		2	6		1
88 Ra	2	8	18	32	2	6	10		2	6		2
89 Ac	2	8	18	32	2	6	10		2	6	1	2
90 Th	2	18	8	32	2	6	10		2	6	2	2
91 Pa	2	18	8	32	2	6	10	2	2	6	1	2
92 U	2	18	8	32	2	6	10	3	2	6	1	2
93 Np	2	18	8	32	2	6	10	4	2	6	1	2
94 Pu	2	18	8	32	2	6	10	5	2	6	1	2

The exact electronic configurations of the later elements are not always certain but the most probable arrangements of the outer electrons are:

95 Am	$(5f)^7(7s)^2$
96 Cm	$(5f)^7(6d)^1(7s)^2$
97 Bk	$(5f)^8(6d)^1(7s)^2$
98 Cf	$(5f)^{10}(7s)^2$
99 Es	$(5f)^{11}(7s)^2$
100 Fm	$(5f)^{12}(7s)^2$
101 Md	$(5f)^{13}(7s)^2$
102 No	$(5f)^{14}(7s)^2$
103 Lw	$(5f)^{14}(6d)^1(7s)^2$
104 —	$(5f)^{14}(6d)^2(7s)^2$

The second short period, from sodium ($Z = 11$) to argon ($Z = 18$), commences with the occupation of the 3*s*-orbital and ends when the 3*p*-orbitals are full (Table 1.3). The long period which follows extends from potassium ($Z = 19$) to krypton ($Z = 36$) and, as mentioned previously, has the unusual feature of the 4*s*-state filling before the 3*d*-state. Thus, potassium has a similarity to sodium and lithium in that the electron of highest energy is in an *s*-state; as a consequence, they have very similar chemical reactivities, forming the group known as the alkali-metal elements. After calcium ($Z = 20$), filling of the 3*d*-state begins.

The 4*s*-state is filled in calcium ($Z = 20$) and the filling of the 3*d*-state becomes energetically favorable to give scandium ($Z = 21$). This belated filling of the five 3*d*-orbitals from scandium to its completion in copper ($Z = 29$) embraces the first series of transition elements. One member of this series, chromium ($Z = 24$), obviously behaves in an unusual manner. Applying Hund's rule, we can reason that maximization of parallel spin is achieved by locating six electrons, of like spin, so that five fill the 3*d*-states and one enters the 4*s*-state. This mode of fully occupying the 3*d*-states reduces the energy of the electrons in this shell considerably. Again, in copper ($Z = 29$), the last member of this transition series, complete filling of all 3*d*-orbitals also produces a significant reduction in energy. It follows from these explanations that the 3*d*- and 4*s*-levels of energy are very close together. After copper, the energy states fill in a straightforward manner and the first long period finishes with krypton ($Z = 36$). It will be noted that lanthanides ($Z = 57$–71) and actinides ($Z = 89$–103), because of their state-filling sequences, have been separated from the main body of Table 1.2. Having demonstrated the manner in which quantum rules are applied to the construction of the Periodic Table for the first 36 elements, we can now examine some general aspects of the classification.

When one considers the small step difference of one electron between adjacent elements in the Periodic Table, it is not really surprising to find that the distinction between metallic and non-metallic elements is imprecise. In fact, there is an intermediate range of elements, the metalloids, which share the properties of both metals and non-metals. However, we can regard the elements which can readily lose an electron, by ionization or bond formation, as strongly metallic in character (e.g. alkali metals). Conversely, elements which have a strong tendency to acquire an electron and thereby form a stable configuration of two or eight electrons in the outermost shell are non-metallic (e.g. the halogens fluorine, chlorine, bromine, iodine). Thus, electropositive metallic elements and the electronegative non-metallic elements lie on the left- and right-hand sides of the Periodic Table, respectively. As will be seen later, these and other aspects of the behavior of the outermost (valence) electrons have a profound and determining effect upon bonding, and therefore upon electrical, magnetic and optical properties.

Prior to the realization that the frequently observed periodicities of chemical behavior could be expressed in terms of electronic configurations, emphasis was placed upon 'atomic weight'. This quantity, which is now referred to as relative atomic mass, increases steadily throughout the Periodic Table as protons and neutrons are added to the nuclei. Atomic mass[2] determines physical properties such as density, specific heat capacity and ability to absorb electromagnetic radiation: it is therefore very relevant to engineering practice. For instance, many ceramics are based upon the light elements aluminum, silicon and oxygen, and consequently have a low density, i.e. $<3000\,\text{kg}\,\text{m}^{-3}$.

1.4 Interatomic bonding in materials

Matter can exist in three states and, as atoms change directly from either the gaseous state (desublimation) or the liquid state (solidification) to the usually denser solid state, the atoms form aggregates in three-dimensional space. Bonding forces develop as atoms are brought into proximity to each other. Generally, there is an attractive force between the atoms but at close range a repulsive force exists.

[2] Atomic mass is now expressed relative to the datum value for carbon (12.01). Thus, a copper atom has 63.55/12.01 or 5.29 times more mass than a carbon atom.

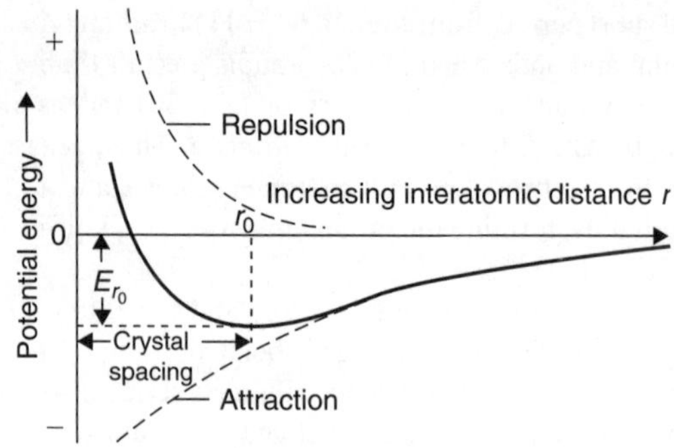

Figure 1.2 *Variation in potential energy with interatomic distance.*

The equilibrium spacing is given when these two forces balance. The energy of interaction decreases as the atoms approach and has its lowest value at the equilibrium spacing, as shown in Figure 1.2. The potential energy U of a pair of atoms can be written as:

$$U = -\frac{A}{r^m} + \frac{B}{r^n} \tag{1.1}$$

where r is the atom separation with $m < n$, and the first term is attractive, the second repulsive. At $r < r_0$, the equilibrium value, the repulsive force dominates and U rises. The force F is given by the rate of change of energy with distance $\mathrm{d}U/\mathrm{d}r$ and is zero at $r = r_0$.

The nature of the bonding forces has a direct effect upon the type of solid structure which develops and therefore upon the physical properties of the material. Melting point provides a useful indication of the amount of thermal energy needed to sever these interatomic (or interionic) bonds. Thus, some solids melt at relatively low temperatures (m.p. of tin = 232°C), whereas many ceramics melt at extremely high temperatures (m.p. of alumina exceeds 2000°C). It is immediately apparent that bond strength has far-reaching implications in all fields of engineering.

Customarily we identify four principal types of bonding in materials, namely metallic bonding, ionic bonding, covalent bonding and the comparatively much weaker van der Waals bonding. However, in many solid materials it is possible for bonding to be mixed, or even intermediate, in character. We will first consider the general chemical features of each type of bonding; in later sections we will examine the resultant disposition of the assembled atoms (ions) in three-dimensional space.

As we have seen, the elements with the most pronounced metallic characteristics are grouped on the left-hand side of the Periodic Table (Table 1.2). In general, they have a few valence electrons, outside the outermost closed shell, which are relatively easy to detach. In a metal, each 'free' valency electron is shared among all atoms, rather than associated with an individual atom, and forms part of the so-called 'electron gas', which circulates at random among the regular array of positively charged electron cores, or cations (Figure 1.3a). Application of an electric potential gradient will cause the 'gas' to drift through the structure with little hindrance, thus explaining the outstanding electrical conductivity of the metallic state. The metallic bond derives from the attraction between the cations and the free electrons and, as would be expected, repulsive components of force develop when cations are brought into close proximity. However, the bonding forces in metallic structures are spatially non-directed and we can readily simulate the packing and space-filling characteristics of the atoms with modeling systems based on equal-sized spheres (polystyrene balls, even soap bubbles). Other properties such as ductility, thermal conductivity and the transmittance of electromagnetic radiation are also directly influenced by the non-directionality and high electron mobility of the metallic bond.

The ionic bond develops when electron(s) are transferred from atoms of active metallic elements to atoms of active non-metallic elements, thereby enabling each of the resultant ions to attain a stable

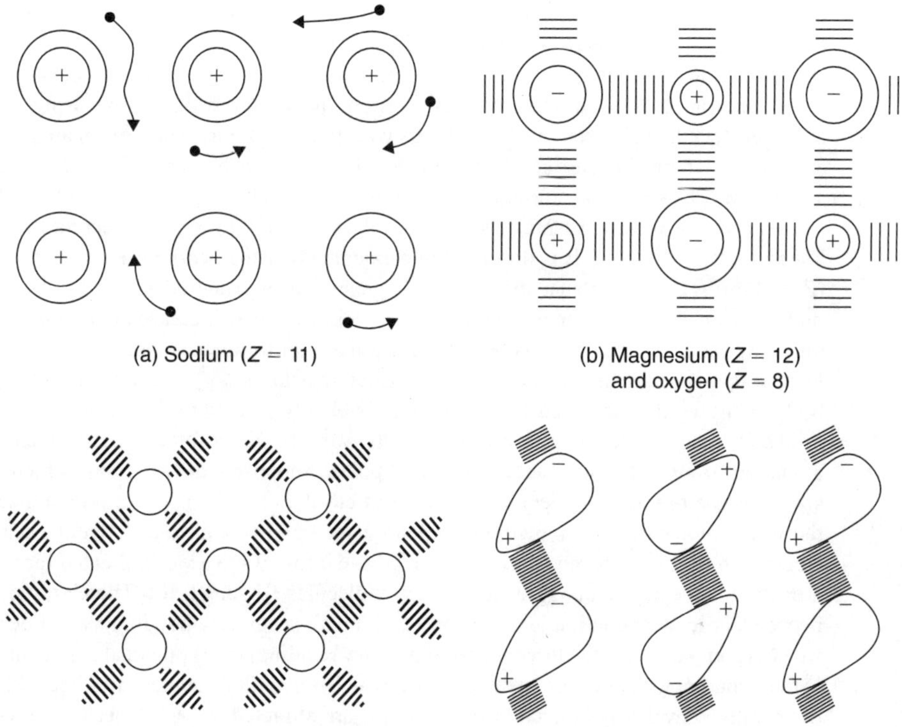

Figure 1.3 *Schematic representation of: (a) metallic bonding, (b) ionic bonding, (c) covalent bonding and (d) van der Waals bonding.*

closed shell. For example, the ionic structure of magnesia (MgO), a ceramic oxide, forms when each magnesium atom ($Z = 12$) loses two electrons from its M-shell ($n = 3$) and these electrons are acquired by an oxygen atom ($Z = 8$), producing a stable octet configuration in its L-shell (Table 1.3). Overall, the ionic charges balance and the structure is electrically neutral (Figure 1.3b). Anions are usually larger than cations. Ionic bonding is omnidirectional, essentially electrostatic in character and can be extremely strong; for instance, magnesia is a very useful refractory oxide (m.p. = 2930°C). At low to moderate temperatures, such structures are electrical insulators but, typically, become conductive at high temperatures when thermal agitation of the ions increases their mobility.

Sharing of valence electrons is the key feature of the third type of strong primary bonding. Covalent bonds form when valence electrons of opposite spin from adjacent atoms are able to pair within overlapping spatially directed orbitals, thereby enabling each atom to attain a stable electronic configuration (Figure 1.3c). Being oriented in three-dimensional space, these localized bonds are unlike metallic and ionic bonds. Furthermore, the electrons participating in the bonds are tightly bound so that covalent solids, in general, have low electrical conductivity and act as insulators, sometimes as semiconductors (e.g. silicon). Carbon in the form of diamond is an interesting prototype for covalent bonding. Its high hardness, low coefficient of thermal expansion and very high melting point (3300°C) bear witness to the inherent strength of the covalent bond. First, using the (8 – N) Rule, in which N is the Group Number[3] in the Periodic Table, we deduce that carbon ($Z = 6$) is tetravalent; that

[3] According to previous IUPAC notation: see top of Table 1.2.

is, four bond-forming electrons are available from the L-shell ($n = 2$). In accordance with Hund's Rule (Figure 1.1), one of the two electrons in the $2s$-state is promoted to a higher $2p$-state to give a maximum spin condition, producing an overall configuration of $1s^2\ 2s^1\ 2p^3$ in the carbon atom. The outermost second shell accordingly has four valency electrons of like spin available for pairing. Thus, each carbon atom can establish electron-sharing orbitals with four neighbors. For a given atom, these four bonds are of equal strength and are set at equal angles (109.5°) to each other and therefore exhibit tetrahedral symmetry. (The structural consequences of this important feature will be discussed in Section 1.9.2.)

This process by which s-orbitals and p-orbitals combine to form projecting hybrid sp-orbitals is known as hybridization. It is observed in elements other than carbon. For instance, trivalent boron ($Z = 5$) forms three co-planar sp^2-orbitals. In general, a large degree of overlap of sp-orbitals and/or a high electron density within the overlap 'cloud' will lead to an increase in the strength of the covalent bond. As indicated earlier, it is possible for a material to possess more than one type of bonding. For example, in calcium silicate (Ca_2SiO_4), calcium cations Ca^{2+} are ionically bonded to tetrahedral SiO_4^{4-} clusters in which each silicon atom is covalently bonded to four oxygen neighbors.

The final type of bonding is attributed to the van der Waals forces which develop when adjacent atoms, or groups of atoms, act as electric dipoles. Suppose that two atoms which differ greatly in size combine to form a molecule as a result of covalent bonding. The resultant electron 'cloud' for the whole molecule can be pictured as pear-shaped and will have an asymmetrical distribution of electron charge. An electric dipole has formed and it follows that weak directed forces of electrostatic attraction can exist in an aggregate of such molecules (Figure 1.3d). There are no 'free' electrons, hence electrical conduction is not favored. Although secondary bonding by van der Waals forces is weak in comparison to the three forms of primary bonding, it has practical significance. For instance, in the technologically important mineral talc, which is hydrated magnesium silicate $Mg_3Si_4O_{10}(OH)_2$, the parallel, covalently bonded layers of atoms are attracted to each other by van der Waals forces. These layers can easily be slid past each other, giving the mineral its characteristically slippery feel. In thermoplastic polymers, van der Waals forces of attraction exist between the extended covalently bonded hydrocarbon chains; a combination of heat and applied shear stress will overcome these forces and cause the molecular chains to glide past each other. To quote a more general case, molecules of water vapor in the atmosphere each have an electric dipole and will accordingly tend to be adsorbed if they strike solid surfaces possessing attractive van der Waals forces (e.g. silica gel).

Worked example

The potential energy U of a pair of atoms in a solid is given by equation (1.1).

(i) Outline the physical significance of the two terms and indicate the values of the two constants m and n.

(ii) Taking the value of $m = 2$ and $n = 10$, calculate the values of A and B for a stable atomic configuration where $r = 3 \times 10^{-10}$ m and $U = -4$ eV. Calculate the force required to break the diatomic configuration.

Solution

(i) $-A/r^m$ is an attractive potential related to the type of bonding in the crystal. B/r^n is a repulsive potential when the ions get close. The value of $m < n$, typically $n \sim 12$ for ionic solids.

(ii) The energy function now reads:

$$U = -\frac{A}{r^2} + \frac{B}{r^{10}}.$$

At equilibrium at $r = r_o$,

$$\text{force } F = \frac{dU}{dr} = \frac{2A}{r^3} - \frac{10B}{r^{11}} = 0, \text{ so } \frac{2A}{r_o^3} = \frac{10B}{r_o^{11}} \text{ or } B = \frac{Ar_o^8}{5}.$$

$$U_o = -\frac{A}{r_o^2} + \frac{Ar_o^8}{5r_o^{10}} = -\frac{4A}{5r_o^2} \text{ or}$$

$$A = -\frac{5r_o^2 U_o}{4}$$

$$= \frac{5 \times 0.3^2 \times 4}{4} = 0.45\,\text{eV nm}^2 = 7.2 \times 10^{-20}\,\text{J nm}^2$$

$$B = \frac{Ar_o^8}{5} = \frac{0.45 \times 0.3^8}{5} = 0.59 \times 10^{-5}\,\text{eV nm}^{10} = 9.4 \times 10^{-25}\,\text{J nm}^{10}.$$

Maximum force is at $d^2U/dr^2 = 0$, at a value for r given by:

$$-\frac{6A}{r^4} + \frac{110B}{r^{12}} = 0.$$

So $r = (110B/6A)^{1/8} = r_o \times (11/3)^{1/8} = 0.352\,\text{nm}$.

Thus, maximum force to break bond:

$$F = \frac{dU}{dr} = \frac{2A}{r^3}\left(1 - \frac{r_o^8}{r^8}\right) = \frac{2 \times 0.45}{0.352^3}\left(1 - \frac{0.3^8}{0.352^8}\right) = 14.9\,\text{eV/nm} = 2.39 \times 10^{-9}\text{N}.$$

1.5 Bonding and energy levels

If one imagines atoms being brought together uniformly to form, for example, a metallic structure, then when the distance between neighboring atoms approaches the interatomic value the outer electrons are no longer localized around individual atoms. Once the outer electrons can no longer be considered to be attached to individual atoms but have become free to move throughout the metal then, because of the Pauli Exclusion Principle, these electrons cannot retain the same set of quantum numbers that they had when they were part of the atoms. As a consequence, the free electrons can no longer have more than two electrons of opposite spin with a particular energy. The energies of the free electrons are distributed over a range which increases as the atoms are brought together to form the metal. If the atoms when brought together are to form a stable metallic structure, it is necessary that the mean energy of the free electrons shall be lower than the energy of the electron level in the free atom from which they are derived. Figure 1.4 shows the broadening of an atomic electron level as the atoms are brought together, and also the attendant lowering of energy of the electrons. It is the extent of the lowering in mean energy of the outer electrons that governs the stability of a metal. The equilibrium spacing between the atoms in a metal is that for which any further decrease in the atomic spacing would lead to an increase in the repulsive interaction of the positive ions as they are forced into closer contact with each other, which would be greater than the attendant decrease in mean electron energy.

In a metallic structure, the free electrons must, therefore, be thought of as occupying a series of discrete energy levels at very close intervals. Each atomic level which splits into a band contains the same number of energy levels as the number N of atoms in the piece of metal. As previously stated, only two electrons of opposite spin can occupy any one level, so that a band can contain a

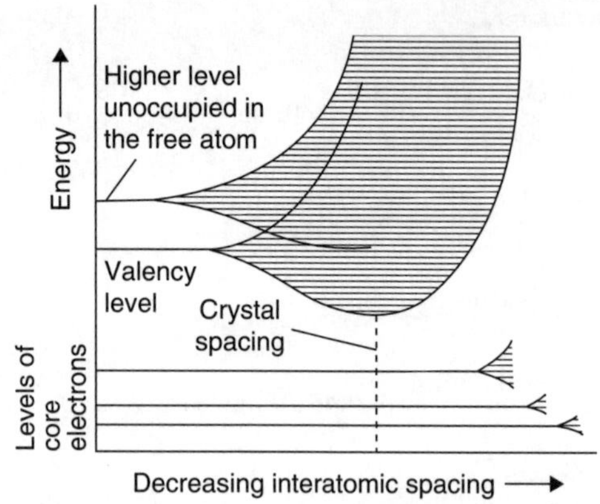

Figure 1.4 *Broadening of atomic energy levels in a metal.*

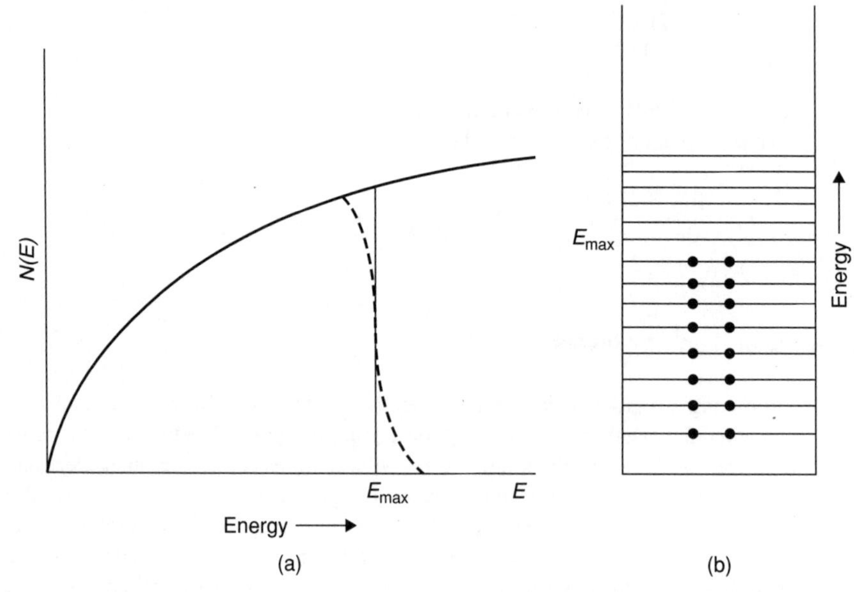

Figure 1.5 *(a) Density of energy levels plotted against energy. (b) Filling of energy levels by electrons at absolute zero. At ordinary temperatures some of the electrons are thermally excited to higher levels than that corresponding to* E_{max}*, as shown by the broken curve in (a).*

maximum of $2N$ electrons. Clearly, in the lowest energy state of the metal all the lower energy levels are occupied.

The energy gap between successive levels is not constant but decreases as the energy of the levels increases. This is usually expressed in terms of the density of electronic states $N(E)$ as a function of the energy E. The quantity $N(E)\mathrm{d}E$ gives the number of energy levels in a small energy interval $\mathrm{d}E$, and for free electrons is a parabolic function of the energy, as shown in Figure 1.5.

Because only two electrons can occupy each level, the energy of an electron occupying a low energy level cannot be increased unless it is given sufficient energy to allow it to jump to an empty level at the top of the band. The energy[4] width of these bands is commonly about 5 or 6 eV and, therefore, considerable energy would have to be put into the metal to excite a low-lying electron. Such energies do not occur at normal temperatures, and only those electrons with energies close to that of the top of the band (known as the Fermi level and surface) can be excited, and therefore only a small number of the free electrons in a metal can take part in thermal processes. The energy of the Fermi level E_F depends on the number of electrons N per unit volume V, and is given by $(\mathbf{h}^2/8m)(3N/\pi V)^{2/3}$.

The electron in a metallic band must be thought of as moving continuously through the structure with an energy depending on which level of the band it occupies. In quantum mechanical terms, this motion of the electron can be considered in terms of a wave with a wavelength which is determined by the energy of the electron according to de Broglie's relationship $\lambda = \mathbf{h}/mv$, where $\mathbf{h}$ is Planck's constant and m and v are, respectively, the mass and velocity of the moving electron. The greater the energy of the electron, the higher will be its momentum mv, and hence the smaller will be the wavelength of the wave function in terms of which its motion can be described. Because the movement of an electron has this wave-like aspect, moving electrons can give rise, like optical waves, to diffraction effects. This property of electrons is used in electron microscopy (Chapter 4).

Worked example

State the equation for the energy of an electron which is confined to a one-dimensional infinite potential well of width L.

Calculate the energy in electron-volts of the ground state and the energy interval ΔE with the first excited state for $L = 0.2$ nm. (Planck's constant $= 6.62 \times 10^{-34}$ J s, electron mass $m = 9.1 \times 10^{-31}$ kg.)

Solution

Electron energy is $E = \dfrac{n^2\mathbf{h}^2}{8mL^2}$, where n is the quantum number.

For $L = 0.2$ nm,

$$E(n = 1) = \frac{1 \times (6.62 \times 10^{-34})^2}{8 \times 9.1 \times 10^{-31} \times (2 \times 10^{-10})^2} = 1.50 \times 10^{-18}\,\text{J} = 9.4\,\text{eV}$$

$$E(n = 2) = 4 \times E(n = 1) = 4 \times 9.4 = 37.6\,\text{eV}$$

Energy interval $\Delta E = E(n = 2) - E(n = 1) = 28.2$ eV.

1.6 Crystal lattices and structures

Having discussed the electronic states and bonding between atoms and so on, we can now focus on the arrangement of atoms to form solids. Except for a few exceptional metallic systems which can form metallic glasses, most metals and alloys usually have crystalline structures. Figure 1.6a shows a three-dimensional lattice in which triple intersections of three families of parallel equidistant lines mark the positions of atoms. In this simple case, three reference axes (x, y, z) are oriented at 90° to each other and atoms are 'shrunk', for convenience. The orthogonal lattice of Figure 1.6a defines eight unit cells, each having a shared atom at every corner. It follows from our recognition of the inherent order of the lattice that we can express the geometrical characteristics of the whole crystal,

[4] An electron-volt is the kinetic energy an electron acquires in falling freely through a potential difference of 1 volt ($1\,\text{eV} = 1.602 \times 10^{-19}$ J; 1 eV per particle $= 23\,050 \times 4.186$ J per mol of particles).

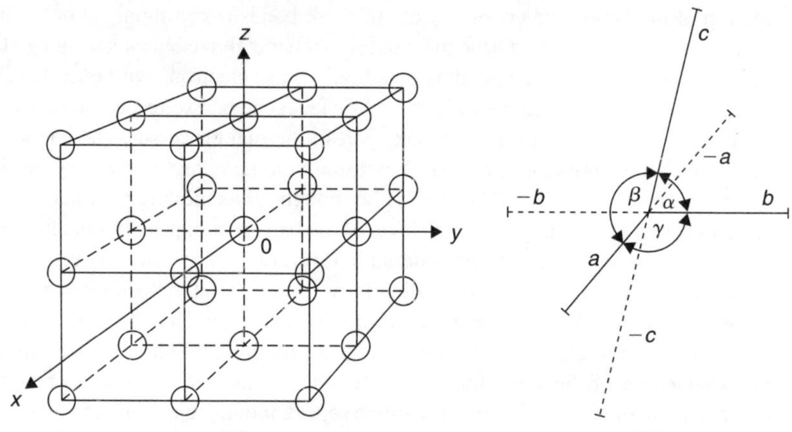

Figure 1.6 *Principles of lattice construction.*

containing millions of atoms, in terms of the size, shape and atomic arrangement of the unit cell, the ultimate repeat unit of structure.[5]

We can assign the lengths of the three cell parameters (a, b, c) to the reference axes, using an internationally accepted notation (Figure 1.6b). Thus, for the simple cubic case portrayed in Figure 1.6a, $\alpha=\beta=\gamma=90°$; $a=b=c$. Economizing in symbols, we only need to quote a single cell parameter (a) for the cubic unit cell. By systematically changing the angles (α, β, γ) between the reference axes, and the cell parameters (a, b, c), and by four skewing operations, we derive the seven crystal systems (Figure 1.7). Any crystal, whether natural or synthetic, belongs to one or other of these systems. From the premise that each point of a space lattice should have identical surroundings, Bravais demonstrated that the maximum possible number of space lattices (and therefore unit cells) is 14. It is accordingly necessary to augment the seven primitive (P) cells shown in Figure 1.7 with seven more non-primitive cells which have additional face-centering, body-centering or end-centering lattice points. Thus, the highly symmetrical cubic system has three possible lattices: primitive (P), body centered (I; from the German word *innenzentrierte*) and face centered (F). We will encounter the latter two again in Section 1.9.1. True primitive space lattices, in which that lattice point has identical surroundings, can sometimes embody awkward angles. In such cases it is common practice to use a simpler orthogonal non-primitive lattice which will accommodate the atoms of the actual crystal structure.[6]

1.7 Crystal directions and planes

In a structurally disordered material, such as fully annealed silica glass, the value of a physical property is independent of the direction of measurement; the material is said to be isotropic. Conversely, in many single crystals, it is often observed that a structurally sensitive property, such as electrical conductivity, is strongly direction dependent because of variations in the periodicity and packing of atoms. Such crystals are anisotropic. We therefore need a precise method for specifying a direction, and equivalent directions, within a crystal. The general method for defining a given direction is

[5] The notion that the striking external appearance of crystals indicates the existence of internal structural units with similar characteristics of shape and orientation was proposed by the French mineralogist Hauy in 1784. Some 130 years elapsed before actual experimental proof was provided by the new technique of X-ray diffraction analysis.

[6] Lattices are imaginary and limited in number; crystal structures are real and virtually unlimited in their variety.

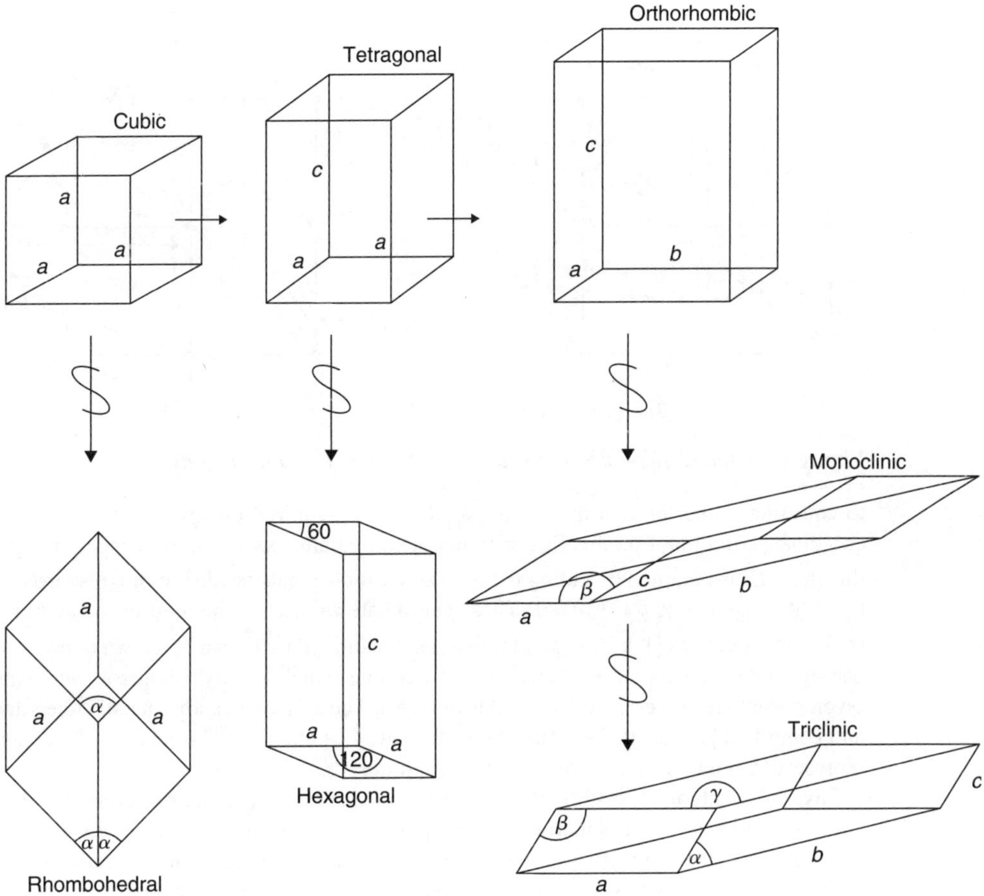

System	Axes	Axial angles
Cubic	$a_1 = a_2 = a_3$	All angles = 90°
Tetragonal	$a_1 = a_2 \neq c$	All angles = 90°
Orthorhombic	$a \neq b \neq c$	All angles = 90°
Monoclinic	$a \neq b \neq c$	Two angles = 90°; 1 angle ≠ 90°
Triclinic	$a \neq b \neq c$	All angles different; none equal 90°
Hexagonal	$a_1 = a_2 = a_3 \neq c$	Angles = 90° and 120°
Rhombohedral	$a_1 = a_2 = a_3$	All angles equal, but not 90°

Figure 1.7 *The seven systems of crystal symmetry (S = skew operation).*

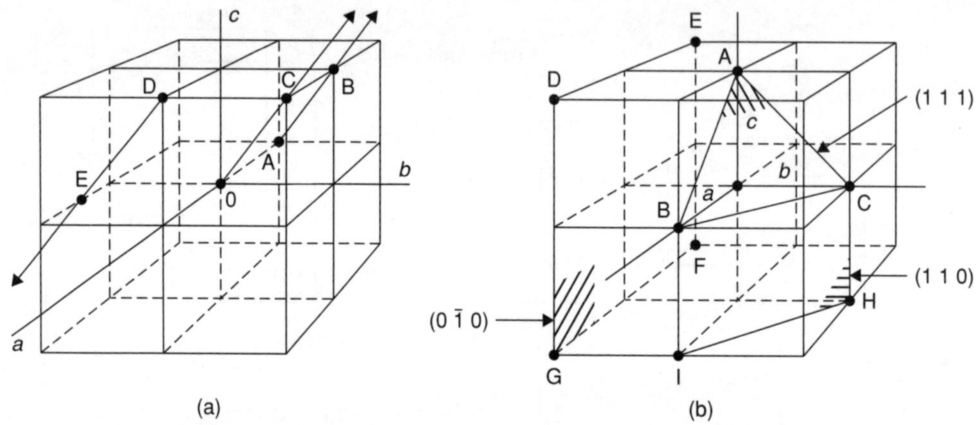

Figure 1.8 *Indexing of directions (a) and planes (b) in cubic crystals.*

to construct a line through the origin parallel to the required direction and then to determine the coordinates of a point on this line in terms of cell parameters (a, b, c). Hence, in Figure 1.8a, the direction $\overrightarrow{AB}$ is obtained by noting the translatory movements needed to progress from the origin O to point C, i.e. $a=1$, $b=1$, $c=1$. These coordinate values are enclosed in square brackets to give the direction indices [1 1 1]. In similar fashion, the direction $\overrightarrow{DE}$ can be shown to be $[\bar{\frac{1}{2}}\ \bar{1}\ \bar{1}]$ with the bar sign indicating use of a negative axis. Directions which are crystallographically equivalent in a given crystal are represented by angular brackets. Thus, $\langle 1\ 0\ 0\rangle$ represents all cube edge directions and comprises [1 0 0], [0 1 0], [0 0 1], $[\bar{1}\ 0\ 0]$, $[0\ \bar{1}\ 0]$ and $[0\ 0\ \bar{1}]$ directions. Directions are often represented in non-specific terms as $[u\,v\,w]$ and $\langle u\,v\,w\rangle$.

Physical events and transformations within crystals often take place on certain families of parallel equidistant planes. The orientation of these planes in three-dimensional space is of prime concern; their size and shape are of lesser consequence. (Similar ideas apply to the corresponding external facets of a single crystal.) In the Miller system for indexing planes, the intercepts of a representative plane upon the three axes (x, y, z) are noted.[7] Intercepts are expressed relatively in terms of a, b, c. Planes parallel to an axis are said to intercept at infinity. Reciprocals of the three intercepts are taken and the indices enclosed by round brackets. Hence, in Figure 1.8b, the procedural steps for indexing the plane ABC are:

	a	b	c
Intercepts	1	1	1
Reciprocals	$\frac{1}{1}$	$\frac{1}{1}$	$\frac{1}{1}$
Miller indices		(111)	

The Miller indices for the planes DEFG and BCHI are $(0\ \bar{1}\ 0)$ and (1 1 0), respectively. Often it is necessary to ignore individual planar orientations and to specify all planes of a given crystallographic type, such as the planes parallel to the six faces of a cube. These planes constitute a crystal form and have the same atomic configurations; they are said to be equivalent and can be represented by a single group of indices enclosed in curly brackets, or braces. Thus, $\{1\ 0\ 0\}$ represents a form of six planar

[7] For mathematical reasons, it is advisable to carry out all indexing operations (translations for directions, intercepts for planes) in the strict sequence a, b, c.

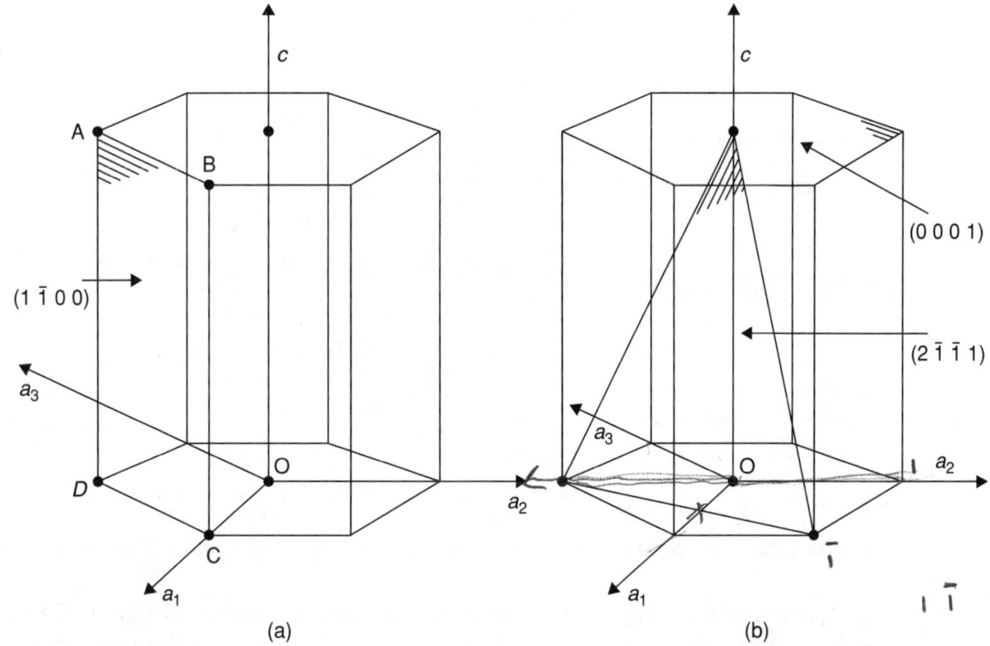

Figure 1.9 *Prismatic, basal and pyramidal planes in hexagonal structures.*

orientations, i.e. (1 0 0), (0 1 0), (0 0 1), ($\bar{1}$ 0 0), (0 $\bar{1}$ 0) and (0 0 $\bar{1}$). Returning to the (1 1 1) plane ABC of Figure 1.8b, it is instructive to derive the other seven equivalent planes, centering on the origin O, which comprise {1 1 1}. It will then be seen why materials belonging to the cubic system often crystallize in an octahedral form in which octahedral {1 1 1} planes are prominent.

It should be borne in mind that the general purpose of the Miller procedure is to define the orientation of a family of parallel equidistant planes; the selection of a convenient representative plane is a means to this end. For this reason, it is permissible to shift the origin provided that the relative disposition of a, b and c is maintained. Miller indices are commonly written in the symbolic form ($h\,k\,l$). Rationalization of indices, either to reduce them to smaller numbers with the same ratio or to eliminate fractions, is unnecessary. This often-recommended step discards information; after all, there is a real difference between the two families of planes (1 0 0) and (2 0 0).

As mentioned previously, it is sometimes convenient to choose a non-primitive cell. The hexagonal structure cell is an important illustrative example. For reasons which will be explained, it is also appropriate to use a four-axis Miller–Bravais notation ($h\,k\,i\,l$) for hexagonal crystals, instead of the three-axis Miller notation ($h\ k\ l$). In this alternative method, three axes (a_1, a_2, a_3) are arranged at 120° to each other in a basal plane and the fourth axis (c) is perpendicular to this plane (Figure 1.9a). Hexagonal structures are often compared in terms of the axial ratio c/a. The indices are determined by taking intercepts upon the axes in strict sequence. Thus, the procedural steps for the plane ABCD, which is one of the six prismatic planes bounding the complete cell, are:

	a_1	a_2	a_3	c
Intercepts	1	-1	∞	∞
Reciprocals	$\frac{1}{1}$	$-\frac{1}{1}$	0	0
Miller–Bravais indices		(1 $\bar{1}$ 0 0)		

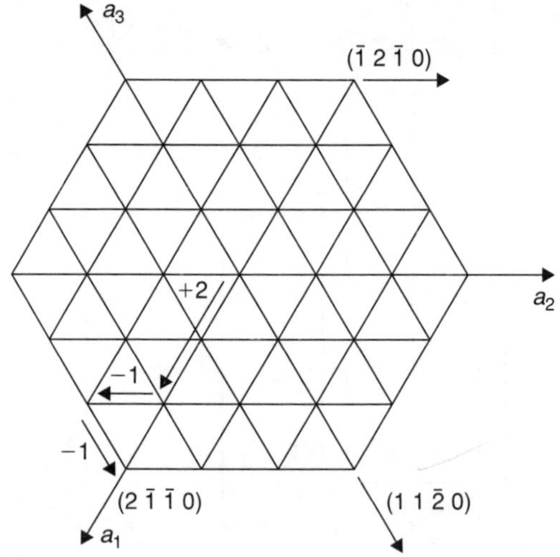

Figure 1.10 *Typical Miller–Bravais directions in (0 0 0 1) basal plane of hexagonal crystal.*

Comparison of these digits with those from other prismatic planes such as $(1\ 0\ \bar{1}\ 0), (0\ 1\ \bar{1}\ 0)$ and $(1\ \bar{1}\ 0\ 0)$ immediately reveals a similarity; that is, they are crystallographically equivalent and belong to the $\{1\ 0\ 1\ 0\}$ form. The three-axis Miller method lacks this advantageous feature when applied to hexagonal structures. For geometrical reasons, it is essential to ensure that the plane indices comply with the condition $(h+k) = -i$. In addition to the prismatic planes, basal planes of (0 0 0 1) type and pyramidal planes of the $(1\ 1\ \bar{2}\ 1)$ type are also important features of hexagonal structures (Figure 1.9b).

The Miller–Bravais system also accommodates directions, producing indices of the form $[u\,v\,t\,w]$. The first three translations in the basal plane must be carefully adjusted so that the geometrical condition $u+v=-t$ applies. This adjustment can be facilitated by subdividing the basal planes into triangles (Figure 1.10). As before, equivalence is immediately revealed; for instance, the close-packed directions in the basal plane have the indices $[2\ \bar{1}\ \bar{1}\ 0]$, $[1\ 1\ \bar{2}\ 0]$, $[\bar{1}\ 2\ \bar{1}\ 0]$, etc. and can be represented by $\langle 2\ 1\ \bar{1}\ 0\rangle$.

1.8 Stereographic projection

Projective geometry makes it possible to represent the relative orientation of crystal planes and directions in three-dimensional space in a more convenient two-dimensional form. The standard stereographic projection is frequently used in the analysis of crystal behavior; X-ray diffraction analyses usually provide the experimental data. Typical applications of the method are the interpretation of strain markings on crystal surfaces, portrayal of symmetrical relationships, determination of the axial orientations in a single crystal and the plotting of property values for anisotropic single crystals. (The basic method can also be adapted to produce a pole figure diagram, which can show preferred orientation effects in polycrystalline aggregates.)

A very small crystal of cubic symmetry is assumed to be located at the center of a reference sphere, as shown in Figure 1.11a, so that the orientation of a crystal plane, such as the (1 1 1) plane marked, may be represented on the surface of the sphere by the point of intersection, or pole, of its normal P. The angle ϕ between the two poles (0 0 1) and (1 1 1), shown in Figure 1.11b, can then be measured in degrees along the arc of the great circle between the poles P and P′. To represent all the planes in a crystal in this three-dimensional way is rather cumbersome; in the stereographic projection,

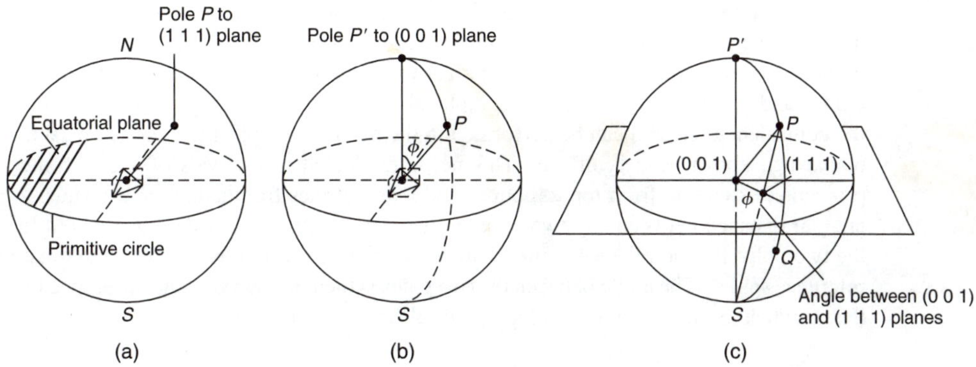

Figure 1.11 *Principles of stereographic projection, illustrating: (a) the pole P to a (1 1 1) plane; (b) the angle between two poles, P, P′; and (c) stereographic projection of P and P′ poles to the (1 1 1) and (0 0 1) planes, respectively.*

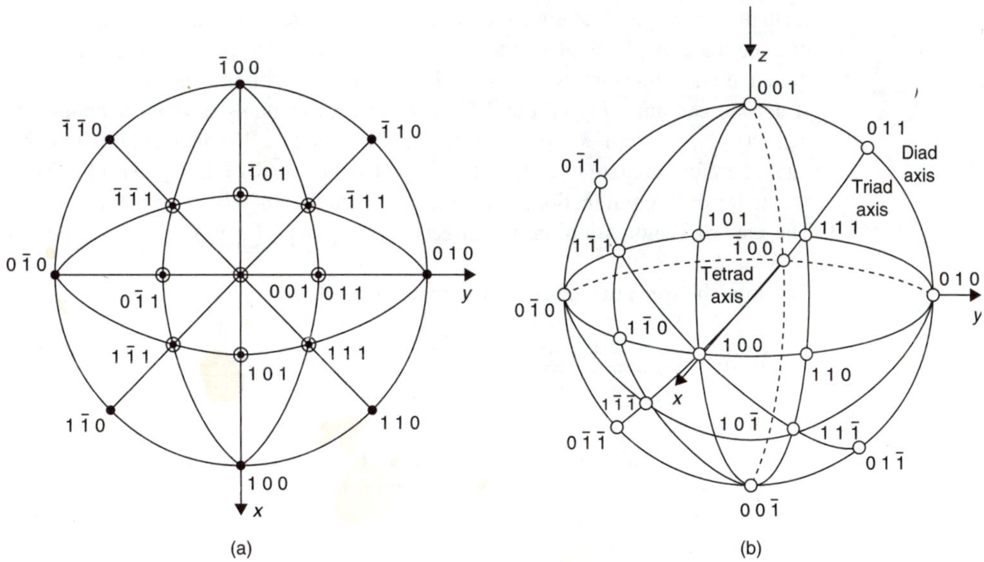

Figure 1.12 *Projections of planes in cubic crystals: (a) standard (0 0 1) stereographic projection and (b) spherical projection.*

the array of poles which represents the various planes in the crystal is projected from the reference sphere onto the equatorial plane. The pattern of poles projected on the equatorial, or primitive, plane then represents the stereographic projection of the crystal. As shown in Figure 1.11c, poles in the northern half of the reference sphere are projected onto the equatorial plane by joining the pole P to the south pole S, while those in the southern half of the reference sphere, such as Q, are projected in the same way in the direction of the north pole N. Figure 1.12a shows the stereographic projection of some simple cubic planes, {1 0 0}, {1 1 0} and {1 1 1}, from which it can be seen that those crystallographic planes which have poles in the southern half of the reference sphere are represented by circles in the stereogram, while those which have poles in the northern half are represented by dots.

As shown in Figure 1.11b, the angle between two poles on the reference sphere is the number of degrees separating them on the great circle passing through them. The angle between P and P′ can be determined by means of a hemispherical transparent cap graduated and marked with meridian circles and latitude circles, as in geographical work. With a stereographic representation of poles, the equivalent operation can be performed in the plane of the primitive circle by using a transparent planar net, known as a Wulff net. This net is graduated in intervals of 2°, with meridians in the projection extending from top to bottom and latitude lines from side to side.[8] Thus, to measure the angular distance between any two poles in the stereogram, the net is rotated about the center until the two poles lie upon the same meridian, which then corresponds to one of the great circles of the reference sphere. The angle between the two poles is then measured as the difference in latitude along the meridian. Some useful crystallographic rules may be summarized:

1. The Weiss Zone Law: the plane $(h\,k\,l)$ is a member of the zone $[u\,v\,w]$ if $hu+kv+lw=0$. A set of planes which all contain a common direction $[u\,v\,w]$ is known as a zone; $[u\,v\,w]$ is the zone axis (rather like the spine of an open book relative to the flat leaves). For example, the three planes $(1\ \bar{1}\ 0)$, $(0\ \bar{1}\ 1)$ and $(\bar{1}\ 0\ 1)$ form a zone about the [1 1 1] direction (Figure 1.12a). The pole of each plane containing $[u\,v\,w]$ must lie at 90° to $[u\,v\,w]$; therefore, these three poles all lie in the same plane and upon the same great circle trace. The latter is known as the zone circle or zone trace. A plane trace is to a plane as a zone circle is to a zone. Uniquely, in the cubic system alone, zone circles and plane traces with the same indices lie on top of one another.
2. If a zone contains $(h_1\,k_1\,l_1)$ and $(h_2\,k_2\,l_2)$ it also contains any linear combination of them, e.g. $m(h_1\,k_1\,l_1)+n(h_2\,k_2\,l_2)$. For example, the zone [1 1 1] contains $(1\ \bar{1}\ 0)$ and $(0\ 1\ \bar{1})$ and it must therefore contain $(1\ \bar{1}\ 0)+(0\ 1\ \bar{1})=(1\ 0\ \bar{1})$, $(1\ \bar{1}\ 0)+2(0\ 1\ \bar{1})=(1\ 1\ \bar{2})$, etc. The same is true for different directions in a zone, provided that the crystal is cubic.
3. The Law of Vector Addition: the direction $[u_1\,v_1\,w_1]+[u_2\,v_2\,w_2]$ lies between $[u_1\,v_1\,w_1]$ and $[u_2\,v_2\,w_2]$.
4. The angle between two directions is given by:

$$\cos\theta=\frac{u_1u_2+v_1v_2+w_1w_2}{\sqrt{(u_1^2+v_1^2+w_1^2)(u_2^2+v_2^2+w_2^2)}},$$

where $u_1\,v_1\,w_1$ and $u_2\,v_2\,w_2$ are the indices for the two directions. Provided that the crystal system is cubic, the angles between planes may be found by substituting the symbols h, k, l for u, v, w in this expression.

When constructing the standard stereogram of any crystal it is advantageous to examine the symmetry elements of that structure. As an illustration, consider a cubic crystal, since this has the highest symmetry of any crystal class. Close scrutiny shows that the cube has 13 axes of symmetry; these axes comprise three fourfold (tetrad) axes, four threefold (triad) axes and six twofold (diad) axes, as indicated in Figure 1.13a. (This diagram shows the standard square, triangular and lens-shaped symbols for the three types of symmetry axis.) An n-fold axis of symmetry operates in such a way that after rotation through an angle $2\pi/n$, the crystal comes into an identical or self-coincident position in space. Thus, a tetrad axis passes through the center of each face of the cube parallel to one of the edges, and a rotation of 90° in either direction about one of these axes turns the cube into a new position which is crystallographically indistinguishable from the old position. Similarly, the cube diagonals form a set of four threefold axes, and each of the lines passing through the center of opposite edges form a set

[8] A less-used alternative to the Wulff net is the polar net, in which the N–S axis of the reference sphere is perpendicular to the equatorial plane of projection.

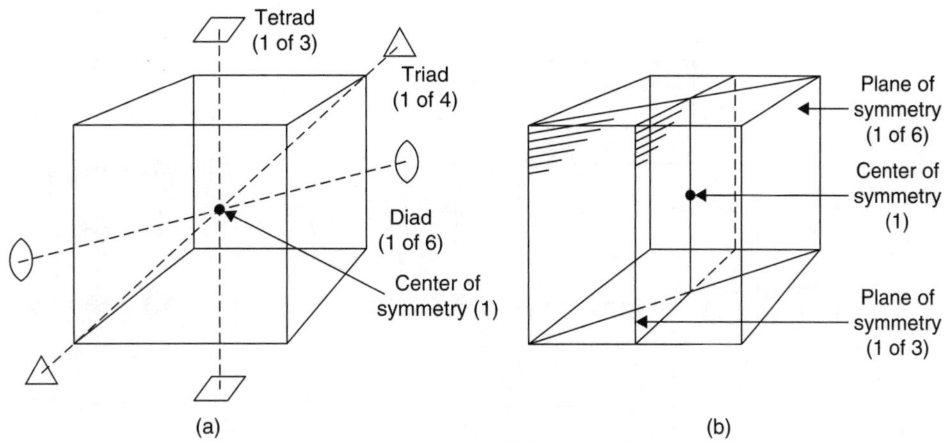

Figure 1.13 *Some elements of symmetry for the cubic system; total number of elements = 23.*

of six twofold symmetry axes. Some tetrad, triad and diad axes are marked on the spherical projection of the cubic crystal shown in Figure 1.12b. The cube also has nine planes of symmetry (Figure 1.13b) and one center of symmetry, giving, together with the axes, a total of 23 elements of symmetry.

In the stereographic projection of Figure 1.12a, planes of symmetry divide the stereogram into 24 equivalent spherical triangles, commonly called unit triangles, which correspond to the 48 (24 on the top and 24 on the bottom) seen in the spherical projection. The two-, three- and fourfold symmetry about the {1 1 0}, {1 1 1} and {1 0 0} poles, respectively, is apparent. It is frequently possible to analyze a problem in terms of a single unit triangle. Finally, reference to a stereogram (Figure 1.12a) confirms rule (2), which states that the indices of any plane can be found merely by adding simple multiples of other planes which lie in the same zone. For example, the (0 1 1) plane lies between the (0 0 1) and (0 1 0) planes and clearly 0 1 1 = 0 0 1 + 0 1 0. Owing to the action of the symmetry elements, it can be reasoned that there must be a total of 12{0 1 1} planes because of the respective three- and fourfold symmetry about the {1 1 1} and {1 0 0} axes. As a further example, it is clear that the (1 1 2) plane lies between the (0 0 1) plane and (1 1 1) plane, since 1 1 2 = 0 0 1 + 1 1 1 and the {1 1 2} form must contain 24 planes, i.e. an icositetrahedron. The plane (1 2 3), which is an example of the most general crystal plane in the cubic system because its hkl indices are all different, lies between the (1 1 2) and (0 1 1) planes; the 48 planes of the {1 2 3} form make up a hexak-isoctahedron.

The tetrahedral form, a direct derivative of the cubic form, is often encountered in materials science (Figure 1.14a). Its symmetry elements comprise four triad axes, three diad axes and six 'mirror' planes, as shown in the stereogram of Figure 1.14b.

Concepts of symmetry, when developed systematically, provide invaluable help in modern structural analysis. As already implied, there are three basic elements, or operations, of symmetry. These operations involve translation (movement along parameters a, b, c), rotation (about axes to give diads, triads, etc.) and reflection (across 'mirror' planes). Commencing with an atom (or group of atoms) at either a lattice point or at a small group of lattice points, a certain combination of symmetry operations will ultimately lead to the three-dimensional development of any type of crystal structure. The procedure provides a unique identifying code for a structure and makes it possible to locate it among 32 point groups and 230 space groups of symmetry. This classification obviously embraces the seven crystal systems. Although many metallic structures can be defined relatively simply in terms of space lattice and one or more lattice constants, complex structures require the key of symmetry theory.

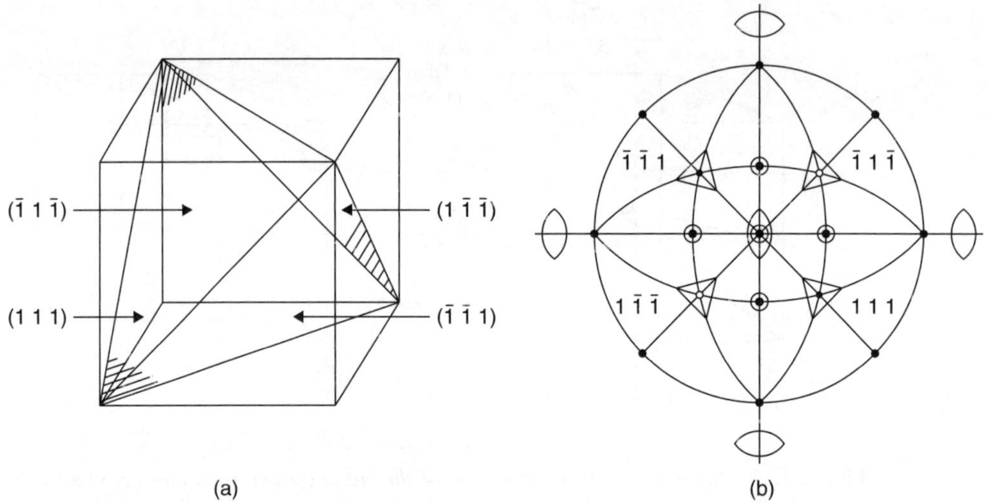

Figure 1.14 *Symmetry of the tetrahedral form.*

1.9 Selected crystal structures

1.9.1 Pure metals

We now examine the crystal structures of various elements (metallic and non-metallic) and compounds, using examples to illustrate important structure-building principles and structure–property relations.[9] Most elements in the Periodic Table are metallic in character; accordingly, we commence with them.

Metal ions are relatively small, with diameters in the order of 0.25 nm. A millimeter cube of metal therefore contains about 10^{20} atoms. The like ions in pure solid metal are packed together in a highly regular manner and, in the majority of metals, are packed so that ions collectively occupy the minimum volume. Metals are normally crystalline and for all of them, irrespective of whether the packing of ions is close or open, it is possible to define and express atomic arrangements in terms of structure cells (Section 1.6). Furthermore, because of the non-directional nature of the metallic bond, it is also possible to simulate these arrangements by simple 'hard-sphere' modeling.

There are two ways of packing spheres of equal size together so that they occupy the minimum volume. The structure cells of the resulting arrangements, face-centered cubic (fcc) and close-packed hexagonal (cph), are shown in Figures 1.15a and b. The other structure cell (Figure 1.15c) has a body-centered cubic (bcc) arrangement; although more 'open' and not based on close packing, it is nevertheless adopted by many metals.

In order to specify the structure of a particular metal completely, it is necessary to give not only the type of crystal structure adopted by the metal, but also the dimensions of the structure cell. In cubic structure cells it is only necessary to give the length of an edge a, whereas in a hexagonal cell the two parameters a and c must be given, as indicated in Figures 1.15a–c. If a hexagonal structure is ideally close packed, the ratio c/a must be 1.633. In hexagonal metal structures, the axial ratio c/a is never exactly 1.633. These structures are therefore never quite ideally close packed, e.g. $c/a(\text{Zn}) = 1.856$, $c/a(\text{Ti}) = 1.587$. As the axial ratio approaches unity, the properties of cph metals begin to show similarities to fcc metals.

[9] Where possible, compound structures of engineering importance have been selected as illustrative examples. Prototype structures, such as NaCl, ZnS, CaF_2, etc., which appear in standard treatments elsewhere, are indicated as appropriate.

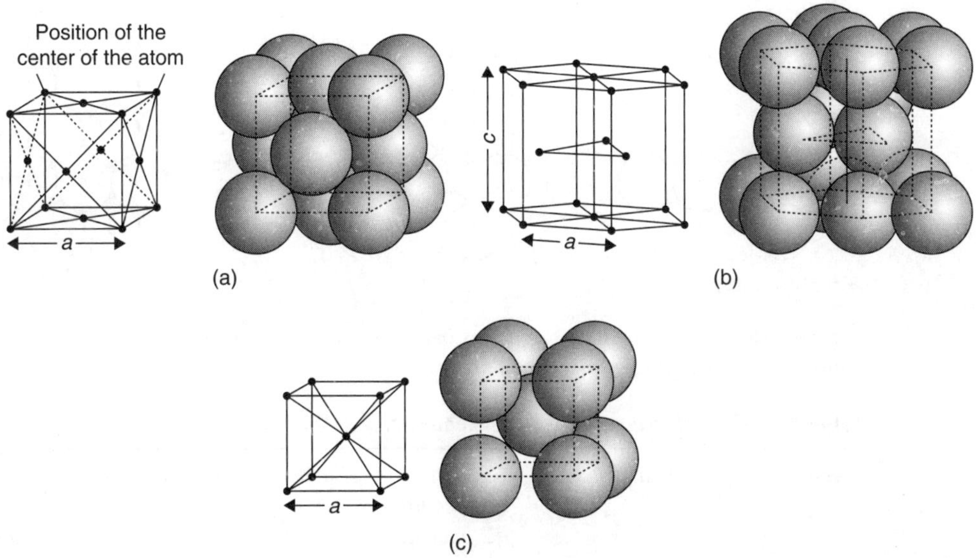

Figure 1.15 *Arrangement of atoms in: (a) face-centered cubic structure, (b) close-packed hexagonal structure and (c) body-centered cubic structure.*

A knowledge of cell parameters permits the atomic radius r of the metal atoms to be calculated on the assumption that they are spherical and that they are in closest possible contact. The reader should verify that in the fcc structure $r = (a\sqrt{2})/4$ and in the bcc structure $r = (a\sqrt{3})/4$, where a is the cell parameter.

The coordination number (CN), an important concept in crystal analysis, is defined as the number of nearest equidistant neighboring atoms around any atom in the crystal structure. Thus, in the bcc structure shown in Figure 1.15c the atom at the center of the cube is surrounded by eight equidistant atoms, i.e. CN = 8. It is perhaps not so readily seen from Figure 1.15a that the coordination number for the fcc structure is 12. Perhaps the easiest method of visualizing this is to place two fcc cells side by side, and then count the neighbors of the common face-centering atom. In the cph structure with ideal packing ($c/a = 1.633$) the coordination number is again 12, as can be seen by once more considering two cells, one stacked on top of the other, and choosing the center atom of the common basal plane. This (0 0 0 1) basal plane has the densest packing of atoms and has the same atomic arrangement as the closest packed plane in the fcc structure.[10]

The cph and fcc structures represent two effective methods of packing spheres closely; the difference between them arises from the different way in which the close-packed planes are stacked. Figure 1.16a shows an arrangement of atoms in **A**-sites of a close-packed plane. When a second plane of close-packed atoms is laid down, its first atom may be placed in either a **B**-site or a **C**-site, which are entirely equivalent. However, once the first atom is placed in one of these two types of site, all other atoms in the second plane must be in similar sites. (This is because neighboring **B**- and **C**-sites are too close together for both to be occupied in the same layer.) At this stage there is no difference between the cph and fcc structure; the difference arises only when the third layer is put in position. In building up the third layer, assuming that sites of type **B** have been used to construct the second layer, as shown in Figure 1.16b, either **A**-sites or **C**-sites may be selected. If **A**-sites are chosen, then the atoms in the third layer will be directly above those in the first layer, and the structure will be cph, whereas if **C**-sites

[10] The Miller indices for the closest packed (octahedral) planes of the fcc structure are {1 1 1}; these planes are best revealed by balancing a ball-and-stick model of the fcc cell on one corner.

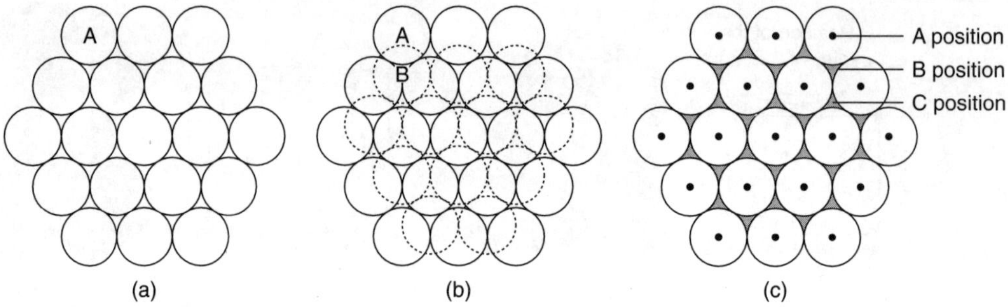

Figure 1.16 *(a) Arrangements of atoms in a close-packed plane. (b) Registry of two close-packed planes. (c) The stacking of successive planes.*

Table 1.4 *Crystal structures of some metals at room temperature.*

Element	*Crystal structure*	*Closest interatomic distance (nm)*	*Element*	*Crystal structure*	*Closest interatomic distance (nm)*
Aluminum	fcc	0.286	Platinum	fcc	0.277
Beryllium	cph ($c/a = 1.568$)	0.223	Potassium	bcc	0.461
Cadmium	cph ($c/a = 1.886$)	0.298	Rhodium	fcc	0.269
Chromium	bcc	0.250	Rubidium	bcc	0.494
Cobalt	cph ($c/a = 1.623$)	0.250	Silver	fcc	0.289
Copper	fcc	0.255	Sodium	bcc	0.372
Gold	fcc	0.288	Tantalum	bcc	0.286
Iron	bcc	0.248	Thorium	fcc	0.360
Lead	fcc	0.350	Titanium	cph ($c/a = 1.587$)	0.299
Lithium	bcc	0.331	Tungsten	bcc	0.274
Magnesium	cph ($c/a = 1.623$)	0.320	Uranium	orthorhombic	0.275
Molybdenum	bcc	0.275	Vanadium	bcc	0.262
Nickel	fcc	0.249	Zinc	cph ($c/a = 1.856$)	0.266
Niobium	bcc	0.286	Zirconium	cph ($c/a = 1.592$)	0.318

are chosen this will not be the case and the structure will be fcc. Thus, a cph structure consists of layers of close-packed atoms stacked in the sequence of **ABABAB** or, of course, equally well, **ACACAC**. An fcc structure has the stacking sequence **ABCABCABC** so that the atoms in the fourth layer lie directly above those in the bottom layer. The density of packing within structures is sometimes expressed as an atomic packing fraction (APF), which is the fraction of the cell volume occupied by atoms. The APF value for a bcc cell is 0.68; it rises to 0.74 for the more closely packed fcc and cph cells.

Table 1.4 gives the crystal structures adopted by some typical metals, the majority of which are either fcc or bcc. As indicated previously, an atom does not have precise dimensions; however, it is convenient to express atomic diameters as the closest distance of approach between atom centers. Table 1.4 lists structures that are stable at room temperature; at other temperatures, some metals undergo transition and the atoms rearrange to form a different crystal structure, each structure being stable over a definite interval of temperature. This phenomenon is known as allotropy. The best-known commercially exploitable example is that of iron, which is bcc at temperatures below 910°C, fcc in the temperature range 910–1400°C, and bcc at temperatures between 1400°C and the melting point (1535°C). Other common examples include titanium and zirconium, which change from cph to bcc at temperatures of 882°C and 815°C, respectively; tin, which changes from cubic (gray) to tetragonal (white) at 13.2°C; and the metals uranium and plutonium. Plutonium is particularly complex in that it has six different allotropes between room temperature and its melting point of 640°C.

These transitions between allotropes are usually reversible and, because they necessitate rearrangement of atoms, are accompanied by volume changes and either the evolution or absorption of thermal energy. The transition can be abrupt but is often sluggish. Fortunately, tetragonal tin can persist in a metastable state at temperatures below the nominal transition temperature. However, the eventual transition to the friable low-density cubic form can be very sudden.[11]

Using the concept of a unit cell, together with data on the atomic mass of constituent atoms, it is possible to derive a theoretical value for the density of a pure single crystal. The parameter a for the bcc cell of pure iron at room temperature is 0.28664 nm. Hence the volume of the unit cell is 0.02355 nm^3. Contrary to first impressions, the bcc cell contains two atoms, i.e. $(8 \times \frac{1}{8}$ atom$) + 1$ atom. Using the Avogadro constant N_A,[12] we can calculate the mass of these two atoms as $2(55.85/N_A)$ or 185.46×10^{-24} kg, where 55.85 is the relative atomic mass of iron. The theoretical density (mass/volume) is thus 7875 kg m^{-3}. The reason for the slight discrepancy between this value and the experimentally determined value of 7870 kg m^{-3} will become evident when we discuss crystal imperfections in Chapter 3.

Worked examples

(1) Calculate what fraction of the structure is occupied by atoms, i.e. atomic packing factor, in (a) fcc and (b) bcc, assuming the atoms behave as hard spheres.
(2) What are the positions (coordinates) of the largest interstices in (a) fcc and (b) bcc? How many are there in each unit cell? Why does carbon not occupy the large interstice in iron?
(3) Draw a cph structure cell and label the $(0\,0\,0\,1)$, $(1\,\bar{1}\,0\,0)$ and $(2\,\bar{1}\,\bar{1}\,1)$ planes. Mark in a $\langle 1\,1\,\bar{2}\,0\rangle$ and $\langle 1\,0\,\bar{1}\,0\rangle$ type direction. Magnesium has almost ideal c/a and has a density 1.74 Mg m^{-3} and relative atomic mass 24.31. Calculate the values of c and a.

Solutions

(1) In the fcc, volume of unit cell $= a^3$, occupied by four atoms.

$$\text{Volume of atoms} = 4 \times \frac{4}{3}\pi r^3$$

$$r = \sqrt{2}a/4$$

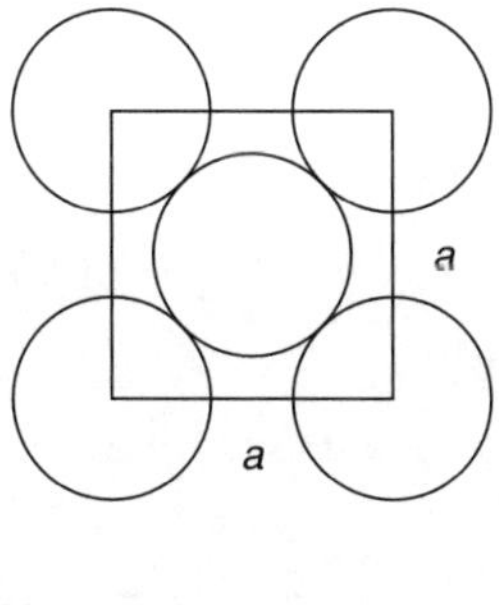

[11] Historical examples of 'tin plague' abound (e.g. buttons, coins, organ pipes, statues).

[12] The Avogadro constant N_A is 0.602217×10^{-24} mol^{-1}. The mole is a basic SI unit. It does not refer to mass and has been likened to terms such as dozen, score, gross, etc. By definition, it is the amount of substance which contains as many elementary units as there are atoms in 0.012 kg of carbon-12. The elementary unit must be specified and may be an atom, a molecule, an ion, an electron, a photon, etc. or a group of such entities.

$$\text{Atomic packing factor} = \frac{16\pi}{3a^3} \times \frac{2\sqrt{2}a^3}{64} = \frac{\pi}{3\sqrt{2}} = 0.74.$$

For bcc, volume of unit cell $= a^3$, occupied by two atoms.

$$\text{Volume of atoms} = 2 \times \frac{4}{3}\pi r^3, \text{ where } r = \sqrt{3}a/4.$$

$$\text{Atomic packing factor} = \frac{8\pi}{3a^3} \times \frac{3\sqrt{3}a^3}{64} = \frac{\pi\sqrt{3}}{8} = 0.68.$$

(2) Fcc has a hole in the center of the cell $[\frac{1}{2}, \frac{1}{2}, \frac{1}{2}]$ and at the middle of each edge $[\frac{1}{2}, 0, 0]$. The number of interstices $= 1 + \frac{1}{4} \times 12 = 4$.

For bcc, the coordinates are $(0, \frac{1}{2}, \frac{1}{4})$ and there are four such sites on each face, which are shared by two unit cells, so number of interstices $= \frac{1}{2} \times 4 \times 6 = 12$.

Carbon does not occupy the tetrahedral site because it has to squeeze out four symmetrically arranged atoms. Instead, it occupies the octahedral site $(0, 0, \frac{1}{2})$ because it pushes out only two atoms.

(3) (i)

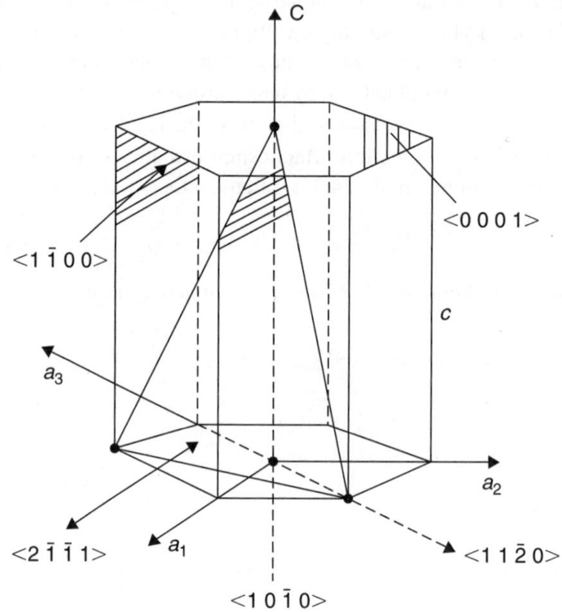

(ii) Volume of cell $= 3 \times a^2 \times \frac{\sqrt{3}}{2} \times c$.

$$\text{Density} = \frac{2 \times 6 \times 24.31}{6.02 \times 10^{23} \times 3\sqrt{3}a^2c} = 1.74\,\text{Mg m}^{-3}, \text{ so } a^2c = 5.36 \times 10^{-29}\,\text{m}^3.$$

Ideal $c/a = 2\sqrt{2/3} = 1.633$.

$a^3 = 5.36 \times 10^{-29} \times 1.633$, so $a = 3.2 \times 10^{-10}$ m, $c = 5.2 \times 10^{-10}$ m.

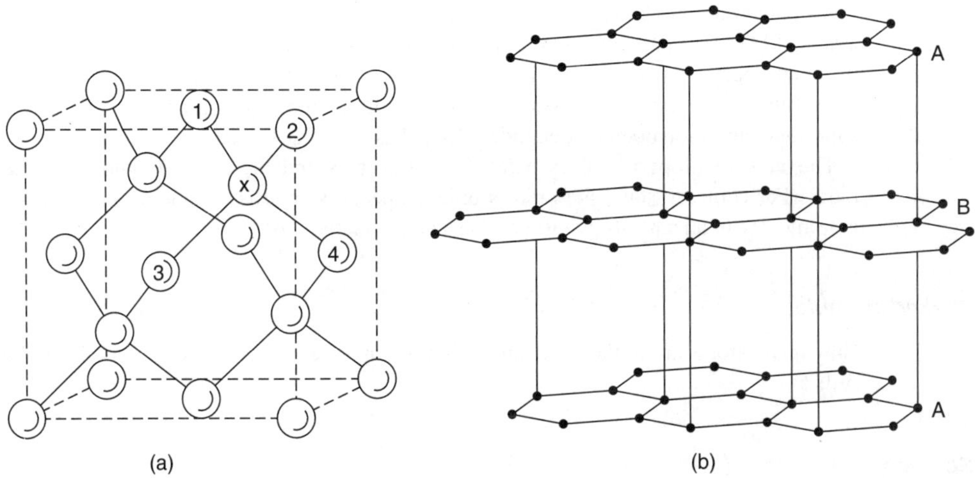

Figure 1.17 *Two crystalline forms of carbon: (a) diamond and (b) graphite (from Kingery, Bowen and Uhlmann, 1976; by permission of Wiley-Interscience).*

1.9.2 Diamond and graphite

It is remarkable that a single element, carbon, can exist in numerous forms (see Section 10.5) including diamond, graphite and fullerene-based structures. Of these, diamond and graphite are most well known and are therefore described here. Diamond is transparent and one of the hardest materials known, finding wide use, notably as an abrasive and cutting medium. Graphite finds general use as a solid lubricant and writing medium (pencil 'lead'). It is now often classed as a highly refractory ceramic because of its strength at high temperatures and excellent resistance to thermal shock.

We can now progress from the earlier representation of the diamond structure (Figure 1.3c) to a more realistic version. Although the structure consists of two interpenetrating fcc substructures, in which one substructure is slightly displaced along the body diagonal of the other, it is sufficient for our purpose to concentrate on a representative structure cell (Figure 1.17a). Each carbon atom is covalently bonded to four equidistant neighbors in regular tetrahedral[13] coordination (CN = 4). For instance, the atom marked X occupies a 'hole', or interstice, at the center of the group formed by atoms marked 1, 2, 3 and 4. There are eight equivalent tetrahedral sites of the X-type, arranged four-square within the fcc cell; however, in the case of diamond, only half of these sites are occupied. Their disposition, which also forms a tetrahedron, maximizes the intervening distances between the four atoms. If the fcc structure of diamond depended solely upon packing efficiency, the coordination number would be 12; actually CN = 4, because only four covalent bonds can form. Silicon ($Z = 14$), germanium ($Z = 32$) and gray tin ($Z = 50$) are fellow members of Group IV in the Periodic Table and are therefore also tetravalent. Their crystal structures are identical in character, but obviously not in dimensions, to the diamond structure of Figure 1.17a.

Graphite is less dense and more stable than diamond. In direct contrast to the cross-braced structure of diamond, graphite has a highly anisotropic layer structure (Figure 1.17b). Adjacent layers in the **ABABAB** sequence are staggered; the structure is not cph. A less stable rhombohedral **ABCABC** sequence has been observed in natural graphite. Charcoal, soot and lampblack have been termed

[13] The stability and strength of a tetrahedral form holds a perennial appeal for military engineers: spiked iron caltrops deterred attackers in the Middle Ages and concrete tetrahedra acted as obstacles on fortified Normandy beaches in World War II.

'amorphous carbon'; actually they are microcrystalline forms of graphite. Covalently bonded carbon atoms, 0.1415 nm apart, are arranged in layers of hexagonal symmetry. These layers are approximately 0.335 nm apart. This distance is relatively large and the interlayer forces are therefore weak. Layers can be readily sheared past each other, thus explaining the lubricity of graphitic carbon. (An alternative solid lubricant, molybdenum disulfide, MoS_2, has a similar layered structure.)

The ratio of property values parallel to the a-axis and the c-axis is known as the anisotropy ratio. (For cubic crystals, the ratio is unity.) Special synthesis techniques can produce near-ideal graphite[14] with an anisotropy ratio of thermal conductivity of 200.

Worked example

How many atoms are in the cubic unit cell of diamond and GaAs? How many are in their primitive cells?

Solution

Diamond is fcc with two atoms per lattice point, the two atoms being separated by $\frac{1}{4}\langle 1\ 1\ 1\rangle$. There are four lattice points per cubic unit cell in fcc and therefore there are eight atoms per cubic unit cell in diamond. GaAs is similar to diamond, but the pair of atoms differ. There are therefore still eight atoms per cubic unit cell.

The primitive cell of fcc contains one lattice point and one atom. That of diamond therefore contains two atoms, as does that of GaAs.

1.9.3 Coordination in ionic crystals

We have seen in the case of diamond how the joining of four carbon atoms outlines a tetrahedron which is smaller than the structure cell (Figure 1.17a). Before examining some selected ionic compounds, it is necessary to develop this aspect of coordination more fully. This approach to structure building concerns packing and is essentially a geometrical exercise. It is subordinate to the more dominant demands of covalent bonding.

In the first of a set of conditional rules, assembled by Pauling, the relative radii of cation (r) and anion (R) are compared. When electrons are stripped from the outer valence shell during ionization, the remaining electrons are more strongly attracted to the nucleus; consequently, cations are usually smaller than anions. **Rule I** states that the coordination of anions around a reference cation is determined by the geometry necessary for the cation to remain in contact with each anion. For instance, in Figure 1.18a, a radius ratio r/R of 0.155 signifies touching contact when three anions are grouped about a cation. This critical value is readily derived by geometry. If the r/R ratio for threefold coordination is less than 0.155 then the cation 'rattles' in the central interstice, or 'hole', and the arrangement is unstable. As r/R exceeds 0.155 then structural distortion begins to develop.

In the next case, that of fourfold coordination, the 'touching' ratio has a value of 0.225 and joining of the anion centers defines a tetrahedron (Figure 1.18b). For example, silicon and oxygen ions have radii of 0.039 and 0.132 nm, respectively, hence $r/R = 0.296$. This value is slightly greater than the critical value of 0.225 and it follows that tetrahedral coordination gives a stable configuration; indeed, the complex anion SiO_4^{4-} is the key structural feature of silica, silicates and silica glasses. The quadruple negative charge is due to the four unsatisfied oxygen bonds which project from the group.

[14] Applications range from rocket nozzles to bowl linings for tobacco pipes.

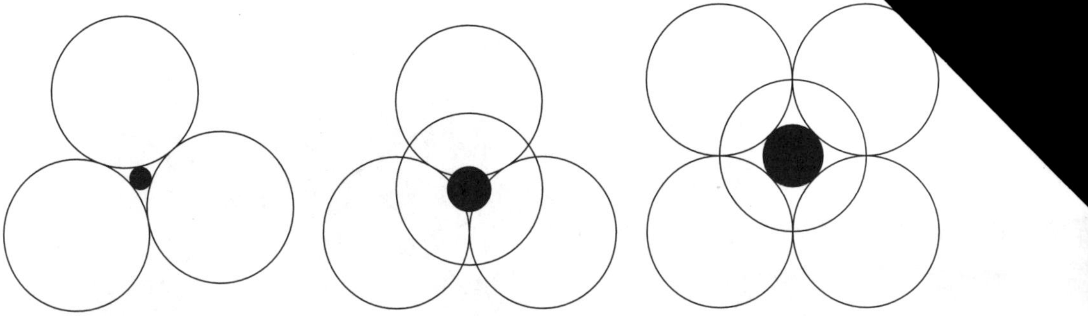

Figure 1.18 *Nesting of cations within anionic groups.*

Table 1.5 *Relation between radius ratio and coordination.*

r/R	*Maximum coordination number (CN)*	*Form of coordination*
<0.155	2	Linear
0.155–0.225	3	Equilateral triangle
0.225–0.414	4	Regular tetrahedron
0.414–0.732	6	Regular octahedron
0.732–1.0	8	Cube
1.00	12	Cuboctahedron

In a feature common to many structures, the tendency for anions to distance themselves from each other as much as possible is balanced by their attraction towards the central cation. Each of the four oxygen anions is only linked by one of its two bonds to the silicon cation, giving an effective silicon/oxygen ratio of 1:2 and thus confirming the stoichiometric chemical formula for silica, SiO_2. Finally, as shown in Figure 1.18c, the next coordination polyhedron is an octahedron for which $r/R = 0.414$. It follows that each degree of coordination is associated with a nominal range of r/R values, as shown in Table 1.5. Caution is necessary in applying these ideas of geometrical packing because (1) range limits are approximative, (2) ionic radii are very dependent upon CN, (3) ions can be non-spherical in anisotropic crystals and (4) considerations of covalent or metallic bonding can be overriding. The other four Pauling rules are as follows.

- **Rule II.** In a stable coordinated structure the total valency of the anion equals the summated bond strengths of the valency bonds which extend to this anion from all neighboring cations. Bond strength is defined as the valency of an ion divided by the actual number of bonds; thus, for Si^{4+} in tetrahedral coordination it is $\frac{4}{4} = 1$. This valuable rule, which expresses the tendency of each ion to achieve localized neutrality by surrounding itself with ions of opposite charge, is useful in deciding the arrangement of cations around an anion. For instance, the important ceramic barium titanate ($BaTiO_3$) has Ba^{2+} and Ti^{4+} cations bonded to a common O^{2-} anion. Given that the coordination numbers of O^{2-} polyhedra centered on Ba^{2+} and Ti^{4+} are 12 and 6, respectively, we calculate the corresponding strengths of the Ba–O and Ti–O bonds as $\frac{2}{12} = \frac{1}{6}$ and $\frac{4}{6} = \frac{2}{3}$. The

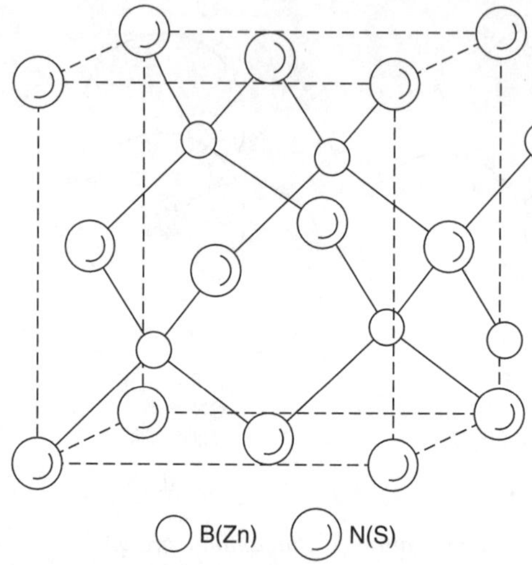

Figure 1.19 *Zinc blende (α-ZnS) structure, prototype for cubic boron nitride (BN) (from Kingery, Bowen and Uhlmann, 1976; by permission of Wiley-Interscience).*

valency of the shared anion is 2, which is numerically equal to $(4 \times \frac{1}{6}) + (2 \times \frac{2}{3})$. Accordingly, coordination of the common oxygen anion with four barium cations and two titanium cations is a viable possibility.

- **Rule III.** An ionic structure tends to have maximum stability when its coordination polyhedra share corners; edge- and face-sharing give less stability. Any arrangement which brings the mutually repelling central cations closer together tends to destabilize the structure. Cations of high valency (charge) and low CN (poor 'shielding' by surrounding anions) aggravate the destabilizing tendency.
- **Rule IV.** In crystals containing different types of cation, cations of high valency and low CN tend to limit the sharing of polyhedra elements; for instance, such cations favor corner sharing rather than edge sharing.
- **Rule V.** If several alternative forms of coordination are possible, one form usually applies throughout the structure. In this way, ions of a given type are more likely to have identical surroundings.

In conclusion, it is emphasized that the Pauling rules are only applicable to structures in which ionic bonding predominates. Conversely, any structure which fails to comply with the rules is extremely unlikely to be ionic.

The structure of the mineral zinc blende (α-ZnS) shown in Figure 1.19 is often quoted as a prototype for other structures. In accord with the radius ratio $r/R = 0.074/0.184 = 0.4$, tetrahedral coordination is a feature of its structure. Coordination tetrahedra share only corners (vertices). Thus, one species of ion occupies four of the eight tetrahedral sites within the cell. These sites have been mentioned previously in connection with diamond (Section 1.9.2); in that case, the directional demands of the covalent bonds between like carbon atoms determined their location. In zinc sulfide, the position of unlike ions is determined by geometrical packing. Replacement of the Zn^{2+} and S^{2-} ions in the prototype cell with boron and nitrogen atoms produces the structure cell of cubic boron nitride (BN). This compound is extremely hard and refractory and, because of the adjacency of boron ($Z = 5$) and nitrogen ($Z = 7$) to carbon ($Z = 6$) in the Periodic Table, is more akin in character to diamond than to zinc sulfide. Its angular crystals serve as an excellent grinding abrasive for hardened steel.

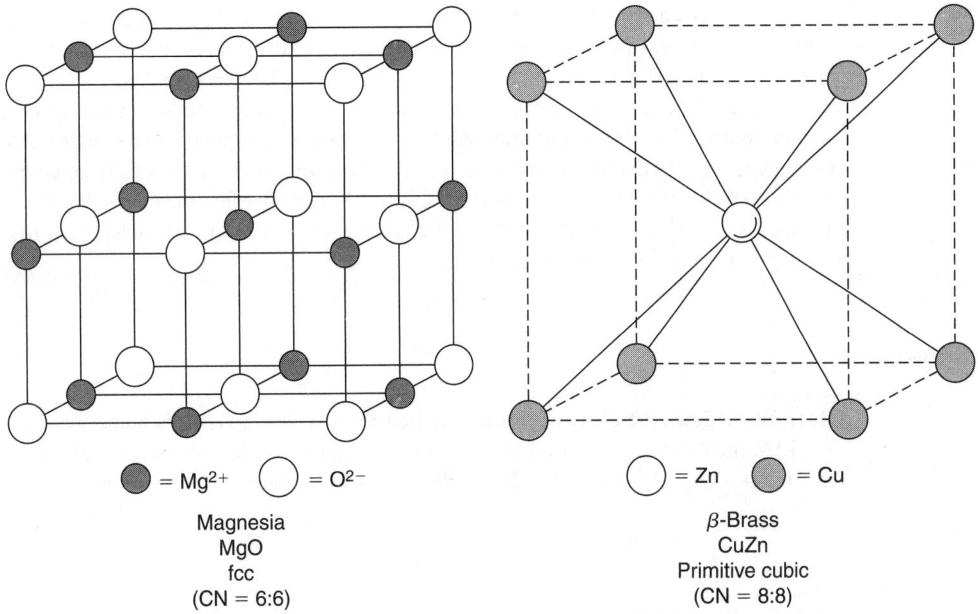

Figure 1.20 *AB-type compounds (from Kingery, Bowen and Uhlmann, 1976; by permission of Wiley-Interscience).*

The precursor for cubic boron nitride is the more common and readily prepared form, hexagonal boron nitride.[15]

This hexagonal form is obtained by replacing the carbon atoms in the layered graphite structure (Figure 1.17b) alternately with boron and nitrogen atoms, and also slightly altering the stacking registry of the layer planes. It feels slippery like graphite and is sometimes called 'white graphite'. Unlike graphite, it is an insulator, having no free electrons.

Another abrasive medium, silicon carbide (SiC), can be represented in one of its several crystalline forms by the zinc blende structure. Silicon and carbon are tetravalent and the coordination is tetrahedral, as would be expected.

1.9.4 AB-type compounds

An earlier diagram (Figure 1.3b) schematically portrayed the ionic bonding within magnesium oxide (periclase). We can now develop a more realistic model of its structure and also apply the ideas of coordination. Generically, MgO is a sodium chloride-type structure (Figure 1.20a), with Mg^{2+} cations and O^{2-} anions occupying two interpenetrating[16] fcc sublattices. Many oxides and halides have this type of structure (e.g. CaO, SrO, BaO, VO, CdO, MnO, FeO, CoO, NiO; NaCl, NaBr, NaI, NaF, KCl, etc.). The ratio of ionic radii $r/R = 0.065/0.140 = 0.46$ and, as indicated by Table 1.5, each Mg^{2+} cation is octahedrally coordinated with six larger O^{2-} anions, and vice versa (CN = 6:6).

[15] The process for converting hexagonal BN to cubic BN (*Borazon*) involves very high temperature and pressure, and was developed by Dr R. H. Wentorf at the General Electric Company, USA (1957).

[16] Sublattices can be discerned by concentrating on each array of like atoms (ions) in turn.

Octahedra of a given type share edges. The 'molecular' formula MgO indicates that there is an exact stoichiometric balance between the numbers of cations and anions; more specifically, the unit cell depicted contains $(8 \times \frac{1}{8}) + (6 \times \frac{1}{2}) = 4$ cations and $(12 \times \frac{1}{4}) + 1 = 4$ anions.

The second example of an AB-type compound is the hard intermetallic compound CuZn (β-brass) shown in Figure 1.20b. It has a cesium chloride-type structure in which two simple cubic sublattices interpenetrate. Copper ($Z = 29$) and zinc ($Z = 30$) have similar atomic radii. Each copper atom is in eightfold coordination with zinc atoms, thus CN = 8:8. The coordination cubes share faces. Each unit cell contains $(8 \times \frac{1}{8}) = 1$ corner atom and 1 central atom; hence the formula CuZn. In other words, this compound contains 50 at.% copper and 50 at.% zinc.

Problems

1.1 (i) Write down the elements for which the energy of the $3d$-level is higher than the $4s$.
(ii) Write down the electronic structure of potassium and bromine in the solid KBr.

1.2 For an $Na^+–Cl^-$ ion pair with separation r nm, the potential energy is given by:

$$U(r) = -\frac{1.436}{r} + \frac{7.32 \times 10^{-6}}{r^8} \text{ eV.}$$

For solid NaCl, estimate the (i) equilibrium ion spacing, (ii) Young's modulus and (iii) the bonding energy per ion pair.

1.3 In an ionically bonded ceramic the force between two ions is given by:

$$F = \frac{b}{r^2} - \frac{c}{r^{10}}$$

(r is the distance between the ions, b and c are constants, $b = 1.2 \times 10^{-28}$ Nm^2 and $c = 1.8 \times 10^{-105}$ Nm^{10}). Estimate in a simple way the equilibrium ion separation in the solid and the energy required to separate the ions in the solid to an infinite distance apart. What physical properties would you measure in order to check your answers and how would you measure them?

1.4 If the density of silicon is 2.55 Mg m^{-3}, what is its lattice parameter? (The relative atomic mass of silicon is 28.09 and Avogadro's number is 6.02×10^{23}.)

1.5 If atoms in silicon are considered to be solid spheres which touch, what fraction of space is occupied by matter? If the lattice parameter is 0.543 nm, what is the atom radius?

1.6 For the atom at $\frac{1}{4}, \frac{1}{4}, \frac{1}{4}$ in the unit cell of diamond, enumerate the vectors to its nearest neighbors.

1.7 (a) Taking the atomic radius of nickel to be 0.1246 nm, calculate the lattice parameter a.
(b) From the relative atomic mass, i.e. atomic weight (58.7 g mol^{-1}) and Avogadro's constant 6.02×10^{23}, calculate the theoretical density of nickel.

1.8 MgO is an ionic solid with Mg^{2+} cations and O^{2-} anions occupying two interpenetrating fcc sub-lattices. The ionic radius of Mg is 0.065 nm and that of oxygen 0.140 nm. Determine the coordination of Mg and O atoms and explain the value in terms of the ionic ratio.

Further reading

Barrett, C. S. and Massalski, T. B. (1966). *Structure of Metals*, 3rd edn. McGraw-Hill, New York.
Brydson, J. A. (1989). *Plastics Materials*, 5th edn. Butterworths, London.
Cottrell, A. H. (1975). *Introduction to Metallurgy*. Edward Arnold, London.

Evans, R. C. (1966). *An Introduction to Crystal Chemistry*, 2nd edn. Cambridge University Press, Cambridge.

Huheey, J. E. (1983). *Inorganic Chemistry*, 3rd edn. Harper & Row, New York.

Hume-Rothery, W., Smallman, R. E. and Haworth, C. W. (1975). *The Structure of Metals and Alloys*, 5th edn (1988 reprint). Institute of Materials, London.

Kelly, A. and Groves, G. W. (1973). *Crystallography and Crystal Defects*. Longmans, Harlow.

Kingery, W. D., Bowen, H. K. and Uhlmann, D. R. (1976). *Introduction to Ceramics*, 2nd edn. John Wiley, Chichester.

Puddephatt, R. J. and Monaghan, P. K. (1986). *The Periodic Table of the Elements*. Clarendon Press, Oxford.

van Vlack, L. H. (1985). *Elements of Materials Science*, 5th edn. Addison-Wesley, Reading, MA.

Schmitt $\cos\lambda \cos\phi = .1749 \quad \times \quad 13.7\ MPa$

$\cos\phi = [321](111)/\sqrt{14 \times 3}$

$\cos\lambda = (321)(1\bar{1}0)/\sqrt{14 \times 2}$

$b = \frac{a\sqrt{3}}{2} = 2.8 \times 10^{-10} \sqrt{3}/2 = 2.4 \times 10^{-10}$

$v = 4.06 \times 10^{-2}\ m/s$

UYP LYP

Tensile strain $\quad v = \left(\frac{\sigma}{\sigma_0}\right)^n$

$0.5 \rho_1 b v_1 = 0.5 \rho_2 b v_2 \Rightarrow \rho_1 \left(\frac{\sigma_1}{\sigma_0}\right)^n = \rho_2 \left(\frac{\sigma_2}{\sigma_0}\right)^n$

$\frac{\rho_1}{\rho_2} = \left(\frac{\sigma_2}{\sigma_1}\right)^n \quad = \quad \frac{\sigma_2}{\sigma_1} = \left(\frac{10^{14}}{10^{11}}\right)^{1/n} = 10^{3/3.5} = 1.2$

$P = T(K)[20 + \log t_r]$

$T(20 + \log t_r) = K$

Chapter 2

Phase equilibria and structure

2.1 Crystallization from the melt

2.1.1 Freezing of a pure metal

At some stage of production the majority of metals and alloys are melted and then allowed to solidify as a casting. The latter may be an intermediate product, such as a large steel ingot suitable for hotworking, or a complex final shape, such as an engine cylinder block of cast iron or a single-crystal gas-turbine blade of superalloy. Solidification conditions determine the structure, homogeneity and soundness of cast products and the governing scientific principles find application over a wide range of fields. For instance, knowledge of the solidification process derived from the study of conventional metal casting is directly relevant to many fusion-welding processes, which may be regarded as 'casting in miniature', and to the fusion casting of oxide refractories. The liquid/solid transition is obviously of great scientific and technological importance.

First, in order to illustrate some basic principles, we will consider the freezing behavior of a melt of like metal atoms. The thermal history of a slowly cooling metal is depicted in Figure 2.1; the plateau on the curve indicates the melting point (m.p.), which is pressure dependent and specific to the metal. Its value relates to the bond strength of the metal. Thus, the drive to develop strong alloys for service at high temperatures has stimulated research into new and improved ways of casting high-m.p. alloys based upon iron, nickel or cobalt.

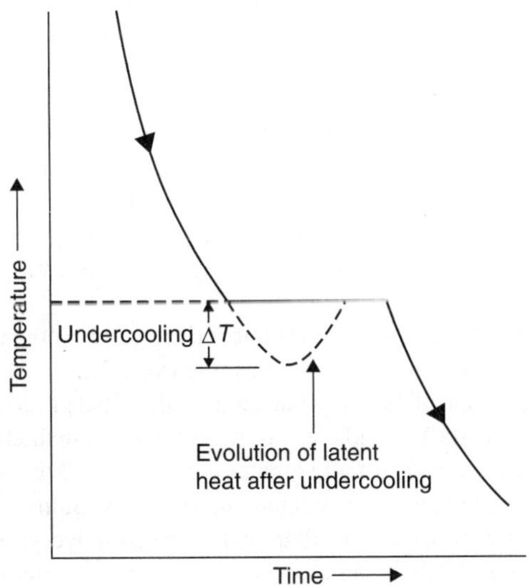

Figure 2.1 *Cooling curve for a pure metal showing possible undercooling.*

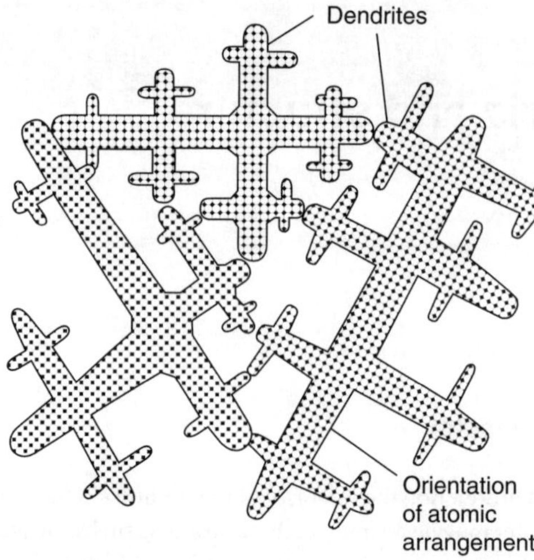

Figure 2.2 *Schematic diagram of three dendrites interlocking.*

The transition from a highly disordered liquid to an ordered solid is accompanied by a lowering in the energy state of the metal and the release of thermal energy (latent heat of solidification), forming the arrest on the cooling curve shown in Figure 2.1. This ordering has a marked and immediate effect upon other structure-sensitive properties of the metal; for instance, the volume typically decreases by 1–6%, the electrical conductivity rises and the diffusivity, or ability of the atoms to migrate, falls.

Solidification is a classic example of a nucleation and growth process. In the general case of freezing within the bulk of pure molten metal, minute crystalline nuclei form independently at random points. After this homogeneous form of nucleation, continued removal of thermal energy from the system causes these small crystalline regions to grow independently at the expense of the surrounding melt. Throughout the freezing process, there is a tendency for bombardment by melt atoms to destroy embryonic crystals; only nuclei which exceed a critical size are able to survive. Rapid cooling of a pure molten metal reduces the time available for nuclei formation and delays the onset of freezing by a temperature interval of ΔT. This thermal undercooling (or supercooling), which is depicted in Figure 2.1, varies in extent, depending upon the metal and conditions, but can be as much as 0.1–$0.3T_m$, where T_m is the absolute melting point. However, commercial melts usually contain suspended insoluble particles of foreign matter (e.g. from the refractory crucible or hearth), which act as seeding nuclei for so-called heterogeneous nucleation. Undercooling is much less likely under these conditions; in fact, very pronounced undercooling is only obtainable when the melt is very pure and extremely small in volume. Homogeneous nucleation is not encountered in normal foundry practice.

The growing crystals steadily consume the melt and eventually impinge upon each other to form a structure of equiaxed (equal-sized) grains (Figures 2.2 and 2.3). Heterogeneous nucleation, by providing a larger population of nuclei, produces a smaller final grain size than homogeneous nucleation. The resultant grain (crystal) boundaries are several atomic diameters wide. The angle of misorientation between adjacent grains is usually greater than 10–15°. Because of this misfit, such high-angle grain boundaries have a higher energy content than the bulk grains and, on reheating, will tend to melt first. (During a grain-contrast etch of diamond-polished polycrystalline metal, the etchant attacks grain boundaries preferentially by an electrochemical process, producing a broad 'canyon' which scatters vertically incident light during normal microscopical examination. The boundary then appears as a black line.)

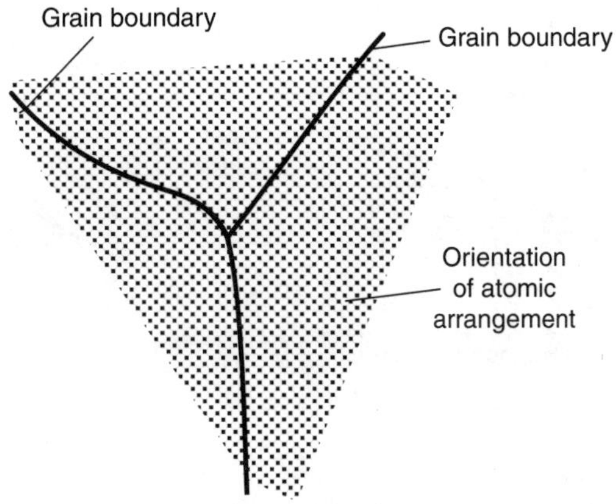

Figure 2.3 *Formation of grains from dendrites of Figure 2.2.*

During the freezing of many metals (and alloys), nucleated crystals grow preferentially in certain directions, causing each growing crystal to assume a distinctive, non-faceted[1] tree-like form, known as a dendrite (Figure 2.2). In cubic crystals, the preferred axes of growth are ⟨1 0 0⟩ directions. As each dendritic spike grows, latent heat is transferred into the surrounding liquid, preventing the formation of other spikes in its immediate vicinity. The spacing of primary dendrites and of dendritic arms therefore tends to be regular. Ultimately, as the various crystals impinge upon each other, it is necessary for the interstices of the dendrites to be well fed with melt if interdendritic shrinkage cavities are to be prevented from forming. Convection currents within the cooling melt are liable to disturb the delicate dendritic branches and produce slight angular misalignments in the final solidified structure (e.g. 5–10°). These low-angle boundaries form a lineage (macromosaic) structure within the final grain, each surface of misfit being equivalent to an array of edge dislocations (Chapter 3). Convection currents can also provide thermal pulses which cause dendritic branch tips to melt off and enter the main body of the melt, where they act as 'kindred nuclei'. Gentle stirring of the melt encourages this process, which is known as dendrite multiplication, and can be used to produce a fine-grained and equiaxed structure (e.g. electromagnetic stirring of molten steel). Dendrite multiplication is now recognized as an important source of crystals in castings and ingots.

2.1.2 Plane-front and dendritic solidification at a cooled surface

The previous section describes random, multidirectional crystallization within a cooling volume of pure molten metal. In practice, freezing often commences at the plane surface of a mold under more complex and constrained conditions, with crystals growing counter to the general direction of heat flow. The morphology of the interface, as well as the final grain structure of the casting, are then decided by thermal conditions at the solid/liquid interface.

Figure 2.4a illustrates the case where all the latent heat evolved at the interface flows into the solid and the temperature gradients in solid and liquid, G_S and G_L, are positive. The solidification front, which moves at a velocity R, is stable, isothermal and planar. Any solid protuberance which chances to

[1] Many metals and a few organic materials grow with non-faceted dendritic morphology, e.g. transparent succinonitrile–6% camphor has proved a valuable means of simulating dendrite growth on a hot-stage optical microscope. Most non-metals grow with a faceted morphology.

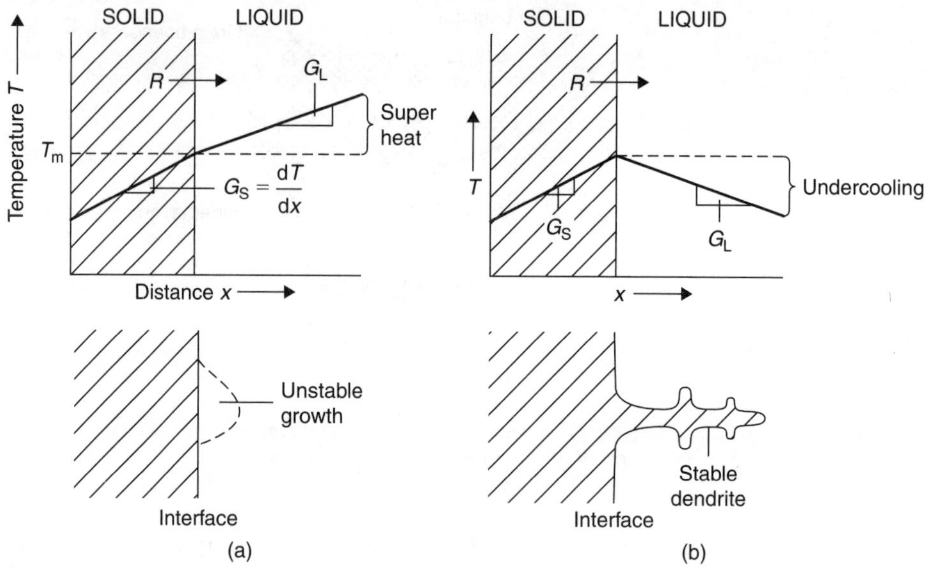

Figure 2.4 *Plane-front solidification (a) and dendritic solidification (b) of a pure metal, as determined by thermal conditions.*

form on this front will project into increasingly hotter, superheated liquid, and will therefore quickly dissolve and be absorbed by the advancing front. Planar-front solidification is characterized by a high G_L/R ratio (e.g. slow cooling). If the solid is polycrystalline, emerging grain boundaries will form grooves in the stable planar front.

In the alternative scenario (Figure 2.4b), for which G_L/R has relatively low values, latent heat flows into both solid and liquid and G_L becomes negative. A planar interface becomes unstable. Dendritic protuberances (spikes) grow rapidly into the undercooled liquid, which quickly absorbs their evolved latent heat. Thermal undercooling is thus an essential prerequisite for dendritic growth; this form of growth becomes more and more likely as the degree of thermal undercooling increases. Melts almost invariably undercool slightly before solidification so that dendritic morphologies are very common. (The ability of dilute alloy melts to produce a cellular morphology as a result of constitutional undercooling will be described in Section 2.2.4.3.)

2.1.3 Forms of cast structure

Because of the interplay of a variety of physical and chemical factors during freezing, the as-cast grain structure is usually not as uniform and straightforward as those discussed in the previous two sections. When solidification commences at the flat surface of a metallic ingot mold there is usually an extreme undercooling or chilling action, which leads to the heterogeneous nucleation of a thin layer of small, randomly oriented chill crystals (Figure 2.5). The size of these equiaxed crystals is strongly influenced by the texture of the mold surface. As the thickness of the zone of chill crystals increases, the temperature gradient G_L becomes less steep and the rate of cooling decreases. Crystal growth rather than the nucleation of new crystals now predominates and, in many metals and alloys, certain favorably oriented crystals at the solid/liquid interface begin to grow into the melt. As in the case of the previously described dendrites, the rapid growth directions are $\langle 1\ 0\ 0 \rangle$ for fcc and bcc crystals, and lie along the direction of heat flow. Sideways growth is progressively hindered so that the

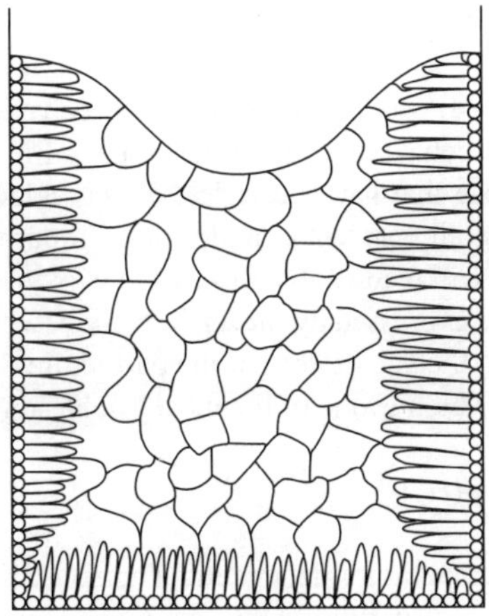

Figure 2.5 *Chill-cast ingot structure.*

crystals develop a preferred orientation and a characteristic columnar form. They therefore introduce directionality into the bulk structure; this effect will be most pronounced if the metal itself is strongly anisotropic (e.g. cph zinc). The preferred growth directions for cph crystals are $\langle 1\ 0\ \bar{1}\ 0 \rangle$. The growth form of the interface between the columnar crystals and the liquid varies from planar to dendritic, depending upon the particular metal (or alloy) and thermal conditions.

As the columnar zone thickens, the temperature gradients within the liquid become more shallow, undercooling more prominent and the presence of kindred nuclei from dendritic multiplication more likely. Under these conditions, independent nucleation (Section 2.1.1) is favored and a central zone of equiaxed, randomly oriented crystals can develop (Figure 2.5). Other factors such as a low pouring temperature (low superheat), molds of low thermal conductivity and the presence of alloying elements also favor the development of this equiaxed zone. There is a related size effect, with the tendency for columnar crystals to form decreasing as the cross-section of the mold cavity decreases. However, in the absence of these influences, growth predominates over nucleation and the columnar zone may extend to the center of the ingot (e.g. pure metals). The balance between the relative proportions of outer columnar crystals and inner equiaxed crystals is important and demands careful control. For some purposes, a completely fine-grained stucture is preferred, being stronger and more ductile. Furthermore, it will not contain the planes of weakness, shown in Figure 2.5, which form when columnar crystals impinge upon each other obliquely. (In certain specialized alloys, however, such as those for high-power magnets and creep-resistant alloys, a coarse grain size is prescribed.)

The addition of various 'foreign' nucleating agents, known as inoculants, is a common and effective method for providing centers for heterogeneous nucleation within the melt, inhibiting undercooling and producing a uniform fine-grained structure. Refining the grain structure disperses impurity elements over a greater area of grain boundary surface, and generally benefits mechanical and founding properties (e.g. ductility, resistance to hot tearing). However, the need for grain refinement during casting operations is often less crucial if the cast structure can be subsequently worked and/or heat treated. Nucleating agents must remain finely dispersed, must survive and must be wetted by the superheated liquid. Examples of inoculants are titanium and/or boron (for aluminum alloys), zirconium or rare earth metals (for magnesium alloys) and aluminum (for steel). Zirconium is an extremely effective grain refiner for magnesium and its alloys. The close similarity in lattice parameters between zirconium and magnesium suggests that the oriented overgrowth (epitaxy) of magnesium upon zirconium is an important factor; however, inoculants have largely been developed empirically.

2.1.4 Gas porosity and segregation

So far we have tended to concentrate upon the behavior of pure metals. It is now appropriate to consider the general behavior of dissimilar types of atoms which, broadly speaking, fall into two main categories: those that have been deliberately added for a specific purpose (i.e. alloying) and those that are accidentally present as undesirable impurities. Most metallic melts, when exposed to a furnace atmosphere, will readily absorb gases (e.g. oxygen, nitrogen, hydrogen). The solubility of gas in liquid metal can be expressed by Sievert's relation, which states that the concentration of dissolved gas is proportional to the square root of the partial pressure of the gas in the contacting atmosphere. Thus, for hydrogen, which is one of the most troublesome gases:

$$[H_{\text{solution}}] = K\{p(H_2)\}^{1/2}. \tag{2.1}$$

The constant K is temperature-dependent. The solubility of gases decreases during the course of freezing, usually quite abruptly, and they are rejected in the form of gas bubbles which may become entrapped within and between the crystals, forming weakening blow-holes. It follows from Sievert's relation that reducing the pressure of the contacting atmosphere will reduce the gas content of the melt; this principle is the basis of vacuum melting and vacuum degassing. Similarly, the passage of numerous bubbles of an inert, low-solubility gas through the melt will also favor gas removal (e.g. scavenging treatment of molten aluminum with chlorine). Conversely, freezing under high applied pressure, as in the die-casting process for light alloys, suppresses the precipitation of dissolved gas and produces a cast shape of high density.

Dissolved gas may precipitate as simple gas bubbles but may, like oxygen, react with melt constituents to form either bubbles of compound gas (e.g. CO_2, CO, SO_2, H_2O_{vap}) or insoluble non-metallic particles. The latter are potential inoculants. Although their presence may be accidental, as indicated previously, their deliberate formation is sometimes sought. Thus, a specific addition of aluminum, an element with a high chemical affinity for oxygen, is used to deoxidize molten steel in the ladle prior to casting; the resultant particles of alumina subsequently act as heterogeneous nucleants, refining the grain size.

Segregation almost invariably occurs during solidification; unfortunately, its complete elimination is impossible. Segregation, in its various forms, can seriously impair the physical, chemical and mechanical properties of a cast material. In normal segregation, atoms different to those which are crystallizing can be rejected into the melt as the solid/liquid interface advances. These atoms may be impurities or, as in the case of a solid solution alloy, solute atoms. Insoluble particles can also be pushed ahead of the interface. Eventually, pronounced macro segregation can be produced in the final regions to solidify, particularly if the volume of the cast mass is large. On a finer scale, micro segregation can occur interdendritically within both equiaxed and columnar grains (coring), and at the surfaces of low- and high-angle grain boundaries. The modern analytical technique of Auger electron spectroscopy (AES) is capable of detecting monolayers of impurity atoms at grain boundary surfaces and has made it possible to study their very significant effect upon properties such as ductility and corrosion resistance (Chapter 4).

In the other main form of separation process,[2] which is known as inverse segregation, thermal contraction of the solidified outer shell forces a residual melt of low melting point outwards along intergranular channels until it freezes on the outside of the casting (e.g. 'tin sweat' on bronzes, 'phosphide sweat' on gray cast iron). The direction of this remarkable migration thus coincides with that of heat flow, in direct contrast to normal macro segregation. Inverse segregation can be prevented by unidirectional solidification. Later, in Section 2.2.4.4, it will be shown how the process of zone

[2] Like all separation processes, even so-called chemical separations, it is essentially a physical process.

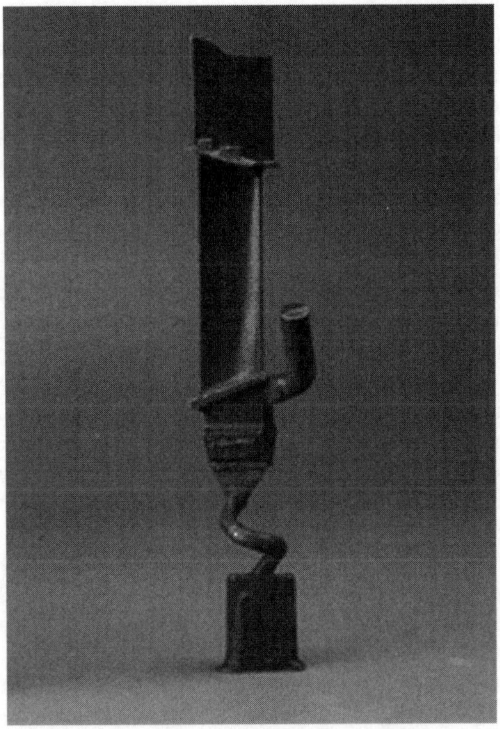

Figure 2.6 *A single-crystal blade embodying a DS starter block and helical constriction, and a test plate facilitating orientation check by XRD (Courtesy of Prof. H. K. D. H. Bhadeshia).*

refining, as used in the production of high-purity materials for the electronics industry, takes positive advantage of segregation.

2.1.5 Directional solidification

The exacting mechanical demands made upon gas-turbine blades have led to the controlled exploitation of columnar crystal growth and the development of directional solidification (DS) techniques for superalloys.[3] As the turbine rotor rotates, the hot blades are subject to extremely large centrifugal forces and to thermal excursions in temperature. Development has proceeded in two stages. First, a wholly columnar grain structure, without grain boundaries transverse to the major axis of the blade, has been produced during precision investment casting by initiating solidification at a water-cooled copper chill plate in the mold base and then slowly withdrawing the vertically positioned mold from the hot zone of an enclosing furnace. Most of the heat is removed by the chill plate. A restricted number of crystals is able to grow parallel to the major axis of the blade. Transverse grain boundaries, which have been the initiation sites for intergranular creep failures in equiaxed blade structures, are virtually eliminated. Grain shape is mainly dependent upon (1) the thermal gradient G_L extending into the melt from the melt/solid interface and (2) the growth rate R at which this interface moves. Graphical plots of G_L versus R have provided a useful means for predicting the grain morphology of DS alloys.

[3] Directional solidification of high-temperature alloys was pioneered by F. L. VerSnyder and R. W. Guard at the General Electric Company, Schenectady, USA, in the late 1950s; by the late 1960s, DS blades were being used in gas turbines of commercial aircraft.

In the second, logical stage of development, a single-crystal (SC) turbine blade is produced by placing a helical constriction between a chilled starter block and the mold proper (Figure 2.6). The constriction acts as a spiral selector which ensures that only one of the upward-growing columnar crystals in the starter block can grow into the blade mold. The orientation of every blade is checked by means of a computerized X-ray diffraction procedure (e.g. the SCORPIO system at Rolls-Royce). In the fcc nickel-based superalloys, the favored [1 0 0] growth direction coincides with the major axis of the blade and fortunately offers the best overall mechanical properties. For instance, the modulus of elasticity is low in the ⟨1 0 0⟩ directions; consequently, thermal stresses are reduced and the resistance to thermal fatigue enhanced. If full three-dimensional control of crystal orientation is required, a seed crystal is precisely located close to the chill plate and gives the desired epitaxial, or oriented, overgrowth. The production of single-crystal turbine blades by directional solidification (DS) techniques has increased blade life substantially and has enabled operating temperatures to be raised by 30°C, thus improving engine efficiency. It has also had a far-reaching effect upon the philosophy of alloy design. Previously, certain elements, such as carbon, boron, hafnium and zirconium, were added in order to strengthen grain boundary surfaces. Unfortunately, their presence lowers the incipient melting point. The DS route dispenses with the need for them and permits beneficially higher temperatures to be used during subsequent heat treatment. The DS approach has to be carefully matched to alloy composition; it is possible for it to lower the creep strength of certain superalloys.

2.1.6 Production of metallic single crystals for research

Development of highly specialized industrial-scale techniques, such as the directional solidification (DS) of turbine blades and the production of silicon, germanium and various compounds for semiconductors, owes much to expertise gained over many years in producing small single crystals for fundamental research. Several methods originally developed for metals have been adapted for ceramics. Experiments with single crystals of metals (and ceramics and polymers) have a special place in the history of materials science. The two basic methods for preparing single crystals involve (1) solidification from the melt and (2) grain growth in the solid state.

In the simplest version of the solidification method, the polycrystalline metal to be converted to a single crystal is supported in a horizontal, non-reactive boat (e.g. graphite) and made to freeze progressively from one end by passing an electric furnace, with its peak temperature set about 10°C above the melting point, over the boat. Although several nuclei may form during initial solidification, the sensitivity of the growth rate to orientation usually results in one of the crystals swamping the others and eventually forming the entire growth front. The method is particularly useful for seeding crystals of a predetermined orientation. A seed crystal is placed adjacent to the polycrystalline sample in the boat and the junction is melted before commencing the melting/solidification process. Wire specimens may be grown in silica or heat-resistant glass tubes internally coated with graphite (*Aquadag*). In a modern development of these methods, the sample is enclosed in an evacuated silica tube and placed in a water-cooled copper boat; passage through a high-frequency heating coil produces a melt zone.

Most solidification techniques for single crystals are derived from the Bridgman and Czochralski methods. In the former, a pure metal sample is loaded in a vertical mold of smooth graphite, tapered to a point at the bottom end. The mold is lowered slowly down a tubular furnace which produces a narrow melt zone. The crystal grows from the point of the mold. In the Czochralski method, often referred to as 'crystal pulling', a seed crystal is withdrawn slowly from the surface of a molten metal, enabling the melt to solidify with the same orientation as the seed. Rotation of the crystal as it is withdrawn produces a cylindrical crystal. This technique is used for the preparation, *in vacuo*, of Si and Ge crystals.

Crystals may also be prepared by a 'floating-zone' technique (e.g. metals of high melting point such as W, Mo and Ta). A pure polycrystalline rod is gripped at the top and bottom in water-cooled grips and rotated in an inert gas or vacuum. A small melt zone, produced by either a water-cooled radio-frequency coil or electron bombardment from a circular filament, is passed up its length. High purity is possible because the specimen has no contact with any source of contamination and also because there is a zone-refining action (Section 2.2.4.4).

Methods involving grain growth in the solid state (2) depend upon the annealing of deformed samples. In the strain-anneal technique, a fine-grained polycrystalline metal is critically strained approximately 1–2% elongation in tension and then annealed in a moving-gradient furnace with a peak temperature set below the melting point or transformation temperature. Light straining produces very few nuclei for crystallization; during annealing, one favored nucleus grows more rapidly than the other potential nuclei, which it consumes. The method has been applied to metals and alloys of high stacking-fault energy, e.g. Al, silicon–iron (see Chapter 3). Single crystals of metals with low stacking-fault energy, such as Au and Ag, are difficult to grow because of the ease of formation of annealing twins which give multiple orientations. Hexagonal metals are also difficult to prepare because deformation twins formed during straining act as effective nucleation sites.

2.2 Principles and applications of phase diagrams

2.2.1 The concept of a phase

The term 'phase' refers to a separate and identifiable state of matter in which a given substance may exist. Being applicable to both crystalline and non-crystalline materials, its use provides a convenient way of expressing a material's structure. Thus, in addition to the three crystalline forms mentioned in Section 1.9.1, the element iron may exist as a liquid or vapor, giving five phases overall. Similarly, the important refractory oxide silica is able to exist as three crystalline phases, quartz, tridymite and cristobalite, as well as a non-crystalline phase, silica glass, and as molten silica. Under certain conditions, it is possible for two or more different phases to coexist. The glass-reinforced polymer (GRP) known as *Fiberglass* is an example of a two-phase structure.

When referring to a particular phase in the structure of a material, we imply a region comprising a large number of atoms (or ions or molecules) and the existence of a bounding surface which separates it from contiguous phases. Local structural disturbances and imperfections are disregarded. Thus, a pure metal or an oxide solid solution are each described, by convention, as single-phase structures even though they may contain grain boundaries, concentration gradients (coring) and microdefects, such as vacancies, dislocations and voids (Chapter 3). Industrial practice understandably favors relatively rapid cooling rates, frequently causing phases to exist in a metastable condition. Some form of 'triggering' process, such as thermal activation, is needed before a metastable phase can adopt the stable, or equilibrium, state of lowest energy (e.g. annealing of metals and alloys). These two features, structural heterogeneity on a micro-scale and non-equilibrium, do not give rise to any untoward scientific difficulty.

The production of the thermal-shock resistant known as *Vycor* provides an interesting example of the potential of phase control.[4] Although glasses of very high silica content are eminently suitable for high-temperature applications, their viscosity is very high, making fabrication by conventional methods difficult and costly. This problem was overcome by taking advantage of phase separation in a workable silica containing a high proportion of boric oxide. After shaping, this glass is heat-treated at a temperature of 500–600°C in order to induce separation of two distinct and interpenetrating glassy phases. Electron microscopy reveals a worm-like boron-rich phase surrounded by a porous siliceous matrix. Leaching in hot acid dissolves away the former phase, leaving a porous silica-rich structure. Consolidation

[4] *Vycor*, developed and patented by Corning Glass Works (USA) just before World War II, has a final composition of $96SiO_2$–$4B_2O_3$. It has been used for laboratory ware and for the outer windows of space vehicles.

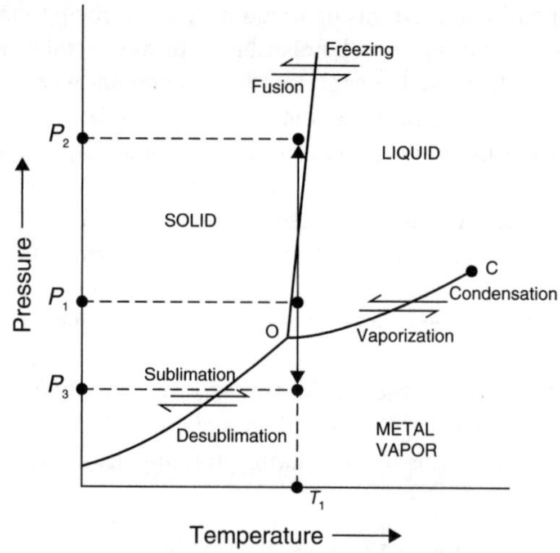

Figure 2.7 *Phase diagram for a typical pure metal.*

heat treatment at a temperature of 1000°C 'shrinks' this skeletal structure by a remarkable 30%. This product has a low linear coefficient of thermal expansion ($\alpha = 0.8 \times 10^{-6}$°$C^{-1}$; 0–300°C) and can withstand quenching into iced water from a temperature of 900°C, its maximum service temperature.

2.2.2 The Phase Rule

For a given metallic or ceramic material, there is a theoretical condition of equilibrium in which each constituent phase is in a reference state of lowest energy. The independent variables determining this energy state, which are manipulated by scientists and technologists, are composition, temperature and pressure. The Phase Rule derived by Willard Gibbs from complex thermodynamical theory provides a device for testing multiphase (heterogeneous) equilibria and deciding the number of variables (degrees of freedom) necessary to define the energy state of a system. Its basic equation, $P + F = C + 2$, relates the number of phases present at equilibrium (P) and the number of degrees of freedom (F) to the number of components (C), which is the smallest number of substances of independently variable composition making up the system. For metallic systems, the components are metallic elements; for ceramics, the components are frequently oxides (e.g. MgO, SiO_2, etc.).

Consider Figure 2.7, which is a single-component (unary) diagram representing the phase relations for a typical pure metal. Transitions such as melting, sublimation and vaporization occur as boundaries between the three single-phase fields are crossed. Suppose that conditions are such that three phases of a pure metal coexist simultaneously, a unique condition represented by the triple-point O. Applying the Phase Rule: $P = 3$, $C = 1$ and $F = 0$. This system is said to be invariant with no degrees of freedom. Having stated that 'three phases coexist', the energy state is fully defined and it is unnecessary to specify values for variables. One or two phases could be caused to disappear, but this would require temperature and/or pressure to be changed. Similarly, at a melting point on the line between single-phase fields, a solid and a liquid phase may coexist (i.e. $P = 2$, $C = 1$ and $F = 1$). Specification of one variable (either temperature or pressure) will suffice to define the energy state of a particular two-phase equilibrium. A point within one of the single-phase fields represents a system that is bivariant ($F = 2$). Both temperature and pressure must be specified if its state is to be fully defined.

Let us consider some practical implications of Figure 2.7. The line for the solid/liquid transition is shown with a slight inclination to the right in this schematic diagram because, theoretically, one may reason from Le Chatelier's principle that a pressure increase on a typical melt, from P_1 to P_2, will favor a constraint-removing contraction and therefore lead to freezing. (Bismuth and gallium are exceptions: like water, they expand on freezing and the solid/liquid line has an opposite slope.) The pressure at the critical point C, beyond which liquid and vapor are indistinguishable, is usually extremely high and considerably greater than one atmosphere. The diagram refers to an enclosed system. For instance, if molten metal is in a vacuum system, and the pressure is reduced isothermally from P_1 to P_3, then vaporization will be favored.

For most metals, the pressure value at the triple point is far below one atmosphere ($1\ \text{atm} = 101.325\ \text{kN m}^{-2} \cong 1\ \text{bar}$). For copper and lead, it is of the order of 10^{-6} atm. The pressures associated with the solid/vapor line are obviously even lower. Thus, for most metals, vapor pressure is disregarded and it is customary to use isobaric (constant pressure) phase diagrams in which the composition (concentration) of component metals is plotted against temperature. As a consequence of this practice, metallurgists use a modified form of the Phase Rule equation, i.e. $P + F = C + 1$.

Nevertheless, it is possible for vapor pressure to be a highly significant factor in alloy systems formed by volatile components such as mercury, zinc, cadmium and magnesium: in such cases, it is advisable to take advantage of full pressure–temperature diagrams. Furthermore, the partial pressures of certain gases in a contacting atmosphere can be highly significant and at elevated temperatures can contribute to serious corrosion of metals (e.g. O_2, S_2 (Chapter 9)). Again, the pressure variable must be taken into account.

2.2.3 Stability of phases

2.2.3.1 The concept of free energy

Every material in a given state has a characteristic heat content or enthalpy, H, and the rate of change of heat content with temperature is equal to the specific heat of the material measured at constant pressure, $C_P = (\text{d}H/\text{d}T)_p$. A knowledge of the quantity H is clearly important to understand reactions but it does not provide a criterion for equilibrium, nor does it determine when a phase change occurs, as shown by the occurrence of both exothermic and endothermic reactions. To provide this criterion it is necessary to consider a second important property of state known as the entropy, S. In statistical terms S may be regarded as a measure of the state of disorder of the structure, but from classical thermodynamics it may be shown that for any material passing through a complete cycle of events:

$$\oint \frac{\text{d}Q}{T} = 0,$$

where $\text{d}Q$ is the heat exchanged between the system and its surroundings during each infinitesimal step and T is the temperature at which the transfer takes place.

It is then convenient to define a quantity S such that $\text{d}S = \text{d}Q/T$, so that $\oint \text{d}S = 0$; entropy so defined is then a state property. At constant pressure, $\text{d}Q = \text{d}H$ and consequently

$$\text{d}S = \text{d}Q/T = C_p \text{d}T/T,$$

which by integration gives

$$S = S_0 + \int_0^T (C_p/T)\text{d}T = S_0 + \int_0^T C_p \delta(\ln T),$$

where S is the entropy at T K, usually measured in J K^{-1}. The integration constant S_0 represents the entropy at absolute zero, which for an ordered crystalline substance is taken to be zero; this is often quoted as the third law of thermodynamics. Clearly, any reaction or transformation within a system will be associated with a characteristic entropy change given by

$$dS = S_\beta - S_\alpha,$$

where dS is the entropy of transformation and S_β and S_α are the entropy values of the new phase β and the old phase α, respectively. It is a consequence of this that any irreversible change which takes place in a system (e.g. the combustion of a metal) must be accompanied by an increase in the total entropy of the system. This is commonly known as the second law of thermodynamics.

The quantity entropy could be used for deciding the equilibrium state of a system, but it is much more convenient to work in terms of energy. Accordingly, it is usual to deal with the quantity TS, which has the units of energy, rather than just S, and to separate the total energy of the system H into two components according to the relation:

$$H = G + TS,$$

where G is that part of the energy of the system which causes the process to occur and TS is the energy requirement resulting from the change involved. The term G is known as the Gibbs free energy and is defined by the equation:

$$G = H - TS.$$

Every material in a given state will have a characteristic value of G. The change of free energy accompanying a change represents the 'driving force' of the change, and is given by the expression:

$$dG = dH - TdS \equiv dE + PdV - TdS.$$

All spontaneous changes in a system must be accompanied by a reduction of the total free energy of that system, and thus for a change to occur the free energy change ΔG must be negative. It also follows that the equilibrium condition of a reaction will correspond to the state where $dG = 0$, i.e. zero driving force.

For solids and liquids at atmospheric pressure the volume change accompanying changes of state is very small and hence PdV is also very small. It is therefore reasonable to neglect this term in the free energy equation and use as the criterion of equilibrium $dE - TdS = 0$. This is equivalent to defining the quantity $(E - TS)$ to be a minimum in the equilibrium state, for by differentiation:

$$\begin{aligned} d(E - TS) &= dE - TdS - SdT \\ &= dE - TdS \text{ (since } T \text{ is constant)} \\ &= 0 \text{ for the equilibrium state.} \end{aligned}$$

The quantity $(E - TS)$ thus defines the equilibrium state at constant temperature and volume, and is given the symbol F, the Helmholtz free energy ($F = E - TS$), to distinguish it from the Gibbs free energy ($G = H - TS$). In considering changes in the solid state it is thus a reasonable approximation to use F in place of G. The enthalpy H is the sum of the internal and external energies, which reduces to $H \simeq E$ when the external energy PV is neglected.

2.2.3.2 *Free energy and temperature*

If a metal undergoes a structural change from phase α to phase β at a temperature T_t then it does so because above this temperature the free energy of the β-phase, G_β, becomes lower than the free

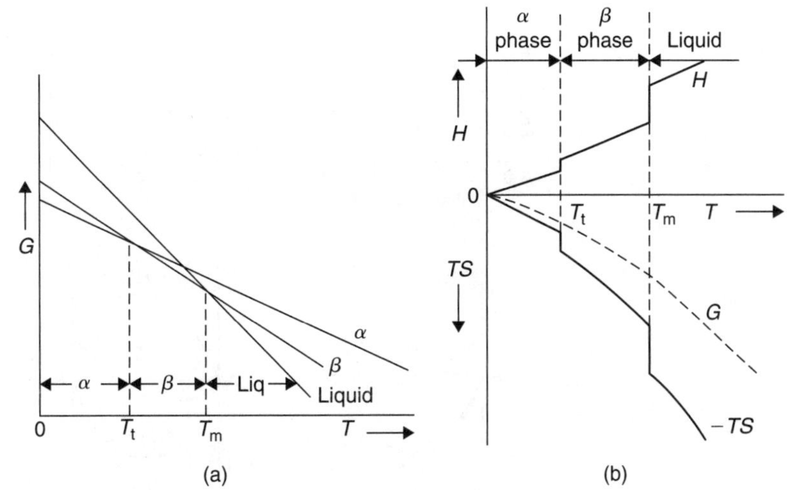

Figure 2.8 *Free energy–temperature curves for α, β and liquid phases.*

energy of the α-phase, G_α. For this to occur the free energy curves must vary with temperature in the manner shown in Figure 2.8a. It can be seen that at T_t the free energy of the α-phase is equal to that of the β-phase so that ΔG is zero; T_t is therefore the equilibrium transformation point.

Figure 2.8a also shows that successive transformations occur in a given temperature range. The way in which the absolute value of the free energy of a crystal varies with temperature is shown in Figure 2.8b, where H and $-TS$ are plotted as a function of temperature. At the transformation temperature, T_t, the change in heat content ΔH is equal to the latent heat L, while the entropy change ΔS is equal to L/T_t. In consequence, a plot of $G = H - TS$ shows no sharp discontinuities at T_t (since $\Delta H = T_t\Delta S$) or T_m but merely a discontinuity of slope. A plot of G versus temperature for each of the three phases considered, α, β and liquid, would then be of the form shown in Figure 2.9a. In any temperature range the phase with the lowest free energy is the stable phase.

2.2.3.3 *Free energy and composition*

Of considerable importance in metallurgical practice is the increase in entropy associated with the formation of a disordered solid solution from the pure components. This arises because, over and above the entropies of the pure components A and B, the solution of B in A has an extra entropy due to the numerous ways in which the two kinds of atoms can be arranged amongst each other. This entropy of disorder or mixing is of the form shown in Figure 2.9a.

As a measure of the disorder of a given state we can, purely from statistics, consider W, the number of distributions which belong to that state. Thus, if the crystal contains N sites, n of which contain A atoms and $(N - n)$ contain B atoms, it can be shown that the total number of ways of distributing the A and B atoms on the N sites is given by

$$W = \frac{N!}{n!(N-n)!}.$$

This is a measure of the extra disorder of solution, since $W = 1$ for the pure state of the crystal because there is only one way of distributing N indistinguishable pure A or pure B atoms on the N sites.

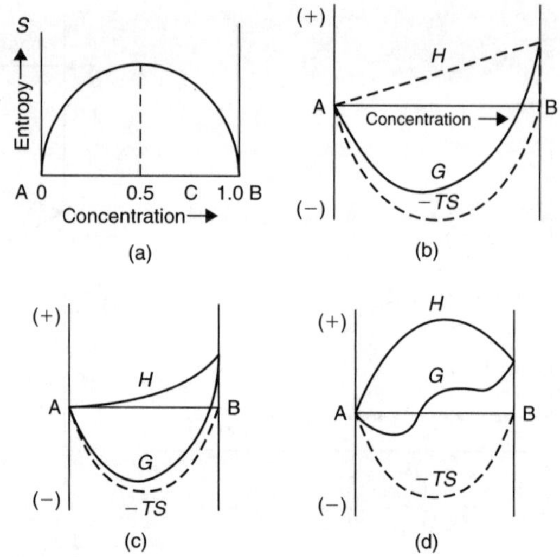

Figure 2.9 *Variation with composition of entropy (a) and free energy (b) for an ideal solid solution, and for non-ideal solid solutions (c and d).*

To ensure that the thermodynamic and statistical definitions of entropy are in agreement the quantity W, which is a measure of the configurational probability of the system, is not used directly, but in the form:

$$S = \mathbf{k} \ln W,$$

where **k** is Boltzmann's constant. From this equation it can be seen that entropy is a property which measures the probability of a configuration, and that the greater the probability, the greater is the entropy. Substituting for W in the statistical equation of entropy and using Stirling's approximation[5] we obtain

$$\begin{aligned} S &= \mathbf{k} \ln [N!/n!(N-n)!] \\ &= \mathbf{k}[N \ln N - n \ln n - (N-n) \ln (N-n)] \end{aligned}$$

for the entropy of disorder, or mixing. The form of this entropy is shown in Figure 2.9a, where $c = n/N$ is the atomic concentration of B in the solution. It is of particular interest to note the sharp increase in entropy for the addition of only a small amount of solute. This fact accounts for the difficulty of producing really pure metals, since the entropy factor, $-T\mathrm{d}S$, associated with impurity addition, usually outweighs the energy term, $\mathrm{d}H$, so that the free energy of the material is almost certainly lowered by contamination.

While Figure 2.9a shows the change in entropy with composition, the corresponding free energy versus composition curve is of the form shown in Figure 2.9b, c or d depending on whether the solid solution is ideal or deviates from ideal behavior. The variation of enthalpy with composition, or heat of mixing, is linear for an ideal solid solution, but if A atoms prefer to be in the vicinity of B atoms rather than A atoms, and B atoms behave similarly, the enthalpy will be lowered by alloying (Figure 2.9c).

---

[5] Stirling's theorem states that if N is large:
$\ln N! = N \ln N - N$.

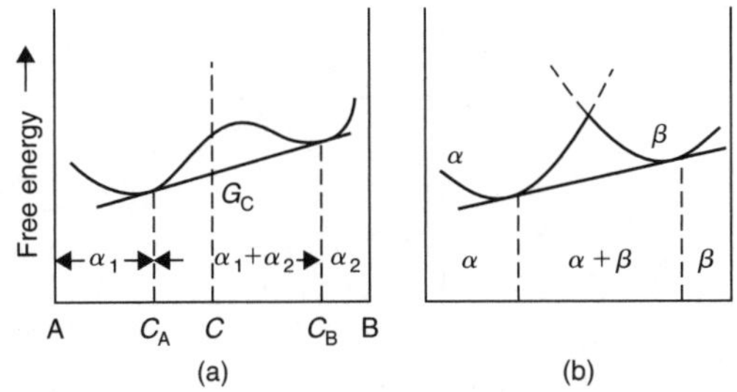

Figure 2.10 *Free energy curves showing extent of phase fields at a given temperature.*

A positive deviation occurs when A and B atoms prefer like atoms as neighbors and the free energy curve takes the form shown in Figure 2.9d. In Figure 2.9b and c the curvature dG^2/dc^2 is always positive, whereas in Figure 2.9d there are two minima and a region of negative curvature between points of inflexion[6] given by $dG^2/dc^2 = 0$. A free energy curve for which d^2G/dc^2 is positive, i.e. simple U-shaped, gives rise to a homogeneous solution. When a region of negative curvature exists, the stable state is a phase mixture rather than a homogeneous solid solution, as shown in Figure 2.10a. An alloy of composition c has a lower free energy G_c when it exists as a mixture of A-rich phase (α_1) of composition c_A and B-rich phase (α_2) of composition c_B in the proportions given by the Lever Rule, i.e. $\alpha_1/\alpha_2 = (c_B - c)/(c - c_A)$. Alloys with composition $c < c_A$ or $c > c_B$ exist as homogeneous solid solutions and are denoted by phases α_1 and α_2 respectively. Partial miscibility in the solid state can also occur when the crystal structures of the component metals are different. The free energy curve then takes the form shown in Figure 2.10b, the phases being denoted by α and β.

2.2.4 Two-phase equilibria

2.2.4.1 Extended and limited solid solubility

Solid solubility is a feature of many metallic and ceramic systems, being favored when the components have similarities in crystal structure and atomic (ionic) diameter. Such solubility may be either extended (continuous) or limited. The former case is illustrated by the binary phase diagram for the nickel–copper system (Figure 2.11) in which the solid solution (α) extends from component to component. In contrast to the abrupt (congruent) melting points of the pure metals, the intervening alloys freeze, or fuse, over a range of temperatures which is associated with a univariant two-phase (α + liquid) field. This 'pasty' zone is located between two lines known as the liquidus and solidus. (The phase diagrams for the Ni–Cu and MgO–FeO systems are similar in form.)

Let us consider the very slow (equilibrating) solidification of a 70Ni–30Cu alloy. A commercial version of this alloy, *Monel*, also contains small amounts of iron and manganese. It is strong, ductile and resists corrosion by all forms of water, including sea water (e.g. chemical and food processing, water treatment). An ordinate is erected from its average composition on the baseline. Freezing starts at a temperature T_1. A horizontal tie-line is drawn to show that the first crystals of solid solution to form have a composition α_1. When the temperature reaches T_2, crystals of composition α_2 are in equilibrium with liquid of composition L_2. Ultimately, at temperature T_3, solidification is completed as the composition α_3 of the crystals coincides with the average composition of the alloy. It will

[6] The composition at which $d^2G/dc^2 = 0$ varies with temperature and the corresponding temperature–composition curves are called spinodal lines.

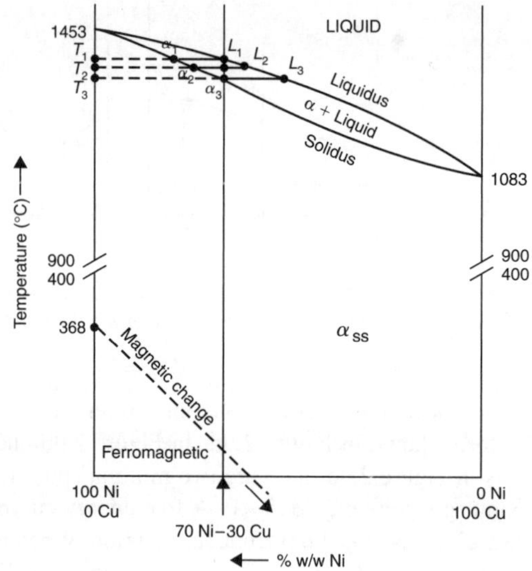

Figure 2.11 *Binary phase diagram for the Ni–Cu system showing extended solid solubility.*

be seen that the compositions of the α-phase and liquid have moved down the solidus and liquidus, respectively, during freezing.

Each tie-line joins two points which represent two phase compositions. One might visualize that a two-phase region in a binary diagram is made up of an infinite number of horizontal (isothermal) tie-lines. Using the average alloy composition as a fulcrum (x) and applying the Lever Rule, it is quickly possible to derive mass ratios and fractions. For instance, for equilibrium at temperature T_2 in Figure 2.11, the mass ratio of solid solution crystals to liquid is $xL_2/\alpha_2 x$. Similarly, the mass fraction of solid in the two-phase mixture at this temperature is $xL_2/L_2\alpha_2$. Clearly, the phase compositions are of greater structural significance than the average composition of the alloy. If the volumetric relations of two phases are required, these being what we naturally assess in microscopy, then the above values must be corrected for phase density.

In most systems, solid solubility is far more restricted and is often confined to the phase field adjacent to the end-component. A portion of a binary phase diagram for the copper–beryllium system, which contains a primary, or terminal, solid solution, is shown in Figure 2.12. Typically, the curving line, known as the solvus, shows an increase in the ability of the solvent copper to dissolve beryllium solute as the temperature is raised. If a typical 'beryllium–copper' containing 2% beryllium is first held at a temperature just below the solidus (solution-treated), water-quenched to preserve the α-phase and then aged at a temperature of 425°C, particles of a second phase (γ) will form within the α-phase matrix because the alloy is equilibrating in the ($\alpha + \gamma$) field of the diagram. This type of treatment, closely controlled, is known as precipitation hardening; the mechanism of this important strengthening process will be discussed in detail in Chapter 7. Precipitation hardening of a typical beryllium–copper, which also contains up to 0.5% cobalt or nickel, can raise the 0.1% proof stress to 1200 MN m^{-2} and the tensile strength to 1400 MN m^{-2}. Apart from being suitable for non-sparking tools, it is a valuable spring material, being principally used for electrically conductive brush springs and contact fingers in electrical switches. A curving solvus is an essential feature of phase diagrams for precipitation-hardenable alloys (e.g. aluminum–copper alloys (*Duralumin*)).

When solid-state precipitation takes place, say of β within a matrix of supersaturated α grains, this precipitation occurs in one or more of the following preferred locations: (1) at grain boundaries, (2) around dislocations and inclusions, and (3) on specific crystallographic planes. The choice of site

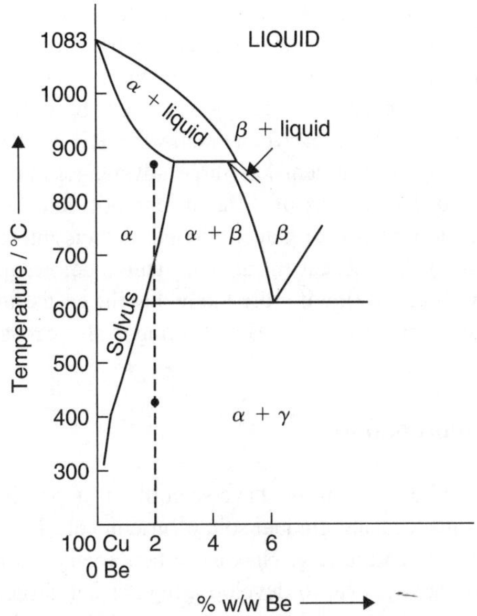

Figure 2.12 *Cu-rich end of the phase diagram for the Cu–Be system, showing field of primary solid solution (α).*

for precipitation depends on several factors, of which grain size and rate of nucleation are particularly important. If the grain size is large, the amount of grain boundary surface is relatively small, and deposition of β-phase within the grains is favored. When this precipitation occurs preferentially on certain sets of crystallographic planes within the grains, the etched structure has a mesh-like appearance which is known as a Widmanstätten-type structure.[7] Widmanstätten structures have been observed in many alloys (e.g. overheated steels).

2.2.4.2 Coring

It is now possible to consider microsegregation, a phenomenon introduced in Section 2.1.4, in more detail. Referring again to the freezing process for a Ni–Cu alloy (Figure 2.11), it is clear that the composition of the α-phase becomes progressively richer in copper and, consequently, if equilibrium is to be maintained in the alloy, the two phases must continuously adjust their compositions by atomic migration. In the liquid phase such diffusion is relatively rapid. Under industrial conditions, the cooling rate of the solid phase is often too rapid to allow complete elimination of differences in composition by diffusion. Each grain of the α-phase will thus contain composition gradients between the core, which will be unduly rich in the metal of higher melting point, and the outer regions, which will be unduly rich in the metal of lower melting point. Such a non-uniform solid solution is said to be cored: etching of a polished specimen can reveal a pattern of dendritic segregation within each cored grain. The faster the rate of cooling, the more pronounced will be the degree of coring. Coring in chill-cast ingots is therefore quite extensive.

---

[7] Named after Count Alois von Widmanstätten, who discovered this morphology within an iron–nickel meteorite sample in 1808.

The physical and chemical hetereogeneity produced by non-equilibrium cooling rates impairs properties. Cored structures can be homogenized by annealing. For instance, an ingot may be heated to a temperature just below the solidus temperature where diffusion is rapid. The temperature must be selected with care because some regions might be rich enough in low-melting-point metal to cause localized fusion. However, when practicable, it is more effective to cold-work a cored structure before annealing. This treatment has three advantages. First, dendritic structures are broken up by deformation so that regions of different composition are intermingled, reducing the distances over which diffusion must take place. Second, defects introduced by deformation accelerate rates of diffusion during the subsequent anneal. Third, deformation promotes recrystallization during subsequent annealing, making it more likely that the cast structure will be completely replaced by a generation of new equiaxed grains. Hot-working is also capable of eliminating coring.

2.2.4.3 Cellular microsegregation

In the case of a solid solution, we have seen that it is possible for solvent atoms to tend to freeze before solute atoms, causing gradual solute enrichment of an alloy melt and, under non-equilibrium conditions, dendritic coring (e.g. Ni–Cu). When a very dilute alloy melt or impure metal freezes, it is possible for each crystal to develop a regular cell structure on a finer scale than coring. The thermal and compositional condition responsible for this cellular microsegregation is referred to as constitutional undercooling.

Suppose that a melt containing a small amount of lower-m.p. solute is freezing. The liquid becomes increasingly enriched in rejected solute atoms, particularly close to the moving solid/liquid interface. The variation of liquid composition with distance from the interface is shown in Figure 2.13a. There is a corresponding variation with distance of the temperature T_L at which the liquid will freeze, since solute atoms lower the freezing temperature. Consequently, for the positive gradient of melt temperature T shown in Figure 2.13b, there is a layer of liquid in which the actual temperature T is below the freezing temperature T_L: this layer is constitutionally undercooled. Clearly, the depth of the undercooled zone, as measured from the point of intersection, will depend upon the slope of the curve for actual temperature, i.e. $G_L = dT/dx$. As G_L decreases, the degree of constitutional undercooling will increase.

Suppose that we visualize a tie-line through the two-phase region of the phase diagram fairly close to the component of higher m.p. Assuming equilibrium, a partition or distribution coefficient k can be defined as the ratio of solute concentration in the solid to that in the liquid, i.e. c_S/c_L. For an alloy of average composition c_0, the solute concentration in the first solid to freeze is kc_0, where $k < 1$, and the liquid adjacent to the solid becomes richer in solute than c_0. The next solid to freeze will have a higher concentration of solute. Eventually, for a constant rate of growth of the solid/liquid interface, a steady state is reached for which the solute concentration at the interface reaches a limiting value of c_0/k and decreases exponentially within the liquid to the bulk composition. This concentration profile is shown in Figure 2.13a.

The following relation can be derived by applying Fick's second law of diffusion (Section 5.4.1):

$$c_L = c_0\left[1 + \frac{1-k}{k}\exp\left(-\frac{Rx}{D}\right)\right], \tag{2.2}$$

where x is the distance into the liquid ahead of the interface, c_L is the solute concentration in the liquid at point x, R is the rate of growth and D is the diffusion coefficient of the solute in the liquid. The temperature distribution in the liquid can be calculated if it is assumed that k is constant and that the liquidus is a straight line of slope m. For the two curves of Figure 2.13b:

$$T = T_0 - mc_0/k + G_L x \tag{2.3}$$

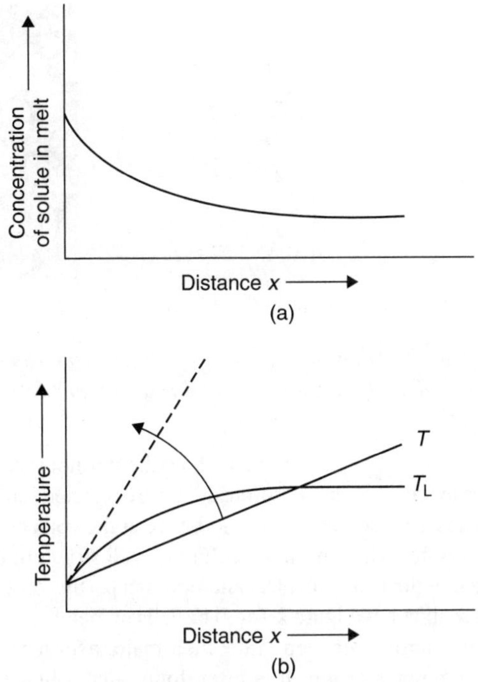

Figure 2.13 *Variation with distance from solid/liquid interface of: (a) melt composition and (b) actual temperature* T *and freezing temperature* T_L.

and

$$T_L = T_0 - mc_0\left[1 + \frac{1-k}{k}\exp\left(-\frac{Rx}{D}\right)\right], \tag{2.4}$$

where T_0 is the freezing temperature of pure solvent, T_L the liquidus temperature for the liquid of composition c_L and T the actual temperature at any point x.

The zone of constitutional undercooling may be eliminated by increasing the temperature gradient G_L, such that:

$$G_L > dT_L/dx. \tag{2.5}$$

Substituting for T_L and putting $[1-(Rx/D)]$ for the exponential gives the critical condition:

$$\frac{G_L}{R} > \frac{mc_0}{D}\left(\frac{1-k}{k}\right). \tag{2.6}$$

This equation summarizes the effect of growth conditions upon the transition from planar to cellular growth and identifies the factors that stabilize a planar interface. Thus, a high G_L, low R and low c_0 will reduce the tendency for cellular (and dendritic) structures to form.

The presence of a zone of undercooled liquid ahead of a macroscopically planar solid/liquid interface (Section 2.1.2) makes it unstable and an interface with cellular morphology develops. The interface grows locally into the liquid from a regular array of points on its surface, forming

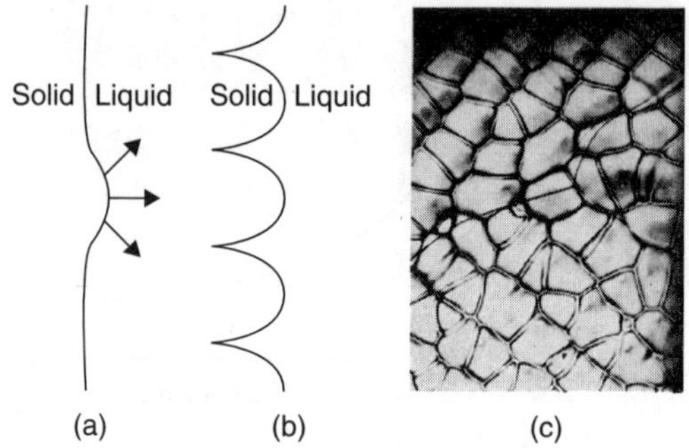

Figure 2.14 *The breakdown of a planar solid/liquid interface (a, b), leading to the formation of a cellular structure of the form shown in (c) for Sn/0.5 at.% Sb × 140.*

dome-shaped cells. Figures 2.14a and b show the development of domes within a metallic melt. As each cell grows by rapid freezing, solute atoms are rejected into the liquid around its base, which thus remains unfrozen. This solute-rich liquid between the cells eventually freezes at a much lower temperature and a crystal with a periodic columnar cell structure is produced. Solute or impurity atoms are concentrated in the cell walls. Decantation of a partly solidified melt will reveal the characteristic surface structure shown in Figure 2.14c. The cells of metals are usually hexagonal in cross-section and about 0.05–1 mm across: for each grain, their major axes have the same crystallographic orientation to within a few minutes of arc. It is often found that a lineage or macromosaic structure (Section 2.1.1) is superimposed on the cellular structure; this other form of substructure is coarser in scale.

Different morphologies of a constitutionally cooled surface, other than cellular, are possible. A typical overall sequence of observed growth forms is planar/cellular/cellular dendritic/dendritic. Substructures produced by constitutional undercooling have been observed in 'doped' single crystals and in ferrous and non-ferrous castings/weldments.[8] When the extent of undercooling into the liquid is increased as, for example, by reducing the temperature gradient G_L, the cellular structure becomes unstable and a few cells grow rapidly as cellular dendrites. The branches of the dendrites are interconnected and are an extreme development of the dome-shaped bulges of the cell structure in directions of rapid growth. The growth of dendrites in a very dilute, constitutionally undercooled alloy is slower than in a pure metal because solute atoms must diffuse away from dendrite/liquid surfaces and also because their growth is limited to the undercooled zone. Cellular impurity-generated substructures have also been observed in 'non-metals' as a result of constitutional undercooling. Unlike the dome-shaped cells produced with metals, non-metals produce faceted projections which relate to crystallographic planes. For instance, cells produced in a germanium crystal containing gallium have been reported in which cell cross-sections are square and the projection tips are pyramid shaped, comprising four octahedral {1 1 1} planes.

2.2.4.4 *Zone refining*

Extreme purification of a metal can radically improve properties such as ductility, strength and corrosion resistance. Zone refining was devised by W. G. Pfann, its development being 'driven' by the demands of the newly invented transistor for homogeneous and ultra-pure metals (e.g. Si, Ge). The method takes advantage of non-equilibrium effects associated with the 'pasty' zone separating

[8] The geological equivalent, formed by very slowly cooling magma, is the hexagonal-columnar structure of the Giant's Causeway, Northern Ireland.

the liquidus and solidus of impure metal. Considering the portion of Figure 2.11 where addition of solute lowers the liquidus temperature, the concentration of solute in the liquid, c_L, will always be greater than its concentration c_S in the solid phase; that is, the distribution coefficient $k = c_S/c_L$ is less than unity. If a bar of impure metal is threaded through a heating coil and the coil is slowly moved, a narrow zone of melt can be made to progress along the bar. The first solid to freeze is purer than the average composition by a factor of k, while that which freezes last, at the trailing interface, is correspondingly enriched in solute. A net movement of impurity atoms to one end of the bar takes place. Repeated traversing of the bar with a set of coils can reduce the impurity content well below the limit of detection (e.g. <1 part in 10^{10} for germanium). Crystal defects are also eliminated: Pfann reduced the dislocation density in metallic and semi-metallic crystals from about $3.5 \times 10^6\ \text{cm}^{-2}$ to almost zero. Zone refining has been used to optimize the ductility of copper, making it possible to cold-draw the fine-gauge threads needed for interconnects in very-large-scale integrated circuits.

Worked example

(i) In a zone-refining process, show that the subsequent impurity concentration along the refining direction x is given by

$$c_S(x) = kc_o(1-x)^{k-1},$$

where c_o is the initial impurity concentration, which is uniform throughout the specimen, and k is the partition coefficient.

(ii) Silicon is purified by a process of zone refining. If it contains an impurity (1 ppm) with a partition coefficient $k = 10^{-5}$, how much would remain in the middle of a bar after one zone pass?

Solution

(i) Consider solidification of the melted zone happening at x (between 0 and 1). The depletion of solute in the purified zone from 0 to x is:

$$\int_0^x [c_o - c_S(x')]\,dx'.$$

The remaining melted zone from x to 1 is enriched by this amount of solute, hence the solute concentration in the melted zone is

$$c_L = c_o + \frac{1}{(1-x)}\int_0^x [c_o - c_S(x')]\,dx' = \frac{1}{(1-x)}\left(c_o - \int_0^x c_S(x')\,dx'\right).$$

At the solidification front at x, partition coefficient k is the ratio of the solute concentrations in solid and liquid, i.e. $k = c_S(x)/c_L(x)$.

So

$$kc_L = \frac{k}{(1-x)}\left(c_o - \int_0^x c_S(x')\,dx'\right) = c_S(x).$$

Rearranging and differentiating gives $(1-k)\,c_S(x) = (1-x)\dfrac{dc_S(x)}{dx}$, the solution of which is $c_S(x) = kc_o(1-x)^{k-1}$.

(ii) The impurity content at $x = 0.5$ is given by

$$c_S = kc_o(1-x)^{k-1} = 10^{-5} \times 10^{-6}(1-0.5)^{-1} = 2 \times 10^{-11}.$$

2.2.5 Three-phase equilibria and reactions

2.2.5.1 The eutectic reaction

In many metallic and ceramic binary systems it is possible for two crystalline phases and a liquid to coexist. The modified Phase Rule reveals that this unique condition is invariant; that is, the temperature and all phase compositions have fixed values. Figure 2.15 shows the phase diagram for the lead–tin system. It will be seen that solid solubility is limited for each of the two component metals, with α and β representing primary solid solutions of different crystal structure. A straight line, the eutectic horizontal, passes through three phase compositions (α_e, L_e and β_e) at the temperature T_e. As will become clear when ternary systems are discussed (Section 2.2.9), this line is a collapsed three-phase triangle: at any point on this line, three phases are in equilibrium. During slow cooling or heating, when the average composition of an alloy lies between its limits, α_e and β_e, a eutectic reaction takes place in accordance with the equation $L_e \leftrightarrow \alpha_e + \beta_e$. The sharply defined minimum in the liquidus, the eutectic (easy-melting) point, is a typical feature of the reaction.

Consider the freezing of a melt, average composition 37Pb–63Sn. At the temperature T_e of approximately 180°C, it freezes abruptly to form a mechanical mixture of two solid phases, i.e. Liquid $L_e \rightarrow \alpha_e + \beta_e$. From the Lever Rule, the α/β mass ratio is approximately 9:11. As the temperature falls further, slow cooling will allow the compositions of the two phases to follow their respective solvus lines. Tie-lines across this ($\alpha + \beta$) field will provide the mass ratio for any temperature. In contrast, a hypo-eutectic alloy melt, say of composition 70Pb–30Sn, will form primary crystals of α over a range of temperature until T_e is reached. Successive tie-lines across the (α + Liquid) field show that the crystals and the liquid become enriched in tin as the temperature falls. When the liquid composition reaches the eutectic value L_e, all of the remaining liquid transforms into a two-phase mixture, as before. However, for this alloy, the final structure will comprise primary grains of α in a eutectic matrix of α and β. Similarly, one may deduce that the structure of a solidified hyper-eutectic alloy containing 30Pb–70Sn will consist of a few primary β grains in a eutectic matrix of α and β.

Low-lead or low-tin alloys, with average compositions beyond the two ends of the eutectic horizontal,[9] freeze by transforming completely over a small range of temperature into a primary phase. (Changes in composition are similar in character to those described for Figure 2.11.) When the temperature 'crosses' the relevant solvus, this primary phase becomes unstable and a small amount of second phase precipitates. Final proportions of the two phases can be obtained by superimposing a tie-line on the central two-phase field: there will be no signs of a eutectic mixture in the microstructure.

The eutectic (37Pb–63Sn) and hypo-eutectic (70Pb–30Sn) alloys chosen for the description of freezing represent two of the numerous types of solder[10] used for joining metals. Eutectic solders containing 60–65% tin are widely used in the electronics industry for making precise, high-integrity

[9] Theoretically, the eutectic horizontal cannot cut the vertical line representing a pure component: some degree of solid solubility, however small, always occurs.

[10] Soft solders for engineering purposes range in composition from 20% to 65% tin; the first standard specifications for solders were produced in 1918 by the ASTM. The USA is currently contemplating the banning of lead-bearing products; lead-free solders are being sought.

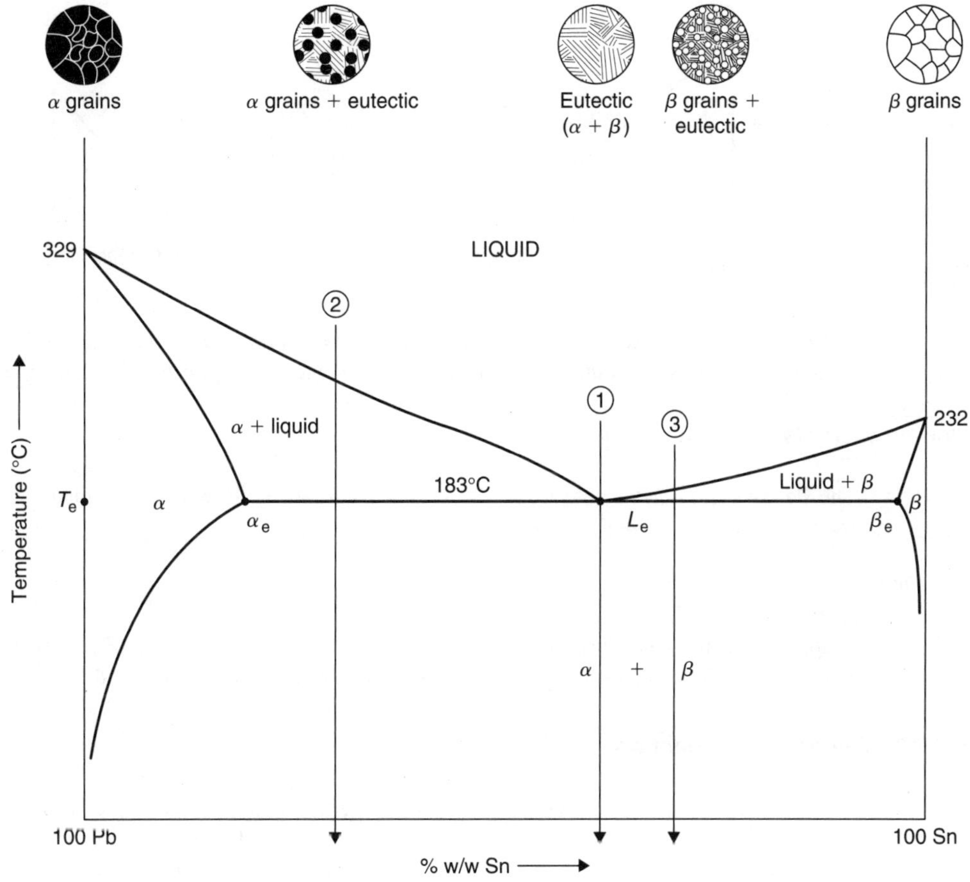

Figure 2.15 *Phase diagram for the Pb–Sn system. Alloy 1: 63Sn–37Pb; Alloy 2: 70Pb–30Sn; Alloy 3: 70Sn–30Pb.*

joints on a mass-production scale without the risk of damaging heat-sensitive components. These solders have excellent 'wetting' properties (contact angle $<10°$), a low liquidus and a negligible freezing range. The long freezing range of the 70Pb–30Sn alloy (plumbers' solder) enables the solder at a joint to be 'wiped' while 'pasty'.

The shear strength of the most widely used solders is relatively low, say 25–55 MN m^{-2}, and mechanically interlocking joints are often used. Fluxes (corrosive zinc chloride, non-corrosive organic resins) facilitate essential 'wetting' of the metal to be joined by dissolving thin oxide films and preventing reoxidation. In electronic applications, minute solder preforms have been used to solve the problems of excess solder and flux.

Figure 2.15 shows the sequence of structures obtained across the breadth of the Pb–Sn system. Cooling curves for typical hypo-eutectic and eutectic alloys are shown schematically in Figure 2.16a. Separation of primary crystals produces a change in slope while heat is being evolved. Much more heat is evolved when the eutectic reaction takes place. The lengths (duration) of the plateaux are proportional to the amounts of eutectic structure formed, as summarized in Figure 2.16b. Although it follows that cooling curves can be used to determine the form of such a simple system, it is usual to confirm details by means of microscopical examination (optical, scanning electron) and X-ray diffraction analysis.

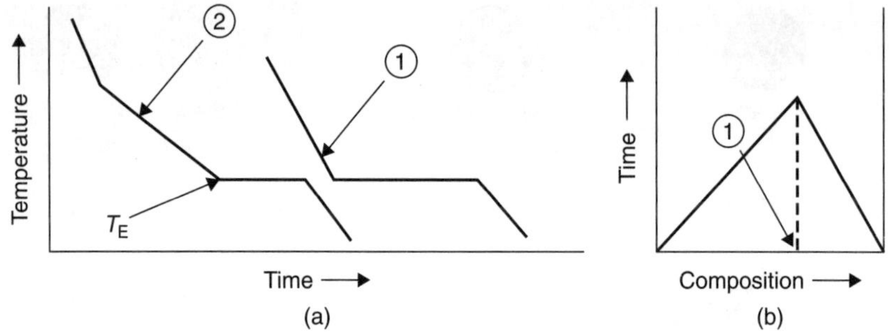

Figure 2.16 *(a) Typical cooling curves for hypo-eutectic alloy 2 and eutectic alloy 1 in Figure 2.15. (b) Dependence of duration of cooling arrest at eutectic temperature T_E on composition.*

Worked example

The Pb–Sn system has a eutectic at 183°C (see Figure 2.15). How does this manifest itself in the modified phase rule $P + F = C + 1$?

Solution

Using the metallurgist's form of the phase rule $F = C + 1 - P$ with $C = 2$, $P = 3$, then $F = 0$. This indicates that there is only one temperature at which the three phases, α, β and liquid, coexist.

2.2.5.2 The peritectic reaction

Whereas eutectic systems often occur when the melting points of the two components are fairly similar, the second important type of invariant three-phase condition, the peritectic reaction, is often found when the components have a large difference in melting points. Usually they occur in the more complicated systems; for instance, there is a cascade of five peritectic reactions in the Cu–Zn system (Figure 2.19).

A simple form of peritectic system is shown in Figure 2.17a; although relatively rare in practice (e.g. Ag–Pt), it can serve to illustrate the basic principles. A horizontal line, the key to the reaction, links three critical phase compositions; that is, α_p, β_p and liquid L_p. A peritectic reaction occurs if the average composition of the alloy crosses this line during either slow heating or cooling. It can be represented by the equation $\alpha_p + L_p \rightleftharpoons \beta_p$. Binary alloys containing less of component B than the point α_p will behave in the manner previously described for solid solutions. A melt of alloy 1, which is of peritectic composition, will freeze over a range of temperature, depositing crystals of primary α-phase. The melt composition will move down the liquidus, becoming richer in component B. At the peritectic temperature T_p, liquid of composition L_p will react with these primary crystals, transforming them completely into a new phase, β, of different crystal structure in accordance with the equation $\alpha_p + L_p \rightarrow \beta_p$. In the system shown, β remains stable during further cooling. Alloy 2 will also deposit primary α, but the reaction at temperature T_p will not consume all these crystals and the final solid will consist of β formed by peritectic reaction and residual α. Initially, the α/β mass ratio will be approximately 2.5 : 1 but both phases will adjust their compositions during subsequent cooling. In the case of alloy 3, fewer primary crystals of α form: later, they are completely destroyed by the peritectic reaction. The amount of β in the resultant mixture of β and liquid increases until the liquid disappears and an entire structure of β is produced.

The above descriptions assume that equilibrium is attained at each stage of cooling. Although very slow cooling is unlikely in practice, the nature of the peritectic reaction introduces a further

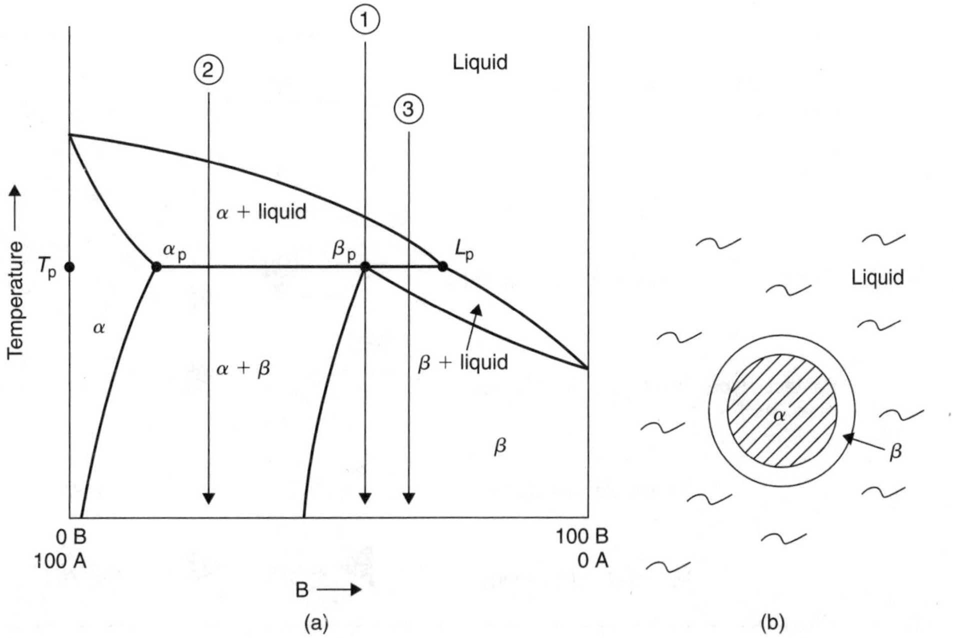

Figure 2.17 *(a) Simple peritectic system. (b) Development of a peritectic 'wall'.*

complication. The reaction product β tends to form a shell around the particles of primary α: its presence obviously inhibits the exchange of atoms by diffusion, which equilibrium demands (Figure 2.17b).

Worked example

The alloy system Ag–Pt has a peritectic form similar to that of Figure 2.17. Draw the free energy curves for the phases which exist at a temperature just above (a) and just below (b) the peritectic temperature.

Solution

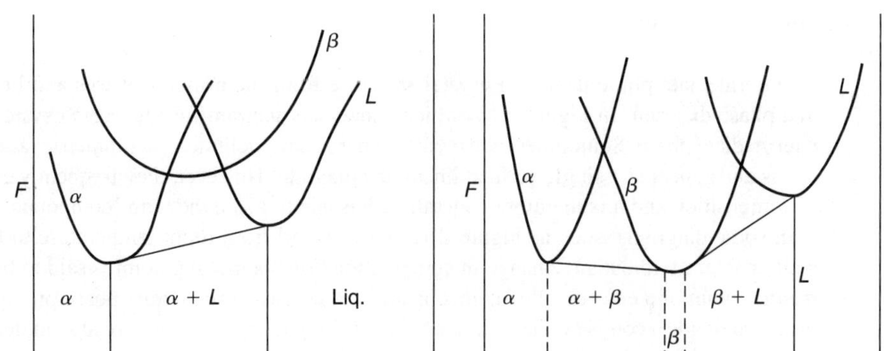

Table 2.1 *Classification of three-phase equilibria.*

Eutectic-type reactions	Eutectic	Liq $\rightleftharpoons \alpha + \beta$	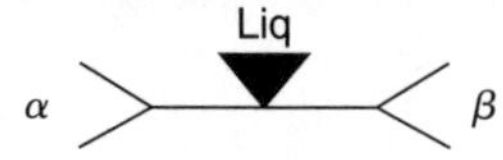	Al–Si, Pb–Sn, Cu–Ag Al_2O_3–SiO_2, Al_2O_3–ZrO_2
	Eutectoid	$\gamma \rightleftharpoons \alpha + \beta$	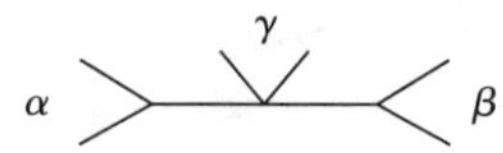	Fe–C, Cu–Zn
	Monotectic	$Liq_1 \rightleftharpoons \alpha + Liq_2$	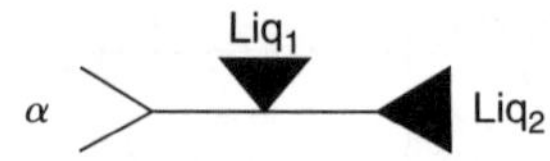	Cu–Pb, Ag–Ni SiO_2–CaO
Peritectic-type reactions	Peritectic	α + Liq $\rightleftharpoons \beta$	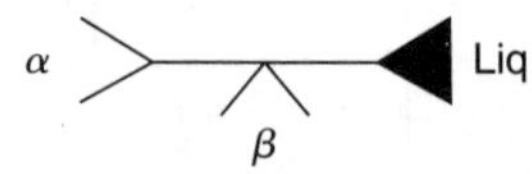	Cu–Zn, Ag–Pt
	Peritectoid	$\alpha + \beta \rightleftharpoons \gamma$	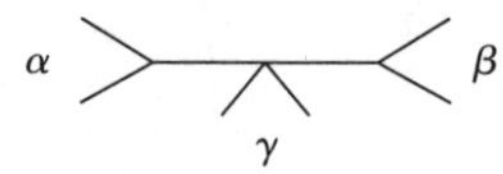	Ag–Al
	Syntectic	$Liq_1 + Liq_2 \rightleftharpoons \alpha$	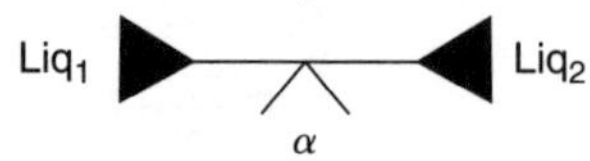	Na–Zn

2.2.5.3 *Classification of three-phase equilibria*

The principal invariant equilibria involving three condensed (solid, liquid) phases can be conveniently divided into eutectic and peritectic types, and classified in the manner shown in Table 2.1. Interpretation of these reactions follows the methodology already set out for the more common eutectic and peritectic reactions.

The inverse relation between eutectic- and peritectic-type reactions is apparent from the line diagrams. Eutectoid and peritectoid reactions occur wholly in the solid state. (The eutectoid reaction $\gamma \leftrightarrow \alpha + Fe_3C$ is the basis of the heat treatment of steels.) In all the systems so far described, the components have been completely miscible in the liquid state. In monotectic and syntectic systems, the liquid-phase field contains a region in which two different liquids (L_1 and L_2) are immiscible.

2.2.6 Intermediate phases

An intermediate phase differs in crystal structure from the primary phases and lies between them in a phase diagram. In Figure 2.18, which shows the diagram for the Mg–Si system, Mg_2Si is the intermediate phase. Sometimes intermediate phases have definite stoichiometric ratios of constituent atoms and appear as a single vertical line in the diagram. However, they frequently exist over a range of composition and it is therefore generally advisable to avoid the term 'compound'.

In some diagrams, such as Figure 2.18, they extend from room temperature to the liquidus and melt or freeze without any change in composition. Such a melting point is said to be congruent; the melting point of a eutectic alloy is incongruent. A congruently melting phase provides a convenient means to divide a complex phase diagram (binary or ternary) into more readily understandable parts. For instance, an ordinate through the vertex of the intermediate phase in Figure 2.18 produces two simple eutectic subsystems. Similarly, an ordinate can be erected to pass through the minimum (or maximum) of the liquidus of a solid solution (Figure 2.31).

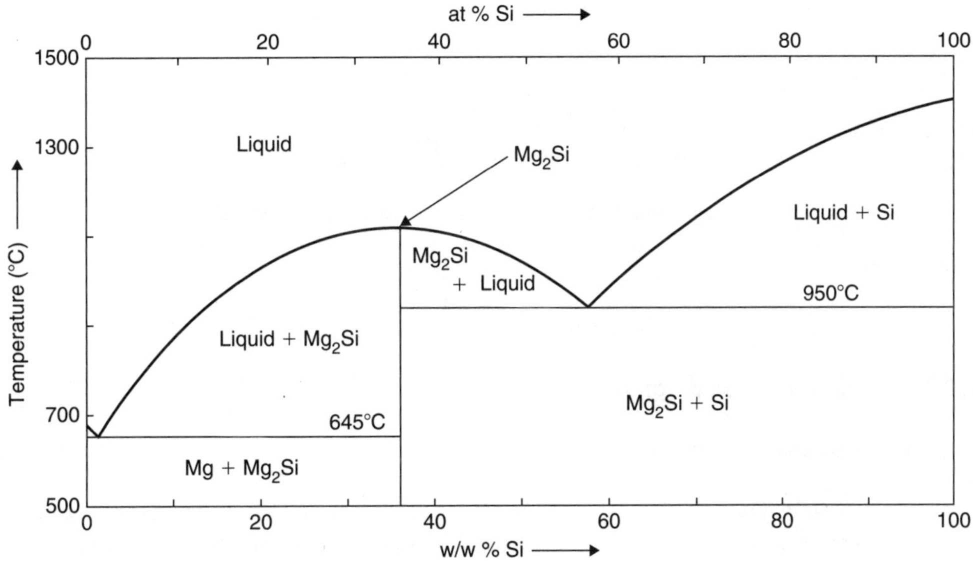

Figure 2.18 *Phase diagram for the Mg–Si system showing intermediate phase Mg_2Si (from Brandes and Brook,* Smithells Metals Reference Book, *1998).*

In general, intermediate phases are hard and brittle, having a complex crystal structure (e.g. Fe_3C, $CuAl_2$ (θ)). For instance, it is advisable to restrict time and temperature when soldering copper alloys, otherwise it is possible for undesirable brittle layers of Cu_3Sn and Cu_6Sn_5 to form at the interface.

2.2.7 Limitations of phase diagrams

Phase diagrams are extremely useful in the interpretation of metallic and ceramic structures but they are subject to several restrictions. Primarily, they identify which phases are likely to be present and provide compositional data. The most serious limitation is that they give no information on the structural form and distribution of phases (e.g. lamellae, spheroids, intergranular films, etc.). This is unfortunate, since these two features, which depend upon the surface energy effects between different phases and strain energy effects due to volume and shape changes during transformations, play an important role in the mechanical behavior of materials. This is understood if we consider a two-phase ($\alpha + \beta$) material containing only a small amount of β-phase. The β-phase may be dispersed evenly as particles throughout the α-grains, in which case the mechanical properties of the material would be largely governed by those of the α-phase. However, if the β-phase is concentrated at grain boundary surfaces of the α-phase, then the mechanical behavior of the material will be largely dictated by the properties of the β-phase. For instance, small amounts of sulfide particles, such as grey manganese sulfide (MnS), are usually tolerable in steels but sulfide films at the grain boundaries cause unacceptable embrittlement.

A second limitation is that phase diagrams portray only equilibrium states. As indicated in previous sections, alloys are rarely cooled or heated at very slow rates. For instance, quenching, as practiced in the heat treatment of steels, can produce metastable phases known as martensite and bainite that will then remain unchanged at room temperature. Neither appears in phase diagrams. In such cases it is necessary to devise methods for expressing the rate at which equilibrium is approached and its temperature dependency.

Whilst it is often convenient to express the composition of an alloy in terms of the percentage by weight (wt%) of the constituent elements, it is sometimes necessary to describe the composition in terms of the percentage of atoms x and y. It is therefore convenient to be able to convert from wt% to at.%, for which it is necessary to know the atomic weight of the constituent atoms. Thus, the conversion of wt% to at.% is given by

$$(\text{at.\%})_x = \frac{(\text{wt\%})_x A_y}{(\text{wt\%})_x A_y + (\text{wt\%})_y A_x} \times 100,$$

where A_x and A_y are the atomic weights of atoms x and y respectively. Correspondingly the conversion of at.% to wt% is

$$(\text{wt\%})_x = \frac{(\text{at.\%})_x A_x}{(\text{at.\%})_x A_x + (\text{at.\%})_y A_y} \times 100.$$

2.2.8 Some key phase diagrams

2.2.8.1 *Copper–zinc system*

Phase diagrams for most systems, metallic and ceramic, are usually more complex than the examples discussed so far. Figure 2.19 for the Cu–Zn system illustrates this point. The structural characteristics and mechanical behavior of the industrial alloys known as brasses can be understood in terms of the

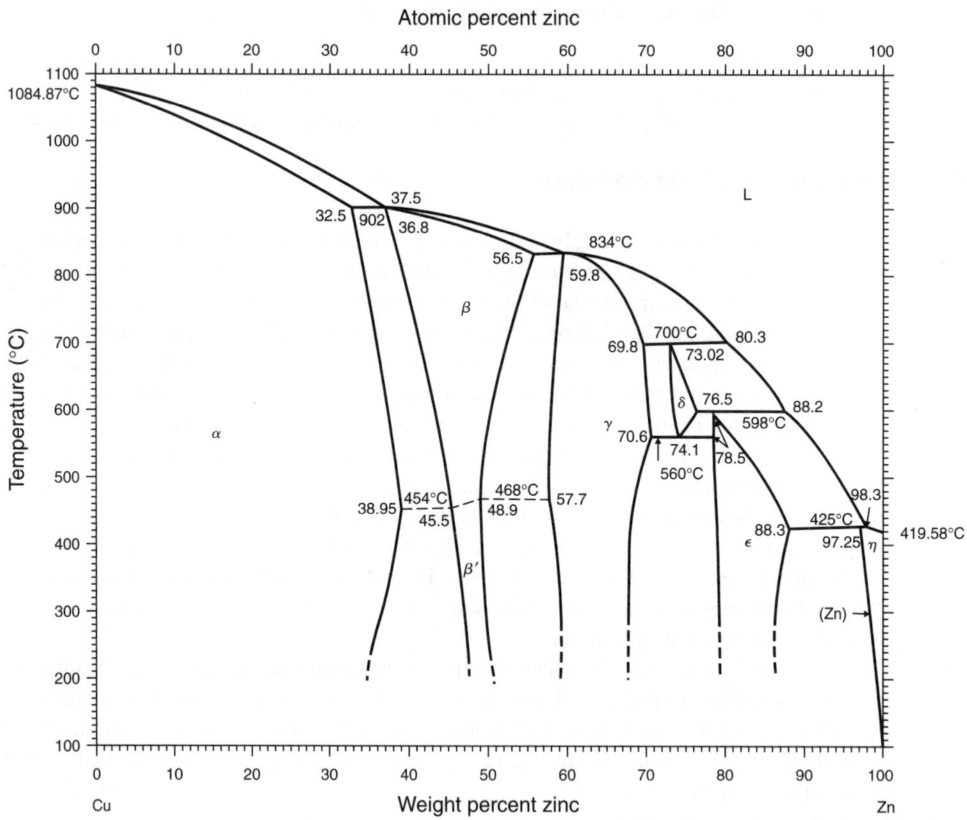

Figure 2.19 *Phase diagram for copper–zinc (from* Binary Alloy Phase Diagrams, *1996, ASM International).*

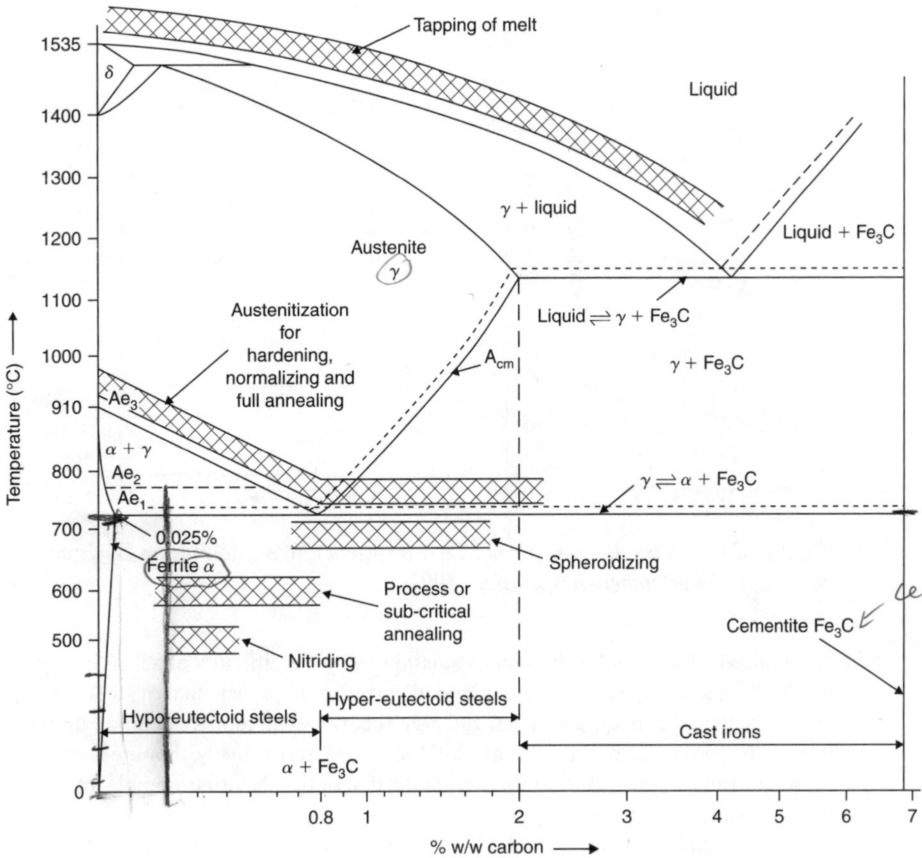

Figure 2.20 *Phase diagram for the Fe–C system (dotted lines represent iron–graphite equilibrium).*

copper-rich end of this diagram. Copper can dissolve up to 40% w/w of zinc and cooling of any alloy in this range will produce an extensive primary solid solution (fcc-α). By contrast the other primary solid solution (η) is extremely limited. A special feature of the diagram is the presence of four intermediate phases (β, γ, δ, ε). Each is formed during freezing by a peritectic reaction and each exists over a range of composition. Another notable feature is the order–disorder transformation which occurs in alloys containing about 50% zinc over the temperature range 450–470°C. Above this temperature range, bcc β-phase exists as a disordered solid solution. At lower temperatures, the zinc atoms are distributed regularly on the bcc lattice: this ordered phase is denoted by β'.

Suppose that two thin plates of copper and zinc are held in very close contact and heated at a temperature of 400°C for several days. Transverse sectioning of the diffusion couple will reveal five phases in the sequence $\alpha/\beta/\gamma/\varepsilon/\eta$, separated from each other by a planar interface. The δ-phase will be absent because it is unstable at temperatures below its eutectoid horizontal (560°C). Continuation of diffusion will eventually produce one or two phases, depending on the original proportions of copper and zinc.

2.2.8.2 Iron–carbon system

The diagram for the part of the Fe–C system shown in Figure 2.20 is the basis for understanding the microstructures of the ferrous alloys known as steels and cast irons. Dissolved carbon clearly has a

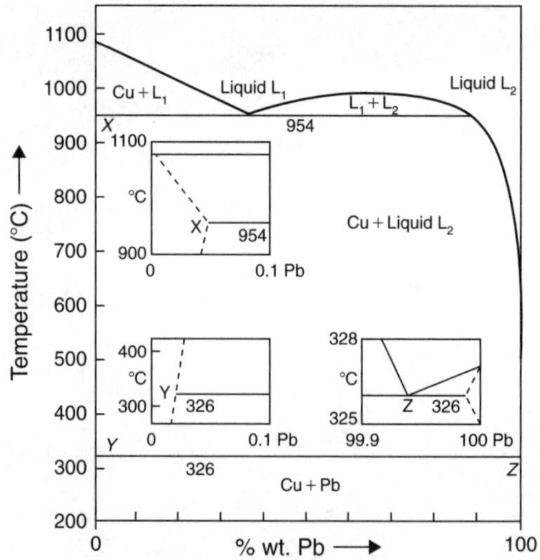

Figure 2.21 *Phase diagram for the Cu–Pb system (from* Equilibrium Diagrams, *by permission of the Copper Development Association, 1993).*

pronounced effect upon the liquidus, explaining why the difficulty of achieving furnace temperatures of 1600°C caused large-scale production of cast irons to predate that of steel. The three allotropes of pure iron are α-Fe (bcc), γ-Fe (fcc) and δ-Fe (bcc).[11] Small atoms of carbon dissolve interstitially in these allotropes to form three primary solid solutions: respectively, they are α-phase (ferrite), γ-phase (austenite) and δ-phase. At the other end of the diagram is the orthorhombic intermediate phase Fe_3C, which is known as cementite.

The large difference in solid solubility of carbon in austenite and ferrite, together with the existence of a eutectoid reaction, are responsible for the versatile behavior of steels during heat treatment. Ae_1, Ae_2, Ae_3 and A_{cm} indicate the temperatures at which phase changes occur: they are arrest points for equilibria detected during thermal analysis. For instance, slow cooling enables austenite (0.8% C) to decompose eutectoidally at the temperature Ae_1 and form the microconstituent pearlite, a lamellar composite of soft, ductile ferrite (initially 0.025% C) and hard, brittle cementite (6.67% C). Quenching of austenite from a temperature above Ae_3 forms a hard metastable phase known as martensite. From the diagram one can see why a medium-carbon (0.4%) steel must be quenched from a higher Ae_3 temperature than a high-carbon (0.8%) steel. Temperature and composition 'windows' for some important heat-treatment operations have been superimposed upon the phase diagram.

2.2.8.3 Copper–lead system

The phase diagram for the Cu–Pb system (Figure 2.21) provides an interesting example of extremely limited solubility in the solid state and partial immiscibility in the liquid state. The two components differ greatly in density and melting point. Solid solutions, α and β, exist at the ends of the diagram. The 'miscibility gap' in the liquid phase takes the form of a dome-shaped two-phase ($L_1 + L_2$) field. At temperatures above the top of the dome, the critical point, liquid miscibility is complete. The upper isothermal represents a monotectic reaction, i.e. $L_1 \rightleftharpoons \alpha + L_2$.

[11] The sequence omits β-Fe, a term once used to denote a non-magnetic form of α-Fe which exists above the Curie point.

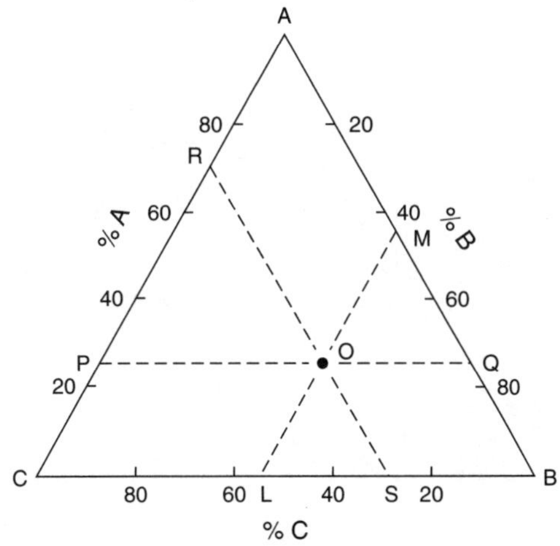

Figure 2.22 *Triangular representation of a ternary alloy system ABC.*

On cooling, a hyper-monotectic 50Cu–50Pb melt will separate into two liquids of different composition. The degree of separation depends on cooling conditions. Like oil and water, the two liquids may form an emulsion of droplets or separate into layers according to density. At a temperature of 954°C, the copper-rich liquid L_1 disappears, forming α crystals and more of the lead-rich liquid L_2. This liquid phase gets richer in lead and eventually decomposes by eutectic reaction, i.e. $L_2 \leftrightarrow \alpha + \beta$. (Tie-lines can be used for all two-phase fields, of course; however, because of density differences, mass ratios may differ greatly from observed volume ratios.)

The hypo-monotectic 70Cu–30Pb alloy, rapidly cast, has been used for steel-backed bearings: dispersed friction-reducing particles of lead-rich β are supported in a supporting matrix of copper-rich α. Binary combinations of conductive metal (Cu, Ag) and refractory arc-resistant metal (W, Mo, Ni) have been used for electrical contacts (e.g. 60Ag–40Ni). These particular monotectic systems, with their liquid immiscibility, are difficult to cast and are therefore made by powder metallurgy techniques.

2.2.9 Ternary phase diagrams

In considering equilibrium diagrams for ternary systems, three independent variables have to be specified, i.e. two to define the composition and the third to define temperature. Consequently a 3-D space is required and an equilateral triangle ABC, with a temperature axis perpendicular to it, is used. The corners of the triangle represent the pure metals, the sides the three appropriate binaries and a point inside the triangle represents a ternary alloy composition. For the ternary alloy O in Figure 2.22, the concentration of the components is given by $C_A = PC$, $C_C = RA$ and $C_B = RP$ and since $C_A + C_B + C_C = 1$, then $PC + RA + RP = 1$. From the triangular representation ABC, all alloys lying on a line (i) parallel to a side must have the same composition of the component opposite to the base, e.g. C_A is constant along PQ, and (ii) through one corner have a constant proportion of the other two components.

2.2.9.1 Ternary diagrams for complete solid miscibility

This is shown in Figure 2.23a. The binaries along each side are similar to that of Figure 2.11. The liquidus and solidus lines of the binaries become liquidus and solidus surfaces in the ternary system,

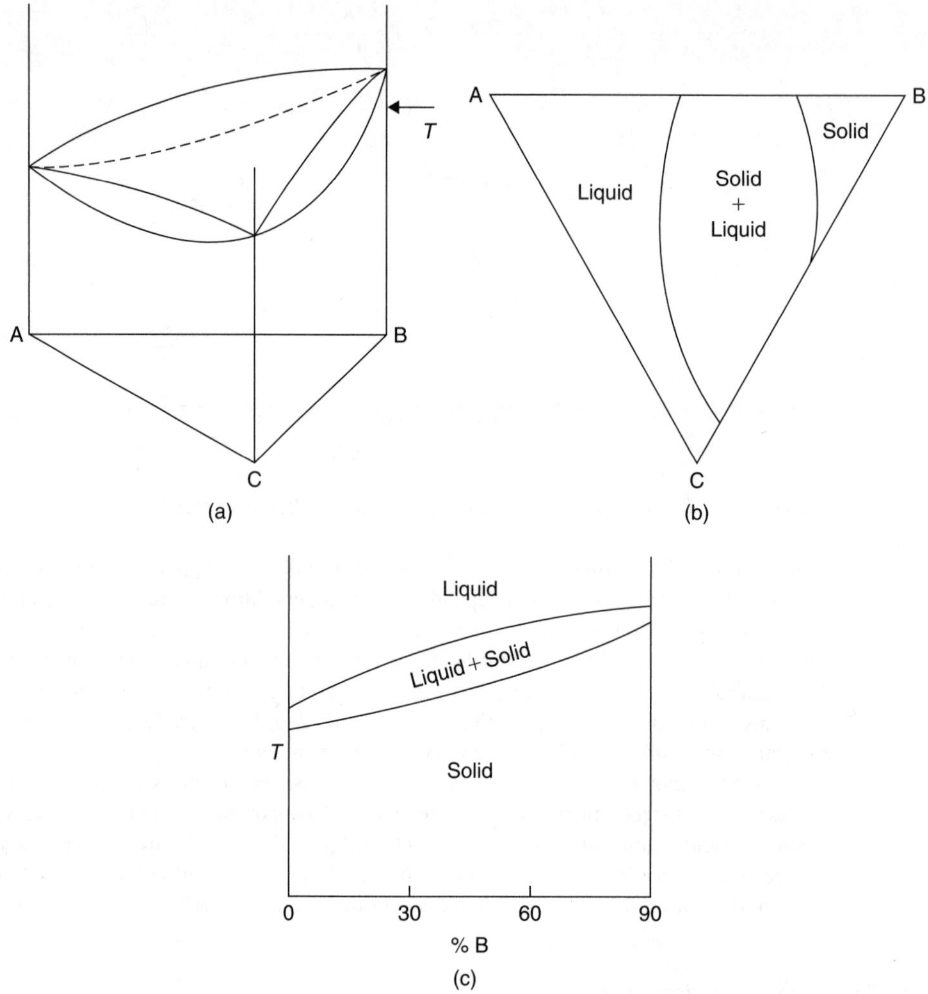

Figure 2.23 *(a) Ternary diagram for a system with complete solid solubility. (b) Horizontal section at temperature T through the diagram shown in (a). (c) Vertical section through diagram shown in (a).*

the (solid + liquid) region forming a convex-lens shape and the solid phase becoming a volume in the ternary bounded by a surface. An example of this system is Ag–Au–Pd.

A series of isothermal or horizontal sections can be examined over the complete temperature range to give a full representation of the equilibrium diagram. For the simple diagram of Figure 2.23a horizontal sections from room temperature up to the solidus surface show no variation, merely solid phase. At temperature T, the horizontal section cuts through the (solid + liquid) phase field and looks like Figure 2.23b; alloys can be liquid, solid or (solid + liquid) depending on the composition. The liquidus surface varies in extent with temperature increasing or decreasing and can be shown as a 'contour' line on the diagram.

Vertical sections may also be useful, particularly if they are taken either (i) parallel to one side of the base triangle, i.e. at constant proportion of one of the components (see Figure 2.23c), or (ii) along a line through a corner of the triangle.

2.2.9.2 Ternary eutectic

For the case when there is complete immiscibility in the solid state, e.g. Pb–Bi–Sn, the three binary eutectics are constructed on the sides of the diagrams (Figure 2.24a). The liquidus of the ternary forms three separate smooth liquidus surfaces, extending inwards from the melting point of each pure metal and sloping downwards to form three valleys which meet at E, the ternary eutectic point. A projection of this liquidus surface on to the base is shown in Figure 2.24b.

Let us consider the solidification of an alloy X which starts at the temperature T_1 when solid of pure A separates out and the liquid becomes richer in B and C. The ratio of B to C in the liquid remains constant as the temperature falls to T_4, when the liquid composition is given by the point X_1 in the valley between the two liquidus surfaces. This liquid is in equilibrium with both solid A (when solid/liquid $= XX_1/AX$) and pure B. On further cooling the composition of the liquid follows the valley X_1E and in the secondary stage of freezing pure B separates out as well as A. The freezing behavior for B-rich or C-rich alloys is similar and the liquid phase eventually reaches either the valley E_1E or E_2E, depositing either A and C or B and C before finally reaching the ternary eutectic point E. Here the liquid is in equilibrium with A, B and C, and at T_E the three-phase mixture freezes into the ternary eutectic structure. Every ternary alloy will completely solidify at the same temperature T_E and so the solidus or tertiary surface is a horizontal plane across the diagram; this can be seen in the vertical section shown in Figure 2.24c. The compositions and proportions of the phases present in an alloy at a given temperature can be found from horizontal sections. Four such sections at temperatures above T_E are shown in Figure 2.25. From these sections it can be seen that boundaries between single- and two-phase regions are curved, between two- and three-phase regions straight lines, and three-phase regions are bounded by three straight lines. In a three-phase field the compositions of the three phases are given by the corners of the triangle, e.g. UVW in Figure 2.26. For any alloy in this field only the proportions of U, V and W change and an alloy X has the proportions $U{:}V{:}W$ given by $MK{:}WH{:}VK$.

2.2.9.3 Ternary diagrams with solid solutions

If primary solid solubility exits, then instead of pure A, B or C separating out, the solid solutions α, β or γ form. In Figure 2.27 the liquidus surface has the same form as before, but beneath it there is a new phase boundary surface near each corner of the diagram, representing the limit of solid solubility in each of the pure metals A, B and C.

Horizontal sections above T_E are shown in Figure 2.28 and below T_E the horizontal section is contained in the base triangle ABC of Figure 2.27. It is readily seen that alloys with compositions near the corners solidify in one stage to primary solid solutions α, β or γ, while alloys near the sides of the diagram do not undergo a tertiary stage of solidification, since no ternary eutectic (which is now a phase mixture of α, β and γ) is formed.

2.2.9.4 Ternary diagrams with a peritectic

There are no new principles involved if other reactions are introduced, but a peritectic system AB is included in Figure 2.29 to illustrate the way in which new features may be considered. The curve EL running into the body of the ternary system represents the path of the liquid composition taking part in the $L+\alpha \rightarrow \beta$ peritectic reaction. The curve DL represents the liquid composition for the eutectic reaction, $L \rightarrow \alpha+\gamma$. The intersection point L is the peritectic point where $L+\alpha \rightarrow \beta+\gamma$. From L the eutectic valley LF runs to the lower eutectic where $L \rightarrow \beta+\gamma$. At the peritectic temperature four

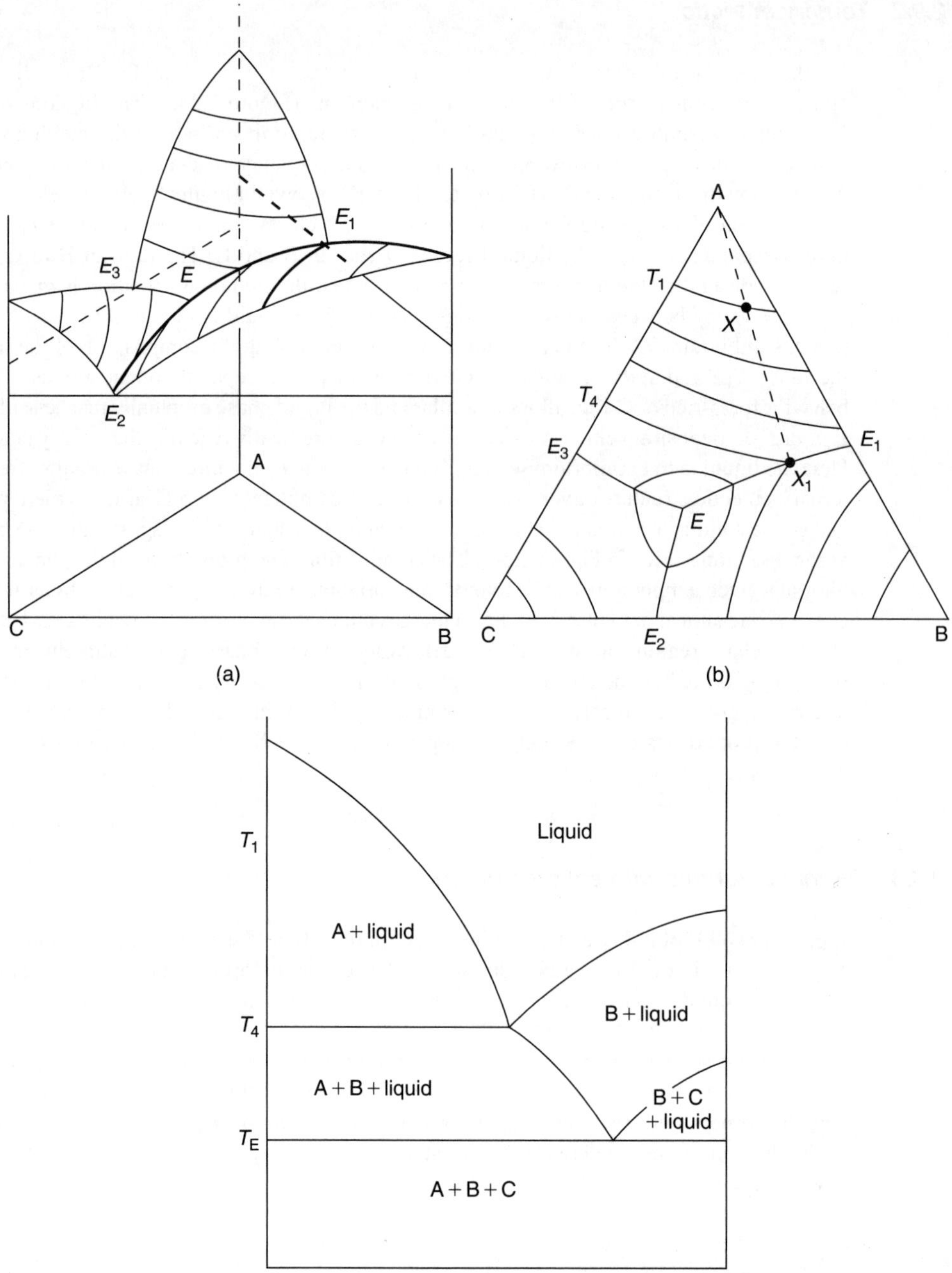

Figure 2.24 *(a) Schematic ternary system for complete immiscibility in the solid state. (b) Schematic ternary system for projection of liquidus surface on to base. (c) Schematic ternary system for a vertical section.*

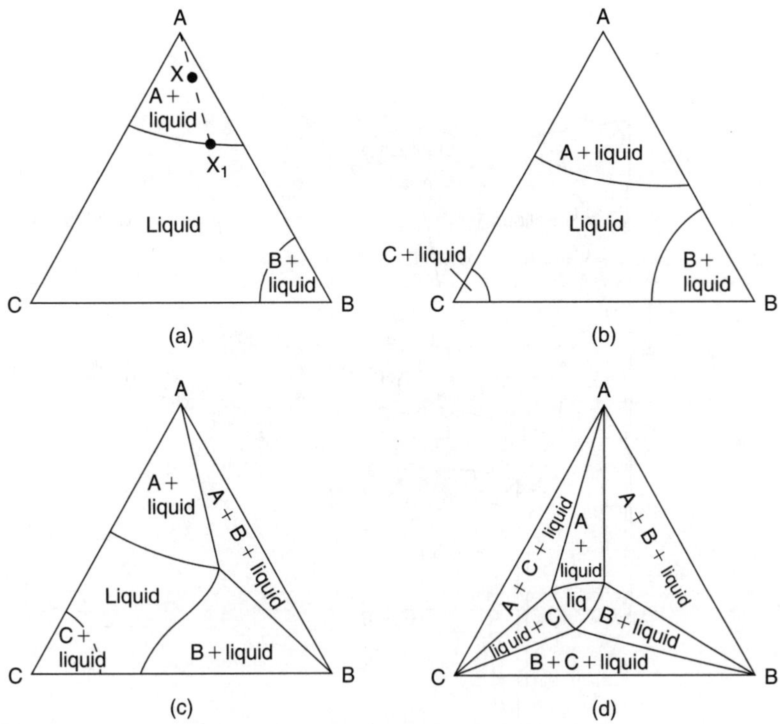

Figure 2.25 *Horizontal sections at four temperatures above T_E through the ternary diagram in Figure 2.24.*

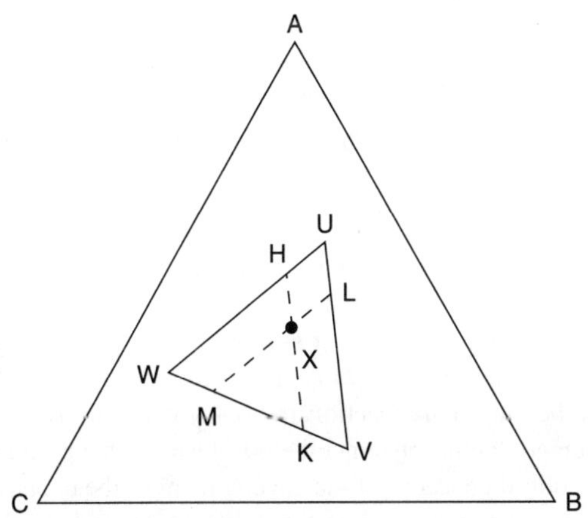

Figure 2.26 *Constitution in a three-phase field UVW.*

phases L, α, β and γ, with compositions lying at the corners of the shaded trapezium, coexist; the trapezium is termed the peritectic plane.

Horizontal sections for five temperatures T_1 to T_5 are given in Figure 2.30 and it is instructive to follow the solidification of four alloys lying in the four different sections of the trapezium (see isotherm T_4). Alloy 1 starts to solidify with a primary separation of α followed by a secondary deposition of

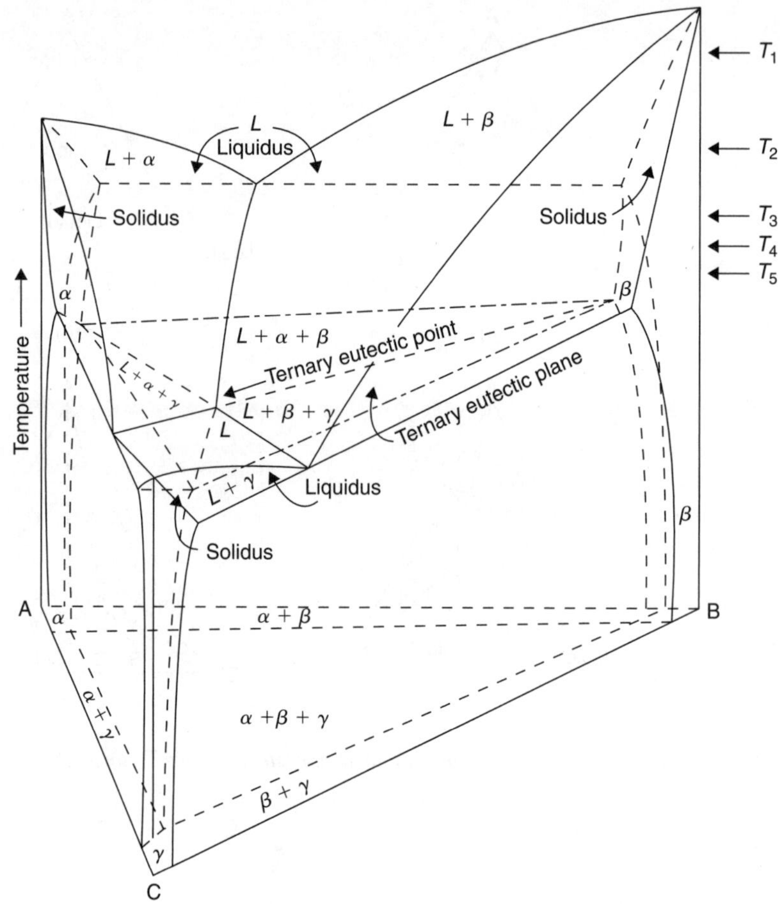

Figure 2.27 *A ternary phase diagram with solid solutions (after Rhines,* Phase Diagrams in Metallurgy, 1956).

$\alpha + \beta$. On cooling through the four-phase reaction temperature $L + \alpha + \beta \rightarrow \alpha + \beta + \gamma$. For alloy 2, primary α also separates initially but the secondary deposition is $\alpha + \gamma$ and in the four-phase reaction $L + \alpha + \gamma \rightarrow \alpha + \beta + \gamma$. For alloys 3 and 4, the primary α is quite small and the secondary deposition is $\alpha + \beta$ and $\alpha + \gamma$, respectively. Both alloys will decrease the α and liquid content to form $\beta + \gamma$ in the four-phase reaction and $\beta + \gamma$ will continue to form at lower temperatures until the liquid is consumed. It must be remembered, however, that peritectic alloys do not often solidify under equilibrium condition because of the envelopment of the α-phase by reaction products which hinder diffusion. An excess amount of α is generally present and also liquid after the four-phase reaction.

2.2.9.5 Ternary diagrams containing intermetallic phases

More complex ternary systems can often be broken down into simpler, basic forms in certain regions of the diagram. A common example of this is of a system containing an intermetallic phase. Figure 2.31 shows such a phase δ in the binary system AB and this forms a quasi-binary with the component C; a vertical section along δC is equivalent to a binary eutectic with solid solubility. This quasi-binary

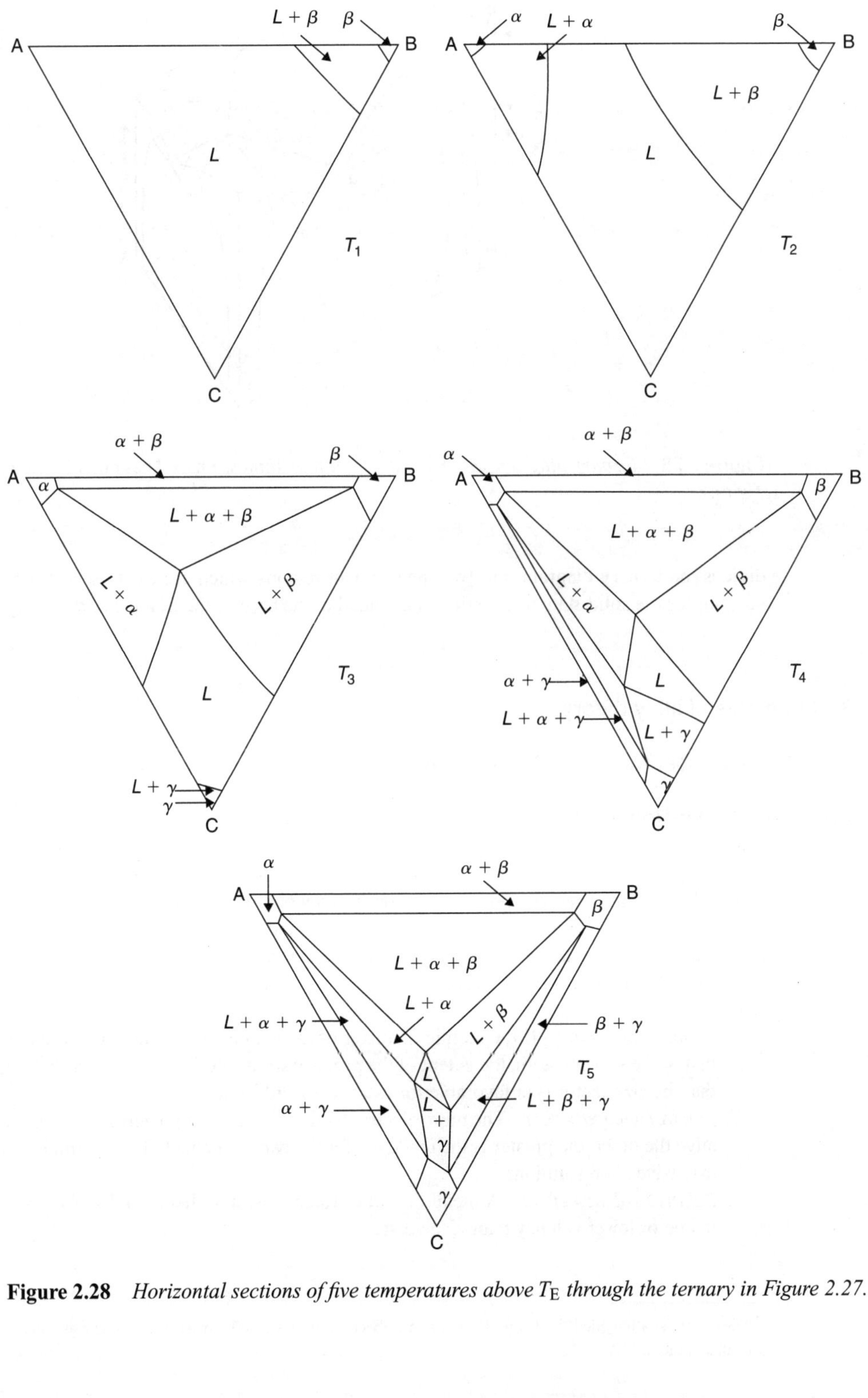

Figure 2.28 *Horizontal sections of five temperatures above* T_E *through the ternary in Figure 2.27.*

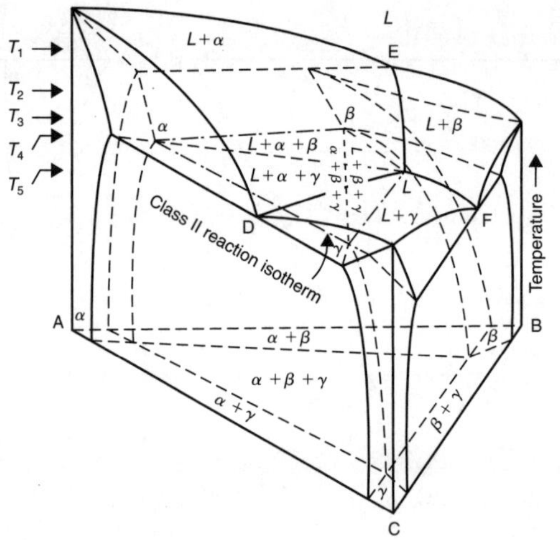

Figure 2.29 *Ternary diagram with a peritectic (after Rhines,* Phase Diagrams in Metallurgy, *1956).*

divides the ternary diagram into two independent regions which are easily seen from a horizontal section; for the solid state this section is included in the base of the triangle ABC of Figure 2.31.

2.3 Principles of alloy theory

2.3.1 Primary substitutional solid solutions

2.3.1.1 The Hume-Rothery rules

The key phase diagrams outlined in Section 2.2.8 exhibit many common features (e.g. primary solid solutions, intermediate phases) and for systems based on simple metals some general rules[12] governing the formation of alloys have been formulated. These rules can form a useful basis for predicting alloying behavior in other more complex systems.

In brief, the rules for primary solid solubility are as follows:

1. *Atomic size factor* – If the atomic diameter of the solute atom differs by more than 15% from that of the solvent atom, the extent of the primary solid solution is small. In such cases it is said that the size factor is unfavorable for extensive solid solution.
2. *Electrochemical effect* – The more electropositive the one component and the more electronegative the other, the greater is the tendency for the two elements to form compounds rather than extensive solid solutions.
3. *Relative valency effect* – A metal of higher valency is more likely to dissolve to a large extent in one of lower valency than vice versa.

[12] These are usually called the Hume-Rothery rules because it was chiefly W. Hume-Rothery and his colleagues who formulated them.

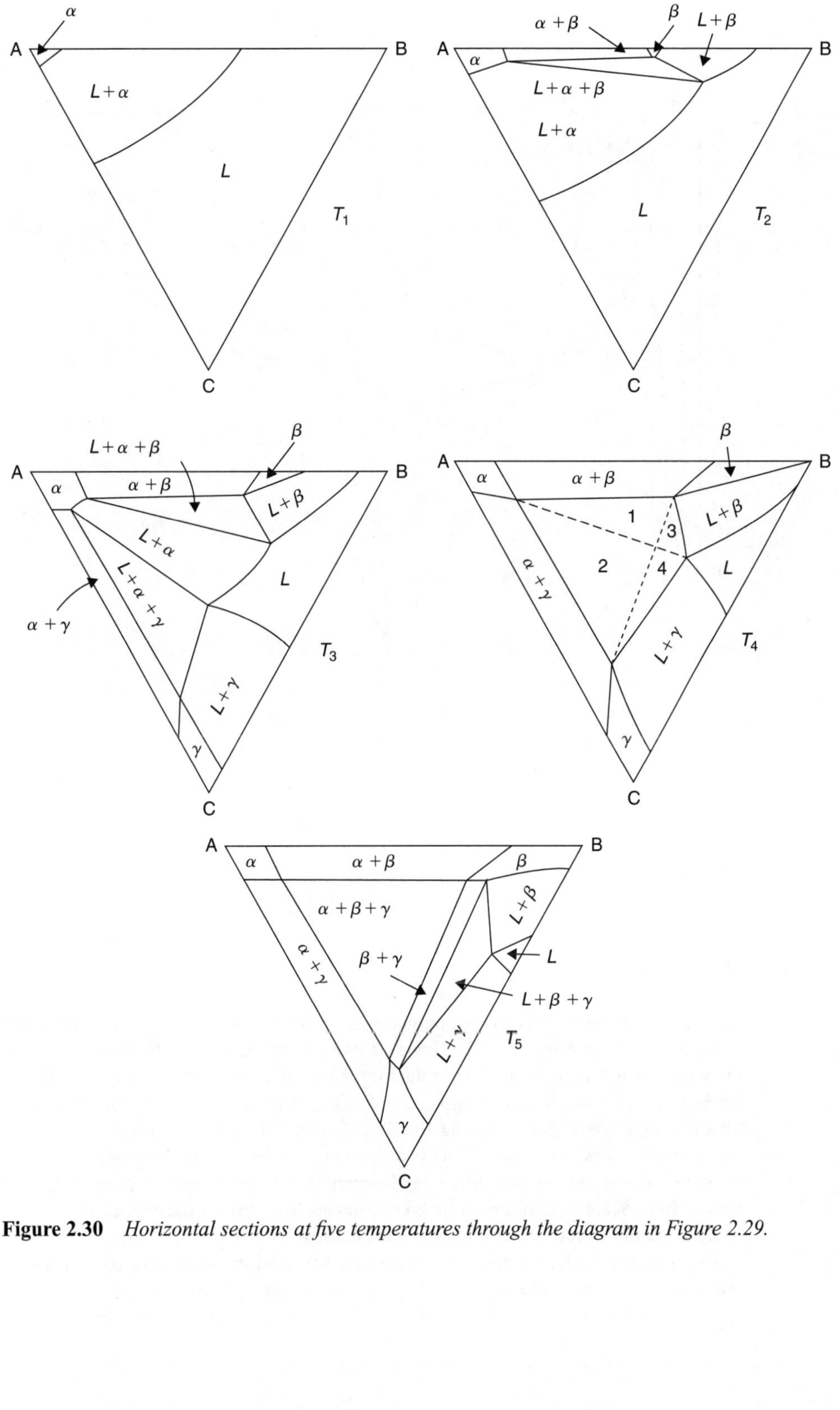

Figure 2.30 *Horizontal sections at five temperatures through the diagram in Figure 2.29.*

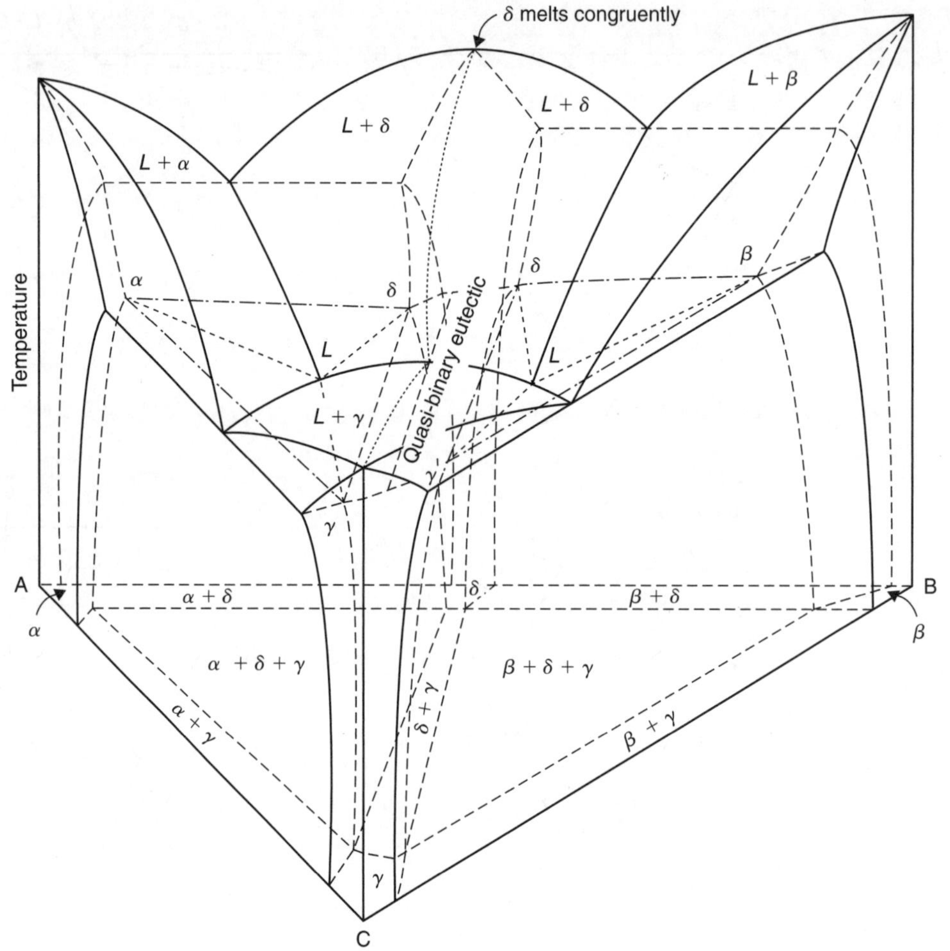

Figure 2.31 *Ternary system containing an intermediate phase δ, which forms a quasi-binary with component C (after Rhines,* Phase Diagrams in Metallurgy, *1956).*

2.3.1.2 Size-factor effect

Two metals are able to form a continuous range of solid solutions only if they have the same crystal structure (e.g. copper and nickel). However, even when the crystal structure of the two elements is the same, the extent of the primary solubility is limited if the atomic size of the two metals, usually taken as the closest distance of approach of atoms in the crystal of the pure metal, is unfavorable. This is demonstrated in Figure 2.32 for alloy systems where rules 2 and 3 have been observed, i.e. the electrochemical properties of the two elements are similar and the solute is dissolved in a metal of lower valency. As the size difference between the atoms of the two component metals A and B approaches 15%, the equilibrium diagram changes from that of the copper–nickel type to one of a eutectic system with limited primary solid solubility.

The size-factor effect is due to the distortion produced in the parent lattice around the dissolved misfitting solute atom. In these localized regions the interatomic distance will differ from that given by the minimum in the *E*–*r* curve of Figure 1.2, so that the internal energy and hence the free energy,

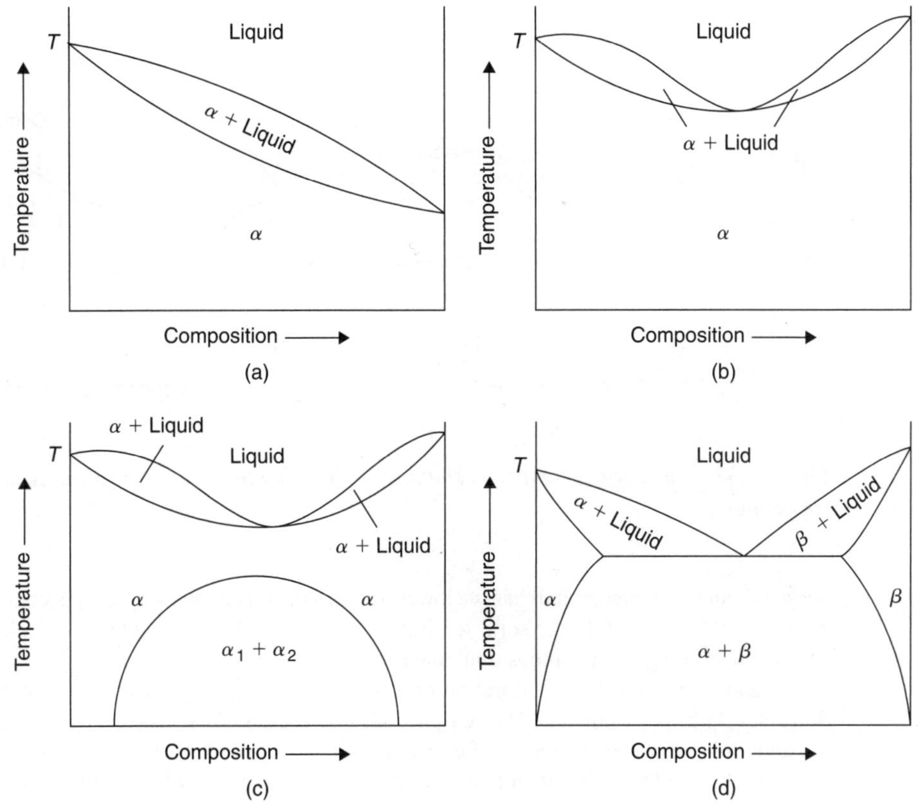

Figure 2.32 *Effect of size factor on the form of the equilibrium diagram. Examples include: (a) Cu–Ni, Au–Pt; (b) Ni–Pt; (c) Au–Ni; and (d) Cu–Ag.*

G, of the system is raised. In the limit when the lattice distortion is greater than some critical value, the primary solid solution becomes thermodynamically unstable relative to some other phase.

2.3.1.3 Electrochemical effect

This effect is best demonstrated by reference to the alloying behavior of an electropositive solvent with solutes of increasing electronegativity. The electronegativity of elements in the Periodic Table increases from left to right in any period and from bottom to top in any group. Thus, if magnesium is alloyed with elements of Group IV the compounds formed, Mg_2 (Si, Sn or Pb), become more stable in the order lead, tin, silicon, as shown by their melting points, 550, 778 and 1085°C, respectively. In accordance with rule 2 the extent of the primary solid solution is small (≈7.75 at.%, 3.35 at.% and negligible, respectively, at the eutectic temperature) and also decreases in the order lead, tin, silicon. Similar effects are also observed with elements of Group V, which includes the elements bismuth, antimony and arsenic, when the compounds Mg_3(Bi, Sb or As)$_2$ are formed.

The importance of compound formation in controlling the extent of the primary solid solution can be appreciated by reference to Figure 2.33, where the curves represent the free energy versus composition relationship between the α-phase and compound at a temperature T. It is clear from Figure 2.33a that at this temperature the α-phase is stable up to a composition c_1, above which the

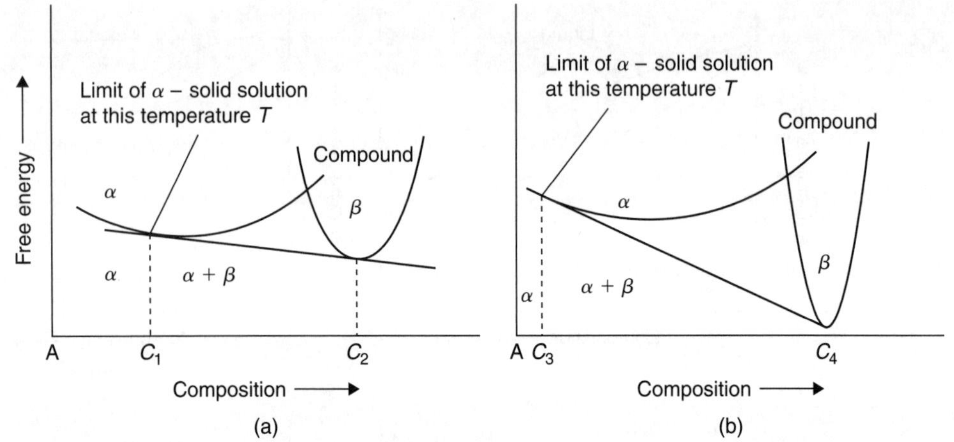

Figure 2.33 *Influence of compound stability on the solubility limit of the α-phase at a given temperature.*

phase mixture (α + compound) has the lower free energy. When the compound becomes more stable, as shown in Figure 2.33b, the solid solubility decreases and correspondingly the phase mixture is now stable over a greater composition range, which extends from c_3 to c_4.

The above example is an illustration of a more general principle that the solubility of a phase decreases with increasing stability, and may also be used to show that the concentration of solute in solution increases as the radius of curvature of the precipitate particle decreases. Small precipitate particles are less stable than large particles and the variation of solubility with particle size is recognized in classical thermodynamics by the Thomson–Freundlich equation:

$$\ln\,[c(r)/c] = 2\gamma\Omega/kTr, \tag{2.7}$$

where $c(r)$ is the concentration of solute in equilibrium with small particles of radius r, c the equilibrium concentration, γ the precipitate/matrix interfacial energy and Ω the atomic volume (see Chapter 7).

2.3.1.4 *Relative valency effect*

This is a general rule for alloys of the univalent metals, copper, silver and gold, with those of higher valency. Thus, for example, copper will dissolve approximately 40% zinc in solid solution but the solution of copper in zinc is limited. For solvent elements of higher valencies the application is not so general, and in fact exceptions, such as that exhibited by the magnesium–indium system, occur.

2.3.1.5 *The primary solid solubility boundary*

It is not yet possible to predict the exact form of the α-solid solubility boundary, but in general terms the boundary may be such that the range of primary solid solution either (1) increases or (2) decreases with rise in temperature. Both forms arise as a result of the increase in entropy which occurs when

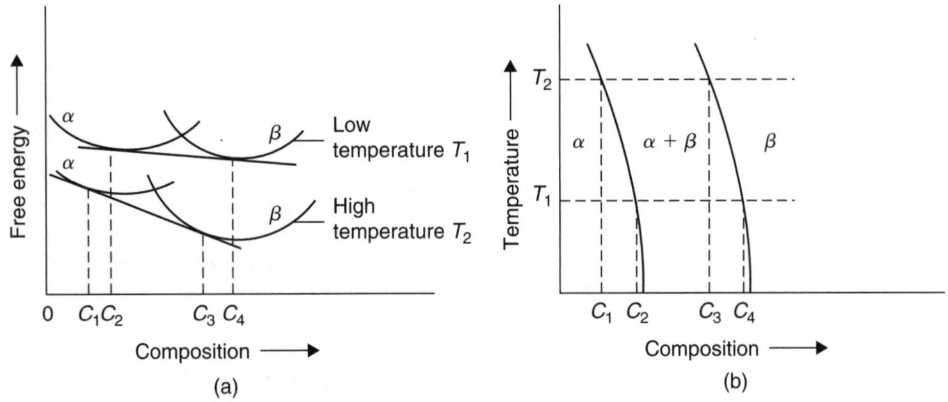

Figure 2.34 *(a) The effect of temperature on the relative positions of the α- and β-phase free energy curves for an alloy system having a primary solid solubility of the form shown in (b).*

solute atoms are added to a solvent. It will be remembered that this entropy of mixing is a measure of the extra disorder of the solution compared with the pure metal.

The most common form of phase boundary is that indicating that the solution of one metal in another increases with rise in temperature. This follows from thermodynamic reasoning, since increasing the temperature favors the structure of highest entropy (because of the $-TS$ term in the relation $G = H - TS$) and in alloy systems of the simple eutectic type an α-solid solution has a higher entropy than a phase mixture $(\alpha + \beta)$. Thus, if the alloy exists as a phase mixture $(\alpha + \beta)$ at the lower temperatures, it does so because the value of H happens to be less for the mixture than for the homogeneous solution at that composition. However, because of its greater entropy term, the solution gradually becomes preferred at high temperatures. In more complex alloy systems, particularly those containing intermediate phases of the secondary solid solution type (e.g. copper–zinc, copper–gallium, copper–aluminum, etc.), the range of primary solid solution decreases with rise in temperature. This is because the β-phase, like the α-phase, is a disordered solid solution. However, since it occurs at a higher composition, it has a higher entropy of mixing, and consequently its free energy will fall more rapidly with rise in temperature. This is shown schematically in Figure 2.34. The point of contact on the free energy curve of the α-phase, determined by drawing the common tangent to the α and β curves, governs the solubility c at a given temperature T. The steep fall with temperature of this common tangent automatically gives rise to a decreasing solubility limit.

Many alloys of copper or silver reach the limit of solubility at an electron-to-atom ratio of about 1.4. The divalent elements zinc, cadmium and mercury have solubilities of approximately 40 at.%[13] (e.g. copper–zinc, silver–cadmium, silver–mercury), the trivalent elements approximately 20 at.% (e.g. copper–aluminum, copper–gallium, silver–aluminum, silver–indium) and the tetravalent elements about 13 at.% (e.g. copper–germanium, copper–silicon, silver–tin), respectively.

The limit of solubility has been explained by Jones in terms of the Brillouin zone structure (see Chapter 5). It is assumed that the density of states–energy curve for the two phases, α (the close-packed phase) and β (the more open phase), is of the form shown in Figure 2.35, where the $N(E)$ curve deviates from the parabolic relationship as the Fermi surface approaches the zone

[13] For example, a copper–zinc alloy containing 40 at.% zinc has an e/a ratio of 1.4, i.e. for every 100 atoms, 60 are copper, each contributing one valency electron, and 40 are zinc, each contributing two valency electrons, so that $e/a = (60 \times 1 + 40 \times 2)/100 = 1.4$.

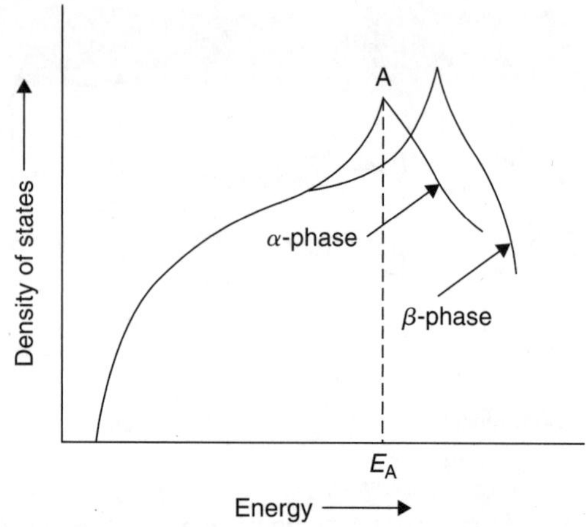

Figure 2.35 *Density of states versus energy diagram.*

boundary.[14] As the solute is added to the solvent lattice and more electrons are added, the top of the Fermi level moves towards A, i.e. where the density of states is high and the total energy E for a given electron concentration is low. Above this point the number of available energy levels decreases so markedly that the introduction of a few more electrons per atom causes a sharp increase in energy. Thus, just above this critical point the α structure becomes unstable relative to the alternative β structure, which can accommodate the electrons within a smaller energy range, i.e. the energy of the Fermi level is lower if the β-phase curve is followed rather than the α-phase curve. The composition for which E_{max} reaches the point E_A is therefore a critical one, since the alloy will adopt that phase which has the lowest energy. It can be shown that this point corresponds to an electron-to-atom ratio of approximately 1.4.

2.3.2 Interstitial solid solutions

Interstitial solid solutions are formed when the solute atoms can fit into the interstices of the lattice of the solvent. However, an examination of the common crystal lattices shows that the size of the available interstices is restricted, and consequently only the small atoms, such as hydrogen, boron, carbon or nitrogen, with atomic radii very much less than one nanometer, form such solutions. The most common examples occur in the transition elements and in particular the solution of carbon or nitrogen in iron is of great practical importance. In fcc iron (austenite) the largest interstice or 'hole' is at the center of the unit cell (coordinates $\frac{1}{2}, \frac{1}{2}, \frac{1}{2}$), where there is space for an atom of radius 52 pm, i.e. $0.41r$ if r is the radius of the solvent atom. A carbon atom (80 pm (0.8 Å) diameter) or a nitrogen atom (70 pm diameter) therefore expands the lattice on solution, but nevertheless dissolves in quantities up to 1.7 and 2.8 wt%, respectively. Although the bcc lattice is the more open structure the largest interstice is smaller than that in the fcc. In bcc iron (ferrite) the largest hole is at the position ($\frac{1}{2}, \frac{1}{4}, 0$) and is a tetrahedral site where four iron atoms are situated symmetrically around it; this can accommodate an atom of radius 36 pm, i.e. $0.29r$, as shown in Figure 2.36a. However,

[14] The shape of the Fermi surface may be determined from measurements of physical properties as a function of orientation in a single crystal. The surface resistance to a high-frequency current at low temperatures (the anomalous skin effect) shows that in copper the Fermi surface is distorted from the spherical shape but becomes more nearly spherical in copper alloys.

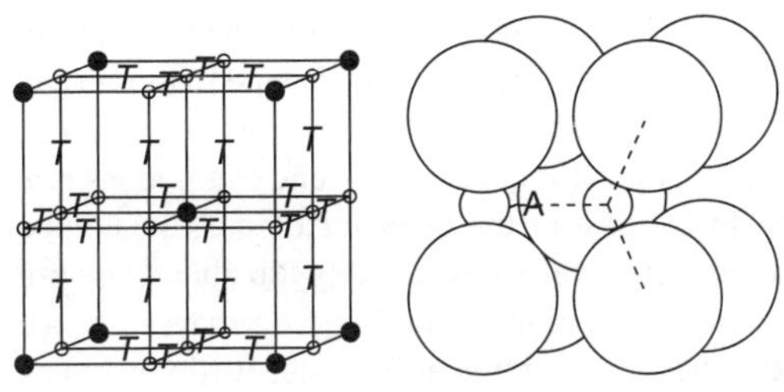

Figure 2.36 *(a) Bcc lattice showing the relative positions of the main lattice sites, the octahedral interstices marked ∘ and the tetrahedral interstices marked T. (b) Structure cell of iron showing the distortions produced by the two different interstitial sites. Only three of the iron atoms surrounding the octahedral sites are shown; the fourth, centered at A, has been omitted for clarity (after Williamson and Smallman, 1953).*

internal friction and X-ray diffraction experiments show that the carbon or nitrogen atoms do not use this site, but instead occupy a smaller site which can accommodate an atom only $0.154r$, or 19 pm. This position $(0, 0, \frac{1}{2})$ at the midpoints of the cell edges is known as the octahedral site since, as can be seen from Figure 2.36b, it has a distorted octahedral symmetry for which two of the iron atoms are nearer to the center of the site than the other four nearest neighbors. The reason for the interstitial atoms preferring this small site is thought to be due to the elastic properties of the bcc lattice. The two iron atoms which lie above and below the interstice, and which are responsible for the smallness of the hole, can be pushed away more easily than the four atoms around the larger interstice. As a result, the solution of carbon in α-iron is extremely limited (0.02 wt%) and the structure becomes distorted into a body-centered tetragonal lattice. The c axis for each interstitial site is, however, disordered, so that this gives rise to a structure which is statistically cubic. The body-centered tetragonal structure forms the basis of martensite (an extremely hard metastable constituent of steel), since the quenching treatment given to steel retains the carbon in supersaturated solution (see Chapter 7).

2.3.3 Types of intermediate phases

2.3.3.1 Electrochemical compounds

The phases which form in the intermediate composition regions of the equilibrium diagram may be (1) electrochemical or full-zone compounds, (2) size-factor compounds or (3) electron compounds. The term 'compound' still persists even though many of these phases do not obey the valency laws of chemistry and often exist over a wide composition range.

We have already seen that a strong tendency for compound formation exists when one element is electropositive and the other is electronegative. The magnesium-based compounds are probably the most common examples, having the formula Mg_2(Pb, Sn, Ge or Si). These have many features in common with salt-like compounds since their compositions satisfy the chemical valency laws, their range of solubility is small, and usually they have high melting points. Moreover, many of these types of compounds have crystal structures identical to definite chemical compounds such as sodium chloride, NaCl, or calcium fluoride, CaF_2. In this respect the Mg_2X series are anti-isomorphous with the CaF_2 fluorspar structure, i.e. the magnesium metal atoms are in the position of the non-metallic fluoride atoms and the metalloid atoms such as tin or silicon take up the position of the metal atoms in calcium fluoride.

Even though these compounds obey all the chemical principles they may often be considered as special electron compounds. For example, the first Brillouin zone[15] of the CaF_2 structure is completely filled at $\frac{8}{3}$ electrons per atom, which significantly is exactly that supplied by the compound Mg_2Pb, Sn, etc. Justification for calling these full-zone compounds is also provided by electrical conductivity measurements. In contrast to the behavior of salt-like compounds which exhibit low conductivity even in the liquid state, the compound Mg_2Pb shows the normal conduction (which indicates the possibility of zone overlapping) while Mg_2Sn behaves like a semiconductor (indicating that a small energy gap exists between the first and second Brillouin zones).

In general, it is probable that both concepts are necessary to describe the complete situation. As we shall see in Section 2.3.3.3, with increasing electrochemical factor even true electron compounds begin to show some of the properties associated with chemical compounds, and the atoms in the structure take up ordered arrangements.

2.3.3.2 *Size-factor compounds*

When the atomic diameters of the two elements differ only slightly, electron compounds are formed, as discussed in the next section. However, when the difference in atomic diameter is appreciable, definite size-factor compounds are formed which may be of the (1) interstitial or (2) substitutional type.

A consideration of several interstitial solid solutions has shown that if the interstitial atom has an atomic radius 0.41 times that of the metal atom then it can fit into the largest available structural interstice without distortion. When the ratio of the radius of the interstitial atom to that of the metal atom is greater than 0.41 but less than 0.59, interstitial compounds are formed; hydrides, borides, carbides and nitrides of the transition metals are common examples. These compounds usually take up a simple structure of either the cubic or hexagonal type, with the metal atoms occupying the normal lattice sites and the non-metal atoms the interstices. In general, the phases occur over a range of composition which is often centered about a simple formula such as M_2X and MX. Common examples are carbides and nitrides of titanium, zirconium, hafnium, vanadium, niobium and tantalum, all of which crystallize in the NaCl structure. It is clear, therefore, that these phases do not form merely as a result of the small atom fitting into the interstices of the solvent structure, since vanadium, niobium and tantalum are bcc, while titanium, zirconium and hafnium are cph. By changing their structure to fcc the transition metals allow the interstitial atom not only a larger 'hole' but also six metallic neighbors. The formation of bonds in three directions at right angles, such as occurs in the sodium chloride arrangement, imparts a condition of great stability to these MX carbides.

When the ratio $r_{(\text{interstitial})}$ to $r_{(\text{metal})}$ exceeds 0.59 the distortion becomes appreciable, and consequently more complicated crystal structures are formed. Thus, iron nitride, where $r_N/r_{Fe} = 0.56$, takes up a structure in which nitrogen lies at the center of six atoms as suggested above, while iron carbide, i.e. cementite, Fe_3C, for which the ratio is 0.63, takes up a more complex structure.

For intermediate atomic size difference, i.e. about 20–30%, an efficient packing of the atoms can be achieved if the crystal structure common to the Laves phases is adopted (Table 2.2). These phases, classified by Laves and his co-workers, have the formula AB_2 and each A atom has 12 B neighbors and four A neighbors, while each B atom is surrounded by six like and six unlike atoms. The average coordination number of the structure (13.33) is higher, therefore, than that achieved by the packing of atoms of equal size. These phases crystallize in one of three closely related structures which are isomorphous with the compounds $MgCu_2$ (cubic), $MgNi_2$ (hexagonal) or $MgZn_2$ (hexagonal). The secret of the close relationship between these structures is that the small atoms are arranged on a space lattice of tetrahedra.

---

[15] Brillouin zones and electrical conductivity are dealt with in Chapter 5.

Table 2.2 *Compounds which exist in a Laves phase structure.*

$MgCu_2$ type	*$MgNi_2$ type*	*$MgZn_2$ type*	
$AgBe_2$			
$BiAu_2$	$BaMg_2$		
$NbCo_2$	Nb(Mn or Fe)$_2$	$NbCo_2$	with excess of B metal
$TaCo_2$	$TaMn_2$	$TaCo_2$	
Ti(Be, Co, or Cr)$_2$	Ti(Mn or Fe)$_2$	$TiCo_2$	
U(Al, Co, Fe or Mn)$_2$	UNi_2	$ZrFe_2$	
Zr(Co, Fe, or W)$_2$	Zr(Cr, Ir, Mn, Re, Ru, Os or V)$_2$		

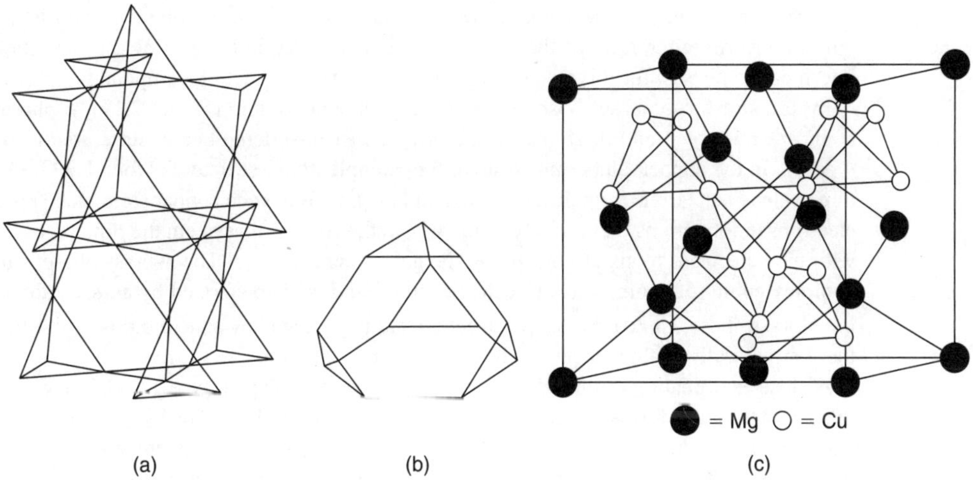

Figure 2.37 *(a) Framework of the $MgCu_2$ structure. (b) Shape of hole in which the large Mg atom is accommodated. (c) Complete $MgCu_2$ structure (after Hume-Rothery, Smallman and Haworth, 1969; by courtesy of the Institute of Materials, Minerals and Mining).*

The different ways of joining such tetrahedra account for the different structures. This may be demonstrated by an examination of the $MgCu_2$ structure. The small B atoms lie at the corners of tetrahedra which are joined point-to point throughout space, as shown in Figure 2.37a. Such an arrangement provides large holes of the type shown in Figure 2.37b and these are best filled when the atomic ratio $r_{(large)}/r_{(small)} = 1.225$. The complete cubic structure of $MgCu_2$ is shown in Figure 2.37c. The $MgZn_2$ structure is hexagonal, and in this case the tetrahedra are joined alternately point-to-point and base-to-base in long chains to form a wurtzite type of structure. The $MgNi_2$ structure is also hexagonal and although very complex it is essentially a mixture of both the $MgCu_2$ and $MgNi_2$ types.

The range of homogeneity of these phases is narrow. This limited range of homogeneity is not due to any ionic nature of the compound, since ionic compounds usually have low coordination numbers, whereas Laves phases have high coordination numbers, but because of the stringent geometrical conditions governing the structure. However, even though the chief reason for their existence is that the ratio of the radius of the large atom to that of the small is about 1.2, there are indications that electronic factors may play some small part. For example, provided the initial size-factor condition is satisfied, then if the e/a ratio is high (e.g. 2), there is a tendency for compounds to crystallize in the $MgZn_2$ structure, while if the e/a ratio is low (e.g. $\frac{4}{3}$), then there is a tendency for the $MgCu_2$ type of structure to be formed. This electronic feature is demonstrated in the magnesium–nickel–zinc ternary system. Thus, even though the binary systems contain both the $MgZn_2$ and $MgNi_2$ phases the ternary compound MgNiZn has the $MgCu_2$ structure, presumably because its e/a ratio is $\frac{4}{3}$. Table 2.2 shows a few common examples of each type of Laves structure, from which it is evident that there is also a general tendency for transition metals to be involved.

2.3.3.3 Electron compounds

Alloys of copper, silver and gold with the B subgroup all possess the sequence α, β, γ, ε of structurally similar phases, and while each phase does not occur at the same composition when this is measured in weight percent or atomic percent, they do so if composition is expressed in terms of electron concentration. Hume-Rothery and co-workers have pointed out that the e/a ratio is important not only in governing the limit of the α-solid solution, but also in controlling the formation of certain intermediate phases; for this reason they have been termed 'electron compounds'.

In terms of those phases observed in the copper–zinc system (Figure 2.19), β-phases are found at an e/a ratio of $\frac{3}{2}$ and these phases are often either disordered bcc in structure or ordered CsCl-type, β'. In the copper–aluminum system, for example, the β-structure is found at Cu_3Al, where the three valency electrons from the aluminum and the one from each copper atom make up a ratio of six electrons to four atoms, i.e. $e/a = \frac{3}{2}$. Similary, in the copper–tin system the β-phase occurs at Cu_5Sn with nine electrons to six atoms, giving the governing e/a ratio. The γ-brass phase, Cu_5Zn_8, has a complex cubic (52 atoms per unit cell) structure, and is characterized by an e/a ratio of $\frac{21}{13}$, while the ε-brass phase, $CuZn_3$, has a cph structure and is governed by an e/a ratio of $\frac{7}{4}$. A list of some of these structurally analogous phases is given in Table 2.3.

A close examination of this table shows that some of these phases, e.g. Cu_5Si and Ag_3Al, exist in different structural forms for the same e/a ratio. Thus, Ag_3Al is basically a $\frac{3}{2}$ bcc phase, but it only exists as such at high temperatures; at intermediate temperatures it is cph and at low temperatures β-Mn. It is also noticeable that to conform with the appropriate electron-to-atom ratio the transition metals are credited with zero valency. The basis for this may be found in their electronic structure, which is characterized by an incomplete d-band below an occupied outermost s-band. The nickel

Table 2.3 *Some selected structurally analogous phases.*

Electron–atom ratio 3:2			*Electron–atom ratio 21:13 γ-brass (complex cubic)*	*Electron–atom ratio 7:4 ε-brass (cph)*
β-Brass (bcc)	*β-Manganese (complex cubic)*	*(cph)*		
(Cu, Ag or Au)Zn		AgZn	(Cu, Ag or Au) (Zn or Cd)$_8$	(Cu, Ag or Au) (Zn or Cd)$_3$
CuBe	(Ag or Au)$_3$Al	AgCd		
	Cu_5Si		Cu_9Al_4	Cu_3Sn
(Ag or Au)Mg	$CoZn_3$	Ag_3Al		Cu_3Si
(Ag or Au)Cd		Au_5Sn	$Cu_{31}Sn_8$	Ag_5Al_3
(Cu or Ag)$_3$Al				
(Cu_5Sn or Si)			(Fe, Co, Ni, Pd or Pt)$_5$Zn$_{21}$	
(Fe, Co or Ni)Al				

atom, for example, has an electronic structure denoted by (2) (8) (16) (2), i.e. two electrons in the first quantum shell, eight in the second, 16 in the third and two in the fourth shells, and while this indicates that the free atom has two valency electrons, it also shows two electrons missing from the third quantum shell. Thus, if nickel contributes valency electrons, it also absorbs an equal number from other atoms to fill up the third quantum shell so that the net effect is zero.

Without doubt the electron concentration is the most important single factor which governs these compounds. However, as for the other intermediate phases, a closer examination shows an interplay of all factors. Thus, in general, the bcc $\frac{3}{2}$ compounds are only formed if the size factor is less than $\pm 18\%$, an increase in the valency of the solute tends to favor cph and β-Mn structures at the expense of the bcc structure, a high electrochemical factor leads to ordering up to the melting point, and an increase in temperature favors the bcc structure in preference to the cph or β-Mn structure.

2.3.4 Order–disorder phenomena

A substitutional solid solution can be one of two types, either ordered in which the A and B atoms are arranged in a regular pattern, or disordered in which the distribution of the A and B atoms is random. From the previous section it is clear that the necessary condition for the formation of a superlattice, i.e. an ordered solid solution, is that dissimilar atoms must attract each other more than similar atoms. In addition, the alloy must exist at or near a composition which can be expressed by a simple formula, such as AB, A_3B or AB_3. The following are common structures:

1. *CuZn*. While the disordered solution is bcc with equal probabilities of having copper or zinc atoms at each lattice point, the ordered lattice has copper atoms and zinc atoms segregated to cube corners (0, 0, 0) and centers $(\frac{1}{2}, \frac{1}{2}, \frac{1}{2})$, respectively. The superlattice in the β-phase therefore takes up the CsCl (also described as B2 or $L2_0$) structure, as illustrated in Figure 2.38a. Other

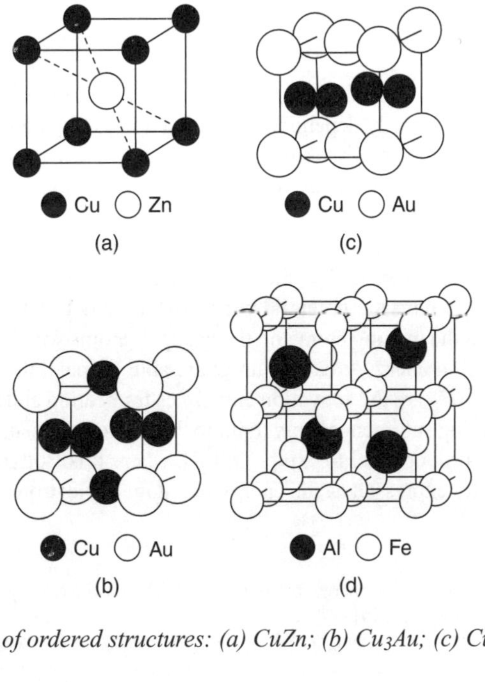

Figure 2.38 *Examples of ordered structures: (a) CuZn; (b) Cu_3Au; (c) CuAu; (d) Fe_3Al.*

examples of the same type, which may be considered as being made up of two interpenetrating simple cubic lattices, are Ag(Mg, Zn or Cd), AuNi, NiAl, FeAl and FeCo.

2. *Cu_3Au*. This structure, which occurs less frequently than the β-brass type, is based on the fcc structure with copper atoms at the centers of the faces $(0, \frac{1}{2}, \frac{1}{2})$ and gold atoms at the corners (0, 0, 0), as shown in Figure 2.38b. Other examples of the $L1_2$ structure include Ni_3Al, Ni_3Ti, Ni_3Si, Pt_3Al, Fe_3Ge, Zr_3Al.
3. *AuCu*. The AuCu structure shown in Figure 2.38c is also based on the fcc lattice, but in this case alternate (0 0 1) layers are made up of copper and gold atoms, respectively. Hence, because the atomic sizes of copper and gold differ, the lattice is distorted into a tetragonal structure having an axial ratio $c/a = 0.93$. Other examples of the $L1_0$ include CoPt, FePt and TiAl.
4. *Fe_3Al*. Like FeAl, the Fe_3Al structure is based on the bcc lattice but, as shown in Figure 2.38d, eight simple cells are necessary to describe the complete ordered arrangement. In this structure any individual atom is surrounded by the maximum number of unlike atoms and the aluminum atoms are arranged tetrahedrally in the cell. Other examples of the $D0_3$ include Fe_3Si and Cu_3Al.
5. *Mg_3Cd*. This ordered structure is based on the cph lattice. Other examples of the $D0_{19}$ structure are Ti_3Al, $MgCd_3$ and Ni_3Sn.

Another important structure which occurs in certain intermetallics is the defect lattice. In the compound NiAl, as the composition deviates from stoichiometry towards pure aluminum, the electron-to-atom ratio becomes greater than $\frac{3}{2}$, but to prevent the compound becoming unstable the lattice takes up a certain proportion of vacancies to maintain the number of electrons per unit cell at a constant value of 3. Such defects obviously increase the entropy of the alloy, but the fact that these phases are stable at low temperatures, where the entropy factor is unimportant, demonstrates that their stability is due to a lowering of internal energy. Such defects produce an anomalous decrease in both the lattice parameter and the density above 50 at.% Al.

2.4 The mechanism of phase changes

2.4.1 Kinetic considerations

Changes of phase in the solid state involve a redistribution of the atoms in that solid and the kinetics of the change necessarily depend upon the rate of atomic migration. The transport of atoms through the crystal is more generally termed diffusion, and is dealt with in Section 5.4. This can occur more easily with the aid of vacancies, since the basic act of diffusion is the movement of an atom to an empty adjacent atomic site.

Let us consider that during a phase change an atom is moved from an α-phase lattice site to a more favorable β-phase lattice site. The energy of the atom should vary with distance as shown in Figure 2.39, where the potential barrier which has to be overcome arises from the interatomic forces between the moving atom and the group of atoms which adjoin it and the new site. Only those atoms (n) with an energy greater than Q are able to make the jump, where $Q_{\alpha\rightarrow\beta} = H_m - H_\alpha$ and $Q_{\beta\rightarrow\alpha} = H_m - H_\beta$ are the activation enthalpies for heating and cooling, respectively. The probability of an atom having sufficient energy to jump the barrier is given, from the Maxwell–Boltzmann distribution law, as proportional to $\exp[-Q/\mathbf{k}T]$, where $\mathbf{k}$ is Boltzmann's constant, T is the temperature and Q is usually expressed as the energy per atom in electron volts.[16]

---

[16] Q may also be given as the energy in J mol^{-1}, in which case the rate equation becomes

Rate of reaction $= A\exp[-Q/\mathbf{R}T]$

where $\mathbf{R} = \mathbf{k}N$ is the gas constant, i.e. 8.314 J mol^{-1} K^{-1} and N is the Avogadro number 6.02×10^{23}.

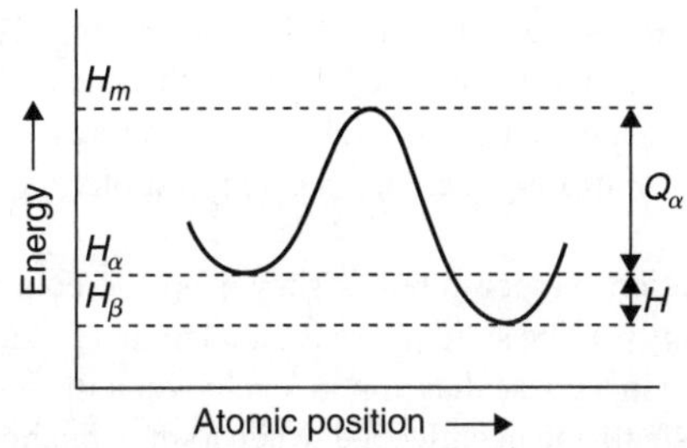

Figure 2.39 *Energy barrier separating structural states.*

The rate of reaction is given by

$$\text{Rate} = A\exp[-Q/\mathbf{k}T], \tag{2.8}$$

where A is a constant involving n and ν, the frequency of vibration. To determine Q experimentally, the reaction velocity is measured at different temperatures and, since

$$\ln(\text{Rate}) = \ln A - Q/\mathbf{k}T, \tag{2.9}$$

the slope of the ln(Rate) versus $1/T$ curve gives $Q/\mathbf{k}$.

In deriving equation (2.8), usually called an Arrhenius equation after the Swedish chemist who first studied reaction kinetics, no account is taken of the entropy of activation, i.e. the change in entropy as a result of the transition. In considering a general reaction the probability expression should be written in terms of the free energy of activation per atom F or G rather than just the internal energy or enthalpy. The rate equation then becomes

$$\begin{aligned}\text{Rate} &= A\exp[-F/\mathbf{k}T]\\ &= A\exp[S/\mathbf{k}]\exp[-E/\mathbf{k}T].\end{aligned} \tag{2.10}$$

The slope of the ln(Rate) versus $1/T$ curve then gives the temperature dependence of the reaction rate, i.e. the activation energy or enthalpy, and the magnitude of the intercept on the ln(Rate) axis depends on the temperature-independent terms, and includes the frequency factor and the entropy term.

During the transformation it is not necessary for the entire system to go from α to β in one jump and, in fact, if this were necessary, phase changes would practically never occur. Instead, most phase changes occur by a process of nucleation and growth (cf. solidification, Section 2.1.1). Chance thermal fluctuations provide a small number of atoms with sufficient activation energy to break away from the matrix (the old structure) and form a small nucleus of the new phase, which then grows at the expense of the matrix until the whole structure is transformed. By this mechanism, the amount of material in the intermediate configuration of higher free energy is kept to a minimum, as it is localized into atomically thin layers at the interface between the phases. Because of this mechanism of transformation, the factors which determine the rate of phase change are: (1) the rate of nucleation, N (i.e. the number of nuclei formed in unit volume in unit time) and (2) the rate of growth, G (i.e. the rate of increase in radius with time). Both processes require activation energies, which in general are not equal, but the values are much smaller than that needed to change the whole structure from α to β in one operation.

Even with such an economical process as nucleation and growth transformation, difficulties occur and it is common to find that the transformation temperature, even under the best experimental conditions, is slightly higher on heating than on cooling. This sluggishness of the transformation is known as hysteresis, and is attributed to the difficulties of nucleation, since diffusion, which controls the growth process, is usually high at temperatures near the transformation temperature and is therefore not rate controlling. Perhaps the simplest phase change to indicate this is the solidification of a liquid metal.

The transformation temperature, as shown on the equilibrium diagram, represents the point at which the free energy of the solid phase is equal to that of the liquid phase. Thus, we may consider the transition, as given in a phase diagram, to occur when the bulk or chemical free energy change, ΔG_V, is infinitesimally small and negative, i.e. when a small but positive driving force exists. However, such a definition ignores the process whereby the bulk liquid is transformed to bulk solid, i.e. nucleation and growth. When the nucleus is formed the atoms which make up the interface between the new and old phase occupy positions of compromise between the old and new structures, and as a result these atoms have rather higher energies than the other atoms. Thus, there will always be a positive free energy term opposing the transformation as a result of the energy required to create the surface of interface. Consequently, the transformation will occur only when the sum $\Delta G_V + \Delta G_S$ becomes negative, where ΔG_S arises from the surface energy of solid/liquid interface. Normally, for the bulk phase change, the number of atoms which form the interface is small and ΔG_S compared with ΔG_V can be ignored. However, during nucleation ΔG_V is small, since it is proportional to the amount transformed, and ΔG_S, the extra free energy of the boundary atoms, becomes important due to the large surface area-to-volume ratio of small nuclei. Therefore, before transformation can take place the negative term ΔG_V must be greater than the positive term ΔG_S and, since ΔG_V is zero at the equilibrium freezing point, it follows that undercooling must result.

Worked example

A thermally activated transformation is complete after 180 seconds at 900°C. How long does it take at the lower temperature of 700°C if the activation energy for the process is 167.5 kJ mol^{-1}?

Solution

Rate of reaction $\propto$ 1/time = 1/t.

$$\frac{\text{Rate}_2}{\text{Rate}_1} = \frac{1/t_2}{1/t_1} = \frac{A\mathrm{e}^{-Q/\mathbf{R}T_2}}{A\mathrm{e}^{-Q/\mathbf{R}T_1}} = \exp\left[-\left(-\frac{1}{T_1}+\frac{1}{T_2}\right)\frac{Q}{\mathbf{R}}\right]$$

$$= \exp\left[-\left(-\frac{1}{1173}+\frac{1}{973}\right)\frac{167.5}{8.314}\right] = \mathrm{e}^{-3.4}.$$

$$\text{So, } \frac{1/t_2}{1/180} = \mathrm{e}^{-3.4} \text{ or } t_2 = 180\,\mathrm{e}^{3.4} = 2220\,\mathrm{s} = 37\,\mathrm{min}.$$

2.4.2 Homogeneous nucleation

Quantitatively, since ΔG_V depends on the volume of the nucleus and ΔG_S is proportional to its surface area, we can write for a spherical nucleus of radius r:

$$\Delta G = (4\pi r^3 \Delta G_V/3) + 4\pi r^2 \gamma, \tag{2.11}$$

where ΔG_V is the bulk free energy change involved in the formation of the nucleus of unit volume and γ is the surface energy of unit area. When the nuclei are small the positive surface energy term

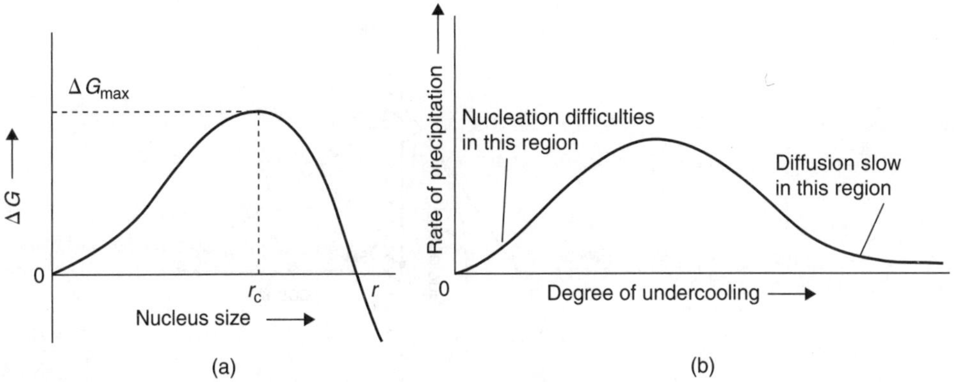

Figure 2.40 *(a) Effect of nucleus size on the free energy of nucleus formation. (b) Effect of undercooling on the rate of precipitation.*

predominates, while when they are large the negative volume term predominates, so that the change in free energy as a function of nucleus size is as shown in Figure 2.40a. This indicates that a critical nucleus size exists below which the free energy increases as the nucleus grows, and above which further growth can proceed with a lowering of free energy; $\Delta G_{\max}$ may be considered as the energy or work of nucleation W. Both r_c and W may be calculated since $d\Delta G/dr = 4\pi r^2 \Delta G_v + 8\pi r\gamma = 0$ when $r = r_c$ and thus $r_c = -2\gamma/\Delta G_v$. Substituting for r_c gives

$$W = 16\pi\gamma^3/3\Delta G_v^2. \tag{2.12}$$

The surface energy factor γ is not strongly dependent on temperature, but the greater the degree of undercooling or supersaturation, the greater is the release of chemical free energy and the smaller the critical nucleus size and energy of nucleation. This can be shown analytically, since $\Delta G_v = \Delta H - T\Delta S$, and at $T = T_e$, $\Delta G_v = 0$, so that $\Delta H = T_e \Delta S$. It therefore follows that

$$\Delta G_v = (T_e - T)\Delta S = \Delta T \Delta S$$

and because $\Delta G_v \propto \Delta T$, then

$$W \propto \gamma^3/\Delta T^2. \tag{2.13}$$

Consequently, since nuclei are formed by thermal fluctuations, the probability of forming a smaller nucleus is greatly improved, and the rate of nucleation increases according to

$$\begin{aligned} \text{Rate} &= A\exp[-Q/\mathbf{k}T]\exp[-\Delta G_{\max}/\mathbf{k}T] \\ &= A\exp[-(Q + \Delta G_{\max})/\mathbf{k}T]. \end{aligned} \tag{2.14}$$

The term exp $[-Q/\mathbf{k}T]$ is introduced to allow for the fact that rate of nucleus formation is in the limit controlled by the rate of atomic migration. Clearly, with very extensive degrees of undercooling, when $\Delta G_{\max} \ll Q$, the rate of nucleation approaches $\exp[-Q/\mathbf{k}T]$ and, because of the slowness of atomic mobility, this becomes small at low temperature (Figure 2.40b). While this range of conditions can be reached for liquid glasses, the nucleation of liquid metals normally occurs at temperatures before this condition is reached. (By splat cooling, small droplets of the metal are cooled very

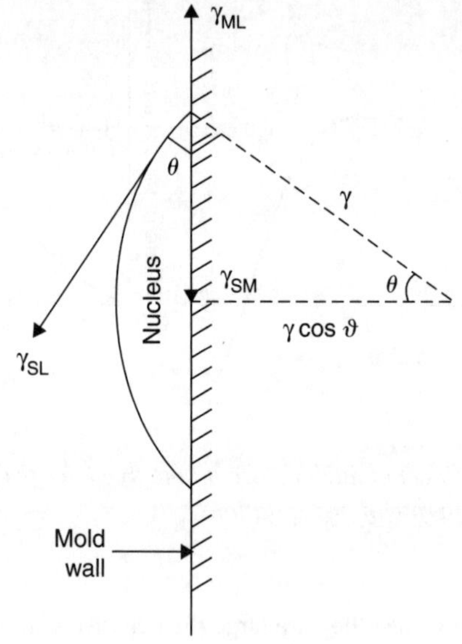

Figure 2.41 *Schematic geometry of heterogeneous nucleation.*

rapidly (10^5 K s^{-1}) and an amorphous solid may be produced.) Nevertheless, the principles are of importance in metallurgy since in the isothermal transformation of eutectoid steel, for example, the rate of transformation initially increases and then decreases with lowering of the transformation temperature (see **TTT** curves, Chapter 7).

2.4.3 Heterogeneous nucleation

In practice, homogeneous nucleation rarely takes place and heterogeneous nucleation occurs either on the mold walls or on insoluble impurity particles. From equation (2.13) it is evident that a reduction in the interfacial energy γ would facilitate nucleation at small values of ΔT. Figure 2.41 shows how this occurs at a mold wall or pre-existing solid particle, where the nucleus has the shape of a spherical cap to minimize the energy and the 'wetting' angle θ is given by the balance of the interfacial tensions in the plane of the mold wall, i.e. $\cos\theta = (\gamma_{ML} - \gamma_{SM})/\gamma_{SL}$.

The formation of the nucleus is associated with an excess free energy given by

$$\begin{aligned}\Delta G &= V\Delta G_v + A_{SL}\gamma_{SL} + A_{SM}\gamma_{SM} - A_{SM}\gamma_{ML} \\ &= \pi/3(2 - 3\cos\theta + \cos^3\theta)r^3\Delta G_v \\ &\quad + 2\pi(1-\cos\theta)r^2\gamma_{SL} \\ &\quad + \pi r^2 \sin^2\theta(\gamma_{SM} - \gamma_{LM}).\end{aligned} \tag{2.15}$$

Differentiation of this expression for the maximum, i.e. $d\Delta G/dr = 0$, gives $r_c = -2\gamma_{SL}/\Delta G_v$ and

$$W = (16\pi\gamma^3/3\Delta G_v^2)[(1-\cos\theta)^2(2+\cos\theta)/4] \tag{2.16}$$

or

$$W_{(\text{heterogeneous})} = W_{(\text{homogeneous})}[S(\theta)].$$

The shape factor $S(\theta) \leq 1$ is dependent on the value of θ and the work of nucleation is therefore less for heterogeneous nucleation. When $\theta = 180°$, no wetting occurs and there is no reduction in W; when $\theta \to 0°$ there is complete wetting and $W \to 0$; and when $0 < \theta < 180°$ there is some wetting and W is reduced.

Worked example

Heterogeneous nucleation of a β-phase in an α-phase occurs on a grain boundary. If $\gamma_{\alpha\beta}$ is 250 mJ m^{-2} and the grain boundary energy is $\frac{1}{2}$ J m^{-2} determine how the nucleation process is aided relative to the homogeneous process.

Solution

Assuming the nucleation is of a spherical nucleus,

$$\Delta G = -A\gamma_{\text{gb}} + 4\pi r^2 \gamma_{\alpha\beta} - \frac{4}{3}\pi r^3 \Delta G_{\text{V}},$$

where A is the area of the grain boundary removed by the nucleus,

$$\Delta G = -\pi r^2 \gamma_{\text{gb}} + 4\pi r^2 \gamma_{\alpha\beta} - \frac{4}{3}\pi r^3 \Delta G_{\text{V}}$$

$$\mathrm{d}(\Delta G)/\mathrm{d}r = -2\pi r \gamma_{\text{gb}} + 8\pi r \gamma_{\alpha\beta} - 4\pi r^2 \Delta G_{\text{V}} = 0$$

$$r_{\text{c}} = \frac{4\gamma_{\alpha\beta} - \gamma_{\text{gb}}}{2\Delta G_{\text{V}}}.$$

For homogeneous nucleation, r_{c} is given by the above formula with $\gamma_{\text{gb}} = 0$ (see also Section 2.4.2), so

$$\frac{r_{\text{c}}\,(\text{heterogeneous})}{r_{\text{c}}\,(\text{homogeneous})} = \frac{(4\gamma_{\alpha\beta} - \gamma_{\text{gb}})/2\Delta G_{\text{V}}}{2\gamma_{\alpha\beta}/\Delta G_{\text{V}}} = \frac{4\gamma_{\alpha\beta} - \gamma_{\text{gb}}}{4\gamma_{\alpha\beta}} = \frac{1000 - 500}{1000} = \frac{1}{2}.$$

So, r_{c} (heterogeneous) $= \frac{1}{2} r_{\text{c}}$ (homogeneous).

2.4.4 Nucleation in solids

When the transformation takes place in the solid state, i.e. between two solid phases, a second factor giving rise to hysteresis operates. The new phase usually has a different parameter and crystal structure from the old so that the transformation is accompanied by dimensional changes. However, the changes in volume and shape cannot occur freely because of the rigidity of the surrounding matrix, and elastic strains are induced. The strain energy and surface energy created by the nuclei of the new phase are positive contributions to the free energy and so tend to oppose the transition.

The total free energy change is

$$\Delta G = V\Delta G_{\text{v}} + A\gamma + V\Delta G_{\text{s}}, \tag{2.17}$$

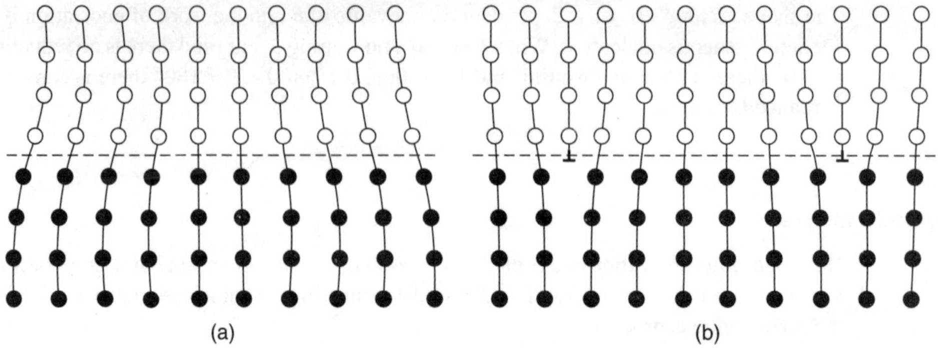

Figure 2.42 *Schematic representation of interface structures. (a) A coherent boundary with misfit strain. (b) A semi-coherent boundary with misfit dislocations.*

where A is the area of interface between the two phases and γ the interfacial energy per unit area, and ΔG_S is the misfit strain energy per unit volume of new phase. For a spherical nucleus of the second phase:

$$\Delta G = \frac{4}{3}\pi r^3(\Delta G_V - \Delta G_S) + 4\pi r^2\gamma \tag{2.18}$$

and the misfit strain energy reduces the effective driving force for the transformation. Differentiation of equation (2.18) gives

$$r_c = -2\gamma/(\Delta G_V - \Delta G_S), \quad \text{and}$$
$$W = 16\pi\gamma^3/3(\Delta G_V - \Delta G_S)^2.$$

The value of γ can vary widely from a few $mJ\,m^{-2}$ to several hundred $mJ\,m^{-2}$ depending on the coherency of the interface. A coherent interface is formed when the two crystals have a good 'match' and the two lattices are continuous across the interface. This happens when the interfacial plane has the same atomic configuration in both phases, e.g. $\{1\ 1\ 1\}$ in fcc and $\{0\ 0\ 0\ 1\}$ in cph. When the 'match' at the interface is not perfect it is still possible to maintain coherency by straining one or both lattices, as shown in Figure 2.42a. These coherency strains increase the energy and for large misfits it becomes energetically more favorable to form a semi-coherent interface (Figure 2.42b) in which the mismatch is periodically taken up by misfit dislocations.[17] The coherency strains can then be relieved by a cross-grid of dislocations in the interface plane, the spacing of which depends on the Burgers vector b of the dislocation and the misfit ε, i.e. b/ε. The interfacial energy for semi-coherent interfaces arises from the change in composition across the interface or chemical contribution as for fully coherent interfaces, plus the energy of the dislocations (see Chapter 3). The energy of a semi-coherent interface is 200–500 $mJ\,m^{-2}$ and increases with decreasing dislocation spacing until the dislocation strain fields overlap. When this occurs, the discrete nature of the dislocations is lost and the interface becomes incoherent. The incoherent interface is somewhat similar to a high-angle grain boundary (see Figure 2.3) with its energy of 0.5–1 $J\,m^{-2}$ relatively independent of the orientation.

The surface and strain energy effects discussed above play an important role in phase separation. When there is coherence in the atomic structure across the interface between precipitate and matrix

[17] A detailed treatment of dislocations and other defects is given in Chapter 3.

the surface energy term is small, and it is the strain energy factor which controls the shape of the particle. A plate-shaped particle is associated with the least strain energy, while a spherical-shaped particle is associated with maximum strain energy but the minimum surface energy. On the other hand, surface energy determines the crystallographic plane of the matrix on which a plate-like precipitate forms. Thus, the habit plane is the one which allows the planes at the interface to fit together with the minimum of disregistry; the frequent occurrence of the Widmanstätten structures may be explained on this basis. It is also observed that precipitation occurs most readily in regions of the structure which are somewhat disarranged, e.g. at grain boundaries, inclusions, dislocations or other positions of high residual stress caused by plastic deformation. Such regions have an unusually high free energy and necessarily are the first areas to become unstable during the transformation. Also, new phases can form there with a minimum increase in surface energy. This behavior is considered again in Chapter 6.

Problems

2.1 Change the following alloy compositions (i) Cu–40 wt% Zn, (ii) Cu–10 wt% Al and (iii) Cu–20 wt% In into (a) atomic percent and (b) electron–atom ratio, e/a. What is the significance of the e/a ratio? Why is Cu–In different from Cu–Zn and Cu–Al?

2.2 A hypothetical alloy transforms from a simple cubic structure to a close packed hexagonal structure. Assuming a hard sphere model, calculate the volume change during the transformation.

2.3 A pressure–temperature diagram for a typical metal is shown in Figure 2.7. How is this diagram modified for the metals Bi and Ga, and why?

2.4 With reference to Figure 2.11, prove the 'Lever Rule' as described in Section 2.2.4.1.

2.5 From the Cu–Ni phase diagram (Figure 2.11) determine for a Cu–40% Ni alloy the composition and amount of each phase at 1250°C.

2.6 From the Pb–Sn phase diagram (Figure 2.15) determine for the eutectic alloy the amount and composition of each phase, just below the eutectic temperature at 182°C.

2.7 (a) The Cu–Ni alloy system is an ideal solid solution system in which entire solubility is observed, while the Cu–Ag system is a typical eutectic system in which limited solubility of Cu into Ag and vice versa is observed. Which one of the the Hume-Rothery alloying rules best accounts for this difference? (b) Which one of the Hume-Rothery rules best explains the fact that in the Cu–Zn alloy system, Cu can dissolve up to ~40 wt% of Zn but Zn can only dissolve only about 2 wt% of Cu?

2.8 Compare the size of the interstitial sites $\left(\frac{1}{2}, \frac{1}{4}, 0\right)$ and $\left(0, 0, \frac{1}{2}\right)$ for a bcc structure containing iron, and explain why the smaller site is occupied by C in low-carbon steel.

2.9 For the Fe–C system, calculate (a) the amount of ferrite and cementite in pearlite formed at 722°C, and (b) the amount of Fe_3C and pearlite for a 0.2% C steel at room temperature.

Further reading

Beeley, P. R. (1972). *Foundry Technology*. Butterworths, London.
Brandes, E. A. and Brook, G. B. (1998). *Smithells Metals Reference Book.* Butterworth-Heinemann, Oxford.
Campbell, J. (1991). *Castings*. Butterworth-Heinemann, London.
Chadwick, G. A. (1972). *Metallography of Phase Transformations*. Butterworths, London.
Davies, G. J. (1973). *Solidification and Casting*. Applied Science, London.
Driver, D. (1985). Aero engine alloy development, Inst. of Metals Conf., Birmingham, 'Materials at their Limits' (25 September 1985).
Flemings, M. C. (1974). *Solidification Processing*. McGraw-Hill, New York.

Hume-Rothery, W., Smallman, R. E. and Haworth, C. (1969). *Structure of Metals and Alloys,* 5th edn. Institute of Metals, London.

Kingery, W. D., Bowen, H. K. and Uhlmann, D. R. (1976). *Introduction to Ceramics,* 2nd edn. Wiley-Interscience, New York.

Massalski, T. B., Okamoto, H. eds (1996). *Binary Alloy Phase Diagrams*, ASM International.

Quets, J. M. and Dresher, W. H. (1969). Thermo-chemistry of hot corrosion of superalloys. *Journal of Materials,* ASTM, JMSLA, **4**, (3), 583–599.

Rhines, F. N. (1956). *Phase Diagrams in Metallurgy: their development and application*. McGraw-Hill, New York.

West, D. R. F. (1982). *Ternary Equilibrium Diagrams,* 2nd edn. Macmillan, London.

Chapter 3

Crystal defects

3.1 Types of imperfection

Real solids invariably contain structural discontinuities and localized regions of disorder. This heterogeneity can exist on both microscopic and macroscopic scales, with defects or imperfections ranging in size from missing or misplaced atoms to features that are visible to the naked eye. The majority of materials used for engineering components and structures are made up from a large number of small interlocking grains or crystals. It is therefore immediately appropriate to regard the grain boundary surfaces of such polycrystalline aggregates as a type of imperfection. Other relatively large defects, such as shrinkage pores, gas bubbles, inclusions of foreign matter and cracks, may be found dispersed throughout the grains of a metal or ceramic material. In general, however, these large-scale defects are very much influenced by the processing of the material and are less fundamental to the basic material. More attention will thus be given to the atomic-scale defects in materials. Within each grain, atoms are regularly arranged according to the basic crystal structure but a variety of imperfections, classified generally as crystal defects, may also occur. A schematic diagram of these basic defects is shown in Figure 3.1. These take the form of:

- Point defects, such as vacant atomic sites (or simply vacancies) and interstitial atoms (or simply interstitials), where an atom sits in an interstice rather than a normal lattice site
- Line defects, such as dislocations
- Planar defects, such as stacking faults and twin boundaries
- Volume defects, such as voids, gas bubbles and cavities.

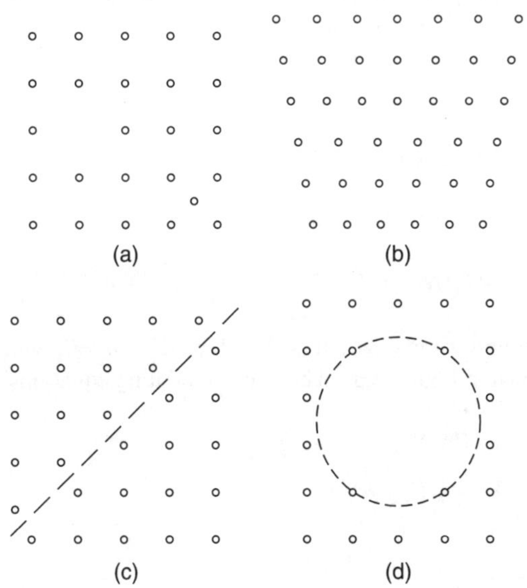

Figure 3.1 *(a) Vacancy–interstitial. (b) Dislocation. (c) Stacking fault. (d) Void.*

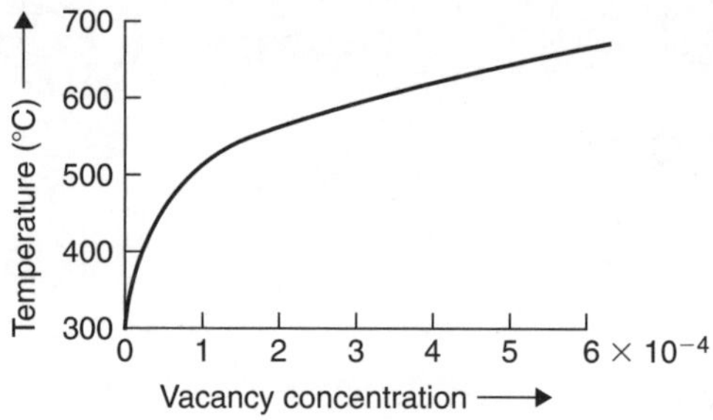

Figure 3.2 *Equilibrium concentration of vacancies as a function of temperature for aluminum (after Bradshaw and Pearson, 1957).*

In the following sections this type of classification will be used to consider the defects which can occur in metallic and ceramic crystals.

3.2 Point defects

3.2.1 Point defects in metals

Of the various lattice defects the vacancy is the only species that is ever present in appreciable concentrations in thermodynamic equilibrium and increases exponentially with rise in temperature, as shown in Figure 3.2. The vacancy is formed by removing an atom from its lattice site and depositing it in a nearby atomic site, where it can be easily accommodated. Favored places are the free surface of the crystal, a grain boundary or the extra half-plane of an edge dislocation. Such sites are termed vacancy sources and the vacancy is created when sufficient energy is available (e.g. thermal activation) to remove the atom. If E_f is the energy required to form one such defect (usually expressed in electron volts per atom), the total energy increase resulting from the formation of n such defects is nE_f. The accompanying entropy increase may be calculated using the relations $S = \mathbf{k} \ln W$, where W is the number of ways of distributing n defects and N atoms on $N + n$ lattice sites, i.e. $(N + n)!/n!N!$ Then the free energy, G, or strictly F, of a crystal of n defects, relative to the free energy of the perfect crystal, is

$$F = nE_f - \mathbf{k}T \ln[(N + n)!/n!N!], \tag{3.1}$$

which by the use of Stirling's theorem[1] simplifies to

$$F = nE_f - \mathbf{k}T[(N + n)\ln(N + n) - n \ln n - N \ln N]. \tag{3.2}$$

The equilibrium value of n is that for which $\mathrm{d}F/\mathrm{d}n = 0$, which defines the state of minimum free energy as shown in Figure 3.3.[2] Thus, differentiating equation (3.2) gives

$$\begin{aligned} 0 &= E_f - \mathbf{k}T[\ln(N + n) - \ln n] \\ &= E_f - \mathbf{k}T \ln[(N + n)/n], \end{aligned}$$

[1] Stirling's approximation states that $\ln N! = N \ln N$.

[2] $\mathrm{d}F/\mathrm{d}n$ or $\mathrm{d}G/\mathrm{d}n$ is known as the chemical potential.

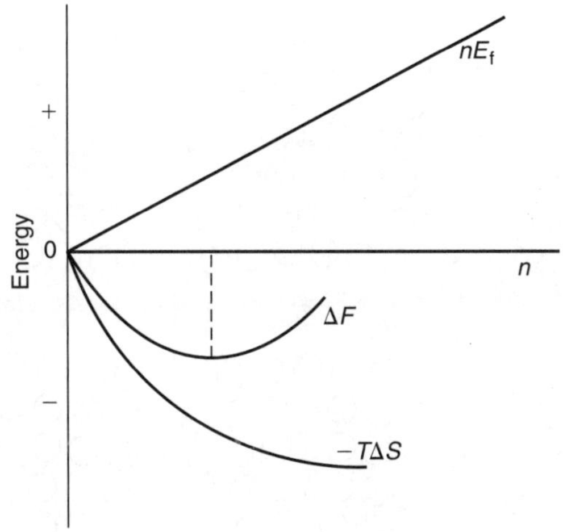

Figure 3.3 *Variation of the energy of a crystal with addition of* n *vacancies.*

so that

$$\frac{n}{N+n} = \exp[-E_f/\mathbf{k}T].$$

Usually N is very large compared with n, so that the expression can be taken to give the atomic concentration, c, of lattice vacancies, $n/N = \exp[-E_f/\mathbf{k}T]$. A more rigorous calculation of the concentration of vacancies in thermal equilibrium in a perfect lattice shows that although c is principally governed by the Boltzmann factor, $\exp[-E_f/\mathbf{k}T]$, the effect of the vacancy on the vibrational properties of the lattice also leads to an entropy term which is independent of temperature and usually written as $\exp[S_f/\mathbf{k}]$. The fractional concentration may thus be written:

$$\begin{aligned} c &= n/N = \exp[S_f/\mathbf{k}]\exp[-E_f/\mathbf{k}T] \\ &= A\exp[-E_f/\mathbf{k}T]. \end{aligned} \tag{3.3}$$

The value of the entropy term is not accurately known but it is usually taken for simplicity to be unity.

The equilibrium number of vacancies rises rapidly with increasing temperature, owing to the exponential form of the expression, and for most common metals has a value of about 10^{-4} near the melting point. For example, $\mathbf{k}T$ at room temperature (300 K) is $\approx 1/40$ eV and for aluminum $E_f = 0.7$ eV, so that at 900 K we have

$$\begin{aligned} c &= \exp\left[-\frac{7}{10} \times \frac{40}{1} \times \frac{300}{900}\right] \\ &= \exp[-9.3] = 10^{-[9.3/2.3]} \approx 10^{-4}. \end{aligned}$$

As the temperature is lowered, c should decrease in order to maintain equilibrium and to do this the vacancies must migrate to positions in the structure where they can be annihilated; these locations are then known as 'vacancy sinks' and include such places as the free surface, grain boundaries and dislocations.

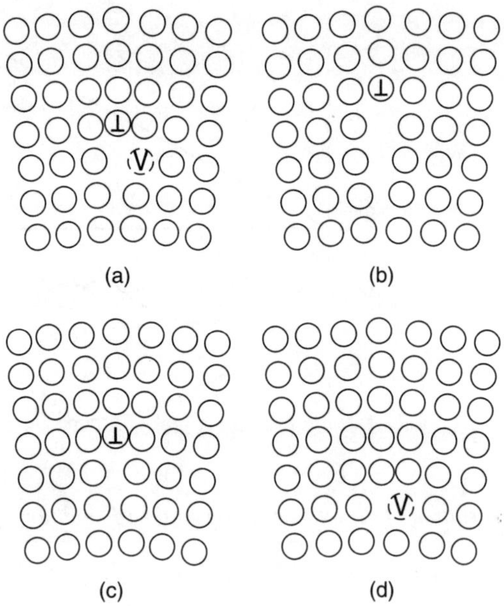

Figure 3.4 *Climb of a dislocation: (a, b) to annihilate; (c, d) to create a vacancy.*

The defect migrates by moving through the energy maxima from one atomic site to the next with a frequency:

$$\nu = \nu_0 \exp\left(\frac{S_m}{\mathbf{k}}\right) \exp\left(-\frac{E_m}{\mathbf{k}T}\right),$$

where ν_0 is the frequency of vibration of the defect in the appropriate direction, S_m is the entropy increase and E_m is the internal energy increase associated with the process. The self-diffusion coefficient in a pure metal is associated with the energy to form a vacancy E_f and the energy to move it E_m, being given by the expression:

$$E_{SD} = E_f + E_m.$$

Clearly the free surface of a sample or the grain boundary interface are a considerable distance, in atomic terms, from the center of a grain and so dislocations in the body of the grain or crystal are the most efficient 'sink' for vacancies. Vacancies are annihilated at the edge of the extra half-plane of atoms of the dislocation, as shown in Figure 3.4a and b. This causes the dislocation to climb, as discussed in Section 3.3.4. The process whereby vacancies are annihilated at vacancy sinks such as surfaces, grain boundaries and dislocations to satisfy the thermodynamic equilibrium concentration at a given temperature is, of course, reversible. When a metal is heated the equilibrium concentration increases and, to produce this additional concentration, the surfaces, grain boundaries and dislocations in the crystal reverse their role and act as vacancy sources and emit vacancies; the extra half-plane of atoms climbs in the opposite sense (see Figure 3.4c and d).

Below a certain temperature, the migration of vacancies will be too slow for equilibrium to be maintained, and at the lower temperatures a concentration of vacancies in excess of the equilibrium number will be retained in the structure. Moreover, if the cooling rate of the metal or alloy is particularly rapid, as, for example, in quenching, the vast majority of the vacancies which exist at high temperatures can be 'frozen-in'.

Vacancies are of considerable importance in governing the kinetics of many physical processes. The industrial processes of annealing, homogenization, precipitation, sintering, surface hardening, as well as oxidation and creep, all involve, to varying degrees, the transport of atoms through the structure with the help of vacancies. Similarly, vacancies enable dislocations to climb, since to move the extra half-plane of a dislocation up or down requires the mass transport of atoms. This mechanism is extremely important in the recovery stage of annealing and also enables dislocations to climb over obstacles lying in their slip plane; in this way materials can soften and lose their resistance to creep at high temperatures.

In metals the energy of formation of an interstitial atom is much higher than that for a vacancy and is of the order of 4 eV. At temperatures just below the melting point, the concentration of such point defects is only about 10^{-15} and therefore interstitials are of little consequence in the normal behavior of metals. They are, however, more important in ceramics because of the more open crystal structure. They are also of importance in the deformation behavior of solids when point defects are produced by the non-conservative motion of jogs in screw dislocation (see Section 3.3.4) and also of particular importance in materials that have been subjected to irradiation by high-energy particles.

Worked example

Resistivity measurements at 300 K show that an annealed metal single crystal has a vacancy concentration $c_v = 10^{-9}$. When this same crystal is quenched from a high temperature the measurements indicate $c_v \approx 2 \times 10^{-3}$. Calculate the temperature of the quench. [Take E_f, the vacancy formation energy $= 51.38\,\text{kJ mol}^{-1}$ and $\mathbf{R} = 8.3\,\text{J mol}^{-1}\,\text{K}^{-1}$.]

Solution

The vacancy concentration $c_v = \exp(-E_f/\mathbf{R}T)$,

$$\text{so that } \frac{2 \times 10^{-3}}{10^{-9}} = \frac{\exp(-E_f/\mathbf{R}T)}{\exp(-E_f/300\mathbf{R})}$$

$$\ln(2 \times 10^6) = \frac{E_f}{\mathbf{R}}\left(\frac{1}{300} - \frac{1}{T}\right)$$

$$\begin{aligned} 14.5 &= 20.7 - 6214.5/T \\ T &= 1001\,\text{K}. \end{aligned}$$

3.2.2 Point defects in non-metallic crystals

Point defects in non-metallic, particularly ionic, structures are associated with additional features (e.g. the requirement to maintain electrical neutrality and the possibility of both anion defects and cation defects existing). An anion vacancy in NaCl, for example, will be a positively charged defect and may trap an electron to become a neutral F-center. Alternatively, an anion vacancy may be associated with either an anion interstitial or a cation vacancy. The vacancy–interstitial pair is called a Frenkel defect and the vacancy pair a Schottky defect, as shown in Figure 3.5. Interstitials are much more common in ionic structures than metallic structures because of the large 'holes' or interstices that are available.

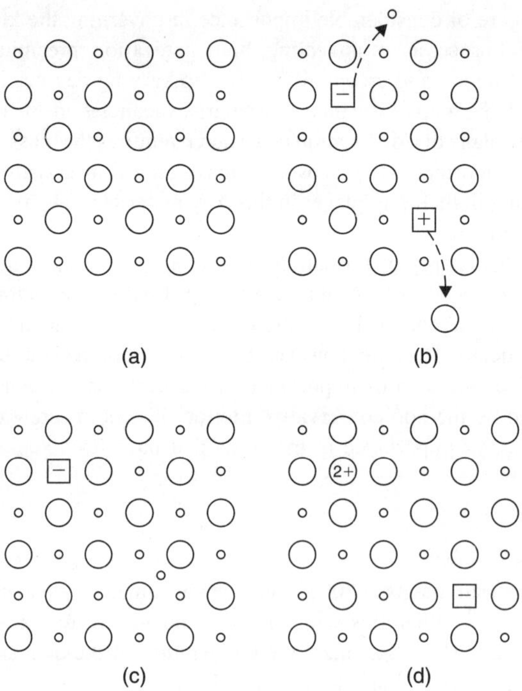

Figure 3.5 *Representation of point defects in two-dimensional ionic structure: (a) perfect structure and monovalent ions, (b) two Schottky defects, (c) Frenkel defect, and (d) substitutional divalent cation impurity and cation vacancy.*

In general, the formation energy of each of these two types of defect is different and this leads to different defect concentrations. With regard to vacancies, when $E_f^- > E_f^+$, the formation will initially produce more cation than anion vacancies from dislocations and boundaries as the temperature is raised. However, the electrical field produced will eventually oppose the production of further cations and promote the formation of anions such that at equilibrium there will be almost equal numbers of both types and the combined or total concentration c of Schottky defects at high temperatures is $\sim 10^{-4}$.

Foreign ions with a valency different from the host cation may also give rise to point defects to maintain charge neutrality. Monovalent sodium ions substituting for divalent magnesium ions in MgO, for example, must be associated with an appropriate number of either cation interstitials or anion vacancies in order to maintain charge neutrality. Deviations from the stoichiometric composition of the non-metallic material as a result of excess (or deficiency) in one (or other) atomic species also results in the formation of point defects.

An example of excess metal due to anion vacancies is found in the oxidation of silicon, which takes place at the metal/oxide interface. Interstitials are more likely to occur in oxides with open crystal structures and when one atom is much smaller than the other as, for example, in ZnO (Figure 3.6a). The oxidation of copper to Cu_2O, shown in Figure 3.6b, is an example of non-stoichiometry involving cation vacancies. Thus, copper vacancies are created at the oxide surface, diffuse through the oxide layer and are eliminated at the oxide/metal interface.

Oxides which contain point defects behave as semiconductors when the electrons associated with the point defects either form positive holes or enter the conduction band of the oxide. If the electrons remain locally associated with the point defects, then charge can only be transferred by the diffusion of the charge carrying defects through the oxide. Both *p*- and *n*-type semiconductors are formed when

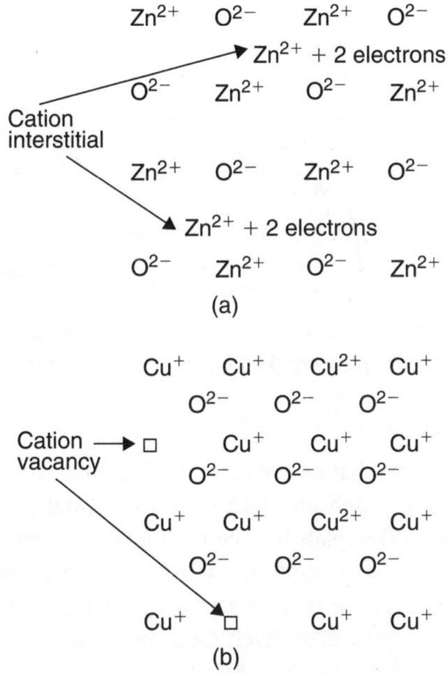

Figure 3.6 *Schematic arrangement of ions in two typical oxides. (a) $Zn_{>1}O$, with excess metal due to cation interstitials. (b) $Cu_{<2}O$, with excess non-metal due to cation vacancies.*

oxides deviate from stoichiometry; the former arises from a deficiency of cations and the latter from an excess of cations.

Examples of *p*-type semiconducting oxides are NiO, PbO and Cu_2O, while the oxides of Zn, Cd and Be are *n*-type semiconductors.

3.2.3 Irradiation of solids

There are many different kinds of high-energy radiation (e.g. neutrons, electrons, α-particles, protons, deuterons, uranium fission fragments, γ-rays, X-rays) and all of them are capable of producing some form of 'radiation damage' in the materials they irradiate. While all are of importance to some aspects of the solid state, of particular interest is the behavior of materials under irradiation in a nuclear reactor. This is because the neutrons produced in a reactor by a fission reaction have extremely high energies of about 2 million electron volts (i.e. 2 MeV), and being electrically uncharged, and consequently unaffected by the electrical fields surrounding an atomic nucleus, can travel large distances through a structure. The resultant damage is therefore not localized, but is distributed throughout the solid in the form of 'damage spikes'.

The fast neutrons (they are given this name because 2 MeV corresponds to a velocity of 2×10^7 ms^{-1}) are slowed down, in order to produce further fission, by the moderator in the pile until they are in thermal equilibrium with their surroundings. The neutrons in a pile will therefore have a spectrum of energies which ranges from about 1/40 eV at room temperature (thermal neutrons) to 2 MeV (fast neutrons). However, when a non-fissile material is placed in a reactor and irradiated, most of the damage is caused by the fast neutrons colliding with the atomic nuclei of the material.

The nucleus of an atom has a small diameter (e.g. 10^{-10} m), and consequently the largest area, or cross-section, which it presents to the neutron for collision is also small. The unit of cross-section is

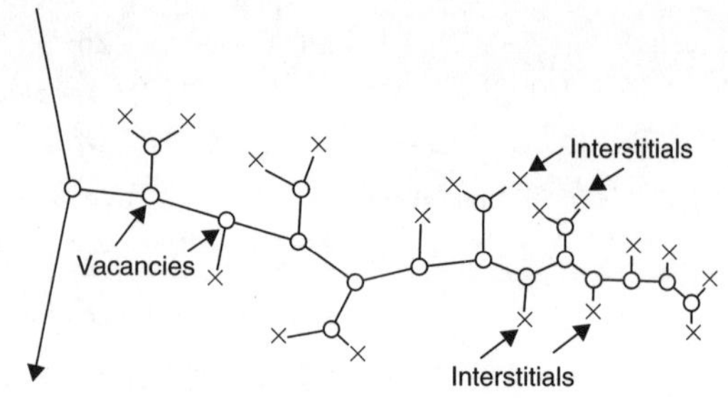

Figure 3.7 *Formation of vacancies and interstitials due to particle bombardment (after Cottrell, 1959; courtesy of the Institute of Mechanical Engineers).*

a barn, i.e. 10^{-28} m^2, so that in a material with a cross-section of 1 barn, an average of 10^9 neutrons would have to pass through an atom (cross-sectional area 10^{-19} m^2) for one to hit the nucleus. Conversely, the mean free path between collisions is about 10^9 atom spacings or about 0.3 m. If a metal such as copper (cross-section, 4 barns) were irradiated for 1 day (10^5 s) in a neutron flux of 10^{17} m^{-2} s^{-1} the number of neutrons passing through unit area, i.e. the integrated flux, would be 10^{22} nm^{-2} and the chance of a given atom being hit (= integrated flux × cross-section) would be 4×10^{-6}, i.e. about 1 atom in 250 000 would have its nucleus struck.

For most metals the collision between an atomic nucleus and a neutron (or other fast particle of mass m) is usually purely elastic, and the struck atom mass M will have equal probability of receiving any kinetic energy between zero and the maximum, $E_{max} = 4E_n Mm/(M+m)^2$, where E_n is the energy of the fast neutron. Thus, the most energetic neutrons can impart an energy of as much as 200 000 eV to a copper atom initially at rest. Such an atom, called a primary 'knock-on', will do much further damage on its subsequent passage through the structure, often producing secondary and tertiary knock-on atoms, so that severe local damage results. The neutron, of course, also continues its passage through the structure, producing further primary displacements until the energy transferred in collisions is less than the energy E_d (≈25 eV for copper) necessary to displace an atom from its lattice site.

The damage produced in irradiation consists largely of interstitials, i.e. atoms knocked into interstitial positions in the lattice, and vacancies, i.e. the holes they leave behind. The damaged region, estimated to contain about 60 000 atoms, is expected to be originally pear-shaped in form, having the vacancies at the center and the interstitials towards the outside. Such a displacement spike or cascade of displaced atoms is shown schematically in Figure 3.7. The number of vacancy–interstitial pairs produced by one primary knock-on is given by $n \simeq E_{max}/4E_d$, and for copper is about 1000. Owing to the thermal motion of the atoms in the lattice, appreciable self-annealing of the damage will take place at all except the lowest temperatures, with most of the vacancies and interstitials annihilating each other by recombination. However, it is expected that some of the interstitials will escape from the surface of the cascade, leaving a corresponding number of vacancies in the center. If this number is assumed to be 100, the local concentration will be 100/60 000 or $\approx 2 \times 10^{-3}$.

Another manifestation of radiation damage concerns the dispersal of the energy of the stopped atom into the vibrational energy of the lattice. The energy is deposited in a small region, and for a very short time the metal may be regarded as locally heated. To distinguish this damage from the 'displacement spike', where the energy is sufficient to displace atoms, this heat-affected zone has been called a 'thermal spike'. To raise the temperature by 1000°C requires about $3R \times 4.2$ kJ mol^{-1} or about 0.25 eV per atom. Consequently, a 25 eV thermal spike could heat about 100 atoms of copper to the melting point, which corresponds to a spherical region of radius about 0.75 nm. It is very doubtful if melting actually takes place, because the duration of the heat pulse is only about

10^{-11} to 10^{-12} s. However, it is not clear to what extent the heat produced gives rise to an annealing of the primary damage, or causes additional quenching damage (e.g. retention of high-temperature phases).

Slow neutrons give rise to transmutation products. Of particular importance is the production of the noble gas elements, e.g. krypton and xenon produced by fission in U and Pu, and helium in the light elements B, Li, Be and Mg. These transmuted atoms can cause severe radiation damage in two ways. First, the inert gas atoms are almost insoluble and hence in association with vacancies collect into gas bubbles, which swell and crack the material. Second, these atoms are often created with very high energies (e.g. as α-particles or fission fragments) and act as primary sources of knock-on damage. The fission of uranium into two new elements is the extreme example when the fission fragments are thrown apart with kinetic energy $\approx$100 MeV. However, because the fragments carry a large charge, their range is short and the damage restricted to the fissile material itself, or in materials which are in close proximity. Heavy ions can be accelerated to kilovolt energies in accelerators to produce heavy ion bombardment of materials being tested for reactor application. These moving particles have a short range and the damage is localized.

Worked example

(i) A 'fast' neutron strikes the nucleus of an iron atom, relative atomic mass 55.847. Estimate (a) the maximum amount of energy that can be transferred to the nucleus, and (b) the number of point defects produced by this amount of energy.

(ii) A certain metal, scattering cross-section 0.75 barn, is exposed to a beam of thermal neutrons. The neutron flux is 10^{12} neutrons cm^{-2} s^{-1} (1 barn = 10^{-24} cm^2). Calculate the proportion of metal atoms which scatter neutrons in a period of 10 days.

Solution

(i) (a) Assuming no energy loss during collision,

$$E_{\mathrm{n}} = E' + E_{\mathrm{max}}, \qquad (1)$$

where E_{n} is the k.e. of the incoming fast neutron, E' is the k.e. of the neutron after collision and E_{max} is the k.e. of the iron atom after collision. Momentum conservation:

$$\sqrt{2mE_{\mathrm{n}}} = \sqrt{2mE'} + \sqrt{2ME_{\mathrm{max}}}. \qquad (2)$$

Eliminating E' between (1) and (2) gives

$$E_{\mathrm{max}} = \frac{4E_{\mathrm{n}}Mm}{(M+m)^2}\ \mathrm{MeV}.$$

For fast neutrons, $E_{\mathrm{n}} = 2$ MeV.

$$E_{\mathrm{max}} = \frac{4 \times 2 \times 55.847 \times 1}{(56.847)^2} = 0.138\ \mathrm{MeV} = 138\,000\ \mathrm{eV}.$$

(b) E_{max} is the *maximum* energy which can be received by a struck atom; the *average* energy taken up by a struck atom must be about one-half of the maximum, i.e. $E_{\mathrm{max}}/2$.

Energy E_d to displace an atom from its site ≈25 eV. Consider a cascade of atoms created by the first collision. The lowest energy of an atom required to displace another, without the former being trapped in the vacancy left behind by the latter, is approximately $2E_d$.

$$\therefore \text{ number of pairs of vacancy/interstitials created in the cascade} \approx \frac{(E_{max}/2)}{2E_d} = \frac{E_{max}}{4E_d}$$

$$= \frac{138\,000}{4 \times 25} = 1380$$

Number of defects $= 2 \times 1380 = 2760$.

(ii) Number $= \sigma\phi t$, where $\sigma =$ scatter cross-section, $\phi =$ flux and $t =$ time.

Number $= (0.75 \times 10^{-24})(10^{12})(10 \times 24 \times 60 \times 60) \approx 7 \times 10^{-7}$.

Proportion $= \dfrac{7}{10^7}$.

3.2.4 Point defect concentration and annealing

Electrical resistivity ρ is one of the simplest and most sensitive properties to investigate the point defect concentration. Point defects are potent scatterers of electrons and the increase in resistivity following quenching ($\Delta\rho$) may be described by the equation:

$$\Delta\rho = A \exp[-E_f/\mathbf{k}T_Q], \tag{3.4}$$

where A is a constant involving the entropy of formation, E_f is the formation energy of a vacancy and T_Q the quenching temperature. Measuring the resistivity after quenching from different temperatures enables E_f to be estimated from a plot of $\Delta\rho$ versus $1/T_Q$. The activation energy, E_m, for the movement of vacancies can be obtained by measuring the rate of annealing of the vacancies at different annealing temperatures. The rate of annealing is inversely proportional to the time to reach a certain value of 'annealed-out' resistivity. Thus, $1/t_1 = A\ \exp[-E_m/\mathbf{k}T_1]$ and $1/t_2 = \exp[-E_m/\mathbf{k}T_2]$ and by eliminating A we obtain $(t_2/t_1) = E_m[(1/T_2) - (1/T_1)]/\mathbf{k}$, where E_m is the only unknown in the expression. Values of E_f and E_m for different materials are given in Table 3.1.

At elevated temperatures the very high equilibrium concentration of vacancies which exists in the structure gives rise to the possible formation of di-vacancy and even tri-vacancy complexes, depending

Table 3.1 *Values of vacancy formation (E_f) and migration (E_m) energies for some metallic materials together with the self-diffusion energy (E_{SD}).*

Energy (eV)	*Cu*	*Al*	*Ni*	*Mg*	*Fe*	*W*	*NiAl*
E_f	1.0–1.1	0.76	1.4	0.9	2.13	3.3	1.05
E_m	1.0–1.1	0.62	1.5	0.5	0.76	1.9	2.4
E_{SD}	2.0–2.2	1.38	2.9	1.4	2.89	5.2	3.45

on the value of the appropriate binding energy. For equilibrium between single and di-vacancies, the total vacancy concentration is given by

$$c_{\text{v}} = c_{1\text{v}} + 2c_{2\text{v}}$$

and the di-vacancy concentration by

$$c_{2\text{v}} = Azc_{1\text{v}}^2 \exp[B_2/\mathbf{k}T],$$

where A is a constant involving the entropy of formation of di-vacancies, B_2 the binding energy for vacancy pairs estimated to be in the range 0.1–0.3 eV and z a configurational factor. The migration of di-vacancies is an easier process and the activation energy for migration is somewhat lower than E_{m} for single vacancies.

Excess point defects are removed from a material when the vacancies and/or interstitials migrate to regions of discontinuity in the structure (e.g. free surfaces, grain boundaries or dislocations) and are annihilated. These sites are termed defect sinks. The average number of atomic jumps made before annihilation is given by

$$n = Az\nu t \exp[-E_{\text{m}}/\mathbf{k}T_{\text{a}}], \tag{3.5}$$

where A is a constant ($\approx$1) involving the entropy of migration, z the coordination around a vacancy, ν the Debye frequency ($\approx 10^{13}$/s), t the annealing time at the ageing temperature T_{a} and E_{m} the migration energy of the defect. For a metal such as aluminum, quenched to give a high concentration of retained vacancies, the annealing process takes place in two stages, as shown in Figure 3.8, Stage I near room temperature with an activation energy $\approx$0.58 eV and $n \approx 10^4$, and Stage II in the range 140–200°C with an activation energy of ~1.3 eV.

Assuming a random walk process, single vacancies would migrate an average distance ($\sqrt{n} \times$ atomic spacing b) $\approx$ 30 nm. This distance is very much less than either the distance to the grain boundary or the spacing of the dislocations in the annealed metal. In this case, the very high supersaturation of vacancies produces a chemical stress, somewhat analogous to an osmotic pressure, which is sufficiently large to create new dislocations in the structure that provide many new 'sinks' to reduce this stress rapidly.

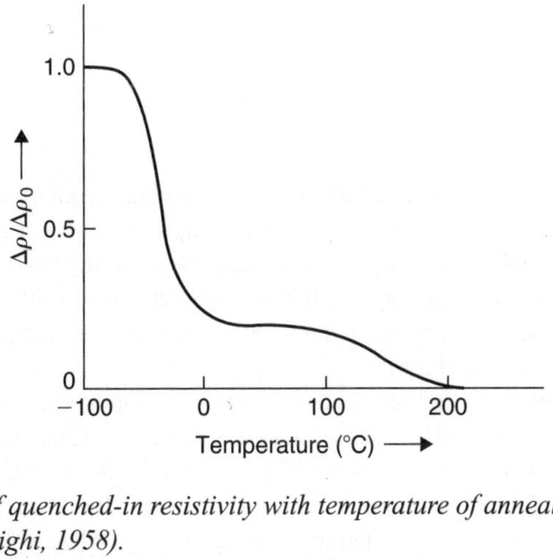

Figure 3.8 *Variation of quenched-in resistivity with temperature of annealing for aluminum (after Panseri and Federighi, 1958).*

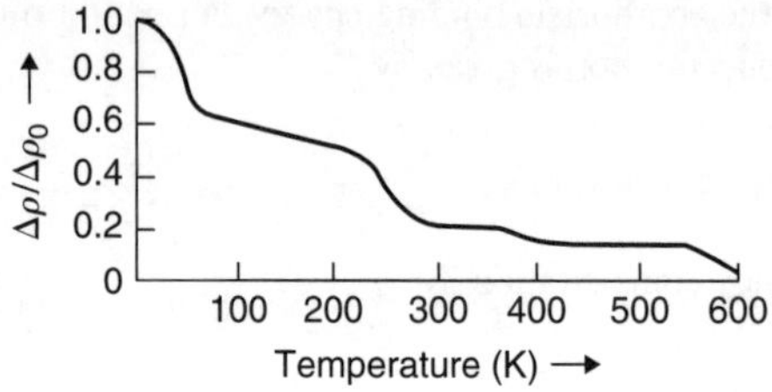

Figure 3.9 *Variation of resistivity with temperature produced by neutron irradiation for copper (after Diehl).*

The magnitude of this chemical stress may be estimated from the chemical potential, if we let dF represent the change of free energy when dn vacancies are added to the system. Then,

$$\begin{aligned} \mathrm{d}F/\mathrm{d}n &= E_\mathrm{f} + \mathbf{k}T\ln(n/N) = -\mathbf{k}T\ln c_0 + \mathbf{k}T\ln c \\ &= \mathbf{k}T\ln(c/c_0), \end{aligned}$$

where c is the actual concentration and c_0 the equilibrium concentration of vacancies. This may be rewritten as

$$\begin{aligned} \mathrm{d}F/\mathrm{d}V &= \text{Energy/volume} \equiv \text{stress} \\ &= (\mathbf{k}T/b^3)[\ln(c/c_0)], \end{aligned} \tag{3.6}$$

where dV is the volume associated with dn vacancies and b^3 is the volume of one vacancy. Inserting typical values, $\mathbf{k}T \cong 1/40$ eV at room temperature and $b = 0.25$ nm, shows $\mathbf{k}T/b^3 \cong 150$ MN/m^2. Thus, even a moderate 1% supersaturation of vacancies, i.e. when $(c/c_0) = 1.01$ and $\ln(c/c_0) = 0.01$, introduces a chemical stress σ_c equivalent to 1.5 MN m^{-2}.

The equilibrium concentration of vacancies at a temperature T_2 will be given by $c_2 = \exp[-E_\mathrm{f}/\mathbf{k}T_2]$ and at T_1 by $c_1 = \exp[-E_\mathrm{f}/\mathbf{k}T_1]$. Then, since

$$\ln(c_2/c_1) = (E_\mathrm{f}/\mathbf{k})\left[\frac{1}{T_1} - \frac{1}{T_2}\right],$$

the chemical stress produced by quenching a metal from a high temperature T_2 to a low temperature T_1 is:

$$\sigma_\mathrm{c} = (\mathbf{k}T_1/b^3)\ln(c_2/c_1) = (E_\mathrm{f}/b^3)\left[1 - \frac{T_1}{T_2}\right].$$

For aluminum, E_f is about 0.7 eV, so that quenching from 900 to 300 K produces a chemical stress of about 3 GN m^{-2}. This stress is extremely high, several times the theoretical yield stress, and must be relieved in some way. Migration of vacancies to grain boundaries and dislocations will occur, of course, but it is not surprising that the point defects form additional vacancy sinks by the spontaneous nucleation of dislocations and other stable lattice defects, such as voids and stacking fault tetrahedra (see Sections 3.5.3 and 3.6).

When the material contains both vacancies and interstitials, the removal of the excess point defect concentration is more complex. Figure 3.9 shows the 'annealing' curve for irradiated copper. The resistivity decreases sharply around 20 K when the interstitials start to migrate, with an activation energy $E_\mathrm{m} \sim 0.1$ eV. In Stage I, therefore, most of the Frenkel (interstitial–vacancy) pairs anneal out. Stage II has been attributed to the release of interstitials from impurity traps as thermal energy

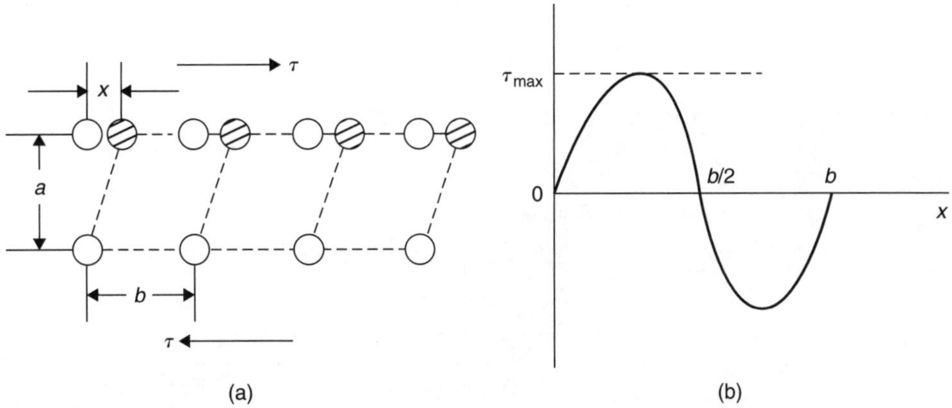

Figure 3.10 *(a) Slip of crystal planes. (b) Shear stress versus displacement curve.*

supplies the necessary activation energy. Stage III is around room temperature and is probably caused by the annihilation of free interstitials with individual vacancies not associated with a Frenkel pair, and also the migration of di-vacancies. Stage IV corresponds to the Stage I annealing of quenched metals arising from vacancy migration and annihilation to form dislocation loops, voids and other defects. Stage V corresponds to the removal of this secondary defect population by self-diffusion.

3.3 Line defects

3.3.1 Concept of a dislocation

All crystalline materials usually contain lines of structural discontinuities running throughout each crystal or grain. These line discontinuities are termed dislocations and there is usually about 10^{10} to 10^{12} m of dislocation line in a meter cube of material.[3] Dislocations enable materials to deform without destroying the basic crystal structure at stresses below that at which the material would break or fracture if they were not present.

A crystal changes its shape during deformation by the slipping of atomic layers over one another. The theoretical shear strength of perfect crystals was first calculated by Frenkel for the simple rectangular-type lattice shown in Figure 3.10 with spacing a between the planes. The shearing force required to move a plane of atoms over the plane below will be periodic, since for displacements $x < b/2$, where b is the spacing of atoms in the shear direction, the lattice resists the applied stress, but for $x > b/2$ the lattice forces assist the applied stress. The simplest function with these properties is a sinusoidal relation of the form:

$$\tau = \tau_m \sin(2\pi x/b) \cong \tau_m 2\pi x/b,$$

where τ_m is the maximum shear stress at a displacement $= b/4$. For small displacements the elastic shear strain given by x/a is equal to τ/μ from Hooke's law, where μ is the shear modulus, so that

$$\tau_m = (\mu/2\pi)b/a \tag{3.7}$$

and since $b \approx a$, the theoretical strength of a perfect crystal is of the order of $\mu/10$.

[3] This is usually expressed as the density of dislocations $\rho = 10^{10}$ to 10^{12} m^{-2}.

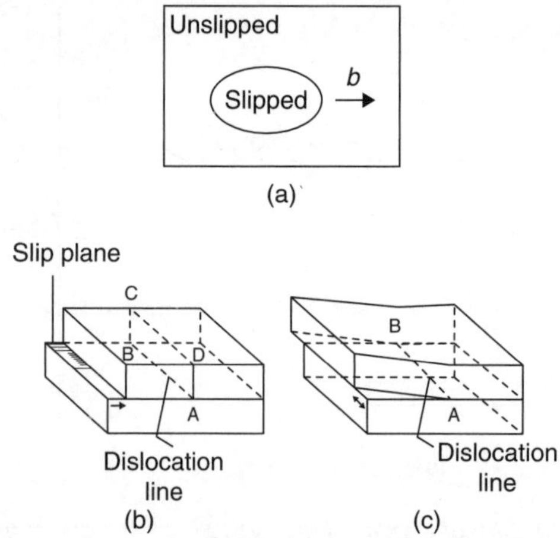

Figure 3.11 *Schematic representation of: (a) a dislocation loop, (b) edge dislocation and (c) screw dislocation.*

This calculation shows that crystals should be rather strong and difficult to deform, but a striking experimental property of single crystals is their softness, which indicates that the critical shear stress to produce slip is very small (about $10^{-5}\,\mu$ or $\approx 50\,\text{gf}\,\text{mm}^{-2}$). This discrepancy between the theoretical and experimental strength of crystals is accounted for if atomic planes do not slip over each other as rigid bodies but instead slip starts at a localized region in the structure and then spreads gradually over the remainder of the plane, somewhat like the disturbance when a pebble is dropped into a pond.

In general, therefore, the slip plane may be divided into two regions, one where slip has occurred and the other which remains unslipped. Between the slipped and unslipped regions the structure will be dislocated (Figure 3.11); this boundary is referred to as a dislocation line, or dislocation. Three simple properties of a dislocation are immediately apparent, namely: (1) it is a line discontinuity, (2) it forms a closed loop in the interior of the crystal or emerges at the surface and (3) the difference in the amount of slip across the dislocation line is constant. The last property is probably the most important, since a dislocation is characterized by the magnitude and direction of the slip movement associated with it. This is called the Burgers vector, b, which for any given dislocation line is the same all along its length.

3.3.2 Edge and screw dislocations

It is evident from Figure 3.11a that some sections of the dislocation line are perpendicular to b, others are parallel to b while the remainder lie at an angle to b. This variation in the orientation of the line with respect to the Burgers vector gives rise to a difference in the structure of the dislocation. When the dislocation line is normal to the slip direction it is called an edge dislocation. In contrast, when the line of the dislocations is parallel to the slip direction the dislocation line is known as a screw dislocation. From the diagram shown in Figure 3.11a it is evident that the dislocation line is rarely pure edge or pure screw, but it is convenient to think of these ideal dislocations since any dislocation can be resolved into edge and screw components. The atomic structures of simple edge and screw dislocations are shown in Figures 3.13 and 3.14.

3.3.3 The Burgers vector

It is evident from the previous sections that the Burgers vector b is an important dislocation parameter. In any deformation situation the Burgers vector is defined by constructing a Burgers circuit in the

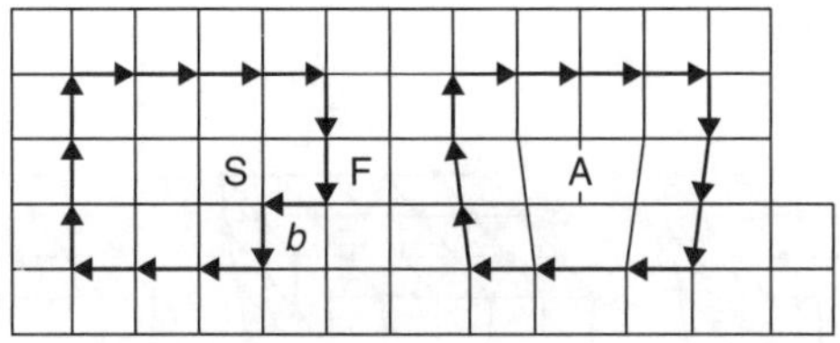

Figure 3.12 *Burgers circuit round a dislocation A fails to close when repeated in a perfect lattice unless completed by a closure vector FS equal to the Burgers vector b.*

dislocated crystal, as shown in Figure 3.12. A sequence of lattice vectors is taken to form a closed clockwise circuit around the dislocation. The same sequence of vectors is then taken in the perfect lattice when it is found that the circuit fails to close. The closure vector FS (finish-start) defines b for the dislocation. With this FS/RH (right-hand) convention it is necessary to choose one direction along the dislocation line as positive. If this direction is reversed the vector b is also reversed. The Burgers vector defines the atomic displacement produced as the dislocation moves across the slip plane. Its value is governed by the crystal structure because during slip it is necessary to retain an identical lattice structure both before and after the passage of the dislocation. This is assured if the dislocation has a Burgers vector equal to one lattice vector and, since the energy of a dislocation depends on the square of the Burgers vector (see Section 3.3.5.2), its Burgers vector is usually the shortest available lattice vector. This vector, by definition, is parallel to the direction of closest packing in the structure, which agrees with experimental observation of the slip direction.

The Burgers vector is conveniently specified by its directional coordinates along the principal crystal axes. In the fcc lattice, the shortest lattice vector is associated with slip from a cube corner to a face center, and has components $a/2$, $a/2$, 0. This is usually written $a/2[1\ 1\ 0]$, where a is the lattice parameter and $[1\ 1\ 0]$ is the slip direction. The magnitude of the vector, or the strength of the dislocation, is then given by $\sqrt{\{a^2(1^2+1^2+0^2)/4\}}=a/\sqrt{2}$. The corresponding slip vectors for the bcc and cph structures are $b=a/2[1\ 1\ 1]$ and $b=a/3[1\ 1\ \bar{2}\ 0]$ respectively.

3.3.4 Mechanisms of slip and climb

The atomic structure of an edge dislocation is shown in Figure 3.13a. Here the extra half-plane of atoms is above the slip plane of the crystal, and consequently the dislocation is called a positive edge dislocation and is often denoted by the symbol $\perp$. When the half-plane is below the slip plane it is termed a negative dislocation. If the resolved shear stress on the slip plane is τ and the Burgers vector of the dislocation b, the force on the dislocation, i.e. force per unit length of dislocation, is $F=\tau b$. This can be seen by reference to Figure 3.13 if the crystal is of side L. The force on the top face (stress × area) is $\tau \times L^2$. Thus, when the two halves of the crystal have slipped the relative amount b, the work done by the applied stress (force × distance) is $\tau L^2 b$. On the other hand, the work done in moving the dislocation (total force on dislocation FL × distance moved) is FL^2, so that equating the work done gives F (force per unit length of dislocation) $=\tau b$. Figure 3.13 indicates how slip is propagated by the movement of a dislocation under the action of such a force. The extra half-plane moves to the right until it produces the slip step shown at the surface of the crystal; the same shear will be produced by a negative dislocation moving from right to left.[4]

[4] An obvious analogy to the slip process is the movement of a caterpillar in the garden or the propagation of a ruck in a carpet to move the carpet into place. In both examples, the effort to move is much reduced by this propagation process.

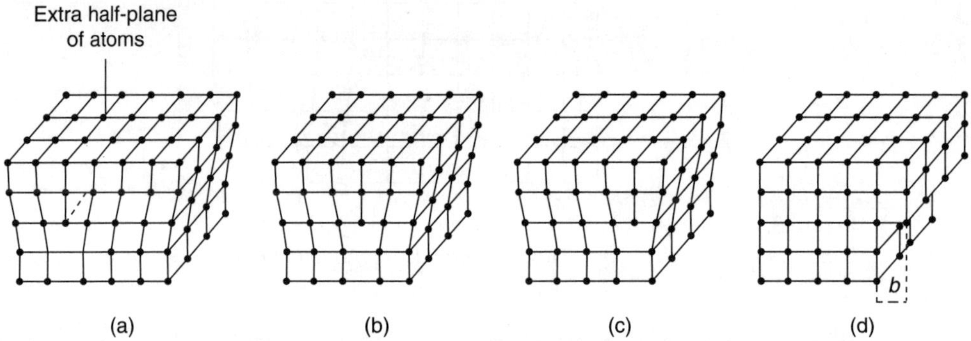

Figure 3.13 *Slip caused by the movement of an edge dislocation.*

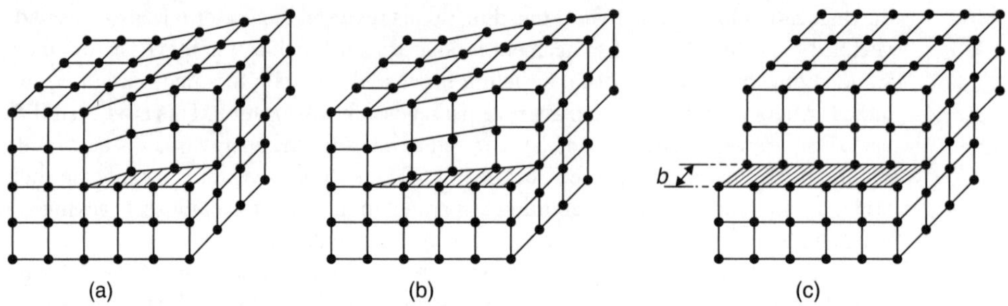

Figure 3.14 *Slip caused by the movement of a screw dislocation.*

The slip process as a result of a screw dislocation is shown in Figure 3.14. It must be recognized, however, that the dislocation is more usually a closed loop and slip occurs by the movement of all parts of the dislocation loop, i.e. edge, screw and mixed components, as shown in Figure 3.15.

A dislocation is able to glide in that slip plane which contains both the line of the dislocation and its Burgers vector. The edge dislocation is confined to glide in one plane only. An important difference between the motion of a screw dislocation and that of an edge dislocation arises from the fact that the screw dislocation is cylindrically symmetrical about its axis with its b parallel to this axis. To a screw dislocation all crystal planes passing through the axis look the same and therefore the motion of the screw dislocation is not restricted to a single slip plane, as is the case for a gliding edge dislocation. The process whereby a screw dislocation glides into another slip plane having a slip direction in common with the original slip plane, as shown in Figure 3.16, is called cross-slip. Usually, the cross-slip plane is also a close-packed plane, e.g. $\{1\ 1\ 1\}$ in fcc crystals.

The mechanism of slip illustrated above shows that the slip or glide motion of an edge dislocation is restricted, since it can only glide in that slip plane which contains both the dislocation line and its Burgers vector. However, movement of the dislocation line in a direction normal to the slip plane can occur under certain circumstances; this is called dislocation climb. To move the extra half-plane either up or down, as is required for climb, requires mass transport by diffusion and is a non-conservative motion. For example, if vacancies diffuse to the dislocation line it climbs up and the extra half-plane will shorten. However, since the vacancies will not necessarily arrive at the dislocation at the same instant, or uniformly, the dislocation climbs one atom at a time and some sections will lie in one

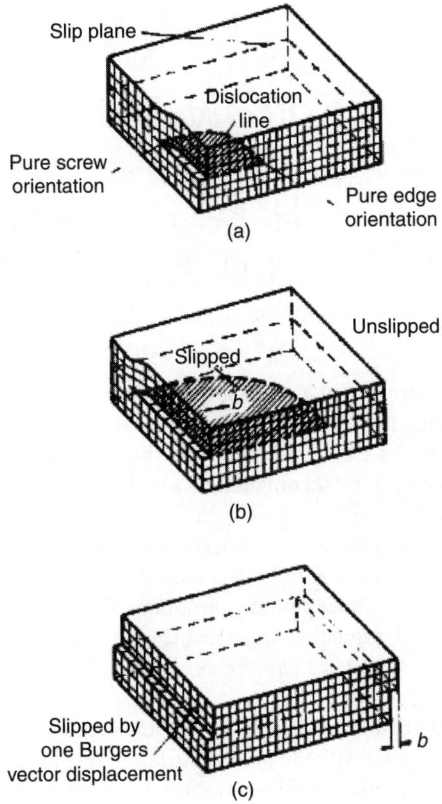

Figure 3.15 *Process of slip by the expansion of a dislocation loop in the slip plane.*

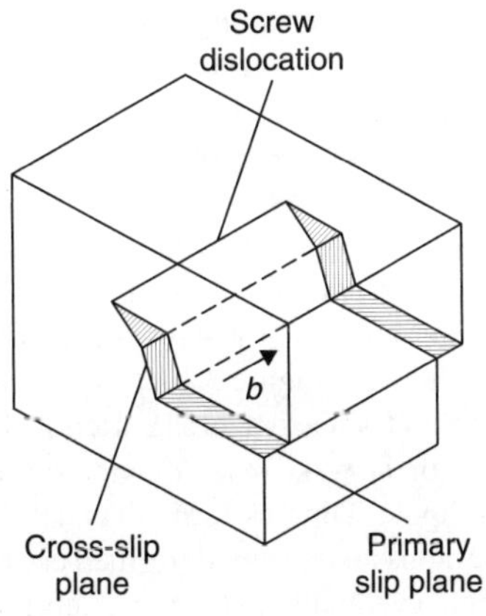

Figure 3.16 *Cross-slip of a screw dislocation in a crystal.*

plane and other sections in parallel neighboring planes. Where the dislocation deviates from one plane to another it is known as a jog, and from the diagrams of Figure 3.17 it is evident that a jog in a dislocation may be regarded as a short length of dislocation not lying in the same slip plane as the main dislocation but having the same Burgers vector.

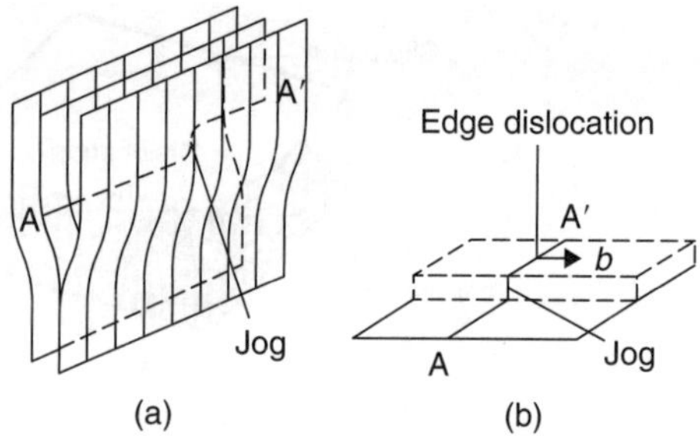

Figure 3.17 *Climb of an edge dislocation in a crystal.*

Jogs may also form when a moving dislocation cuts through intersecting dislocations, i.e. forest[5] dislocations, during its glide motion. In the lower range of temperature this will be the major source of jogs. Two examples of jogs formed from the crossings of dislocations are shown in Figure 3.18. Figure 3.18a shows a crystal containing a screw dislocation running from top to bottom, which has the effect of 'ramping' all the planes in the crystal. If an edge dislocation moves through the crystal on a horizontal plane then the screw dislocation becomes jogged as the top half of the crystal is sheared relative to the bottom. In addition, the screw dislocation becomes jogged since one part has to take the upper ramp and the other part the lower ramp. The result is shown schematically in Figure 3.18b. Figure 3.18c shows the situation for a moving screw cutting through the vertical screw; the jog formed in each dislocation is edge in character, since it is perpendicular to its Burgers vector, which lies along the screw axis.

A jog in an edge dislocation will not impede the motion of the dislocation in its slip plane because it can, in general, move with the main dislocation line by glide, not in the same slip plane (see Figure 3.17b) but in an intersecting slip plane that does contain the line of the jog and the Burgers vector. In the case of a jog in a screw dislocation the situation is not so clear, since there are two ways in which the jog can move. Since the jog is merely a small piece of edge dislocation it may move sideways, i.e. conservatively, along the screw dislocation and attach itself to an edge component of the dislocation line. Conversely, the jog may be dragged along with the screw dislocation. This latter process requires the jog to climb and, because it is a non-conservative process, must give rise to the creation of a row of point defects, i.e. either vacancies or interstitials depending on which way the jog is forced to climb. Clearly, such a movement is difficult but, nevertheless, may be necessary to give the dislocation sufficient mobility. The 'frictional' drag of jogs will make a contribution to the work hardening[6] of the material.

Apart from elementary jogs, or those having a height equal to one atomic plane spacing, it is possible to have multiple jogs where the jog height is several atomic plane spacings. Such jogs can be produced, for example, by part of a screw dislocation cross-slipping from the primary plane to the cross-slip plane, as shown in Figure 3.19a. In this case, as the screw dislocation glides forward it trails the multiple jog behind, since it acts as a frictional drag. As a result, two parallel dislocations of opposite sign are created in the wake of the moving screw, as shown in Figure 3.19b; this arrangement is called a dislocation dipole. Dipoles formed as debris behind moving dislocations are frequently seen in electron micrographs taken from deformed crystals (see Chapter 6). As the dipole gets longer the screw dislocation will eventually jettison the debris by cross-slipping and pinching off the dipole to form a prismatic loop, as shown in Figure 3.20. The loop is capable of gliding on the surface of a prism, the cross-sectional area of which is that of the loop.

[5] A number of dislocation lines may project from the slip plane like a forest, hence the term 'forest dislocation'.

[6] When material is deformed by straining or working the flow stress increases with increase in strain (i.e. it is harder to deform a material which has been strained already). This is called strain or work hardening.

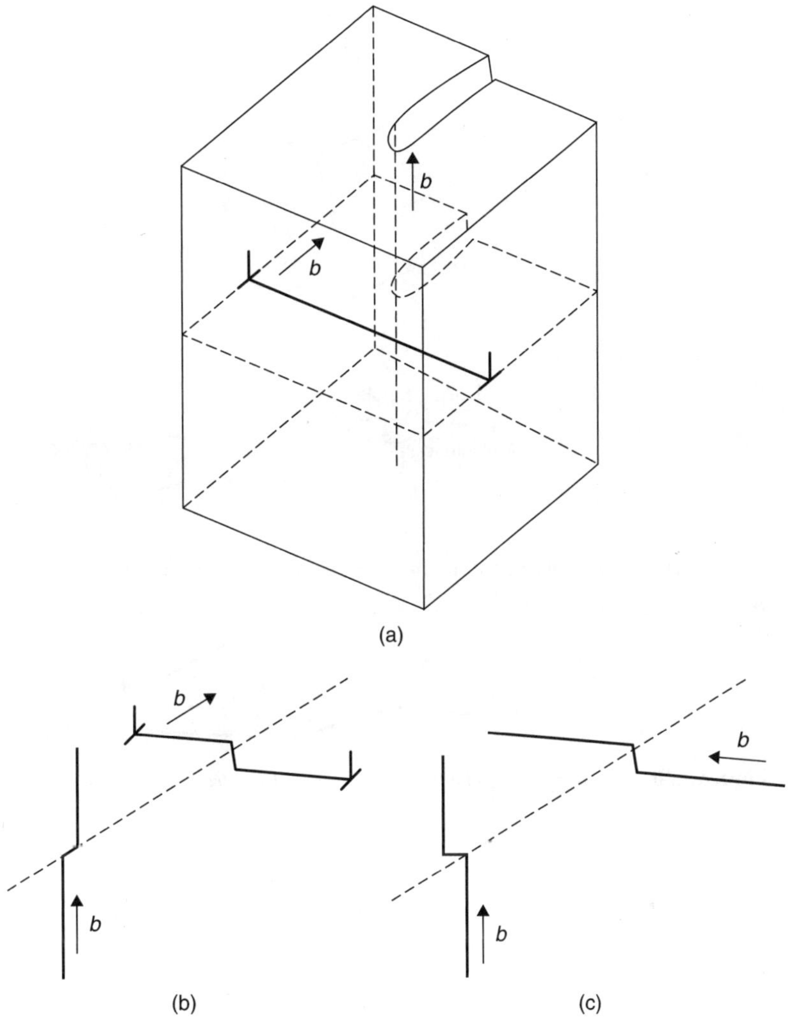

Figure 3.18 *Dislocation intersections. (a, b) Screw–edge. (c) Screw–screw.*

3.3.5 Strain energy associated with dislocations

3.3.5.1 Stress fields of screw and edge dislocations

The distortion around a dislocation line is evident from Figures 3.1 and 3.13. At the center of the dislocation the strains are too large to be treated by elasticity theory, but beyond a distance r_0, equal to a few atom spacings, Hooke's law can be applied. It is therefore necessary to define a core to the dislocation at a cut-off radius r_0 ($\approx b$) inside which elasticity theory is no longer applicable. A screw dislocation can then be considered as a cylindrical shell of length l and radius r contained in an elastically isotropic medium (Figure 3.21). A discontinuity in displacement exists only in the z-direction, i.e. parallel to the dislocation, such that $u = v = 0$, $w = b$. The elastic strain thus has to accommodate a displacement $w = b$ around a length $2\pi r$. In an elastically isotropic crystal the accommodation must occur equally all round the shell and indicates the simple relation $w = b\theta/2\pi$

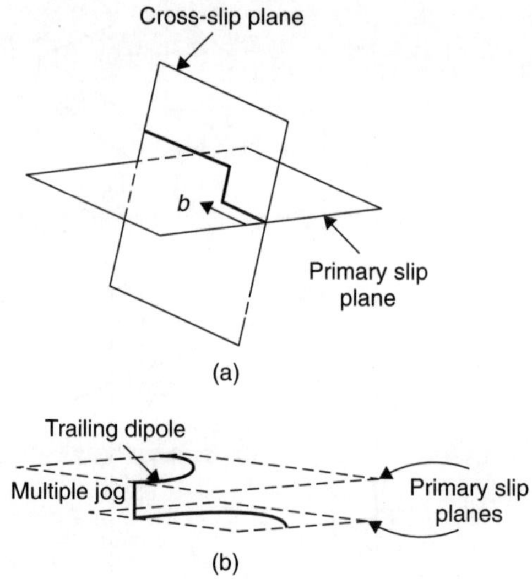

Figure 3.19 *(a) Formation of a multiple jog by cross-slip. (b) Motion of jog to produce a dipole.*

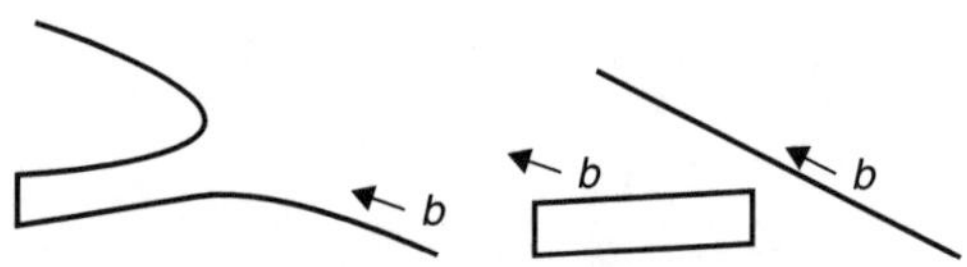

Figure 3.20 *Formation of prismatic dislocation loop from screw dislocation trailing a dipole.*

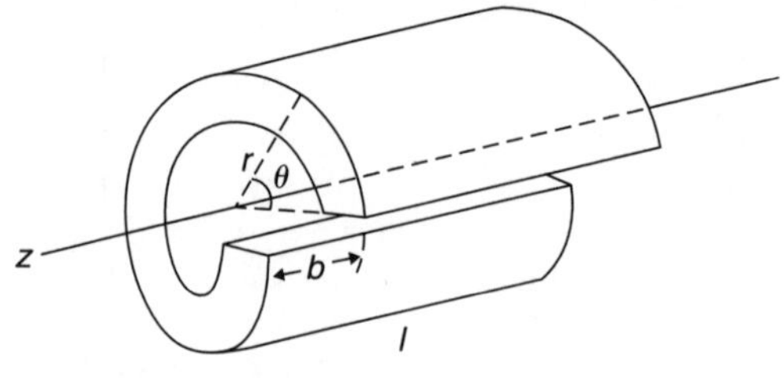

Figure 3.21 *Screw dislocation in an elastic continuum.*

in polar (r, θ, z) coordinates. The corresponding shear strain $\gamma_{\theta z}$ $(=\gamma_{z\theta})=b/2\pi r$ and shear stress $\tau_{\theta z}(=\tau_{z\theta})=\mu b/2\pi r$, which acts on the end faces of the cylinder with σ_{rr} and $\tau_{r\theta}$ equal to zero.[7] Alternatively, the stresses are given in cartesian coordinates (x, y, z):

$$\begin{aligned}\tau_{xz}(=\tau_{zx}) &= -\mu by/2\pi(x^2+y^2)\\ \tau_{yz}(=\tau_{zy}) &= -\mu bx/2\pi(x^2+y^2),\end{aligned} \tag{3.8}$$

[7] This subscript notation $z\theta$ indicates that the stress is in the θ-direction on an element perpendicular to the z-direction. The stress with subscript rr is thus a normal stress and denoted by σ_{rr} and the subscript $r\theta$ a shear stress and denoted by $\tau_{r\theta}$.

with all other stresses equal to zero. The field of a screw dislocation is therefore purely one of shear, having radial symmetry with no dependence on θ. This mathematical description is related to the structure of a screw which has no extra half-plane of atoms and cannot be identified with a particular slip plane.

An edge dislocation has a more complicated stress and strain field than a screw. The distortion associated with the edge dislocation is one of plane strain, since there are no displacements along the z-axis, i.e. $w = 0$. In plane deformation the only stresses to be determined are the normal stresses σ_{xx}, σ_{yy} along the x- and y-axes respectively, and the shear stress τ_{xy} which acts in the direction of the y-axis on planes perpendicular to the x-axis. The third normal stress $\sigma_{zz} = \nu(\sigma_{xx} + \sigma_{yy})$ where ν is Poisson's ratio, and the other shear stresses τ_{yz} and τ_{zx} are zero. In polar coordinates r, θ and z, the stresses are σ_{rr}, $\sigma_{\theta\theta}$, and $\tau_{r\theta}$.

Even in the case of the edge dislocation the displacement b has to be accommodated round a ring of length $2\pi r$, so that the strains and the stresses must contain a term in $b/2\pi r$. Moreover, because the atoms in the region $0 < \theta < \pi$ are under compression and for $\pi < \theta < 2\pi$ in tension, the strain field must be of the form $(b/2\pi r) f(\theta)$, where $f(\theta)$ is a function such as $\sin\theta$ which changes sign when θ changes from 0 to 2π. It can be shown that the stresses are given by

$$\begin{aligned}\sigma_{rr} &= \sigma_{\theta\theta} = -D\sin\theta/r; \quad \sigma_{r\theta} = D\cos\theta/r;\\ \sigma_{xz} &= -D\frac{y(3x^2 + y^2)}{(x^2 + y^2)^2}; \quad \sigma_{yy} = D\frac{y(x^2 - y^2)}{(x^2 + y^2)^2}\\ \sigma_{xy} &= D\frac{x(x^2 - y^2)}{(x^2 + y^2)^2},\end{aligned} \tag{3.9}$$

where $D = \mu b/2\pi(1 - \nu)$. These equations show that the stresses around dislocations fall off as $1/r$ and hence the stress field is long range in nature.

3.3.5.2 *Strain energy of a dislocation*

A dislocation is a line defect extending over large distances in the crystal and, since it has a strain energy per unit length ($\mathrm{J\,m^{-1}}$), it possesses a total strain energy. An estimate of the elastic strain energy of screw dislocation can be obtained by taking the strain energy (i.e. $\frac{1}{2} \times$ stress $\times$ strain per unit volume) in an annular ring around the dislocation of radius r and thickness $\mathrm{d}r$ to be $\frac{1}{2} \times (\mu b/2\pi r) \times (b/2\pi r) \times 2\pi r \mathrm{d}r$. The total strain energy per unit length of dislocation is then obtained by integrating from r_0 (the core radius) to r (the outer radius of the strain field) and is:

$$E = \frac{\mu b^2}{4\pi}\int_{r_0}^{r}\frac{\mathrm{d}r}{r} = \frac{\mu b^2}{4\pi}\ln\left[\frac{r}{r_0}\right]. \tag{3.10}$$

With an edge dislocation this energy is modified by the term $(1 - \nu)$ and hence is about 50% greater than a screw. For a unit dislocation in a typical crystal $r_0 \cong 0.25$ nm, $r \cong 2.5$ µm and $\ln[r/r_0] \cong 9.2$, so that the energy is approximately μb^2 per unit length of dislocation, which for copper (taking $\mu = 40\,\mathrm{GN\,m^{-2}}$, $b = 0.25$ nm and 1 eV $= 1.6 \times 10^{-19}$ J) is about 4 eV for every atom plane threaded by the dislocation.[8] If the reader prefers to think in terms of 1 meter of dislocation line, then this length is associated with about 2×10^{10} electron volts. We shall see later that heavily deformed metals contain approximately $10^{16}\,\mathrm{m\,m^{-3}}$ of dislocation line, which leads to a large amount of energy stored

[8] The energy of the core must be added to this estimate. The core energy is about $\mu b^2/10$ or 0.5 eV per atom length.

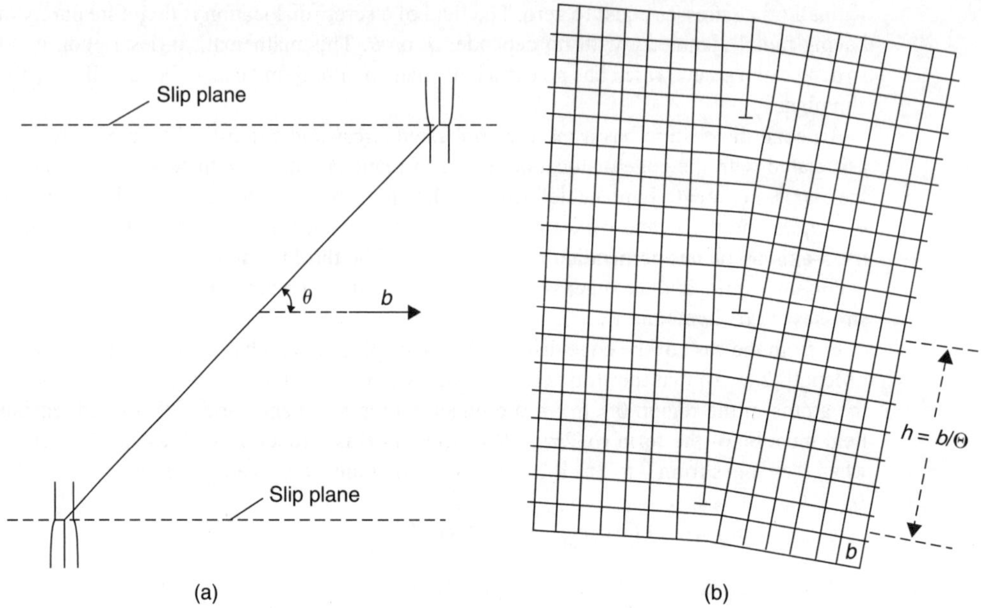

Figure 3.22 *Interaction between dislocations not on the same slip plane: (a) unlike dislocations;* (b) *like dislocations. The arrangement in (b) constitutes a small-angle boundary.*

in the lattice (i.e. $\approx 4\,\text{J}\,\text{g}^{-1}$ for Cu). Clearly, because of this high line energy a dislocation line will always tend to shorten its length as much as possible, and from this point of view it may be considered to possess a line tension, $T \approx \alpha\mu b^2$, analogous to the surface energy of a soap film, where $\alpha \approx \frac{1}{2}$.

3.3.5.3 Interaction of dislocations

The strain field around a dislocation, because of its long-range nature, is also important in influencing the behavior of other dislocations in the crystal. Thus, it is not difficult to imagine that a positive dislocation will attract a negative dislocation lying on the same slip plane in order that their respective strain fields should cancel. Moreover, as a general rule it can be said that the dislocations in a crystal will interact with each other to take up positions of minimum energy to reduce the total strain energy of the lattice.

Two dislocations of the same sign will repel each other, because the strain energy of two dislocations on moving apart would be $2 \times b^2$, whereas if they combined to form one dislocation of Burgers vector $2b$, the strain energy would then be $(2b)^2 = 4b^2$; a force of repulsion exists between them. The force is, by definition, equal to the change of energy with position ($\text{d}E/\text{d}r$) and for screw dislocations is simply $F = \mu b^2/2\pi r$, where r is the distance between the two dislocations. Since the stress field around screw dislocations has cylindrical symmetry the force of interaction depends only on the distance apart, and the above expression for F applies equally well to parallel screw dislocations on neighboring slip planes. For parallel edge dislocations the force–distance relationship is less simple. When the two edge dislocations lie in the same slip plane the relation is similar to that for two screws and has the form $F = \mu b^2/(1-\nu)2\pi r$, but for edge dislocations with the same Burgers vector but not on the same slip plane the force also depends on the angle θ between the Burgers vector and the line joining the two dislocations (Figure 3.22a).

Edge dislocations of the same sign repel and opposite sign attract along the line between them, but the component of force in the direction of slip, which governs the motion of a dislocation, varies with the angle θ. With unlike dislocations an attractive force is experienced for $\theta > 45°$ but a repulsive force for $\theta < 45°$, and in equilibrium the dislocations remain at an angle of 45° to each other. For like dislocations the converse applies and the position $\theta = 45°$ is now one of unstable equilibrium. Thus, edge dislocations which have the same Burgers vector but which do not lie on the same slip plane will be in equilibrium when $\theta = 90°$, and consequently they will arrange themselves in a plane normal to the slip plane, one above the other a distance h apart. Such a wall of dislocations constitutes a small-angle grain boundary, as shown in Figure 3.22b, where the angle across the boundary is given by $\theta = b/h$. This type of dislocation array is also called a sub-grain or low-angle boundary, and is important in the annealing of deformed metals.

By this arrangement the long-range stresses from the individual dislocations are cancelled out beyond a distance of the order of h from the boundary. It then follows that the energy of the crystal boundary will be given approximately by the sum of the individual energies, each equal to $\{\mu b^2/4\pi(1-\nu)\}\ \ln(h/r_0)$ per unit length. There are $1/h$ or θ/b dislocations in a unit length, vertically, and hence, in terms of the misorientation across the boundary $\theta = b/h$, the energy γ_{gb} per unit area of boundary is

$$\begin{aligned}\gamma_{gb} &= \frac{\mu b^2}{4\pi(1-\nu)} \ln\left(\frac{h}{r_0}\right) \times \frac{\theta}{b} \\ &= \left[\frac{\mu b}{4\pi(1-\nu)}\right]\theta \ln\left(\frac{b}{\theta r_0}\right) \\ &= E_0\theta[A - \ln\theta],\end{aligned} \tag{3.11}$$

where $E_0 = \mu b/4\pi(1-\nu)$ and $A = \ln(b/r_0)$; this is known as the Read–Shockley formula. Values from it give good agreement with experimental estimates, even up to relatively large angles. For $\theta \sim 25°$, $\gamma_{gb} \sim \mu b/25$ or $\sim 0.4\ \mathrm{J\,m^{-2}}$, which surprisingly is close to the value for the energy per unit area of a general large-angle grain boundary

3.3.6 Dislocations in ionic structures

The slip system which operates in materials with NaCl structure is predominantly $a/2\langle 1\ \bar{1}\ 0\rangle\{1\ 1\ 0\}$. The closest packed plane $\{1\ 0\ 0\}$ is not usually the preferred slip plane because of the strong electrostatic interaction that would occur across the slip plane during slip; like ions are brought into neighboring positions across the slip plane for (1 0 0) but not for (1 1 0). Dislocations in these materials are therefore simpler than fcc metals, but they may carry an electric charge (the edge dislocation on $\{1\ 1\ 0\}$, for example). Figure 3.23a has an extra 'half-plane' made up of a sheet of Na^+ ions and one of Cl^- ions. The line as a whole can be charged up to a maximum of $e/2$ per atom length by acting as a source or sink for point defects. Figure 3.23b shows different jogs in the line which may either carry a charge or be uncharged. The jogs at B and C would be of charge $+e/2$ because the section BC has a net charge equal to e. The jog at D is uncharged.

3.4 Planar defects

3.4.1 Grain boundaries

The small-angle boundary described in Section 3.3.5.3 is a particular example of a planar defect interface in a crystal. Many such planar defects occur in materials ranging from the large-angle grain

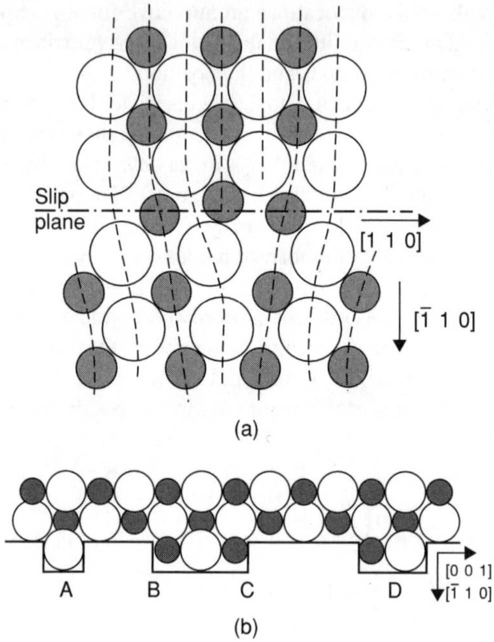

Figure 3.23 *Edge dislocation in NaCl, showing: (a) two extra half-sheets of ions; anions are open circles, cations are shaded; (b) charged and uncharged jogs (after Kelly, Groves and Kidd, 2000).*

boundary, which is an incoherent interface with a relatively high energy of $\sim$0.5 J m^{-2}, to atomic planes in the crystal across which there is a mis-stacking of the atoms, i.e. twin interfaces and stacking faults that retain the coherency of the packing and have much lower energies (≤ 0.1 J/m^2). Generally, all these planar defects are associated with dislocations in an extended form.

A small-angle tilt boundary can be described adequately by a vertical wall of dislocations. Rotation of one crystal relative to another, i.e. a twist boundary, can be produced by a crossed grid of two sets of screw dislocations, as shown in Figure 3.24. These boundaries are of a particularly simple kind, separating two crystals which have a small difference in orientation, whereas a general grain boundary usually separates crystals which differ in orientation by large angles. In this case, the boundary has five degrees of freedom, three of which arise from the fact that the adjoining crystals may be rotated with respect to each other about the three perpendicular axes, and the other two from the degree of freedom of the orientation of the boundary surface itself with respect to the crystals. Such a large-angle (30–40°) grain boundary may simply be regarded as a narrow region, about two atoms thick, across which the atoms change from the lattice orientation of one grain to that of the other. Nevertheless, such a grain boundary may be described by an arrangement of dislocations, but their arrangement will be complex and the individual dislocations are not easily recognized or analyzed.

The simplest extension of the dislocation model for low-angle boundaries to high-angle grain boundaries is to consider that there are islands of good atomic fit surrounded by non-coherent regions. In a low-angle boundary the 'good fit' is a perfect crystal and the 'bad fit' is accommodated by lattice dislocations, whereas for high-angle boundaries the 'good fit' could be an interfacial structure with low energy and the 'bad fit' accommodated by dislocations which are not necessarily lattice dislocations. These dislocations are often termed intrinsic secondary grain boundary dislocations (gbds) and are essential to maintain the boundary at that misorientation.

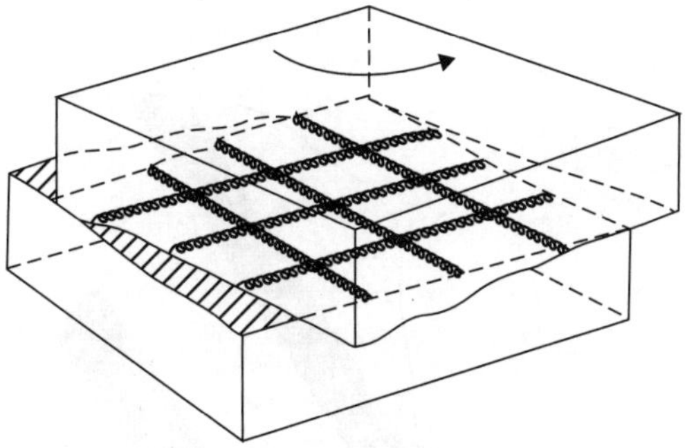

Figure 3.24 *Representation of a twist boundary produced by a cross-grid of screw dislocations.*

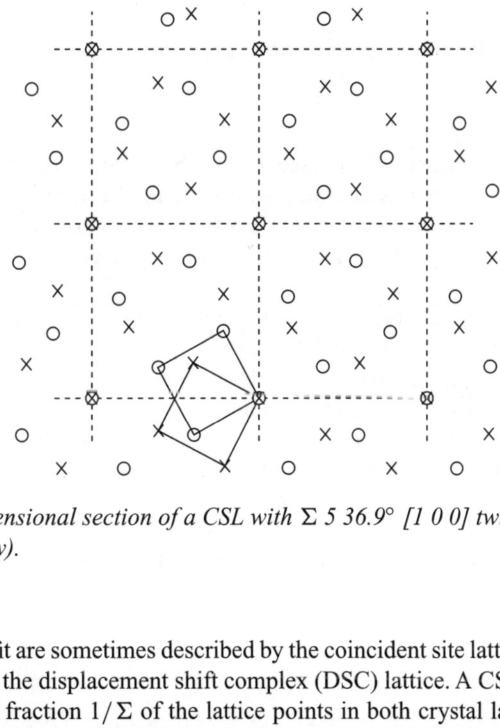

Figure 3.25 *Two-dimensional section of a CSL with Σ 5 36.9° [1 0 0] twist orientation (courtesy of P. Goodhew).*

The regions of good fit are sometimes described by the coincident site lattice (CSL) model, with its development to include the displacement shift complex (DSC) lattice. A CSL is a three-dimensional superlattice on which a fraction $1/\Sigma$ of the lattice points in both crystal lattices lie; for the simple structures there will be many such CSLs, each existing at a particular misorientation. One CSL is illustrated in Figure 3.25, but it must be remembered that the CSL is three-dimensional, infinite and interpenetrates both crystals; it does not in itself define an interface. However, an interface is likely to have a low energy if it lies between two crystals oriented such that they share a high proportion of lattice sites, i.e. preferred misorientations will be those with CSLs having low Σ values. Such misorientations can be predicted from the expression:

$$\theta = 2\tan^{-1}\left(\frac{b}{a}\sqrt{N}\right),$$

where b and a are integers and $N = h^2 + k^2 + l^2$; the Σ value is then given by $a^2 + Nb^2$, divided by 2 until an odd number is obtained.

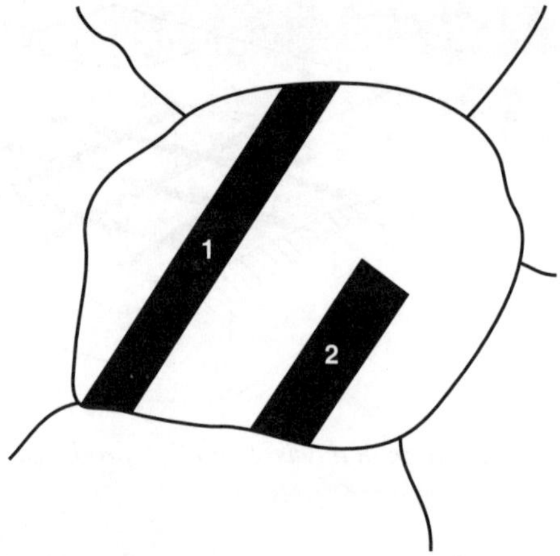

Figure 3.26 *Twinned regions within a single etched grain, produced by deformation and annealing.*

The CSL model can only be used to describe certain specific boundary misorientations but it can be extended to other misorientations by allowing the presence of arrays of dislocations which act to preserve a special orientation between them. Such intrinsic secondary dislocations must conserve the boundary structure and, generally, will have Burgers vectors smaller than those of the lattice dislocations.

When a polycrystalline specimen is examined in TEM other structural features apart from intrinsic grain boundary dislocations (GBDS) may be observed in a grain boundary, such as 'extrinsic' dislocations which have probably run-in from a neighboring grain, and interface ledges or steps which curve the boundary. At low temperatures the run-in lattice dislocation tends to retain its character while trapped in the interface, whereas at high temperatures it may dissociate into several intrinsic GBDS, resulting in a small change in misorientation across the boundary. The analysis of GBDS in TEM is not easy, but information about them will eventually further our understanding of important boundary phenomena (e.g. migration of boundaries during recrystallization and grain growth, the sliding of grains during creep and superplastic flow, and the way grain boundaries act as sources and sinks for point defects).

3.4.2 Twin boundaries

Annealing of cold-worked fcc metals and alloys, such as copper, α-brass and austenitic stainless steels usually causes many of the constituent crystals to form annealing twins. The lattice orientation changes at the twin boundary surface, producing a structure in which one part of the crystal or grain is the mirror image of the other, the amount of atomic displacement being proportional to the distance from the twin boundary.

The surfaces of a sample within and outside an annealing twin have different surface energies, because of their different lattice orientations, and hence respond quite differently when etched with a chemical etchant (Figure 3.26). In this diagram, twins 1 and 2 both have two straight parallel sides which are coherent low-energy interfaces. The short end face of twin 2 is non-coherent and therefore has a higher energy content per unit surface. Stacking faults are also coherent and low in energy content; consequently, because of this similarity in character, we find that crystalline materials which twin readily are also likely to contain many stacking faults (e.g. copper, α-brass).

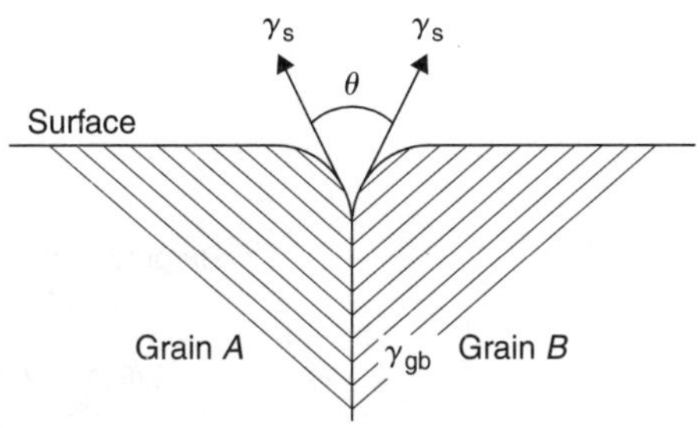

Figure 3.27 *Grain boundary/surface triple junction.*

Although less important and less common than slip, another type of twinning can take place during plastic deformation. These so-called deformation twins sometimes form very easily, e.g. during mechanical polishing of metallographic samples of pure zinc; this process is discussed in Chapter 6.

The free energies of interfaces can be determined from the equilibrium form of the triple junction where three interfaces, such as surfaces, grain boundaries or twins, meet. For the case of a grain boundary intersecting a free surface, shown in Figure 3.27,

$$\gamma_{gb} = 2\gamma_s \cos(\theta/2), \tag{3.12}$$

and hence γ_{gb} can be obtained by measuring the dihedral angle θ and knowing γ_s. Similarly, measurements can be made of the ratio of twin boundary energy to the average grain boundary energy and, knowing either γ_s or γ_{gb} gives an estimate of γ_T.

3.4.3 Extended dislocations and stacking faults in close-packed crystals

3.4.3.1 Stacking faults

Stacking faults associated with dislocations can be an extremely significant feature of the structure of many materials, particularly those with fcc and cph structures. They arise because, to a first approximation, there is little to choose electrostatically between the stacking sequence of the close-packed planes in the fcc metals **ABCABC** ... and that in the cph metals **ABABAB** .... Thus, in a metal like copper or gold, the atoms in a part of one of the close-packed layers may fall into the 'wrong' position relative to the atoms of the layers above and below, so that a mistake in the stacking sequence occurs (e.g. **ABCBCABC** ...). Such an arrangement will be reasonably stable, but because some work will have to be done to produce it, stacking faults are more frequently found in deformed metals than annealed metals.

3.4.3.2 Dissociation into Shockley partials

The relationship between the two close-packed structures cph and fcc has been discussed in Chapter 1, where it was seen that both structures may be built up from stacking close-packed planes of spheres. The shortest lattice vector in the fcc structure joins a cube corner atom to a neighboring face center atom and defines the observed slip direction; one such slip vector $a/2[1\ 0\ \bar{1}]$ is shown as b_1 in Figure 3.28a, which is for glide in the $(1\ 1\ 1)$ plane. However, an atom which sits in a **B** position

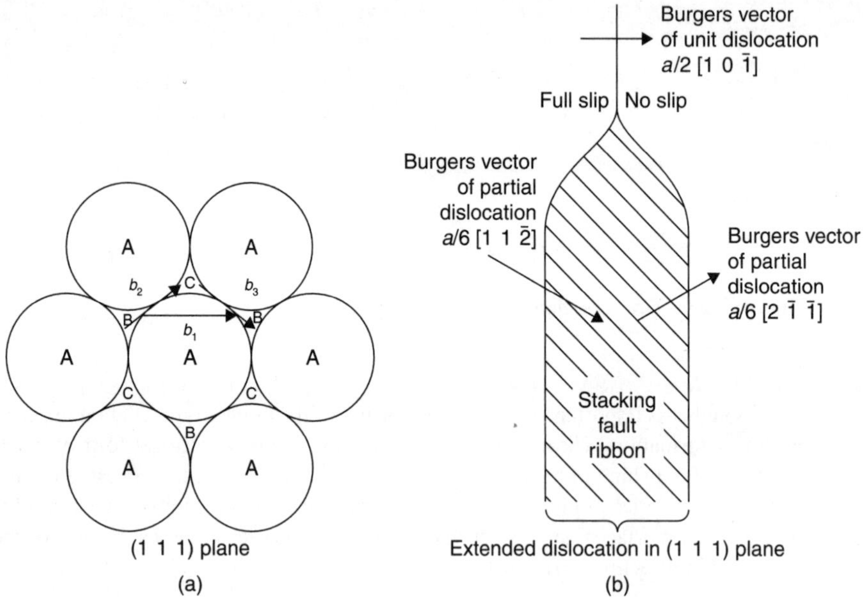

Figure 3.28 *Schematic representation of slip in a (1 1 1) plane of a fcc crystal.*

on top of the A plane would move most easily initially towards a **C** position and, consequently, to produce a macroscopical slip movement along [1 0 $\bar{1}$] the atoms might be expected to take a zigzag path of the type **B** → **C** → **B** following the vectors $b_2 = a/6[2\ \bar{1}\ \bar{1}]$ and $b_3 = a/6[1\ 1\ \bar{2}]$ alternately. It will be evident, of course, that during the initial part of the slip process when the atoms change from **B** positions to **C** positions, a stacking fault in the (1 1 1) layers is produced and the stacking sequence changes from **ABCABC** ... to **ABCACABC** .... During the second part of the slip process the correct stacking sequence is restored.

To describe the atoms' movement during slip, discussed above, Heidenreich and Shockley have pointed out that the unit dislocation must dissociate into two half-dislocations,[9] which for the case of glide in the (1 1 1) plane would be according to the reaction:

$$a/2[1\ 0\ \bar{1}] \rightarrow a/6[2\ \bar{1}\ \bar{1}] + a/6[1\ 1\ \bar{2}].$$

Such a dissociation process is: (1) algebraically correct, since the sum of the Burgers vector components of the two partial dislocations, i.e. $a/6[2+1]$, $a/6[-1+1]$, $a/6[-1-2]$, are equal to the components of the Burgers vector of the unit dislocation, i.e. $a/2,\ 0, \bar{a}/2$; and (2) energetically favorable, since the sum of the strain energy values for the pair of half-dislocations is less than the strain energy value of the single unit dislocation, where the initial dislocation energy is proportional to $b_1^2 (= a^2/2)$ and the energy of the resultant partials to $b_2^2 + b_3^2 = a^2/3$. These half dislocations, or Shockley partial dislocations, repel each other by a force that is approximately $F = (\mu b_2 b_3 \cos 60°)/2\pi d$ and separate, as shown in Figure 3.28b. A sheet of stacking faults is then formed in the slip plane between the partials, and it is the creation of this faulted region, which has a higher energy than the normal lattice, that prevents the partials from separating too far. Thus, if γ J m^{-2} is the energy per unit

[9] The correct indices for the vectors involved in such dislocation reactions can be obtained from Figure 3.35.

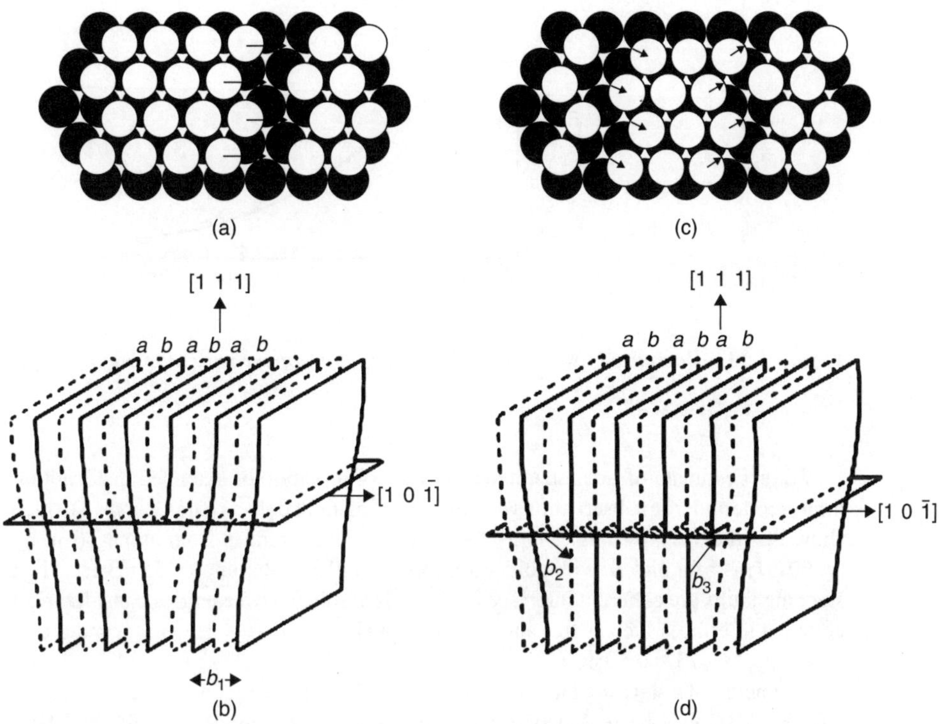

Figure 3.29 *Edge dislocation structure in the fcc lattice. (a, b) Undissociated. (c, d) Dissociated. (a) and (c) are viewed normal to the (1 1 1) plane (from Seeger, 1957; courtesy of John Wiley).*

area of the fault, the force per unit length exerted on the dislocations by the fault is γ N m^{-1} and the equilibrium separation d is given by equating the repulsive force F between the two half-dislocations to the force exerted by the fault, γ. The equilibrium separation of two partial dislocations is then given by

$$d = \left(\frac{\mu b_2 b_3 \cos 60^\circ}{2\pi\gamma}\right) = \left(\frac{\mu \frac{a}{\sqrt{6}}\frac{a}{\sqrt{6}}\frac{1}{2}}{2\pi\gamma}\right) = \frac{\mu a^2}{24\pi\gamma}, \tag{3.13}$$

from which it can be seen that the width of the stacking fault 'ribbon' is inversely proportional to the value of the stacking fault energy γ and also depends on the value of the shear modulus μ.

Figure 3.29a shows that the undissociated edge dislocation has its extra half-plane corrugated, which may be considered as two $(1\ 0\ \bar{1})$ planes displaced relative to each other (labeled a and b in Figure 3.29b). On dissociation, planes a and b are separated by a region of crystal in which, across the slip plane, the atoms are in the wrong sites (see Figure 3.29c). Thus, the high strain energy along a line through the crystal associated with an undissociated dislocation is spread over a plane in the crystal for a dissociated dislocation (see Figure 3.29d), thereby lowering its energy.

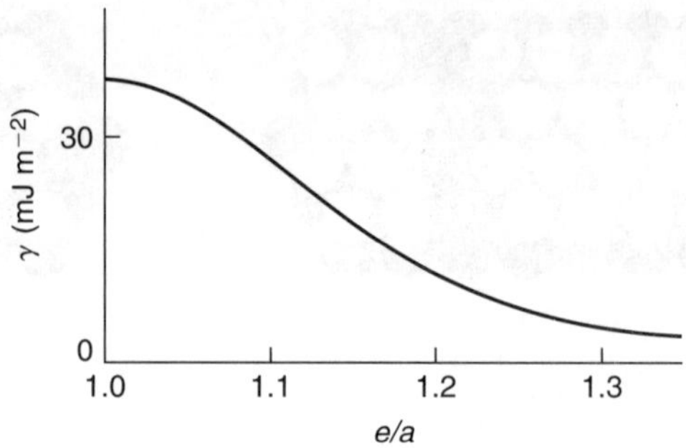

Figure 3.30 *Decrease in stacking-fault energy γ for copper with alloying addition (e/a).*

A direct estimate of γ can be made from the observation of extended dislocations in the electron microscope and from observations on other stacking fault defects (see Chapter 4). Such measurements show that the stacking fault energy for pure fcc metals ranges from about 16 mJ m^{-2} for silver to $\approx$200 mJ m^{-2} for nickel, with gold $\approx$30, copper $\approx$40 and aluminum 135 mJ m^{-2}, respectively. Since stacking faults are coherent interfaces or boundaries they have energies considerably lower than non-coherent interfaces such as free surfaces, for which $\gamma_s \approx \mu b/8 \approx 1.5$ J m^{-2}, and grain boundaries, for which $\gamma_{gb} \approx \gamma_s/3 \approx 0.5$ J m^{-2}.

The energy of a stacking fault can be estimated from twin boundary energies, since a stacking fault **ABCBCABC** may be regarded as two overlapping twin boundaries **CBC** and **BCB** across which the next nearest neighboring plane is wrongly stacked. In fcc crystals any sequence of three atomic planes not in the **ABC** or **CBA** order is a stacking violation and is accompanied by an increased energy contribution. A twin has one pair of second nearest neighbor planes in the wrong sequence, two third neighbors, one fourth neighbor and so on; an intrinsic stacking fault two second nearest neighbors, three third and no fourth nearest neighbor violations. Thus, if next-next nearest neighbor interactions are considered to make a relatively small contribution to the energy, then an approximate relation $\gamma \approx 2\gamma_T$ is expected.

The frequency of occurrence of annealing twins generally confirms the above classification of stacking fault energy and it is interesting to note that in aluminum, a metal with a relatively high value of γ, annealing twins are rarely, if ever, observed, while they are seen in copper, which has a lower stacking fault energy. Electron microscope measurements of γ show that the stacking fault energy is lowered by solid solution alloying and is influenced by those factors which affect the limit of primary solubility. The reason for this is that, on alloying, the free energies of the α-phase and its neighboring phase become more nearly equal, i.e. the stability of the α-phase is decreased relative to some other phase, and hence can more readily tolerate mis-stacking. Figure 3.30 shows the reduction of γ for copper with addition of solutes such as Zn, Al, Sn and Ge, and is consistent with the observation that annealing twins occur more frequently in α-brass or Cu–Sn than pure copper. Substituting the appropriate values for μ, a and γ in equation (3.13) indicates that in silver and copper the partials are separated to about 12 and six atom spacings, respectively. For nickel the width is about $2b$, since although nickel has a high γ its shear modulus is also very high. In contrast, aluminum has a lower $\gamma \approx 135$ mJ m^{-2} but also a considerably lower value for μ, and hence the partial separation is limited to about $1b$ and may be considered to be unextended. Alloying significantly reduces γ and very wide dislocations are produced, as found in the brasses, bronzes and austenitic stainless steels. However, no matter how narrow or how wide the partials are separated, the two half-dislocations are bound together by the stacking fault and, consequently, they must move together as a unit across the slip plane.

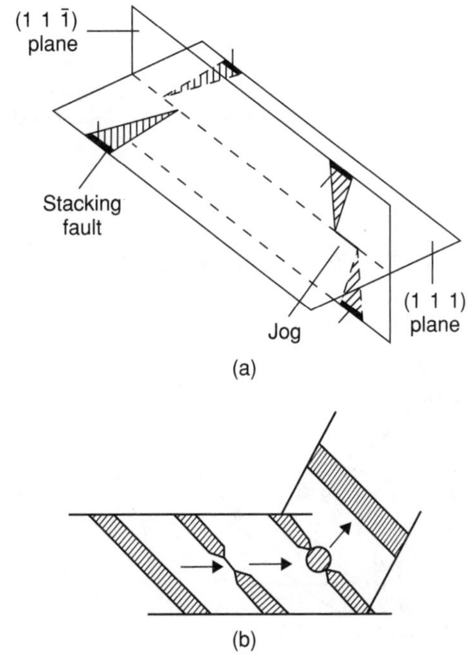

Figure 3.31 *(a) The crossing of extended dislocations. (b) Various stages in the cross-slip of a dissociated screw dislocation.*

The width of the stacking-fault ribbon is of importance in many aspects of plasticity because at some stage of deformation it becomes necessary for dislocations to intersect each other; the difficulty which dislocations have in intersecting each other gives rise to one source of work hardening. With extended dislocations the intersecting process is particularly difficult, since the crossing of stacking faults would lead to a complex fault in the plane of intersection. The complexity may be reduced, however, if the half-dislocations coalesce at the crossing point, so that they intersect as perfect dislocations: the partials then are constricted together at their jogs, as shown in Figure 3.31a.

The width of the stacking-fault ribbon is also important to the phenomenon of cross-slip in which a dislocation changes from one slip plane to another intersecting slip plane. As discussed previously, for glide to occur the slip plane must contain both the Burgers vector and the line of the dislocation and, consequently, for cross-slip to take place a dislocation must be in an exact screw orientation. If the dislocation is extended, however, the partials have first to be brought together to form an unextended dislocation, as shown in Figure 3.31b, before the dislocation can spread into the cross-slip plane. The constriction process will be aided by thermal activation and hence the cross-slip tendency increases with increasing temperature. The constriction process is also more difficult the wider the separation of the partials. In aluminum, where the dislocations are relatively unextended, the frequent occurrence of cross-slip is expected, but for low stacking-fault energy metals (e.g. copper or gold) the activation energy for the process will be high. Nevertheless, cross-slip may still occur in those regions where a high concentration of stress exists, as, for example, when dislocations pile up against some obstacle, where the width of the extended dislocation may be reduced below the equilibrium separation. Often, screw dislocations escape from the piled-up group by cross-slipping but then after moving a certain distance in this cross-slip plane return to a plane parallel to the original slip plane because the resolved shear stress is higher. This is a common method of circumventing obstacles in the structure.

Worked example

Consider dissociation and slip on the $(\bar{1}\,\bar{1}\,1)$ plane of an fcc crystal with the aid of a stereogram. Complete the vectorial reactions, so that each represents dissociation of a unit dislocation on this plane into Shockley partials in one of the six <1 1 0>-type slip directions.

Solution

A quick and simple way is to look at the stereogram and draw in the trace of the $(\bar{1}\,\bar{1}\,1)$.

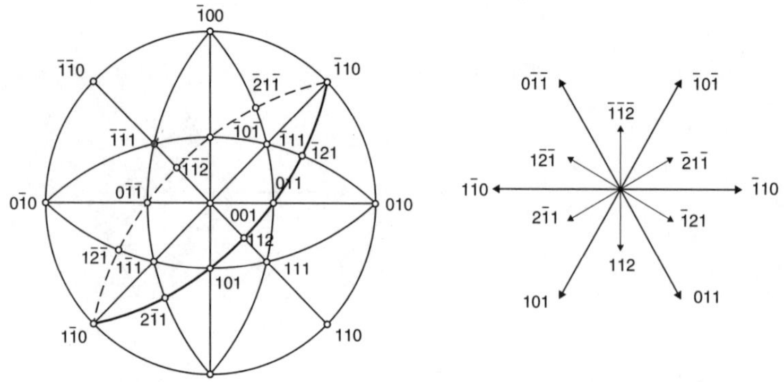

The six <1 1 0> directions can be read off the stereogram. The <1 1 2> directions can be drawn in using adding rule or from the stereogram. Hence the dissociation reactions are:

$$\frac{a}{2}[\bar{1}\,1\,0] \rightarrow \frac{a}{6}[\bar{2}\,1\,\bar{1}] + \frac{a}{6}[\bar{1}\,2\,1] \qquad \frac{a}{2}[1\,\bar{1}\,0] \rightarrow \frac{a}{6}[1\,\bar{2}\,\bar{1}] + \frac{a}{6}[2\,\bar{1}\,1]$$

$$\frac{a}{2}[0\,1\,1] \rightarrow \frac{a}{6}[1\,1\,2] + \frac{a}{6}[\bar{1}\,2\,1] \qquad \frac{a}{2}[0\,\bar{1}\,\bar{1}] \rightarrow \frac{a}{6}[1\,\bar{2}\,\bar{1}] + \frac{a}{6}[\bar{1}\,\bar{1}\,\bar{2}]$$

$$\frac{a}{2}[1\,0\,1] \rightarrow \frac{a}{6}[1\,1\,2] + \frac{a}{6}[2\,\bar{1}\,1] \qquad \frac{a}{2}[\bar{1}\,0\,\bar{1}] \rightarrow \frac{a}{6}[\bar{2}\,1\,\bar{1}] + \frac{a}{6}[\bar{1}\,\bar{1}\,\bar{2}]$$

3.4.3.3 Sessile dislocations

The Shockley partial dislocation has its Burgers vector lying in the plane of the fault and hence is glissile. Some dislocations, however, have their Burgers vector not lying in the plane of the fault with which they are associated, and are incapable of gliding, i.e. they are sessile. The simplest of these, the Frank sessile dislocation loop, is shown in Figure 3.32a. This dislocation is believed to form as a result of the collapse of the lattice surrounding a cavity which has been produced by the aggregation of vacancies on to a (1 1 1) plane. As shown in Figure 3.32a, if the vacancies aggregate on the central A-plane the adjoining parts of the neighboring B- and C-planes collapse to fit in close-packed formation. The Burgers vector of the dislocation line bounding the collapsed sheet is normal to the plane with $b = \frac{a}{3}[1\,1\,1]$, where a is the lattice parameter, and such a dislocation is sessile

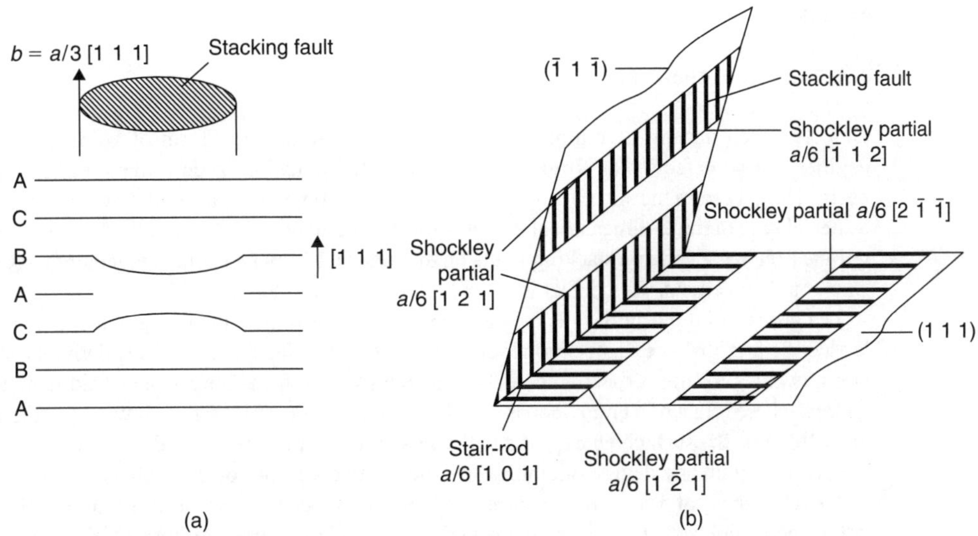

Figure 3.32 *Sessile dislocations: (a) a Frank sessile dislocation; (b) stair-rod dislocation as part of a Lomer–Cottrell barrier.*

since it encloses an area of stacking fault which cannot move with the dislocation. A Frank sessile dislocation loop can also be produced by inserting an extra layer of atoms between two normal planes of atoms, as occurs when interstitial atoms aggregate following high-energy particle irradiation. For the loop formed from vacancies the stacking sequence changes from the normal **ABCABCA** ... to **ABCBCA** ..., whereas inserting a layer of atoms, e.g. an A-layer between B and C, the sequence becomes **ABCABACA** .... The former type of fault with one violation in the stacking sequence is called an intrinsic fault, the latter with two violations is called an extrinsic fault. The stacking sequence violations are conveniently shown by using the symbol Δ to denote any normal stacking sequence **AB, BC, CA** but ∇ for the reverse sequence **AC, CB, BA.** The normal fcc stacking sequence is then given by $\Delta\Delta\Delta\Delta$..., the intrinsic fault by $\Delta\Delta\nabla\Delta\Delta$... and the extrinsic fault by $\Delta\Delta\nabla\nabla$ $\Delta\Delta$.... The reader may verify that the fault discussed in the previous section is also an intrinsic fault, and that a series of intrinsic stacking faults on neighboring planes gives rise to a twinned structure **ABCABACBA** or $\Delta\Delta\Delta\Delta\nabla\nabla\nabla\nabla$. Electron micrographs of Frank sessile dislocation loops are shown in Figures 3.36 and 3.37.

Another common obstacle is that formed between extended dislocations on intersecting {1 1 1} slip planes, as shown in Figure 3.32b. Here, the combination of the leading partial dislocation lying in the (1 1 1) plane with that which lies in the $(\bar{1}\ 1\ \bar{1})$ plane forms another partial dislocation, often referred to as a 'stair-rod' dislocation, at the junction of the two stacking fault ribbons by the reaction:

$$\frac{a}{6}[\bar{1}\ 1\ 2] + \frac{a}{6}[2\ \bar{1}\ \bar{1}] \rightarrow \frac{a}{6}[1\ 0\ 1].$$

The indices for this reaction can be obtained from Figure 3.35 and it is seen that there is a reduction in energy from $[(a^2/6) + (a^2/6)]$ to $a^2/18$. This triangular group of partial dislocations, which bounds the wedge-shaped stacking-fault ribbon lying in a $\langle 1\ 0\ 1 \rangle$ direction, is obviously incapable of gliding and such an obstacle, first considered by Lomer and Cottrell, is known as a Lomer–Cottrell barrier. Such a barrier impedes the motion of dislocations and leads to work hardening, as discussed in Chapter 6.

3.5 Volume defects

3.5.1 Void formation and annealing

Defects which occupy a volume within the crystal may take the form of voids, gas bubbles and cavities. These defects may form by heat treatment, irradiation or deformation and their energy is derived largely from the surface energy (1–3 $J\,m^{-2}$). In some materials with low stacking-fault energy a special type of three-dimensional defect is formed, namely the defect tetrahedron. This consists of a tetrahedron made up from stacking faults on the four {1 1 1} planes joined together by six low-energy stair-rod dislocations. This defect is discussed more fully in Section 3.6.2.3.

The growth of the original vacancy cluster into a three-dimensional aggregate, i.e. void, or a collapsed vacancy disk, i.e. dislocation loop, should, in principle, depend on the relative surface to strain energy values for the respective defects. The energy of a three-dimensional void is mainly surface energy, whereas that of a Frank loop is mainly strain energy at small sizes. However, without a detailed knowledge of the surface energy of small voids and the core energy of dislocations it is impossible to calculate, with any degree of confidence, the relative stability of these clustered vacancy defects.

The clustering of vacancies to form voids has now been observed in a number of metals with either fcc or cph structure. In as-quenched specimens the voids are not spherical but bounded by crystallographic faces (see Figure 3.54) and usually are about 50 nm radius in size. In fcc metals they are octahedral in shape with sides along ⟨1 1 0⟩, sometimes truncated by {1 0 0} planes, and in cph metals bounded by prism and pyramidal planes. Void formation is favored by slow quenching rates and high ageing temperatures, and the density of voids increases when gas is present in solid solution (e.g. hydrogen in copper, and either hydrogen or oxygen in silver). In aluminum and magnesium, void formation is favored by quenching from a wet atmosphere, probably as a result of hydrogen production due to the oxidation reactions. It has been postulated that small clustered vacancy groups are stabilized by the presence of gas atoms and prevented from collapsing to a planar disk, so that some critical size for collapse can be exceeded. The voids are not conventional gas bubbles, however, since only a few gas atoms are required to nucleate the void, after which it grows by vacancy adsorption.

3.5.2 Irradiation and voiding

Irradiation produces both interstitials and vacancies in excess of the equilibrium concentration. Both species initially cluster to form dislocation loops, but it is the interstitial loops formed from clustering of interstitials which eventually develop into a dislocation structure.

In general, interstitial loops grow during irradiation because the large elastic misfit associated with an interstitial causes dislocations to attract interstitials more strongly than vacancies (see Chapter 6). Interstitial loops are therefore intrinsically stable defects, whereas vacancy loops are basically unstable defects during irradiation. Thus, interstitials attracted to a vacancy loop, i.e. a loop formed by clustering vacancies, will cause it to shrink as the interstitials are annihilated. Increasing the irradiation temperature results in vacancies aggregating to form voids. Voids are formed in an intermediate temperature range $\approx 0.3\text{–}0.6T_m$, above that for long-range single-vacancy migration and below that for thermal vacancy emission from voids. To create the excess vacancy concentration it is also necessary to build up a critical dislocation density from loop growth to bias the interstitial flow.

There are two important factors contributing to void formation. The first is the degree of bias the dislocation density (developed from the growth of interstitial loops) has for attracting interstitials, which suppresses the interstitial content compared to vacancies. The second factor is the important role played in void nucleation by gases, both surface-active gases such as oxygen, nitrogen and hydrogen frequently present as residual impurities, and inert gases such as helium, which may be generated continuously during irradiation due to transmutation reactions. The surface-active gases

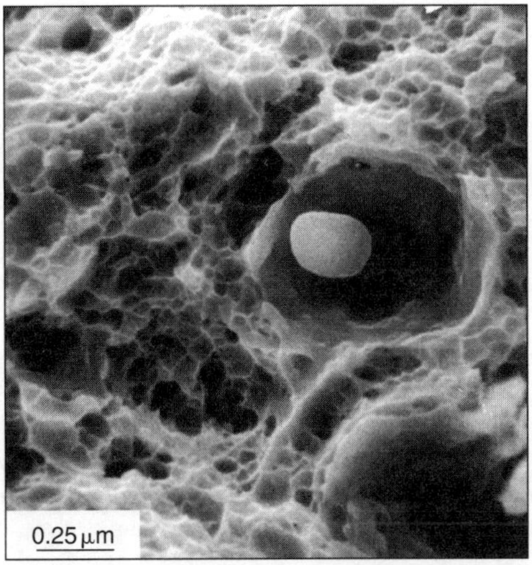

Figure 3.33 *Scanning electron micrograph of a medium-carbon (0.4%) steel with a quenched and tempered martensite structure, showing large dimples associated with oxide inclusions and small dimples associated with small carbide precipitates (courtesy Dr L. Sidjanin).*

such as oxygen in copper can migrate to embryo vacancy clusters and reduce the surface energy. The inert gas atoms can acquire vacancies to become gas molecules inside voids (when the gas pressure is not in equilibrium with the void surface tension) or gas bubbles when the gas pressure is considerable ($P \gtrsim 2\gamma_s/r$). Voids and bubbles can give rise to irradiation swelling and embrittlement of materials.

3.5.3 Voiding and fracture

The formation of voids is an important feature in the ductile failure of materials. The fracture process involves three stages. First, small holes or cavities nucleate, usually at weak internal interfaces (e.g. particle/matrix interfaces). These cavities then expand by plastic deformation and finally coalesce by localized necking of the metal between adjacent cavities to form a fibrous fracture. A scanning electron micrograph illustrating the characteristics of a typical ductile failure is shown in Figure 3.33. This type of fracture may be regarded as taking place by the nucleation of an internal plastic cavity, rather than a crack, which grows outwards to meet the external neck that is growing inwards. Experimental evidence suggests that nucleation occurs at foreign particles. For example, oxygen-free high conductivity (OFHC) copper necks down to over 90% reduction in area, whereas tough-pitch copper shows only 70% reduction in area; similar behavior is noted for super-pure and commercial purity aluminum. Thus, if no inclusions were present, failure should occur by the specimen pulling apart entirely by the inward growth of the external neck, giving nearly 100% reduction in area. Dispersion-hardened materials in general fail with a ductile fracture, the fibrous region often consisting of many dimples arising from the dispersed particles nucleating holes and causing local ductile failure. Ductile failure is discussed further in Chapter 7.

3.6 Defect behavior in common crystal structures

3.6.1 Dislocation vector diagrams and the Thompson tetrahedron

The classification of defects into point, line, planar and volume is somewhat restrictive in presenting an overview of defect behavior in materials, since it is clear, even from the discussion so far, that these

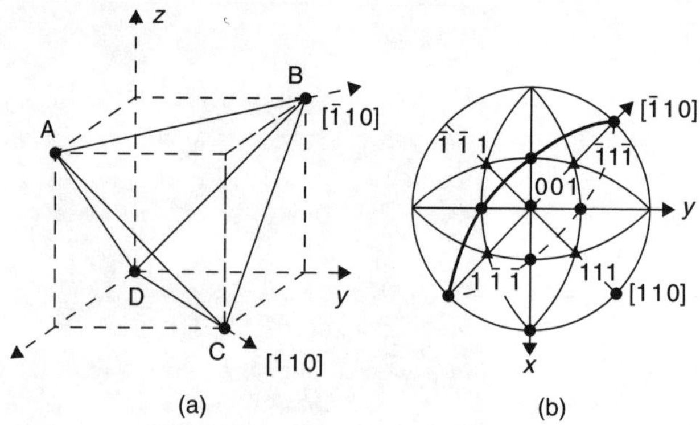

Figure 3.34 *Construction (a) and orientation (b) of the Thompson tetrahedron ABCD. The slip directions in a given {1 1 1} plane may be obtained from the trace of that plane, as shown for the (1 1 1) plane in (b).*

defects are interrelated and interdependent. In the following sections these features will be brought out as well as those which relate to specific structures.

In dealing with dislocation interactions and defects in real material it is often convenient to work with a vector notation rather than use the more conventional Miller indices notation. This may be illustrated by reference to the fcc structure and the Thompson tetrahedron.

All the dislocations common to the fcc structure, discussed in the previous sections, can be represented conveniently by means of the Thompson reference tetrahedron (Figure 3.34a), formed by joining the three nearest face-centering atoms to the origin D. Here ABCD is made up of four $\{1\ 1\ 1\}$ planes $(1\ 1\ 1), (\bar{1}\ \bar{1}\ 1), (\bar{1}\ 1\ \bar{1})$ and $(1\ \bar{1}\ \bar{1})$ as shown by the stereogram given in Figure 3.34b, and the edges **AB**, **BC**, **CA** ... correspond to the $\langle 1\ 1\ 0\rangle$ directions in these planes. Then, if the midpoints of the faces are labeled α, β, γ, δ, as shown in Figure 3.35a, all the dislocation Burgers vectors are represented. Thus, the edges (**AB**, **BC** ...) correspond to the normal slip vectors, $a/2\langle 1\ 1\ 0\rangle$. The half-dislocations, or Shockley partials, into which these are dissociated have Burgers vectors of the $a/6\langle 1\ 1\ 2\rangle$ type and are represented by the Roman–Greek symbols Aγ, Bγ, Dγ, Aδ, Bδ, etc., or Greek–Roman symbols γA, γB, γD, δA, δB, etc. The dissociation reaction given in the first reaction in Section 3.4.3.2 is then simply written:

$$\mathrm{BC} \rightarrow \mathrm{B}\delta + \delta\mathrm{C}$$

and there are six such dissociation reactions in each of the four $\{1\ 1\ 1\}$ planes (see Figure 3.35). It is conventional to view the slip plane from outside the tetrahedron along the positive direction of the unit dislocation BC, and on dissociation to produce an intrinsic stacking fault arrangement; the Roman–Greek partial Bδ is on the right and the Greek–Roman partial δC on the left. A screw dislocation with Burgers vector BC, which is normally dissociated in the δ-plane, is capable of cross-slipping into the α-plane by first constricting Bδ + δC → BC and then redissociating in the α-plane BC → Bα + αC.

3.6.2 Dislocations and stacking faults in fcc structures

3.6.2.1 Frank loops

A powerful illustration of the use of the Thompson tetrahedron can be made if we look at simple Frank loops in fcc metals (see Figure 3.32a). The Frank partial dislocation has a Burgers vector perpendicular

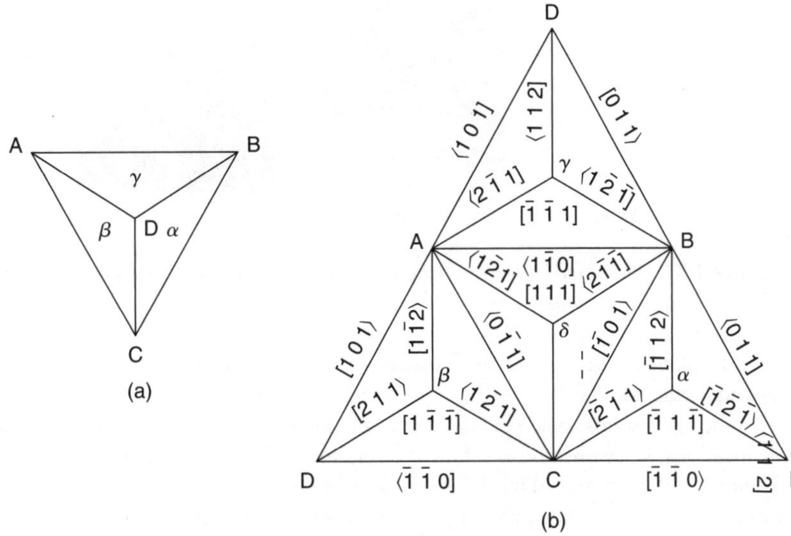

Figure 3.35 *A Thompson tetrahedron closed (a) and opened out (b). In (b) the notation [1 1 0⟩ is used in place of the usual notation [1 1 0] to indicate the sense of the vector direction.*

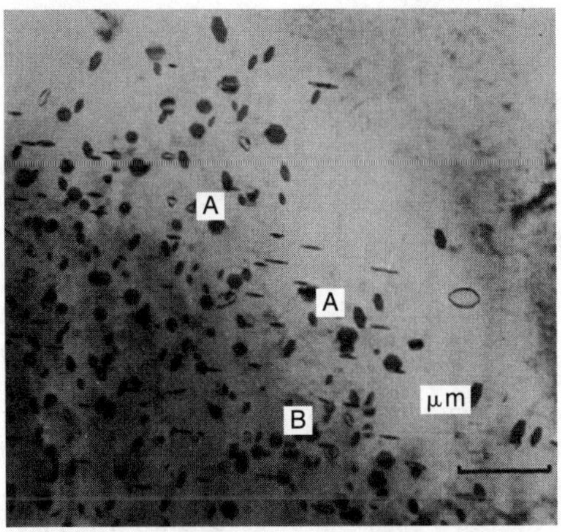

Figure 3.36 *Single-faulted, double-faulted (A) and unfaulted (B) dislocation loops in quenched aluminum (after Edington and Smallman, 1965; courtesy of Taylor & Francis).*

to the (1 1 1) plane on which it lies and is represented by Aα, Bβ, Cγ, Dδ, αA, etc. Such loops shown in the electron micrograph of Figure 3.36 have been produced in aluminum by quenching from about 600°C. Each loop arises from the clustering of vacancies into a disk-shaped cavity, which then forms a dislocation loop. To reduce their energy, the loops take up regular crystallographic forms with their edges parallel to the ⟨1 1 0⟩ directions in the loop plane. Along a ⟨1 1 0⟩ direction it can reduce its energy by dissociating on an intersecting {1 1 1} plane, forming a stair-rod at the junction of the two

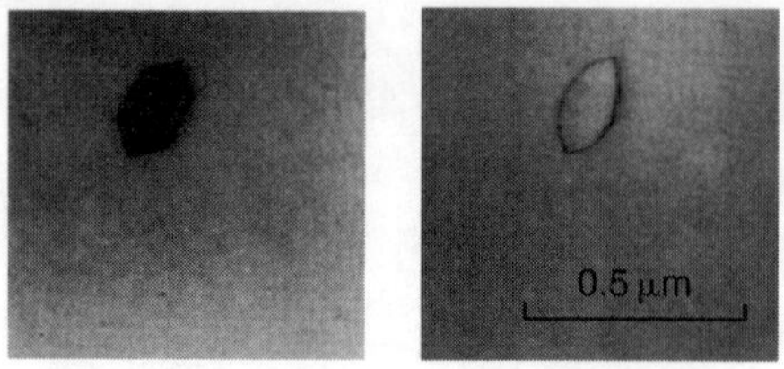

Figure 3.37 *Removal of the stacking fault from a Frank sessile dislocation by stress (after Goodhew and Smallman).*

{1 1 1} planes, e.g. $A\alpha \rightarrow A\delta + \delta\alpha$ when the Frank dislocation lies along $[\bar{1}\ 0\ 1]$ common to both α- and δ-planes.

Some of the loops shown in Figure 3.36 are not Frank sessile dislocations as expected, but prismatic dislocations, since no contrast of the type arising from stacking faults can be seen within the defects. The fault will be removed by shear if it has a high stacking fault energy, thereby changing the sessile Frank loop into a glissile prismatic loop according to the reaction:

$$a/3[1\ 1\ 1] + a/6[1\ 1\ \bar{2}] \rightarrow a/2[1\ 1\ 0].$$

Stressing the foil while it is under observation in the microscope allows the unfaulting process to be observed directly (see Figure 3.37). This reaction is more easily followed with the aid of the Thompson tetrahedron and rewritten as:

$$D\delta + \delta C \rightarrow DC.$$

Physically, this means that the disk of vacancies aggregated on a (1 1 1) plane of a metal with high stacking fault energy, besides collapsing, also undergoes a shear movement. The dislocation loops shown in Figure 3.37b are therefore unit dislocations with their Burgers vector $a/2[1\ 1\ 0]$ inclined at an angle to the original (1 1 1) plane. A prismatic dislocation loop lies on the surface of a cylinder, the cross-section of which is determined by the dislocation loop, and the axis of which is parallel to the [1 1 0] direction. Such a dislocation is not sessile, and under the action of a shear stress it is capable of movement by prismatic slip in the [1 1 0] direction.

Many of the large Frank loops in Figure 3.36 (for example, marked A) contain additional triangular-shaped loop contrast within the outer hexagonal loop. The stacking fault fringes within the triangle are usually displaced relative to those between the triangle and the hexagon by half the fringe spacing, which is the contrast expected from overlapping intrinsic stacking faults. The structural arrangement of those double-faulted loops is shown schematically in Figure 3.38, from which it can be seen that two intrinsic faults on next neighboring planes are equivalent to an extrinsic fault. The observation of double-faulted loops in aluminum indicates that it is energetically more favorable to nucleate a Frank sessile loop on an existing intrinsic fault than randomly in the perfect lattice, and it therefore follows that the energy of a double or extrinsic fault is less than twice that of the intrinsic fault, i.e. $\gamma_E < 2\gamma_I$. The double loops marked **B** have the outer intrinsic fault removed by stress.

The addition of a third overlapping intrinsic fault would change the stacking sequence from the perfect **ABCABCABC** to **ABC ↓ B ↓ A ↓ CABC**, where the arrows indicate missing planes of atoms, and produce a coherent twinned structure with two coherent twin boundaries. This structure would be energetically favorable to form, since $\gamma_{twin} < \gamma_I < \gamma_E$. It is possible, however, to reduce the energy of the crystal even further by aggregating the third layer of vacancies between the two previously formed neighboring intrinsic faults to change the structure from an extrinsically faulted **ABC ↓ B ↓ ABC** to the perfect **ABC ↓↓↓ ABC** structure. Such a triple-layer dislocation loop is shown in Figure 3.39.

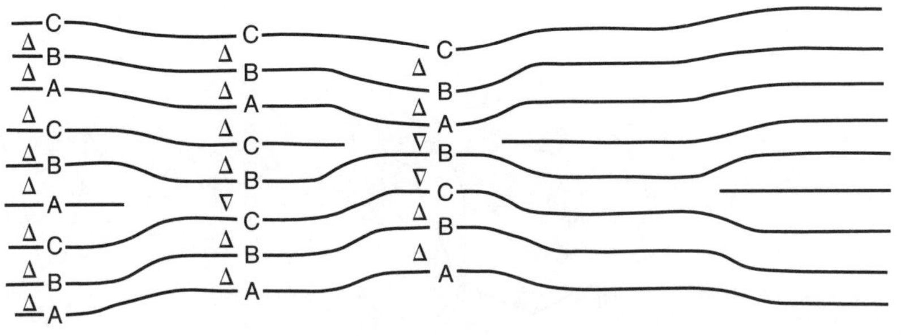

Figure 3.38 *The structure of a double dislocation loop in quenched aluminum (after Edington and Smallman, 1965; courtesy of Taylor & Francis).*

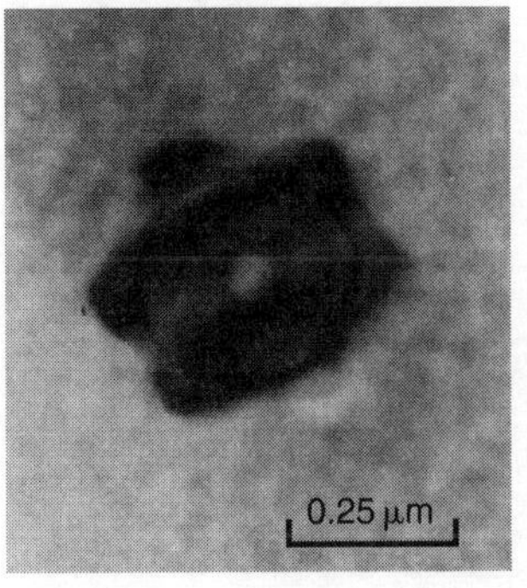

Figure 3.39 *Triple-loop and Frank sessile loop in Al–0.65% Mg (after Kritzinger, Smallman and Dobson, 1969).*

3.6.2.2 Stair-rod dislocations

The stair-rod dislocation formed at the apex of a Lomer–Cottrell barrier can also be represented by the Thompson notation. As an example, let us take the interaction between dislocations on the δ- and α-planes. Two unit dislocations BA and DB, respectively, are dissociated according to:

$$\text{BA} \rightarrow \text{B}\delta + \delta\text{A} \text{ (on the } \delta\text{-plane)}$$

$$\text{and DB} \rightarrow \text{D}\alpha + \alpha\text{B} \text{ (on the } \alpha\text{-plane)}$$

and when the two Shockley partials αB and Bδ interact, a stair-rod dislocation $\alpha\delta = a/6[1\ 0\ 1]$ is formed. This low-energy dislocation is pure edge and therefore sessile. If the other pair of partials interact then the resultant Burgers vector is $(\delta\text{A} + \text{D}\alpha) = a/3[1\ 0\ 1]$ and of higher energy. This vector is written in Thompson's notation as $\delta D/\text{A}\alpha$ and is a vector equal to twice the length joining the midpoints of δA and Dα.

3.6.2.3 Stacking-fault tetrahedra

In fcc metals and alloys, the vacancies may also cluster into a three-dimensional defect, forming a tetrahedral arrangement of stacking faults on the four $\{1\ 1\ 1\}$ planes with the six $\langle 1\ 1\ 0\rangle$ edges of the tetrahedron, where the stacking faults bend from one $\{1\ 1\ 1\}$ plane to another, consisting of stair-rod dislocations. The crystal structure is perfect inside and outside the tetrahedron, and the

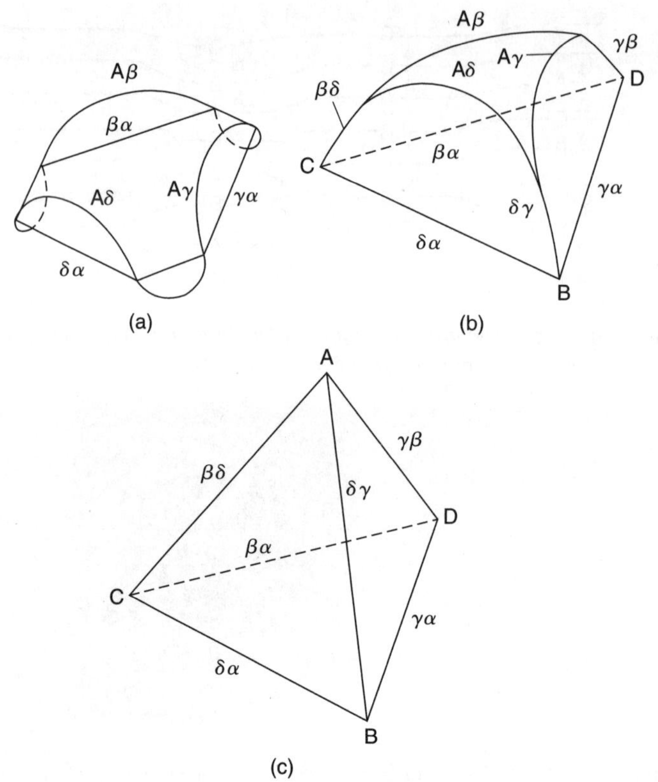

Figure 3.40 *Formation of defect tetrahedron: (a) dissociation of Frank dislocations; (b) formation of new stair-rod dislocations; (c) arrangement of the six stair-rod dislocations.*

three-dimensional array of faults exhibits characteristic projected shape and contrast when seen in transmission electron micrographs, as shown in Figure 3.42. This defect was observed originally in quenched gold but occurs in other materials with low stacking-fault energy. One mechanism for the formation of the defect tetrahedron by the dissociation of a Frank dislocation loop (see Figure 3.40) was first explained by Silcox and Hirsch. The Frank partial dislocation bounding a stacking fault has, because of its large Burgers vector, a high strain energy, and hence can lower its energy by dissociation according to a reaction of the type:

$$\underset{\left(\frac{1}{3}\right)}{a/3[1\ 1\ 1]} \rightarrow \underset{\left(\frac{1}{6}\right)}{a/6[1\ 2\ 1]} + \underset{\left(\frac{1}{18}\right)}{a/6[1\ 0\ 1]}.$$

The figures underneath the reaction represent the energies of the dislocations, since they are proportional to the squares of the Burgers vectors. This reaction is therefore energetically favorable. This reaction can be seen with the aid of the Thompson tetrahedron, which shows that the Frank partial dislocation Aα can dissociate into a Shockley partial dislocation (Aβ, Aδ or Aγ) and a low energy stair-rod dislocation ($\beta\alpha$, $\delta\alpha$ or $\gamma\alpha$), for example A$\alpha \rightarrow$ A$\gamma + \gamma\alpha$.

The formation of the defect tetrahedron of stacking faults may be envisaged as follows. The collapse of a vacancy disk will, in the first instance, lead to the formation of a Frank sessile loop bounding a stacking fault, with edges parallel to the $\langle 1\ 1\ 0\rangle$ directions. Each side of the loop then dissociates according to the above reaction into the appropriate stair-rod and partial dislocations, and, as shown in Figure 3.40a, the Shockley dislocations formed by dissociation will lie on intersecting {1 1 1} planes, above and below the plane of the hexagonal loop; the decrease in energy accompanying the dissociation will give rise to forces which tend to pull any rounded part of the loop into ⟨1 1 0⟩. Moreover, because the loop will not in general be a regular hexagon, the short sides will be eliminated by the preferential addition of vacancies at the constricted site, and a triangular-shaped loop will form (Figure 3.40b). The partials Aβ, Aγ and Aδ bow out on their slip plane as they are repelled by the stair-rods. Taking into account the fact that adjacent ends of the bowing loops are of opposite sign, the partials attract each other in pairs to form stair-rod dislocations along *DA, BA* and *CA*, according to the reactions:

$$\gamma\text{A} + \text{A}\beta \rightarrow \gamma\beta,\ \delta\text{A} + \text{A}\gamma \rightarrow \delta\gamma,\ \beta\text{A} + \text{A}\delta \rightarrow \beta\delta.$$

In vector notation the reactions are of the type:

$$\begin{array}{ccccc} a/6[\bar{1}\ \bar{1}\ \bar{2}] & + & a/6[1\ 2\ 1] & \rightarrow & a/6[0\ 1\ \bar{1}] \\ \left(\frac{1}{6}\right) & & \left(\frac{1}{6}\right) & & \left(\frac{1}{6}\right) \end{array}$$

(the reader may deduce the appropriate indices from Figure 3.35), and from the addition of the squares of the Burgers vectors underneath it is clear that this reaction is also energetically favorable. The final defect will therefore be a tetrahedron made up from the intersection of stacking faults on the four {1 1 1} planes, so that the ⟨1 1 0⟩ edges of the tetrahedron will consist of low-energy stair-rod dislocations (Figure 3.40c).

The tetrahedron of stacking faults formed by the above sequence of events is essentially symmetrical, and the same configuration would have been obtained if collapse had taken place originally on any other (1 1 1) plane. The energy of the system of stair-rod dislocations in the final configuration is proportional to $6 \times \frac{1}{18} = \frac{1}{3}$, compared with $3 \times \frac{1}{3} = 1$ for the original stacking fault triangle bounded by Frank partials. Considering the dislocation energies alone, the dissociation leads to a lowering of energy to one-third of the original value. However, three additional stacking-fault areas, with energies of γ per unit area, have been newly created and if there is to be no net rise in energy these areas will impose an upper limit on the size of the tetrahedron formed of about 50 nm.

Tetrahedra may also form by the nucleation and growth of a three-dimensional vacancy cluster. The smallest cluster that is able to collapse to a tetrahedron and subsequently grow by the absorption of vacancies is a hexa-vacancy cluster. Growth would then occur by the nucleation and propagation of jog lines across the faces of the tetrahedron, as shown in Figure 3.41. The hexa-vacancy cluster may form by clustering di-vacancies and is aided by impurities which have excess positive charge relative to the matrix (e.g. Mg, Cd or Al in Au). Hydrogen in solution is also a potent nucleating agent because the di-vacancy/proton complex is mobile and attracted to 'free' di-vacancies. Figure 3.42 shows the increase in tetrahedra nucleation after pre-annealing gold in hydrogen.

Worked example

By considering the energy of a triangular-shaped dislocation loop and that of a tetrahedron into which it could dissociate, estimate the maximum size of the tetrahedron. (Assume that the dislocation energy per unit length is μb^2, where μ, the shear modulus, is $2.7 \times 10^{10}\ \text{N m}^{-2}$, b is the Burgers vector, a the lattice parameter is 0.4 nm and the stacking-fault energy γ is $33\ \text{mJ m}^{-2}$.)

Solution

Let $\ell =$ side length of triangular loop or tetrahedron.

Energy E_L of loop $= 3\ell \times \mu b_F^2 + (\text{Area}) \times \gamma$, where $b_F = \frac{a}{3}$<1 1 1> (Frank dislocation)

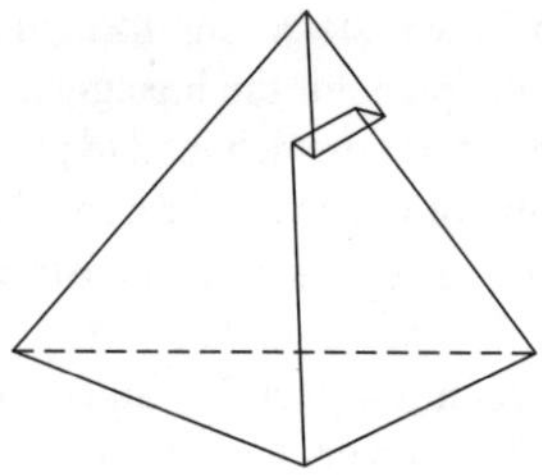

Figure 3.41 *Jog line forming a ledge on the face of a tetrahedron.*

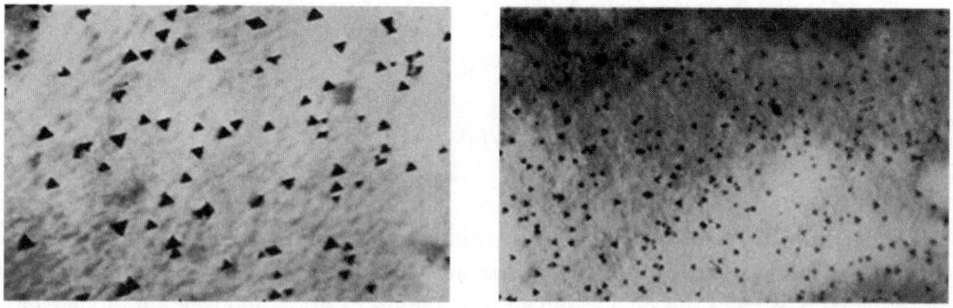

Figure 3.42 *Tetrahedra in gold quenched (a) and preannealed in $H_2 - N_2$ gas (b) (after Johnston, Dobson and Smallman).*

Energy E_T of tetrahedron $= 6\ell \times \mu b_{SR}^2 + 4A\gamma$, where $b_{SR} = \frac{a}{6}$<1 1 0> (stair-rod dislocation)

$$E_L = \mu a^2\ell + \frac{\sqrt{3}}{4}\ell^2\gamma; \quad E_T = \frac{\mu a^2}{3}\ell + \sqrt{3}\ell^2\gamma$$

$$E_L = E_T \quad \text{for} \quad \mu a^2\ell + \frac{\sqrt{3}}{4}\ell^2\gamma = \frac{\mu a^2}{3}\ell + \sqrt{3}\ell^2\gamma$$

$$\frac{2}{3}\mu a^2\ell = \frac{3\sqrt{3}}{4}\ell^2\gamma \quad \text{or} \quad \ell = \frac{8}{9\sqrt{3}}\frac{\mu a^2}{\gamma} = 0.51\frac{\mu a^2}{\gamma}$$

$$\ell = 0.51 \times \frac{2.7 \times 10^{10} \times (0.4 \times 10^{-9})^2}{0.033} = 6.68 \times 10^{-8}\,\text{m} = 668\,\text{Å}.$$

When $\ell < 668$Å, $E_T < E_L$ so the tetrahedron is more stable. Hence maximum size of the tetrahedron is 668 Å, beyond which the loop is more stable.

Worked example

Explain with well-labeled sketches the dislocation reactions in the fcc structure written in the Thompson tetrahedron notation:

(i) $B\delta + \delta C \rightarrow BC \rightarrow B\alpha + \alpha C$
(ii) $A\alpha + \alpha D \rightarrow AD$
(iii) $C\alpha + \delta C \rightarrow \alpha\delta$
(iv) $A\alpha \rightarrow A\delta + \delta\alpha$.

Solution

The Thompson tetrahedron ABCD is made up of the four operative {111} planes in the fcc structure. These are labeled α-, β-, γ- and δ-planes; A opposite to α, etc.

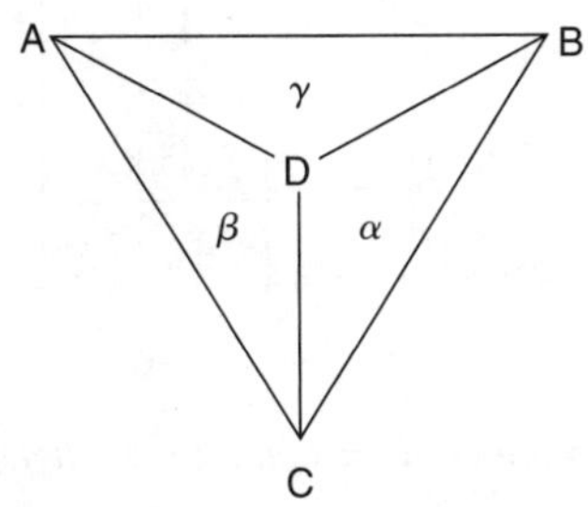

(i) This is a cross-slip reaction of a dissociated dislocation $B\delta + \delta C$ in the δ-plane, which recombines to BC and then dissociates to $B\alpha + \alpha C$ on moving into the α-plane.

(ii) This is the removal of the stacking fault from the Frank dislocation $A\alpha$ by the sweeping cross (shear) of a Shockley partial αD in the α-plane to produce a perfect dislocation AD:

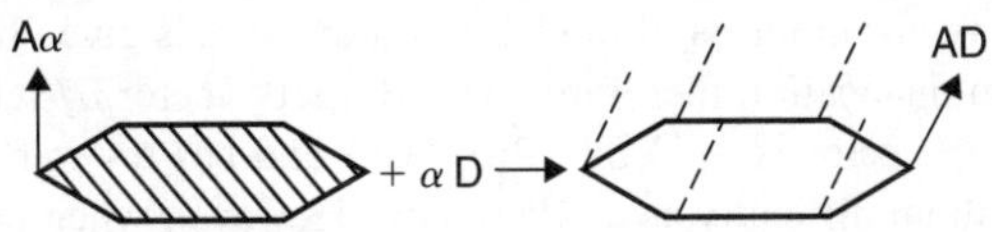

(iii) The formation of a stair-rod dislocation at the intersection of two {111} planes α and δ by the coming together (interaction) of Shockley partials $C\alpha$ in the α-plane and δC in the δ-plane to form $\alpha\delta$.

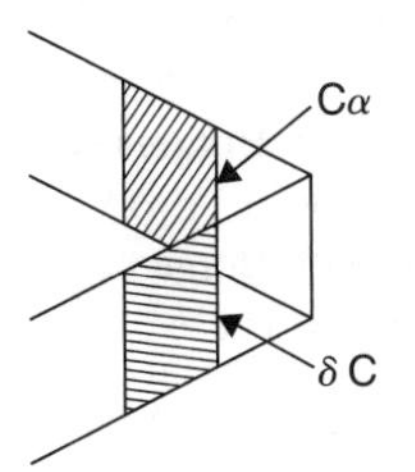

(iv) This indicates the dissociation of a Frank dislocation $A\alpha$ lying in the α-plane into a Shockley partial $A\delta$ on the intersecting δ-plane and leaving a stair-rod $\delta\alpha$ at the junction of the α-plane and the intersecting δ-plane.

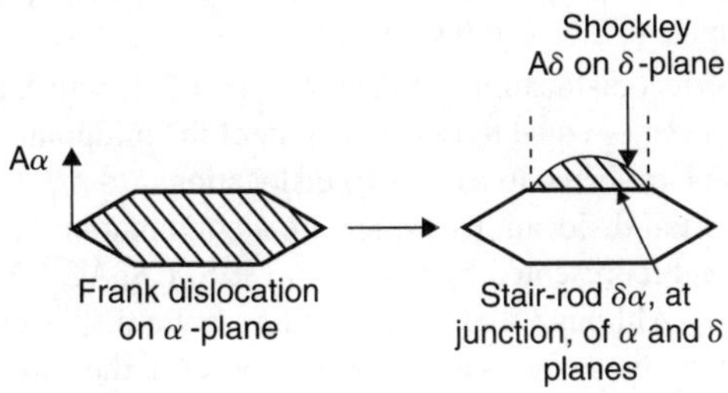

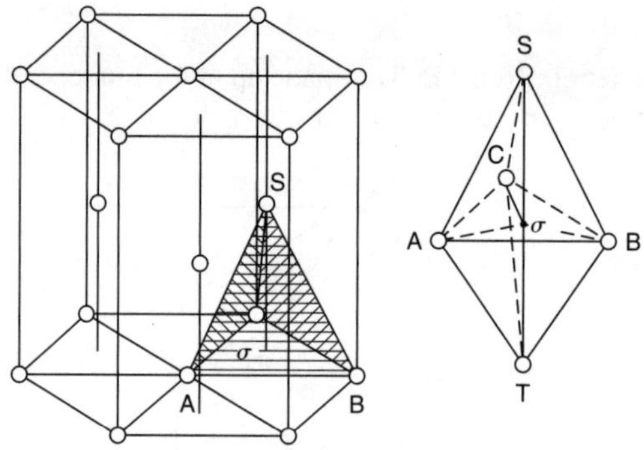

Figure 3.43 *Burgers vectors in the cph lattice (after Berghezan, Fourdeux and Amelinckx, 1961).*

3.6.3 Dislocations and stacking faults in cph structures

In a cph structure with axial ratio c/a, the most closely packed plane of atoms is the basal plane (0 0 0 1) and the most closely packed directions $\langle 1\ 1\ \bar{2}\ 0\rangle$. The smallest unit lattice vector is a, but to indicate the direction of the vector $\langle u, v, w\rangle$ in Miller–Bravais indices it is written as $a/3\langle 1\ 1\ \bar{2}\ 0\rangle$, where the magnitude of the vector in terms of the lattice parameters is given by $a[3(u^2+uv+v^2)+(c/a)^2w^2]^{1/2}$. The usual slip dislocation therefore has a Burgers vector $a/3\langle 1\ 1\ \bar{2}\ 0\rangle$ and glides in the (0 0 0 1) plane. This slip vector is $\langle a/3, a/3, \bar{2}(a/3), 0\rangle$ and has no component along the c-axis, and so can be written without difficulty as $a/3\langle 1\ 1\ \bar{2}\ 0\rangle$. However, when the vector has a component along the c-axis, as for example $\langle a/3, a/3, \bar{2}(a/3), 3c\rangle$, difficulty arises and to avoid confusion the vectors are referred to unit distances (a, a, a, c) along the respective axes (e.g. $1/3\langle 1\ 1\ \bar{2}\ 0\rangle$ and $1/3\langle 1\ 1\ \bar{2}\ 3\rangle$). Other dislocations can be represented in a notation similar to that for the fcc structure, but using a double-tetrahedron or bipyramid instead of the single tetrahedron previously adopted, as shown in Figure 3.43. An examination leads to the following simple types of dislocation:

1. Six perfect dislocations with Burgers vectors in the basal plane along the sides of the triangular base ABC. They are AB, BC, CA, BA, CB and AC, and are denoted by a or $1/3\langle 1\ 1\ \bar{2}\ 0\rangle$.
2. Six partial dislocations with Burgers vectors in the basal plane represented by the vectors Aσ, Bσ, Cσ and their negatives. These dislocations arise from dissociation reactions of the type:

$$\text{AB} \rightarrow \text{A}\sigma + \sigma\text{B}$$

   and may also be written as p or $\frac{1}{3}\langle 1\ 0\ \bar{1}\ 0\rangle$.
3. Two perfect dislocations perpendicular to the basal plane represented by the vectors ST and TS of magnitude equal to the cell height c or $\langle 0\ 0\ 0\ 1\rangle$.
4. Partial dislocations perpendicular to the basal plane represented by the vectors σS, σT, Sσ, Tσ of magnitude $c/2$ or $\frac{1}{2}\langle 0\ 0\ 0\ 1\rangle$.
5. Twelve perfect dislocations of the type $\frac{1}{3}\langle 1\ 1\ \bar{2}\ 3\rangle$ with a Burgers vector represented by SA/TB, which is a vector equal to twice the join of the midpoints of SA and TB. These dislocations are more simply referred to as $(c+a)$ dislocations.
6. Twelve partial dislocations, which are a combination of the partial basal and non-basal dislocations, and represented by vectors AS, BS, CS, AT, BT and CT or simply $(c/2)+p$ equal to $\frac{1}{6}\langle 2\ 0\ \bar{2}\ 3\rangle$. Although these vectors represent a displacement from one atomic site to another, the resultant dislocations are imperfect because the two sites are not identical.

Table 3.2 *Dislocations in hcp structures.*

Type	*AB, BC*	*Aσ, Bσ*	*ST, TS*	*σS, σT*	*AS, BS*	*SA/TB*
Vector	$\frac{1}{3}\langle 1\,1\,\bar{2}\,0\rangle$	$\frac{1}{3}\langle 1\,0\,\bar{1}\,0\rangle$	$\langle 0\,0\,0\,1\rangle$	$\frac{1}{2}\langle 0\,0\,0\,1\rangle$	$\frac{1}{6}\langle 2\,0\,\bar{2}\,3\rangle$	$\frac{1}{3}\langle 1\,1\,\bar{2}\,3\rangle$
Energy	a^2	$a^2/3$	$c^2 = 8a^2/3$	$2a^2/3$	a^2	$11a^2/3$

The energies of the different dislocations are given in a relative scale in Table 3.2, assuming c/a is ideal.

There are many similarities between the dislocations in the cph and fcc structure, and thus it is not necessary to discuss them in great detail. It is, however, of interest to consider the two basic processes of glide and climb.

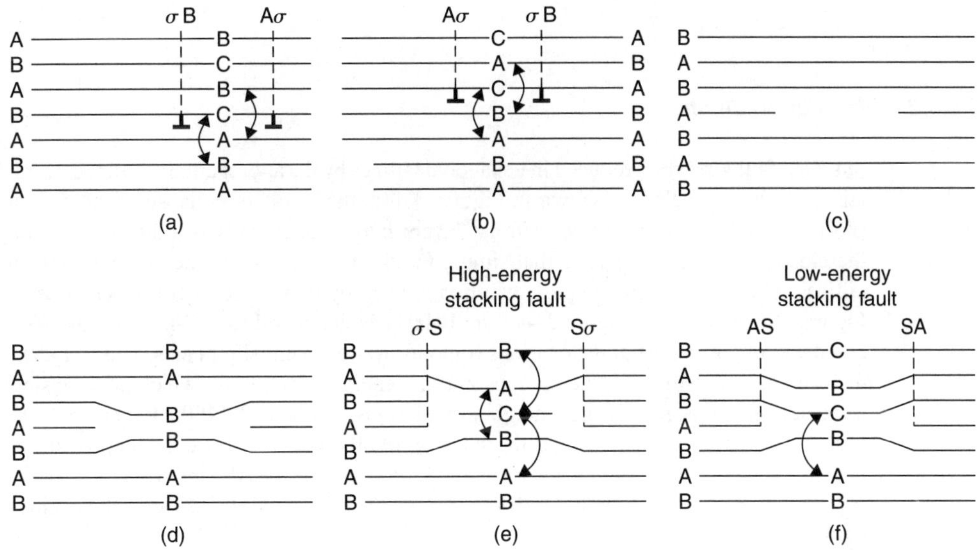

Figure 3.44 *Stacking faults in the cph lattice (after Partridge, 1967; by courtesy of the American Society for Metals).*

3.6.3.1 *Dislocation glide*

A perfect slip dislocation in the basal plane $\text{AB} = \frac{1}{3}[\bar{1}\;2\;\bar{1}\;0]$ may dissociate into two Shockley partial dislocations separating a ribbon of intrinsic stacking fault which violates the two next-nearest neighbors in the stacking sequence. There are actually two possible slip sequences: either a B-layer slides over an A-layer, i.e. Aσ followed by σB (see Figure 3.44a) or an A-layer slides over a B-layer by the passage of a σB partial followed by an Aσ (see Figure 3.44b). The dissociation given by

$$\text{AB} \rightarrow \text{A}\sigma + \sigma\text{B}$$

may be written in Miller–Bravais indices as:

$$\tfrac{1}{3}[\bar{1}\,2\,\bar{1}\,0] \rightarrow \tfrac{1}{3}[0\,1\,\bar{1}\,0] + \tfrac{1}{3}[\bar{1}\,1\,0\,0].$$

This reaction is similar to that in the fcc lattice and the width of the ribbon is again inversely proportional to the stacking fault energy γ. Dislocations dissociated in the basal planes have been observed in cobalt, which undergoes a phase transformation and for which γ is considered to be low ($\approx$25 mJ m^{-2}). For the other common cph metals Zn, Cd, Mg, Ti, Be, etc., γ is high (250–300 mJ m^{-2}). No measurements of intrinsic faults with two next-nearest neighbor violations have been made, but intrinsic faults with one next-nearest neighbor violation have been measured (see 3.7.1) and show that Mg $\approx$ 125 mJ m^{-2}, Zn $\approx$140 mJ m^{-2}, and Cd $\approx$150–175 mJ m^{-2}. It is thus reasonable to conclude that intrinsic faults associated with Shockley partials have somewhat higher energy. Dislocations in these metals are therefore not very widely dissociated. A screw dislocation lying along a $[\bar{1}\ 2\ \bar{1}\ 0]$ direction is capable of gliding in three different glide planes, but the small extension in the basal plane will be sufficient to make basal glide easier than in either the pyramidal $(1\ 0\ \bar{1}\ 1)$ or prismatic $(1\ 0\ \bar{1}\ 0)$ glide (see Figure 6.18). Pyramidal and prismatic glide will be more favored at high temperatures in metals with high stacking-fault energy, when thermal activation aids the constriction of the dissociated dislocations.

3.6.3.2 Dislocation climb

Stacking faults may be produced in hexagonal lattices by the aggregation of point defects. If vacancies aggregate as a platelet, as shown in Figure 3.44c, the resultant collapse of the disk-shaped cavity (Figure 3.44d) would bring two similar layers into contact. This is a situation incompatible with the close packing and suggests that simple Frank dislocations are energetically unfavorable in cph lattices. This unfavorable situation can be removed by either one of two mechanisms, as shown in Figures 3.44e and f. In Figure 3.44e the B-layer is converted to a C-position by passing a pair of equal and opposite partial dislocations (dipole) over adjacent slip planes. The Burgers vector of the dislocation loop will be of the σS type and the energy of the fault, which is extrinsic, will be high because of the three next-nearest neighbor violations. In Figure 3.44f the loop is swept by an Aσ-type partial dislocation, which changes the stacking of all the layers above the loop according to the rule A $\rightarrow$ B $\rightarrow$ C $\rightarrow$ A. The Burgers vector of the loop is of the type AS, and from the dislocation reaction Aσ + σS $\rightarrow$ AS or $\frac{1}{3}[1\ 0\ \bar{1}\ 0] + \frac{1}{2}[0\ 0\ 0\ 1] \rightarrow \frac{1}{6}[2\ 0\ \bar{2}\ 3]$ and the associated stacking fault, which is intrinsic, will have a lower energy because there is only one next-nearest neighbor violation in the stacking sequence. Faulted loops with b = AS or $(\frac{1}{2}c + p)$ have been observed in Zn, Mg and Cd (see Figure 3.45). Double-dislocation loops have also been observed when the inner dislocation loop encloses a central region of perfect crystal and the outer loop an annulus of stacking fault. The structure of such a double loop is shown in Figure 3.46. The vacancy loops on adjacent atomic planes are bounded by dislocations with non-parallel Burgers vectors, i.e. $b = (\frac{1}{2}c + p)$ and $b = (\frac{1}{2}c - p)$, respectively; the shear component of the second loop acts in such a direction as to eliminate the fault introduced by the first loop. There are six partial vectors in the basal plane, p_1, p_2, p_3 and the negatives, and if one side of the loop is sheared by either p_1, p_2 or p_3 the stacking sequence is changed according to A $\rightarrow$ B $\rightarrow$ C $\rightarrow$ A, whereas reverse shearing A $\rightarrow$ C $\rightarrow$ B $\rightarrow$ A, results from either $-p_1$, $-p_2$ or $-p_3$. It is clear that the fault introduced by a positive partial shear can be eliminated by a subsequent shear brought about by any of the three negative partials. Three, four and more layered loops have also been observed in addition to the more common double loop. The addition of each layer of vacancies alternately introduces or removes stacking faults, no matter whether the loops precipitate one above the other or on opposite sides of the original defect.

As in fcc metals, interstitials may be aggregated into platelets on close-packed planes and the resultant structure, shown in Figure 3.47a, is a dislocation loop with Burgers vector Sσ, containing a high-energy stacking fault. This high-energy fault can be changed to one with lower energy by having the loop swept by a partial, as shown in Figure 3.47b.

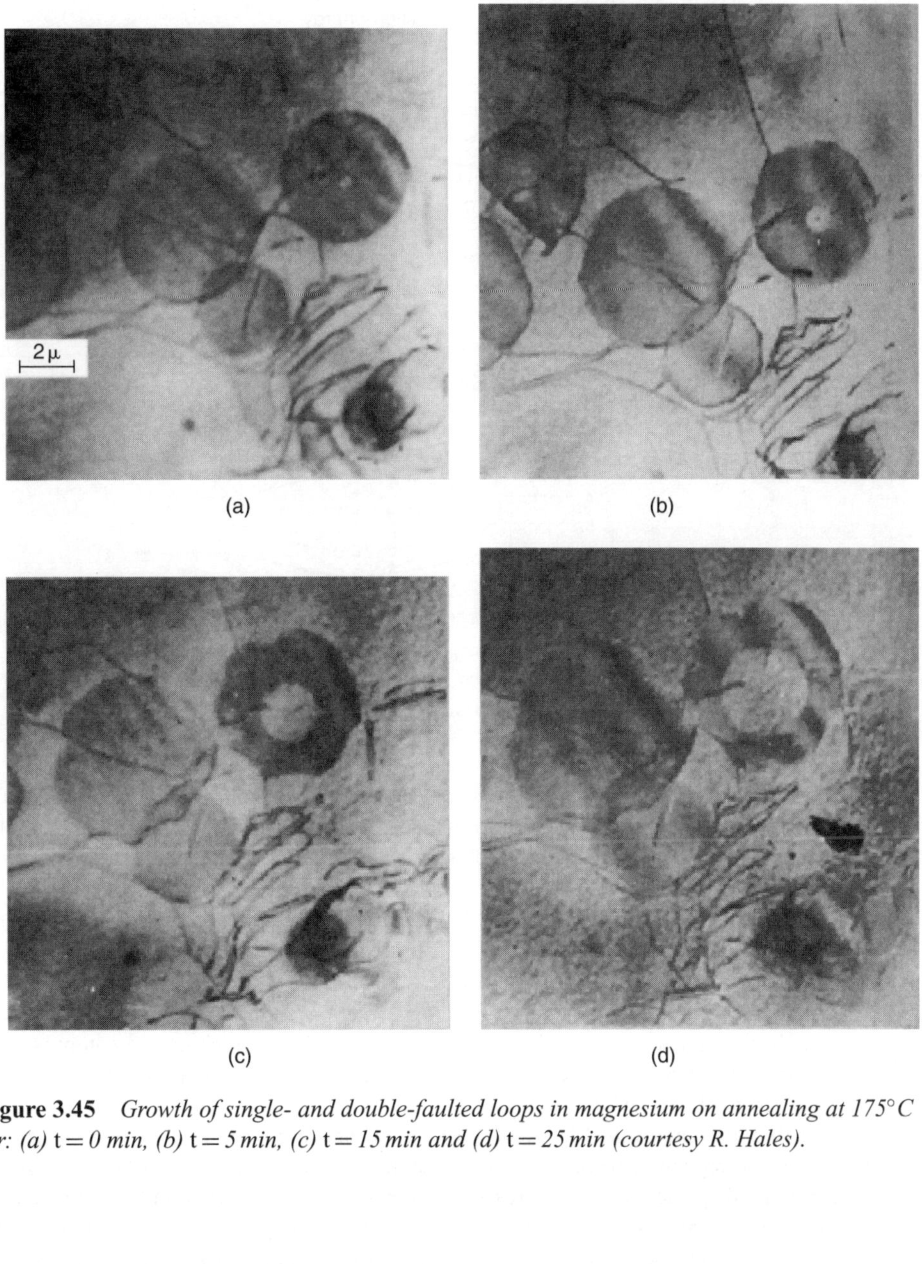

Figure 3.45 *Growth of single- and double-faulted loops in magnesium on annealing at 175°C for: (a)* t = *0 min, (b)* t = *5 min, (c)* t = *15 min and (d)* t = *25 min (courtesy R. Hales).*

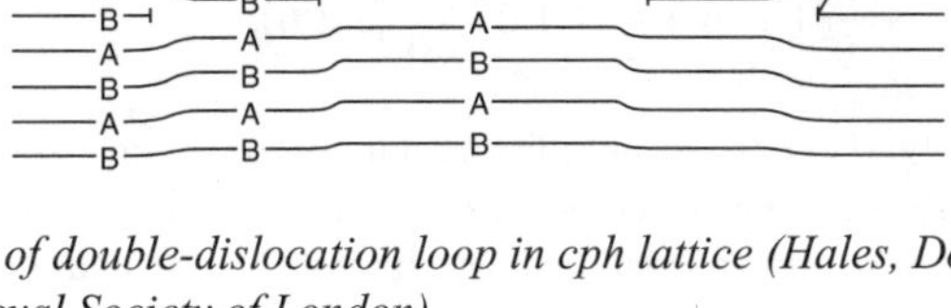

Figure 3.46 *Structure of double-dislocation loop in cph lattice (Hales, Dobson and Smallman, 1968; courtesy of the Royal Society of London).*

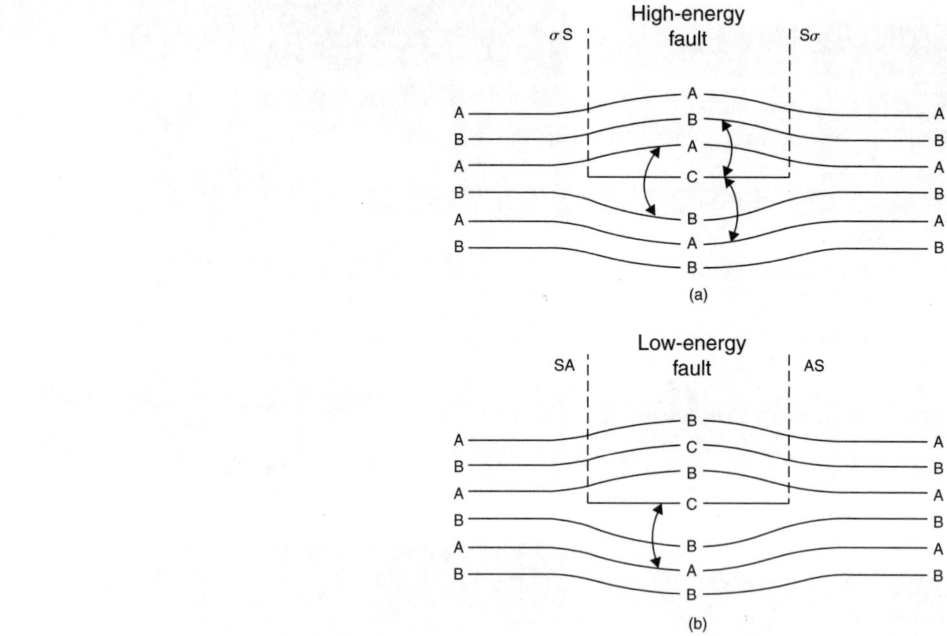

Figure 3.47 *Dislocation loop formed by aggregation of interstitials in a cph lattice with high-energy (a) and low-energy (b) stacking faults.*

All these faulted dislocation loops are capable of climbing by the addition or removal of point defects to the dislocation line. The shrinkage and growth of vacancy loops has been studied in some detail in Zn, Mg and Cd, and examples, together with the climb analysis, are discussed in Section 3.7.1.

3.6.4 Dislocations and stacking faults in bcc structures

The shortest lattice vector in the bcc lattice is $a/2[1\ 1\ 1]$, which joins an atom at a cube corner to the one at the center of the cube; this is the observed slip direction. The slip plane most commonly observed is (1 1 0) which, as shown in Figure 3.48, has a distorted close-packed structure. The (1 1 0) planes are packed in an **ABABAB** sequence and three {1 1 0} type planes intersect along a ⟨1 1 1⟩ direction. It therefore follows that screw dislocations are capable of moving in any of the three {1 1 0} planes and for this reason the slip lines are often wavy and ill-defined. By analogy with the fcc structure it is seen that in moving the B-layer along the $[\bar{1}\ \bar{1}\ 1]$ direction it is easier to shear in the directions indicated by the three vectors b_1, b_2 and b_3. These three vectors define a possible dissociation reaction:

$$\frac{a}{2}[\bar{1}\ \bar{1}\ 1] \rightarrow \frac{a}{8}[\bar{1}\ \bar{1}\ 0] + \frac{a}{4}[\bar{1}\ \bar{1}\ 2] + \frac{a}{8}[\bar{1}\ \bar{1}\ 0].$$

The stacking-fault energy of pure bcc metals is considered to be very high, however, and hence no faults have been observed directly. Because of the stacking sequence **ABABAB** of the (1 1 0) planes the formation of a Frank partial dislocation in the bcc structure gives rise to a situation similar to that for the cph structure, i.e. the aggregation of vacancies or interstitials will bring either two A-layers or

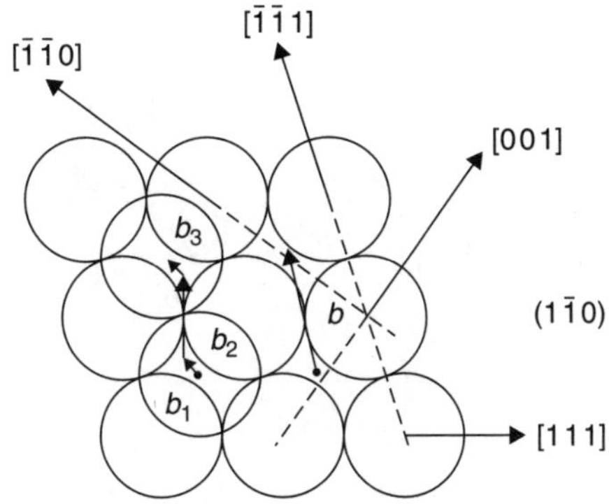

Figure 3.48 *The (1 1 0) plane of the bcc lattice (after Weertman; by courtesy of Oxford University Press).*

two B-layers into contact with each other. The correct stacking sequence can be restored by shearing the planes to produce perfect dislocations $a/2[1\ 1\ 1]$ or $a/2[1\ 1\ \bar{1}]$.

Slip has also been observed on planes indexed as (1 1 2) and (1 2 3) planes, and although some workers attribute this latter observation to varying amounts of slip on different (1 1 0) planes, there is evidence to indicate that (1 1 2) and (1 2 3) are definite slip planes. The packing of atoms in a (1 1 2) plane conforms to a rectangular pattern, the rows and columns parallel to the $[1\ \bar{1}\ 0]$ and $[1\ 1\ \bar{1}]$ directions, respectively, with the closest distance of approach along the $[1\ 1\ \bar{1}]$ direction. The stacking sequence of the (1 1 2) planes is **ABCDEFAB**... and the spacing between the planes $a/\sqrt{6}$. It has often been suggested that the unit dislocation can dissociate in the (1 1 2) plane according to the reaction:

$$\frac{a}{2}[1\ 1\ \bar{1}] \rightarrow \frac{a}{3}[1\ 1\ \bar{1}] + \frac{a}{6}[1\ 1\ \bar{1}]$$

because the homogeneous shear necessary to twin the structure is $1/\sqrt{2}$ in a $\langle 1\ 1\ 1\rangle$ on a (1 1 2) and this shear can be produced by a displacement $a/6[1\ 1\ \bar{1}]$ on every successive (1 1 2) plane. It is therefore believed that twinning takes place by the movement of partial dislocations. However, it is generally recognized that the stacking-fault energy is very high in bcc metals so that dissociation must be limited. Moreover, because the Burgers vectors of the partial dislocations are parallel, it is not possible to separate the partials by an applied stress unless one of them is anchored by some obstacle in the crystal.

When the dislocation line lies along the $[1\ 1\ \bar{1}]$ direction it is capable of dissociating in any of the three $\{1\ 1\ 2\}$ planes, i.e. (1 1 2), ($\bar{1}$ 2 1) and (2 $\bar{1}$ 1), which intersect along $[1\ 1\ \bar{1}]$. Furthermore, the $a/2[1\ 1\ \bar{1}]$ screw dislocation could dissociate according to:

$$\frac{a}{2}[1\ 1\ \bar{1}] \rightarrow \frac{a}{6}[1\ 1\ \bar{1}] + \frac{a}{6}[1\ 1\ \bar{1}] + \frac{a}{6}[1\ 1\ \bar{1}]$$

to form the symmetrical fault shown in Figure 3.49.

The symmetrical configuration may be unstable, and the equilibrium configuration is one partial dislocation at the intersection of two $\{1\ 1\ 2\}$ planes and the other two lying equidistant, one in each of the other two planes. At larger stresses this unsymmetrical configuration can be broken up and the partial dislocations induced to move on three neighboring parallel planes, to produce a three-layer twin. In recent years an asymmetry of slip has been confirmed in many bcc single crystals, i.e. the preferred slip plane may differ in tension and compression. A yield stress asymmetry has also been

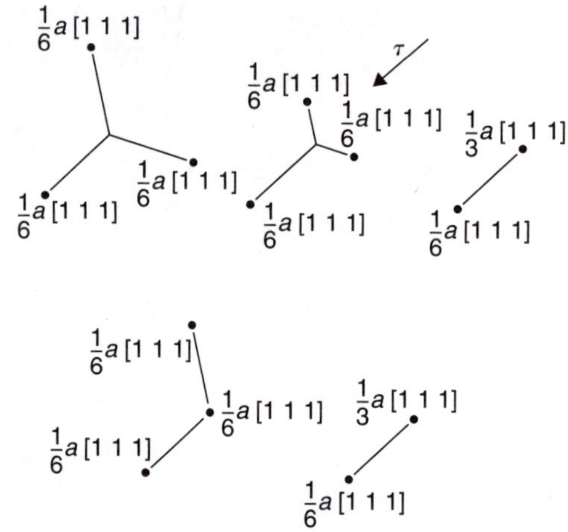

Figure 3.49 *Dissociated* a/2 *[1 1 1] dislocation in the bcc lattice (after Mitchell, Foxall and Hirsch, 1963; courtesy of Taylor & Francis).*

noted and has been related to asymmetric glide resistance of screw dislocations arising from their 'core' structure.

An alternative dissociation of the slip dislocation proposed by Cottrell is

$$\frac{a}{2}[1\ 1\ 1] \rightarrow \frac{a}{3}[1\ 1\ 2] + \frac{a}{6}[1\ 1\ \bar{1}]$$

The dissociation results in a twinning dislocation $a/6[1\ 1\ \bar{1}]$ lying in the (1 1 2) plane and an $a/3[1\ 1\ 2]$ partial dislocation with Burgers vector normal to the twin fault, and hence is sessile. There is no reduction in energy by this reaction and is therefore not likely to occur except under favorable stress conditions.

Another unit dislocation can exist in the bcc structure, namely $a[0\ 0\ 1]$, but it will normally be immobile. This dislocation can form at the intersection of normal slip bands by the reaction:

$$\frac{a}{2}[\bar{1}\ \bar{1}\ 1] + \frac{a}{2}[1\ 1\ 1] \rightarrow a[0\ 0\ 1]$$

with a reduction of strain energy from $3a^2/2$ to a^2. The new $a[0\ 0\ 1]$ dislocation lies in the (0 0 1) plane and is pure edge in character and may be considered as a wedge, one lattice constant thick, inserted between the (0 0 1) and hence has been considered as a crack nucleus. $a[0\ 0\ 1]$ dislocations can also form in networks of $a/2\langle 1\ 1\ 1\rangle$-type dislocations.

3.6.5 Dislocations and stacking faults in ordered structures

When the alloy orders, a unit dislocation in a disordered alloy becomes a partial dislocation in the superlattice with its attached anti-phase boundary interface, as shown in Figure 3.50a. Thus, when this dislocation moves through the lattice it will completely destroy the order across its slip plane. However, in an ordered alloy, any given atom prefers to have unlike atoms as its neighbors, and consequently such a process of slip would require a very high stress. To move a dislocation against the force γ exerted on it by the fault requires a shear stress $\tau = \gamma/b$, where b is the Burgers vector; in β-brass, where γ is about 0.07 N m^{-1}, this stress is 300 MN m^{-2}. In practice, the critical shear stress of β-brass is an order of magnitude less than this value, and thus one must conclude that slip occurs by an easier process than

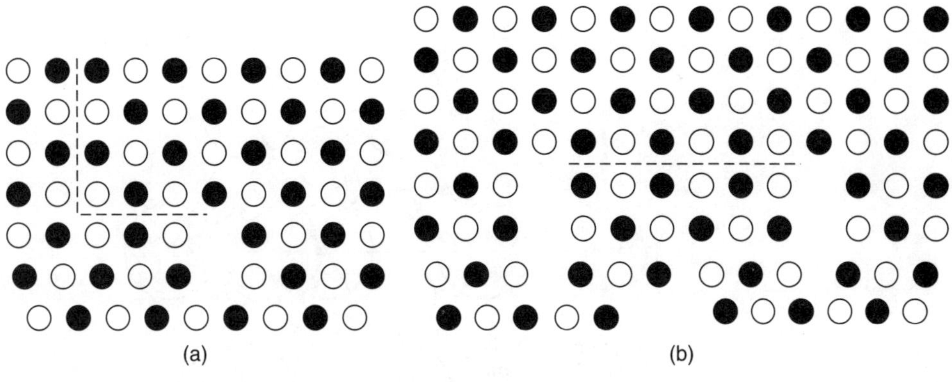

Figure 3.50 *Dislocations in ordered structures.*

the movement of unit dislocations. In consequence, by analogy with the slip process in fcc crystals, where the leading partial dislocation of an extended dislocation trails a stacking fault, it is believed that the dislocations which cause slip in an ordered lattice are not single dislocations but coupled pairs of dislocations, as shown in Figure 3.50b. The first dislocation of the pair, on moving across the slip plane, destroys the order and the second half of the couple completely restores it again, the third dislocation destroys it once more, and so on. In β-brass[10] and similar weakly ordered alloys such as AgMg and FeCo, the crystal structure is ordered bcc (or CsCl-type) and, consequently, deformation is believed to occur by the movement of coupled pairs of $a/2[1\ 1\ 1]$-type dislocations. The combined slip vector of the coupled pair of dislocations, sometimes called a super-dislocation, is then equivalent to $a[1\ 1\ 1]$ and, since this vector connects like atoms in the structure, long-range order will be maintained.

The separation of the super-partial dislocations may be calculated, as for Shockley partials, by equating the repulsive force between the two like $a/2\langle 1\ 1\ 1\rangle$ dislocations to the surface tension of the anti-phase boundary. The values obtained for β-brass and FeCo are about 70 and 50 nm, respectively, and thus superdislocations can be detected in the electron microscope using the weak beam technique (see Chapter 4). The separation is inversely proportional to the square of the ordering parameter and superdislocation pairs $\approx$ 12.5 nm width have been observed more readily in partly ordered FeCo (S = 0.59).

In alloys with high ordering energies the anti-phase boundaries associated with superdislocations cannot be tolerated and dislocations with a Burgers vector equal to the unit lattice vector $a\langle 1\ 0\ 0\rangle$ operate to produce slip in $\langle 1\ 0\ 0\rangle$ directions. The extreme case of this is in ionic-bonded crystals such as CsBr, but strongly ordered intermetallic compounds such as NiAl are also observed to slip in the $\langle 1\ 0\ 0\rangle$ direction with dislocations having $b = a\langle 1\ 0\ 0\rangle$.

Ordered A_3B-type alloys also give rise to superdislocations. Figure 3.51a illustrates three (1 1 1) layers of the $L1_2$ structure, with different size atoms for each layer. The three vectors shown give rise to the formation of different planar faults; $a/2[\bar{1}\ 0\ 1]$ is a super-partial producing apb, $a/6[\bar{2}\ 1\ 1]$ produces the familiar stacking fault and $a/3[\bar{1}\ \bar{1}\ 2]$ produces a superlattice intrinsic stacking fault (SISF). A $[\bar{1}\ 0\ 1]$ superdislocation can therefore be composed of either

$$[\bar{1}\ 0\ 1] \rightarrow \frac{a}{2}[\bar{1}\ 0\ 1] + \text{apb on } (1\ 1\ 1) + \frac{a}{2}[\bar{1}\ 0\ 1]$$

[10] Chapter 1, Figure 1.20, shows the CsCl or β_2 structure. When disordered, the slip vector is $a/2[1\ 1\ 1]$, but this vector in the ordered structure moves an A atom to a B site. The slip vector to move an A atom to an A site is twice the length and equal to $a[1\ 1\ 1]$.

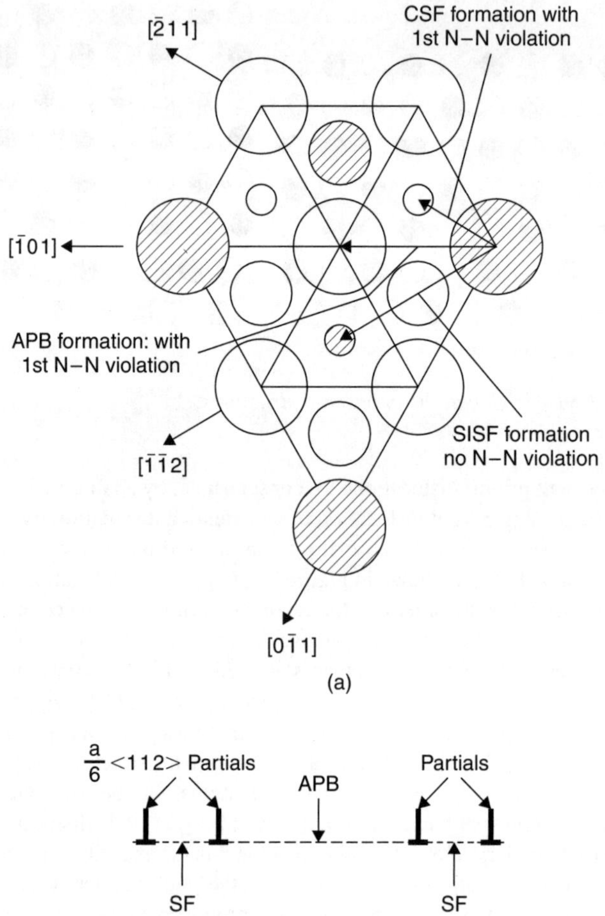

Figure 3.51 *(a) Stacking of (1 1 1) planes of the $L1_2$ structure, illustrating the apb and fault vectors. (b) Schematic representation of superdislocation structure.*

or

$$[\bar{1}\ 0\ 1] \rightarrow \frac{a}{3}[\bar{1}\ \bar{1}\ 2] + \text{SISF on}(1\ 1\ 1) + \frac{a}{3}[\bar{2}\ 1\ 1].$$

Each of the $a/2[\bar{1}\ 0\ 1]$ super-partials may also dissociate, as for fcc, according to:

$$\frac{a}{2}[\bar{1}\ 0\ 1] \rightarrow \frac{a}{6}[\bar{2}\ 1\ 1] + \frac{a}{6}[\bar{1}\ \bar{1}\ 2].$$

The resultant superdislocation is shown schematically in Figure 3.51b. In alloys such as Cu_3Au, Ni_3Mn, Ni_3Al, etc., the stacking-fault ribbon is too small to be observed experimentally but superdislocations have been observed. It is evident, however, that the cross-slip of these superdislocations will be an extremely difficult process. This can lead to a high work-hardening rate in these alloys, as discussed in Chapter 6.

In an alloy possessing short-range order, slip will not occur by the motion of superdislocations since there are no long-range faults to couple the dislocations together in pairs. However, because the distribution of neighboring atoms is not random the passage of a dislocation will destroy the short-range order between the atoms, across the slip plane. As before, the stress to do this will be large, but in this case there is no mechanism, such as coupling two dislocations together, to make the process easier. The fact that, for instance, a crystal of $AuCu_3$ in the quenched state (short-range order) has nearly double the yield strength of the annealed state (long-range order) may be explained on this basis. The maximum strength is exhibited by a partially ordered alloy with a critical domain size of about 6 nm. The transition from deformation by unit dislocations in the disordered state to deformation by superdislocations in the ordered condition gives rise to a peak in the flow stress with change in degree of order (see Chapter 5).

3.7 Stability of defects

3.7.1 Dislocation loops

During annealing, defects such as dislocation loops, stacking-fault tetrahedra and voids may shrink in size. This may be strikingly demonstrated by observing a heated specimen in the microscope. On heating, the dislocation loops and voids act as vacancy sources and shrink. This process occurs in the temperature range where self-diffusion is rapid, and confirms that the removal of the residual resistivity associated with Stage II is due to the dispersal of the loops, voids, etc.

The driving force for the emission of vacancies from a vacancy defect arises in the case of (1) a prismatic loop from the line tension of the dislocation, (2) a Frank loop from the force due to the stacking fault on the dislocation line, since in intermediate and high γ-metals this force far outweighs the line tension contribution, and (3) a void from the surface energy γ_s. The annealing of Frank loops and voids in quenched aluminum is shown in Figures 3.52 and 3.54, respectively. In a thin metal foil the rate of annealing is generally controlled by the rate of diffusion of vacancies away from the defect to any nearby sinks, usually the foil surfaces, rather than the emission of vacancies at the defect itself. To derive the rate equation governing the annealing, the vacancy concentration at the surface of the defect is used as one boundary condition of a diffusion-controlled problem and the second boundary condition is obtained by assuming that the surfaces of a thin foil act as ideal sinks for vacancies. The rate then depends on the vacancy concentration gradient developed between the defect, where the vacancy concentration is given by

$$c = c_0 \exp\{(\mathrm{d}F/\mathrm{d}n)/\mathbf{k}T\}, \tag{3.14}$$

with $(\mathrm{d}F/\mathrm{d}n)$ the change of free energy of the defect configuration per vacancy emitted at the temperature T, and the foil surface where the concentration is the equilibrium value c_0.

For a single, intrinsically faulted circular dislocation loop of radius r the total energy of the defect F is given by the sum of the line energy and the fault energy, i.e.

$$F \cong 2\pi r\{[\mu b^2/4\pi(1-\nu)]\ln(r/r_0)\} + \pi r^2\gamma.$$

In the case of a large loop ($r > 50$ nm) in a material of intermediate or high stacking-fault energy ($\gamma \gtrsim 60\,\mathrm{mJ\,m^{-2}}$) the term involving the dislocation line energy is negligible compared with the stacking-fault energy term and thus, since $(\mathrm{d}F/\mathrm{d}n) = (\mathrm{d}F/\mathrm{d}r) \times (\mathrm{d}r/\mathrm{d}n)$, is given simply by γB^2, where B^2 is the cross-sectional area of a vacancy in the (1 1 1) plane. For large loops the diffusion

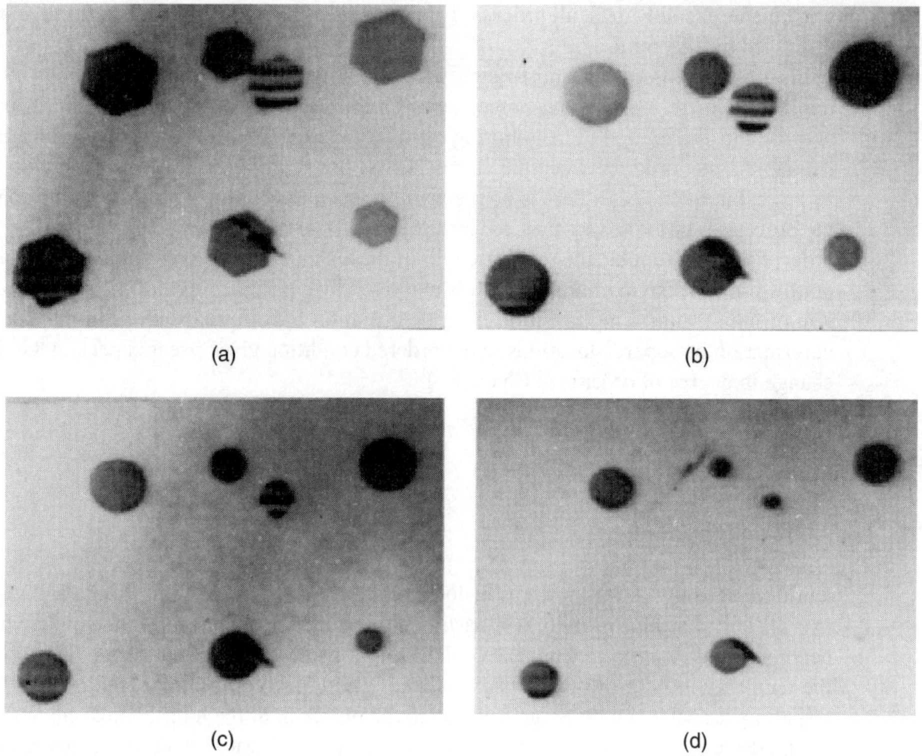

Figure 3.52 *Climb of faulted loops in aluminum at 140°C. (a)* t = *0 min, (b)* t = *12 min, (c)* t = *24 min, (d)* t = *30 min (after Dobson, Goodhew and Smallman, 1967; courtesy of Taylor & Francis).*

geometry approximates to cylindrical diffusion[11] and a solution of the time-independent diffusion equation gives for the annealing rate:

$$\begin{aligned} \mathrm{d}r/\mathrm{d}t &= -[2\pi D/b \ln(L/b)][\exp(\gamma B^2/\mathbf{k}T) - 1] \\ &= \text{const.}[\exp(\gamma B^2/\mathbf{k}T) - 1], \end{aligned} \tag{3.15}$$

where $D = D_0 \exp(-U_\mathrm{D}/\mathbf{k}T)$ is the coefficient of self-diffusion and L is half the foil thickness. The annealing rate of a prismatic dislocation loop can be similarly determined; in this case $\mathrm{d}F/\mathrm{d}r$ is determined solely by the line energy, and then

$$\begin{aligned} \mathrm{d}r/\mathrm{d}t &= -[2\pi D/b \ln(L/b)](\alpha b/r) \\ &= \text{const.}[\alpha b/r], \end{aligned} \tag{3.16}$$

where the term containing the dislocation line energy can be approximated to $\alpha b/r$. The annealing of Frank loops obeys the linear relation given by equation (3.15) at large r (Figure 3.53); at small r the curve deviates from linearity because the line tension term can no longer be neglected and also

[11] For spherical diffusion geometry the pre-exponential constant is D/b.

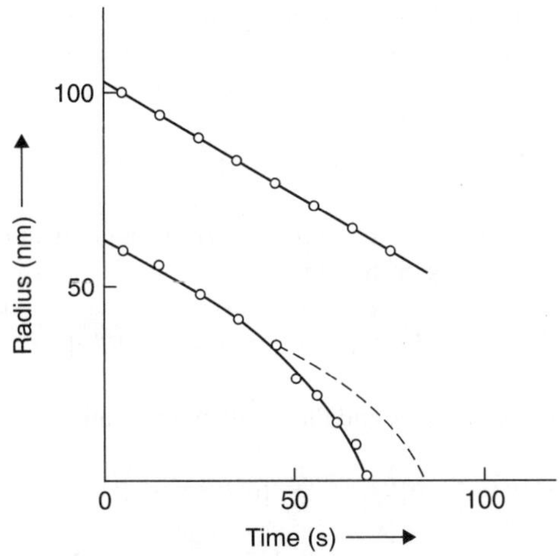

Figure 3.53 *Variation of loop radius with time of annealing for Frank dislocations in Al showing the deviation from linearity at small* r.

because the diffusion geometry changes from cylindrical to spherical symmetry. The annealing of prismatic loops is much slower, because only the line tension term is involved, and obeys an r^2 versus t relationship.

In principle, equation (3.15) affords a direct determination of the stacking fault energy γ by substitution, but since U_D is usually much bigger than γB^2 this method is unduly sensitive to small errors in U_D. This difficulty may be eliminated, however, by a comparative method in which the annealing rate of a faulted loop is compared to that of a prismatic one at the same temperature. The intrinsic stacking-fault energy of aluminum has been shown to be $135\,\mathrm{mJ\,m^{-2}}$ by this technique.

In addition to prismatic and single-faulted (Frank) dislocation loops, double-faulted loops have also been annealed in a number of quenched fcc metals. It is observed that, on annealing, the intrinsic loop first shrinks until it meets the inner, extrinsically faulted region, following which the two loops shrink together as one extrinsically faulted loop. The rate of annealing of this extrinsic fault may be derived in a way similar to equation (3.15) and is given by

$$\begin{aligned} \mathrm{d}r/\mathrm{d}t &= -[\pi D/b\ln(L/b)][\exp(\gamma_E B^2/\mathbf{k}T) - 1] \\ &= \text{const.}\{\exp(\gamma_E B^2/2\mathbf{k}T) - 1\}, \end{aligned} \tag{3.17}$$

from which the extrinsic stacking-fault energy may be determined. Generally, γ_E is about 10–30% higher in value than the intrinsic energy γ.

Loop growth can occur when the direction of the vacancy flux is towards the loop rather than away from it, as in the case of loop shrinkage. This condition can arise when the foil surface becomes a vacancy source, as, for example, during the growth of a surface oxide film. Loop growth is thus commonly found in Zn, Mg, Cd, although loop shrinkage is occasionally observed, presumably due to the formation of local cracks in the oxide film, at which vacancies can be annihilated. Figure 3.45 shows loops growing in Mg as a result of the vacancy supersaturation produced by oxidation. For the double loops, it is observed that a stacking fault is created by vacancy absorption at the growing outer perimeter of the loop and is destroyed at the growing inner perfect loop. The perfect regions expand faster than the outer stacking fault, since the addition of a vacancy to the inner loop decreases the energy of the defect by γB^2, whereas the addition of a vacancy to the outer loop increases the energy by the same amount. This effect is further enhanced as the two loops approach each other due to vacancy transfer from the outer to inner loops. Eventually the two loops coalesce to give a perfect prismatic loop of Burgers vector $c = [0\ 0\ 0\ 1]$, which continues to grow under the vacancy supersaturation. The outer loop growth rate is thus given by

$$\dot{r}_0 = [2\pi D/B\ln(L/b)][(c_s/c_0) - \exp(\gamma B^2/\mathbf{k}T)], \tag{3.18}$$

when the vacancy supersaturation term (c_s/c_0) is larger than the elastic force term tending to shrink the loop. The inner loop growth rate is

$$\dot{r}_i = -[2\pi D/B\ln(L/b)][(c_s/c_0) - \exp(-\gamma B^2/\mathbf{k}T)], \tag{3.19}$$

where $\exp(-\gamma B^2/\mathbf{k}T) << 1$, and the resultant prismatic loop growth rate is

$$\dot{r}_p = -[\pi D/B\ln(L/b)]\{(c_s/c_0) - [(\alpha b/r) + 1]\} \tag{3.20}$$

where $(\alpha b/r) < 1$ and can be neglected. By measuring these three growth rates, values for γ, (c_s/c_0) and D may be determined; Mg has been shown to have $\gamma = 125\ \mathrm{mJ\,m^{-2}}$ from such measurements.

3.7.2 Voids

Voids will sinter on annealing at a temperature where self-diffusion is appreciable. The driving force for sintering arises from the reduction in surface energy as the emission of vacancies takes place from the void surface. In a thin metal foil the rate of annealing is generally controlled by the rate of diffusion of vacancies away from the defect to any nearby sinks, usually the foil surfaces. The rate then depends on the vacancy concentration gradient developed between the defect (where the vacancy concentration is given by

$$c = c_0 \exp\{(\mathrm{d}F/\mathrm{d}n)/\mathbf{k}T\}, \tag{3.21}$$

with $(\mathrm{d}F/\mathrm{d}n)$ the change in free energy of the defect configuration per vacancy emitted at the temperature T and the foil surface where the concentration is the equilibrium value c_0.

For a void in equilibrium with its surroundings the free energy $F \cong 4\pi r^2\gamma_s$, and since $(\mathrm{d}F/\mathrm{d}n) = (\mathrm{d}F/\mathrm{d}r)(\mathrm{d}r/\mathrm{d}n) = (8\pi r\gamma_s)(\Omega/4\pi r^2)$, where Ω is the atomic volume and n the number of vacancies in the void, the concentration of vacancies in equilibrium with the void is

$$c_v = c_0 \exp[(\mathrm{d}F/\mathrm{d}n)/\mathbf{k}T] = c_0 \exp(2\gamma_s\Omega/r\mathbf{k}T).$$

Assuming spherical diffusion geometry, the diffusion equation may be solved to give the rate of shrinkage of a void as

$$\mathrm{d}r/\mathrm{d}t = -(D/r)\{\exp(2\Omega\gamma_s/r\mathbf{k}T) - 1\}. \tag{3.22}$$

For large r (>50 nm) the exponential term can be approximated to the first two terms of the series expansion and equation (3.22) may then be integrated to give

$$r^3 = r_i^3 - (6D\Omega\gamma_s/\mathbf{k}T)t, \tag{3.23}$$

where r_i is the initial void radius at $t = 0$. By observing the shrinkage of voids as a function of annealing time at a given temperature (see Figure 3.54), it is possible to obtain either the diffusivity D or the surface energy γ_s. From such observations, γ_s for aluminum is shown to be $1.14\ \mathrm{J\,m^{-2}}$ in the

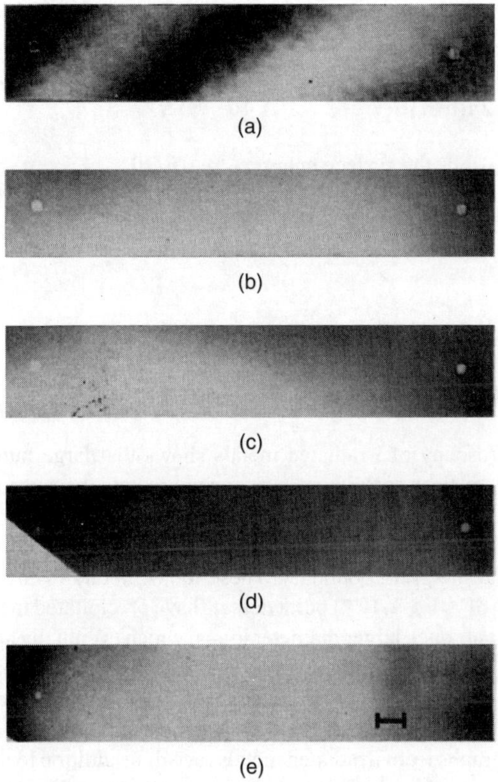

Figure 3.54 *Sequence of micrographs showing the shrinkage of voids in quenched aluminum during isothermal annealing at 170°C. (a) t = 3 min, (b) t = 8 min, (c) t = 21 min, (d) t = 46 min, (e) t = 98 min. In all micrographs the scale corresponds to 0.1 μm (after Westmacott, Smallman and Dobson, 1968; courtesy of the Institute of Materials, Minerals and Mining).*

temperature range 150–200°C, and $D = 0.176 \times \exp(-1.31\ \text{eV}/\mathbf{k}T)$. It is difficult to determine γ_S for Al by zero creep measurements because of the oxide. This method of obtaining γ_S has been applied to other metals and is particularly useful since it gives a value of γ_S in the self-diffusion temperature range rather than near the melting point.

Worked example

Excess vacancies in a crystal may aggregate into dislocation loops or voids. Show that the void is the lower energy configuration at small sizes but above a critical radius has a higher energy than a dislocation loop.

Assume that the surface energy $\gamma_S = \mu b/10$, where μ is the shear modulus.

Solution

The energy of the loop $E_L \approx 2\pi R_L(\mu b^2)$, where μb^2 is the dislocation line energy per unit length. This becomes $E_L = 2\pi(n/\pi)^{1/2}\mu b^3$, since the area of the loop $\pi R_L^2 = nb^2$, where n is the number of vacancies and b^2 is the cross-sectional area of a vacancy.

The energy of a spherical void is $E_V = 4\pi R_V^2\gamma_S = 4\pi(3n/4\pi)^{2/3}b^2\gamma_S$ since the volume of the sphere $4\pi R_V^3/3 = nb^3$, where b^3 is the atomic volume.

Thus, the ratio

$$\frac{E_V}{E_L} = \frac{4\pi(3n/4\pi)^{2/3}b^2\gamma_s}{2\pi(n/\pi)^{1/2}\mu b^3} = 2\left(\frac{3}{4}\right)^{2/3}\left(\frac{n}{\pi}\right)^{1/6}\frac{\gamma_s}{\mu b} = 1.36(n)^{1/6}\frac{\gamma_s}{\mu b} \approx (n)^{1/6}\frac{\gamma_s}{\mu b}.$$

Typically for solids the surface energy $\gamma_s \sim \mu b/10$.

Thus, $E_V/E_L \approx n^{1/6}/10$, and $E_V > E_L$ when $n^{1/6} > 10$, i.e. when the radius of the sphere is bigger than about 60 atom spacings.

3.7.3 Nuclear irradiation effects

3.7.3.1 *Behavior of point defects and dislocation loops*

Electron microscopy of irradiated metals shows that large numbers of small point defect clusters are formed on a finer scale than in quenched metals, because of the high supersaturation and low diffusion distance. Bombardment of copper foils with 1.4×10^{21} 38 MeV α-particles m^{-2} produces about $10^{21}\,m^{-3}$ dislocation loops, as shown in Figure 3.55a; a denuded region 0.8 μm wide can also be seen at the grain boundary. These loops, about 40 nm in diameter, indicate that an atomic concentration of $\approx 1.5 \times 10^{-4}$ point defects have precipitated in this form. Heavier doses of α-particle bombardment produce larger diameter loops, which eventually appear as dislocation tangles. Neutron bombardment produces similar effects to α-particle bombardment, but unless the dose is greater than 10^{21} neutrons m^{-2} the loops are difficult to resolve. In copper irradiated at pile temperature the density of loops increases with dose and can be as high as $10^{14}\,m^{-2}$ in heavily bombarded metals.

The micrographs from irradiated metals reveal, in addition to the dislocation loops, numerous small centers of strain in the form of black dots somewhat less than 5 nm in diameter, which are difficult to resolve (see Figure 3.55a). Because the two kinds of clusters differ in size and distribution, and also in their behavior on annealing, it is reasonable to attribute the presence of one type of defect, i.e. the large loops, to the aggregation of interstitials and the other, i.e. the small dots, to the aggregation of vacancies. This general conclusion has been confirmed by detailed contrast analysis of the defects.

The addition of an extra (1 1 1) plane in a crystal with fcc structure (see Figure 3.56) introduces two faults in the stacking sequence and not one, as is the case when a plane of atoms is removed. In consequence, to eliminate the fault it is necessary for two partial dislocations to slip across the loop, one above the layer and one below, according to a reaction of the form:

$$\frac{a}{3}[\bar{1}\,\bar{1}\,\bar{1}] + \frac{a}{6}[1\,1\,\bar{2}] + \frac{a}{6}[1\,\bar{2}\,1] \rightarrow \frac{a}{2}[0\,\bar{1}\,\bar{1}].$$

The resultant dislocation loop formed is identical to the prismatic loop produced by a vacancy cluster but has a Burgers vector of opposite sign. The size of the loops formed from interstitials increases with the irradiation dose and temperature, which suggests that small interstitial clusters initially form and subsequently grow by a diffusion process. In contrast, the vacancy clusters are much more numerous, and although their size increases slightly with dose, their number is approximately proportional to the dose and equal to the number of primary collisions which occur. This observation supports the suggestion that vacancy clusters are formed by the redistribution of vacancies created in the cascade.

Changing the type of irradiation from electron to light charged particles such as protons, to heavy ions such as self-ions, to neutrons, results in a progressive increase in the mean recoil energy. This results in an increasingly non-uniform point defect generation due to the production of displacement cascades by primary knock-ons. During the creation of cascades, the interstitials are transported outwards (see Figure 3.7), most probably by focused collision sequences, i.e. along a close-packed row

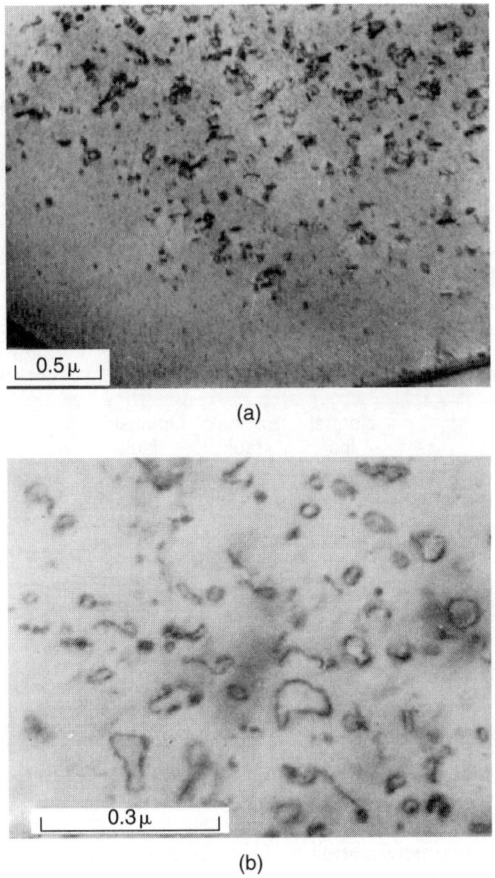

Figure 3.55 *A thin film of copper after bombardment with 1.4×10^{21} α-particles m^{-2}. (a) Dislocation loops (∼40 nm dia.) and small centers of strain (∼4 nm dia.). (b) After a 2-hour anneal at 350°C showing large prismatic loops (after Barnes and Mazey, 1960).*

of atoms by a sequence of replacement collisions, to displace the last atom in this same crystallographic direction, leaving a vacancy-rich region at the center of the cascade which can collapse to form vacancy loops. As the irradiation temperature increases, vacancies can also aggregate to form voids.

Frank sessile dislocation loops, double faulted loops, tetrahedra and voids have all been observed in irradiated metals, but usually under different irradiation conditions. Results from Cu, Ag and Au show that cascades collapse to form Frank loops, some of which dissociate towards stacking fault tetrahedra. The fraction of cascades collapsing to form visible loops, defined as the defect yield, is high, ≈0.5 in Cu to 1.0 in Au irradiated with self-ions. Moreover, the fraction of vacancies taking part in the collapse process, expressed as the cascade efficiency, is also high (≈0.3–0.5). Vacancy loops have been observed on irradiation at R.T. in some bcc metals (e.g. Mo, Nb, W, α-Fe). Generally, the loops are perfect with $b = a/2\langle 1\ 1\ 1\rangle$, although they are thought to nucleate as $a/2\langle 1\ 1\ 0\rangle$ faulted loops on $\{1\ 1\ 0\}$ but unfault at an early stage because of the high stacking-fault energy. Vacancy loops have also been observed in some cph metals (e.g. Zr and Ti).

Interstitial defects in the form of loops are commonly observed in all metals. In fcc metals Frank loops containing extrinsic faults occur in Cu, Ag, Au, Ni, Al and austenitic steels. Clustering of

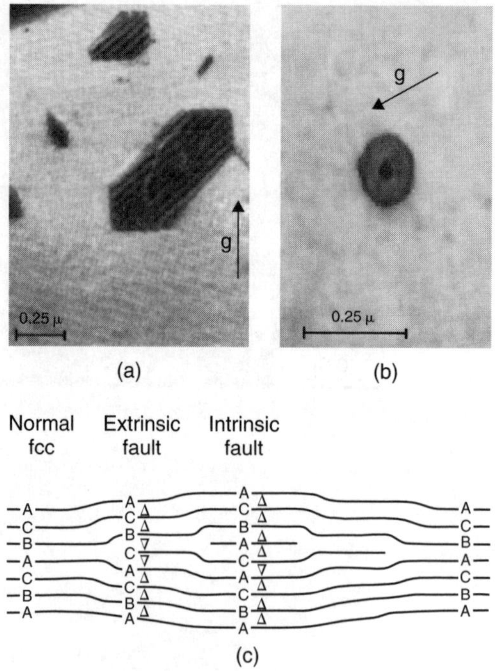

Figure 3.56 *Single (near top) and double (near middle) dislocation loops (a) and a small double-faulted loop (b), in proton-irradiated copper. (c) Structure of a double-dislocation loop (after Mazey and Barnes, 1968; courtesy of Taylor & Francis).*

interstitials on two neighboring (1 1 1) planes to produce an intrinsically faulted defect may also occur, as shown in Figure 3.56. In bcc metals they are predominantly perfect $a/2\langle 1\ 1\ 1\rangle$.

The damage produced in cph metals by electron irradiation is very complex and for Zn and Cd ($c/a > 1.633$) several types of dislocation loops, interstitial in nature, nucleate and grow; thus, $c/2$ loops, i.e. with $b = [c/2]$, c loops, $(c/2+p)$ loops, i.e. with $b = \frac{1}{6}\langle 2\ 0\ \bar{2}\ 3\rangle$, $[c/2]+[c/2]$ loops and $\langle c/2+p\rangle + \langle c/2-p\rangle$ loops are all formed; in the very early stages of irradiation most of the loops consist of $[c/2]$ dislocations, but as they grow a second loop of $b=[c/2]$ forms in the center, resulting in the formation of a $[c/2]+[c/2]$ loop. The $\langle c/2+p\rangle + \langle c/2-p\rangle$ loops form either from the nucleation of a $\langle c/2+p\rangle$ loop inside a $\langle c/2-p\rangle$ loop or when a $[c/2]+[c/2]$ loop shears. At low dose rates and low temperatures many of the loops facet along $\langle 1\ 1\ \bar{2}\ 0\rangle$ directions.

In magnesium, with c/a almost ideal, the nature of the loops is very sensitive to impurities, and interstitial loops with either $b = \frac{1}{3}\langle 1\ 1\ \bar{2}\ 0\rangle$ on non-basal planes or basal loops with $b = (c/2+p)$ have been observed in samples with different purity. Double loops with $b = (c/2+p)+(c/2-p)$ also form but no $c/2$ loops have been observed.

In Zr and Ti ($c/a < 1.633$) irradiated with either electrons or neutrons both vacancy and interstitial loops form on non-basal planes with $b = \frac{1}{3}\langle 1\ 1\ \bar{2}\ 0\rangle$. Loops with a c-component, namely $b = \frac{1}{3}\langle 1\ 1\ \bar{2}\ 3\rangle$ on $\{1\ 0\ \bar{1}\ 0\}$ planes and $b = c/2$ on basal planes have also been observed; voids also form in the temperature range 0.3–0.6T_m. The fact that vacancy loops are formed on electron irradiation indicates that cascades are not essential for the formation of vacancy loops. Several factors can give rise to the increased stability of vacancy loops in these metals. One factor is the possibility of stresses arising from oxidation or anisotropic thermal expansion, i.e. interstitial loops are favored perpendicular to a tensile axis and vacancy loops parallel. A second possibility is impurities segregating to dislocations and reducing the interstitial bias.

3.7.3.2 Radiation growth and swelling

In non-cubic materials, partitioning of the loops on to specific habit planes can lead to an anisotropic dimensional change, known as irradiation growth. The aggregation of vacancies into a disk-shaped cavity which collapses to form a dislocation loop will give rise to a contraction of the material in the direction of the Burgers vector. Conversely, the precipitation of a plane of interstitials will result in the growth of the material. Such behavior could account for the growth which takes place in α-uranium single crystals during neutron irradiation, since electron micrographs from thin films of irradiated uranium show the presence of clusters of point defects.

The energy of a fission fragment is extremely high ($\approx$200 MeV) so that a high concentration of both vacancies and interstitials might be expected. A dose of 10^{24} n m^{-2} at room temperature causes uranium to grow about 30% in the [0 1 0] direction and contract in the [1 0 0] direction. However, a similar dose at the temperature of liquid nitrogen produces 10 times this growth, which suggests the preservation of about 10^4 interstitials in clusters for each fission event that occurs. Growth also occurs in textured polycrystalline α-uranium and to avoid the problem a random texture has to be produced during fabrication. Similar effects can be produced in graphite.

During irradiation, vacancies may aggregate to form voids and the interstitials form dislocation loops. The voids can grow by acquiring vacancies which can be provided by the climb of the dislocation loops. However, because these loops are formed from interstitial atoms they grow, not shrink, during the climb process and eventually become a tangled dislocation network.

Interstitial point defects have two properties important in both interstitial loop and void growth. First, the elastic size interaction (see Chapter 6) causes dislocations to attract interstitials more strongly than vacancies and, secondly, the formation energy of an interstitial E_f^i is greater than that of a vacancy E_f^v so that the dominant process at elevated temperatures is vacancy emission. The importance of these factors to loop stability is shown by the spherical diffusion-controlled rate equation:

$$\frac{dr}{dt} = \frac{1}{b}\left\{D_v c_v - Z_i D_i c_i - D_v c_0 \exp\left[\frac{(F_{el} + \gamma)b^2}{\mathbf{k}T}\right]\right\}. \tag{3.24}$$

For the growth of voids during irradiation the spherical diffusion equation

$$\frac{dr}{dt} = \frac{1}{r}\left\{[1 + (\rho r)^{1/2}]D_v c_v - [1 + (Z_i \rho r)^{1/2}]D_i c_i - [1 + (\rho r)^{1/2}]D_v c_0 \exp\left[\frac{[(2\gamma_s/r) - P]\Omega}{\mathbf{k}T}\right]\right\}$$

has been developed, where c_i and c_v are the interstitial and vacancy concentrations, respectively, D_v and D_i their diffusivities, γ_s the surface energy and Z_i is a bias term defining the preferred attraction of the loops for interstitials.

At low temperatures, voids undergo bias-driven growth in the presence of biased sinks, i.e. dislocation loops or network of density ρ. At higher temperatures, when the thermal emission of vacancies becomes important, whether voids grow or shrink depends on the sign of $[(2\gamma_s/r) - P]$. During neutron irradiation, when gas is being created continuously, the gas pressure $P > \gamma_s/r$ and a flux of gas atoms can arrive at the voids, causing gas-driven growth.

The formation of voids leads to the phenomenon of void swelling and is of practical importance in the dimensional stability of reactor core components. Voids are formed in an intermediate temperature range $\approx$0.3–0.6T_m, above that for long-range single vacancy migration and below that for thermal vacancy emission from voids. To create the excess vacancy concentration it is also necessary to build up a critical dislocation density from loop growth to bias the interstitial flow. The sink strength of the dislocations, i.e. the effectiveness of annihilating point defects, is given by $K_i^2 = Z_i \rho$ for interstitials and $K_v^2 = Z_v\rho$ for vacancies, where $(Z_i - Z_v)$ is the dislocation bias for interstitials ≈10% and ρ is the dislocation density. As voids form they also act as sinks, and are considered neutral to vacancies and interstitials, so that $K_i^2 = K_v^2 = 4\pi r_v C_v$, where r_v and C_v are the void radius and concentration, respectively.

The rate theory of void swelling takes all these factors into account and (1) for moderate dislocation densities as the dislocation structure is evolving, swelling is predicted to increase linearly with irradiation dose, (2) when ρ reaches a quasi-steady state the rate should increase as $(\text{dose})^{3/2}$, and (3) when the void density is very high, i.e. the sink strength of the voids is greater than the sink strength of the dislocations ($K_v^2 >> K_d^2$), the rate of swelling should again decrease. Results from electron irradiation of stainless steel show that the swelling rate is linear with dose up to 40 dpa (displacements per atom) and there is no tendency to a $(\text{dose})^{3/2}$ law, which is consistent with dislocation structure continuing to evolve over the dose and temperature range examined.

In the fuel element itself, fission gas swelling can occur since uranium produces one atom of gas (Kr and Ze) for every five uranium atoms destroyed. This leads to ≈2 m^3 of gas (STP) per m^3 of U after a 'burn-up' of only 0.3% of the uranium atoms.

In practice, it is necessary to keep the swelling small and also to prevent nucleation at grain boundaries when embrittlement can result. In general, variables which can affect void swelling include alloying elements together with specific impurities, and microstructural features such as precipitates, grain size and dislocation density. In ferritic steels, the interstitial solutes carbon and nitrogen are particularly effective in (1) trapping the radiation-induced vacancies and thereby enhancing recombination with interstitials, and (2) interacting strongly with dislocations and therefore reducing the dislocation bias for preferential annihilation of interstitials, and also inhibiting the climb rate of dislocations. Substitutional alloying elements with a positive misfit such as Cr, V and Mn with an affinity for C or N can interact with dislocations in combination with interstitials and are considered to have a greater influence than C and N alone.

These mechanisms can operate in fcc alloys with specific solute atoms trapping vacancies and also elastically interacting with dislocations. Indeed, the inhibition of climb has been advanced to explain the low swelling of *Nimonic PE16* nickel-based alloys. In this case precipitates were considered to restrict dislocation climb. Such a mechanism of dislocation pinning is likely to be less effective than solute atoms, since pinning will only occur at intervals along the dislocation line. Precipitates in the matrix which are coherent in nature can also aid swelling resistance by acting as regions of enhanced vacancy–interstitial recombination. TEM observations on θ' precipitates in Al–Cu alloys have confirmed that as these precipitates lose coherency during irradiation, the swelling resistance decreases.

3.7.3.3 Radiation-induced segregation, diffusion and precipitation

Radiation-induced segregation is the segregation under irradiation of different chemical species in an alloy towards or away from defect sinks (free surfaces, grain boundaries, dislocations, etc.). The segregation is caused by the coupling of the different types of atom with the defect fluxes towards the sinks. There are four different possible mechanisms, which fall into two pairs, one pair connected with size effects and the other with the Kirkendall effect.[12] With size effects, the point defects drag the solute atoms to the sinks because the size of the solute atoms differs from the other types of atom present (solvent atoms). Thus interstitials drag small solute atoms to sinks and vacancies drag large solute atoms to sinks. With Kirkendall effects, the faster diffusing species move in the opposite direction to the vacancy current, but in the same direction as the interstitial current. The former case is usually called the 'inverse Kirkendall effect', although it is still the Kirkendall effect, but solute atoms rather than the vacancies are of interest. The most important of these mechanisms, which are

[12] The Kirkendall effect is discussed in Chapter 5, Section 5.4.2.

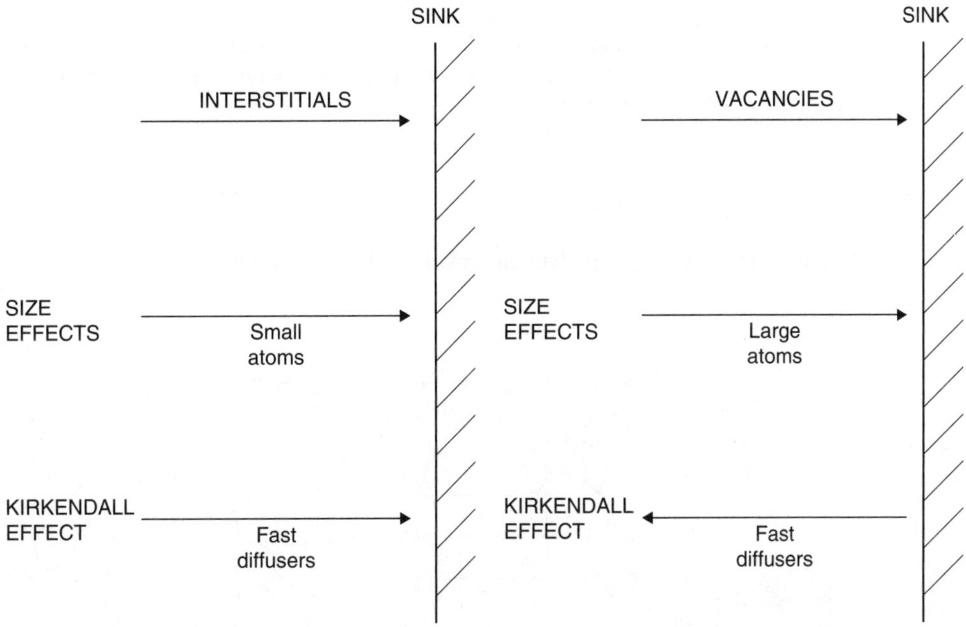

Figure 3.57 *Schematic representation of radiation-induced segregation produced by interstitial and vacancy flow to defect sinks.*

summarized in Figure 3.57, appear to be (1) the interstitial size effect mechanism – the dragging of small solute atoms to sinks by interstitials – and (2) the vacancy Kirkendall effect – the migration away from sinks of fast-diffusing atoms.

Radiation-induced segregation is technologically important in fast breeder reactors, where the high radiation levels and high temperatures cause large effects. Thus, for example, in Type 316 stainless steels, at temperatures in the range 350–650°C (depending on the position in the reactor), silicon and nickel segregate strongly to sinks. The small silicon atoms are dragged there by interstitials and the slow-diffusing nickel stays there in increasing concentration as the other elements diffuse away by the vacancy inverse Kirkendall effect. Such diffusion (1) denudes the matrix of void-inhibiting silicon and (2) can cause precipitation of brittle phases at grain boundaries, etc.

Diffusion rates may be raised by several orders of magnitude because of the increased concentration of point defects under irradiation. Thus, phases expected from phase diagrams may appear at temperatures where kinetics are far too slow under normal circumstances. Many precipitates of this type have been seen in stainless steels that have been in reactors. Two totally new phases have definitely been produced and identified in alloy systems (e.g. Pd_8W and Pd_8V) and others appear likely (e.g. Cu–Ni miscibility gap).

3.7.3.4 Irradiation of ordering alloys

Ordering alloys have a particularly interesting response to the influence of point defects in excess of the equilibrium concentration. Irradiation introduces point defects and their effect on the behavior of ordered alloys depends on two competitive processes, i.e. radiation-induced ordering and radiation-induced disordering, which can occur simultaneously. The interstitials do not contribute significantly to ordering but the radiation-induced vacancies give rise to ordering by migrating through the crystal. Disordering is assumed to take place athermally by displacements. The final state of the alloy at any irradiation temperature is independent of the initial condition. At 323 K, Cu_3Au is fully ordered on irradiation, whether it is initially ordered or not, but at low temperatures it becomes largely disordered because of the inability of the vacancies to migrate and develop order; the interstitials ($E_m^I \approx 0.1$ eV) can migrate at low temperatures.

Problems

In each of the following four diagrams, mark the Burgers vectors:

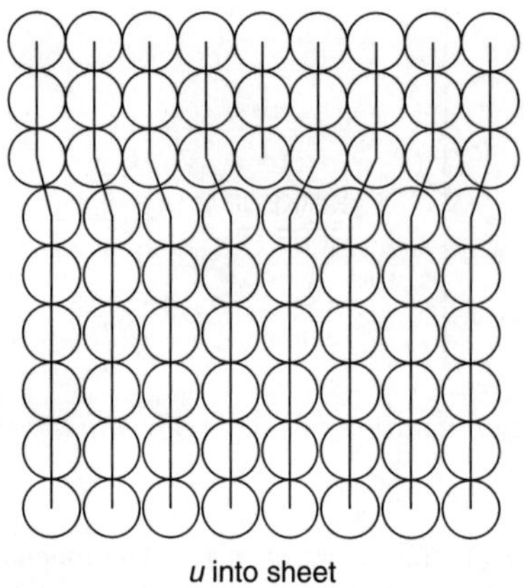

3.1

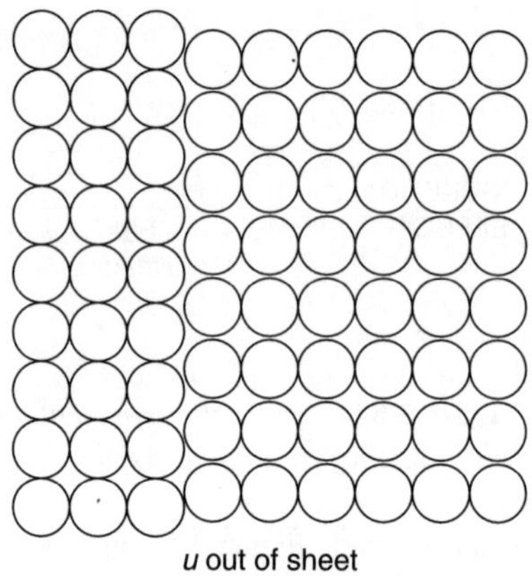

3.2

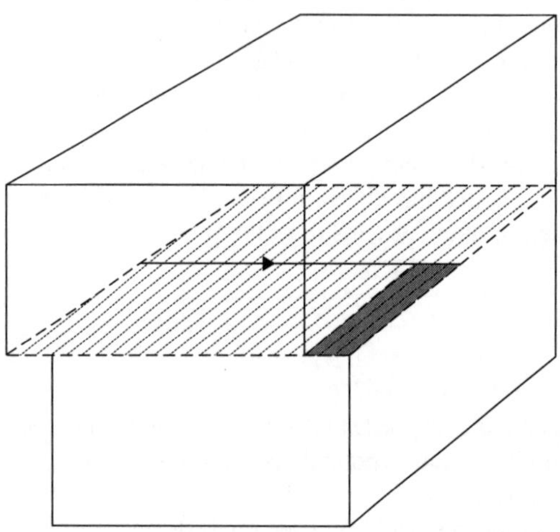

3.3

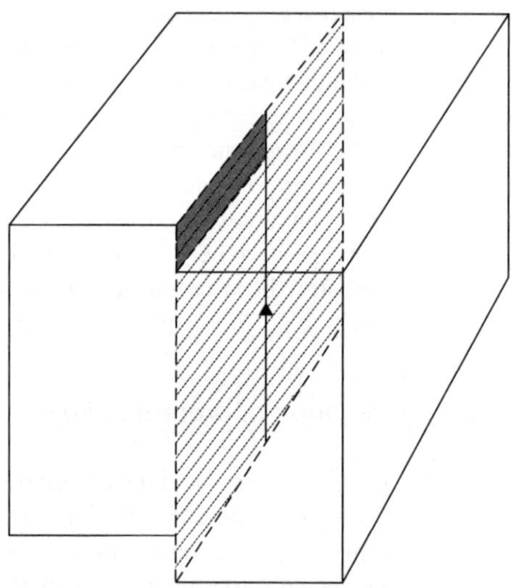

3.4

Fcc metals

3.5 Write down the possible Burgers vectors for glide dislocations in an fcc metal.

3.6 For a $1/2[1\ \bar{1}\ 0]$ dislocation moving on a $(1\ 1\ \bar{1})$ plane, what are the line directions of the screw and edge segments respectively?

3.7 On what planes can a dislocation with Burgers vector $\frac{1}{2}[1\ 0\ 1]$ glide?

Bcc metals

3.8 What is the line direction of an edge dislocation lying on $(1\ 0\ 1)$ whose Burgers vector is $1/2[1\ \bar{1}\ \bar{1}]$?

3.9 In what planes can a dislocation with Burgers vector $\frac{1}{2}[1\ 1\ 1]$ glide?

General

3.10 Complete the following dislocation reactions:

(a) $a/2[1\ 1\ 1] + a/2[1\ \bar{1}\ \bar{1}] \rightarrow ?$

(b) $a/6[1\ 1\ \bar{1}] + a/3[1\ 1\ 2] \rightarrow ?$

(c) $1/6[\bar{2}\ 0\ 2\ 3] + 1/6[2\ 0\ \bar{2}\ 3] \rightarrow ?$

3.11 What are the magnitudes of the following Burgers vectors?

(a) $a/2[1\ 1\ 1]$

(b) $a/6[1\ 1\ 2]$.

3.12 What is the angle between the Burgers vectors $a/6[1\ 1\ \bar{1}]$ and $a/3[1\ 1\ \bar{2}]$ (in a cubic crystal), and what is the pole of the plane containing their two directions?

3.13 A sample of nickel is quenched from a temperature of 1500 K to a temperature of 300 K. Estimate the 'chemical stress' developed in the nickel by this treatment (GPa). The energy of formation for a vacancy in nickel is 1.4 eV.

3.14 (i) State the Read–Shockley formula for the energy of a tilt boundary.

(ii) Show, by calculation, why it is feasible to disregard the term $(A - \ln\theta)$ when applying the formula to high-angle boundary surfaces. (Assume that the Burgers vector is 1.3 times larger than the core radius of the constituent dislocations.)

(iii) How does this energy vary with angle of misorientation? Estimate the energy for a large-angle boundary.

(iv) Comment on the value of the tilt boundary energy and how it relates to (a) grain boundary energy, (b) surface energy and (c) stacking-fault energy.

3.15 The sketch shows a diagramatric representation of a dissociated edge dislocation, in plan view, on the $(1\ 1\ 1)$ plane of a face-centered cubic crystal. Two Shockley partial dislocations (b_2 and b_3) are separated by a stacking fault.

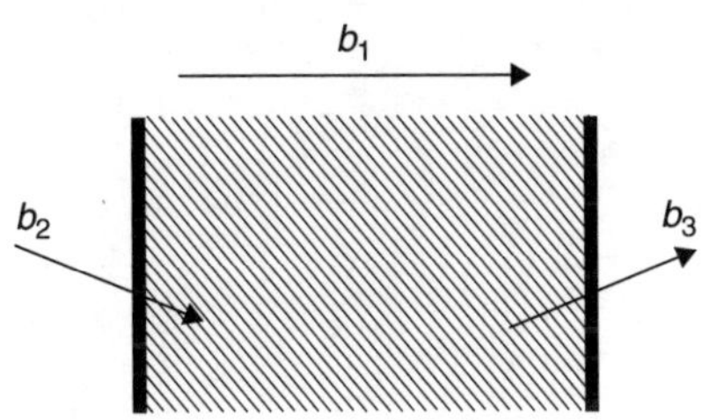

(a) Given that the separation is four atom spacings calculate the stacking-fault energy of the material. (Assume the lattice parameter of the material is 0.35 nm and the shear modulus is $40\,\mathrm{GN\,m^{-2}}$.)
(b) If the material is assumed to be pure nickel, how would you increase the dissociated width? Justify your conclusion.
(c) Write down a possible dissociation reaction in Miller index terms.
(d) What would this reaction be in the Thompson tetrahedron notation?
(e) It is possible to separate the two partials completely by applying a stress in the plane of the fault. Indicate whether this stress needs to be parallel or perpendicular to the b_1 direction and show that this separation will occur if the stress exceeds $5\gamma/a$.

3.16 Given that the anti-phase boundary energy of Ni_3Al is $175\,\mathrm{mJ\,m^{-2}}$, calculate the width of the superdislocation in atom spacings. (Shear modulus $= 7.5 \times 10^{10}\,\mathrm{N\,m^{-2}}$, $a = 3.5 \times 10^{-10}$m.)

Further reading

Hirth, J. P. and Lothe, J. (1984). *Theory of Dislocations*. McGraw-Hill, New York.
Hume-Rothery, W., Smallman, R. E. and Haworth, C. W. (1969). *Structure of Metals and Alloys*, Monograph No. 1. Institute of Metals.
Kelly, A., Groves, G. W. and Kidd, P. (2000). *Crystallography and Crystal Defects*. Wiley, Chichester.
Loretto, M. H. (ed.) (1985). *Dislocations and Properties of Real Materials*. Institute of Metals, London.
Smallman, R. E. and Harris, J. E. (eds) (1976). *Vacancies*. The Metals Society, London.
Sprackling, M. T. (1976). *The Plastic Deformation of Simple Ionic Solids*. Academic Press, London.
Thompson, N. (1953). Dislocation nodes in fcc lattices. *Proc. Phys. Soc.*, **B66**, 481.

Chapter 4

Characterization and analysis

4.1 Tools of characterization

Determination of the structural character of a material, whether massive in form or particulate, crystalline or glassy, is a central activity of materials science. The general approach adopted in most techniques is to probe the material with a beam of radiation or high-energy particles. The radiation is electromagnetic in character and may be monochromatic or polychromatic: the electromagnetic spectrum (Figure 4.1) conveniently indicates the wide choice of energy which is available. Wavelengths (λ) range from heat, through the visible range ($\lambda = 700$–400 nm) to penetrating X-radiation. Using de Broglie's important relation $\lambda = \mathbf{h}/mv$, which expresses the duality of radiation frequency and particle momentum, it is possible to apply the idea of wavelength to a stream of electrons.

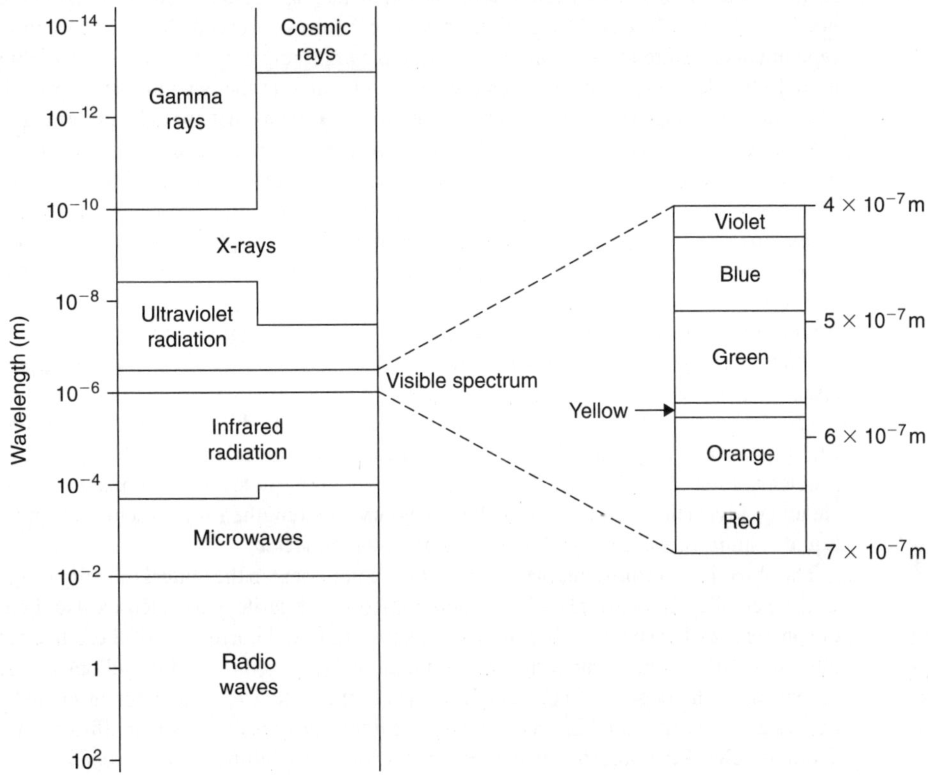

Figure 4.1 *The electromagnetic spectrum of radiation (from Askeland, 1994, p. 732; by permission of Chapman & Hall).*

The microscope, in its various forms, is the principal tool of the materials scientist. The magnification of the image produced by an electron microscope can be extremely high; however, on occasion, the modest magnification produced by a light stereomicroscope can be sufficient to solve a problem. In practical terms, the microscopist attaches more importance to resolution than magnification – that is, the ability of the microscope to distinguish fine detail. In a given microscope, increasing the magnification beyond a certain limit will fail to reveal further structural detail; such magnification is said to be 'empty'. Unaided, the human eye has a resolution of about 0.1 mm: the resolution of light microscopes and electron microscopes are, respectively, about 200 and 0.5 nm. In order to perceive or image a structural feature it is necessary that the wavelength of the probing radiation should be similar in size to that of the feature. In other words, and as will be enlarged upon later, resolution is a function of wavelength.

In this chapter we examine the principal ways in which light, X-rays, electrons and neutrons are used to explore the structure of metals. Some degree of selectivity has been unavoidable. Although the prime purpose of microscopy is to provide qualitative information on structure, many complementary techniques are available that provide quantitative data on the chemical and physical attributes of a material.

4.2 Light microscopy

4.2.1 Basic principles

The light microscope provides two-dimensional representation of structure over a total magnification range of roughly ×40 to ×1250. Interpretation of such images is a matter of skill and experience and needs to allow for the three-dimensional nature of features observed. The main components of a bench-type microscope are (1) an illumination system comprising a light source and variable apertures, (2) an objective lens and an ocular lens (eyepiece) mounted at the ends of a cylindrical body tube, and (3) a specimen stage (fixed or rotatable). Metallic specimens that are to be examined at high magnifications are successively polished with 6, 1 and sometimes 0.25 μm diamond grit. Examination in the as-polished condition, which is generally advisable, will reveal structural features such as shrinkage or gas porosity, cracks and inclusions of foreign matter. Etching with an appropriate chemical reagent is used to reveal the arrangement and size of grains, phase morphology, compositional gradients (coring), orientation-related etch pits and the effects of plastic deformation. Although actually only a few atomic diameters wide, grain boundaries are preferentially and grossly attacked by many etchants. In bright-field illumination, light is reflected back towards the objective from reflective surfaces, causing them to appear bright. Dark-field illumination reverses this effect, causing grain boundaries to appear bright. The degree of chemical attack is sensitive to crystal orientation and an etched polycrystalline aggregate will often display its grain structure clearly (Figure 4.2a). Preparation techniques for ceramics are essentially similar to those for metals and alloys. However, their porosity can cause two problems. First, there is a risk of entrapping diamond particles during polishing, making ultrasonic cleaning advisable. Second, it may be necessary to strengthen the structure by impregnating with liquid resin *in vacuo*, provided that pores are interconnected.

The objective, the most important and critical component in the optical train of the light microscope, is made up of a number of glass lenses and, sometimes, fluorite (CaF_2) lenses also. Lenses are subject to spherical and chromatic aberrations. Minimization and correction of these undesirable physical effects, greatly aided by modern computational techniques, is possible and objectives are classified according to the degree of correction, i.e. achromats, fluorites (semi-apochromats), apochromats. Lenses are usually coated in order to increase light transmission. As magnification is increased, the depth of field of the objective becomes smaller, typically falling from 250 μm at ×15 to 0.08 μm at ×1200, so that specimen flatness becomes more critical. The focal length and the working distance (separating its front lens from the specimen) of an objective differ. For instance, an *f* 2 mm objective may have a working distance of 0.15 mm.

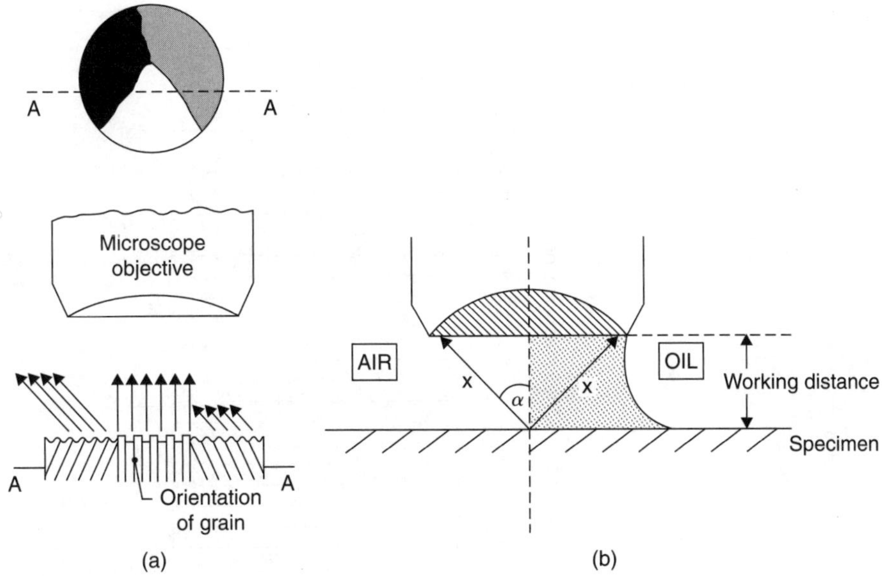

Figure 4.2 *(a) Reflection of light from etched specimen. (b) Use of oil to improve numerical aperture of objective.*

Resolution, rather than magnification, is usually the prime concern of the skilled microscopist. It is the smallest separating distance (δ) that can be discerned between two lines in the image. The unaided eye, at the least distance of comfortable vision (about 250 mm), can resolve 0.1 mm. Confusingly, the resolution value for a lens with a so-called high resolving power is small. Resolution is determined by (1) the wavelength (λ) of the radiation and (2) the numerical aperture (NA) of the objective, and is expressed by the Abbe formula $\delta = \lambda/2\text{NA}$.

The numerical aperture value, which is engraved upon the side of the objective, indicates the light-gathering power of the compound lens system and is obtained from the relation $\text{NA} = n \sin \alpha$, where n is the refractive index of the medium between the front lens face of the objective and the specimen and α is the semi-apex angle of the light cone defined by the most oblique rays collected by the lens. Numerical apertures range in typical value from 0.08 to 1.25. Despite focusing difficulties and the need for costly lenses, efforts have been made to use short-wavelength ultraviolet radiation: developments in electron microscopy have undermined the feasibility of this approach. Oil-immersion objectives enable the refractive index term to be increased (Figure 4.2b). Thus, by replacing air ($n = 1$) with a layer of cedar wood oil ($n = 1.5$) or monobromonaphthalene ($n = 1.66$), the number of rays of reflected light accepted by the front lens of the objective is increased, and resolution and contrast are improved. The range of wavelengths for visible light is approximately 400–700 nm; consequently, using the Abbe formula, it can readily be shown that the resolution limit of the light microscope is of the order of 200 nm. The 'useful' range of magnification is approximately 500–1000 NA. The lower end of the range can be tiring to the eyes; at the top end, oil-immersion objectives are useful.

Magnification is a subjective term; for instance, it varies with the distance of an image or object from the eye. Hence, microscopists sometimes indicate this difficulty by using the more readily defined term 'scale of reproduction', which is the lineal size ratio of an image (on a viewing screen or photomicrograph) to the original object. Thus, strictly speaking, a statement such as $\times 500$ beneath a photomicrograph gives the scale of reproduction, not the magnification.

The ocular magnifies the image formed by the objective: the finally observed image is virtual. It can also correct for certain objective faults and, in photomicrography, be used to project a real image.

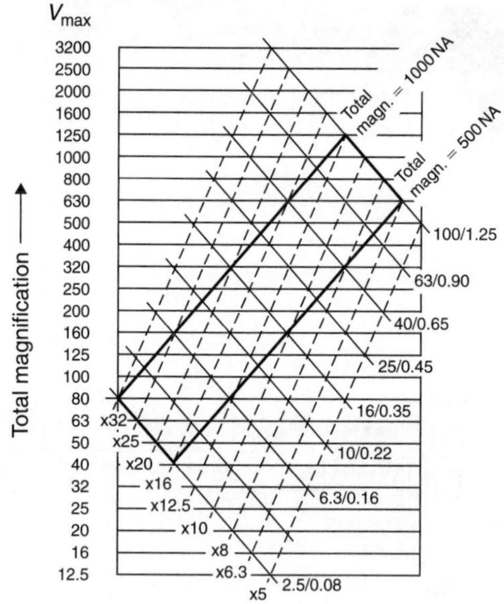

Figure 4.3 *Range of 'useful' magnification in light microscope (from Optical Systems for the Microscope, 1967, p. 15; by courtesy of Carl Zeiss, Germany).*

The ocular cannot improve the resolution of the system but, if inferior in quality, can worsen it. The most widely used magnifications for oculars are ×8 and ×12.5.

Two-dimensional features of a standard bench microscope, the mechanical tube length t_m and optical tube length t_o, are of special significance. The former is the fixed distance between the top of the body tube, on which the ocular rests, and the shoulder of the rotatable nosepiece into which several objectives are screwed. Objectives are designed for a certain t_m value. A value of 160 mm is commonly used. (In Victorian times, it was 250 mm, giving a rather unwieldy instrument.) The optical tube length t_o is the distance between the front focal point of the ocular and the rear focal plane of the objective. Parfocalization, using matched parfocal objectives and oculars, enables the specimen to remain in focus when objectives are step-changed by rotating the nosepiece. With each change, t_o changes but the image produced by the objective always forms in the fixed focal phase of the ocular. Thus, the distance between the specimen and the aerial image is kept constant. Some manufacturers base their sequences of objective and ocular magnifications upon preferred numbers[1] rather than upon a decimal series. This device facilitates the selection of a basic set of lenses that is comprehensive and 'useful' (exempt from 'empty' magnification). For example, the Michel series of ×6.3, ×8, ×10, ×12.5, ×16, ×20, ×25, etc., a geometrical progression with a common ratio of approximately 1.25, provides a basis for magnification values for objectives and oculars. This rational approach is illustrated in Figure 4.3. Dashed lines represent oculars and thin solid lines represent objectives. The bold lines outline a box within which objective/ocular combinations give 'useful' magnifications. Thus, pairing of a ×12.5 ocular with a ×40 objective (NA = 0.65) gives a 'useful' magnification of ×500.

[1] The valuable concept of preferred numbers/sizes, currently described in document PD 6481 of the British Standards Institution, was devised by a French military engineer, Colonel Charles Renard (1847–1905). In 1879, during the development of dirigible (steerable) balloons, he used a geometrical progression to classify cable diameters. A typical Renard series is 1.25, 1.6, 2.0, 2.5, 3.2, 4.0, 5.0, 6.4, 8.0, etc.

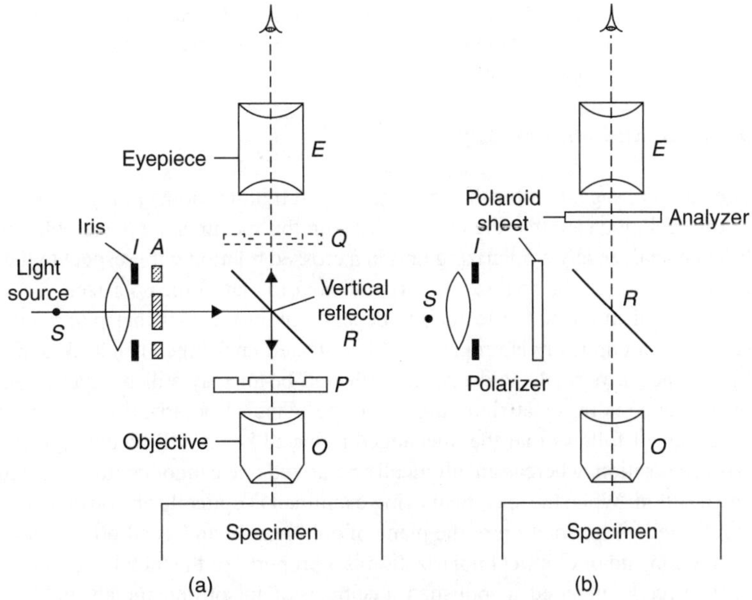

Figure 4.4 *Schematic arrangement of microscope system for phase-contrast (a) and polarized light (b) microscopy.*

4.2.2 Selected microscopical techniques

4.2.2.1 Phase-contrast microscopy

Phase-contrast microscopy is a technique that enables special surface features to be studied even when there is no color or reflectivity contrast. The light reflected from a small depression in a metallographic specimen will be retarded in phase by a fraction of a light wavelength relative to that reflected from the surrounding matrix and, whereas in ordinary microscopy a phase difference in the light collected by the objective will not contribute to contrast in the final image, in phase-contrast microscopy small differences in phases are transformed into differences in brightness which the eye can detect.

General uses of the technique include the examination of multi-phased alloys after light etching, the detection of the early stages of precipitation, and the study of cleavage faces, twins and other deformation characteristics. The optimum range of differences in surface level is about 20–50 nm, although under favorable conditions these limits may be extended. A schematic diagram of the basic arrangement for phase contrast in the metallurgical microscope is shown in Figure 4.4a. A hollow cone of light produced by an annulus A is reflected by the specimen and brought to an image in the back focal plane of the objective. A phase plate of suitable size should, strictly, be positioned in this plane but, for the ease of interchangeability of phase plates, the position Q in front of the eyepiece E is often preferred. This phase plate has an annulus, formed either by etching or deposition, such that the light it transmits is either advanced or retarded by a quarter of a wavelength relative to the light transmitted by the rest of the plate and, because the light reflected from a surface feature is also advanced or retarded by approximately $\lambda/4$, the beam is either in phase or approximately $\lambda/2$ or π out of phase with that diffracted by the surface features of the specimen. Consequently, reinforcement or cancellation occurs, and the image intensity at any point depends on the phase difference produced at the corresponding point on the specimen surface, and this in turn depends upon the height of this point relative to the adjacent parts of the surface. When the light passing through the annulus is advanced in phase, positive phase contrast results and areas of the specimen which are proud of the matrix appear bright and depressions dark; when the phase is retarded, negative contrast is produced and 'pits' appear bright and 'hills' dark.

4.2.2.2 Polarized-light microscopy

The basic arrangement for the use of polarized light is shown in Figure 4.4b. The only requirements of this technique are that the incident light on the specimen be plane polarized and that the reflected light be analyzed by a polarizing unit in a crossed relation with respect to the polarizer, i.e. the plane of polarization of the analyzer is perpendicular to that of the polarizer.

The application of the technique depends upon the fact that plane-polarized light striking the surface of an optically isotropic metal is reflected unchanged if it strikes at normal incidence. If the light is not at normal incidence the reflected beam may still be unchanged, but only if the angle of incidence is in, or at right angles to, the plane of polarization, otherwise it will be elliptically polarized. It follows that the unchanged reflected beam will be extinguished by an analyzer in the crossed position, whereas an elliptically polarized one cannot be fully extinguished by an analyzer in any position. When the specimen being examined is optically anisotropic, the light incident normally is reflected with a rotation of the plane of polarization and as elliptically polarized light; the amount of rotation and of elliptical polarization is a property of the metal and of the crystal orientation.

If correctly prepared, as-polished specimens of anisotropic metals will 'respond' to polarized light and a grain-contrast effect is observed under crossed polars as a variation of brightness with crystal orientation. Metals which have cubic structure, on the other hand, will appear uniformly dark under crossed polars, unless etched to invoke artificial anisotropy, by producing anisotropic surface films or well-defined pits. An etch pit will reflect the light at oblique incidence and elliptically polarized light will be produced. However, because such a beam cannot be fully extinguished by the analyzer in any position, it will produce a background illumination in the image which tends to mask the grain-contrast effect.

Clearly, one of the main uses of polarized light is to distinguish between areas of varying orientation, since these are revealed as differences of intensity under crossed polars. The technique is therefore very useful for studying the effects of deformation, particularly the production of preferred orientation, but information on cleavage faces, twin bands and sub-grain boundaries can also be obtained. If a 'sensitive tint' plate is inserted between the vertical illuminator and the analyzer, each grain of a sample may be identified by a characteristic color which changes as the specimen is rotated on the stage. This application is useful in the assessment of the degree of preferred orientation and in recrystallization studies. Other uses of polarized light include distinguishing and identifying phases in multi-phase alloys.

Near-perfect extinction occurs when the polars of a transmission microscope are crossed. If a thin section or slice of ceramic, mineral or rock is introduced and the stage slowly rotated, optically anisotropic crystals will produce polarization colors, developing maximum brilliance at 45° to any of the four symmetrical positions of extinction. The color of a crystal depends upon its birefringence, or capacity for double-refraction, and thickness. By standardizing the thickness of the section at 30–50 μm and using a Michel–Lévy color chart, it is possible to identify crystalline species. In refractory materials, it is relatively easy to identify periclase (MgO), chromite ($FeCrO_4$), tridymite (SiO_2) and zircon ($ZrSiO_4$) by their characteristic form and color.

As birefringence occurs within the crystal, each incident ray forms ordinary and extraordinary rays which are polarized in different planes and travel through the crystal at different velocities. On leaving the analyzer, these out-of-phase 'fast' and 'slow' rays combine to produce the polarization color. This color is complementary to color cancelled by interference and follows Newton's sequence: yellow, orange, red, violet, blue, green. More delicate, higher-order colors are produced as the phase difference between the emergent rays increases. Anisotropic crystals are either uniaxial or biaxial, having one or two optic axes, respectively, along which birefringence does not occur. (Optic axes do not necessarily correspond with crystallographic axes.) It is therefore possible for quartz (uniaxial) and mica (biaxial) crystals to appear black because of their orientation in the slice. Uniaxial (tetragonal and hexagonal systems) can be distinguished from biaxial crystals (orthorhombic, triclinic and monoclinic systems) by introducing a Bertrand lens into the light train of the microscope to give a convergent beam, rotating the stage and comparing their interference figures: uniaxial crystals give a moving 'ring and brush' pattern, biaxial crystals give two static 'eyes'. Cubic crystals are isotropic, being highly symmetrical. Glassy phases are isotropic and also appear black between crossed polars; however, glass containing residual stresses from rapid cooling produces fringe patterns and polarization colors. The stress-anisotropic properties of plastics are utilized in photoelastic analyses of transparent models of engineering structures or components made from standard sheets of constant thickness and stress-optic coefficient (e.g. clear *Bakelite*, epoxy resin). The fringe patterns produced by monochromatic light and crossed polars in a polariscope reveal the magnitude and direction of the principal stresses that are developed when typical working loads are applied.

4.2.2.3 Hot-stage microscopy

The ability to observe and photograph phase transformations and structural changes in metals, ceramics and polymers at high magnifications while being heated holds an obvious attraction. Various designs of microfurnace cell are available for mounting in light microscope systems.

For studies at moderate temperatures, such as spherulite formation in a cooling melt of polypropylene, the sample can be placed on a glass slide, heated in a stage fitment and viewed through crossed polars with transmitted light. For metals, which have an increasing tendency to vaporize as the temperature is raised, the polished sample is enclosed in a resistance-heated microfurnace and viewed by reflected light through an optically worked window of fused silica. The chamber can be either evacuated (10^{-6} torr) or slowly purged with inert gas (argon). The latter must be dry and oxygen free. A Pt.Pt–10Rh thermocouple is inserted in the specimen. The furnace should have a low thermal inertia and be capable of heating or cooling the specimen at controlled rates; temperatures of up to 1800°C are possible in some designs. The presence of a window, and possibly cooling devices, drastically reduces the available working distance for the objective lens, particularly when a large numerical aperture or high magnification are desired. One common solution is to use a Burch-type reflecting objective with an internal mirror system, which gives a useful working distance of 13–14 mm. The type of stage unit described has been used for studies of grain growth in austenite and the formation of bainite and martensite in steels, allotropic transformations, and sintering mechanisms in powder compacts.

When polished polycrystalline material is heated, individual grains tend to reduce their volume as a result of surface tension and grain boundaries appear as black lines, an effect referred to as thermal etching or grooving. If a grain boundary migrates, as in the grain growth stage of annealing, ghost images of former grooves act as useful markers. As the melting point is approached, there is often a noticeable tendency for grain boundary regions to fuse before the bulk grains; this liquation effect is due to the presence of impurities and the atomic misfit across the grain boundary surface. When interpreting the visible results of hot-stage microscopy, it is important to bear in mind that surface effects do not necessarily reflect what is happening within the bulk material beneath the surface. The technique can produce artefacts; the choice between evacuation and gas purging can be crucial. For instance, heating *in vacuo* can favor decarburization and grain coarsening in steel specimens.

The classic method for studying high-temperature phases and their equilibria in oxide systems was based upon rapid quenching (e.g. silicates). This indirect method is slow and does not always preserve the high-temperature phase(s). A direct microscopical technique uses the U-shaped notch of a thermocouple hot junction as the support for a small non-metallic sample. In the original design,[2] the junction was alternately connected by high-speed relay at a frequency of 50 Hz to a power circuit and a temperature-measuring circuit. The sample could be heated to temperatures of up to 2150°C and it was possible to observe crystallization from a melt directly through crossed polars. Although unsuitable for metals and highly volatile materials, the technique has been applied to glasses, slags, refractories, Portland cements, etc., providing information on phase changes, devitrification, sintering shrinkage, grain growth and the 'wetting' of solids by melts.

4.2.2.4 *Microhardness testing*

The measurement of hardness with a microscope attachment, comprising a diamond indenter and means for applying small loads, dates back more than 50 years. Initially used for small components (watch gears, thin wire, foils), microhardness testing was extended to research studies of individual phases, orientation effects in single crystals, diffusion gradients, ageing phenomena, etc. in metallic and ceramic materials. Nowadays, testing at temperatures up to 1000°C is possible. In Europe, the pyramidal Vickers-type (interfacial angle 136°) indenter, which produces a square impression, is generally favored. Its counterpart in general engineering employs test loads of 5–100 kgf: in microhardness testing, typical test loads are in the range 1–100 gf (1 gf = 1 pond = 1 p = 9.81 mN). A rhombic-based Knoop indenter of American origin has been recommended for brittle and/or anisotropic material (e.g. carbides, oxides, glass) and for thin foils and coatings where a shallow depth of impression is desired. The kite-shaped Knoop impression is elongated, with a 7:1 axial ratio.

Microhardness tests need to be very carefully controlled and replicated, using as large a load as possible. The surface of the specimen should be strain-free (e.g. electropolished), plane and perpendicular to the indenter axis. The indenter is lowered slowly at a rate of $<1\ \text{mm min}^{-1}$ under vibration-free conditions, eventually deforming the test surface in a manner analogous to steady-state creep. This condition is achieved within 15 s, a test period commonly used.

The equations for Vickers hardness (H_V) and Knoop hardness (H_K) take the following forms:

$$H_V = 1854.4(P/d^2)\quad \text{kgf mm}^{-2} \tag{4.1}$$

$$H_K = 14228(P/d^2)\quad \text{kgf mm}^{-2} \tag{4.2}$$

In these equations, which have the dimensions of stress, load P and diagonal length d are measured in gf and μm, respectively. The first equation is based upon the surface area of the impression; the second is based upon its projected area and the length of the long diagonal.

The main potential difficulty concerns the possible dependence of microhardness values (H_m) upon test load. As the test load is reduced below a certain threshold, the measured microhardness value may tend to decrease or increase, depending upon the material. In these circumstances, when measuring absolute hardness rather than relative hardness, it is useful to consider the material's behavior in terms of the Meyer equation, which relates indenting force P to the diagonal length of the Vickers-type impression produced, d, as follows:

$$P = kd^n. \tag{4.3}$$

The Meyer exponent n expresses the strain-hardening characteristics of the material as it deforms plastically during the test; it increases in value with the degree of strain hardening. For simple comparisons of relative microhardness, hardness values at a fixed load can be compared without allowance for

[2] Developed by W. Gutt and co-workers at the Building Research Station, Watford.

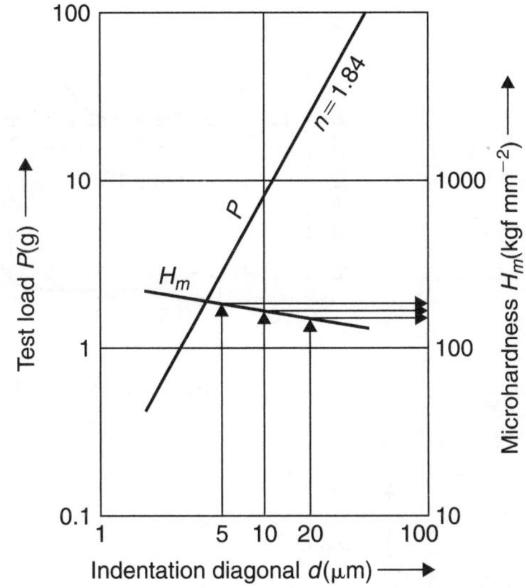

Figure 4.5 *Meyer line for material with load-independent hardness (by courtesy of Carl Zeiss, Germany).*

load dependence. On the other hand, if absolute values of hardness are required from low-load tests, it is advisable to determine the Meyer line for the particular material over a comparatively small load range by plotting P values against the corresponding d values, using log–log graph paper. (Extrapolations beyond the chosen load range are unwise because the profile of the Meyer line may change.) Figure 4.5 shows the Meyer line, slope n, for a material giving load-dependent microhardness values. The slope n is less than 2, which is usual. The H_m curve has a negative slope and microhardness values increase as the load increases. One way of reporting load-dependent microhardness results is to state three hardness numbers in terms of a standard set of three diagonal d-values, as shown in Figure 4.5. The approximate values for the set shown are $H_{5\,\mu m} = 160$, $H_{10\,\mu m} = 140$, $H_{20\,\mu m} = 120$. When the anisotropy ratio for elastic moduli is high, microhardness values can vary greatly from grain to grain in polycrystalline material.

Combination of the Vickers equation with the Meyer equation gives the following expression:

$$H_V = \text{constant} \times d^{n-2}. \tag{4.4}$$

Accordingly, if $n = 2$, which is true for the conventional Vickers macrohardness test, the gradient of the H_m line becomes zero and hardness values are conveniently load independent.

4.3 X-ray diffraction analysis

4.3.1 Production and absorption of X-rays

The use of diffraction methods is of great importance in the analysis of crystalline solids. Not only can they reveal the main features of the structure, i.e. the lattice parameter and type of structure, but also other details such as the arrangement of different kinds of atoms in crystals, the presence of imperfections, the orientation, sub-grain and grain size, the size and density of precipitates.

X-rays are a form of electromagnetic radiation differing from light waves ($\lambda = 400$–800 nm) in that they have a shorter wavelength ($\lambda \approx 0.1$ nm). These rays are produced when a metal target is

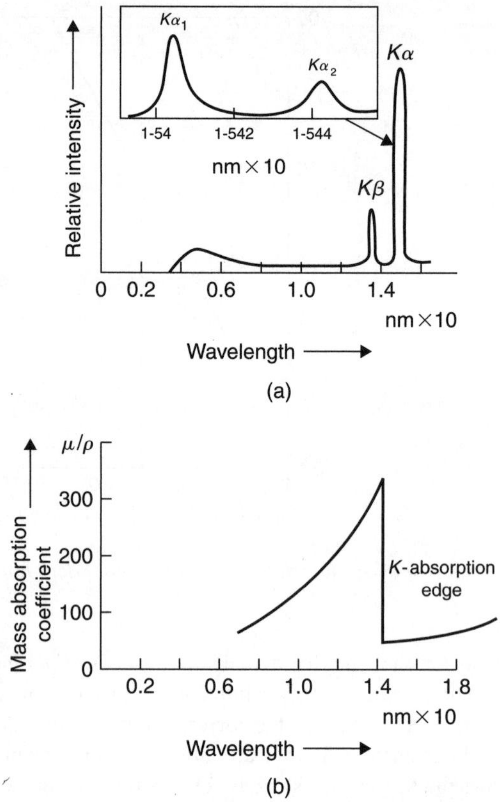

Figure 4.6 *(a) Distribution of X-ray intensity from a copper target. (b) Dependence of mass absorption coefficient on X-ray wavelength for nickel.*

bombarded with fast electrons in a vacuum tube. The radiation emitted, as shown in Figure 4.6a, can be separated into two components, a continuous spectrum which is spread over a wide range of wavelengths and a superimposed line spectrum characteristic of the metal being bombarded. The energy of the 'white' radiation, as the continuous spectrum is called, increases as the atomic number of the target and approximately as the square of the applied voltage, while the characteristic radiation is excited only when a certain critical voltage is exceeded. The characteristic radiation is produced when the accelerated electrons have sufficient energy to eject one of the inner electrons (1*s*-level, for example) from its shell. The vacant 1*s*-level is then occupied by one of the other electrons from a higher energy level, and during the transition an emission of X-radiation takes place. If the electron falls from an adjacent shell then the radiation emitted is known as $K\alpha$-radiation, since the vacancy in the first K-shell $n = 1$ is filled by an electron from the second L-shell and the wavelength can be derived from the relation

$$\mathbf{h}\nu = E_L - E_K. \tag{4.5}$$

However, if the K-shell vacancy is filled by an electron from an M-shell (i.e. the next highest quantum shell) then $K\beta$-radiation is emitted. Figure 4.6 shows that, in fact, one cannot be excited without the other, and the characteristic K-radiation emitted from a copper target is in detail composed of a strong $K\alpha$-doublet and a weaker $K\beta$-line.

In transversing a specimen, an X-ray beam loses intensity according to the equation:

$$I = I_0 \exp[-\mu x], \tag{4.6}$$

where I_0 and I are the values of the initial and final intensities respectively, μ is a constant, known as the linear absorption coefficient, which depends on the wavelength of the X-rays and the nature of the absorber, and x is the thickness of the specimen.[3] The variation of mass absorption coefficient,i.e. linear absorption coefficient divided by density, μ/ρ, with wavelength is of particular interest, as shown in Figure 4.6b, which is the curve for nickel. It varies approximately as λ^3 until a critical value of λ (=0.148 nm) is reached, when the absorption decreases precipitously. The critical wavelength λ_K at which this decrease occurs is known as the K absorption edge, and is the value at which the X-ray beam has acquired just sufficient energy to eject an electron from the K-shell of the absorbing material. The value of λ_K is characteristic of the absorbing material, and similar L and M absorption edges occur at higher wavelengths.

This sharp variation in absorption with wavelength has many applications in X-ray practice, but its most common use is in filtering out unwanted $K\beta$-radiation. For example, if a thin piece of nickel foil is placed in a beam of X-rays from a copper target, absorption of some of the short-wavelength 'white' radiation and most of the $K\beta$-radiation will result, but the strong $K\alpha$-radiation will be only slightly attenuated. This filtered radiation is sufficiently monochromatic for many X-ray techniques, but for more specialized studies when a pure monochromatic beam is required, crystal monochromators are used. The X-ray beam is then reflected from a crystal, such as quartz or lithium fluoride, which is oriented so that only the desired wavelength is reflected according to the Bragg law (see below).

4.3.2 Diffraction of X-rays by crystals

The phenomena of interference and diffraction are commonplace in the field of light. The standard school physics laboratory experiment is to determine the spacing of a grating, knowing the wavelength of the light impinging on it, by measuring the angles of the diffracted beam. The only conditions imposed on the experiment are that (1) the grating be periodic and (2) the wavelength of the light is of the same order of magnitude as the spacing to be determined. This experiment immediately points to the application of X-rays in determining the spacing and interatomic distances in crystals, since both are about 0.1–0.4 nm in dimension. Rigorous consideration of diffraction from a crystal in terms of a three-dimensional diffraction grating is complex, but Bragg simplified the problem by showing that diffraction is equivalent to symmetrical reflection from the various crystal planes, provided certain conditions are fulfilled. Figure 4.7a shows a beam of X-rays of wavelength λ, impinging at an angle θ on a set of crystal planes of spacing d. The beam reflected at the angle θ can be real only if the rays from each successive plane reinforce each other. For this to be the case, the extra distance a ray, scattered from each successive plane, has to travel, i.e. the path difference, must be equal to an integral number of wavelengths, $n\lambda$. For example, the second ray shown in Figure 4.7a has to travel further than the first ray by the distance $PO + OQ$. The condition for reflection and reinforcement is then given by:

$$n\lambda = PO + OQ = 2ON \sin\theta = 2d \sin\theta. \tag{4.7}$$

This is the well-known Bragg law and the critical angular values of θ for which the law is satisfied are known as Bragg angles.

The directions of the reflected beams are determined entirely by the geometry of the lattice, which in turn is governed by the orientation and spacing of the crystal planes. If, for a crystal of cubic

[3] This absorption equation is the basis of radiography, since a cavity, crack or similar defect will have a much lower μ-value than the sound metal. Such defects can be detected by the appearance of an intensity difference registered on a photographic film placed behind the X-irradiated object.

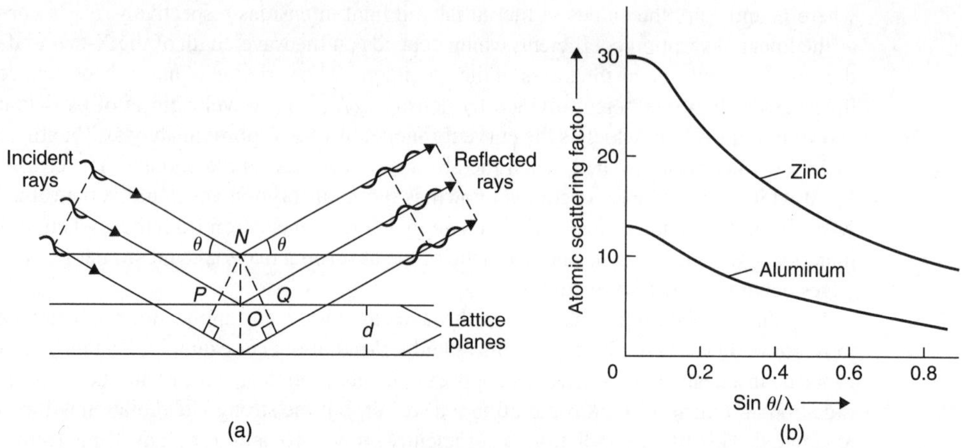

Figure 4.7 *(a) Diffraction from crystal planes. (b) Form of the atomic scattering curves for aluminum and zinc.*

symmetry, we are given the size of the structure cell, a, the angles at which the beam is diffracted from the crystal planes $(h\ k\ l)$ can easily be calculated from the interplanar spacing relationship:

$$d_{(h\ k\ l)} = a/\sqrt{(h^2 + k^2 + l^2)}. \tag{4.8}$$

It is conventional to incorporate the order of reflection, n, with the Miller index, and when this is done the Bragg law becomes:

$$\begin{aligned} \lambda &= 2a \sin\theta/\sqrt{(n^2h^2 + n^2k^2 + n^2l^2)} \\ &= 2a \sin\theta/\sqrt{N}, \end{aligned} \tag{4.9}$$

where N is known as the reflection or line number. To illustrate this let us take as an example the second-order reflection from (1 0 0) planes. Then, since $n=2$, $h=1$, $k=0$ and $l=0$ this reflection is referred to either as the 2 0 0 reflection or as line 4. The lattice planes which give rise to a reflection at the smallest Bragg angle are those which are most widely spaced, i.e. those with a spacing equal to the cell edge, $d_{1\ 0\ 0}$. The next planes in order of decreased spacing will be $\{1\ 1\ 0\}$ planes, for which $d_{1\ 1\ 0} = a/\sqrt{2}$, while the octahedral $\{1\ 1\ 1\}$ planes will have a spacing equal to $a/\sqrt{3}$. The angle at which any of these planes in a crystal reflect an X-ray beam of wavelength λ may be calculated by inserting the appropriate value of d into the Bragg equation.

To ensure that Bragg's law is satisfied and that reflections from various crystal planes can occur, it is necessary to provide a range of either θ or λ values. The various ways in which this can be done lead to the standard methods of X-ray diffraction, namely: (1) the Laue method and (2) the powder method.

4.3.3 X-ray diffraction methods

4.3.3.1 Laue method

In the Laue method, a stationary single crystal is bathed in a beam of 'white' radiation. Then, because the specimen is a fixed single crystal, the variable necessary to ensure that the Bragg law is satisfied

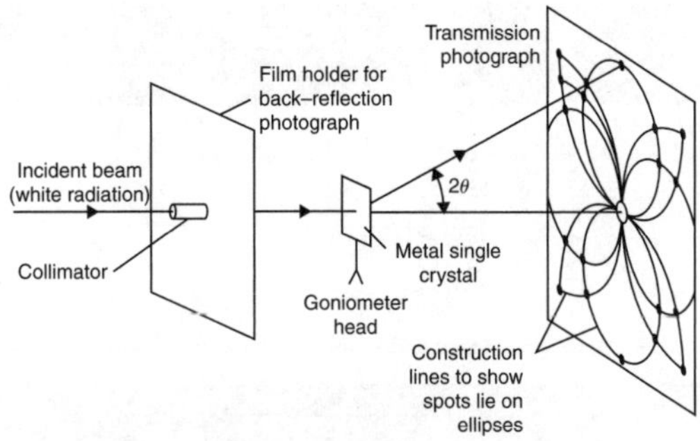

Figure 4.8 *Laue method of X-ray diffraction.*

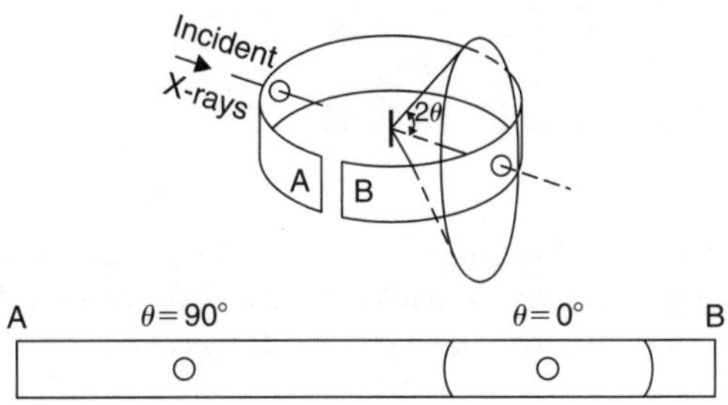

Figure 4.9 *Powder method of X-ray diffraction.*

for all the planes in the crystal has to be provided by the range of wavelengths in the beam, i.e. each set of crystal planes chooses the appropriate λ from the 'white' spectrum to give a Bragg reflection. Radiation from a target metal having a high atomic number (e.g. tungsten) is often used, but almost any form of 'white' radiation is suitable. In the experimental arrangement shown in Figure 4.8, either a transmission photograph or a back-reflection photograph may be taken, and the pattern of spots which are produced lie on ellipses in the transmission case or hyperbolae in the back-reflection case. All spots on any ellipse or hyperbola are reflections from planes of a single zone (i.e. where all the lattice planes are parallel to a common direction, the zone axis) and, consequently, the Laue pattern is able to indicate the symmetry of the crystal. For example, if the beam is directed along a [1 1 1] or [1 0 0] direction in the crystal, the Laue pattern will show three- or fourfold symmetry, respectively. The Laue method is used extensively for the determination of the orientation of single crystals and, while charts are available to facilitate this determination, the method consists essentially of plotting the zones taken from the film on to a stereogram, and comparing the angles between them with a standard projection of that crystal structure. In recent years the use of the Laue technique has been extended to the study of imperfections resulting from crystal growth or deformation, because it is found that the Laue spots from perfect crystals are sharp, while those from deformed crystals are, as shown in Figure 6.49(c), elongated. This elongated appearance of the diffraction spots is known as asterism and it arises in an analogous way to the reflection of light from curved mirrors.

4.3.3.2 Powder method

The powder method, devised independently by Debye and Scherrer, is probably the most generally useful of all the X-ray techniques. It employs monochromatic radiation and a finely powdered, or fine-grained polycrystalline, wire specimen. In this case, θ is the variable, since the collection of randomly

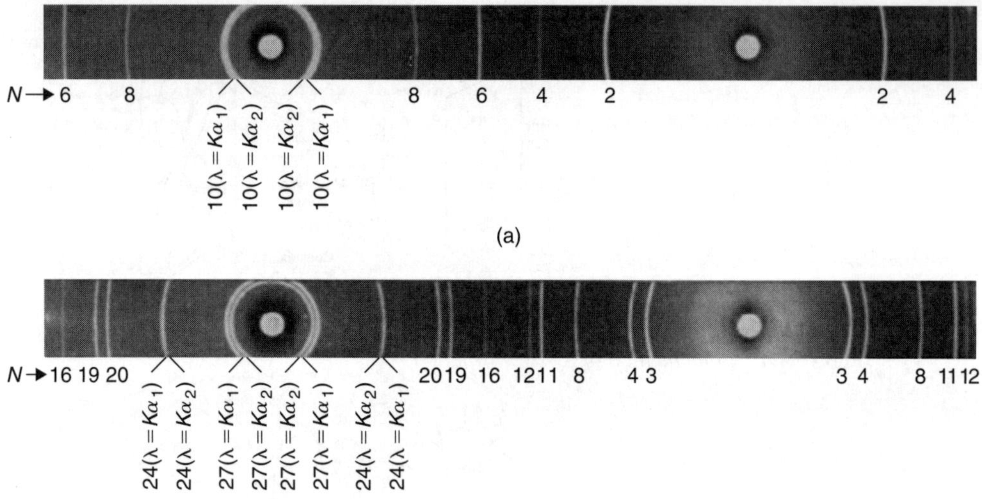

Figure 4.10 *Powder photographs taken in a Philips camera (114 mm radius) of iron with cobalt radiation using an iron filter (a) and aluminum with copper radiation using a nickel filter (b). The high-angle lines are resolved and the separate reflections for $\lambda = K\alpha_1$ and $\lambda = K\alpha_2$ are observable.*

oriented crystals will contain sufficient particles with the correct orientation to allow reflection from each of the possible reflecting planes, i.e. the powder pattern results from a series of superimposed rotating crystal patterns. The angle between the direct X-ray beam and the reflected ray is 2θ, and consequently each set of crystal planes gives rise to a cone of reflected rays of semi-angle 2θ, where θ is the Bragg angle for that particular set of reflecting planes producing the cone. Thus, if a film is placed around the specimen, as shown in Figure 4.9, the successive diffracted cones, which consist of rays from hundreds of grains, intersect the film to produce concentric curves around the entrance and exit holes. Figure 4.10 shows examples of patterns from bcc and fcc materials, respectively.

Precise measurement of the pattern of diffraction lines is required for many applications of the powder method, but a good deal of information can readily be obtained merely by inspection. One example of this is in the study of deformed metals, since after deformation the individual spots on the diffraction rings are blurred so much that line broadening occurs, especially at high Bragg angles. On low-temperature annealing, the cold-worked material will tend to recover and this is indicated on the photograph by a sharpening of the broad diffraction lines. At higher annealing temperatures the metal will completely regain its softness by a process known as recrystallization (see Chapter 6) and this phenomenon is accompanied by the completion of the line-sharpening process. With continued annealing, the grains absorb each other to produce a structure with an overall coarser grain size and, because fewer reflections are available to contribute to the diffraction cones, the lines on the powder photograph take on a spotty appearance. This latter behavior is sometimes used as a means of determining the grain size of a polycrystalline sample. In practice, an X-ray photograph is taken for each of a series of known grain sizes to form a set of standards, and with them an unknown grain size can be determined quite quickly by comparing the corresponding photograph with the set of standards. Yet a third use of the powder method as an inspection technique is in the detection of a preferred orientation of the grains of a polycrystalline aggregate. This is because a random orientation of the grains will produce a uniformly intense diffraction ring, while a preferred orientation, or texture, will concentrate the intensity at certain positions on the ring. The details of the texture require considerable interpretation and are discussed in Chapter 6.

Worked example

A Debye–Scherrer pattern of an unknown powder showed diffraction peaks using copper $K\alpha$ X-rays ($\lambda = 1.54$ Å) at the following Bragg angles: 19.25°, 22.45°, 32.7°, 39.3° and 41.4°. (i) Confirm that the powder is cubic in structure and (ii) determine its lattice parameter. Suggest what the powder might be.

Solution

Bragg's law: $\lambda = \frac{2a}{\sqrt{N}} \sin\theta$, where $N = \sqrt{h^2 + k^2 + l^2}$.

$$\therefore \quad \sin^2\theta = \left(\frac{\lambda}{2a}\right)^2 N;$$

θ (°)	$\sin\theta$	$\sin^2\theta$	$\frac{\sin^2\theta}{0.1087}$	Ratio × 3
19.25	0.3297	0.1087	1	3
22.45	0.3819	0.1458	1.3416	4.02
32.7	0.5402	0.2919	2.6851	8.06
39.3	0.6334	0.4012	3.6908	11.07
41.4	0.6613	0.4373	4.0235	12.07

(i) Lines correspond to $N = 3, 4, 8, 11, 12$, which corresponds to an fcc pattern (1 1 1, 2 0 0, 2 2 0, 1 1 3, 2 2 2).
(ii) Average value of ($\sin^2\theta/N$), where $N = 3, 4, 8, 11, 12$, is 0.03642.

$$\text{So } \frac{\sin^2\theta}{N} = \left(\frac{\lambda}{2a}\right)^2 = 0.03642,$$

$$a = \frac{\lambda}{2 \times \sqrt{0.03642}} = \frac{1.54}{2 \times \sqrt{0.03642}} = 4.03 \text{ Å}.$$

This is close to the lattice parameter of aluminum (4.05 Å), so the powder could be Al.

4.3.3.3 X-ray diffractometry

In addition to photographic recording, the diffracted X-ray beam may be detected directly using a counter tube (either Geiger, proportional or scintillation type) with associated electrical circuitry. The geometrical arrangement of such an X-ray diffractometer is shown in Figure 4.11a. A divergent beam of filtered or monochromatized radiation impinges on the flat face of a powder specimen. This specimen is rotated at precisely one-half of the angular speed of the receiving slit so that a constant angle between the incident and reflected beams is maintained. The receiving slit is mounted in front of the counter on the counter tube arm, and behind it is usually fixed a scatter slit to ensure that the counter receives radiation only from the portion of the specimen illuminated by the primary beam. The intensity diffracted at the various angles is recorded automatically on a chart of the form shown in Figure 4.11c, and this can quickly be analyzed for the appropriate θ and d values.

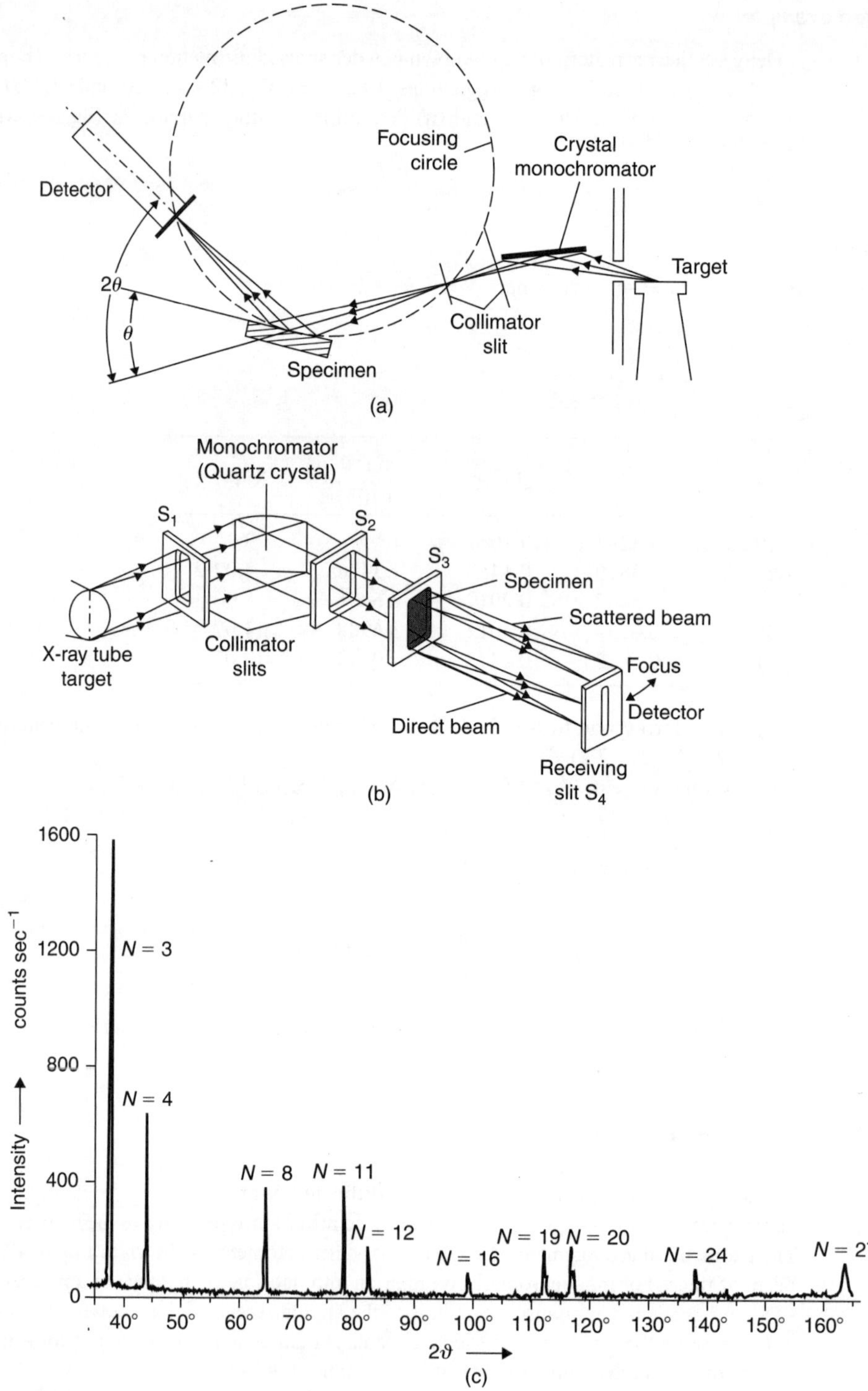

Figure 4.11 *Geometry of conventional diffractometer (a) and small-angle scattering diffractometer (b). (c) Chart record of diffraction pattern from aluminum powder with copper radiation using nickel filter.*

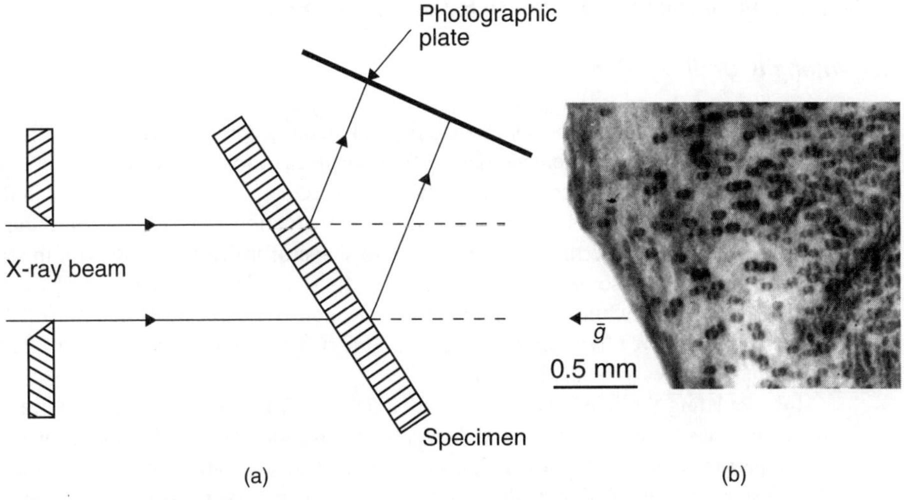

Figure 4.12 *(a) Geometry of X-ray topographic technique. (b) Topograph from a magnesium single crystal showing dislocation loops, g* = $0\,1\,\bar{1}\,0$ *(after Vale and Smallman, 1977).*

The technique is widely used in routine chemical analysis, since accurate intensity measurements allow a quantitative estimate of the various elements in the sample to be made. In research, the technique has been applied to problems such as the degree of order in alloys, the density of stacking faults in deformed alloys, elastic constant determination, the study of imperfections and preferred orientation.

4.3.3.4 X-ray topography

With X-rays it is possible to study individual crystal defects by detecting the differences in intensity diffracted by regions of the crystal near dislocations, for example, and more nearly perfect regions of the crystal. Figure 4.12a shows the experimental arrangement schematically in which collimated monochromatic $K\alpha$-radiation and photographic recording is used.

Any imperfections give rise to local changes in diffracted or transmitted X-ray intensities and, consequently, dislocations show up as bands of contrast, some 5–50 μm wide. No magnification is used in recording the diffraction image, but subsequent magnification of up to 500 times may be achieved with high-resolution X-ray emulsions. Large areas of the crystal to thicknesses of 10–100 μm can be mapped using scanning techniques provided the dislocation density is not too high ($\not> 10^{10}\,m^{-2}$).

The X-ray method of detecting lattice defects suffers from the general limitations that the resolution is low and exposure times are long (12 h), although very high intensity X-ray sources are now available from synchrotrons and are being used increasingly with very short exposure times (the order of minutes). By comparison, the thin-film electron microscopy method (see Section 4.4.2) is capable of revealing dislocations with a much higher resolution because the dislocation image width is 10 nm or less and magnifications up to 100 000 times are possible. The X-ray method does, however, have the great advantage of being able to reveal dislocations in crystals which are comparatively thick (~1 mm; cf. 0.1 μm in foils suitable for transmission electron microscopy). The technique has been used for studying in detail the nature of dislocations in thick single crystals with very low dislocation densities, such as found in semiconducting materials; Figure 4.12b shows an example of an X-ray topograph revealing dislocations in magnesium by this technique.

4.3.4 Typical interpretative procedures for diffraction patterns

4.3.4.1 Intensity of diffraction

Many applications of the powder method depend on the accurate measurement of either line position or line intensity. The arrangement of the diffraction lines in any pattern is characteristic of the material being examined and, consequently, an important practical use of the method is in the identification of unknown phases. Thus, it will be evident that equation (4.9) can indicate the position of the reflected beams, as determined by the size and shape of the unit cell, but not the intensities of the reflected beams. These are determined not by the size of the unit cell, but by the distribution of atoms within it, and while cubic lattices give reflections for every possible value of $(h^2+k^2+l^2)$, all other structures give characteristic absences. Studying the indices of the 'absent' reflections enables different structures to be distinguished.

In calculating the intensity scattered by a given atomic structure, we have first to consider the intensity scattered by one atom, and then go on to consider the contribution from all the other atoms in the particular arrangement which make up that structure. The efficiency of an atom in scattering X-rays is usually denoted by f, the atomic scattering factor, which is the ratio of amplitude scattered by an atom A_a to that by a single electron A_e. If atoms were merely points, their scattering factors would be equal to the number of electrons they contain, i.e. to their atomic numbers, and the relation $I_a = Z^2 I_e$ would hold since intensity is proportional to the square of amplitude. However, because the size of the atom is comparable to the wavelength of X-rays, scattering from different parts of the atom is not in phase, and the result is that $I_a \leq Z^2 I_e$. The scattering factor therefore depends both on the angle θ and on the wavelength of X-rays used, as shown in Figure 4.7b, because the path difference for the individual waves scattered from the various electrons in the atom is zero when $\theta = 0$ and increases with increasing θ. Thus, to consider the intensity scattered by a given structure, it is necessary to sum up the waves which come from all the atoms of one unit cell of that structure, since each wave has a different amplitude and a different phase angle due to the fact that it comes from a different part of the structure. The square of the amplitude of the resultant wave, F, then gives the intensity, and this may be calculated by using the f-values and the atomic coordinates of each atom in the unit cell. It can be shown that a general formula for the intensity is

$$\begin{aligned} I \propto |F|^2 &= [f_1 \cos 2\pi(hx_1 + ky_1 + lz_1) \\ &\quad + f_2 \cos 2\pi(hx_2 + ky_2 + lz_2) + \cdots]^2 \\ &\quad + [f_1 \sin 2\pi(hx_1 + ky_1 + lz_1) \\ &\quad + f_2 \sin 2\pi(hx_2 + ky_2 + lz_2) + \cdots]^2, \end{aligned} \tag{4.10}$$

where $x_1, y_1, z_1; x_2, y_2, z_2$, etc., are the coordinates of those atoms having scattering factors f_1, f_2, etc., respectively, and hkl are the indices of the reflection being computed. For structures having a center of symmetry, which includes most metals, the expression is much simpler because the sine terms vanish.

This equation may be applied to any structure, but to illustrate its use let us examine a pure metal crystallizing in the bcc structure. From Figure 1.15c it is clear that the structure has identical atoms (i.e. $f_1 = f_2$) at the coordinates (0 0 0) and $\left\{\frac{1}{2}\ \frac{1}{2}\ \frac{1}{2}\right\}$, so that equation (4.10) becomes:

$$\begin{aligned} I &\propto f^2[\cos 2\pi \cdot 0 + \cos 2\pi(h/2 + k/2 + l/2)]^2 \\ &= f^2[1 + \cos \pi(h + k + l)]^2. \end{aligned} \tag{4.11}$$

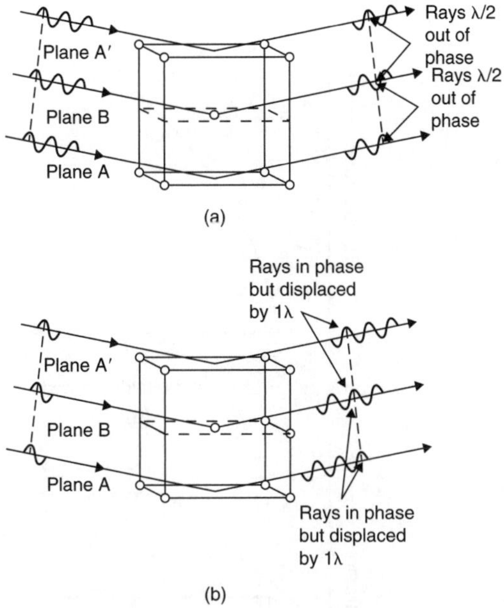

Figure 4.13 *(a) 1 0 0 reflection from bcc cell showing interference of diffracted rays. (b) 2 0 0 reflection showing reinforcement (after Barrett and Massalski, 1980).*

It then follows that I is equal to zero for every reflection where $(h+k+l)$ is an odd number. The significance of this is made clear if we consider in a qualitative way the 1 0 0 reflection shown in Figure 4.13a. To describe a reflection as the first-order reflection from (1 0 0) planes implies that there is 1λ phase difference between the rays reflected from planes A and those reflected from planes A′. However, the reflection from the plane B situated halfway between A and A′ will be $\lambda/2$ out of phase with that from plane A, so that complete cancellation of the 1 0 0 reflected ray will occur. The 1 0 0 reflection is therefore absent, which agrees with the prediction made from equation (4.11) that the reflection is missing when $(h+k+l)$ is an odd number. A similar analysis shows that the 2 0 0 reflection will be present (Figure 4.13b), since the ray from the B plane is now exactly 1λ out of phase with the rays from A and A′. In consequence, if a diffraction pattern is taken from a material having a bcc structure, because of the rule governing the sum of the indices, the film will show diffraction lines almost equally spaced with indices $N=2$, (1 1 0); 4, (2 0 0); 6, (2 1 1); 8, (2 2 0); ..., as shown in Figure 4.10a. Application of equation (4.10) to a pure metal with fcc structure shows that 'absent' reflections will occur when the indices of that reflection are mixed, i.e. when they are neither all odd nor all even. Thus, the corresponding diffraction pattern will contain lines according to $N=3$, 4, 8, 11, 12, 16, 19, 20, etc., and the characteristic feature of the arrangement is a sequence of two lines close together and one line separated, as shown in Figure 4.10b.

Equation (4.10) is the basic equation used for determining unknown structures, since the determination of the atomic positions in a crystal is based on this relation between the coordinates of an atom in a unit cell and the intensity with which it will scatter X-rays.

4.3.4.2 Determination of lattice parameters

Perhaps the most common use of the powder method is in the accurate determination of lattice parameters. From the Bragg law we have the relation $a=\lambda\sqrt{N}/2\sin\theta$, which, because both λ and N

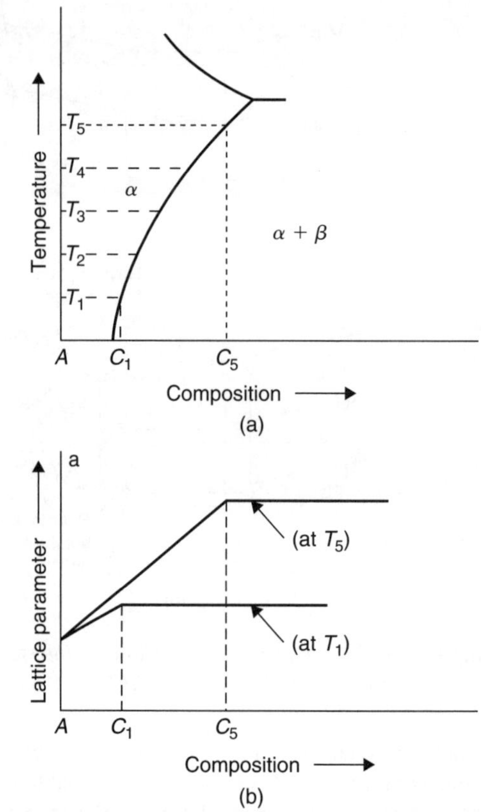

Figure 4.14 *(a, b) Phase-boundary determination using lattice parameter measurements.*

are known and θ can be measured for the appropriate reflection, can be used to determine the lattice parameter of a material. Several errors are inherent in the method, however, and the most common include shrinkage of the film during processing, eccentricity of the specimen and the camera, and absorption of the X-rays in the sample. These errors affect the high-angle diffraction lines least and, consequently, the most accurate parameter value is given by determining a value of a from each diffraction line, plotting it on a graph against an angular function[4] of the $\cos^2\theta$ type and then extrapolating the curve to $\theta = 90°$.

The determination of precision lattice parameters is of importance in many fields of materials science, particularly in the study of thermal expansion coefficients, density determinations, the variation of properties with composition, precipitation from solid solution, and thermal stresses. At this stage it is instructive to consider the application of lattice parameter measurements to the determination of phase boundaries in equilibrium diagrams, since this illustrates the general usefulness of the technique. The diagrams shown in Figure 4.14a and b indicate the principle of the method. A variation of alloy composition within the single-phase field, α, produces a variation in the lattice parameter, a,

[4] Nelson and Riley suggest the function:

$$\left(\frac{\cos^2\theta}{\sin\theta} + \frac{\cos^2\theta}{\theta}\right).$$

since solute B, which has a different atomic size to the solvent A, is being taken into solution. However, at the phase boundary solvus this variation in a ceases, because at a given temperature the composition of the α-phase remains constant in the two-phase field, and the marked discontinuity in the plot of lattice parameter versus composition indicates the position of the phase boundary at that temperature. The change in solid solubility with temperature may then be obtained, either by taking diffraction photographs in a high-temperature camera at various temperatures or by quenching the powder sample from the high temperature to room temperature (in order to retain the high temperature state of solid solution down to room temperature) and then taking a powder photograph at room temperature.

4.3.4.3 Line broadening

Diffraction lines are not always sharp because of various instrumental factors such as slit size, specimen condition and spread of wavelengths, but in addition the lines may be broadened as a result of lattice strain in the region of the crystal diffracting and also its limited dimension. Strain gives rise to a variation of the interplanar spacing Δd and hence diffraction occurs over a range $\Delta\theta$ and the breadth due to strain is then

$$\beta_s = \eta \tan\theta, \tag{4.12}$$

where η is the strain distribution. If the dimension of the crystal diffracting the X-rays is small,[5] then this also gives rise to an appreciable 'particle-size' broadening given by the Scherrer formula:

$$\beta_p = \lambda / t \cos\theta, \tag{4.13}$$

where t is the effective particle size. In practice, this size is the region over which there is coherent diffraction and is usually defined by boundaries such as dislocation walls. It is possible to separate the two effects by plotting the experimentally measured broadening $\beta \cos\theta/\lambda$ against $\sin\theta/\lambda$, when the intercept gives a measure of t and the slope η.

4.3.4.4 Small-angle scattering

The scattering of intensity into the low-angle region ($\varepsilon = 2\theta < 10°$) arises from the presence of inhomogeneities within the material being examined (such as small clusters of solute atoms), where these inhomogeneities have dimensions only 10–100 times the wavelength of the incident radiation. The origin of the scattering can be attributed to the differences in electron density between the heterogeneous regions and the surrounding matrix,[6] so that precipitated particles afford the most common source of scattering; other heterogeneities such as dislocations, vacancies and cavities must also give rise to some small-angle scattering, but the intensity of the scattered beam will be much weaker than that from precipitated particles. The experimental arrangement suitable for this type of study is shown in Figure 4.11b.

---

[5] The optical analog of this effect is the broadening of diffraction lines from a grating with a limited number of lines.

[6] The halo around the moon seen on a clear frosty night is the best example, obtained without special apparatus, of the scattering of light at small angles by small particles.

Interpretation of much of the small-angle scatter data is based on the approximate formula derived by Guinier,

$$I = Mn^2 I_e \exp[-4\pi^2\varepsilon^2 R^2/3\lambda^2], \tag{4.14}$$

where M is the number of scattering aggregates, or particles, in the sample, n represents the difference in number of electrons between the particle and an equal volume of the surrounding matrix, R is the radius of gyration of the particle, I_e is the intensity scattered by an electron, ε is the angle of scattering and λ is the wavelength of X-rays. From this equation it can be seen that the intensity of small-angle scattering is zero if the inhomogeneity, or cluster, has an electron density equivalent to that of the surrounding matrix, even if it has quite different crystal structure. On a plot of $\log_{10} I$ as a function of ε^2, the slope near the origin, $\varepsilon = 0$, is given by

$$P = -(4\pi^2/3\lambda^2)R^2 \log_{10} e,$$

which for Cu $K\alpha$ radiation gives the radius of gyration of the scattering aggregate to be

$$R = 0.0645 \times P^{1/2}\text{nm}. \tag{4.15}$$

It is clear that the technique is ideal for studying regions of the structure where segregation on too fine a scale to be observable in the light microscope has occurred, e.g. the early stages of phase precipitation (see Chapter 7), and the aggregation of lattice defects (see Chapter 3).

4.3.4.5 The reciprocal lattice concept

The Bragg law shows that the conditions for diffraction depend on the geometry of sets of crystal planes. To simplify the more complex diffraction problems, use is made of the reciprocal lattice concept in which the sets of lattice planes are replaced by a set of points, this being geometrically simpler.

The reciprocal lattice is constructed from the real lattice by drawing a line from the origin normal to the lattice plane $h\,k\,l$ under consideration of length, d^*, equal to the reciprocal of the interplanar spacing d_{hkl}. The construction of part of the reciprocal lattice from a face-centered cubic crystal lattice is shown in Figure 4.15.

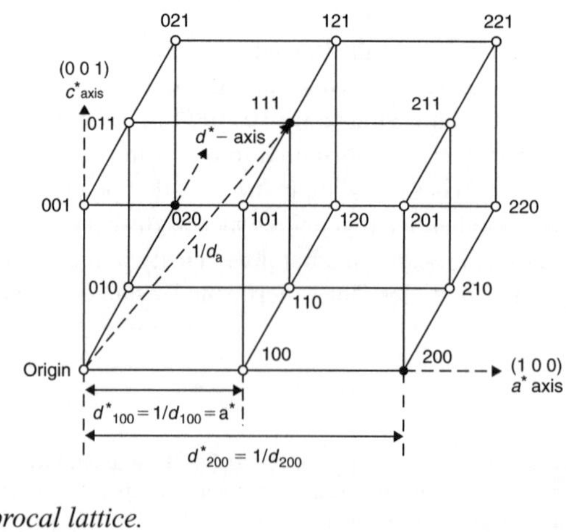

Figure 4.15 *Fcc reciprocal lattice.*

Included in the reciprocal lattice are the points which correspond not only to the true lattice planes with Miller indices ($h\ k\ l$), but also to the fictitious planes ($nh,\ nk,\ nl$) which give possible X-ray reflections. The reciprocal lattice therefore corresponds to the diffraction spectrum possible from a particular crystal lattice and, since a particular lattice type is characterized by 'absent' reflections the corresponding spots in the reciprocal lattice will also be missing. It can be deduced that an fcc Bravais lattice is equivalent to a bcc reciprocal lattice, and vice versa.

A simple geometrical construction using the reciprocal lattice gives the conditions that correspond to Bragg reflection. Thus, if a beam of wavelength λ is incident on the origin of the reciprocal lattice, then a sphere of radius $1/\lambda$ drawn through the origin will intersect those points which correspond to the reflecting planes of a stationary crystal. This can be seen from Figure 4.16, in which the reflecting plane AB has a reciprocal point at d^*. If d^* lies on the surface of the sphere of radius $1/\lambda$ then:

$$d^* = 1/d_{hkl} = 2\sin\theta/\lambda \tag{4.16}$$

and the Bragg law is satisfied; the line joining the origin to the operating reciprocal lattice spot is usually referred to as the g-vector. It will be evident that at any one setting of the crystal, few, if any, points will touch the sphere of reflection. This is the condition for a stationary single crystal and a monochromatic beam of X-rays, when the Bragg law is not obeyed except by chance. To ensure that the Bragg law is satisfied the crystal has to be rotated in the beam, since this corresponds to a rotation of the reciprocal lattice about the origin when each point must pass through the reflection surface. The corresponding reflecting plane reflects twice per revolution.

To illustrate this feature let us re-examine the powder method. In the powder specimen, the number of crystals is sufficiently large that all possible orientations are present and, in terms of the reciprocal lattice construction, we may suppose that the reciprocal lattice is rotated about the origin in all possible directions. The locus of any one lattice point during such a rotation is, of course, a sphere. This locus-sphere will intersect the sphere of reflection in a small circle about the axis of the incident beam, as shown in Figure 4.17, and any line joining the center of the reflection sphere to a point on this small circle is a possible direction for a diffraction maximum. This small circle corresponds to the powder halo discussed previously. From Figure 4.17 it can be seen that the radius of the sphere describing the locus of the reciprocal lattice point ($h\ k\ l$) is $1/d_{(hkl)}$ and that the angle of deviation of the diffracted beam 2θ is given by the relation:

$$(2/\lambda)\sin\theta = 1/d_{(hkl)},$$

which is the Bragg condition.

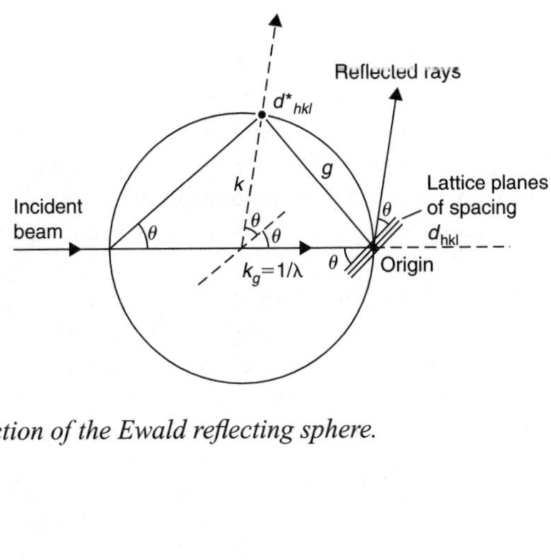

Figure 4.16 *Construction of the Ewald reflecting sphere.*

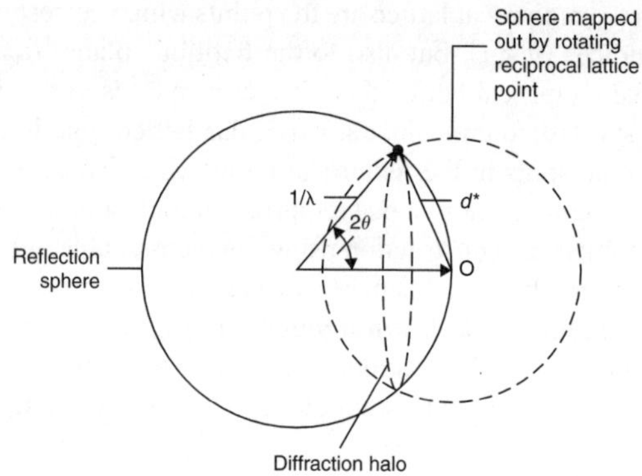

Figure 4.17 *Principle of the power method.*

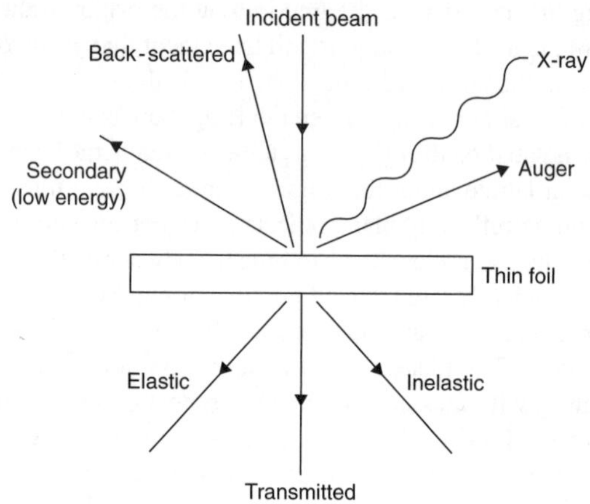

Figure 4.18 *Scattering of incident electrons by thin foil. With a bulk specimen the transmitted, elastic and inelastic scattered beams are absorbed.*

4.4 Analytical electron microscopy

4.4.1 Interaction of an electron beam with a solid

When an electron beam is incident on a solid specimen a number of interactions take place which generate useful structural information. Figure 4.18 illustrates these interactions schematically. Some of the incident beam is back-scattered and some penetrates the sample. If the specimen is thin enough a significant amount is transmitted, with some electrons elastically scattered without loss of energy and some inelastically scattered. Interaction with the atoms in the specimen leads to the ejection of low-energy electrons and the creation of X-ray photons and Auger electrons, all of which can be used to characterize the material.

The two inelastic scattering mechanisms important in chemical analysis are (1) excitation of the electron gas plasmon scattering and (2) single-electron scattering. In *plasmon scattering* the fast electron excites a ripple in the plasma of free electrons in the solid. The energy of this 'plasmon' depends only on the volume concentration of free electrons n in the solid and is given by $E_p = [ne^2/m]^{1/2}$. Typically, E_p, the energy loss suffered by the fast electron, is ≈15 eV and the scattering intensity/unit solid angle has an angular half-width given by $\theta_E = E_p/2E_0$, where E_0 is the incident voltage; θ_E is therefore $\approx 10^{-4}$ radian. The energy of the plasmon is converted very quickly into atom vibrations (heat) and the mean free path for plasmon excitation is small, $\approx$50–150 nm. With *single-electron scattering* energy may be transferred to single electrons (rather than to the large number $\approx 10^5$ involved in plasmon excitation) by the incident fast electrons. Lightly bound valency electrons may be ejected, and these electrons can be used to form secondary images in SEM; a very large number of electrons with energies up to $\approx$50 eV are ejected when a high-energy electron beam strikes a solid. The useful collisions are those where the single electron is bound. There is a minimum energy required to remove the single electron, i.e. ionization, but provided the fast electron gives the bound electron more than this minimum amount, it can give the bound electron any amount of energy, up to its own energy (e.g. 100 keV). Thus, instead of the single-electron excitation process turning up in the energy loss spectrum of the fast electron as a peak, as happens with plasmon excitation, it turns up as an edge. Typically, the mean free path for inner shell ionization is several micrometers and the energy loss can be several keV. The angular half-width of scattering is given by $\Delta E/2E_0$. Since the energy loss ΔE can vary from $\approx$10 eV to tens of keV, the angle can vary upwards from 10^{-4} radian (see Figure 4.34).

A plasmon, once excited, decays to give heat, which is not at all useful. In contrast, an atom which has had an electron removed from it decays in one of two ways, both of which turn out to be very useful in chemical analysis, leading to the creation of X-rays and Auger electrons. The first step is the same for both cases. An electron from an outer shell, which therefore has more energy than the removed electron, drops down to fill the hole left by the removal of the bound electron. Its extra energy, ΔE, equal to the difference in energy between the two levels involved and therefore absolutely characteristic of the atom, must be dissipated. This may happen in two ways. (1) By the creation of a photon whose energy, $\mathbf{h}\nu$, equals the energy difference ΔE. For electron transitions of interest, ΔE, and therefore $\mathbf{h}\nu$, is such that the photon is an X-ray. (2) By transferring the energy to a neighboring electron, which is then ejected from the atom. This is an 'Auger' electron. Its energy when detected will depend on the original energy difference ΔE minus the binding energy of the ejected electron. Thus, the energy of the Auger electron depends on three atomic levels rather than two as for emitted photons. The energies of the Auger electrons are sufficiently low that they escape from within only about 5 nm of the surface. This is therefore a surface analysis technique. The ratio of photon–Auger yield is called the fluorescence ratio ω, and depends on the atom and the shells involved. For the K-shell, ω is given by $\omega_K = X_K/(A_K + X_K)$, where X_K and A_K are, respectively, the number of X-ray photons and Auger electrons emitted. A_K is independent of atomic number Z, and X_K is proportional to Z^4, so that $\omega_K Z^4/(a + Z^4)$, where $a = 1.12 \times 10^6$. Light elements and outer shells (L-lines) have lower yields; for K-series transitions ω_K varies from a few percent for carbon up to $\geq$90% for gold.

4.4.2 The transmission electron microscope (TEM)

Section 4.2.1 shows that to increase the resolving power of a microscope it is necessary to employ shorter wavelengths. For this reason the electron microscope has been developed to allow the observation of structures which have dimensions down to less than 1 nm. An electron microscope consists of an electron gun and an assembly of lenses all enclosed in an evacuated column. A very basic system for a transmission electron microscope is shown schematically in Figure 4.19. The optical arrangement is similar to that of the glass lenses in a projection-type light microscope, although it is customary to use several stages of magnification in the electron microscope. The lenses are usually of the magnetic type, i.e. current-carrying coils which are completely surrounded by a soft iron shroud except for a narrow gap in the bore, energized by d.c. and, unlike the lenses in a light microscope, which have fixed focal lengths, the focal length can be controlled by regulating the current through the coils of the lens. This facility compensates for the fact that it is difficult to move the large magnetic

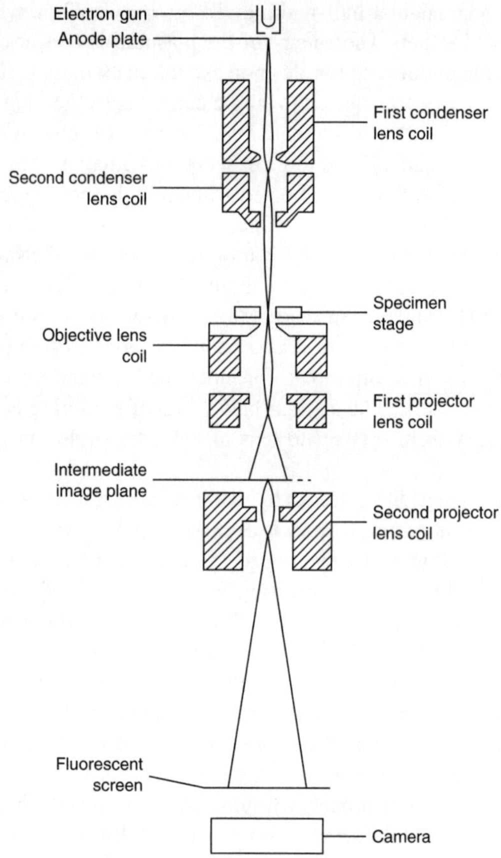

Figure 4.19 *Schematic arrangement of a basic transmission electron microscope system.*

lenses in the evacuated column of the electron microscope in an analogous manner to the glass lenses in a light microscope.

The condenser lenses are concerned with collimating the electron beam and illuminating the specimen which is placed in the bore of the objective lens. The function of the objective lens is to form a magnified image of up to about ×40 in the object plane of the intermediate, or first projector, lens. A small part of this image then forms the object for the first projector lens, which gives a second image, again magnified in the object plane of the second projector lens. The second projector lens is capable of enlarging this image further to form a final image on the fluorescent viewing screen. This image, magnified up to ×100 000, may be recorded on a photographic film beneath the viewing screen. A stream of electrons can be assigned a wavelength λ given by the equation $\lambda = \mathbf{h}/mv$, where $\mathbf{h}$ is Planck's constant and mv is the momentum of the electron. The electron wavelength is inversely proportional to the velocity, and hence to the voltage applied to the electron gun, according to the approximate relation

$$\lambda = \sqrt{(1.5/V)}\ \text{nm} \tag{4.17}$$

and, since normal operating voltages are between 50 and 100 kV, the value of λ used varies from 0.0054 to 0.0035 nm. With a wavelength of 0.005 nm, if one could obtain a value of $(\mu \sin \alpha)$ for electron lenses comparable to that for optical lenses, i.e. 1.4, it would be possible to see the orbital electrons. However, magnetic lenses are more prone to spherical and chromatic aberration than glass lenses and, in consequence, small apertures, which correspond to α-values of about 0.002 radian, must be used. As a result, the resolution of the electron microscope is limited to about 0.2 nm. It will be appreciated, of course, that a variable magnification is possible in the electron microscope without relative movement of the lenses, as in a light microscope, because the depth of focus of each image, being inversely proportional to the square of the numerical aperture, is so great.

Worked example

What is the wavelength of 200 keV electrons?

Solution

Non-relativistically, $\frac{1}{2}mv^2 = 200\,000 \times 1.6 \times 10^{-19}$

$$\therefore v = 2.652 \times 10^{8}\,\text{m s}^{-1}.$$

$$mv = 2.416 \times 10^{-22}\,\text{m kg s}^{-1}.$$

$$\lambda = \frac{\mathbf{h}}{mv} = 2.74\,\text{pm}.$$

4.4.3 The scanning electron microscope

The surface structure of a metal can be studied in the TEM by the use of thin transparent replicas of the surface topography. Three different types of replica are in use: (1) oxide, (2) plastic and (3) carbon replicas. However, since the development of the scanning electron microscope (SEM) it is very much easier to study the surface structure directly.

A diagram of the SEM is shown in Figure 4.20. The electron beam is focused to a spot $\approx$10 nm in diameter and made to scan the surface in a raster. Electrons from the specimen are focused with an electrostatic electrode on to a biased scintillator. The light produced is transmitted via a *Perspex* light pipe to a photomultiplier and the signal generated is used to modulate the brightness of an oscilloscope spot, which traverses a raster in exact synchronism with the electron beam at the specimen surface. The image observed on the oscilloscope screen is similar to the optical image and the specimen is usually tilted towards the collector at a low angle ($<30°$) to the horizontal, for general viewing.

As initially conceived, the SEM used back-scattered electrons (with $E \approx 30$ kV, which is the incident energy) and secondary electrons ($E \approx 100$ eV) which are ejected from the specimen. Since the secondary electrons are of low energy they can be bent round corners and give rise to the topographic contrast. The intensity of back-scattered (BS) electrons is proportional to atomic number, but contrast from these electrons tends to be swamped because, being of higher energy, they are not so easily collected by the normal collector system used in SEMs. If the secondary electrons are to be collected a positive bias of $\approx$200 V is applied to the grid in front of the detector; if only the back-scattered electrons are to be collected the grid is biased negatively to $\approx$200 V.

Perhaps the most significant development in recent years has been the gathering of information relating to chemical composition. As discussed in Section 4.4.1, materials bombarded with high-energy electrons can give rise to the emissions of X-rays characteristic of the material being bombarded. The X-rays emitted when the beam is stopped on a particular region of the specimen may be detected either with a solid-state (Li-drifted silicon) detector which produces a voltage pulse proportional to the energy of the incident photons (energy-dispersive method) or with an X-ray spectrometer to measure the wavelength and intensity (wavelength-dispersive method). The microanalysis of materials is

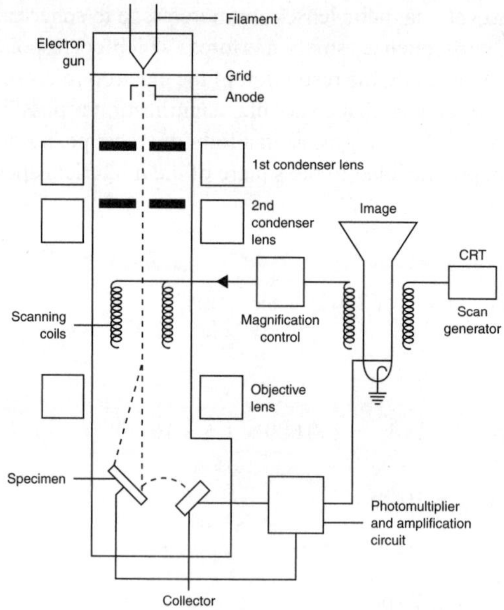

Figure 4.20 *Schematic diagram of a basic scanning electron microscope.*

presented in Section 4.4.5. Alternatively, if the beam is scanned as usual and the intensity of the X-ray emission, characteristic of a particular element, is used to modulate the CRT, an image showing the distribution of that element in the sample will result. X-ray images are usually very 'noisy' because the X-ray production efficiency is low, necessitating exposures 1000 times greater than electron images.

Collection of the back-scattered (BS) electrons with a specially located detector on the bottom of the lens system gives rise to some exciting applications and opens up a completely new dimension for SEM from bulk samples. The BS electrons are very sensitive to atomic number Z and hence are particularly important in showing contrast from changes of composition, as illustrated by the image from a silver alloy in Figure 4.21. This atomic number contrast is particularly effective in studying alloys which normally are difficult to study because they cannot be etched. The intensity of back-scattered electrons is also sensitive to the orientation of the incident beam relative to the crystal. This effect will give rise to 'orientation' contrast from grain to grain in a polycrystalline specimen as the scan crosses several grains. In addition, the effect is also able to provide crystallographic information from bulk specimens by a process known as electron channeling. As the name implies, the electrons are channeled between crystal planes and the amount of channeling per plane depends on its packing and spacing. If the electron beam impinging on a crystal is rocked through a large angle then the amount of channeling will vary with angle and hence the BS image will exhibit contrast in the form of electron channeling patterns which can be used to provide crystallographic information. Figure 4.22 shows the 'orientation' or channeling contrast exhibited by an Fe–3% Si specimen during secondary recrystallization (a process used for transformer lamination production) and the channeling pattern can be analyzed to show that the new grain possesses the Goss texture. Electron channeling occurs only in relatively perfect crystals and hence the degradation of electron channeling patterns may be used to monitor the level of plastic strain, for example to map out the plastic zone around a fatigue crack as it develops in an alloy.

The electron beam may also induce electrical effects which are of importance particularly in semiconductor materials. Thus, a 30 kV electron beam can generate some thousand excess free electrons and the equivalent number of ions ('holes'), the vast majority of which recombine.

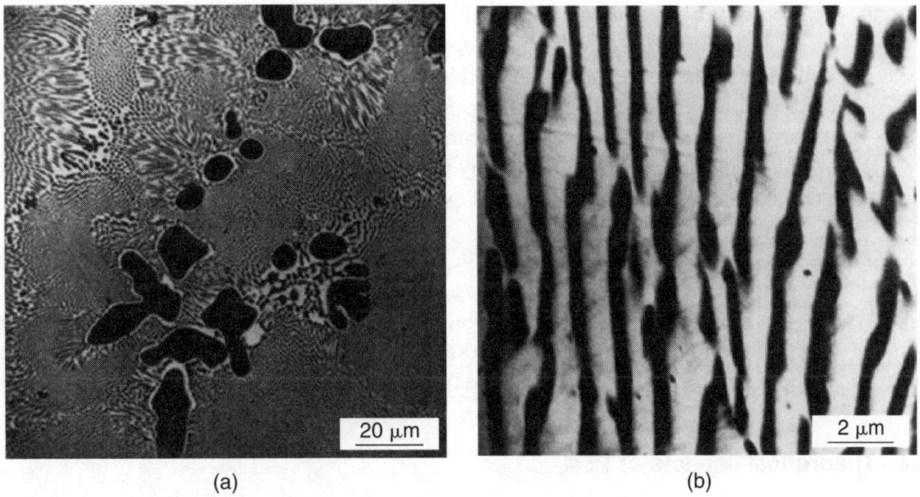

Figure 4.21 *Back-scattered electron image by atomic number contrast from 70Ag–30Cu alloy showing α-dendrites + eutectic (a) and eutectic (b) (courtesy of B. W. Hutchinson).*

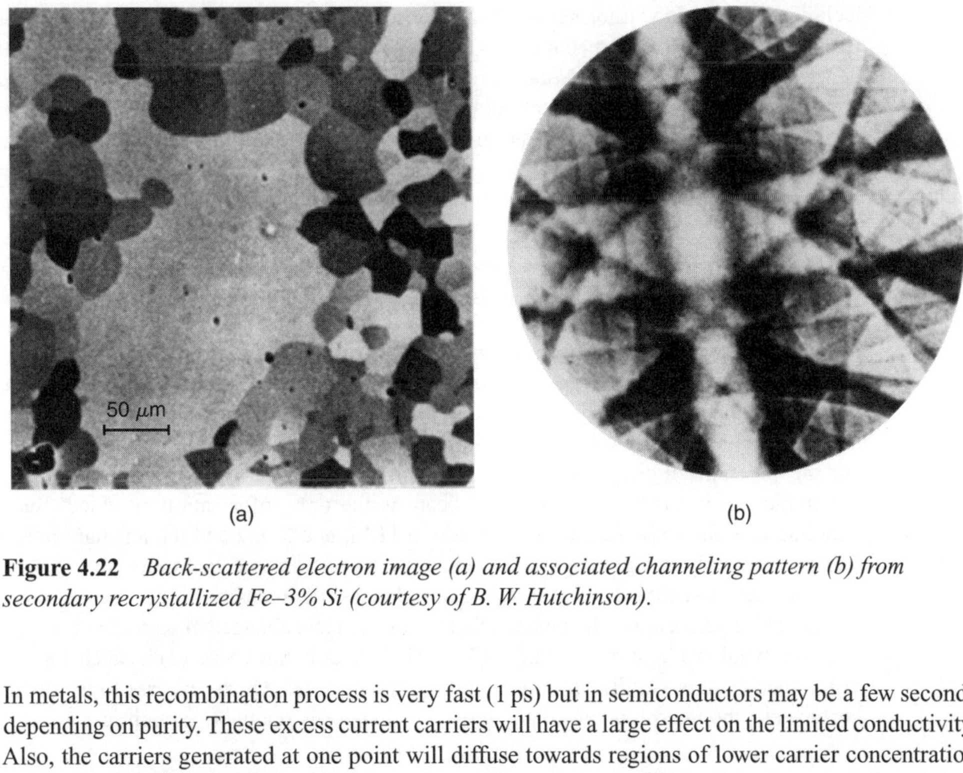

Figure 4.22 *Back-scattered electron image (a) and associated channeling pattern (b) from secondary recrystallized Fe–3% Si (courtesy of B. W. Hutchinson).*

In metals, this recombination process is very fast (1 ps) but in semiconductors may be a few seconds depending on purity. These excess current carriers will have a large effect on the limited conductivity. Also, the carriers generated at one point will diffuse towards regions of lower carrier concentration and voltages will be established whenever the carriers encounter regions of different chemical composition (e.g. impurities around dislocations). The conductivity effect can be monitored by applying a potential difference across the specimen from an external battery and using the magnitude of the resulting current to modulate the CRT brightness to give an image of conductivity variation.

The voltage effect arising from different carrier concentrations or from accumulation of charge on an insulator surface or from the application of an external electromotive force can modify the collection of the emitted electrons and hence give rise to voltage contrast. Similarly, a magnetic field arising from ferromagnetic domains, for example, will affect the collection efficiency of emitted electrons and lead to magnetic field contrast.

The secondary electrons, i.e. lightly bound electrons ejected from the specimen which give topographical information, are generated by the incident electrons, by the back-scattered electrons and by X-rays. The resolution is typically ≈10 nm at 20 kV for medium-atomic-weight elements and is limited by spreading of electrons as they penetrate into the specimen. The back-scattered electrons are also influenced by beam spreading and for a material of medium-atomic-weight the resolution is ≈100 nm. The specimen current mode is limited both by spreading of the beam and the noise of electronic amplification to a spatial resolution of 500 nm and somewhat greater values ≈1 μm apply to the beam-induced conductivity and X-ray modes.

4.4.4 Theoretical aspects of TEM

4.4.4.1 Imaging and diffraction

Although the examination of materials may be carried out with the electron beam impinging on the surface at a 'glancing incidence', most electron microscopes are aligned for the use of a transmission technique, since added information on the interior of the specimen may be obtained. In consequence, the thickness of the metal specimen has to be limited to below a micrometer, because of the restricted penetration power of the electrons. Three methods now in general use for preparing such thin films are (1) chemical thinning, (2) electropolishing and (3) bombarding with a beam of ions at a potential of about 3 kV. Chemical thinning has the disadvantage of preferentially attacking either the matrix or the precipitated phases, and so the electropolishing technique is used extensively to prepare thin metal foils. Ion beam thinning is quite slow but is the only way of preparing thin ceramic and semiconducting specimens.

Transmission electron microscopy provides both image and diffraction information from the same small volume down to 1 μm in diameter. Ray diagrams for the two modes of operation, imaging and diffraction, are shown in Figure 4.23. Diffraction contrast[7] is the most common technique used and, as shown in Figure 4.24a, involves the insertion of an objective aperture in the back focal plane, i.e. in the plane in which the diffraction pattern is formed, to select either the directly transmitted beam or a strong diffracted beam. Images obtained in this way cannot possibly contain information concerning the periodicity of the crystal, since this information is contained in the spacing of diffraction maxima and the directions of diffracted beams, information excluded by the objective aperture.

Variation in intensity of the selected beam is the only information provided. Such a mode of imaging, carried out by selecting one beam in TEM, is unusual and the resultant images cannot be interpreted simply as high-magnification images of periodic objects. In formulating a suitable theory it is necessary to consider what factors can influence the intensity of the directly transmitted beam and the diffracted beams. The obvious factors are (1) local changes in scattering factor, e.g. particles of heavy metal in a light metal matrix, (2) local changes in thickness, (3) local changes in orientation of the specimen, or (4) discontinuities in the crystal planes which give rise to the diffracted beams. Fortunately, the interpretation of any intensity changes is relatively straightforward if it is assumed

[7] Another imaging mode does allow more than one beam to interfere in the image plane and hence crystal periodicity can be observed; the larger the collection angle, which is generally limited by lens aberrations, the smaller the periodicity that can be resolved. Interpretation of this direct imaging mode, while apparently straightforward, is still controversial and will not be covered here.

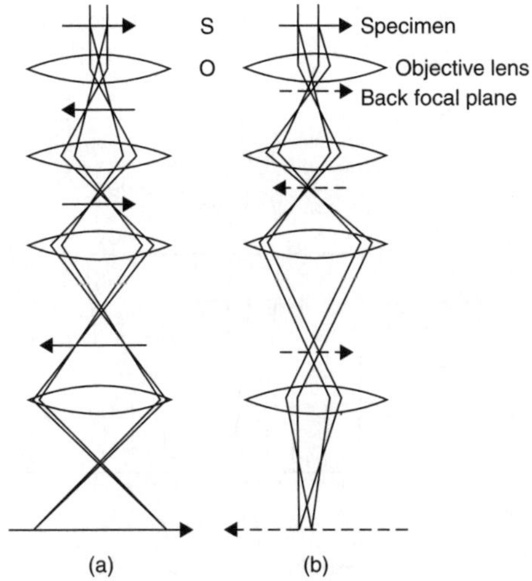

Figure 4.23 *Schematic ray diagrams for imaging (a) and diffraction (b).*

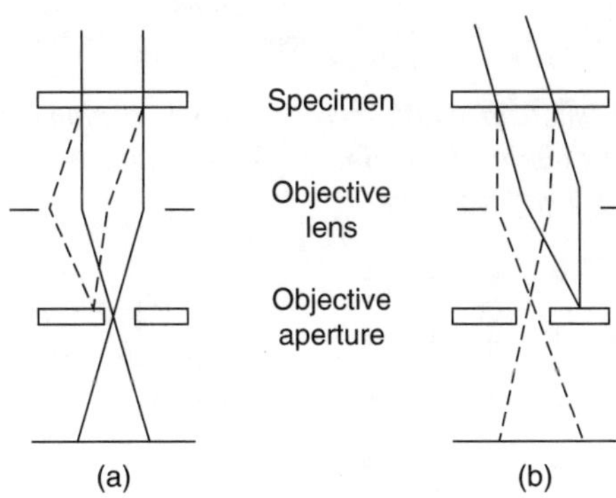

Figure 4.24 *Schematic diagram illustrating bright-field (a) and dark-field (b) image formation.*

that there is only one strong diffracted beam excited. Moreover, since this can be achieved quite easily experimentally, by orienting the crystal such that strong diffraction occurs from only one set of crystal planes, virtually all TEM is carried out with a two-beam condition: a direct and a diffracted beam. When the direct, or transmitted, beam only is allowed to contribute to the final image by inserting a small aperture in the back focal plane to block the strongly diffracted ray, then contrast is shown on a bright background and is known as bright-field imaging. If the diffracted ray only is allowed through the aperture by tilting the incident beam then contrast on a dark background is observed and is known as dark-field imaging. These two arrangements are shown in Figure 4.24.

A dislocation can be seen in the electron microscope because it locally changes the orientation of the crystal, thereby altering the diffracted intensity. This is illustrated in Figure 4.25. Any region of a grain or crystal which is not oriented at the Bragg angle, i.e. $\theta > \theta_B$, is not strongly diffracting electrons. However, in the vicinity of the dislocation the lattice planes are tilted such that, locally, the Bragg law is satisfied and then strong diffraction arises from near the defect. These diffracted rays are blocked by the objective aperture and prevented from contributing to the final image. The dislocation therefore appears as a dark line (where electrons have been removed) on a bright background in the bright-field picture.

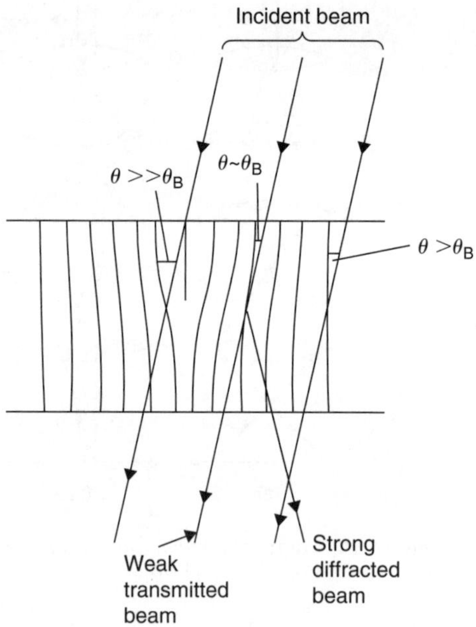

Figure 4.25 *Mechanism of diffraction contrast: the planes to the RHS of the dislocation are bent so that they closely approach the Bragg condition and the intensity of the direct beam emerging from the crystal is therefore reduced.*

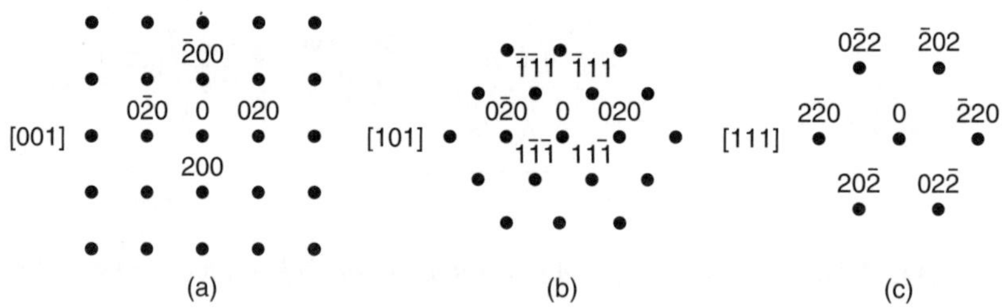

Figure 4.26 *Fcc cross-grating patterns: (a)* [*0 0 1*]*, (b)* [*1 0 1*] *and (c)* [*1 1 1*].

The success of transmission electron microscopy (TEM) is due, to a great extent, to the fact that it is possible to define the diffraction conditions which give rise to the dislocation contrast by obtaining a diffraction pattern from the same small volume of crystal (as small as 1 μm diameter) as that from which the electron micrograph is taken. Thus, it is possible to obtain the crystallographic and associated diffraction information necessary to interpret electron micrographs. To obtain a selected area diffraction pattern (SAD) an aperture is inserted in the plane of the first image so that only that part of the specimen which is imaged within the aperture can contribute to the diffraction pattern. The power of the diffraction lens is then reduced so that the back focal plane of the objective is imaged, and then the diffraction pattern, which is focused in this plane, can be seen after the objective aperture is removed.

The usual type of transmission electron diffraction pattern from a single crystal region is a cross-grating pattern of the form shown in Figure 4.26. The simple explanation of the pattern can be given by considering the reciprocal lattice and reflecting sphere construction commonly used in X-ray diffraction. In electron diffraction, the electron wavelength is extremely short ($\lambda = 0.0037$ nm

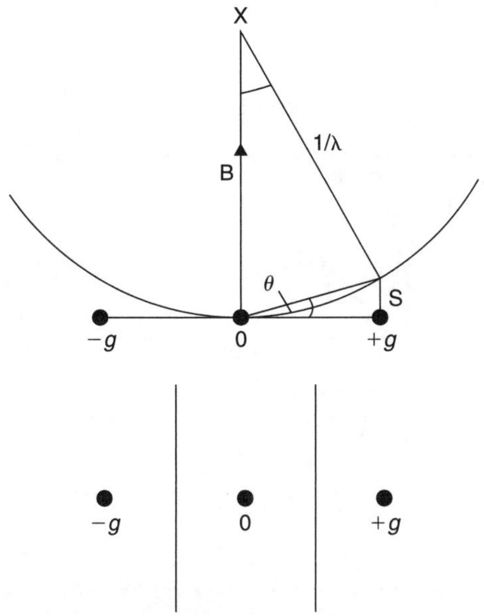

Figure 4.27 *Schematic diagram to illustrate the determination of s at the symmetry position, together with associated diffraction pattern.*

at 100 kV) so that the radius of the Ewald reflecting sphere is about 2.5 nm^{-1}, which is about 50 times greater than g, the reciprocal lattice vector. Moreover, because λ is small the Bragg angles are also small (about 10^{-2} radian or $\frac{1}{2}^\circ$ for low-order reflections) and hence the reflection sphere may be considered as almost planar in this vicinity. If the electron beam is closely parallel to a prominent zone axis of the crystal, then several reciprocal points (somewhat extended because of the limited thickness of the foil) will intersect the reflecting sphere, and a projection of the prominent zone in the reciprocal lattice is obtained, i.e. the SAD pattern is really a photograph of a reciprocal lattice section. Figure 4.26 shows some standard cross-gratings for fcc crystals. Because the Bragg angle for reflection is small ($\approx \frac{1}{2}^\circ$), only those lattice planes which are almost vertical, i.e. almost parallel to the direction of the incident electron beam, are capable of Bragg-diffracting the electrons out of the objective aperture and giving rise to image contrast. Moreover, because the foil is buckled or purposely tilted, only one family of the various sets of approximately vertical lattice planes will diffract strongly and the SAD pattern will then show only the direct beam spot and one strongly diffracted spot (see insert, Figure 4.38). The indices g of the crystal planes $(h\,k\,l)$ which are set at the Bragg angle can be obtained from the SAD. Often the planes are near to, but not exactly at, the Bragg angle and it is necessary to determine the precise deviation which is usually represented by the parameter s, as shown in the Ewald sphere construction in Figure 4.27. The deviation parameter s is determined from Kikuchi lines, observed in diffraction patterns obtained from somewhat thicker areas of the specimen, which form a pair of bright and dark lines associated with each reflection, spaced $|g|$ apart.

The Kikuchi lines arise from inelastically scattered rays, originating at some point P in the specimen (see Figure 4.28), being subsequently Bragg-diffracted. Thus, for the set of planes in Figure 4.28a, those electrons traveling in the directions PQ and PR will be Bragg-diffracted at Q and R and give rise to rays in the directions QQ′ and RR′. Since the electrons in the beam RR′ originate from the scattered ray PR, this beam will be less intense than QQ′, which contains electrons scattered through a smaller angle at P. Because P is a spherical source this rediffraction at points such as Q and R gives rise to cones of rays which, when they intersect the film, approximate to straight lines.

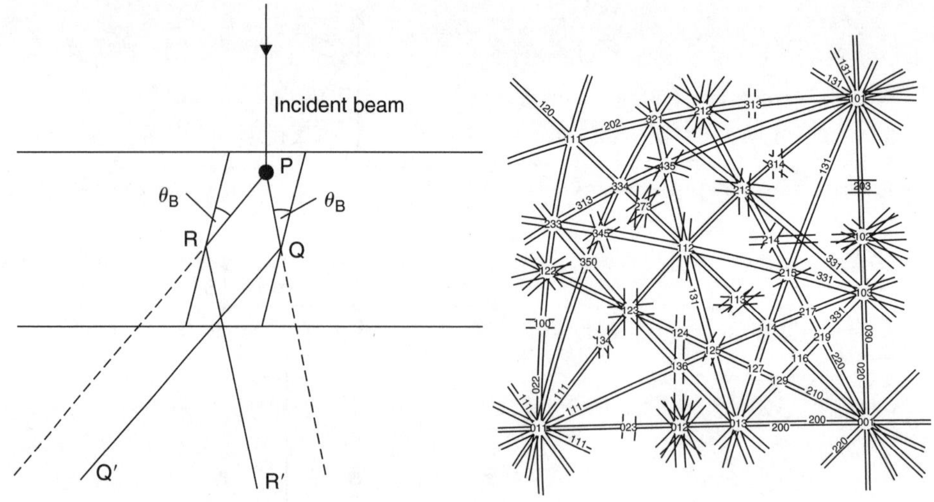

(a) (b)

Figure 4.28 *Kikuchi lines. (a) Formation. (b) From fcc crystal, forming a Kikuchi map.*

The selection of the diffracting conditions used to image the crystal defects can be controlled using Kikuchi lines. Thus, the planes $(h\,k\,l)$ are at the Bragg angle when the corresponding pair of Kikuchi lines passes through 0 0 0 and g_{hkl}, i.e. $s = 0$. Tilting of the specimen so that this condition is maintained (which can be done quite simply, using modern double-tilt specimen stages) enables the operator to select a specimen orientation with a close approximation to two-beam conditions. Tilting the specimen to a particular orientation, i.e. electron beam direction, can also be selected using the Kikuchi lines as a 'navigational' aid. The series of Kikuchi lines make up a Kikuchi map, as shown in Figure 4.28b, which can be used to tilt from one pole to another (as one would use an Underground map).

4.4.4.2 Convergent beam diffraction patterns

When a selected area diffraction pattern is taken with a convergent beam of electrons, the resultant pattern contains additional structural information. A ray diagram illustrating the formation of a convergent beam diffraction pattern (CBDP) is shown in Figure 4.29a. The disks of intensity which are formed in the back focal plane contain information which is of three types:

1. Fringes within disks formed by strongly diffracted beams. If the crystal is tilted to two-beam conditions, these fringes can be used to determine the specimen thickness very accurately.
2. High-angle information in the form of fine lines (somewhat like Kikuchi lines) which are visible in the direct beam and in the higher-order Laue zones (HOLZ). These HOLZ are visible in a pattern covering a large enough angle in reciprocal space. The fine line pattern can be used to measure the lattice parameter to 1 in 10^4. Figure 4.29b shows an example of HOLZ lines for a silicon crystal centered [1 1 1]. Pairing a dark line through the zero-order disk with its corresponding bright line through the higher-order disk allows the lattice parameter to be determined, the distance between the pair being sensitive to the temperature, etc.
3. Detailed structure both within the direct beam and within the diffracted beams which show certain well-defined symmetries when the diffraction pattern is taken precisely along an important zone axis. The patterns can therefore be used to give crystal structure information, particularly the point group and space group. This information, together with the chemical composition

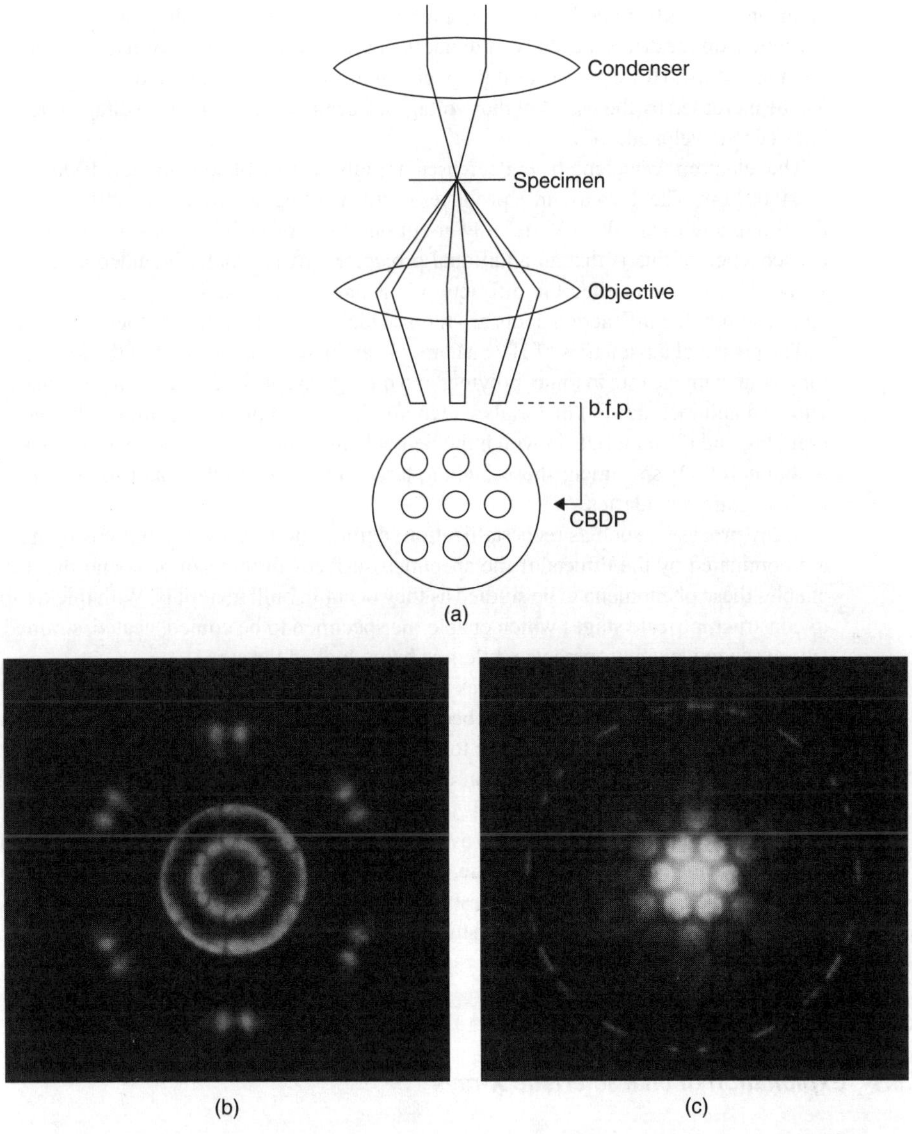

Figure 4.29 *(a) Schematic formation of convergent beam diffraction pattern in the back focal plane of the objective lens. (b, c) ⟨1 1 1⟩ CBDPs from Si: (b) zero layer and HOLZ (Higher Order Laue Zones) in direct beam and (c) zero layer + FOLZ (First Order Laue Zones).*

from EELS or EDX, and the size of the unit cell from the indexed diffraction patterns can be used to define the specific crystal structure, i.e. the atomic positions. Figure 4.29c indicates the threefold symmetry in a CBDP from silicon taken along the [1 1 1] axis.

4.4.4.3 Higher-voltage electron microscopy

The most serious limitation of conventional transmission electron microscopes (CTEM) is the limited thickness of specimens examined (50–500 nm). This makes preparation of samples from heavy elements difficult, gives limited containment of particles and other structural features within the specimen, and restricts the study of dynamical processes such as deformation, annealing, etc., within the microscope. However, the usable specimen thickness is a function of the accelerating voltage and can be increased by the use of higher voltages. Because of this, higher-voltage microscopes (HVEM) have been developed.

The electron wavelength λ decreases rapidly with voltage and at 1000 kV the wavelength $\lambda \approx 0.001$ nm. The decrease in λ produces corresponding decreases in the Bragg angles θ, and hence the Bragg angles at 1000 kV are only about one-third of their corresponding values at 100 kV. One consequence of this is that an additional projector lens is usually included in a high-voltage microscope. This is often called the diffraction lens and its purpose is to increase the diffraction camera length so that the diffraction spots are more widely spaced on the photographic plate.

The principal advantages of HVEM are: (1) an increase in usable foil thickness, and (2) a reduced ionization damage rate in ionic, polymer and biological specimens. The range of materials is therefore widened and includes (1) materials which are difficult to prepare as thin foils, such as tungsten and uranium, and (2) materials in which the defect being studied is too large to be conveniently included within a 100 kV specimen; these include large voids, precipitates and some dislocation structures such as grain boundaries.

Many processes such as recrystallization, deformation, recovery, martensitic transformation, etc. are dominated by the effects of the specimen surfaces in thin samples, and the use of thicker foils enables these phenomena to be studied as they occur in bulk materials. With thicker foils it is possible to construct intricate stages which enable the specimen to be cooled, heated, strained and exposed to various chemical environments while it is being looked through.

A disadvantage of HVEM is that as the beam voltage is raised, the energy transferred to the atom by the fast electron increases until it becomes sufficient to eject the atom from its site. The amount of energy transferred from one particle to another in a collision depends on the ratio of the two masses (see Chapter 3). Because the electron is very light compared with an atom, the transfer of energy is very inefficient and the electron needs to have several hundred keV before it can transmit the 25 eV or so necessary to displace an atom. To avoid radiation damage it is necessary to keep the beam voltage below the critical displacement value, which is $\approx$100 kV for Mg and $\approx$1300 kV for Au. There is, however, much basic scientific interest in radiation damage for technological reasons and an HVEM enables the damage processes to be studied directly.

4.4.5 Chemical microanalysis

4.4.5.1 Exploitation of characteristic X-rays

Electron probe microanalysis (EPMA) of bulk samples is now a routine technique for obtaining rapid, accurate analysis of alloys. A small electron probe ($\approx$100 nm diameter) is used to generate X-rays from a defined area of a polished specimen and the intensity of the various characteristic X-rays measured using either wavelength-dispersive spectrometry (WDS) or energy-dispersive spectrometry (EDS). Typically the accuracy of the analysis is $\pm$0.1%. One of the limitations of EPMA of bulk samples is that the volume of the sample which contributes to the X-ray signal is relatively independent of the size of the electron probe, because high-angle elastic scattering of electrons within the sample generates X-rays (see Figure 4.30). The consequence of this is that the spatial resolution of EPMA is no better than $\sim$2 μm. In the last few years EDX detectors have been interfaced to transmission electron microscopes which are capable of operating with an electron probe as small as 2 nm. The combination of electron-transparent samples, in which high-angle elastic scattering is limited, and a small electron probe leads to a significant improvement in the potential spatial resolution of X-ray microanalysis. In addition, interfacing of energy loss spectrometers has enabled light elements to

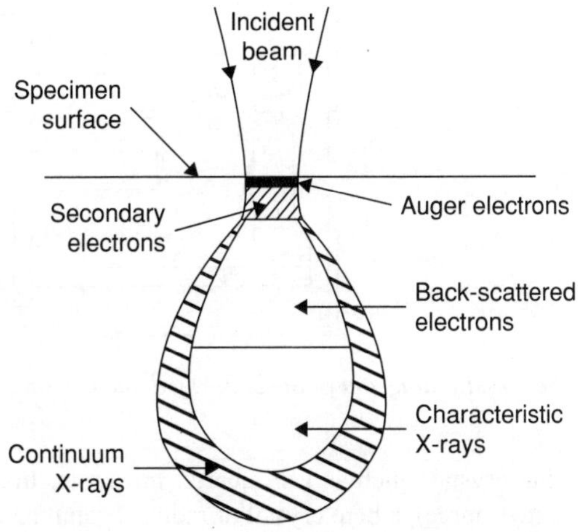

Figure 4.30 *Schematic diagram showing the generation of electrons and X-rays within the specimen.*

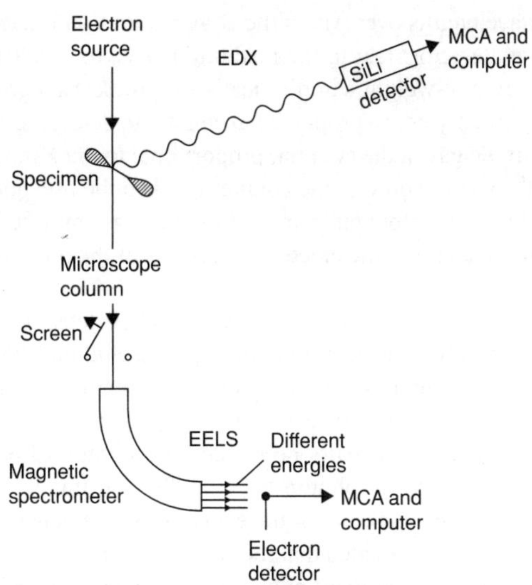

Figure 4.31 *Schematic diagram of EDX and EELS in TEM.*

be detected and measured, so that electron microchemical analysis is now a powerful tool in the characterization of materials. With electron beam instrumentation it is required to measure (1) the wavelength or energies of emitted X-rays (WDX and EDX), (2) the energy losses of the fast electrons (EELS) and (3) the energies of emitted electrons (AES). Nowadays (1) and (2) can be carried out on the modern TEM using special detector systems, as shown schematically in Figure 4.31.

In a WDX spectrometer a crystal of known d-spacing is used which diffracts X-rays of a specific wavelength, λ, at an angle θ, given by the Bragg equation, $n\lambda = 2d \sin\theta$. Different wavelengths are selected by changing θ and thus, to cover the necessary range of wavelengths, several crystals of different d-spacings are used successively in a spectrometer. The range of wavelength is 0.1–2.5 nm and the corresponding d-spacing for practicable values of θ, which lie between $\approx 15°$ and 65°, is

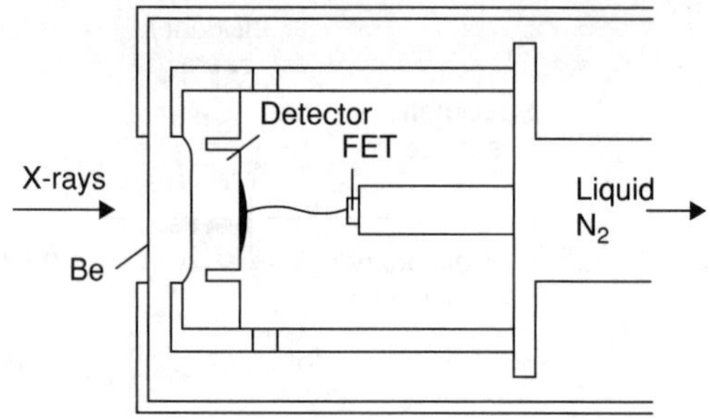

Figure 4.32 *Schematic diagram of Si–Li X-ray detector.*

achieved by using crystals such as LiF, quartz, mica, etc. In a WDX spectrometer, the specimen (which is the X-ray source), a bent crystal of radius $2r$ and the detector all lie on the focusing circle radius r and different wavelength X-rays are collected by the detector by setting the crystal at different angles, θ. The operation of the spectrometer is very time-consuming since only one particular X-ray wavelength can be focused on to the detector at any one time.

The resolution of WDX spectrometers is controlled by the perfection of the crystal, which influences the range of wavelengths over which the Bragg condition is satisfied, and by the size of the entrance slit to the X-ray detector; taking the resolution ($\Delta\lambda$) to $\sim$0.001 nm then $\lambda/\Delta\lambda$ is about 300 which, for a medium-atomic-weight sample, leads to a peak–background ratio of about 250. The crystal spectrometer normally uses a proportional counter to detect the X-rays, producing an electrical signal, by ionization of the gas in the counter, proportional to the X-ray energy, i.e. inversely proportional to the wavelength. The window of the counter needs to be thin and of low atomic number to minimize X-ray absorption. The output pulse from the counter is amplified and differentiated to produce a short pulse. The time constant of the electrical circuit is of the order of 1 μs, which leads to possible count rates of at least 10^5/s.

In recent years EDX detectors have replaced WDX detectors on transmission microscopes and are used together with WDX detectors on microprobes and on SEMs. A schematic diagram of an Si–Li detector is shown in Figure 4.32. X-rays enter through the thin Be window and produce electron–hole pairs in the Si–Li. Each electron–hole pair requires 3.8 eV, at the operating temperature of the detector, and the number of pairs produced by a photon of energy E_p is thus $E_p/3.8$. The charge produced by a typical X-ray photon is $\approx 10^{-16}$C and this is amplified to give a shaped pulse, the height of which is then a measure of the energy of the incident X-ray photon. The data are stored in a multi-channel analyzer. Provided that the X-ray photons arrive with a sufficient time interval between them, the energy of each incident photon can be measured and the output presented as an intensity versus energy display. The amplification and pulse shaping takes about 50 μs and if a second pulse arrives before the preceding pulse is processed, both pulses are rejected. This results in significant dead time for count rates $\geq$4000/s.

The number of electron–hole pairs generated by an X-ray of a given energy is subject to normal statistical fluctuations and this, taken together with electronic noise, limits the energy resolution of an Si–Li detector to about a few hundred eV, which worsens with increase in photon energy. The main advantage of EDX detectors is that simultaneous collection of the whole range of X-rays is possible and the energy characteristics of all the elements with $Z > 11$ in the Periodic Table can be obtained in a matter of seconds. The main disadvantages are the relatively poor resolution, which leads to a peak–background ratio of about 50, and the limited count rate.

The variation in efficiency of an Si–Li detector must be allowed for when quantifying X-ray analysis. At low energies ($\leq$1 kV) the X-rays are mostly absorbed in the Be window and at high energies ($\geq$20 kV) the X-rays pass through the detector, so that the decreasing cross-section for electron–hole pair generation results in a reduction in efficiency. The Si–Li detector thus has optimum detection efficiency between about 1 and 20 kV.

Worked example

In the output from a WDX spectrometer, where the analyzing crystal was LiF, with the (2 0 0) planes ($d = 0.2013$ nm) parallel to the surface, X-ray peaks were detected when θ (the Bragg angle for the LiF analyzer) was 28.8° and 34.8°. From the following table of characteristic X-ray energies, determine which elements were present in the specimen.

Table of characteristic X-ray energies (keV)

		M			*L*			*K*				
Z	*El*	M_α	M_β	M_γ	$L_{\alpha 1}$	$L_{\beta 1}$	$L_{\beta 2}$	$K_{\alpha 2}$	$K_{\alpha 1}$	$K_{\beta 1}$	*El*	*Z*
21	Sc				0.395	0.400		4.085	4.090	4.460	Sc	21
22	Ti				0.452	0.458		4.504	4.510	4.931	Ti	22
23	V				0.511	0.519		4.944	4.951	5.426	V	23
24	Cr				0.573	0.583		5.405	5.414	5.946	Cr	24
25	Mn				0.637	0.649		5.887	5.898	6.489	Mn	25
26	Fe				0.705	0.718		6.390	6.403	7.057	Fe	26
27	Co				0.776	0.791		6.914	6.929	7.648	Co	27
28	Ni				0.851	0.869		7.460	7.477	8.263	Ni	28
29	Cu				0.930	0.950		8.026	8.046	8.904	Cu	29
30	Zn				1.012	1.034		8.614	8.637	9.570	Zn	30
31	Ga				1.098	1.125		9.223	9.250	10.263	Ga	31
32	Ge				1.188	1.218		9.854	9.885	10.980	Ge	32
33	As				1.282	1.317		10.506	10.542	11.724	As	33
34	Se				1.379	1.419		11.179	11.220	12.494	Se	34
35	Br				1.480	1.526		11.876	11.922	13.289	Br	35
36	Kr				1.586	1.636		12.596	12.648	14.110	Kr	36
37	Rb				1.694	1.752		13.333	12.393	14.959	Rb	37
38	Sr				1.806	1.871		14.095	14.163	15.833	Sr	38
39	Y				1.922	1.995		14.880	14.956	16.735	Y	39
40	Zr				2.042	2.124	(2.219)	15.688	15.772	17.665	Zr	40

Solution

If $\theta = 28.8°$, then if $\lambda = 2d \sin\theta$ (Bragg's law)

$$\lambda = 2 \times 0.2013 \times \sin 28.8° = 0.194 \text{ nm}.$$

$$c = \lambda\nu$$

$$\text{so } \nu = \frac{2.998 \times 10^8}{0.194 \times 10^{-9}} = 1.546 \times 10^{18} \text{s}^{-1}$$

$$E = \mathbf{h}\nu \text{ (Planck's equation)}$$

$$\therefore\ E = 6.626 \times 10^{-34} \times 1.546 \times 10^{18} = 1.024 \times 10^{-15} \text{J} = 6.393 \text{ keV}.$$

From the table of characteristic X-ray energies, this energy corresponds to FeK_α, so iron is present.

Similarly, if $\theta = 34.8°$, $\lambda = 0.2298$ nm and $E = 5.398$ keV, which corresponds to CrK_α. Therefore, iron and chromium are present.

4.4.5.2 *Electron microanalysis of thin foils*

There are several simplifications which arise from the use of thin foils in microanalysis. The most important of these arises from the fact that the average energy loss which electrons suffer on passing through a thin foil is only about 2%, and this small average loss means that the ionization cross-section can be taken as a constant. Thus, the number of characteristic X-ray photons generated from a thin sample is given simply by the product of the electron path length and the appropriate cross-section Q, i.e. the probability of ejecting the electron, and the fluorescent yield ω. The intensity generated by element A is then given by

$$I_A = iQ\omega n,$$

where Q is the cross-section per cm^2 for the particular ionization event, ω the fluorescent yield, n the number of atoms in the excited volume and i the current incident on the specimen. Microanalysis is usually carried out under conditions where the current is unknown and interpretation of the analysis simply requires that the ratio of the X-ray intensities from the various elements be obtained. For the simple case of a very thin specimen for which absorption and X-ray fluorescence can be neglected, then the measured X-ray intensity from element A is given by

$$I_A \propto n_A Q_A \omega_A a_A \eta_A$$

and for element B by

$$I_B \propto n_B Q_B \omega_B a_B \eta_B,$$

where n, Q, ω, a and η represent the number of atoms, the ionization cross-sections, the fluorescent yields, the fraction of the K line (or L and M) which is collected and the detector efficiencies, respectively, for elements A and B. Thus, in the alloy made up of elements A and B:

$$\frac{n_A}{n_B} \propto \frac{I_A Q_B \omega_B a_B \eta_B}{I_B Q_A \omega_A a_A \eta_A} = K_{AB} \frac{I_A}{I_B}.$$

This equation forms the basis for X-ray microanalysis of thin foils, where the constant K_{AB} contains all the factors needed to correct for atomic number differences, and is known as the Z-correction. Thus, from the measured intensities, the ratio of the number of atoms A to the number of atoms B, i.e. the concentrations of A and B in an alloy, can be calculated using the computed values for Q, ω, η, etc. A simple spectrum for stoichiometric Ni–Al is shown in Figure 4.33 and the values of I_K^{Al} and I_K^{Ni}, obtained after stripping the background, are given in Table 4.1, together with the final analysis. The absolute accuracy of any X-ray analysis depends either on the accuracy and the constants Q, ω, etc. or on the standards used to calibrate the measured intensities.

If the foil is too thick then an absorption correction (A) may have to be made to the measured intensities, since in traversing a given path length to emerge from the surface of the specimen, the X-rays of different energies will be absorbed differently. This correction involves a knowledge of the specimen thickness, which has to be determined by one of various techniques but usually from CBDPs. Occasionally a fluorescence (F) correction is also needed, since elements Z between Ti and

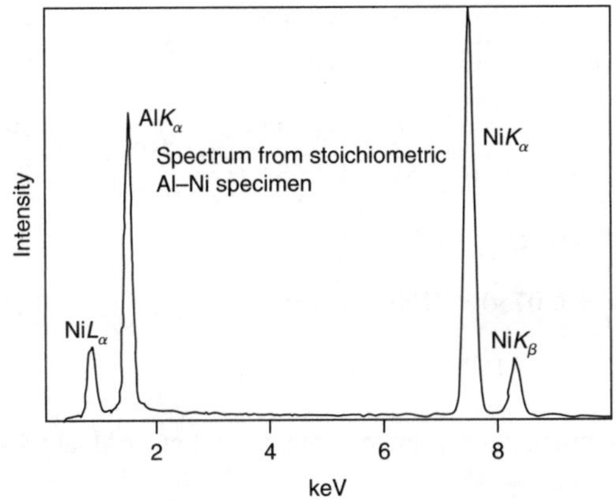

Figure 4.33 *EDX spectrum from a stoichiometric Ni–Al specimen.*

Table 4.1 *Relationships between measured intensities and composition for a Ni–Al alloy.*

	Measured intensities	*Cross-section Q, (10^{-24} cm^2)*	*Fluorescent yield ω*	*Detector efficiency η*	*Analysis at.%*
NiK_α	16250	297	0.392	0.985	50.6
AlK_α	7981	2935	0.026	0.725	49.4

Ni are fluoresced by the element $Z+2$. This 'no standards' $Z(AF)$ analysis can given an overall accuracy of ≈2% and can be carried out online with laboratory computers.

Worked example

In a TEM/EDX experiment the X-ray counts from a Ti – x at.% Al – y at.% Nb specimen and a Ti – 45 at.% Al – 4 at.% Nb standard were as follows:

	TiK_α	AlK	NbK_α
Ti – x at.% Al – y at.% Nb	14 391	5132	831
Ti – 45 at.% Al – 4 at.% Nb	15 228	6335	918

What was the composition of the unknown specimen?

Solution

Ratioing to titanium, for example:

	AlK/TiK_α	NbK_α/TiK_α
Ti – x at.% Al – y at.% Nb	0.3566	0.0577
Ti – 45 at.% Al – 4 at.% Nb	0.4160	0.0603

$$\text{Thus,} \quad \frac{x}{100-x-y} = \frac{45}{51} \times \frac{0.3566}{0.4160} = 0.7564 \qquad \left(\frac{\text{Al}}{\text{Ti}}\right)$$

$$\frac{y}{100-x-y} = \frac{4}{51} \times \frac{0.0577}{0.0603} = 0.0750 \qquad \left(\frac{\text{Nb}}{\text{Ti}}\right)$$

$$\text{Then,} \quad x = 0.7564 \times (100-x-y)$$

$$y = 0.0750 \times (100-x-y)$$

$$\therefore \quad x = 41.3\% \qquad y = 4.1\%.$$

Therefore, the unknown specimen contains 41.3 at.% Al and 4.1 at.% Nb.

4.4.6 Electron energy-loss spectroscopy (EELS)

A disadvantage of EDX is that the X-rays from the light elements are absorbed in the detector window. Windowless detectors can be used but have some disadvantages, such as the overlapping of spectrum lines, which have led to the development of EELS.

EELS is possible only on transmission specimens, and so electron spectrometers have been interfaced to TEMs to collect all the transmitted electrons lying within a cone of width α. The intensity of the various electrons, i.e. those transmitted without loss of energy and those that have been inelastically scattered and lost energy, is then obtained by dispersing the electrons with a magnetic prism, which separates spatially the electrons of different energies.

A typical EELS spectrum, illustrated in Figure 4.34, shows three distinct regions. The zero-loss peak is made up from those electrons which have (1) not been scattered by the specimen, (2) suffered

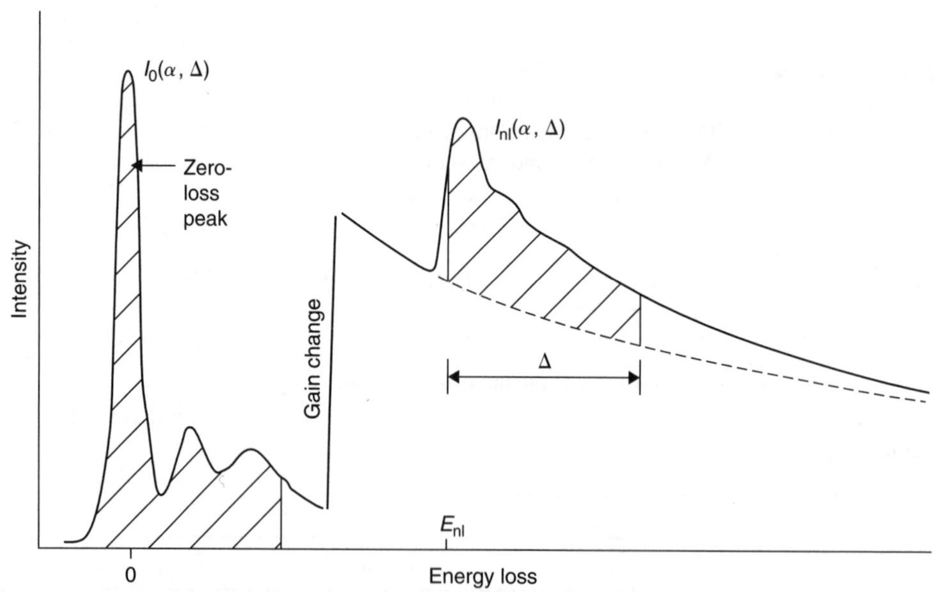

Figure 4.34 *Schematic energy-loss spectrum, showing the zero-loss and plasmon regions together with the characteristic ionization edge, energy E_{nl} and intensity I_{nl}.*

photon scattering ($\approx 1/40\,\text{eV}$) and (3) been elastically scattered. The energy width of the zero-loss peak is caused by the energy spread of the electron source (up to $\approx$2 eV for a thermionic W filament) and the energy resolution of the spectrometer (typically a few eV). The second region of the spectrum extends up to about 50 eV loss and is associated with plasmon excitations corresponding to electrons which have suffered one, two or more plasmon interactions. Since the typical mean free path for the generation of a plasmon is about 50 nm, many electrons suffer single-plasmon losses and only in specimens which are too thick for electron loss analysis will there be a significant third plasmon peak. The relative size of the plasmon loss peak and the zero-loss peak can also be used to measure the foil thickness. Thus, the ratio of the probability of exciting a plasmon loss, P_1, to not exciting a plasmon, P_0, is given by $P_1/P_0 = t/L$, where t is the thickness, L the mean free path for plasmon excitation, and P_1 and P_0 are given by the relative intensities of the zero loss and the first plasmon peak. If the second plasmon peak is a significant fraction of the first peak, this indicates that the specimen will be too thick for accurate microanalysis.

The third region is made up of a continuous background on which the characteristic ionization losses are superimposed. Qualitative elemental analysis can be carried out simply by measuring the energy of the edges and comparing them with tabulated energies. The actual shape of the edge can also help to define the chemical state of the element. Quantitative analysis requires the measurement of the ratios of the intensities of the electrons from elements A and B which have suffered ionization losses. In principle, this allows the ratio of the number of A atoms, N_A, and B atoms, N_B, to be obtained simply from the appropriate ionization cross-sections, Q_K. Thus, the number of A atoms will be given by

$$N_\text{A} = (1/Q_K^\text{A})[I_K^\text{A}/I_0]$$

and the number of B atoms by a similar expression, so that

$$N_\text{A}/N_\text{B} = I_K^\text{A} Q_K^\text{B} / I_K^\text{B} Q_K^\text{A},$$

where I_K^A is the measured intensity of the K edge for element A, similarly for I_K^B and I_0 is the measured intensity of the zero-loss peak. This expression is similar to the thin foil EDX equation.

To obtain I_K the background has to be removed so that only loss electrons remain. Because of the presence of other edges there is a maximum energy range over which I_K can be measured, which is about 50–100 eV. The value of Q_K must therefore be replaced by $Q_K(\Delta)$, which is a partial cross-section calculated for atomic transition within an energy range Δ of the ionization threshold. Furthermore, only the loss electrons arising from an angular range of scatter α at the specimen are collected by the spectrometer so that a double partial cross-section $Q(\Delta, \alpha)$ is appropriate. Thus, analysis of a binary alloy is carried out using the equation:

$$\frac{N_\text{A}}{N_\text{B}} = \frac{Q_K^\text{B}(\Delta,\alpha) I_K^\text{A}(\Delta,\alpha)}{Q_K^\text{A}(\Delta,\alpha) I_K^\text{B}(\Delta,\alpha)}.$$

Values of $Q(\Delta, \alpha)$ may be calculated from data in the literature for the specific value of ionization edge, Δ, α and incident accelerating voltage, but give an analysis accurate to only about 5%; a greater accuracy might be possible if standards are used.

4.4.7 Auger electron spectroscopy (AES)

Auger electrons originate from a surface layer a few atoms thick and therefore AES is a technique for studying the composition of the surface of a solid. It is obviously an important method for studying oxidation, catalysis and other surface chemical reactions, but has also been used successfully to determine the chemistry of fractured interfaces and grain boundaries (e.g. temper embrittlement of steels).

The basic instrumentation involves a focusable electron gun, an electron analyzer, and a sample support and manipulation system, all in an ultra-high-vacuum environment to minimize adsorption of gases onto the surface during analysis. Two types of analyzer are in use, a cylindrical mirror analyzer (CMA) and a hemispherical analyzer (HSA), both of which are of the energy-dispersive type as for EELS, with the difference that the electron energies are much lower, and electrostatic rather than magnetic 'lenses' are used to separate out the electrons of different energies.

In the normal distribution the Auger electron peaks appear small on a large and often sloping background, which gives problems in detecting weak peaks, since amplification enlarges the background slope as well as the peak. It is therefore customary to differentiate the spectrum so that the Auger peaks are emphasized as doublet peaks with a positive and negative displacement against a nearly flat background. This is achieved by electronic differentiation by applying a small a.c. signal of a particular frequency in the detected signal. Chemical analysis through the outer surface layers can be carried out by depth profiling with an argon-ion gun.

4.5 Observation of defects

4.5.1 Etch pitting

Since dislocations are regions of high energy, their presence can be revealed by the use of an etchant which chemically attacks such sites preferentially. This method has been applied successfully in studying metals, alloys and compounds, and there are many fine examples in existence of etch-pit patterns showing small-angle boundaries and pile-ups. Figure 4.35a shows an etch-pit pattern from an array of piled-up dislocations in a zinc crystal. The dislocations are much closer together at the head of the pile-up, and an analysis of the array, made by Gilman, shows that their spacing is in reasonable

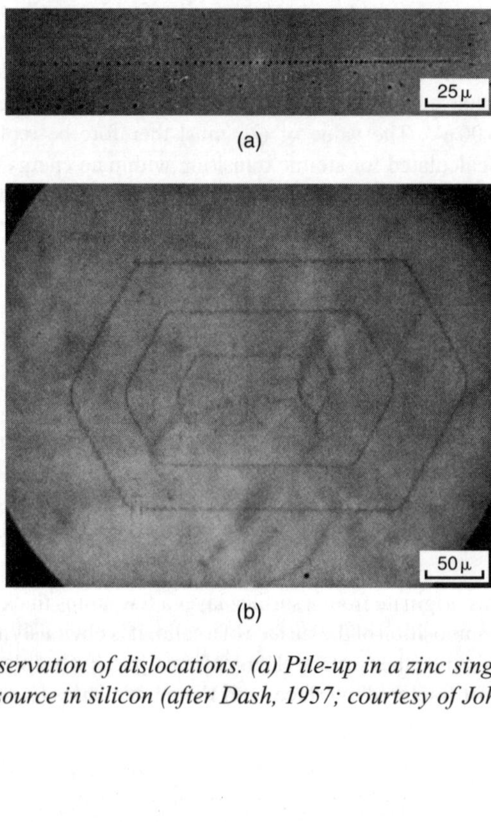

Figure 4.35 *Direct observation of dislocations. (a) Pile-up in a zinc single crystal (after Gilman, 1956). (b) Frank–Read source in silicon (after Dash, 1957; courtesy of John Wiley and Sons).*

agreement with the theory of Eshelby, Frank and Nabarro, who have shown that the number of dislocations n that can be packed into a length L of slip plane is $n = 2L\tau/\mu b$, where τ is the applied stress. The main disadvantage of the technique is its inability to reveal networks or other arrangements in the interior of the crystal, although some information can be obtained by taking sections through the crystal. Its use is also limited to materials with low dislocation contents ($<10^4$ mm^{-2}) because of the limited resolution. In recent years it has been successfully used to determine the velocity v of dislocations as a function of temperature and stress by measuring the distance traveled by a dislocation after the application of a stress for a known time (see Chapter 6).

4.5.2 Dislocation decoration

It is well known that there is a tendency for solute atoms to segregate to grain boundaries and, since these may be considered as made up of dislocations, it is clear that particular arrangements of dislocations and sub-boundaries can be revealed by preferential precipitation. Most of the studies in metals have been carried out on aluminum–copper alloys, to reveal the dislocations at the surface, but recently several decoration techniques have been devised to reveal internal structures. The original experiments were made by Hedges and Mitchell, in which they made visible the dislocations in AgBr crystals with photographic silver. After a critical annealing treatment and exposure to light, the colloidal silver separates along dislocation lines. The technique has since been extended to other halides, and to silicon, where the decoration is produced by diffusing copper into the crystal at 900°C so that, on cooling the crystal to room temperature, the copper precipitates. When the silicon crystal is examined optically, using infrared illumination, the dislocation-free areas transmit the infrared radiation, but the dislocations decorated with copper are opaque. A fine example of dislocations observed using this technique is shown in Figure 4.35b.

The technique of dislocation decoration has the advantage of revealing internal dislocation networks but, when used to study the effect of cold-work on the dislocation arrangement, suffers the disadvantage of requiring some high-temperature heat treatment, during which the dislocation configuration may become modified.

4.5.3 Dislocation strain contrast in TEM

The most notable advance in the direct observation of dislocations in materials has been made by the application of transmission techniques to thin specimens. The technique has been used widely because the dislocation arrangements inside the specimen can be studied. It is possible, therefore, to investigate the effects of plastic deformation, irradiation, heat treatment, etc. on the dislocation distribution and to record the movement of dislocations by taking cine-films of the images on the fluorescent screen of the electron microscope. One disadvantage of the technique is that the materials have to be thinned before examination and, because the surface-to-volume ratio of the resultant specimen is high, it is possible that some rearrangement of dislocations may occur.

A theory of image contrast has been developed which agrees well with experimental observations. The basic idea is that the presence of a defect in the lattice causes displacements of the atoms from their position in the perfect crystal and these lead to phase changes in the electron waves scattered by the atoms so that the amplitude diffracted by a crystal is altered. The image seen in the microscope represents the electron intensity distribution at the lower surface of the specimen. This intensity distribution has been calculated by a dynamical theory (see Section 4.5.7) which considers the coupling between the diffracted and direct beams, but it is possible to obtain an explanation of many observed contrast effects using a simpler (kinematical) theory in which the interactions between the transmitted and scattered waves are neglected. Thus, if an electron wave, represented by

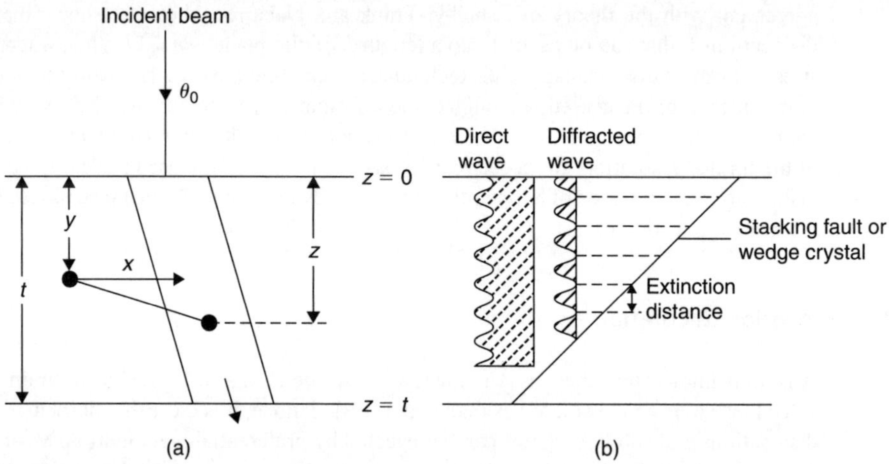

Figure 4.36 *(a) Column approximation used to calculate the amplitude of the diffracted beam ϕ_g on the bottom surface of the crystal. The dislocation is at a depth* y *and a distance* x *from the column. (b) Variation of intensity with depth in a crystal.*

the function $\exp(2\pi i k_0 \cdot r)$, where k_0 is the wave vector of magnitude $1/\lambda$, is incident on an atom at position r there will be an elastically scattered wave $\exp(2\pi i k_1 \cdot r)$ with a phase difference equal to $2\pi r(k_1 - k_0)$ when k_1 is the wave vector of the diffracted wave. If the crystal is not oriented exactly at the Bragg angle the reciprocal lattice point will be either inside or outside the reflecting sphere and the phase difference is then $2\pi r(g + s)$, where g is the reciprocal lattice vector of the lattice plane giving rise to reflection and s is the vector indicating the deviation of the reciprocal lattice point from the reflection sphere (see Figure 4.37). To obtain the total scattered amplitude from a crystal it is necessary to sum all the scattered amplitudes from all the atoms in the crystal, i.e. take account of all the different path lengths for rays scattered by different atoms. Since most of the intensity is concentrated near the reciprocal lattice point it is only necessary to calculate the amplitude diffracted by a column of crystal in the direction of the diffracted beam and not the whole crystal, as shown in Figure 4.36. The amplitude of the diffracted beam ϕ_g for an incident amplitude $\phi_0 = 1$, is then

$$\phi_g = (\pi i/\xi_g)\int_0^t \exp[-2\pi i(g+s)\cdot r]\mathrm{d}r$$

and since $r \cdot s$ is small and $g \cdot r$ is an integer this reduces to

$$\begin{aligned}\phi_g &= (\pi i/\xi_g)\textstyle\int_0^t \exp[-2\pi i s\cdot r]\mathrm{d}r\\ &= (\pi i/\xi_g)\textstyle\int_0^t \exp[-2\pi i s z]\mathrm{d}z,\end{aligned}$$

where z is taken along the column. The intensity from such a column is

$$|\phi_g|^2 = I_g = [\pi^2/\xi_g^2](\sin^2 \pi t s/(\pi s)^2),$$

from which it is evident that the diffracted intensity oscillates with depth z in the crystal with a periodicity equal to $1/s$. The maximum wavelength of this oscillation is known as the extinction[8] distance ξ_g since the diffracted intensity is essentially extinguished at such positions in the crystal.

[8] $\xi_g = \pi V \cos\theta/\lambda F$, where V is the volume of the unit cell, θ the Bragg angle and F the structure factor.

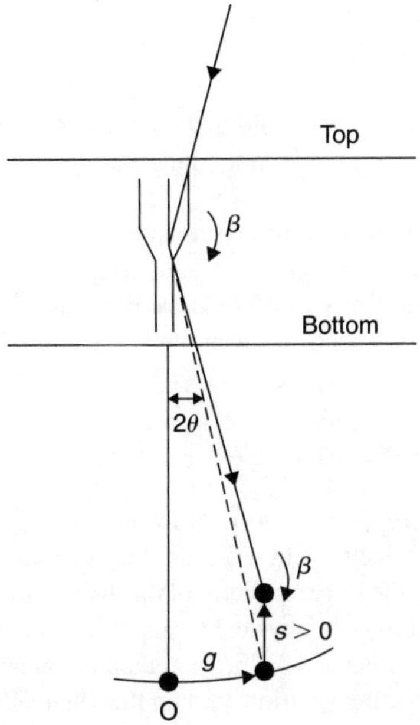

Figure 4.37 *Schematic diagram showing the dependence of the dislocation image position on diffraction conditions.*

This sinusoidal variation of intensity gives rise to fringes in the electron-optical image of boundaries and defects inclined to the foil surface, e.g. a stacking fault on an inclined plane is generally visible on an electron micrograph as a set of parallel fringes running parallel to the intersection of the fault plane with the plane of the foil (see Figure 4.41).

In an imperfect crystal, atoms are displaced from their true lattice positions. Consequently, if an atom at r_n is displaced by a vector $\mathbf{R}$, the amplitude of the wave diffracted by the atom is multiplied by an additional phase factor $\exp[2\pi i(k_1 - k_0) \cdot \mathbf{R}]$. Then, since $(k_1 - k_0) = g + s$, the resultant amplitude is

$$\phi_g = (\pi i/\xi_g) \int_0^t \exp[-2\pi i(g + s) \cdot (r + \mathbf{R})]\mathrm{d}r.$$

If we neglect $s \cdot \mathbf{R}$, which is small in comparison with $g \cdot \mathbf{R}$, and $g \cdot r$, which gives an integer, then in terms of the column approximation:

$$\phi_g = (\pi i/\xi_g) \int_0^t \exp(-2\pi isz) \exp(-2\pi ig \cdot \mathbf{R})\mathrm{d}z.$$

The amplitude, and hence the intensity, therefore may differ from that scattered by a perfect crystal, depending on whether the phase factor $\alpha = 2\pi g \cdot \mathbf{R}$ is finite or not, and image contrast is obtained when $g \cdot \mathbf{R} \neq 0$.

4.5.4 Contrast from crystals

In general, crystals observed in the microscope appear light because of the good transmission of electrons. In detail, however, the foils are usually slightly buckled so that the orientation of the crystal relative to the electron beam varies from place to place, and if one part of the crystal is oriented at the Bragg angle, strong diffraction occurs. Such a local area of the crystal then appears dark under bright-field illumination, and is known as a bend or extinction contour. If the specimen is tilted while under observation, the angular conditions for strong Bragg diffraction are altered, and the extinction contours, which appear as thick dark bands, can be made to move across the specimen. To interpret micrographs correctly, it is essential to know the correct sense of both g and s. The g-vector is the line joining the origin of the diffraction pattern to the strong diffraction spot and it is essential that its sense is correct with respect to the micrograph, i.e. to allow for any image inversion or rotation by the electron optics. The sign of s can be determined from the position of the Kikuchi lines with respect to the diffraction spots, as discussed in Section 4.4.1.

4.5.5 Imaging of dislocations

Image contrast from imperfections arises from the additional phase factor $\alpha = 2\pi g \cdot \mathbf{R}$ in the equation for the diffraction of electrons by crystals. In the case of dislocations the displacement vector $\mathbf{R}$ is essentially equal to b, the Burgers vector of the dislocation, since atoms near the core of the dislocation are displaced parallel to b. In physical terms, it is easily seen that if a crystal, oriented off the Bragg condition, i.e. $s \neq 0$, contains a dislocation then on one side of the dislocation core the lattice planes are tilted into the reflecting position, and on the other side of the dislocation the crystal is tilted away from the reflecting position. On the side of the dislocation in the reflecting position the transmitted intensity, i.e. passing through the objective aperture, will be less and hence the dislocation will appear as a line in dark contrast. It follows that the image of the dislocation will lie slightly to one or other side of the dislocation core, depending on the sign of $(g \cdot b)s$. This is shown in Figure 4.37 for the case where the crystal is oriented in such a way that the incident beam makes an angle greater than the Bragg angle with the reflecting planes, i.e. $s > 0$. The image occurs on that side of the dislocation where the lattice rotation brings the crystal into the Bragg position, i.e. rotates the reciprocal lattice point onto the reflection sphere. Clearly, if the diffracting conditions change, i.e. g or s change sign, then the image will be displaced to the other side of the dislocation core.

The phase angle introduced by a lattice defect is zero when $g \cdot \mathbf{R} = 0$, and hence there is no contrast, i.e. the defect is invisible when this condition is satisfied. Since the scalar product $g \cdot \mathbf{R}$ is equal to $g\mathbf{R} \cos\theta$, where θ is the angle between g and $\mathbf{R}$, then $g \cdot \mathbf{R} = 0$ when the displacement vector $\mathbf{R}$ is normal to g, i.e. parallel to the reflecting plane producing the image. If we think of the lattice planes which reflect the electrons as mirrors, it is easy to understand that no contrast results when $g \cdot \mathbf{R} = 0$, because the displacement vector $\mathbf{R}$ merely moves the reflecting planes parallel to themselves without altering the intensity scattered from them. Only displacements which have a component perpendicular to the reflecting plane, i.e. tilting the planes, will produce contrast.

A screw dislocation only produces atomic displacements in the direction of its Burgers vector, and hence because $\mathbf{R} = b$ such a dislocation will be completely 'invisible' when b lies in the reflecting plane producing the image. A pure edge dislocation, however, produces some minor atomic displacements perpendicular to b, as discussed in Chapter 3, and the displacements give rise to a slight curvature of the lattice planes. An edge dislocation is therefore not completely invisible when b lies in the reflecting planes, but usually shows some evidence of faint residual contrast. In general, however, a dislocation goes out of contrast when the reflecting plane operating contains its Burgers vector, and this fact is commonly used to determine the Burgers vector. To establish b uniquely, it is necessary to tilt the foil so that the dislocation disappears on at least two different reflections. The Burgers vector must then be parallel to the direction which is common to these two reflecting planes. The magnitude of b is usually the repeat distance in this direction.

The use of the $g \cdot b = 0$ criterion is illustrated in Figure 4.38. The helices shown in this micrograph have formed by the condensation of vacancies on to screw dislocations having their Burgers vector b

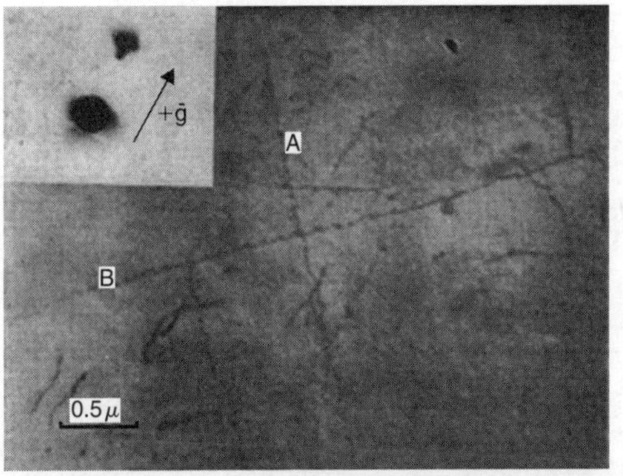

(a)

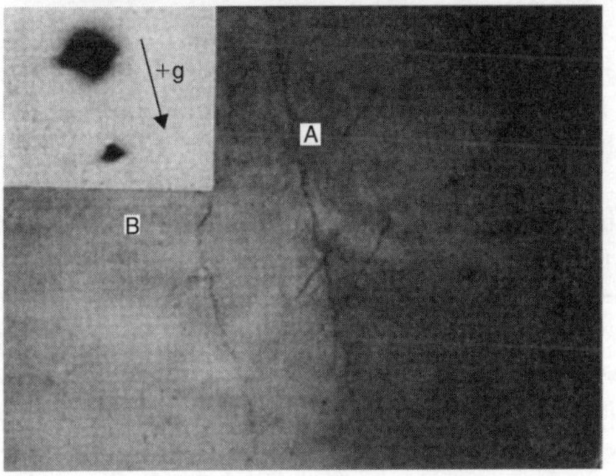

(b)

Figure 4.38 *(a) Application of the $g \cdot b = 0$ criterion. The effect of changing the diffraction condition (see diffraction pattern inserts) makes the long helical dislocation B in (a) disappear in (b) (after Hirsch, Howie and Whelan, 1960; courtesy of the Royal Society).*

parallel to the axis of the helix. Comparison of the two pictures in (a) and (b) shows that the effect of tilting the specimen, and hence changing the reflecting plane, is to make the long helix B in (a) disappear in (b). In detail, the foil has a [0 0 1] orientation and the long screws lying in this plane are $1/2[\bar{1}\ 1\ 0]$ and $1/2[1\ 1\ 0]$. In Figure 4.38a the insert shows the 0 2 0 reflection is operating and so $g \cdot b \neq 0$ for either A or B, but in Figure 4.38b the insert shows that the $2\ \bar{2}\ 0$ reflection is operating and the dislocation B is invisible, since its Burgers vector b is normal to the g-vector, i.e. $g \cdot b = 2\ \bar{2}\ 0 \cdot 1/2[1\ 1\ 0] = (\frac{1}{2} \times 1 \times 2) + (\frac{1}{2} \times 1 \times 2) + 0 = 0$ for the dislocation B, and is therefore invisible.

4.5.6 Imaging of stacking faults

Contrast at a stacking fault arises because such a defect displaces the reflecting planes relative to each other, above and below the fault plane, as illustrated in Figure 4.39a. In general, the contrast from a stacking fault will not be uniformly bright or dark as would be the case if it were parallel to the foil surface, but in the form of interference fringes running parallel to the intersection of the foil surface with the plane containing the fault. These appear because the diffracted intensity oscillates with depth in the crystal as discussed. The stacking-fault displacement vector **R**, defined as the shear parallel to the fault of the portion of crystal below the fault relative to that above the fault which is

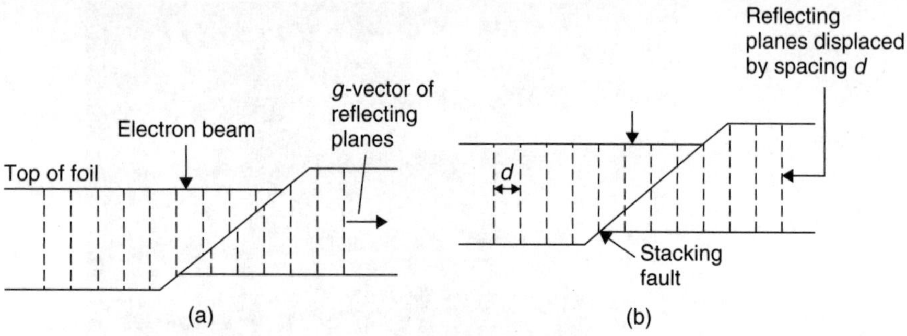

Figure 4.39 *Schematic diagram showing displacement of reflecting planes by a stacking fault (a) and the condition for $g \cdot \boldsymbol{R} = n$ when the fault would be invisible (b).*

as fixed, gives rise to a phase difference $\alpha = 2\pi g \cdot \mathbf{R}$ in the electron waves diffracted from either side of the fault. It then follows that stacking-fault contrast is absent with reflections for which $\alpha = 2\pi$, i.e. for which $g \cdot \mathbf{R} = n$. This is equivalent to the $g \cdot b = 0$ criterion for dislocations and can be used to deduce **R**.

The invisibility of stacking-fault contrast when $g \cdot \mathbf{R} = 0$ is exactly analogous to that of a dislocation when $g \cdot b = 0$, namely that the displacement vector is parallel to the reflecting planes. The invisibility when $g \cdot \mathbf{R} = 1, 2, 3, \ldots$ occurs because in these cases the vector **R** moves the imaging reflecting planes normal to themselves by a distance equal to a multiple of the spacing between the planes. From Figure 4.39b it can be seen that for this condition the reflecting planes are once again in register on either side of the fault and, as a consequence, there is no interference between waves from the crystal above and below the fault.

Worked example

A bright-field micrograph of stacking faults in silicon shows characteristic fringes on $(\bar{1}\ 1\ 1)$ when imaged with $g = 2\ \bar{2}\ 0$ which go out of contrast when the imaging condition is changed to $g = 0\ \bar{2}\ 2$. What information about the stacking fault can be obtained from this observation?

Solution

The invisibility criterion for a stacking fault is $g \cdot \mathbf{R} = n$, where $n = 0, 1, 2, \ldots$. Silicon is diamond cubic and faults on $\{1\ 1\ 1\}$ planes could be formed (a) by Frank sessile formation, i.e. $a/3 \langle 1\ 1\ 1 \rangle$ or (b) by movement of a Shockley partial $a/6 \langle 1\ 1\ 2 \rangle$.

(a) For fringes on $(\bar{1}\ 1\ 1)$ the Frank would be $\frac{a}{3}[\bar{1}\ 1\ 1]$ when imaged with $g = 2\ \bar{2}\ 0$, i.e. $g \cdot \mathbf{R} = (1/3)(-2 - 2 + 0) = -4/3$, and would be visible. When imaged with $g = 0\ \bar{2}\ 2$, $g \cdot \mathbf{R} = (1/3)(0 - 2 + 2) = 0$ and would be invisible.

(b) If formed by Shockley partial $a/6[2\ 1\ 1]$, which lies in $(\bar{1}\ 1\ 1)$, $g \cdot R = (1/6)(4 - 2 + 0) = 1/3$ when $g = 2\ \bar{2}\ 0$, i.e. in contrast, but $g \cdot R = (1/6)(0 - 2 + 2) = 0$ or out of contrast when $g = 0\ \bar{2}\ 2$.

4.5.7 Application of dynamical theory

The kinematical theory, although very useful, has limitations. The equations are only valid when the crystal is oriented far from the exact Bragg condition, i.e. when s is large. The theory is also only strictly

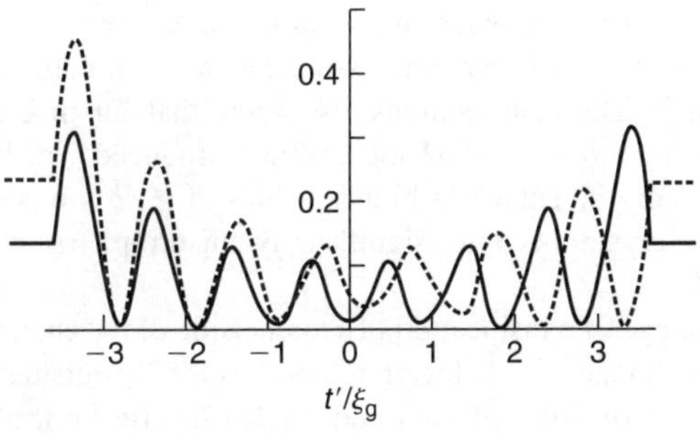

Figure 4.40 *Computed intensity profiles about the foil center for a stacking fault with $\alpha = +2\pi/3$. The full curve is the bright-field and the broken curve the dark-field image (from Hirsch et al., 1965, by courtesy of The Royal Society).*

applicable for foils whose thickness is less than about half an extinction distance ($\frac{1}{2}\xi_g$) and no account is taken of absorption. The dynamical theory has been developed to overcome these limitations.

The object of the dynamical theory is to take into account the interactions between the diffracted and transmitted waves. Again only two beams are considered, i.e. the transmitted and one diffracted beam, and experimentally it is usual to orient the specimen in a double-tilt stage so that only one strong diffracted beam is excited. The electron wave function is then considered to be made up of two plane waves – an incident or transmitted wave and a reflected or diffracted wave:

$$\psi(r) = \phi_0 \exp(2\pi i k_0 \cdot r) + \phi_g \exp(2\pi i k_1 \cdot r).$$

The two waves can be considered to propagate together down a column through the crystal, since the Bragg angle is small. Moreover, the amplitudes ϕ_0 and ϕ_g of the two waves are continually changing with depth z in the column because of the reflection of electrons from one wave to another. This is described by a pair of coupled first-order differential equations linking the wave amplitudes ϕ_0 and ϕ_g. Displacement of an atom **R** causes a phase change $\alpha = 2\pi g \cdot \mathbf{R}$ in the scattered wave, as before, and the two differential equations describing the dynamical equilibrium between incident and diffracted waves are

$$\frac{\mathrm{d}\phi_0}{\mathrm{d}z} = \frac{\pi i}{\xi_g}\phi_g$$

$$\frac{\mathrm{d}\phi_g}{\mathrm{d}z} = \frac{\pi i}{\xi_g}\phi_0 + 2\pi\phi_g\left(s + g \cdot \frac{\mathrm{d}\mathbf{R}}{\mathrm{d}z}\right).$$

These describe the change in reflected amplitude ϕ_g because electrons are reflected from the transmitted wave (this change is proportional to ϕ_0, the transmitted wave amplitude, and contains the phase factor) and the reflection in the reverse direction.

These equations show that the effect of a displacement **R** is to modify s locally, by an amount proportional to the derivative of the displacement, i.e. $\mathrm{d}\mathbf{R}/\mathrm{d}z$, which is the variation of displacement with depth z in the crystal. This was noted in the kinematical theory, where $\mathrm{d}\mathbf{R}/\mathrm{d}z$ is equivalent to a local tilt of the lattice planes. The variation of the intensities $|\phi_0|^2$ and $|\phi_g|^2$ for different positions of the column in the crystal, relative to the defect, then gives the bright- and dark-field images respectively. Figure 4.40 shows the bright- and dark-field intensity profiles from a stacking fault on an inclined plane, in full and broken lines, respectively. A wide variety of defects have been computed, some of which are summarized below:

1. *Dislocations*. In elastically isotropic crystals, perfect screw dislocations show no contrast if the condition $g \cdot b = 0$ is satisfied. Similarly, an edge dislocation will be invisible if $g \cdot b = 0$

and if $g \cdot b \times \mathbf{u} = 0$, where $\mathbf{u}$ is a unit vector along the dislocation line, and $b \times \mathbf{u}$ describes the secondary displacements associated with an edge dislocation normal to the dislocation line and b. The computations also show that for mixed dislocations and edge dislocations for which $g \cdot b \times \mathbf{u} < 0.64$ the contrast produced will be so weak as to render the dislocation virtually invisible. At higher values of $g \cdot b \times \mathbf{u}$ some contrast is expected. In addition, when the crystal becomes significantly anisotropic residual contrast can be observed even for $g \cdot b = 0$.

The image of a dislocation lies to one side of the core, the side being determined by $(g \cdot b)s$. Thus, the image of a dislocation loop lies totally outside the core when (using the appropriate convention) $(g \cdot b)s$ is positive and inside when $(g \cdot b)s$ is negative. Vacancy and interstitial loops can thus be distinguished by examining their size after changing from $+g$ to $-g$, since these loops differ only in the sign of b.

2. *Partial dislocations*. Partials for which $g \cdot b = \pm\frac{1}{3}$ (e.g. partial $a/6[\bar{1}\ \bar{1}\ 2]$ on (1 1 1) observed with 2 0 0 reflection) will be invisible at both small and large deviations from the Bragg condition. Partials examined under conditions for which $g \cdot b = \pm\frac{2}{3}$ (i.e. partial $a/6[\bar{2}\ 1\ 1]$ on (1 1 1) with 2 0 0 reflection) are visible except at large deviations from the Bragg condition. A partial dislocation lying above a similar stacking fault is visible for $g \cdot b = \pm\frac{1}{3}$ and invisible for $g \cdot b = \pm\frac{2}{3}$.
3. *Stacking faults*. For stacking faults running from top to bottom of the foil, the bright-field image is symmetrical about the center, whereas the dark-field image is asymmetrical (see Figure 4.40). The top of the foil can thus be determined from the non-complementary nature of the fringes by comparing bright- and dark-field images. Moreover, the intensity of the first fringe is determined by the sign of the phase factor α, such that when α is positive the first fringe is bright (corresponding to a higher transmitted intensity) and vice versa on a positive photographic print.

It is thus possible to distinguish between intrinsic and extrinsic faults and an example is shown in Figure 4.41 for an intrinsic fault on (1 1 1). The foil orientation is [1 1 0] and the non-complementary nature of the first fringe between bright- and dark-field indicates the top of the foil, marked T. Furthermore, from the bright-field images the first fringe is bright with $\bar{1}\ 1\ 1$, and dark with $1\ \bar{1}\ \bar{1}$ and $\bar{1}\ 1\ \bar{1}$.

4.5.8 Weak-beam microscopy

One of the limiting factors in the analysis of defects is the fact that dislocation images have widths of $\xi_g/3$, i.e. typically >10.0 nm. It therefore follows that dislocations closer together than about 20.0 nm are not generally resolved. With normal imaging techniques, the detail that can be observed is limited to a value about 50–100 times greater than the resolution of the microscope. This limitation can be overcome by application of the weak-beam technique, in which crystals are imaged in dark field using a very large deviation parameter s. Under these conditions the background intensity is very low so that weak images are seen in very high contrast and the dislocation images are narrow ($\approx$1.5 nm), as shown in Figure 4.42. At the large value of s used in weak beam, the transfer of energy from the direct to the diffracted beam is very small, i.e. the crystal is a long way from the Bragg condition and there is negligible diffraction. Moreover, it is only very near the core of the dislocation that the crystal planes are sufficiently bent to cause the Bragg condition to be locally satisfied, i.e. $g \cdot (\mathrm{d}R/\mathrm{d}z)$ is large enough to satisfy the condition $[s + g \cdot (\mathrm{d}R/\mathrm{d}z)] = 0$. Therefore, diffraction takes place from only a small volume near the center of the dislocation, giving rise to narrow images. The absolute intensity of these images is, however, very small, even though the signal-to-background ratio is high, and hence long exposures are necessary to record them.

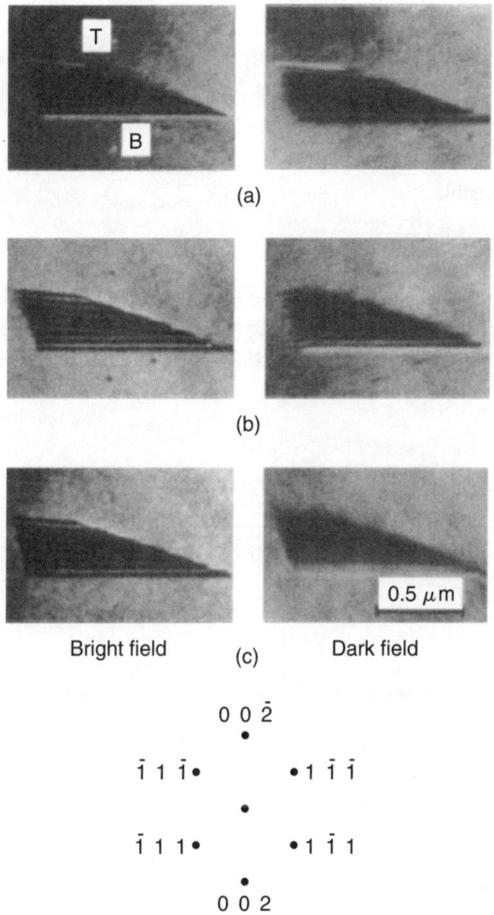

Figure 4.41 *Bright-field and dark-field micrographs of an intrinsic stacking fault in a copper–aluminum alloy: the operating diffraction vectors are: (a)* $\bar{1}\,1\,1$ *(b)* $1\,\bar{1}\,\bar{1}$ *and (c)* $\bar{1}\,1\,\bar{1}$ *(after Howie and Valdre, 1963; courtesy of Taylor & Francis).*

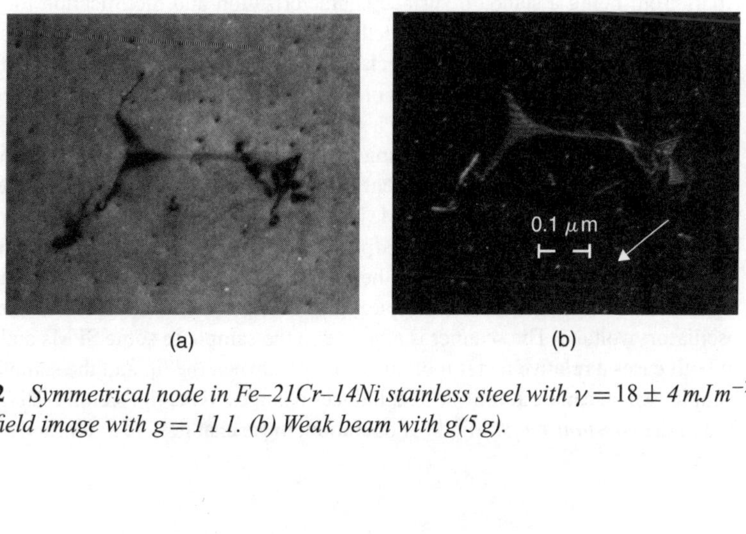

Figure 4.42 *Symmetrical node in Fe–21Cr–14Ni stainless steel with* $\gamma = 18 \pm 4\,mJ\,m^{-2}$. *(a) Bright-field image with* $g = 1\,1\,1$. *(b) Weak beam with* $g(5\,g)$.

Worked example

A dislocation in a specimen of slightly deformed Ni_3Al was imaged in a TEM in bright field at the Bragg position for various reflections. Visibilities and invisibilities were as follows:

$\bar{1}\ 1\ 1$ visible

$1\ 1\ \bar{1}$ visible

$0\ 2\ 0$ invisible

On changing to weak beam dark field ($\bar{1}\ 1\ 1$; $g/5g$), it could be seen that the dislocation consisted of two partial dislocations of similar appearance. When g was reversed to $1\ \bar{1}\ \bar{1}$ (still $g/5g$), the separation and appearance of the partials were unchanged. Explain these observations.

Solution

Glide dislocations in Ni_3Al have Burgers vector <1 1 0>. Thus,

	[1 1 0]	[1 $\bar{1}$ 0]	[1 0 1]	[1 0 $\bar{1}$]	[0 1 $\bar{1}$]	[0 1 1]
$\bar{1}$ 1 1	i	v	i	v	i	v
1 1 $\bar{1}$	v	i	i	v	v	i
0 2 0	v	v	i	i	v	v

The Burgers vector is [1 0 $\bar{1}$].

Weak beam shows the dislocation dissociated into two partials. Probably these are 1/2[1 0 $\bar{1}$], especially since the appearance and separation do not change on reversing g (i.e. not a dipole).

4.6 Scanning probe microscopy

Scanning probe microscopy (SPM) was invented as a microscopic technique capable of producing topographical images of samples with atomic resolution, but subsequently it has also been exploited as a tool for modifying the atomic structure of surfaces. The introduction of SPM not only witnessed but also contributed significantly to the rapid development of nanotechnology over the last two decades. Apart from being a standard surface characterization and modification tool for basic nanoscience research, the development of SPM led to the maturation of a range of techniques for the accurate positioning of submicron-sized objects down to sub-angstrom displacement resolution. Experience learnt from the development of such precision positioning techniques has benefitted enormously the development of nanotechnologies in a wide context.

SPM includes (i) scanning tunneling microscopy (STM) and (ii) atomic force microscopy (AFM). Figure 4.43 shows the basic components in an SPM. A sharp tip, which can be electrically conducting as in STM, or insulating as in AFM, is used to probe the surface. The tip usually has submicron but not necessarily atomic scale sharpness. All SPMs have a computer-controlled mechanism for approaching the tip to the surface at the beginning of an experiment. Scanning is usually provided by a piezoelectric scanner, which is a piezoelectric tube that bends back and forth when excited by an oscillatory voltage. The scanner is attached to the sample in some SPMs and to the tip in others, and in both cases a relative raster motion is set up between the tip and the sample. The roughness on the sample, which can be atomic scaled, induces changes to the tip–sample interactions during scanning, and in the *constant-height* mode of operation, these changes are magnified electronically to generate

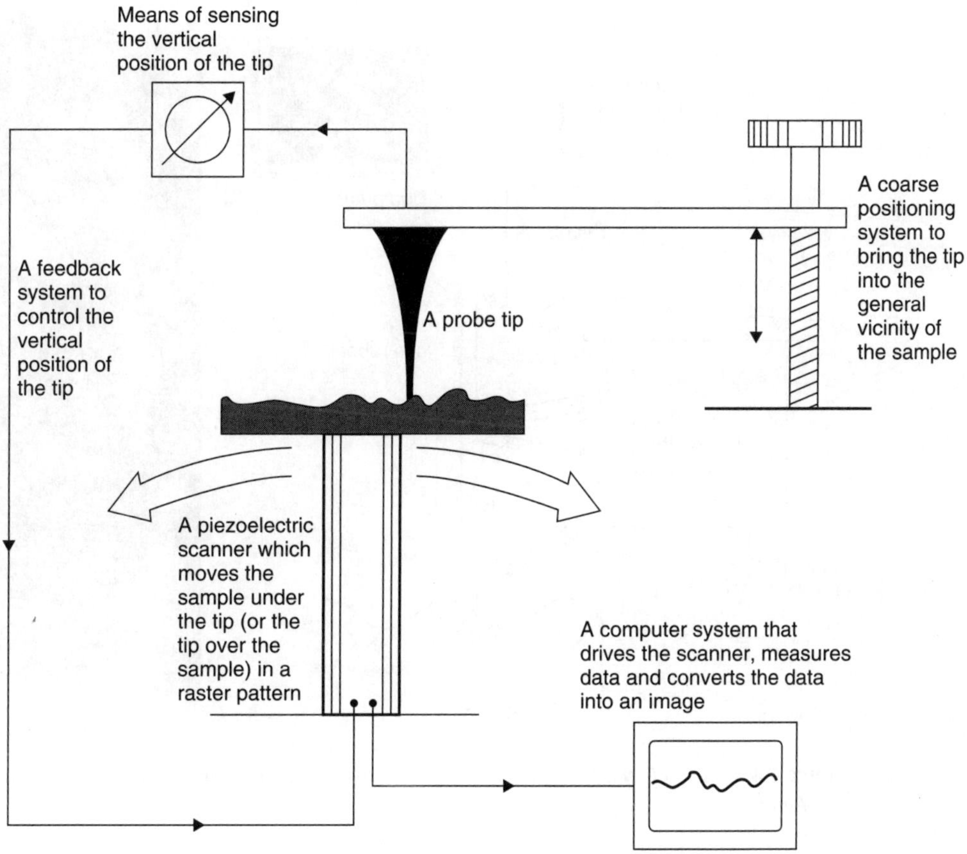

Figure 4.43 *Basic construction of an SPM (courtesy of Park Scientific Instruments).*

a topographical image of the sample. The scanner is also capable of lengthening and shortening by an applied voltage in the appropriate direction, and in the *constant-signal* mode of operation, a feedback control loop is used to continuously adjust the height of the scanner to maintain an approximately constant tip–sample interaction during scanning. In this case, the signal driving the height change of the scanner is used to generate the topographical image of the sample.

4.6.1 Scanning tunneling microscopy (STM)

The STM was invented by G. Binnig and H. Rohrer at IBM Zürich Research Laboratory in 1981, and in 1986 both were awarded the Nobel Prize for their invention. Only electrically conducting samples can be imaged by an STM, and the tip used is also made of a conductor such as tungsten or Pt–Ir alloy. The tip is brought into a small distance d from the sample surface, as shown in Figure 4.44a, and a voltage applied between the two produces a tunneling current to flow.

Electron tunneling is a wave-mechanical phenomenon predictable by the Schrödinger equation, the time-independent version of which can be written as

$$-\frac{\hbar^2}{2m}\nabla^2\psi = (E - V)\psi. \tag{4.18}$$

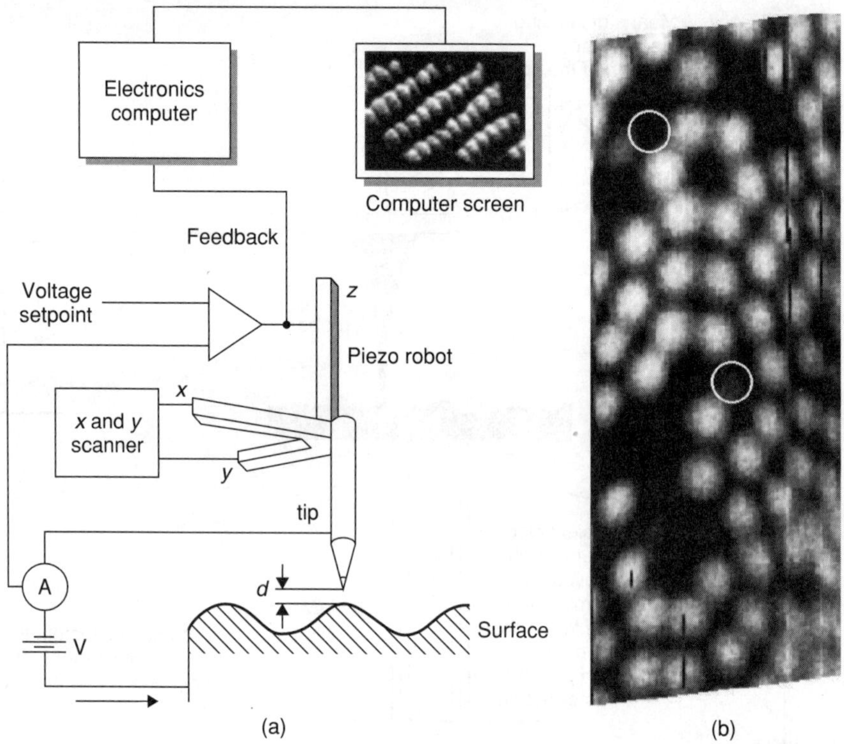

Figure 4.44 *(a) Schematic of an STM. (b) An image revealing atomic locations (courtesy J. B. Pethica).*

Here ψ (*x*,*y*,*z*) is the wave function such that $|\psi(x,y,z)|^2$ is the probability of finding an electron at the point (*x*,*y*,*z*), *m* is the mass of an electron, $\hbar = 2\pi \times$ Planck's constant, *E* the total energy and *V* the potential energy of the electron. The STM configuration can be idealized as shown in Figure 4.45a. Here, the gap *d* constitutes a potential barrier (Region 2) that separates the tip and the sample, which are modeled as two low-potential regions 1 and 3 respectively. Electrons are now injected in Region 1 towards the barrier at 2. Applying equation (4.18) to the three regions yields the following:

Region 1: $-\dfrac{\hbar^2}{2m}\dfrac{d^2\psi_1}{dz^2} = E\psi_1$, giving $\psi_1 = e^{ikz} + Ae^{-ikz}$,

where $k^2 = 2mE/\hbar^2$, e^{ikz} is the incident wave, and Ae^{-ikz} is the reflected wave;

Region 2: $-\dfrac{\hbar^2}{2m}\dfrac{d^2\psi_2}{dz^2} + (V_o - E)\psi_2 = 0$, giving $\psi_2 = Be^{-\kappa z} + Ce^{\kappa z}$,
where V_o is the potential height, and $\kappa^2 = 2m(V_o - E)/\hbar^2$; and

Region 3: $-\dfrac{\hbar^2}{2m}\dfrac{d^2\psi_3}{dz^2} = E\psi_3$, giving $\psi_3 = De^{ikz}$ as the transmitted wave.

Matching the values of ψ and $d\psi/dz$ in the above solutions at $z = 0$ and $z = d$ gives the wave amplitude in Region 3 as:

$$D = \frac{-4ik\kappa}{e^{ikd}[e^{-\kappa d}(\kappa + ik)^2 - e^{\kappa d}(\kappa - ik)^2]}.$$

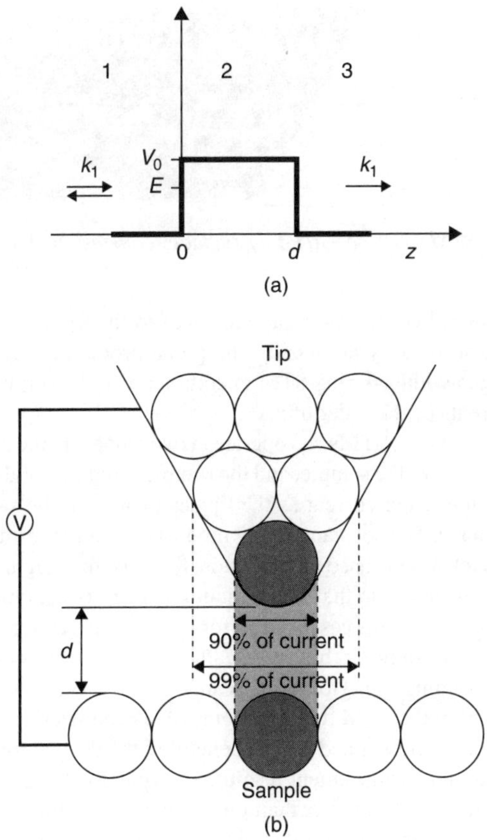

Figure 4.45 *(a) Potential barrier through which tunneling occurs. (b) Tunneling between STM tip and sample.*

Since in the above, the incident electron wave e^{ik} in Region 1 has unit amplitude, a transmission coefficient T can be defined as:

$$T = |D|^2 = \frac{1}{1 + [(k^2 + \kappa^2)^2/(4k^2\kappa^2)] \sinh^2(\kappa d)},$$

which tends to the value $[16k^2\kappa^2/(k^2 + \kappa^2)^2]\, e^{-2kd}$ for $\kappa d \gg 1$. A non-zero T therefore means that electrons incident on the barrier at Region 2 can leak through it, and this phenomenon is known as tunneling. The tunneling current I is proportional to T, and so for $\kappa d \gg 1$,

$$I \propto e^{-2\kappa d}, \tag{4.19}$$

where $\kappa = \sqrt{2m(V_o - E)}/\hbar$ is a constant that increases with the barrier height V_o. The exponential dependence in equation (4.19) means that tunneling current is very sensitive to the gap distance d. In an STM tip, although many atoms can form tunneling current with the sample, as shown in Figure 4.45b, the few atoms that are closest to the sample contribute to the majority of the overall

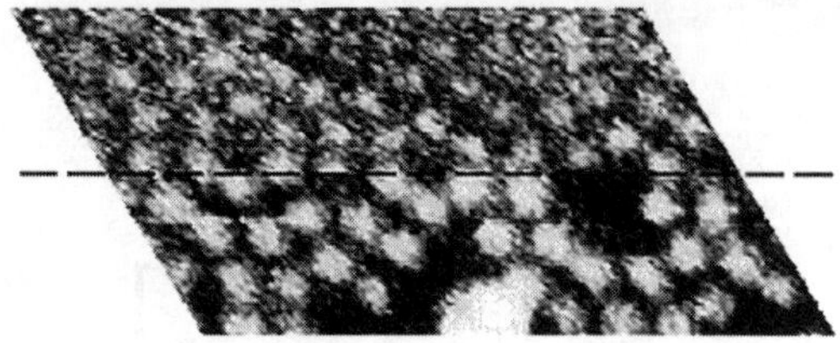

Figure 4.46 *STM image of Si(1 1 1) surface (courtesy of J. B. Pethica).*

tunneling current, because their gap distances to the sample are shortest. For this reason, an STM tip that is not necessarily atomically sharp can produce topographical images with atomic resolution. Typical gap widths during STM operation are at the sub-nanometer level, and these give rise to tunneling currents of the order of nA.

As mentioned above, STMs can operate in two modes. In the *constant-height* mode, the tip travels at constant height above the sample, and the sample roughness induces changes in the tunneling current. The changing tunneling current as the tip scans over on the sample surface is used to generate the topographic image. The constant-height mode works only for smooth samples. In the *constant-current* mode, a feedback loop is used to continuously adjust the height of either the sample or the tip to keep a constant gap distance so that the tunneling current is maintained constant. The signal driving the sample or tip height in this case is used for image construction. The constant-current mode can cope with more irregular samples, but the usually rather slow response time of the feedback loop makes it more time-consuming to perform one scan.

The invention of the STM required tremendous engineering efforts in solving a number of critical problems which, as mentioned above, benefitted the development of nanotechnology in many ways. First, STM requires displacement resolutions typically down to ~0.01 Å or 1 pm, which are at least six orders of magnitude smaller than the typical floor vibration amplitudes of 0.1–1 μm. STMs are therefore equipped with special damping systems involving clever mechanical designs plus a wise choice of damping materials to provide the necessary vibration insulation. Secondly, techniques had to be developed to position objects to very high precisions. During the 'approach' operation before the start of a scanning experiment, the tip has to be brought to a suitable interacting distance from the sample surface. This is often the most difficult step in high-resolution work. During a 'coarse' approach, the sample is moved from millimeter to submicron distances towards the tip, and this is usually accomplished by servo-motors. The 'fine' approach is done by a piezoelectric drive of the tip.

Technologies also had to be available to fabricate suitable STM tips, but fortunately, before the birth of STM, sharp tips were already frequently used in field-ion microscopy, and so the technologies used to fabricate them were relatively mature. For ultra-high-vacuum work, STM tips are usually made of tungsten. Tungsten tips are usually made by electrolytic etching, and focused-ion beam milling can be applied afterwards to further increase the sharpness if necessary. Special tungsten tips which can handle extremely rough surfaces with deep grooves can also be made by electron beam-induced chemical vapor deposition of tungsten on top of a substrate tip. Tips with diameters as small as ~10 nm and with high aspect ratios of ~10 can be fabricated this way. For topographic STM work in air or moderate vacuum conditions, platinum–iridium (Pt–Ir) alloy tips are used. Pt–Ir tips can be simply prepared by mechanically cutting a piece of Pt–Ir wire at an oblique angle to its length. The severe deformation can, by chance, produce very sharp severed sections suitable for STM imaging. During repeated scanning, it is possible that the tip can be changed by interaction with the sample. Tip change can lead to scanning artefacts, as illustrated by the example in Figure 4.46. Here, the change in the image quality across the dashed line, which is along the scanning direction, is likely to be due to tip modification.

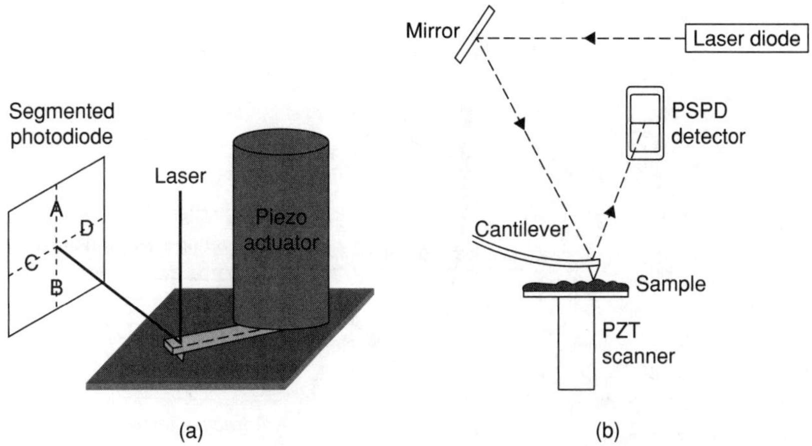

Figure 4.47 *Two designs of AFM (courtesy of Park Scientific Instruments).*

4.6.2 Atomic force microscopy (AFM)

In an AFM, the mechanical interaction between the tip and the sample is exploited to yield surface topography. The probing tip is typically a sharp tip which is a few microns in length and less than 10 nm in diameter. The tip is attached to the free end of a micro-cantilever that is typically 100–200 μm long. During scanning, the mechanical interaction with the sample surface causes the cantilever to deflect, just like a soundtrack induces vibrations on a stylus in an old record player. In an AFM, a detector measures the cantilever deflection as the tip is scanned over the sample, as shown in Figure 4.47a, or the sample is scanned under the tip, as in Figure 4.47b. The cantilever deflection is usually measured by shining a focused laser beam on the back side of the cantilever, and by either detecting the reflected position of the laser by a position-sensitive photodiode, as shown in Figure 4.47, or by measuring the path difference between the incident and the reflected beams, by collecting the reflected beam in the same optical fiber that delivers the incident beam. The measured cantilever deflection is used to generate the topographical image of the sample surface. Unlike STM, which can only be used to study conductors, both insulators and conductors can be studied by AFM.

AFM tips are usually made of a rigid ceramic such as Si_3N_4 by lithographic means. Apart from being chemically inert, ceramics are chosen because they have high elastic modulus-to-density ratio, which means that the natural frequency of the cantilever can be made very high if this is required. Also, ceramics are typically low damping materials due to the difficulty of operation of dislocation processes, and so ceramic cantilevers have high sensitivity.

An AFM can operate in either the *contact mode* or *non-contact mode*. In the contact mode, the tip makes ‘mechanical contact’ with the sample, meaning that the tip is held less than a few angstroms from the sample surface, so that the tip–sample interaction is within the repulsive regime, as shown in Figure 4.48. In this case, the tip–sample interaction stiffness is much higher than the stiffness of the cantilever to which the tip is attached. Typical microfabricated AFM cantilevers have stiffness of the order of 0.1–1 $N\,m^{-1}$, but the ‘stiffness’ of an atomic bond in metallic samples is much higher, of the order of $10^2\,N\,m^{-1}$. The cantilever therefore deflects rather than having the tip penetrating into the sample, and so, during scanning, the tip’s deflection follows closely the sample topography. Either *constant-height* or *constant-force* images can be generated by fixing the height of the scanner or the tip–sample interaction force through a feedback loop, as explained above.

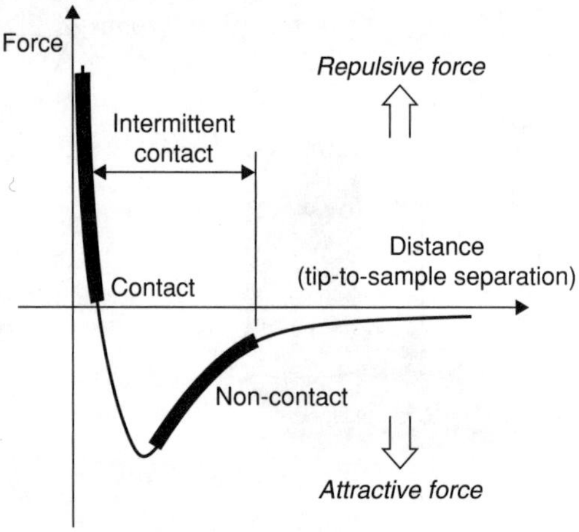

Figure 4.48 *Tip–sample interaction force curve (courtesy of Park Scientific Instruments).*

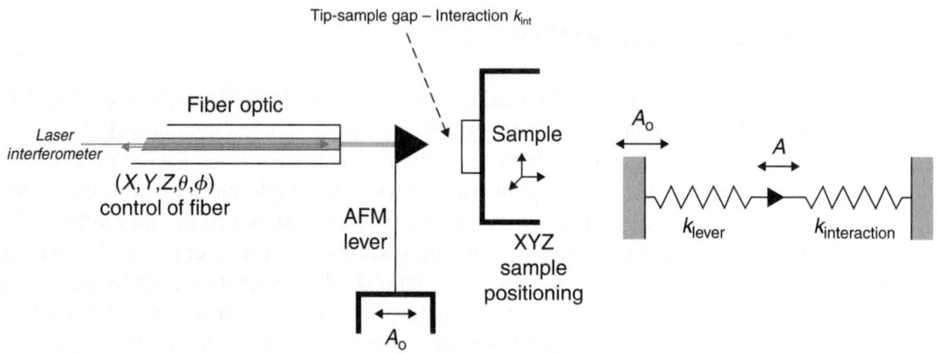

Figure 4.49 *Tip–sample interaction in NC-AFM.*

In *non-contact* AFM (NC-AFM), the cantilever is vibrated near the sample surface, while keeping a distance of the order of tens to hundreds of angstroms away from it. The tip is vibrated about its resonant frequency, which is typically between 100 and 400 Hz, with an amplitude a few tens to hundreds of angstroms. The tip–sample interaction can be described by the model shown in Figure 4.49. Neglecting inertia effects, force balance on the tip gives $k_{int}A = k_{lev}(A_o - A)$ or

$$A/A_o = 1/(1 + k_{int}/k_{lev}), \tag{4.20}$$

where A_o is the driving amplitude on the lever base, A is the responding amplitude of the tip, k_{int} is the force gradient of the tip–sample potential (positive if repulsive) and k_{lev} is the stiffness of the cantilever. Since NC-AFM is usually operated within the attractive force region, $k_{int} < 0$ and so $A/A_o > 1$, but in any case, the response amplitude ratio A/A_o is a direct measure of the tip–sample interaction k_{int}. Apart from the vibrating amplitude, the resonant frequency ω is also a direct measure of k_{int}, because

$$\omega = \sqrt{(k_{lev} + k_{int})/M}, \tag{4.21}$$

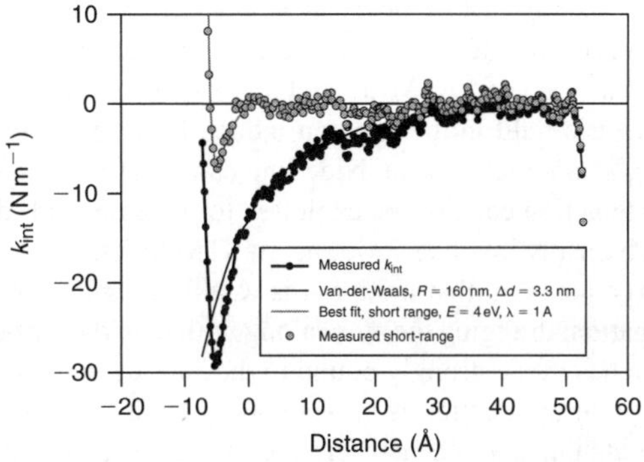

Figure 4.50 *Measuring short-range interactions with sub-angstrom amplitude AFM (Pethica and co-workers, 2001).*

where M is the effective mass of the cantilever. Therefore, by monitoring either the vibrational amplitude or the resonating frequency of the cantilever, and by keeping this constant using a feedback loop that continuously adjusts the height of the scanner, a topographical image of the sample can be obtained from the signal that drives the motion of the scanner. Understandably, tip and sample degradation is in general more rapid in contact-AFM than in NC-AFM. Also, NC-AFM is better than contact-AFM in measuring soft samples, including liquid surfaces.

A hybrid mode between the contact and non-contact modes can also be performed. This is known as intermittent-contact AFM or, more commonly, the 'tapping' mode. The tapping mode is very similar to NC-AFM, except that the vibrating tip is brought closer to the sample, so that it barely hits, or 'taps', the sample. As in NC-AFM, the image is generated by monitoring and controlling the cantilever's oscillation amplitude or resonant frequency changes in response to the tip–sample spacing. The tapping mode can overcome some of the limitations of both contact and non-contact AFM. For example, the tapping mode is less likely to cause damage to both the tip and the sample than contact-AFM, and also it is more effective than NC-AFM in coping with topographies with greater variations in height.

Another newer technique is small-amplitude AFM. The usual vibrating-tip technique as in NC- or tapping AFM is seldom performed within or even close to the repulsive force region for which $k_{int} > 0$. To apply the technique to the repulsive regime, for example, in an attempt to measure the short-range tip–sample interaction, because the tip is already very close to the sample surface, the tip amplitude has to be in the sub-angstrom regime. This is achievable by using a very stiff cantilever (for example, a tungsten lever with $k_{lev} = 100\,\mathrm{N\,m^{-1}}$) vibrating at a sub-angstrom amplitude A_o at a sub-resonance frequency. The resultant amplitude A is also sub-angstrom, because $A/A_o < 1$ in this case. To achieve the high sensitivity required ($\sim 4 \times 10^{-4}\,\text{Å}/\sqrt{\mathrm{Hz}}$), the end of the optical fiber and the back of the cantilever (see Figure 4.49) have to be coated with Au to enhance the reflectivity, so as to encourage multiple reflections which will greatly improve the measured sensitivity. After subtracting the van der Waals component from the measured k_{int}, the short-range atomic interaction can be revealed, as shown in the example given in Figure 4.50.

4.6.3 Applications of SPM

Apart from a surface microscopy technique, SPMs are also being continuously and rapidly developed into various tools for surface modification. Simple ideas, such as modifying an AFM tip into a nano-scale 'ball pen' construction which can transfer organic molecules as 'ink' onto a surface, are being put into practice. More novel ideas involve exploiting the chemical, electrostatic, magnetic and other means of interaction between the tip and the sample. For example, by applying an electric voltage across a Pt-coated AFM tip and a metal or semiconductor sample in a moist environment, the intense electric field in the gap can induce local oxidation of the sample, and so by traversing the tip slowly, a nanometer-wide oxide line can be written on the sample surface. Material-specific tip–sample interaction can also be exploited to 'pick up' and 'drag' individual atoms along a sample surface. This typically involves lowering an STM tip onto a surface atom to within the attractive van der Waals force range so that the atom is weakly bound to the tip. The tip is then translated to the desired destination, dragging the atom along with it on the surface. Retracting the tip will then release the atom, which is more strongly bound to the surface in the normal direction.

Another type of application is to develop AFMs as nano-scale mechanical testing devices. Truly nano-scale scratching and indentation experiments are possible with an AFM tip, albeit the very rapid tip damage in this case due to the much smaller tip size than a Berkovich diamond tip as in usual nanoindentation, for example. In more novel applications, tensile tests on single organic molecules such as DNA can be performed. These are done by dipping an AFM tip into a solution containing the molecules and, by chance, several of these molecules will be adhered to the tip surface. Lifting the tip out of the solution will then stretch these molecules in succession, first the shortest one and, when this breaks or its junction with the tip slips, the second shortest one, and so on. The tensile force exerted by the tip is measurable from the cantilever deflection, and because the molecules are stretched in succession, their individual tensile properties may be revealed from the recorded force data.

An STM also allows the electronic properties of a nano-sized structure to be measured. In particular, the electron density of states (DOS) is related to the current (I) and voltage (V) characteristics by:

$$\mathrm{DOS} \propto \left(\frac{\mathrm{d}I}{\mathrm{d}V}\right)\left(\frac{V}{I}\right). \tag{4.22}$$

For example, the I–V curve of a semiconductor exhibits a plateau, over which $\mathrm{d}I/\mathrm{d}V = 0$, between two conducting regions for which $\mathrm{d}I/\mathrm{d}V > 0$. The DOS is therefore positive on either side of the plateau and is zero within it, corresponding to the band structure expected from a semiconductor, with the zero DOS region representing the band gap. The I–V characteristics of a nano-sized structure such as a nanoparticle or a nanowire can be conveniently measured by an STM, by holding the tip over a fixed location on the structure, and by recording the tunneling current at different applied voltages. Figure 4.51 shows the tunneling I–V curves measured from silicon nanowires with different diameters in the nano-scale. The resultant band structures on the right reveal a remarkable dependence of the band gap on the sample size, a phenomenon which is known as the quantum confinement effect. The enlarged band gap at very small sample sizes allows blue-shift of the florescence from the nanowires from the infrared regime at large sizes to the ultraviolet regime at small sizes.

4.6.4 Nanoindentation

Traditional microhardness tests involve applied loads of at least a few gf, and for most materials the resultant indents are usually large enough to be observable by an optical microscope (Section 4.2.2.4). However, the typical resolution of the optical microscope imposes a limit of about 1 μm for the smallest indent size to be measurable, and this restricts the application of the technique to very small samples such as fine conduction lines or thin films in microelectronics. In the 1980s, in a development closely related to SPM, a new technique known as nanoindentation was established. Nanoindentation involves applied loads in the sub-millinewton (<1 mN) range, so that the usual indent size is in the sub-micron regime. Unlike traditional microindentation, nanoindentation allows

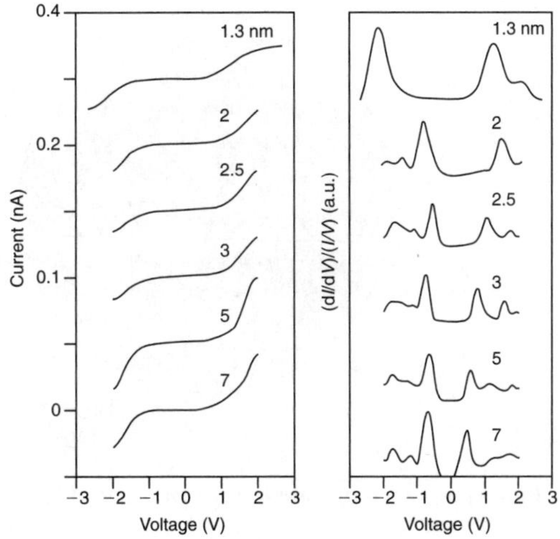

Figure 4.51 *STM current (I) vs voltage (V) curves measured from Si nanowires with different diameters (left). The (dI/dV)(V/I) vs V curves revealing an increase in the band gap as size decreases (right) (C. S. Lee and co-workers).*

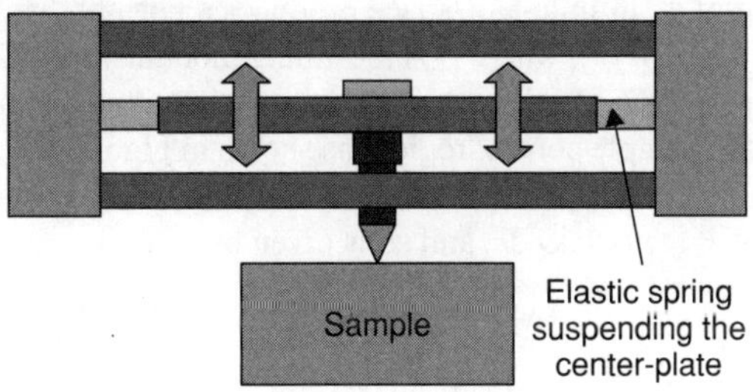

Figure 4.52 *Schematic of a capacitive nanoindenter transducer (courtesy of Hysitron Inc.).*

the load and displacement to be continuously controlled and recorded during the indentation process, and so the technique is also known as depth-sensing indentation. The load–displacement data are analyzed afterwards using standard contact mechanics equations to yield quantities such as hardness and elastic modulus from sub-micron-sized indents, without the need to image the indents. Newer nanoindenters that were manufactured after 2000 are also commonly equipped with an AFM objective for direct imaging of the indents, and this provides independent measurement of the tip–sample contact size in the sub-micron regime.

A state-of-the-art nanoindenter has sub-angstrom displacement resolution and nano-Newton force resolution. These high resolutions are achievable by electromagnetic or capacitive principles in the load–displacement transducer. In one design, the nanoindenter tip is attached to the center plate of a three-plate capacitor, as shown in Figure 4.52. Application of a d.c. voltage of up to several hundred volts results in an electrostatic force to drive the tip down to the sample.

The Vickers and Knoop tip geometries used in microhardness testing are four-sided pyramids, but because four planes in general do not meet at a point, these are usually not sharp enough for nanoindentation purposes. Common nanoindentation tips are therefore three-sided pyramids, which are usually sharper than the four-sided geometry. One standard shape is the Berkovich geometry, which has an included apex angle of 142°, giving the same projected area vs depth relation as the conventional Vickers geometry.

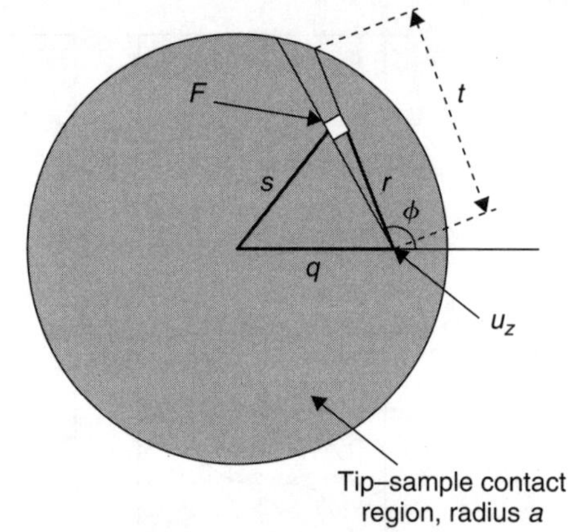

Figure 4.53 *Plan view of the tip–sample contact region.*

A standard procedure to analyze the nanoindentation data was proposed by Oliver and Pharr in 1992, which makes use of the specimen's unloading response, assumed to be purely elastic, to calculate the tip–sample contact size at the onset of unloading. This method is based on the Hertz theory for elastic contact and is briefly outlined here. From the theory of stress potential, a point force F acting normal on the surface of an infinite half-space produces a normal displacement u_z on the surface, given by $u_z = [(1-\nu^2)/\pi E](F/r)$, where E is the Young modulus, ν the Poisson ratio and r the radial distance from the position of F. In the nanoindentation situation, the indenter exerts a pressure distribution $p(s)$ over the tip–sample contact region, as shown in Figure 4.53. By the principle of superposition in elasticity, the normal displacement of the sample surface is due to contributions from infinitesimal point forces of $F = p(s)r\mathrm{d}\phi\ \mathrm{d}r$, and so is given by

$$u_z(q) = \frac{(1-\nu^2)}{\pi E} \int_{\phi=-\pi}^{\pi} \int_{r=0}^{t} p(s)\mathrm{d}r\mathrm{d}\phi, \quad q < a. \tag{4.23}$$

Here a is the radius of the tip–sample contact region, which is assumed to be circular: $s^2 = r^2 + q^2 + 2rq\cos\phi$ and $t = -q\cos\phi + \sqrt{a^2 - q^2\sin^2\phi}$.

Assuming that before deformation, the specimen surface and the tip are both spherical with radii of curvature R_1 and R_2 respectively ($R_1 \to \infty$ if the specimen surface is flat). For a small radial distance q from the indentation axis (see Figure. 4.53), the specimen and the tip surfaces are well approximated by $z_1 \approx q^2/(2R_1)$ and $z_2 \approx q^2/(2R_2)$ respectively. The deformation requires that $z_1 + u_{z1} + z_2 + u_{z2} = h$, where h is the relative displacement of the tip into the sample surface, and u_{z1} and u_{z2} are deformations of the specimen surface and the tip according to equation (4.23). Therefore, equation (4.23) becomes

$$h - \frac{q^2}{2R} = \frac{1}{\pi E_r} \int_{\phi=-\pi}^{\pi} \int_{r=0}^{t} p(s)\mathrm{d}r\mathrm{d}\phi, \tag{4.24}$$

where R and E_r, given by $1/R = (1/R_1) + (1/R_2)$ and $1/E_r = [(1-\nu_1^2)/E_1] + [(1-\nu_2^2)/E_2]$, are respectively an effective radius and elastic modulus. E_r is known as the reduced modulus of the tip–sample contact. A new Berkovich diamond indenter should have a tip radius R_2 as small as ~50 nm, but a worn-out tip may have an R_2 value as large as a micron. Diamond tips have $E_2 = 1140$ GPa and $\nu_2 = 0.07$, and so if E_r is known, the quantity $(1-\nu_1^2)/E_1$ for the sample can be obtained.

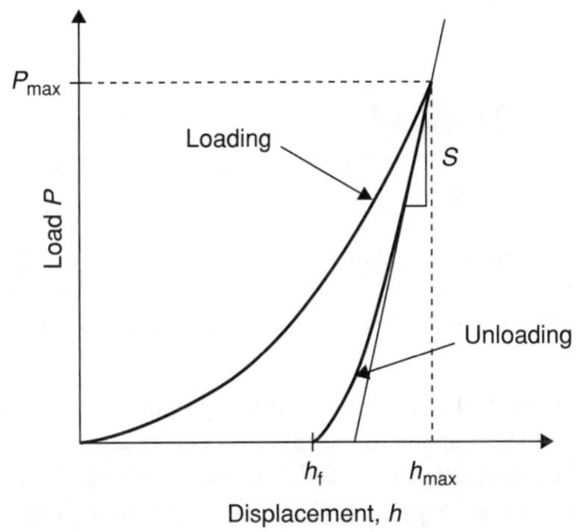

Figure 4.54 *Schematic load–displacement graph during nanoindentation.*

Hertz showed that the solution to equation (4.24) is $p(s) = p_o\sqrt{1-(s/a)^2}$, where $a = (\pi R p_o)/(2E_r)$ and $h = (\pi a p_o)/(2E_r)$. The total indentation load is given by $P = \int_0^a 2\pi s p(s) ds = (2\pi p_o a^2/3)$. Key results of the Hertzian theory are therefore as follows:

$$a = \left(\frac{3PR}{4E_r}\right)^{1/3}; \quad h = \frac{a^2}{R} = \left(\frac{9P^2}{16RE_r^2}\right)^{1/3}; \quad p_o = \frac{3P}{2\pi a^2} = \left(\frac{6PE_r^2}{\pi^3 R^2}\right)^{1/3}. \tag{4.25}$$

The contact stiffness S between two elastic spheres is defined as $S = dP/dh$, and from the first two equations in (4.25), this is given by

$$E_r = \frac{\sqrt{\pi}}{2}\frac{S}{\sqrt{A_c}}, \tag{4.26}$$

where $A_c = \pi a^2$ is the projected area of the tip–sample contact circle. Equation (4.26) suggests that the elastic modulus E_r can be estimated if S and A_c are measured. In addition, knowledge of A_c enables hardness to be evaluated as

$$\text{Hardness} = \frac{P_{max}}{A_c}. \tag{4.27}$$

Note that the 'nanohardness' defined in equation (4.27) is based on the projected area at full load. Ductile metals usually exhibit negligible elastic or viscoelastic recovery on unloading and so the projected indent area at full load is about the same as the residual area after load removal. In this case, the Vickers hardness defined in equation (4.1), which is based on the surface area, is 0.927 times the projected-area hardness defined in equation (4.27). In polymers, significant recovery can take place upon load removal and, in this case, the relation between the hardness defined in equation (4.27) and the Vickers hardness is by and large unclear.

Figure 4.54 shows a schematic load–displacement curve during a typical loading–unloading nanoindentation experiment. The unloading curve does not follow the loading curve because of the occurrence of plasticity. In the Oliver–Pharr method, the contact stiffness S is measured as the slope

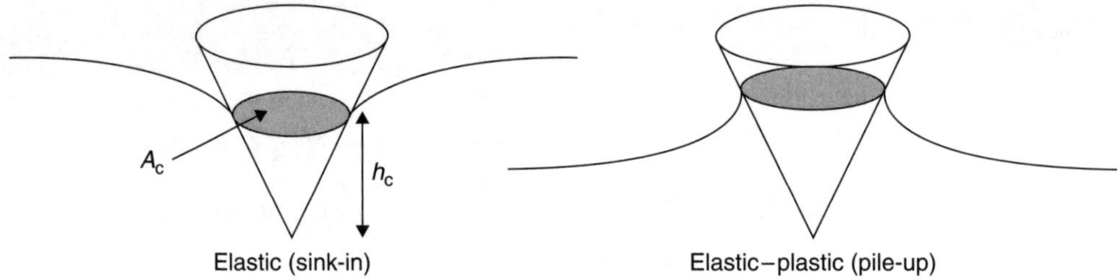

Figure 4.55 *Pile-up and sink-in morphologies of indentation.*

of the unloading part of a nanoindentation curve at the maximum load P_{max}. Oliver and Pharr also proposed that if the tip-shape function $A_c = f(h_c)$ is known (see Figure 4.55), A_c can be obtained by measuring the contact depth h_c at full load. For a spherical tip, $h_c \approx a^2/2R_2$, so $A_c = \pi a^2 = 2\pi R_2 h_c$, i.e. $A_c \propto h_c$. A conical or pyramidal tip has a quadratic shape function, i.e. $A_c \propto h_c^2$; for example, the tip-shape function for an ideal Berkovich tip is $A_c = 24.5h_c^2$.

For an unknown sample, Oliver and Pharr proposed using the following formula to calculate the contact depth h_c from the load–displacement data:

$$h_c = h_{max} - \varepsilon \frac{P_{max}}{S}. \tag{4.28}$$

Here h_{max} is the total indenter displacement at full load P_{max} (see Figure 4.54) and ε is a constant for a given indenter. For a spherical indenter, we have seen that $h_c \approx a^2/2R_2$. If the sample surface is initially flat, $R = R_2$, and so $h_{max} \approx a^2/2R_2$ from the second equation in (4.25). Therefore, $h_c \approx h_{max}/2$. However, this relation was derived assuming pure elasticity. In plastic indentation situations, the corresponding plastic depth, h_f, should be deducted from the h data, i.e. instead of $h_c = h_{max}/2$, we have

$$(h_c - h_f) = \frac{(h_{max} - h_f)}{2}. \tag{4.29}$$

Also, from the second equation in (4.25), $P \propto (h - h_f)^{3/2}$, and from this, the contact stiffness is given by

$$\frac{dP}{dh} = S = \frac{3}{2}\frac{P_{max}}{(h_{max} - h_f)}. \tag{4.30}$$

Combining equations (4.29) and (4.30), we have $h_c = h_{max} - (3/4)P_{max}/S$, so that equation (4.28) is verified with $\varepsilon = \frac{3}{4} = 0.75$, for the case of a spherical indenter on an initially flat specimen. Table 4.2 gives the values of ε for other indenter geometries.

In applying the Oliver–Pharr method to analyze nanoindentation data, several points require special attention:

(i) *Pile-up and sink-in around the indent.* The Oliver–Pharr method originates from elastic contact mechanics equations which predict that the indent can only be of the 'sink-in' type. However, pile-up around an indent can also occur, as shown in Figure 4.55. If a specimen has a low work-hardening rate (e.g. a metal which has received severe deformation before indentation, so that further deformation will cause less stress increment), it is easier for the material underneath the indenter to flow sideways to make room for the indenter, rather than to flow downward to cause the underlying material to deform. Material will then pile up around the indenter in this

Table 4.2 *Values of ε in equation (4.28).*

Indenter shape	ε
Spherical and paraboloid	$\frac{3}{4}=0.75$
Flat ended	1
Conical	$\frac{2}{\pi}(\pi-2)=0.73$

case. A specimen with a high work-hardening rate (e.g. an annealed metal) is likely to have a sink-in morphology for the indent, because the material just underneath the indent will be quickly hardened and so further penetration of the tip can only proceed by deformation of the underlying material, i.e. by spreading the deformation front deeper into the specimen. If the pile-up situation occurs, the Oliver–Pharr method may give rise to erroneous results.

(ii) *Thermal and electronic drifts*. In addition to electronic drifts, the major cause for drift in a nanoindentation experiment is thermal expansion or contraction of machine parts. Drifts are detrimental to nanoindentation analyses. As a typical example, a drift rate of 0.1 nm s^{-1} can occur during a test lasting for, say, 5 minutes. The total displacement drifted would be $0.1\times300=30$ nm. This may not be negligible compared to h_{max}, which may be as small as 100 nm for hard materials under a small load. In very stable situations, drift rates of the order of ~0.01 nm s^{-1} are achievable, but drift rates of the order of ~0.1 nm s^{-1} are more commonly encountered. The average drift rate during a test can be measured by inserting a holding period at a very low load in the load schedule, and then to monitor the displacement change. The measured drift rate $\dot{h}$ can be used to correct the displacement data h according to $h_{corrected}(t)=h_{raw}(t)-\dot{h}\times t$.

(iii) *Creep*. The Oliver–Pharr method assumes purely elastic recovery of the sample during unloading. However, under nanoindentation situations, even high-melting solids can exhibit a certain extent of viscoelastic deformation, or creep, during the initial stage of unloading. A sample can creep more severely if: (a) the unloading speed is slow, (b) there is no or insufficient holding of the load before unloading and (c) the peak load is high. Also, soft materials such as polymers and biotissues creep much faster than hard materials like metals and ceramics. Creep is also more significant if the nanoindentation test is performed at an elevated temperature. In extreme cases, the creep during unloading may be so severe that the unloading curve exhibits a 'nose' (see Figure 11.14), so that the indenter tip keeps sinking into the sample even though the force on it is reducing. In this case the apparent contact stiffness S will be negative, and the application of the Oliver–Pharr method (equation (4.26)) will lead to a negative estimate of the elastic modulus, which is obviously erroneous. Creep effects, however, can be corrected for by the following equation:

$$\frac{1}{S_e}=\frac{1}{S}+\frac{\dot{h}_h}{|\dot{P}|}, \tag{4.31}$$

where $\dot{h}_h$ is the indenter displacement rate just before the unloading, $\dot{P}$ is the unloading rate and $S=dP/dh$ is the apparent unloading stiffness. $\dot{h}_h$, $\dot{P}$ and S are either measurable or known, and using S_e calculated from equation (4.31) to replace S in equations (4.26) and (4.28), more accurate estimates of the elastic modulus and hardness can be obtained. Figure 4.56 shows the application of equation (4.31) to correct for creep effects in the Young modulus measurement of an aluminum sample.

An alternative way to overcome thermal drifts and specimen creep problems is to run the nanoindentation test in dynamic mode at a high frequency. In the continuous stiffness measurement technique,

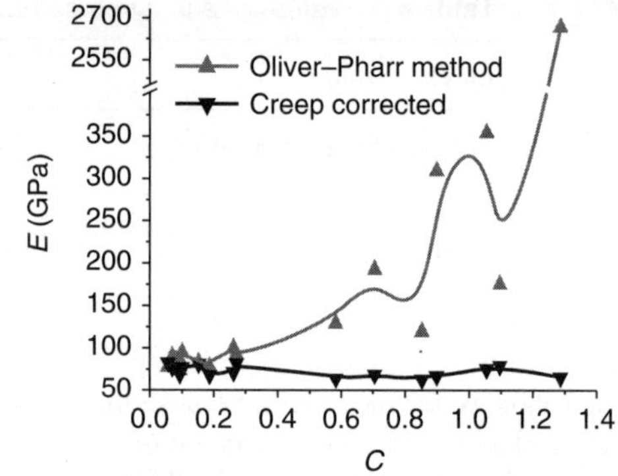

Figure 4.56 *Youngs modulus data in aluminum measured by nanoindentation with and without creep correction using equation (4.31).* $C = \dot{h}_h S/|\dot{P}|$ *is a measurement for the severity of creep. (Feng and Ngan, 2002, by permission of the Materials Research Society).*

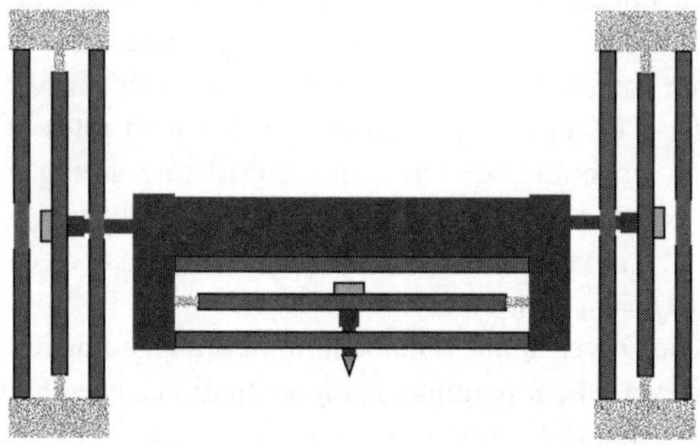

Figure 4.57 *Schematic of a capacitive 2-D (lateral force) transducer system (courtesy of Hysitron Inc.).*

a small a.c. signal (~40 Hz) is superimposed onto the static force applied on the tip, and the stiffness S is measured as the response ratio $S = \Delta P/\Delta h$, where ΔP and Δh are the amplitudes of the load and displacement modulations. Any drift or creep is unlikely to affect the measurement during a few periods of the oscillation. For example, at 40 Hz, a drift rate of 0.1 nm s^{-1} produces a drift of only 0.0025 nm during one period, and this is insignificant. Equation (4.26) can be rewritten as $S = 2aE_r$, and so the measured stiffness is a direct measure of the tip–sample contact size a.

Nanoindentation has quickly become a standard laboratory tool to study the mechanical properties of fine structures such as grain boundaries, thin films and coatings, and small second-phase precipitates. It is also an indispensable technique for the mechanical characterization of materials and structures for micro/nano devices. Nanoindentation is also becoming increasingly used in biomedical research, in areas such as the development of implant materials in orthopedic surgery, where local mechanical properties and interactions due to biological reactions have to be characterized.

Apart from uniaxial, depth-sensing indentation, the newest type of nanoindenters can also perform scratching experiments, through the addition of lateral-force transducers, as shown in Figure 4.57. This is a useful technique in nanotribology research. For example, the coefficient of friction can be evaluated as the ratio between the recorded lateral force and the normal force.

Figure 4.58 shows a sub-micron scratch produced on an Ni_3Al surface by a nanoindenter operating in the scratching mode. From the directions of the slip traces decorating the scratch, deformation

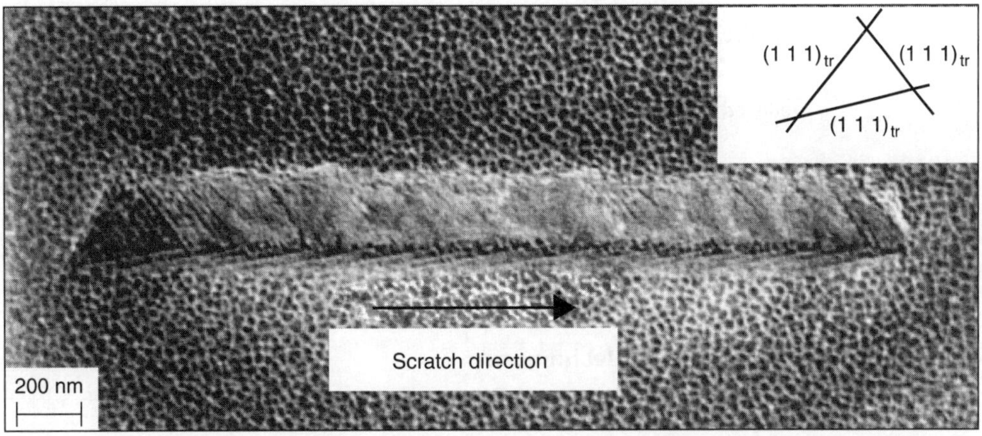

Figure 4.58 *Low-energy (8 keV) secondary-electron SEM micrograph of a sub-micron scratch on Ni_3Al (Wo and Ngan, 2004, by permission of Taylor & Francis).*

took place on {1 1 1} planes inclined to the specimen surface. The coefficient of friction between the diamond tip and the specimen evaluated in this case was about 0.24.

Worked example

The following figure shows the load–displacement curve measured during nanoindentation on a Ni_3Al sample by a cube-corner tip. Use the Oliver–Pharr method to calculate the hardness and the value of $E/(1-\nu^2)$ for Ni_3Al.

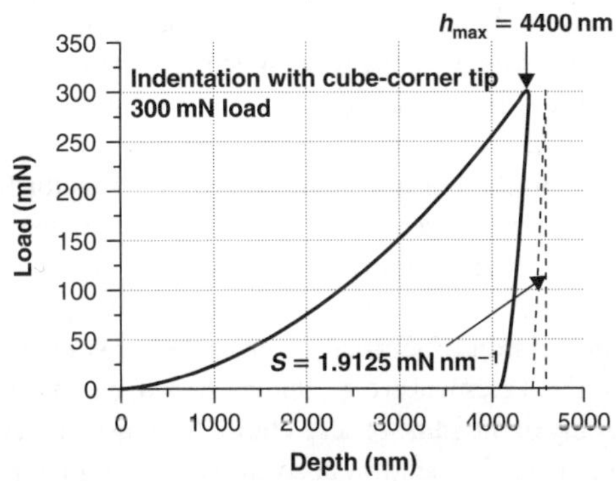

Solution

From the graph, $S = 1.9125\,\text{mN nm}^{-1}$; $h_{max} = 4400\,\text{nm}$; $P_{max} = 300\,\text{mN}$.

For a cube-corner tip, $\varepsilon = 0.73$, and so contact depth $h_c = 4400 - 0.73 \times \dfrac{300}{1.9125} = 4285.5\,\text{nm}$.

The shape function of a cube-corner tip is $A_c = (3\sqrt{3}/2)h_c^2$, i.e.

$$A_c = (3\sqrt{3}/2) \times (4285.5 \times 10^{-9})^2 = 4.7715 \times 10^{-11}\text{m}^2.$$

$$\text{Therefore, hardness } H = \frac{300 \times 10^{-3}}{4.7715 \times 10^{-11}} = 6.287 \text{ GPa, and}$$

$$\text{reduced modulus } E_r = \frac{\sqrt{\pi}}{2} \frac{1.9125 \times 10^{-3} \times 10^9}{\sqrt{4.7715 \times 10^{-11}}} = 245.37 \text{ GPa.}$$

For the diamond tip, $E = 1140$ GPa and $\nu = 0.07$, and so the reduced modulus for the sample is

$$\left(\frac{E}{1-\nu^2}\right)_{\text{sample}} = 1/\left(\frac{1}{245.37} - \frac{1-0.07^2}{1140}\right) = 312.25 \text{ GPa.}$$

4.7 Specialized bombardment techniques

4.7.1 Neutron diffraction

The advent of nuclear reactors stimulated the application of neutron diffraction to those problems of materials science which could not be solved satisfactorily by other diffraction techniques. In a conventional pile the fast neutrons produced by fission are slowed down by repeated collisions with a 'moderator' of graphite or heavy water until they are slow enough to produce further fission. If a collimator is inserted into the pile, some of these slow neutrons[9] will emerge from it in the form of a beam, and the equivalent wavelength λ of this neutron beam of energy E in electron-volts is given by $\lambda = 0.0081/E$. The equilibrium temperature in a pile is usually in the range 0–100°C, which corresponds to a peak energy of several hundredths of an electron-volt. The corresponding wavelength of the neutron beam is about 0.15 nm and, since this is very similar to the wavelength of X-rays, it is to be expected that thermal neutrons will be diffracted by crystals.

The properties of X-ray and neutron beams differ in many respects. The distribution of energy among the neutrons in the beam approximately follows the Maxwellian curve appropriate to the equilibrium temperature and, consequently, there is nothing which corresponds to characteristic radiation. The neutron beam is analogous to a beam of 'white' X-rays, and as a result it has to be monochromatized before it can be used in neutron crystallography. Then, because only about one in 10^3 of the neutrons in the originally weak collimated beam are reflected from the monochromator, it is necessary to employ very wide beams several inches in cross-section to achieve a sufficiently high counting rate on the boron trifluoride counter detector (photographic detection is possible but not generally useful). In consequence, neutron spectrometers, although similar in principle to X-ray diffractometers, have to be constructed on a massive scale.

Neutron beams do, however, have advantages over X-rays or electrons, and one of these is the extremely low absorption of thermal neutrons by most elements. Table 4.3 shows that, even in the most highly absorbent elements (e.g. lithium, boron, cadmium and gadolinium), the mass absorption coefficients are only of the same order as those for most elements for a comparable X-ray wavelength, and for other elements the neutron absorption is very much less indeed. This penetrative property of the neutron beam presents a wide scope for neutron crystallography, since the whole body of a specimen may be examined and not merely its surface. Problems concerned with preferred orientation, residual stresses, cavitation and structural defects are but a few of the possible applications, some of which are discussed more fully later.

Another difference concerns the intensity of scattering per atom, I_a. For X-rays, where the scattering is by electrons, the intensity I_a increases with atomic number and is proportional to the square of the atomic-form factor. For neutrons, where the scattering is chiefly by the nucleus, I_a appears to be quite

[9] These may be called 'thermal' neutrons because they are in thermal equilibrium with their surroundings.

Table 4.3 *X-ray and neutron mass absorption coefficients.*

Element	*At. no.*	*X-rays ($\lambda = 0.19\,nm$)*	*Neutrons ($\lambda = 0.18\,nm$)*
Li	3	1.5	5.8
B	5	5.8	38.4
C	6	10.7	0.002
Al	13	92.8	0.005
Fe	26	72.8	0.026
Cu	29	98.8	0.03
Ag	47	402	0.3
Cd	48	417	13.0
Gd	61	199	183.0
Au	79	390	0.29
Pb	82	429	0.0006

Table 4.4 *Scattering amplitudes for X-rays and thermal neutrons.*

Element	*At. no.*	*Scattering amplitudes $\times 10^{-12}$*		
		X-rays for $\sin\theta/\lambda = 0.5$	*Neutrons**	
H	1	0.02		−0.4
Li	3	0.28	Li^6	0.7
			Li^7	−0.25
C	6	0.48		0.64
N	7	0.54		0.85
O	8	0.62		0.58
Al	13	1.55		0.35
Ti	22	2.68		−0.38
Fe	26	3.27	Fe^{56}	1.0
			Fe^{57}	0.23
Co	27	3.42		0.28
Cu	29	3.75		0.76
Zn	30	3.92		0.59
Ag	47	6.71	Ag^{107}	0.83
			Ag^{109}	0.43
Au	79	12.37		0.75

* The negative sign indicates that the scattered and incident waves are in phase for certain isotopes and hence for certain elements. Usually the scattered wave from an atom is 180° out of phase with the incident wave.

unpredictable. The scattering power per atom varies not only apparently at random from atom to atom, but also from isotope to isotope of the same atom. Moreover, the nuclear component to the scattering does not decrease with increasing angle, as it does with X-rays, because the nucleus which causes the scattering is about 10^{-12} mm in size compared with 10^{-7} mm, which is the size of the electron cloud that scatters X-rays. Table 4.4 gives some of the scattering amplitudes for X-rays and thermal neutrons.

The fundamental difference in the origin of scattering between X-rays and neutrons affords a method of studying structures, such as hydrides and carbides, which contain both heavy and light

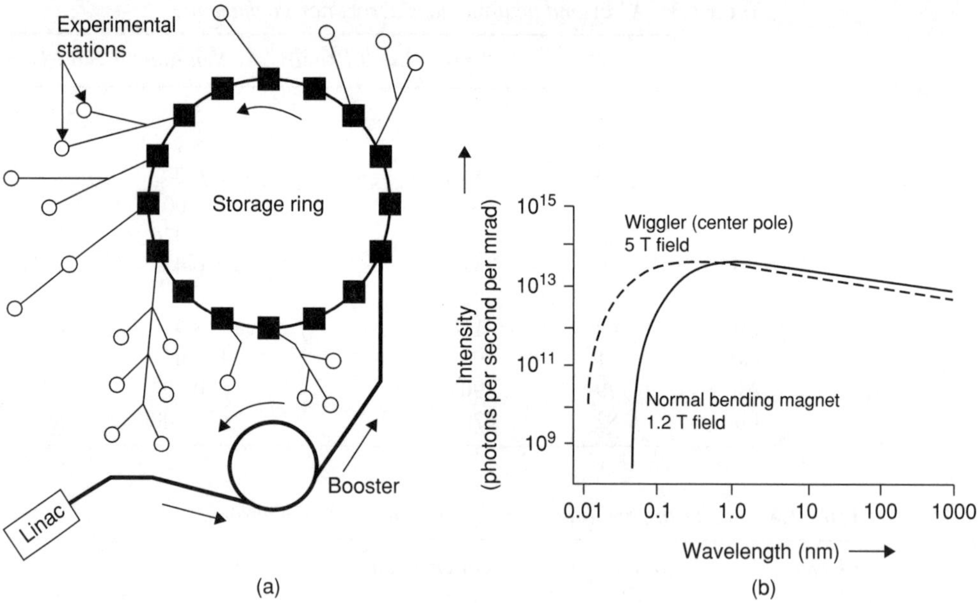

Figure 4.59 *(a) Layout of SRS, Daresbury. (b) Wavelength spectrum of synchrotron radiation (after Barnes, 1990, by permission of the Institute of Materials, Minerals and Mining).*

atoms. When X-rays are used, the weak intensity contributions of the light atoms are swamped by those from the heavy atoms, but when neutrons are used, the scattering power of all atoms is roughly of the same order. Similarly, structures made up of atoms whose atomic numbers are nearly the same (e.g. iron and cobalt, or copper and zinc) can be studied more easily by using neutrons. This aspect is discussed later in relation to the behavior of ordered alloy phases.

The major contribution to the scattering power arises from the nuclear component, but there is also an electronic (magnetic spin) component to the scattering. This arises from the interaction between the magnetic moment of the neutron and any resultant magnetic moment which the atom might possess. As a result, the neutron diffraction pattern from paramagnetic materials, where the atomic moments are randomly directed (see Chapter 5), shows a broad diffuse background, due to incoherent (magnetic) scattering, superimposed on the sharp peaks which arise from coherent (nuclear) scattering. In ferromagnetic metals the atomic moments are in parallel alignment throughout a domain, so that this cause of incoherent scattering is absent. In some materials (e.g. NiO or FeO) an alignment of the spins takes place, but in this case the magnetization directions of neighboring pairs of atoms in the structure are opposed and, in consequence, cancel each other out. For these materials, termed anti-ferromagnetic, there is no net spontaneous magnetization, and neutron diffraction is a necessary and important tool for investigating their behavior (see Chapter 5).

4.7.2 Synchrotron radiation studies

Very large electrical machines known as synchrotron radiation sources (SRS) provide a unique source of electromagnetic radiation for materials characterization.[10] Electrons from a hot cathode

[10] In 1980, the world's first totally radiation-dedicated SRS came into operation at Daresbury, England. Electrons are 'stored' in the main ring for 10–20 h, traversing its 96 m periphery more than 3×10^6 times per second.

are accelerated in three stages by a linear accelerator (Linac), a booster synchrotron and an evacuated storage ring (Figure 4.59a). As bunches of electrons travel around the periphery of the storage ring they attain energies of up to 2 GeV and velocities approaching that of light. At these relativistic velocities, electron mass becomes 4000 times greater than the rest mass. Dipole and quadrupole magnets constrain the bunches into an approximately circular orbit and, by accelerating them centripetally, cause electromagnetic radiation to be produced. The spectrum of this synchrotron radiation is very wide, extending from short-wavelength ('hard') X-rays to the infrared range (Figure 4.59b). A wiggler magnet produces a strong (5 tesla) field and can extend the spectrum to even shorter wavelengths. Compared with more orthodox sources of electromagnetic radiation, the synchrotron offers the advantages of very high intensity, short wavelengths, precise collimation of the beam and a smooth, continuous spectrum. The high radiation intensity permits exposure times that are often several orders of magnitude shorter than those for comparable laboratory methods. The risk of beam damage to specimens by the flashes of radiation is accordingly lessened. Specimens of metals, ceramics, polymers, semiconductors, catalysts, etc. are placed in independent experimental stations located around the periphery of the ring chamber and irradiated in order to produce spectroscopic, diffraction or imaging effects.

In the technique known as extended X-ray absorption fine-structure spectroscopy (EXAFS), attention is directed to the small discontinuities on the higher-energy flank beyond each vertical, characteristic 'edge' which appears in a plot of mass absorption versus X-ray wavelength. These 'fine structure' (FS) features derive from interference effects between electron waves from excited atoms and waves back-scattered from surrounding atoms. Mathematical treatment (using a Fourier transform) of the EXAFS spectra yields a radial distribution plot of surrounding atomic density versus distance from the excited atom. By selecting the 'edge' for a particular type of atom/ion and studying its fine structure, it is thus possible to obtain information on its local environment and coordination. This type of information is of great value in structural studies of materials, such as glasses, which only exhibit short-range order. For instance, the EXAFS technique has demonstrated that the network structure of SiO_2–Na_2O–CaO glass is threaded by percolation channels of modifier (sodium) cations.

4.7.3 Secondary ion mass spectrometry (SIMS)

This important and rapidly developing technique, which enables material surfaces to be analyzed with great chemical sensitivity and excellent resolution in depth, is based upon the well-known phenomenon of sputtering. The target surface is bombarded with a focused beam of primary ions that has been accelerated with a potential of 1–30 kV within a high-vacuum chamber (10^{-5}–10^{-10} torr). These ions generate a series of collision cascades in a shallow surface layer, 0.5–5 nm deep, causing neutral atoms and, to a much smaller extent, secondary ions to be ejected (sputtered) from the specimen surface. Thus, a metallic oxide (MO) sample may act as a source of M, O, M^+, O^+, M^-, O^-, MO^+ and MO^- species. The secondary ions, which are thus either monatomic or clustered, positive or negative, are directed into a mass spectrometer (analyzer), wherein they are sorted and identified according to their mass/charge ratio. Exceptionally high elemental sensitivities, expressed in parts per million and even parts per billion, are achievable. All elements in the Periodic Table can be analyzed and it is possible to distinguish between individual isotopes. Studies of the self-diffusion of oxygen and nitrogen have been hindered because these light elements have no isotopes that can be used as radioactive tracers. SIMS based on the stable isotope ^{18}O provides a rapid method for determining self-diffusion coefficients. The physical process whereby ions are ejected is difficult to express in rigorous theoretical terms; consequently, SIMS is usually semiquantitative, with dependence upon calibration with standard samples of known composition. SIMS is a valuable complement to other methods of surface analysis.

The available range of beam diameter is 1 μm to several millimeters. Although various types of ion beam are available (e.g. Ar^-, $^{32}O_2^+$, $^{16}O^-$, Cs^+, etc.) positively charged beams are a common choice. However, if the sample is insulating, positive charge tends to accumulate in the bombarded region, changing the effective value of the beam voltage and degrading the quality of signals. One partial remedy, applicable at low beam voltages, is to 'flood' the ion-bombarded area with a high-intensity electron beam. In some variants of SIMS laser beams are used instead of ion beams.

Of the large and growing variety of methods covered by the term SIMS, the dynamic, static and imaging modes are especially useful. Materials being investigated include metals, ceramics, polymers, catalysts, semiconductors and composites. Dynamic SIMS, which uses a relatively high beam current, is an important method for determining the distribution and very low concentration of dopants in semiconductors. The beam scans a raster, 100–500 μm in size, and slowly erodes the surface of the sample. Secondary ions from the central region of the crater are analyzed to produce a precise depth profile of concentration. Static SIMS uses a much smaller beam current and the final spectra tend to be more informative, providing chemical data on the top few atomic layers of the sample surface. Little surface damage occurs and the method has been applied to polymers. The imaging version of SIMS has a resolution comparable to SEM and provides 'maps' that show the lateral distribution of elements at grain boundaries and precipitated particles and hydrogen segregation in alloys. Imaging SIMS has been applied to transverse sections through the complex scale layers which form when alloys are exposed to hot oxidizing gases (e.g. O_2, CO_2). Its sensitivity is greater than that obtainable with conventional EDX in SEM analysis and has provided a better understanding of growth mechanisms and the special role of trace elements such as yttrium.

4.8 Thermal analysis

4.8.1 General capabilities of thermal analysis

Heating a material at a steady rate can produce chemical changes, such as oxidation and degradation, and/or physical changes, such as the glass transition in polymers, conversions/inversions in ceramics and phase changes in metals. Thermal analysis is used to complement X-ray diffraction analysis, optical and electron microscopy during the development of new materials and in production control. Sometimes it is used to define the temperature and energy change associated with a structural change; at other times it is used qualitatively to provide a characteristic 'fingerprint' trace of a particular material. The various techniques of thermal analysis measure one or more physical properties of a sample as a function of temperature. Figure 4.60 illustrates three basic methods of thermal analysis, namely thermogravimetric analysis (TGA), differential thermal analysis (DTA) and differential scanning calorimetry (DSC). Respectively, they measure change in mass (TGA) and energy flow (DTA, DSC). They can apply programmed heating and cooling, but usually operate with a slowly rising temperature. The sample chamber may contain air, oxygen, nitrogen, argon, etc. or be evacuated. A sample of a few tens of milligrams will often suffice.

Recently developed methods have extended the range of thermal analysis and other aspects of behavior can now be studied. For instance, using dynamic mechanical thermal analysis (DMTA), mechanical as well as structural information can be obtained on the viscoelastic response of a polymeric sample to tensile, bend or shear stresses during heating.

4.8.2 Thermogravimetric analysis

In a thermobalance the mass of a sample is continuously determined and recorded while the sample is being slowly heated (Figure 4.60a). Temperatures up to at least 1000°C are available. It has been

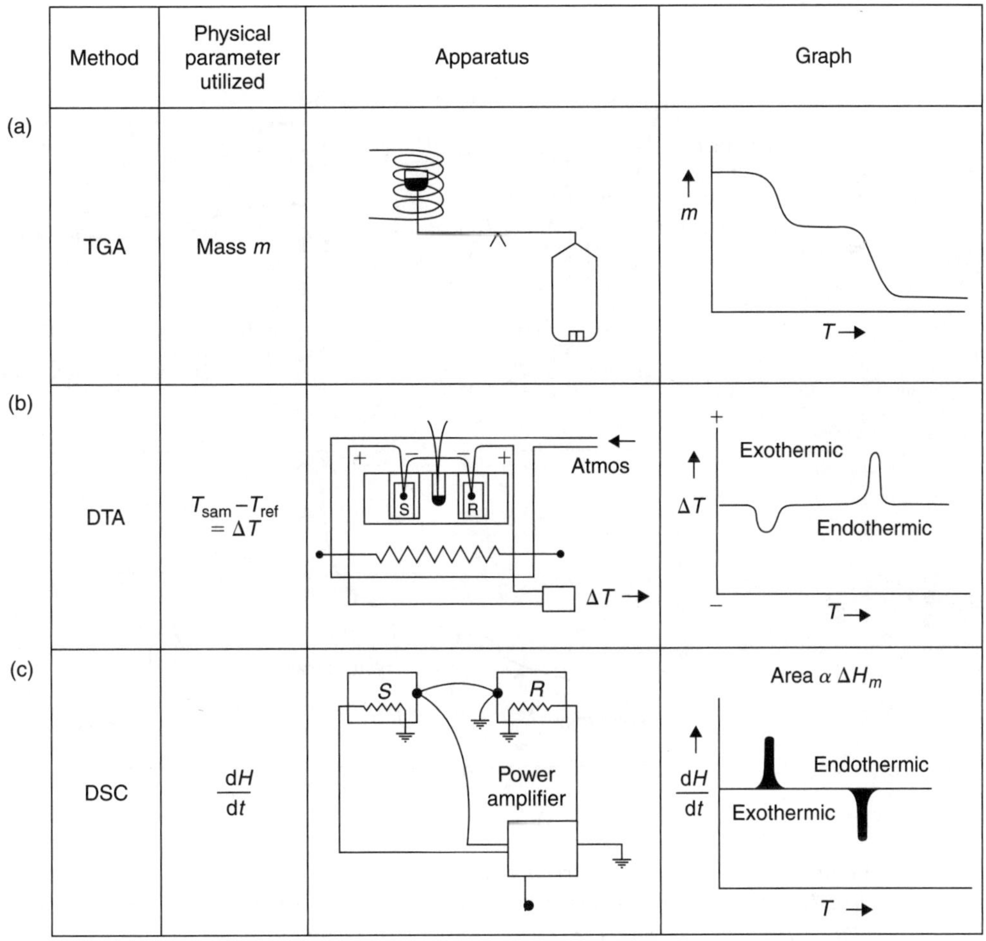

Figure 4.60 *Basic methods of thermal analysis. (a) Thermogravimetric analysis (TGA). (b) Differential thermal analysis (DTA). (c) Differential scanning calorimetry (DSC).*

applied to the decomposition of rubbers (Figure 4.61a), kinetic studies of metallic oxidation, glass transitions and softening in polymers. Equilibrium is not attained within the sample and the method is insensitive to the more subtle solid-state changes. When changes overlap, it can be helpful to plot the first derivative, $\delta m/\delta t$, of the graphical trace in a procedure known as derivative thermogravimetric analysis (DTGA).

4.8.3 Differential thermal analysis

DTA[11] reveals changes during the heating of a sample which involve evolution or absorption of energy. As shown diagrammatically in Figure 4.60b, a sample S and a chemically and thermally inert reference material R (sintered alumina or precipitated silica) are mounted in a recessed heating block

[11] Usually accredited to H. Le Chatelier (1887): improved version and forerunner of modern DTA used by W. C. Roberts-Austen (1899) in metallurgical studies of alloys.

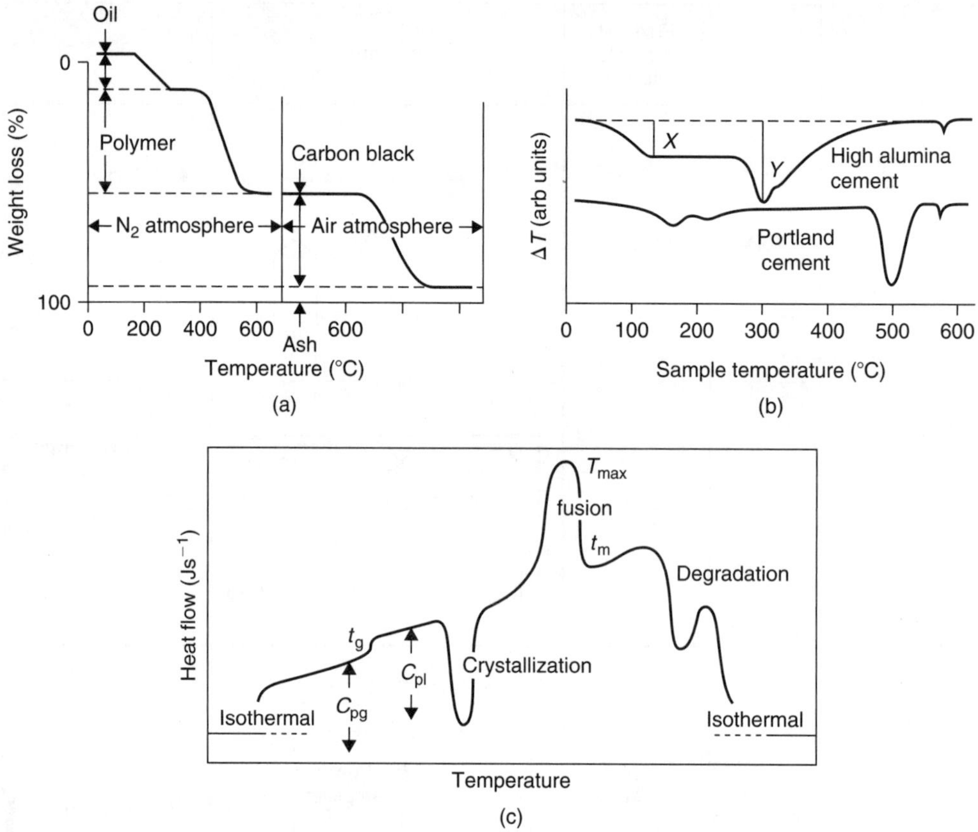

Figure 4.61 *Examples of thermal analysis. (a) TGA curve for decomposition of rubber, showing decomposition of oil and polymer in N_2 up to 600°C and oxidation of carbon black in air above 600°C (Hill and Nicholas, 1989). (b) DTA curve for a high-alumina cement and Portland cement (Hill and Nicholas, 1989). (c) DTA curve for a quenched glassy polymer (Hay, 1982).*

and slowly heated. The thermocouples in S and R are connected in opposition; their temperature difference ΔT is amplified and plotted against temperature. Peak area on this trace is a function of the change in enthalpy (ΔH) as well as the mass and thermal characteristics of the sample S. Small samples can be used to give sharper, narrower peaks, provided that they are fully representative of the source material. Ideally, the specific heat capacities of S and R should be similar. DTA is generally regarded as a semiquantitative or qualitative method. It has been used in studies of devitrification in oxide glasses and the glass transition in polymers. Figure 4.61b shows a comparison of the thermal response of high-alumina cement (HAC) and Portland cement. The amount of an undesirable weakening phase can be derived from the relative lengths of the ordinates X and Y in the HAC trace.

4.8.4 Differential scanning calorimetry

In this method, unlike DTA, the sample and reference body have separate resistive heaters (Figure 4.60c). When a difference in temperature develops between sample S and reference R, an automatic control loop heats the cooler of the two until the difference is eliminated. The electrical power needed to accomplish this equalizer is plotted against temperature. An endothermic change signifies that an enthalpy increase has occurred in S; accordingly, its peak is plotted upwards (unlike DTA traces). Differences in thermal conductivity and specific heat capacity have no effect and peak areas can be expressed as energy per unit mass. DSC has proved particularly valuable in polymer research, often being used in combination with other techniques, such as evolved gas analysis (EGA). DSC has been used in studies of the curing characteristics of rubbers and thermoset resins, transitions in liquid crystals and isothermal crystallization rates in thermoplastics. Figure 4.61c is a trace obtained for a quenched amorphous polymer. DSC has also been used in studies of the exothermic behavior of cold-worked metals as they release 'stored energy' during annealing, energy absorption during eutectic melting of alloys, precipitation in aluminum-based alloys, relaxation transformations in metallic glasses and drying/firing transitions in clay minerals.

Problems

4.1 What is the angle between the diffracted beam and the incident beam when X-rays of wavelength 0.1 nm are reflected by {1 1 1} planes in copper ($a = 0.362$ nm)?

4.2 The energy of CuK_α X-rays is 8.04 keV. What is their wavelength?

4.3 What different types of planes in aluminum have the three smallest Bragg angles?

4.4 In a 200 kV transmission electron microscope, electromagnetic lenses make the *effective* camera length 1 m. What is the distance on the screen of a {1 1 0} diffracted beam spot from the main (incident) beam for an iron specimen at room temperature ($a = 0.287$ nm)?

4.5 In an X-ray powder camera, CuK_α X-rays are directed at a titanium powder sample. What is the distance apart of the $\{1\ \bar{1}\ 0\ 1\}$ lines? (For Ti, $a = 0.295$ nm; $c = 0.468$ nm; diameter of camera = 115 mm.)

4.6 In an X-ray powder camera experiment the diffracted beams emerge at the following angles to the incident X-ray beam: 57.34°, 67.28°, 103.20°, 133.59°, 147.36°. The specimen has an fcc crystal structure and the wavelength of the X-rays is 0.2 nm. What is the lattice parameter?

4.7 In an EDX spectrum a strong peak was detected at 8.04 keV. What other peaks would you expect necessarily to see?

4.8 In an XRF experiment, the X-ray intensities from the specimen of unknown composition and from a standard containing 50% of magnesium and 50% oxygen were as follows:

	$Mg_x\ O_{1-x}$	MgO
MgK	823	746
OK	605	423

What was the composition of the specimen?

4.9 In an SEM analysis, the following X-ray intensities were measured:

	Specimen	*Elemental standard*
Iron FeK	28 231	38 113
Chromium CrK	6189	36 627
Nickel NiK	3527	40 077

What was the composition of the specimen? What material was it?

4.10 In a thin foil of aluminum with (0 0 1) orientation a screw dislocation is in contrast when 0 2 0 and $2\ \bar{2}\ 0$ reflections operate. It is 'invisible' when the 2 2 0 reflection operates. When the

specimen is tilted to excite the $1\ 1\ \bar{1}$ reflection, the dislocation is also invisible. (i) Determine the possible Burgers vector of the dislocation and
(ii) comment on the slip plane it is capable of gliding in.

4.11 Stacking faults A, B, C are observed in a (1 1 1)-oriented TEM specimen of silicon on $(\bar{1}\ 1\ 1)$, $(1\ 1\ \bar{1})$ and $(1\ \bar{1}\ 1)$ planes respectively. Determine which fault will be 'invisible' using the $g = 0\ \bar{2}\ 2$ reflection and which with the $g = 2\ \bar{2}\ 0$ reflection.

4.12 Rod precipitates lying along [1 0 0] in a cubic crystal are examined in the electron microscope with the beam pointing along [1 1 3]. Predict the direction of the precipitate images.

Further reading

Barnes, P. (1990). Synchrotron radiation for materials science research. *Metals and Materials*, November, pp. 708–715. Institute of Materials.

Barrett, C. S. and Massalski, T. B. (1980). *Structure of Metals: Crystallographic Methods, Principles and Data*. Pergamon, Oxford.

Cullity, B. D. (1978). *Elements of X-ray Diffraction*. Addison-Wesley, Reading, MA.

Dehoff, R. T. and Rhines, F. N. (eds) (1968). *Quantitative Microscopy*. McGraw-Hill, New York.

Fischer-Cripps, A.C. (2004). *Nanoindentation*. Springer, New York.

Gifkins, R. C. (1970). *Optical Microscopy of Metals*. Pitman, Melbourne.

Hay, J. N. (1982). Thermal methods of analysis of polymers. In *Analysis of Polymer Systems* (edited by L. S. Bark and N. S. Allen), Chap. 6. Applied Science, London.

Hill, M. and Nicholas, P. (1989). Thermal analysis in materials development. *Metals and Materials*, November, pp. 639–642. Institute of Materials.

Johnson, K. L. (1985). *Contact Mechanics*. Cambridge University Press, Cambridge.

Jones, I. P. (1992). *Chemical Microanalysis using Electron Beams*. Institute of Materials, London.

Loretto, M. H. (1984). *Electron Beam Analysis of Materials*. Chapman & Hall, London.

Loretto, M. H. and Smallman, R. E. (1975). *Defect Analysis in Electron Microscopy*. Chapman & Hall, London.

Modin, H. and Modin, S. (1973). *Metallurgical Microscopy*. Butterworths, London.

Patzelt, W. J. (1974). *Polarised Light Microscopy: Principles, Instruments, Applications*. Ernst Leitz Wetzlar GmbH, Lahn-Wetzlar.

Pickering, F. B. (1976). *The Basis of Quantitative Metallography*, Institute of Metallurgical Technicians Monograph No. 1.

Richardson, J. H. (1971). *Optical Microscopy for the Materials Sciences*. Marcel Dekker, New York.

Tabor, D. (1951). *The Hardness of Metals*. Oxford University Press, Oxford.

Vickerman, J. C., Brown, A. and Reed, N. M. (eds) (1990). *Secondary Ion Mass Spectrometry: Principles and Applications*. Clarendon Press, Oxford.

Wendlandt, W. W. (1986). *Thermal Analysis*, 3rd edn. Wiley, New York.

Williams, D. B. and Carter, C. B. (1996). *Transmission Electron Microscopy – A Textbook for Materials Science, I Basic, II Diffraction, III Imaging*. Plenum Press, New York.

Chapter 5

Physical properties

5.1 Introduction

The ways in which any material interacts and responds to various forms of energy are of prime interest to scientists and, in the context of engineering, provide the essential base for design and innovation. The energy acting on a material may derive from force fields (gravitational, electric, magnetic), electromagnetic radiation (heat, light, X-rays), high-energy particles, etc. The responses of a material, generally referred to as its physical properties, are governed by the structural arrangement of atoms/ions/molecules in the material. The theme of the structure–property relation which has run through previous chapters is developed further. Special attention will be given to the diffusion of atoms/ions within materials because of the importance of thermal behavior during manufacture and service. In this brief examination, which will range from density to superconductivity, the most important physical properties of materials are considered.

5.2 Density

This property, defined as the mass per unit volume of a material, increases regularly with increasing atomic numbers in each subgroup. The reciprocal of the density is the specific volume v, while the product of v and the relative atomic mass W is known as the atomic volume Ω. The density may be determined by the usual 'immersion' method, but it is instructive to show how X-rays can be used. For example, a powder photograph may give the lattice parameter of an fcc metal, say copper, as 0.36 nm. Then $1/(3.6 \times 10^{-10})^3$ or 2.14×10^{28} cells of this size (0.36 nm) are found in a cube of 1 m edge length. The total number of atoms in 1 m^3 is then $4 \times 2.14 \times 10^{28} = 8.56 \times 10^{28}$ since an fcc cell contains four atoms. Furthermore, the mass of a copper atom is 63.57 times the mass of a hydrogen atom (which is 1.63×10^{-24} g) so that the mass of 1 m^3 of copper, i.e. the density, is $8.56 \times 10^{28} \times 63.57 \times 1.63 \times 10^{-24} = 8900\,\text{kg}\,\text{m}^{-3}$.

On alloying, the density of a metal changes. This is because the mass of the solute atom differs from that of the solvent, and also because the lattice parameter usually changes on alloying. The parameter change may often be deduced from Vegard's law, which assumes that the lattice parameter of a solid solution varies linearly with atomic concentration, but numerous deviations from this ideal behavior do exist.

The density clearly depends on the mass of the atoms, their size and the way they are packed. Metals are dense because they have heavy atoms and close packing; ceramics have lower densities than metals because they contain light atoms, either C, N or O; polymers have low densities because they consist of light atoms in chains. Figure 5.1 shows the spread in density values for the different

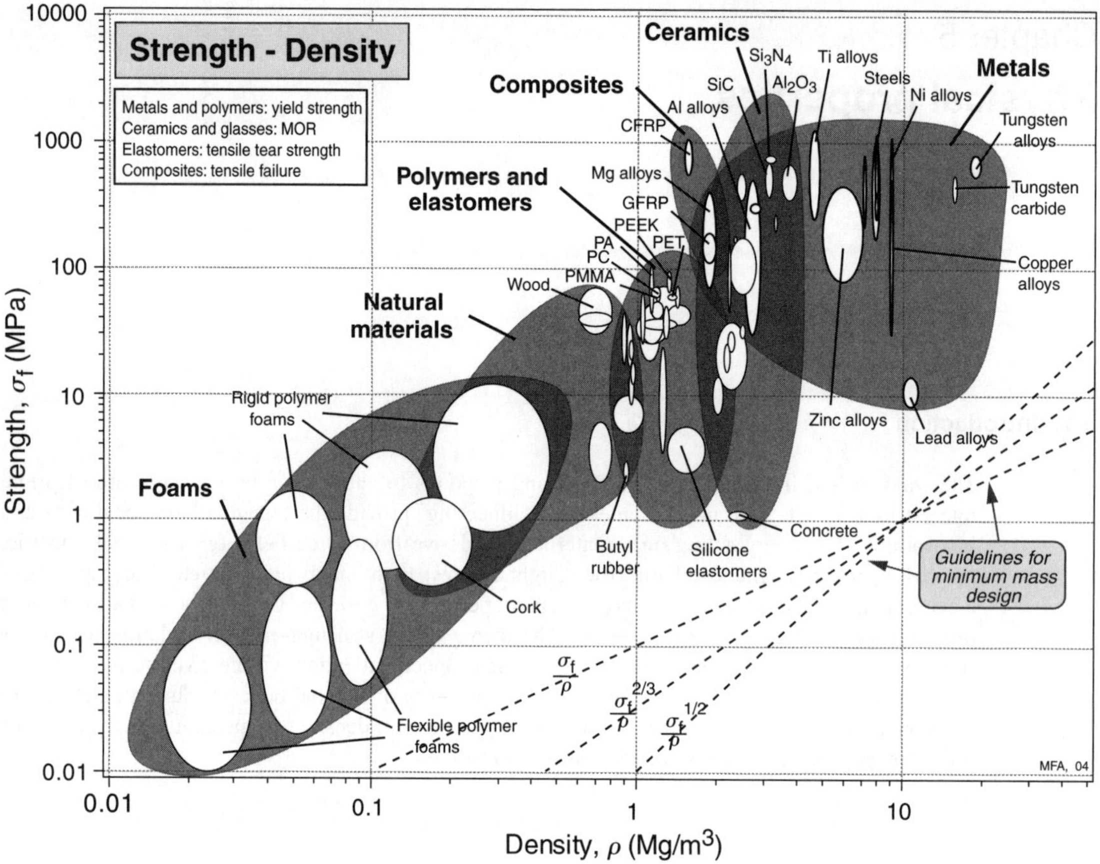

Figure 5.1 *Strength σ, plotted against density, ρ (yield strength for metals and polymers, compressive strength for ceramics, tear strength for elastomers and tensile strength for composites). The guide lines of constant σ/ρ, $\sigma^{2/3}/\rho$ and $\sigma^{1/2}/\rho$ are used in minimum weight, yield-limited, design (after Ashby, 1992).*

material classes. Such 'material property charts', as developed by Ashby, are useful when selecting materials during engineering design.

5.3 Thermal properties

5.3.1 Thermal expansion

If we consider a crystal at absolute zero temperature, the ions sit in a potential well of depth E_{r_0} below the energy of a free atom (Figure 1.3). The effect of raising the temperature of the crystal is to cause the ions to oscillate in this asymmetrical potential well about their mean positions. As a consequence, this motion causes the energy of the system to rise, increasing with increasing amplitude of vibration. The increasing amplitude of vibration also causes an expansion of the crystal, since as a result of the sharp rise in energy below r_0 the ions as they vibrate to and fro do not approach much closer than the equilibrium separation, r_0, but separate more widely when moving apart. When the distance r is such that the atoms are no longer interacting, the material is transformed to the gaseous phase, and the energy to bring this about is the energy of evaporation.

The change in dimensions with temperature is usually expressed in terms of the linear coefficient of expansion α, given by $\alpha = (1/l)(dl/dT)$, where l is the original length of the specimen and T is the absolute temperature. Because of the anisotropic nature of crystals, the value of α usually varies with the direction of measurement and even in a particular crystallographic direction the dimensional change with temperature may not always be uniform.

Phase changes in the solid state are usually studied by *dilatometry*. The change in dimensions of a specimen can be transmitted to a sensitive dial gauge or electrical transducer by means of a fused silica rod. When a phase transformation takes place, because the new phase usually occupies a different volume to the old phase, discontinuities are observed in the coefficient of thermal expansion α–T curve. Some of the 'nuclear metals' which exist in many allotropic forms, such as uranium and plutonium, show a negative coefficient along one of the crystallographic axes in certain of their allotropic modifications.

The change in volume with temperature is important in many metallurgical operations such as casting, welding and heat treatment. Of particular importance is the volume change associated with the melting or, alternatively, the freezing phenomenon, since this is responsible for many of the defects, both of a macroscopic and microscopic size, which exist in crystals. Most metals increase their volume by about 3% on melting, although those metals which have crystal structures of lower coordination, such as bismuth, antimony or gallium, contract on melting. This volume change is quite small, and while the liquid structure is more open than the solid structure, it is clear that the liquid state resembles the solid state more closely than it does the gaseous phase. For the simple metals the latent heat of melting, which is merely the work done in separating the atoms from the close-packed structure of the solid to the more open liquid structure, is only about one-thirtieth of the latent heat of evaporation, while the electrical and thermal conductivities are reduced only to three-quarters to one-half of the solid-state values.

Worked example

In copper, what percentage of the volume change which occurs as the specimen is heated from room temperature to its melting point is due to the increased vacancy concentration, assuming that the vacancy concentration at the melting point (1083°C) is $\sim 10^{-4}$? (Linear thermal expansion coefficient of copper α is $16.5 \times 10^{-6} \text{K}^{-1}$.)

Solution

At the melting point (1083°C) vacancy concentration $= 10^{-4}$, i.e. one vacancy every 10^4 atom sites.

On heating to melting point, Δ Volume/initial volume $= 3\alpha\Delta T$

$$= 3 \times 16.5 \times 10^{-6} \times (1083 - rT)$$

$$= 49.5 \times 10^{-6} \times 1060$$

$$= 5.25 \times 10^{-2}.$$

$$\text{Fractional change due to vacancies} = \frac{10^{-4}}{5.25 \times 10^{-2}} = 0.19 \times 10^{-2} \approx 0.2\%.$$

5.3.2 Specific heat capacity

The *specific heat* is another thermal property important in the processing operations of casting or heat treatment, since it determines the amount of heat required in the process. Thus, the specific heat (denoted by C_p, when dealing with the specific heat at constant pressure) controls the increase in temperature, $\mathrm{d}T$, produced by the addition of a given quantity of heat, $\mathrm{d}Q$, to one gram of matter, so that $\mathrm{d}Q = C_p\mathrm{d}T$.

The specific heat of a metal is due almost entirely to the vibrational motion of the ions. However, a small part of the specific heat is due to the motion of the free electrons, which becomes important at high temperatures, especially in transition metals with electrons in incomplete shells.

The classical theory of specific heat assumes that an atom can oscillate in any one of three directions, and hence a crystal of N atoms can vibrate in $3N$ independent normal modes, each with its characteristic frequency. Furthermore, the mean energy of each normal mode will be $\mathbf{k}T$, so that the total vibrational thermal energy of the metal is $E = 3N\mathbf{k}T$. In solid and liquid metals, the volume changes on heating are very small and, consequently, it is customary to consider the specific heat at constant volume. If N, the number of atoms in the crystal, is equal to the number of atoms in a gram-atom (i.e. Avogadro number), the heat capacity per gram-atom, i.e. the atomic heat, at constant volume is given by

$$C_v\left(\frac{\mathrm{d}Q}{\mathrm{d}T}\right)_v = \frac{\mathrm{d}E}{\mathrm{d}T} = 3N\mathbf{k} = 24.95\,\mathrm{J\,K^{-1}}.$$

In practice, of course, when the specific heat is experimentally determined, it is the specific heat at constant pressure, C_p, which is measured, not C_v, and this is given by

$$C_p\left(\frac{\mathrm{d}E + P\mathrm{d}V}{\mathrm{d}T}\right)_p = \frac{\mathrm{d}H}{\mathrm{d}T},$$

where $H = E + PV$ is known as the heat content or enthalpy, C_p is greater than C_v by a few percent because some work is done against interatomic forces when the crystal expands, and it can be shown that

$$C_p - C_v = 9\alpha^2 VT/\beta,$$

where α is the coefficient of linear thermal expansion, V is the volume per gram-atom and β is the compressibility.

Dulong and Petit were the first to point out that the specific heat of most materials, when determined at sufficiently high temperatures and corrected to apply to constant volume, is approximately equal to $3\mathbf{R}$, where $\mathbf{R}$ is the gas constant. However, deviations from the 'classical' value of the atomic heat occur at low temperatures, as shown in Figure 5.2a. This deviation is readily accounted for by the quantum theory, since the vibrational energy must then be quantized in multiples of $\mathbf{h}\nu$, where $\mathbf{h}$ is Planck's constant and ν is the characteristic frequency of the normal mode of vibration.

According to the quantum theory, the mean energy of a normal mode of the crystal is

$$E(\nu) = \tfrac{1}{2}\mathbf{h}\nu + \{\mathbf{h}\nu/\exp(\mathbf{h}\nu/\mathbf{k}T) - 1\},$$

where $\frac{1}{2}\mathbf{h}\nu$ represents the energy a vibrator will have at the absolute zero of temperature, i.e. the zero-point energy. Using the assumption made by Einstein (1907) that all vibrations have the same

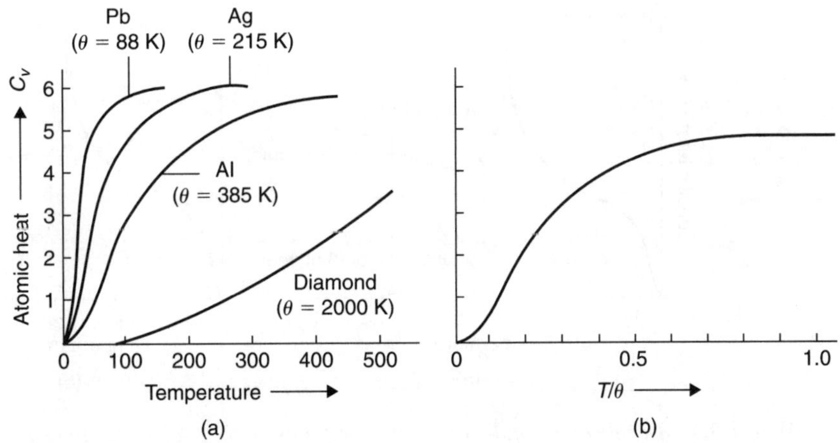

Figure 5.2 *The variation of atomic heat with temperature.*

frequency (i.e. all atoms vibrate independently), the heat capacity is

$$\begin{aligned} C_v &= (\mathrm{d}E/\mathrm{d}T)_v \\ &= 3N\mathbf{k}(\mathbf{h}\nu/\mathbf{k}T)^2[\exp(\mathbf{h}\nu/\mathbf{k}T)/\{\exp(\mathbf{h}\nu/\mathbf{k}T)-1\}^2] \end{aligned}$$

This equation is rarely written in such a form because most materials have different values of ν. It is more usual to express ν as an equivalent temperature defined by $\Theta_E = \mathbf{h}\nu/\mathbf{k}$, where Θ_E is known as the Einstein characteristic temperature. Consequently, when C_v is plotted against T/Θ_E, the specific heat curves of all pure metals coincide and the value approaches zero at very low temperatures and rises to the classical value of $3N\mathbf{k} = 3\mathbf{R} \cong 25.2\,\mathrm{J\,g^{-1}}$ at high temperatures.

Einstein's formula for the specific heat is in good agreement with experiment for $T \gtrsim \Theta_E$, but is poor for low temperatures where the practical curve falls off less rapidly than that given by the Einstein relationship. However, the discrepancy can be accounted for, as shown by Debye, by taking account of the fact that the atomic vibrations are not independent of each other. This modification to the theory gives rise to a Debye characteristic temperature Θ_D, which is defined by

$$\mathbf{k}\Theta_D = \mathbf{h}\nu_D,$$

where ν_D is Debye's maximum frequency. Figure 5.2b shows the atomic heat curves of Figure 5.2a plotted against T/Θ_D; in most metals for low temperatures ($T/\Theta_D \ll 1$) a T^3 law is obeyed, but at high temperatures the free electrons make a contribution to the atomic heat which is proportional to T and this causes a rise of C above the classical value.

5.3.3 The specific heat curve and transformations

The specific heat of a metal varies smoothly with temperature, as shown in Figure 5.2a, provided that no phase change occurs. On the other hand, if the metal undergoes a structural transformation the specific heat curve exhibits a discontinuity, as shown in Figure 5.3. If the phase change occurs at a fixed temperature, the metal undergoes what is known as a first-order transformation; for example, the α to γ, γ to δ and δ to liquid phase changes in iron shown in Figure 5.3a. At the transformation

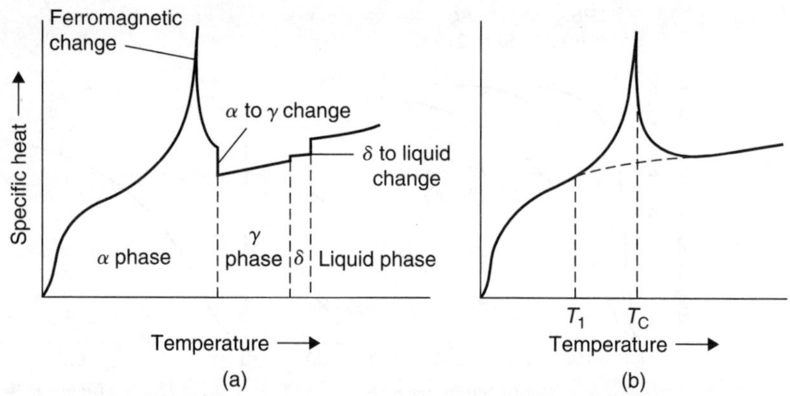

Figure 5.3 *The effect of solid-state transformations on the specific heat–temperature curve.*

temperature the latent heat is absorbed without a rise in temperature, so that the specific heat (dQ/dT) at the transformation temperature is infinite. In some cases, known as transformations of the second order, the phase transition occurs over a range of temperature (e.g. the order–disorder transformation in alloys), and is associated with a specific heat peak of the form shown in Figure 5.3b. Obviously the narrower the temperature range T_1–T_c, the sharper is the specific heat peak, and in the limit when the total change occurs at a single temperature, i.e. $T_1 = T_c$, the specific heat becomes infinite and equal to the latent heat of transformation. A second-order transformation also occurs in iron (see Figure 5.3a), and in this case is due to a change in ferromagnetic properties with temperature.

5.3.4 Free energy of transformation

In Section 2.2.3.2 it was shown that any structural changes of a phase could be accounted for in terms of the variation of free energy with temperature. The relative magnitude of the free energy value governs the stability of any phase, and from Figure 3.9a it can be seen that the free energy G at any temperature is in turn governed by two factors: (1) the value of G at 0 K, G_0, and (2) the slope of the G–T curve, i.e. the temperature dependence of free energy. Both of these terms are influenced by the vibrational frequency, and consequently the specific heat of the atoms, as can be shown mathematically. For example, if the temperature of the system is raised from T to $T + dT$ the change in free energy of the system dG is

$$\begin{aligned} dG &= dH - TdS - SdT \\ &= C_p dT - T(C_p dT/T) - SdT \\ &= -SdT, \end{aligned}$$

so that the free energy of the system at a temperature T is

$$G = G_0 - \int_0^T S dT.$$

At the absolute zero of temperature, the free energy G_0 is equal to H_0, and then

$$G = H_0 - \int_0^T S dT,$$

which if S is replaced by $\int_0^T (C_p/T)\mathrm{d}T$ becomes

$$G = H_0 - \int_0^T \left[\int_0^T (C_p/T)\mathrm{d}T \right] \mathrm{d}T. \tag{5.1}$$

Equation (5.1) indicates that the free energy of a given phase decreases more rapidly with rise in temperature the larger its specific heat. The intersection of the free energy–temperature curves, shown in Figure 2.8a, therefore takes place because the low-temperature phase has a smaller specific heat than the higher-temperature phase.

At low temperatures the second term in equation (5.1) is relatively unimportant, and the phase that is stable is the one which has the lowest value of H_0, i.e. the most close-packed phase which is associated with a strong bonding of the atoms. However, the more strongly bound the phase, the higher is its elastic constant, the higher the vibrational frequency and consequently the smaller the specific heat (see Figure 5.2a). Thus, the more weakly bound structure, i.e. the phase with the higher H_0 at low temperature, is likely to appear as the stable phase at higher temperatures. This is because the second term in equation (5.1) now becomes important and G decreases more rapidly with increasing temperature, for the phase with the largest value of $\int (C_p/T)\mathrm{d}T$. From Figure 5.2b it is clear that a large $\int (C_p/T)\mathrm{d}T$ is associated with a low characteristic temperature and hence with a low vibrational frequency, such as is displayed by a metal with a more open structure and small elastic strength. In general, therefore, when phase, changes occur the more close-packed structure usually exists at the low temperatures and the more open structures at the high temperatures. From this viewpoint a liquid, which possesses no long-range structure, has a higher entropy than any solid phase, so that ultimately all metals must melt at a sufficiently high temperature, i.e. when the TS term outweighs the H term in the free energy equation.

The sequence of phase changes in such metals as titanium, zirconium, etc. is in agreement with this prediction and, moreover, the alkali metals, lithium and sodium, which are normally bcc at ordinary temperatures, can be transformed to fcc at sub-zero temperatures. It is interesting to note that iron, being bcc (α-iron) even at low temperatures and fcc (γ-iron) at high temperatures, is an exception to this rule. In this case, the stability of the bcc structure is thought to be associated with its ferromagnetic properties. By having a bcc structure the interatomic distances are of the correct value for the exchange interaction to allow the electrons to adopt parallel spins (this is a condition for magnetism). While this state is one of low entropy it is also one of minimum internal energy, and in the lower temperature ranges this is the factor which governs the phase stability, so that the bcc structure is preferred.

Iron is also of interest because the bcc structure, which is replaced by the fcc structure at temperatures above 910°C, reappears as the δ-phase above 1400°C. This behavior is attributed to the large electronic specific heat of iron, which is a characteristic feature of most transition metals. Thus, the Debye characteristic temperature of γ-iron is lower than that of α-iron and this is mainly responsible for the α to γ transformation. However, the electronic specific heat of the α-phase becomes greater than that of the γ-phase above about 300°C and eventually at higher temperatures becomes sufficient to bring about the return to the bcc structure at 1400°C.

5.4 Diffusion

5.4.1 Diffusion laws

Some knowledge of diffusion is essential in understanding the behavior of materials, particularly at elevated temperatures. A few examples include such commercially important processes as annealing, heat treatment, the age hardening of alloys, sintering, surface hardening, oxidation and creep. Apart

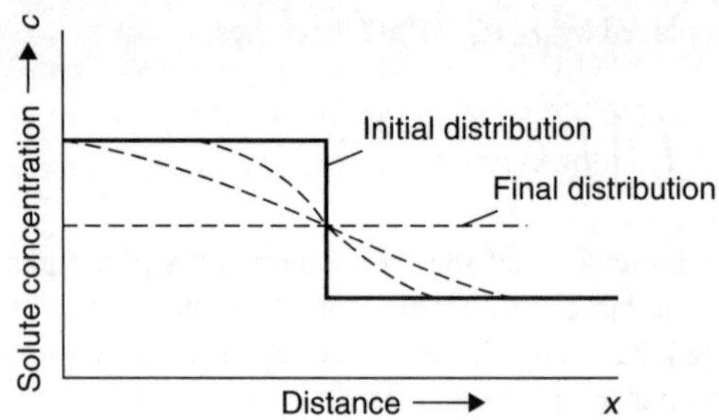

Figure 5.4 *Effect of diffusion on the distribution of solute in an alloy.*

from the specialized diffusion processes, such as grain boundary diffusion and diffusion down dislocation channels, a distinction is frequently drawn between diffusion in pure metals, homogeneous alloys and inhomogeneous alloys. In a pure material self-diffusion can be observed by using radioactive tracer atoms. In a homogeneous alloy diffusion of each component can also be measured by a tracer method, but in an inhomogeneous alloy diffusion can be determined by chemical analysis merely from the broadening of the interface between the two metals as a function of time. Inhomogeneous alloys are common in metallurgical practice (e.g. cored solid solutions) and in such cases diffusion always occurs in such a way as to produce a macroscopic flow of solute atoms down the concentration gradient. Thus, if a bar of an alloy, along which there is a concentration gradient (Figure 5.4), is heated for a few hours at a temperature where atomic migration is fast, i.e. near the melting point, the solute atoms are redistributed until the bar becomes uniform in composition. This occurs even though the individual atomic movements are random, simply because there are more solute atoms to move down the concentration gradient than there are to move up. This fact forms the basis of Fick's law of diffusion, which is

$$dn/dt = -D\,dc/dx. \tag{5.2}$$

Here the number of atoms diffusing in unit time across unit area through a unit concentration gradient is known as the diffusivity or diffusion coefficient,[1] D. It is usually expressed as units of $cm^2\,s^{-1}$ or $m^2\,s^{-1}$ and depends on the concentration and temperature of the alloy.

To illustrate, we may consider the flow of atoms in one direction x, by taking two atomic planes A and B of unit area separated by a distance b, as shown in Figure 5.5. If c_1 and c_2 are the concentrations of diffusing atoms in these two planes ($c_1 > c_2$), the corresponding number of such atoms in the respective planes is $n_1 = c_1 b$ and $n_2 = c_2 b$. If the probability that any one jump in the $+x$ direction is p_x, then the number of jumps per unit time made by one atom is $p_x \nu$, where ν is the mean frequency with which an atom leaves a site irrespective of directions. The number of diffusing atoms leaving A and arriving at B in unit time is $(p_x \nu c_1 b)$ and the number making the reverse transition is $(p_x \nu c_2 b)$ so that the net gain of atoms at B is

$$p_x \nu b(c_1 - c_2) = J_x,$$

with J_x the flux of diffusing atoms. Setting $c_1 - c_2 = -b(dc/dx)$, this flux becomes:

$$\begin{aligned} J_x &= -p_x \nu_\nu b^2(dc/dx) = -\tfrac{1}{2}\nu b^2(dc/dx) \\ &= -D(dc/dx). \end{aligned} \tag{5.3}$$

[1] The conduction of heat in a still medium also follows the same laws as diffusion.

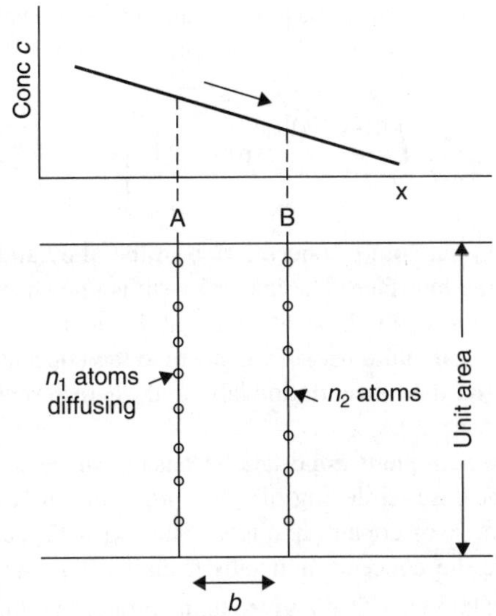

Figure 5.5 *Diffusion of atoms down a concentration gradient.*

In cubic lattices, diffusion is isotropic and hence all six orthogonal directions are equally likely, so that $p_x = \frac{1}{6}$. For simple cubic structures $b = a$ and thus

$$D_x = D_y = D_z = \tfrac{1}{6}va^2 = D, \tag{5.4}$$

whereas in fcc structures $b = a/\sqrt{2}$ and $D = \frac{1}{12}va^2$, and in bcc structures $D = \frac{1}{24}va^2$.

Fick's first law only applies if a steady state exists in which the concentration at every point is invariant, i.e. $(\mathrm{d}c/\mathrm{d}t) = 0$ for all x. To deal with non-stationary flow in which the concentration at a point changes with time, we take two planes A and B, as before, separated by unit distance and consider the rate of increase of the number of atoms $(\mathrm{d}c/\mathrm{d}t)$ in a unit volume of the specimen; this is equal to the difference between the flux into and that out of the volume element. The flux across one plane is J_x and across the other $J_x + \mathrm{d}J_x$, the difference being $-\mathrm{d}J_x$. We thus obtain Fick's second law of diffusion:

$$\frac{\mathrm{d}c}{\mathrm{d}t} = -\frac{\mathrm{d}J_x}{\mathrm{d}x} = \frac{\mathrm{d}}{\mathrm{d}x}\left(D_x\frac{\mathrm{d}c}{\mathrm{d}x}\right). \tag{5.5}$$

When D is independent of concentration this reduces to

$$\frac{\mathrm{d}c}{\mathrm{d}t} = D_x\frac{\mathrm{d}^2c}{\mathrm{d}x^2} \tag{5.6}$$

and in three dimensions becomes

$$\frac{\mathrm{d}c}{\mathrm{d}t} = \frac{\mathrm{d}}{\mathrm{d}x}\left(D_x\frac{\mathrm{d}c}{\mathrm{d}x}\right) + \frac{\mathrm{d}}{\mathrm{d}y}\left(D_y\frac{\mathrm{d}c}{\mathrm{d}y}\right) + \frac{\mathrm{d}}{\mathrm{d}z}\left(D_z\frac{\mathrm{d}c}{\mathrm{d}z}\right).$$

An illustration of the use of the diffusion equations is the behavior of a diffusion couple, where there is a sharp interface between pure metal and an alloy. Figure 5.4 can be used for this example and as the solute moves from alloy to the pure metal, the way in which the concentration varies is shown by the dotted lines. The solution to Fick's second law is given by

$$c = \frac{c_0}{2}\left[1 - \frac{2}{\sqrt{\pi}}\int_0^{x/[2\sqrt{(Dt)}]} \exp(-y^2)\mathrm{d}y\right], \tag{5.7}$$

where c_0 is the initial solute concentration in the alloy and c is the concentration at a time t at a distance x from the interface. The integral term is known as the Gauss error function (erf (y)) and as $y \to \infty$, erf $(y) \to 1$. It will be noted that at the interface where $x = 0$, then $c = c_0/2$, and in those regions where the curvature $\partial^2 c/\partial x^2$ is positive the concentration rises, in those regions where the curvature is negative the concentration falls, and where the curvature is zero the concentration remains constant.

This particular example is important because it can be used to model the depth of diffusion after time t, e.g. in the case hardening of steel, providing the concentration profile of the carbon after a carburizing time t, or dopant in silicon. Starting with a constant composition at the surface, the value of x where the concentration falls to half the initial value, i.e. $1 - \mathrm{erf}(y) = \frac{1}{2}$, is given by $x = \sqrt{(Dt)}$. Thus, knowing D at a given temperature, the time to produce a given depth of diffusion can be estimated.

The diffusion equations developed above can also be transformed to apply to particular diffusion geometries. If the concentration gradient has spherical symmetry about a point, c varies with the radial distance r and, for constant D,

$$\frac{\mathrm{d}c}{\mathrm{d}t} = D\left(\frac{\mathrm{d}^2 c}{\mathrm{d}r^2} + \frac{2}{r}\frac{\mathrm{d}c}{\mathrm{d}r}\right). \tag{5.8}$$

When the diffusion field has radial symmetry about a cylindrical axis, the equation becomes

$$\frac{\mathrm{d}c}{\mathrm{d}t} = D\left(\frac{\mathrm{d}^2 c}{\mathrm{d}r^2} + \frac{1}{r}\frac{\mathrm{d}c}{\mathrm{d}r}\right) \tag{5.9}$$

and the steady-state condition $(\mathrm{d}c/\mathrm{d}t) = 0$ is given by

$$\frac{\mathrm{d}^2 c}{\mathrm{d}r^2} + \frac{1}{r}\frac{\mathrm{d}c}{\mathrm{d}r} = 0, \tag{5.10}$$

which has a solution $c = A\ln(r) + B$. The constants A and B may be found by introducing the appropriate boundary conditions and for $c = c_0$ at $r = r_0$ and $c = c_1$ at $r = r_1$ the solution becomes

$$c = \frac{c_0 \ln(r_1/r) + c_1 \ln(r/r_0)}{\ln(r_1/r_0)}.$$

The flux through any shell of radius r is $-2\pi rD(\mathrm{d}c/\mathrm{d}r)$ or

$$J = -\frac{2\pi D}{\ln(r_1/r_0)}(c_1 - c_0). \tag{5.11}$$

Diffusion equations are of importance in many diverse problems and in Chapter 3 are applied to the diffusion of vacancies from dislocation loops and the sintering of voids.

5.4.2 Mechanisms of diffusion

The transport of atoms through the lattice may conceivably occur in many ways. The term 'interstitial diffusion' describes the situation when the moving atom does not lie on the crystal lattice, but instead occupies an interstitial position. Such a process is likely in interstitial alloys where the migrating atom is very small (e.g. carbon, nitrogen or hydrogen in iron). In this case, the diffusion process for the atoms to move from one interstitial position to the next in a perfect lattice is not defect-controlled. A possible variant of this type of diffusion has been suggested for substitutional solutions in which the diffusing atoms are only temporarily interstitial and are in dynamic equilibrium with others in substitutional positions. However, the energy to form such an interstitial is many times that to produce a vacancy and, consequently, the most likely mechanism is that of the continual migration of vacancies. With vacancy diffusion, the probability that an atom may jump to the next site will depend on: (1) the probability that the site is vacant (which in turn is proportional to the fraction of vacancies in the crystal) and (2) the probability that it has the required activation energy to make the transition. For self-diffusion, where no complications exist, the diffusion coefficient is therefore given by

$$\begin{aligned} D &= \tfrac{1}{6}a^2 f\nu \exp[(S_f + S_m)/k] \times \exp[-E_f/\mathbf{k}T]\exp[-E_m/\mathbf{k}T] \\ &= D_0 \exp[-(E_f + E_m)/\mathbf{k}T]. \end{aligned}$$

The factor f appearing in D_0 is known as a correlation factor and arises from the fact that any particular diffusion jump is influenced by the direction of the previous jump. Thus, when an atom and a vacancy exchange places in the lattice there is a greater probability of the atom returning to its original site than moving to another site, because of the presence there of a vacancy; f is 0.80 and 0.78 for fcc and bcc lattices, respectively. Values for E_f and E_m are discussed in Chapter 3: E_f is the energy of formation of a vacancy, E_m the energy of migration, and the sum of the two energies, $Q = E_f + E_m$, is the activation energy for self-diffusion[2] E_d.

In alloys, the problem is not so simple and it is found that the self-diffusion energy is smaller than in pure metals. This observation has led to the suggestion that in alloys the vacancies associate preferentially with solute atoms in solution; the binding of vacancies to the impurity atoms increases the effective vacancy concentration near those atoms so that the mean jump rate of the solute atoms is much increased. This association helps the solute atom on its way through the lattice but, conversely, the speed of vacancy migration is reduced because it lingers in the neighborhood of the solute atoms, as shown in Figure 5.6. The phenomenon of association is of fundamental importance in all kinetic studies, since the mobility of a vacancy through the lattice to a vacancy sink will be governed by its ability to escape from the impurity atoms which trap it. This problem has been mentioned in Chapter 3.

When considering diffusion in alloys it is important to realize that in a binary solution of A and B the diffusion coefficients D_A and D_B are generally not equal. This inequality of diffusion was first demonstrated by Kirkendall using an α-brass/copper couple (Figure 5.7). He noted that if the position of the interfaces of the couple were marked (e.g. with fine W or Mo wires), during diffusion the markers move towards each other, showing that the zinc atoms diffuse out of the alloy more rapidly than copper atoms diffuse in. This being the case, it is not surprising that several workers have shown that porosity develops in such systems on that side of the interface from which there is a net loss of atoms.

The Kirkendall effect is of considerable theoretical importance, since it confirms the vacancy mechanism of diffusion. This is because the observations cannot easily be accounted for by any other

---

[2] The entropy factor $\exp[(S_f + S_m)/k]$ is usually taken to be unity.

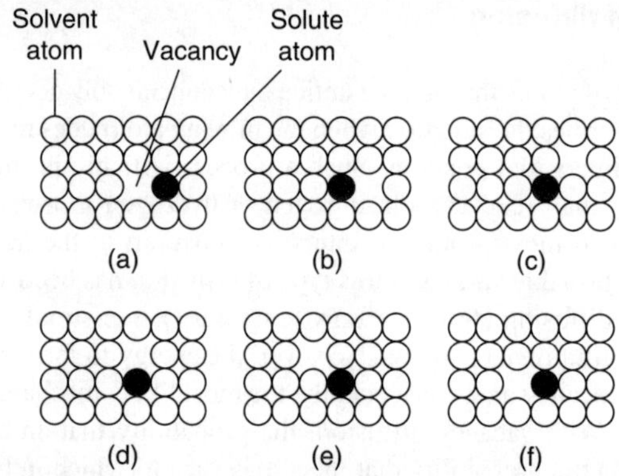

Figure 5.6 *Solute atom–vacancy association during diffusion.*

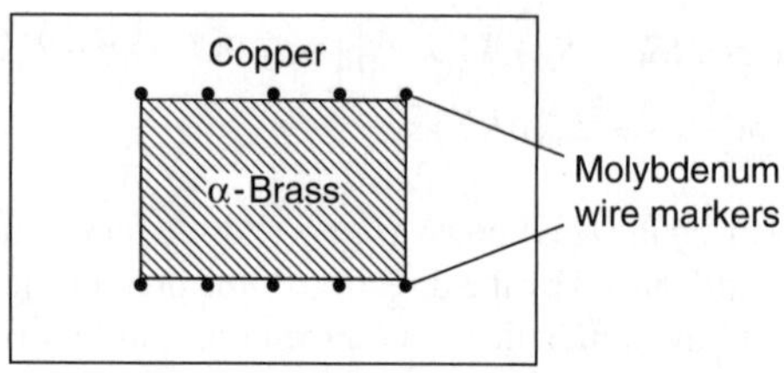

Figure 5.7 *α-Brass–copper couple for demonstrating the Kirkendall effect.*

postulated mechanisms of diffusion, such as direct place exchange, i.e. where neighboring atoms merely change place with each other. The Kirkendall effect is readily explained in terms of vacancies, since the lattice defect may interchange places more frequently with one atom than the other. The effect is also of some practical importance, especially in the fields of metal-to-metal bonding, sintering and creep.

5.4.3 Factors affecting diffusion

The two most important factors affecting the diffusion coefficient D are temperature and composition. Because of the activation energy term the rate of diffusion increases with temperature according to equation (5.12), while each of the quantities D, D_0 and Q varies with concentration; for a metal at high temperatures $Q \approx 20\mathbf{R}T_m$, D_0 is 10^{-5} to 10^{-3} m^2 s^{-1}, and $D \cong 10^{-12}$ m^2 s^{-1}. Because of this variation of diffusion coefficient with concentration, the most reliable investigations into the effect of other variables necessarily concern self-diffusion in pure metals.

Diffusion is a structure-sensitive property and therefore D is expected to increase with increasing lattice irregularity. In general, this is found experimentally. In metals quenched from a high temperature the excess vacancy concentration $\approx 10^9$ leads to enhanced diffusion at low temperatures, since $D = D_0 c_v \exp(-E_m/\mathbf{k}T)$ where c_v is the vacancy concentration. Grain boundaries and dislocations are particularly important in this respect and produce enhanced diffusion. Diffusion is faster in the cold-worked state than in the annealed state, although recrystallization may take place and tend to mask the effect. The enhanced transport of material along dislocation channels has been demonstrated in aluminum, where voids connected to a free surface by dislocations anneal out at appreciably higher rates than isolated voids. Measurements show that surface and grain boundary forms of diffusion also obey Arrhenius equations, with lower activation energies than for volume diffusion, i.e. $Q_{\text{vol}} \geq 2Q_{g \cdot b} \geq 2Q_{\text{surface}}$. This behavior is understandable in view of the progressively more open atomic structure found at grain boundaries and external surfaces. It will be remembered, however, that the relative importance of the various forms of diffusion does not entirely depend on the relative activation energy or diffusion coefficient values. The amount of material transported by any diffusion process is given by Fick's law and for a given composition gradient also depends on the effective area through which the atoms diffuse. Consequently, since the surface area (or grain boundary area) to volume ratio of any polycrystalline solid is usually very small, it is only in particular phenomena (e.g. sintering, oxidation, etc.) that grain boundaries and surfaces become important. It is also apparent that grain boundary diffusion becomes more competitive the finer the grain and the lower the temperature. The lattice feature follows from the lower activation energy, which makes it less sensitive to temperature change. As the temperature is lowered, the diffusion rate along grain boundaries (and also surfaces) decreases less rapidly than the diffusion rate through the lattice. The importance of grain boundary diffusion and dislocation pipe diffusion is discussed again in Chapter 6 in relation to deformation at elevated temperatures, and is demonstrated convincingly in the deformation maps (see Figure 6.67), where the creep field is extended to lower temperatures when grain boundary (Coble creep) rather than lattice diffusion (Herring–Nabarro creep) operates.

Because of the strong binding between atoms, pressure has little or no effect, but it is observed that with extremely high pressure on soft metals (e.g. sodium) an increase in Q may result. The rate of diffusion also increases with decreasing density of atomic packing. For example, self-diffusion is slower in fcc iron or thallium than in bcc iron or thallium when the results are compared by extrapolation to the transformation temperature. This is further emphasized by the anisotropic nature of D in metals of open structure. Bismuth (rhombohedral) is an example of a metal in which D varies by 10^6 for different directions in the lattice; in cubic crystals D is isotropic.

5.5 Anelasticity and internal friction

For an elastic solid it is generally assumed that stress and strain are directly proportional to one another, but in practice the elastic strain is usually dependent on time as well as stress, so that the strain lags behind the stress; this is an anelastic effect. On applying a stress at a level below the conventional elastic limit, a specimen will show an initial elastic strain ε_e followed by a gradual increase in strain until it reaches an essentially constant value, $\varepsilon_e + \varepsilon_{an}$, as shown in Figure 5.8. When the stress is removed the strain will decrease, but a small amount remains, which decreases slowly with time. At any time t the decreasing anelastic strain is given by the relation $\varepsilon = \varepsilon_{an} \exp(-t/\tau)$, where τ is known as the relaxation time, and is the time taken for the anelastic strain to decrease to $1/e \cong 36.79\%$ of its initial value. Clearly, if τ is large, the strain relaxes very slowly, while if small the strain relaxes quickly.

In materials under cyclic loading this anelastic effect leads to a decay in amplitude of vibration and therefore a dissipation of energy by internal friction. Internal friction is defined in several different but related ways. Perhaps the most common uses the logarithmic decrement $\delta = \ln(A_n/A_{n+1})$, the natural logarithm of successive amplitudes of vibration. In a forced vibration experiment near a resonance, the factor $(\omega_2 - \omega_1)/\omega_0$ is often used, where ω_1 and ω_2 are the frequencies on the two sides of the resonant frequency ω_0 at which the amplitude of oscillation is $1/\sqrt{2}$ of the resonant amplitude. Also used is the specific damping capacity $\Delta E/E$, where ΔE is the energy dissipated per cycle of

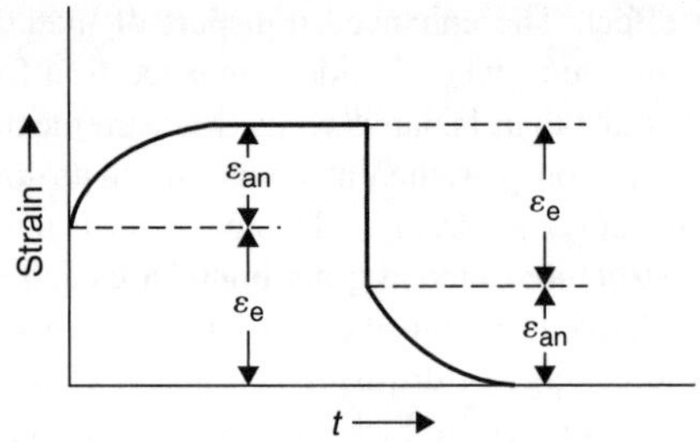

Figure 5.8 *Anelastic behavior.*

vibrational energy E, i.e. the area contained in a stress–strain loop. Yet another method uses the phase angle α by which the strain lags behind the stress, and if the damping is small it can be shown that

$$\tan\alpha = \frac{\delta}{\pi} = \frac{1}{2\pi}\frac{\Delta E}{E} = \frac{\omega_2 - \omega_1}{\omega_0} = Q^{-1}. \tag{5.12}$$

By analogy with damping in electrical systems, $\tan\alpha$ is often written equal to Q^{-1}.

There are many causes of internal friction arising from the fact that the migration of atoms, lattice defects and thermal energy are all time-dependent processes. The latter gives rise to thermoelasticity and occurs when an elastic stress is applied to a specimen too fast for the specimen to exchange heat with its surroundings and so cools slightly. As the sample warms back to the surrounding temperature it expands thermally, and hence the dilatation strain continues to increase after the stress has become constant.

The diffusion of atoms can also give rise to anelastic effects in an analogous way to the diffusion of thermal energy giving thermoelastic effects. A particular example is the stress-induced diffusion of carbon or nitrogen in iron. A carbon atom occupies the interstitial site along one of the cell edges, slightly distorting the lattice tetragonally. Thus, when iron is stretched by a mechanical stress, the crystal axis oriented in the direction of the stress develops favored sites for the occupation of the interstitial atoms relative to the other two axes. Then if the stress is oscillated, such that first one axis and then another is stretched, the carbon atoms will want to jump from one favored site to the other. Mechanical work is therefore done repeatedly, dissipating the vibrational energy and damping out the mechanical oscillations. The maximum energy is dissipated when the time per cycle is of the same order as the time required for the diffusional jump of the carbon atom.

The simplest and most convenient way of studying this form of internal friction is by means of a Kê torsion pendulum, shown schematically in Figure 5.9. The specimen can be oscillated at a given frequency by adjusting the moment of inertia of the torsion bar. The energy loss per cycle $\Delta E/E$ varies smoothly with the frequency according to the relation:

$$\frac{\Delta E}{E} = 2\left(\frac{\Delta E}{E}\right)_{\max}\left[\frac{\omega\tau}{1+(\omega\tau)^2}\right]$$

and has a maximum value when the angular frequency of the pendulum equals the relaxation time of the process; at low temperatures around room temperature this is interstitial diffusion. In practice, it is difficult to vary the angular frequency over a wide range and thus it is easier to keep ω constant and vary the relaxation time. Since the migration of atoms depends strongly on temperature according to an Arrhenius-type equation, the relaxation time $\tau_1 = 1/\omega_1$ and the peak occurs at a temperature T_1. For a different frequency value ω_2 the peak occurs at a different temperature T_2, and so on

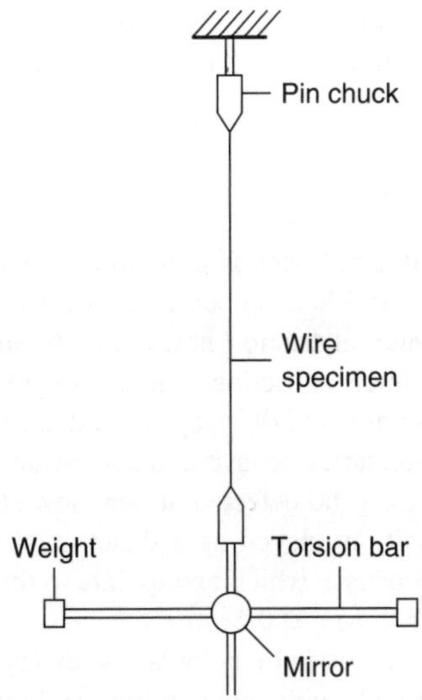

Figure 5.9 *Schematic diagram of a Kê torsion pendulum.*

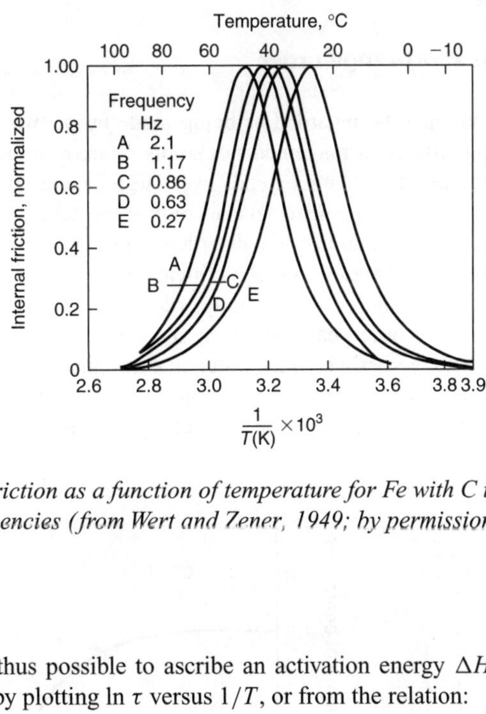

Figure 5.10 *Internal friction as a function of temperature for Fe with C in solid solution at five different pendulum frequencies (from Wert and Zener, 1949; by permission of the American Physical Society).*

(see Figure 5.10). It is thus possible to ascribe an activation energy ΔH for the internal process producing the damping by plotting $\ln \tau$ versus $1/T$, or from the relation:

$$\Delta H = R\frac{\ln(\omega_2/\omega_1)}{1/T_1 - 1/T_2}.$$

In the case of iron the activation energy is found to coincide with that for the diffusion of carbon in iron. Similar studies have been made for other metals. In addition, if the relaxation time is τ the mean time an atom stays in an interstitial position is $(\frac{3}{2})\tau$, and from the relation $D = \frac{1}{24}a^2\nu$ for bcc lattices derived previously the diffusion coefficient may be calculated directly from

$$D = \frac{1}{36}\left(\frac{a^2}{\tau}\right).$$

Many other forms of internal friction exist in metals arising from different relaxation processes to those discussed above, and hence occurring in different frequency and temperature regions. One important source of internal friction is that due to stress relaxation across grain boundaries. The occurrence of a strong internal friction peak due to grain boundary relaxation was first demonstrated on polycrystalline aluminum at 300°C by Kê and has since been found in numerous other metals. It indicates that grain boundaries behave in a somewhat viscous manner at elevated temperatures and grain boundary sliding can be detected at very low stresses by internal friction studies. The grain boundary sliding velocity produced by a shear stress τ is given by $\nu = \tau d/\eta$ and its measurement gives values of the viscosity η which extrapolate to that of the liquid at the melting point, assuming the boundary thickness to be $d \cong 0.5$ nm.

Movement of low-energy twin boundaries in crystals, domain boundaries in ferromagnetic materials, and dislocation bowing and unpinning all give rise to internal friction and damping.

5.6 Ordering in alloys

5.6.1 Long-range and short-range order

An ordered alloy may be regarded as being made up of two or more interpenetrating sublattices, each containing different arrangements of atoms. Moreover, the term 'superlattice' would imply that such a coherent atomic scheme extends over large distances, i.e. the crystal possesses long-range order. Such a perfect arrangement can exist only at low temperatures, since the entropy of an ordered structure is much lower than that of a disordered one, and with increasing temperature the degree of long-range order, S, decreases until at a critical temperature T_c it becomes zero; the general form of the curve is shown in Figure 5.11. Partially ordered structures are achieved by the formation of small regions (domains) of order, each of which are separated from each other by domain or anti-phase domain boundaries, across which the order changes phase (Figure 5.12). However, even when long-range order is destroyed, the tendency for unlike atoms to be neighbors still exists, and short-range order results above T_c. The transition from complete disorder to complete order is a nucleation and

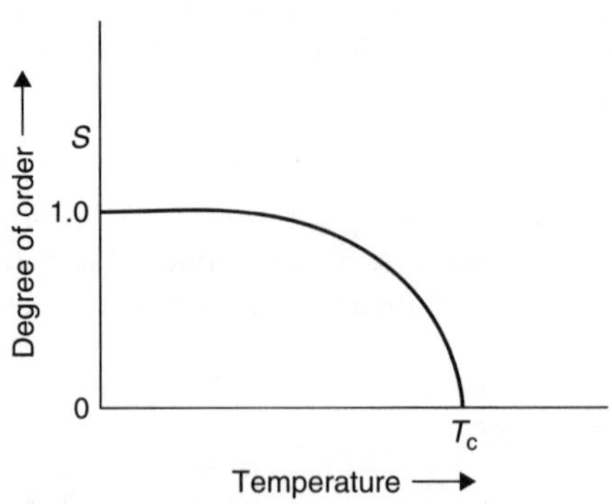

Figure 5.11 *Influence of temperature on the degree of order.*

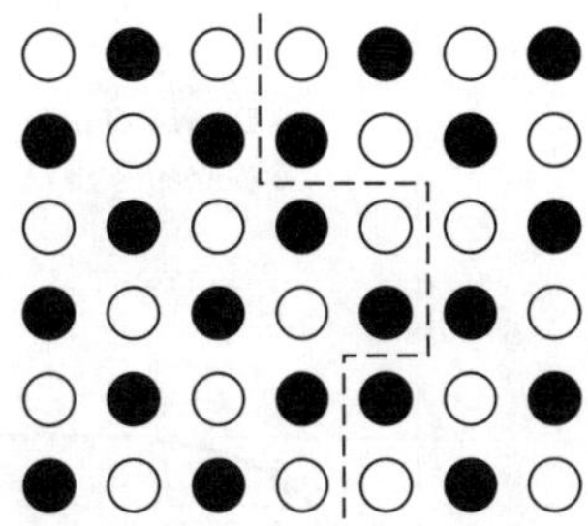

Figure 5.12 *An anti-phase domain boundary.*

growth process and may be likened to the annealing of a cold-worked structure. At high temperatures well above T_c, there are more than the random number of AB atom pairs, and with the lowering of temperature small nuclei of order continually form and disperse in an otherwise disordered matrix. As the temperature, and hence thermal agitation, is lowered these regions of order become more extensive, until at T_c they begin to link together and the alloy consists of an interlocking mesh of small ordered regions. Below T_c these domains absorb each other (cf. grain growth) as a result of anti-phase domain boundary mobility until long-range order is established.

Some order–disorder alloys can be retained in a state of disorder by quenching to room temperature, while in others (e.g. β-brass) the ordering process occurs almost instantaneously. Clearly, changes in the degree of order will depend on atomic migration, so that the rate of approach to the equilibrium configuration will be governed by an exponential factor of the usual form, i.e. Rate $= Ae^{-Q/\mathbf{R}T}$. However, Bragg has pointed out that the ease with which interlocking domains can absorb each other to develop a scheme of long-range order will also depend on the number of possible ordered schemes the alloy possesses. Thus, in β-brass only two different schemes of order are possible, while in fcc lattices such as Cu_3Au four different schemes are possible and the approach to complete order is less rapid.

5.6.2 Detection of ordering

The determination of an ordered superlattice is usually done by means of the X-ray powder technique. In a disordered solution every plane of atoms is statistically identical and, as discussed in Chapter 4, there are reflections missing in the powder pattern of the material. In an ordered lattice, on the other hand, alternate planes become A-rich and B-rich, respectively, so that these 'absent' reflections are no longer missing but appear as extra superlattice lines. This can be seen from Figure 5.13: while

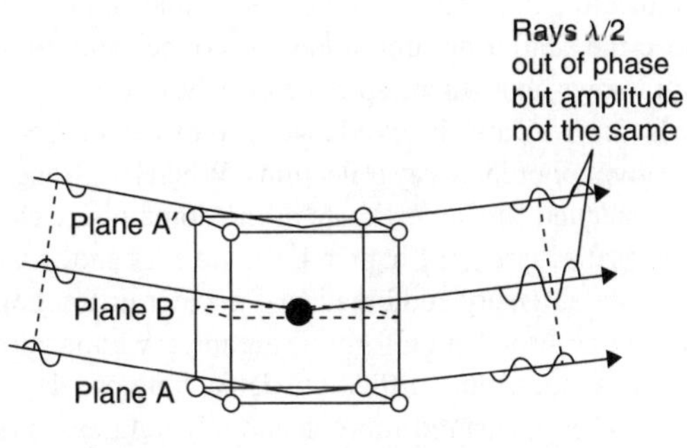

Figure 5.13 *Formation of a weak 1 0 0 reflection from an ordered lattice by the interference of diffracted rays of unequal amplitude.*

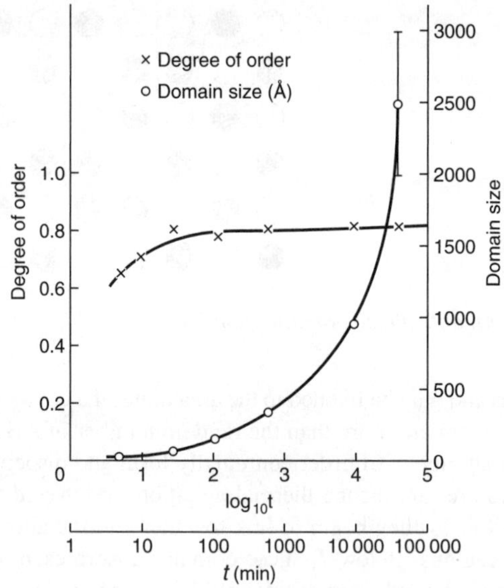

Figure 5.14 *Degree of order (×) and domain size (o) during isothermal annealing at 350°C after quenching from 465°C (after Morris, Besag and Smallman, 1974; courtesy of Taylor & Francis).*

the diffracted rays from the A planes are completely out of phase with those from the B planes their intensities are not identical, so that a weak reflection results.

Application of the structure factor equation indicates that the intensity of the superlattice lines is proportional to $|F^2| = S^2(f_A - f_B)^2$, from which it can be seen that in the fully disordered alloy, where $S = 0$, the superlattice lines must vanish. In some alloys, such as copper–gold, the scattering factor difference $(f_A - f_B)$ is appreciable and the superlattice lines are therefore quite intense and easily detectable. In other alloys, however, such as iron–cobalt, nickel–manganese and copper–zinc, the term $(f_A - f_B)$ is negligible for X-rays and the superlattice lines are very weak; in copper–zinc, for example, the ratio of the intensity of the superlattice lines to that of the main lines is only about 1:3500. In some cases special X-ray techniques can enhance this intensity ratio; one method is to use an X-ray wavelength near to the absorption edge when an anomalous depression of the f-factor occurs which is greater for one element than for the other. As a result, the difference between f_A and f_B is increased. A more general technique, however, is to use neutron diffraction, since the scattering factors for neighboring elements in the Periodic Table can be substantially different. Conversely, as Table 4.4 indicates, neutron diffraction is unable to show the existence of superlattice lines in Cu_3Au, because the scattering amplitudes of copper and gold for neutrons are approximately the same, although X-rays show them up quite clearly.

Sharp superlattice lines are observed as long as order persists over lattice regions of about 10^{-3} mm, large enough to give coherent X-ray reflections. When long-range order is not complete the superlattice lines become broadened, and an estimate of the domain size can be obtained from a measurement of the line breadth, as discussed in Chapter 4. Figure 5.14 shows variation of order S and domain size as determined from the intensity and breadth of powder diffraction lines. The domain sizes determined from the Scherrer line-broadening formula are in very good agreement with those observed by TEM. Short-range order is much more difficult to detect but nowadays direct measuring devices allow weak X-ray intensities to be measured more accurately, and as a result considerable information on the nature of short-range order has been obtained by studying the intensity of the diffuse background between the main lattice lines.

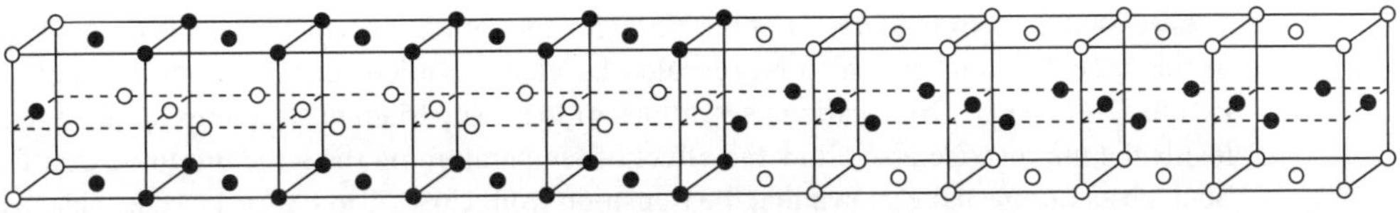

Figure 5.15 *One unit cell of the orthorhombic superlattice of CuAu, i.e. CuAu 11 (from Pashley and Presland, 1958–9; courtesy of the Institute of Materials, Minerals and Mining).*

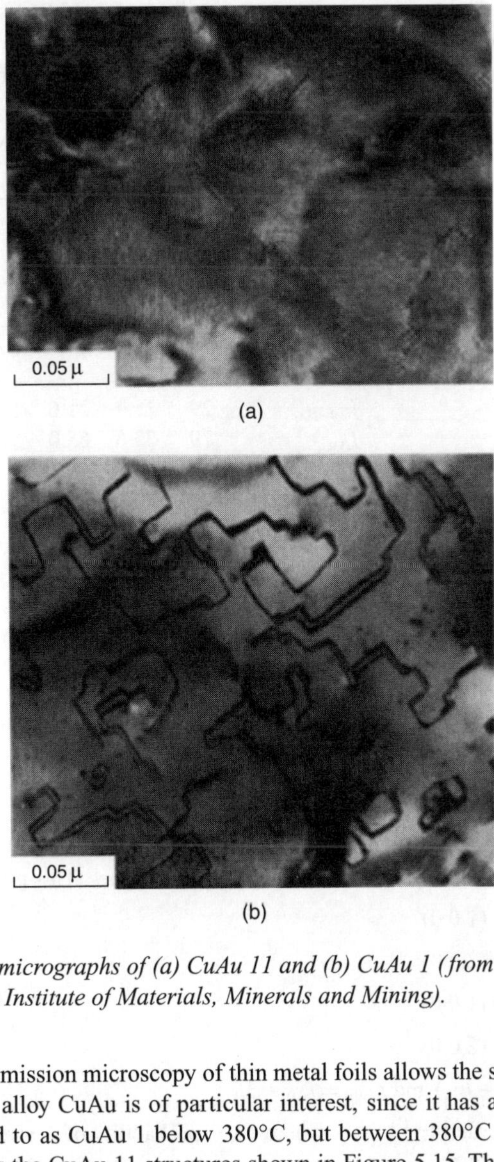

Figure 5.16 *Electron micrographs of (a) CuAu 11 and (b) CuAu 1 (from Pashley and Presland, 1958–9; courtesy of the Institute of Materials, Minerals and Mining).*

High-resolution transmission microscopy of thin metal foils allows the structure of domains to be examined directly. The alloy CuAu is of particular interest, since it has a face-centered tetragonal structure, often referred to as CuAu 1 below 380°C, but between 380°C and the disordering temperature of 410°C it has the CuAu 11 structures shown in Figure 5.15. The (0 0 2) planes are again alternately gold and copper, but halfway along the *a*-axis of the unit cell the copper atoms switch to gold planes and vice versa. The spacing between such periodic anti-phase domain boundaries is five unit cells or about 2 nm, so that the domains are easily resolvable in TEM, as seen in Figure 5.16a.

The isolated domain boundaries in the simpler superlattice structures such as CuAu 1, although not in this case periodic, can also be revealed by electron microscopy, and an example is shown in Figure 5.16b. Apart from static observations of these superlattice structures, annealing experiments inside the microscope also allow the effect of temperature on the structure to be examined directly. Such observations have shown that the transition from CuAu 1 to CuAu 11 takes place, as predicted, by the nucleation and growth of anti-phase domains.

Worked example

The X-ray diffractometer data given below were obtained from a partially ordered 75 at.% Cu/25 at.% Au alloy (a-spacing = 0.3743 nm), using CuK_α radiation ($\lambda_{\text{average}} = 0.15418$ nm). Using this and the other information provided, calculate the ordering parameter S for this alloy.

Diffraction Peak	*Integrated intensity (counts × 10^3)*
{1 0 0}	715
{2 0 0}	1660

$\frac{\sin\theta}{\lambda}$ (nm^{-1})	0.0	1.0	2.0	3.0	4.0
f_{Cu}	29	25.9	21.6	17.9	15.2
f_{Au}	79	73.6	65.0	57.0	49.7

$$\text{Lorentz polarization factor} = \frac{1+\cos^2 2\theta}{\sin^2\theta\cos\theta}.$$

Solution

Bragg's law: $\lambda = 2d\sin\theta$, $\lambda = 0.15418$ nm

$$\frac{\sin\theta}{\lambda} = \frac{1}{2d_{100}} = \frac{1}{2\times 0.3743} = 1.335, \text{ so } \theta_{100} = 11.886°.$$

$$\frac{\sin\theta}{\lambda} = \frac{1}{2d_{200}} = \frac{1}{0.3743} = 2.672, \text{ so } \theta_{200} = 24.325°$$

$(\text{LPF})_{100} = 44.28$; $(\text{LPF})_{200} = 9.29$.

Plot f_{Cu} and f_{Au} vs $\frac{\sin\theta}{\lambda}$ and read off appropriate values. This gives:

$f_{\text{Cu}} = 24.6$ for (1 0 0)

$f_{\text{Cu}} = 19.4$ for (2 0 0)

$f_{\text{Au}} = 71.0$ for (1 0 0)

$f_{\text{Au}} = 60.2$ for (2 0 0)

$F_{100} = S(f_{\text{Au}} - f_{\text{Cu}})$ and $F_{200} = f_{\text{Au}} + 3f_{\text{Cu}}$.

S (degree of order) $= \frac{P_A - C_A}{1-C_A}$, where P_A = probability of A sites filled by A atoms, and C_A = atom fraction of A atoms.

$$\frac{I_{100}}{I_{200}} = \frac{S^2(f_{\text{Au}} - f_{\text{Cu}})^2 \times (\text{LPF})_{100}}{(f_{\text{Au}} + 3f_{\text{Cu}})^2 \times (\text{LPF})_{200}}.$$

Substitute in expression:

$$\frac{715}{1660} = \frac{S^2 \times (46.4)^2 \times 44.28}{(118.4)^2 \times 9.29},$$

so $S = 0.77$.

5.6.3 Influence of ordering on properties

5.6.3.1 *Specific heat*

The order–disorder transformation has a marked effect on the specific heat, since energy is necessary to change atoms from one configuration to another. However, because the change in lattice arrangement takes place over a range of temperature, the specific heat–temperature curve will be of the form shown in Figure 5.3b. In practice the excess specific heat, above that given by Dulong and Petit's law, does not fall sharply to zero at T_c owing to the existence of short-range order, which also requires extra energy to destroy it as the temperature is increased above T_c.

5.6.3.2 *Electrical resistivity*

As discussed in Chapter 3, any form of disorder in a metallic structure (e.g. impurities, dislocations or point defects) will make a large contribution to the electrical resistance. Accordingly, superlattices below T_c have a low electrical resistance, but on raising the temperature the resistivity increases, as shown in Figure 5.17a for ordered Cu_3Au. The influence of order on resistivity is further demonstrated by the measurement of resistivity as a function of composition in the copper–gold alloy system. As shown in Figure 5.17b, at composition near Cu_3Au and CuAu, where ordering is most complete, the resistivity is extremely low, while away from these stoichiometric compositions the resistivity increases; the quenched (disordered) alloys given by the dotted curve also have high resistivity values.

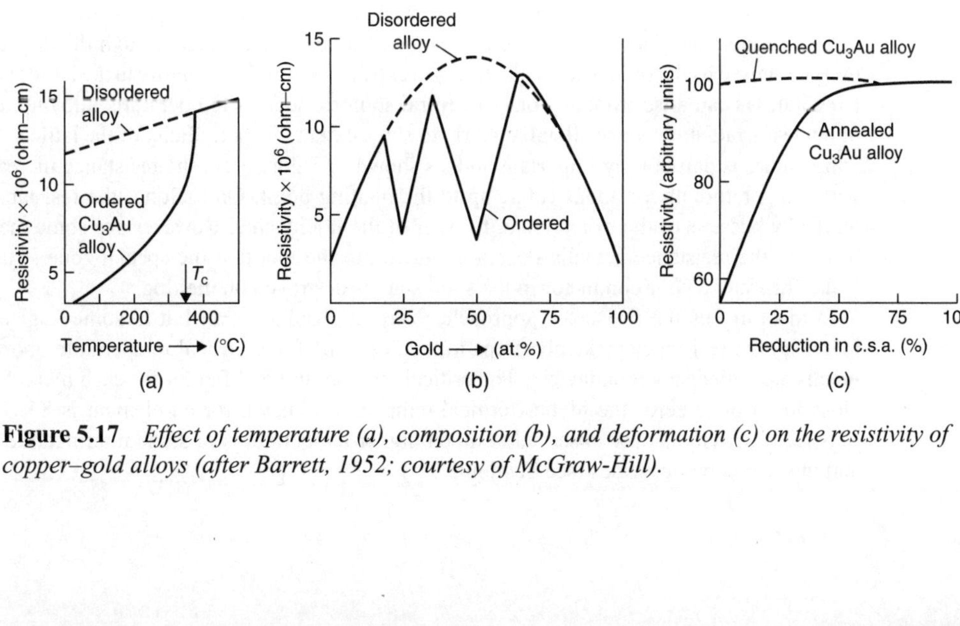

Figure 5.17 *Effect of temperature (a), composition (b), and deformation (c) on the resistivity of copper–gold alloys (after Barrett, 1952; courtesy of McGraw-Hill).*

5.6.3.3 Mechanical properties

The mechanical properties are altered when ordering occurs. The change in yield stress is not directly related to the degree of ordering, however, and in fact Cu_3Au crystals have a lower yield stress when well ordered than when only partially ordered. Experiments show that such effects can be accounted for if the maximum strength as a result of ordering is associated with critical domain size. In the alloy Cu_3Au, the maximum yield strength is exhibited by quenched samples after an annealing treatment of 5 min at 350°C, which gives a domain size of 6 nm (see Figure 5.14). However, if the alloy is well ordered and the domain size larger, the hardening is insignificant. In some alloys such as CuAu or CuPt, ordering produces a change of crystal structure and the resultant lattice strains can also lead to hardening. Thermal agitation is the most common means of destroying long-range order, but other methods (e.g. deformation) are equally effective. Figure 5.17c shows that cold work has a negligible effect upon the resistivity of the quenched (disordered) alloy but considerable influence on the well-annealed (ordered) alloy. Irradiation by neutrons or electrons also markedly affects the ordering (see Chapter 3).

5.6.3.4 Magnetic properties

The order–disorder phenomenon is of considerable importance in the application of magnetic materials. The kind and degree of order affects the magnetic hardness, since small ordered regions in an otherwise disordered lattice induce strains which affect the mobility of magnetic domain boundaries (see Section 5.8.4).

5.7 Electrical properties

5.7.1 Electrical conductivity

One of the most important electronic properties of metals is the electrical conductivity, κ, and the reciprocal of the conductivity (known as the resistivity, ρ) is defined by the relation $R = \rho l/A$, where R is the resistance of the specimen, l is the length and A is the cross-sectional area.

A characteristic feature of a metal is its high electrical conductivity, which arises from the ease with which the electrons can migrate through the lattice. The high thermal conduction of metals also has a similar explanation, and the Wiedmann–Franz law shows that the ratio of the electrical and thermal conductivities is nearly the same for all metals at the same temperature.

Since conductivity arises from the motion of conduction electrons through the lattice, resistance must be caused by the scattering of electron waves by any kind of irregularity in the lattice arrangement. Irregularities can arise from any one of several sources, such as temperature, alloying, deformation or nuclear irradiation, since all will disturb, to some extent, the periodicity of the lattice. The effect of temperature is particularly important and, as shown in Figure 5.18, the resistance increases linearly with temperature above about 100 K up to the melting point. On melting, the resistance increases markedly because of the exceptional disorder of the liquid state. However, for some metals such as bismuth, the resistance actually decreases, owing to the fact that the special zone structure which makes bismuth a poor conductor in the solid state is destroyed on melting.

In most metals the resistance approaches zero at absolute zero, but in some (e.g. lead, tin and mercury) the resistance suddenly drops to zero at some finite critical temperature above 0 K. Such metals are called superconductors. The critical temperature is different for each metal but is always close to absolute zero; the highest critical temperature known for an element is 8 K for niobium. Superconductivity is now observed at much higher temperatures in some intermetallic compounds and in some ceramic oxides (see Section 5.7.5).

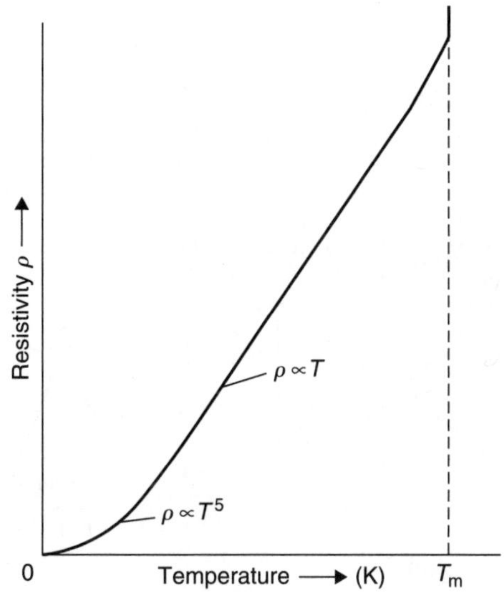

Figure 5.18 *Variation of resistivity with temperature.*

An explanation of electrical and magnetic properties requires a more detailed consideration of electronic structure than that briefly outlined in Chapter 1. There the concept of band structure was introduced and the electron can be thought of as moving continuously through the structure with an energy depending on the energy level of the band it occupies. The wave-like properties of the electron were also mentioned. For the electrons the regular array of atoms on the metallic lattice can behave as a three-dimensional diffraction grating, since the atoms are positively charged and interact with moving electrons. At certain wavelengths, governed by the spacing of the atoms on the metallic lattice, the electrons will experience strong diffraction effects, the results of which are that electrons having energies corresponding to such wavelengths will be unable to move freely through the structure. As a consequence, in the bands of electrons, certain energy levels cannot be occupied and therefore there will be energy gaps in the otherwise effectively continuous energy spectrum within a band.

The interaction of moving electrons with the metal ions distributed on a lattice depends on the wavelength of the electrons and the spacing of the ions in the direction of movement of the electrons. Since the ionic spacing will depend on the direction in the lattice, the wavelength of the electrons suffering diffraction by the ions will depend on their direction. The kinetic energy of a moving electron is a function of the wavelength according to the relationship:

$$E = \mathbf{h}^2/2m\lambda^2. \tag{5.13}$$

Since we are concerned with electron energies, it is more convenient to discuss interaction effects in terms of the reciprocal of the wavelength. This quantity is called the wave number and is denoted by k.

In describing electron–lattice interactions it is usual to make use of a vector diagram in which the direction of the vector is the direction of motion of the moving electron and its magnitude is the wave number of the electron. The vectors representing electrons having energies which, because of diffraction effects, cannot penetrate the lattice, trace out a three-dimensional surface known as a Brillouin zone. Figure 5.19a shows such a zone for a face-centered cubic lattice. It is made up of plane faces which are, in fact, parallel to the most widely spaced planes in the lattice, i.e. in this case the {1 1 1} and {2 0 0} planes. This is a general feature of Brillouin zones in all lattices.

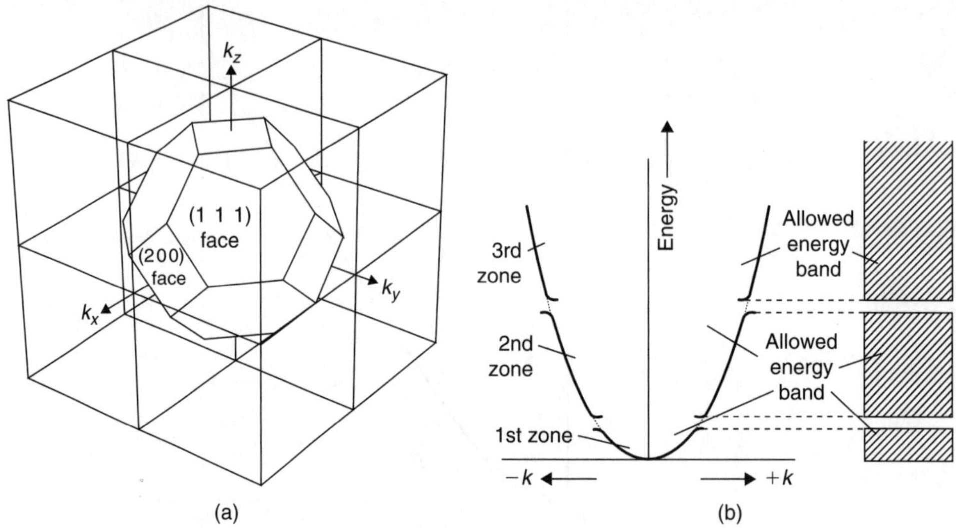

Figure 5.19 *Schematic representation of a Brillouin zone in a metal.*

For a given direction in the lattice, it is possible to consider the form of the electron energies as a function of wave number. The relationship between the two quantities as given from equation (5.13) is

$$E = \mathbf{h}^2k^2/2m, \tag{5.14}$$

which leads to the parabolic relationship shown as a broken line in Figure 5.19b. Because of the existence of a Brillouin zone at a certain value of k, depending on the lattice direction, there exists a range of energy values which the electrons cannot assume. This produces a distortion in the form of the $E-k$ curve in the neighborhood of the critical value of k and leads to the existence of a series of energy gaps, which cannot be occupied by electrons. The E–k curve showing this effect is given as a continuous line in Figure 5.19b.

The existence of this distortion in the E–k curve, due to a Brillouin zone, is reflected in the density of states–energy curve for the free electrons. As previously stated, the density of states–energy curve is parabolic in shape, but it departs from this form at energies for which Brillouin zone interactions occur. The result of such interactions is shown in Figure 5.20a, in which the broken line represents the $N(E)$–E curve for free electrons in the absence of zone effects and the full line is the curve where a zone exists. The total number of electrons needed to fill the zone of electrons delineated by the full line in Figure 5.20a is $2N$, where N is the total number of atoms in the metal. Thus, a Brillouin zone would be filled if the metal atoms each contributed two electrons to the band. If the metal atoms contribute more than two per atom, the excess electrons must be accommodated in the second or higher zones.

In Figure 5.20a the two zones are separated by an energy gap, but in real metals this is not necessarily the case, and two zones can overlap in energy in the $N(E)$–E curves so that no such energy gaps appear. This overlap arises from the fact that the energy of the forbidden region varies with direction in the lattice, and often the energy level at the top of the first zone has a higher value in one direction than the lowest energy level at the bottom of the next zone in some other direction. The energy gap in the $N(E)$–E curves, which represent the summation of electronic levels in all directions, is then closed (Figure 5.20b).

For electrical conduction to occur, it is necessary that the electrons at the top of a band should be able to increase their energy when an electric field is applied to materials so that a net flow of

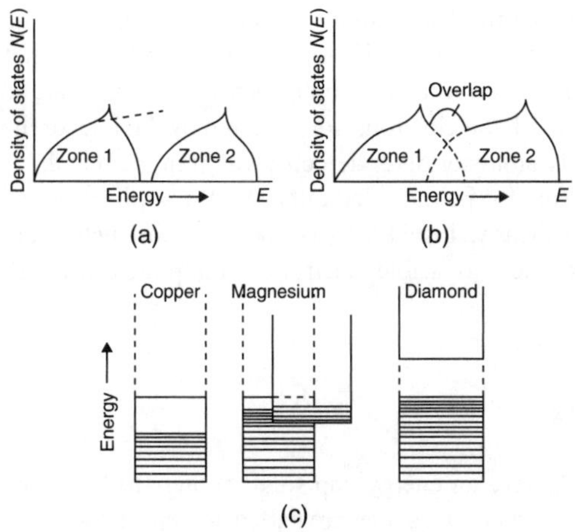

Figure 5.20 *Schematic representation of Brillouin zones.*

electrons in the direction of the applied potential, which manifests itself as an electric current, can take place. If an energy gap between two zones of the type shown in Figure 5.20a occurs, and if the lower zone is just filled with electrons, then it is impossible for any electrons to increase their energy by jumping into vacant levels under the influence of an applied electric field, unless the field strength is sufficiently great to supply the electrons at the top of the filled band with enough energy to jump the energy gap. Thus, metallic conduction is due to the fact that in metals the number of electrons per atom is insufficient to fill the band up to the point where an energy gap occurs. In copper, for example, the 4*s* valency electrons fill only one-half of the outer *s*-band. In other metals (e.g. Mg) the valency band overlaps a higher energy band and the electrons near the Fermi level are thus free to move into the empty states of a higher band. When the valency band is completely filled and the next higher band, separated by an energy gap, is completely empty, the material is either an insulator or a semiconductor. If the gap is several electron-volts wide, such as in diamond, where it is 7 eV, extremely high electric fields would be necessary to raise electrons to the higher band and the material is an insulator. If the gap is small enough, such as 1–2 eV as in silicon, then thermal energy may be sufficient to excite some electrons into the higher band and also create vacancies in the valency band; such a material is a semiconductor. In general, the lowest energy band which is not completely filled with electrons is called a conduction band and the band containing the valency electrons the valency band. For a conductor the valency band is also the conduction band. The electronic state of a selection of materials of different valencies is presented in Figure 5.20c. Although all metals are relatively good conductors of electricity, they exhibit among themselves a range of values for their resistivities. There are a number of reasons for this variability. The resistivity of a metal depends on the density of states of the most energetic electrons at the top of the band, and the shape of the $N(E)$–E curve at this point.

In the transition metals, for example, apart from producing the strong magnetic properties, great strength and high melting point, the *d*-band is also responsible for the poor electrical conductivity and high electronic specific heat. When an electron is scattered by a lattice irregularity it jumps into a different quantum state, and it will be evident that the more vacant quantum states there are available in the same energy range, the more likely will be the electron to deflect at the irregularity. The high resistivities of the transition metals may therefore be explained by the ease with which electrons can be deflected into vacant *d*-states. Phonon-assisted *s*–*d* scattering gives rise to the non-linear variation of ρ with temperature observed at high temperatures. The high electronic specific heat is also due to the high density of states in the unfilled *d*-band, since this gives rise to a considerable number of electrons at the top of the Fermi distribution which can be excited by thermal activation. In copper, of course, there are no unfilled levels at the top of the d-band into which electrons can go, and consequently both the electronic specific heat and electrical resistance are low. The conductivity also depends on the degree to which the electrons are scattered by the ions of the metal which are thermally vibrating, and by impurity atoms or other defects present in the metal.

Insulators can also be modified either by the application of high temperatures or by the addition of impurities. Clearly, insulators may become conductors at elevated temperatures if the thermal agitation is sufficient to enable electrons to jump the energy gap into the unfilled zone above.

5.7.2 Semiconductors

Some materials have an energy gap small enough to be surmounted by thermal excitation. In such intrinsic semiconductors, as they are called, the current carriers are electrons in the conduction band and holes in the valency band in equal numbers. The relative position of the two bands is as shown in Figure 5.21. The motion of a hole in the valency band is equivalent to the motion of an electron in the opposite direction. Alternatively, conduction may be produced by the presence of impurities, which either add a few electrons to an empty zone or remove a few from a full one. Materials which have their conductivity developed in this way are commonly known as semiconductors. Silicon and germanium containing small amounts of impurity have semiconducting properties at ambient temperatures and, as a consequence, they are frequently used in electronic transistor devices. Silicon normally has completely filled zones, but becomes conducting if some of the silicon atoms, which have four valency electrons, are replaced by phosphorus, arsenic or antimony atoms, which have five valency electrons. The extra electrons go into empty zones, and as a result silicon becomes an n-type semiconductor, since conduction occurs by negative carriers. On the other hand, the addition of elements of lower valency than silicon, such as aluminum, removes electrons from the filled zones, leaving behind 'holes' in the valency band structure. In this case silicon becomes a p-type semiconductor, since the movement of electrons in one direction of the zone is accompanied by a movement of 'holes' in the other, and consequently they act as if they were positive carriers. The conductivity may be expressed as the product of (1) the number of charge carriers, n, (2) the charge carried by each (i.e. $e = 1.6 \times 10^{-19}$C) and (3) the mobility of the carrier, μ.

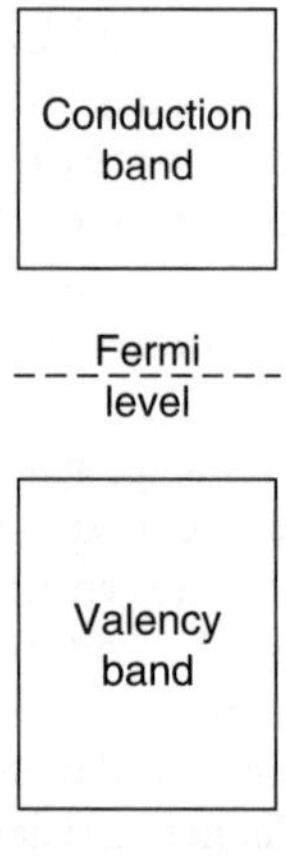

Figure 5.21 *Schematic diagram of an intrinsic semiconductor, showing the relative positions of the conduction and valency bands.*

A pentavalent impurity which donates conduction electrons without producing holes in the valency band is called a donor. The spare electrons of the impurity atoms are bound in the vicinity of the impurity atoms in energy levels known as the donor levels, which are near the conduction band. If the impurity exists in an otherwise intrinsic semiconductor the number of electrons in the conduction band becomes greater than the number of holes in the valency band and, hence, the electrons are the majority carriers and the holes the minority carriers. Such a material is an n-type extrinsic semiconductor (see Figure 5.22a).

Trivalent impurities in Si or Ge show the opposite behavior, leaving an empty electron state, or hole, in the valency band. If the hole separates from the so-called acceptor atom an electron is excited from the valency band to an acceptor level $\Delta E \approx 0.01$ eV. Thus, with impurity elements such as Al, Ga or In creating holes in the valency band in addition to those created thermally, the majority carriers are holes and the semiconductor is of the p-type extrinsic form (see Figure 5.22b). For a semiconductor where both electrons and holes carry current the conductivity is given by

$$\kappa = n_e e \mu_e + n_h e \mu_h, \tag{5.15}$$

where n_e and n_h are, respectively, the volume concentration of electrons and holes, and μ_e and μ_h the mobilities of the carriers, i.e. electrons and holes.

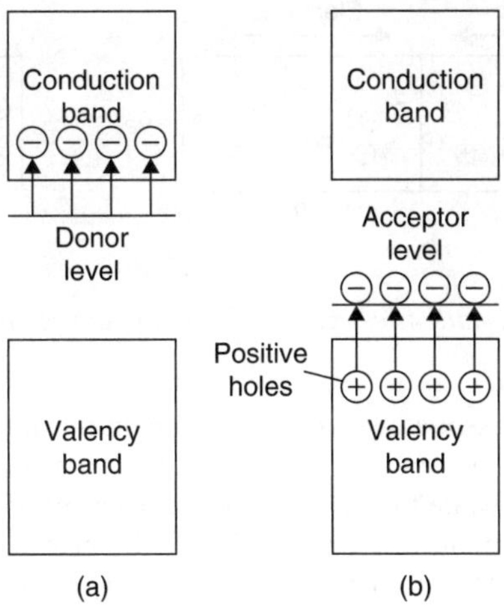

Figure 5.22 *Schematic energy band structure of n-type (a) and p-type (b) semiconductors.*

Worked example

The conductivity of a semiconductor is given by $\kappa = n_e e \mu_e + n_h e \mu_h$. What values of the carrier concentration n_e and n_h give minimum conductivity at a given temperature? Determine n_e/n_h if $\mu_e/\mu_h = 3$.

Solution

$\kappa = n_e e \mu_e + n_h e \mu_h$ and $n_e n_h = n_i^2$

$\therefore \kappa = n_e e \mu_e + n_i^2 e \mu_h / n_e.$

For minimum κ, $d\kappa/dn_e = 0 = e\mu_e - n_i^2 e\mu_h/n_e^2$

$\therefore n_e^2 = n_i^2\mu_h/\mu_e$ or $n_e = n_i\sqrt{\mu_h/\mu_e}$.

$n_h = n_i^2/n_e = n_i\sqrt{\mu_e/\mu_h}$; $n_e/n_h = \mu_h/\mu_e = 1/3$.

Semiconductor materials are extensively used in electronic devices such as the *p–n* rectifying junction, transistor (a double-junction device) and the tunnel diode. Semiconductor regions of either *p-* or *n*-type can be produced by carefully controlling the distribution and impurity content of Si or Ge single crystals, and the boundary between *p-* and *n*-type extrinsic semiconductor materials is called a *p–n* junction. Such a junction conducts a large current when the voltage is applied in one direction, but only a very small current when the voltage is reversed. The action of a *p–n* junction as a rectifier is shown schematically in Figure 5.23. The junction presents no barrier to the flow of minority carriers from either side, but since the concentration of minority carriers is low, it is the flow of majority carriers which must be considered. When the junction is biased in the forward direction, i.e. *n*-type made negative and the *p*-type positive, the energy barrier opposing the flow of majority carriers from both sides of the junction is reduced. Excess majority carriers enter the *p* and *n* regions, and these recombine continuously at or near the junction to allow large currents to flow. When the junction is reverse biased, the energy barrier opposing the flow of majority carriers is raised, few carriers move and little current flows.

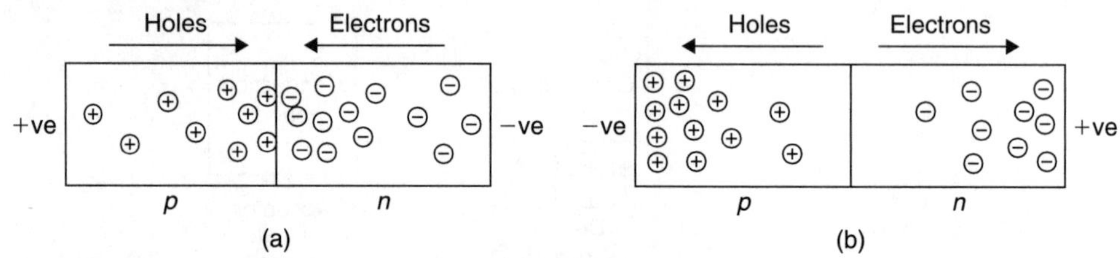

Figure 5.23 *Schematic illustration of p–n junction rectification with forward bias (a) and reverse bias (b).*

A transistor is essentially a single crystal with two *p–n* junctions arranged back to back to give either a *p–n–p* or *n–p–n* two-junction device. For a *p–n–p* device the main current flow is provided by the positive holes, while for an *n–p–n* device the electrons carry the current. Connections are made to the individual regions of the *p–n–p* device, designated emitter, base and collector respectively, as shown in Figure 5.24, and the base is made slightly negative and the collector more negative relative to the emitter. The emitter–base junction is therefore forward biased and a strong current of holes passes through the junction into the *n*-layer which, because it is thin (10^{-2} mm), largely reach the

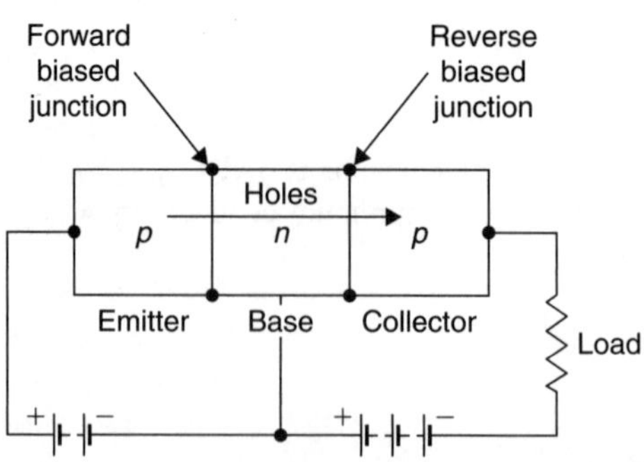

Figure 5.24 *Schematic diagram of a p–n–p transistor.*

collector base junction without recombining with electrons. The collector–base junction is reverse biased and the junction is no barrier to the passage of holes; the current through the second junction is thus controlled by the current through the first junction. A small increase in voltage across the emitter–base junction produces a large injection of holes into the base and a large increase in current in the collector, to give the amplifying action of the transistor.

Many varied semiconductor materials such as InSb and GaAs have been developed apart from Si and Ge. However, in all cases very high purity and crystal perfection is necessary for efficient semiconducting operations and, to produce the material, zone-refining techniques are used. Semiconductor integrated circuits are extensively used in microelectronic equipment and these are produced by vapor deposition through masks on to a single Si-slice, followed by diffusion of the deposits into the base crystal.

Doped ceramic materials are used in the construction of *thermistors*, which are semiconductor devices with a marked dependence of electrical resistivity upon temperature. The change in resistance can be quite significant at the critical temperature. Positive temperature coefficient (PTC) thermistors are used as switching devices, operating when a control temperature is reached during a heating process. PTC thermistors are commonly based on barium titanate. Conversely, negative temperature coefficient (NTC) thermistors are based on oxide ceramics and can be used to signal a desired temperature change during cooling; the change in resistance is much more gradual and does not have the step characteristic of the PTC types.

Doped zinc oxide does not exhibit the linear voltage–current relation that one expects from Ohm's law. At low voltage, the resistivity is high and only a small current flows. When the voltage increases there is a sudden decrease in resistance, allowing a heavier current to flow. This principle is adopted in the *varistor*, a voltage-sensitive on/off switch. It is wired in parallel with high-voltage equipment and can protect it from transient voltage 'spikes' or overload.

5.7.3 Hall effect

When dealing with electrical properties it is often useful to know the type of charge carriers, their concentrations and mobility. Electrical conductivity measurements do not provide such information but they may be given by means of Hall effect measurements. The Hall effect is produced when a magnetic field is applied perpendicular to the direction of motion of a charge carrier whereby a force is exerted on the carrier particle perpendicular to both the magnetic field and the direction of the particle motion.

Electrons (or holes) moving in a particular direction, x, producing a current I_x will be deflected in a direction y by a magnetic field B_z. Thus, a Hall voltage V_H will be established in the y-direction, the magnitude of which depends on I_x, B_z and the specimen thickness d according to

$$V_H = R_H I_x B_z / d,$$

where R_H is the Hall coefficient for a given material. When conduction is by electrons, R_H is negative and equal to $1/ne$. The mobility may be determined if the conductivity κ is known from the equation:

$$\mu_e = R_H \kappa.$$

Worked example

An n-type germanium sample is 2 mm wide and 0.2 mm thick. A current of 10 mA is passed longitudinally through the sample and a magnetic field of 0.1 Wb m^{-2} is directed parallel to the thickness. The magnitude of the Hall voltage developed is 1.0 mV. Calculate the magnitude of the Hall constant and the concentration of electrons.

Solution

Hall effect is to have:

(Hall electric field E) $\propto$ (applied magnetic field B) $\times$ (current density J)

where E, B and J are mutually perpendicular.

Hall constant is defined as $R_H = \frac{E}{BJ}$.

Magnetic field $B = 0.1\,\text{Wb}\,\text{m}^{-2}$

$$\text{Electric field } E = \frac{\text{voltage}}{\text{gap}} = \frac{1.0 \times 10^{-3}}{0.2 \times 10^{-3}} = 5\,\text{V}\,\text{m}^{-1}.$$

$$\text{Current density } J = \frac{\text{current}}{\text{area}} = \frac{10 \times 10^{-3}}{2 \times 10^{-3} \times 0.2 \times 10^{-3}} = 2.5 \times 10^4\,\text{A}\,\text{m}^{-2}$$

$$\therefore R_H = \frac{5}{0.1 \times 2.5 \times 10^4} = 2 \times 10^{-3}\,\text{m}^3\text{C}^{-1}$$

$$\text{Electron concentration } n = \frac{1}{R_H e} = \frac{1}{2 \times 10^{-3} \times 1.6 \times 10^{-19}} = 3.1 \times 10^{21}\,\text{m}^{-3}.$$

Worked example

A silicon crystal is doped with indium for which the electron acceptor level is 0.16 eV above the top of the valence band. The energy gap of silicon is 1.10 eV and the effective masses of electrons and holes are $0.26m_0$ and $0.39m_0$ respectively ($m_0 = 9.1 \times 10^{-31}$ kg is the rest mass of an electron). What impurity concentration would cause the Fermi level to coincide with the impurity level at 300 K and what fraction of the acceptor levels will be filled? What are the majority and minority carrier concentrations in the crystal?

Solution

Putting energy level at top of valence band $E_v = 0$, then bottom of conduction band $E_c = 1.10$ eV (band gap) and the required Fermi level $E_f = 0.16$ eV.

Concentration (number per m^3) of electrons (minority carriers) is given by:

$$n = 2\left(\frac{2\pi m_e^* \mathbf{k}T}{h^2}\right)^{3/2} \exp\left(\frac{E_f - E_c}{\mathbf{k}T}\right), \text{ where } m_e^* = \text{effective mass of electrons} = 0.26\,m_o.$$

Concentration of holes (majority carriers) is given by:

$$p = 2\left(\frac{2\pi m_h^* \mathbf{k}T}{h^2}\right)^{3/2} \exp\left(\frac{E_v - E_f}{\mathbf{k}T}\right), \text{ where } m_h^* = \text{effective mass of holes} = 0.39\,m_o.$$

$$\therefore n = 2 \times \left(\frac{2\pi \times 0.26 \times 9.1 \times 10^{-31} \times 1.38 \times 10^{-23} \times 300}{(6.62 \times 10^{-34})^2}\right)^{3/2}$$

$$\times \exp\left(\frac{-0.94 \times 1.6 \times 10^{-19}}{1.38 \times 10^{-23} \times 300}\right)^{-3}$$

$$= 5.4 \times 10^8\,\text{m}^{-3}$$

$$p = 2 \times \left(\frac{2\pi \times 0.39 \times 9.1 \times 10^{-31} \times 1.38 \times 10^{-23} \times 300}{(6.62 \times 10^{-34})^2} \right)^{3/2}$$

$$\times \exp\left(\frac{-0.16 \times 1.6 \times 10^{-19}}{1.38 \times 10^{-23} \times 300} \right)^{-3}$$

$$= 1.25 \times 10^{22}\,\text{m}^{-3}.$$

Since E_f coincides with the acceptor level, half the acceptors are ionized and therefore the total acceptor concentration $= 2p = 2.5 \times 10^{22}\,\text{m}^{-3}$.

5.7.4 Superconductivity

At low temperatures (<20 K) some metals have zero electrical resistivity and become superconductors. This superconductivity disappears if the temperature of the metal is raised above a critical temperature T_c, if a sufficiently strong magnetic field is applied or when a high current density flows. The critical field strength H_c, current density J_c and temperature T_c are interdependent. Figure 5.25 shows the dependence of H_c on temperature for a number of metals; metals with high T_c and H_c values, which include the transition elements, are known as hard superconductors, those with low values such as Al, Zn, Cd, Hg and white-Sn are soft superconductors. The curves are roughly parabolic and approximate to the relation $H_c = H_0[1 - (T/T_c)^2]$, where H_0 is the critical field at 0 K; H_0 is about $1.6 \times 10^5\,\text{A m}^{-1}$ for Nb.

Superconductivity arises from conduction electron–electron attraction resulting from a distortion of the lattice through which the electrons are traveling; this is clearly a weak interaction, since for most metals it is destroyed by thermal activation at very low temperatures. As the electron moves through the lattice it attracts nearby positive ions, thereby locally causing a slightly higher positive charge density. A nearby electron may in turn be attracted by the net positive charge, the magnitude of the attraction depending on the electron density, ionic charge and lattice vibrational frequencies, such that under favorable conditions the effect is slightly stronger than the electrostatic repulsion between electrons. The importance of the lattice ions in superconductivity is supported by the observation that different isotopes of the same metal (e.g. Sn and Hg) have different T_c values proportional to

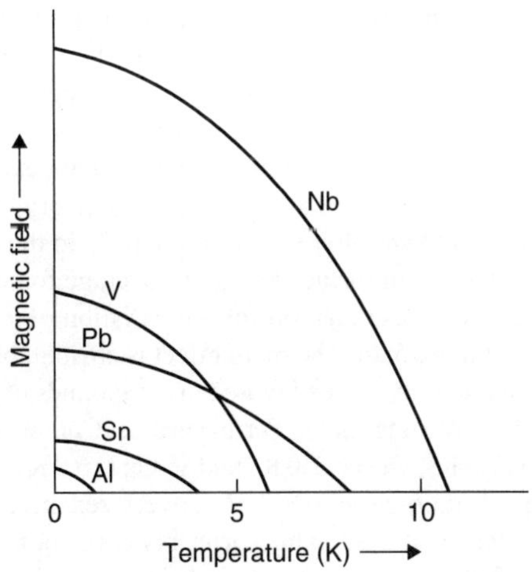

Figure 5.25 *Variation of critical field* H_c *as a function of temperature for several pure metal superconductors.*

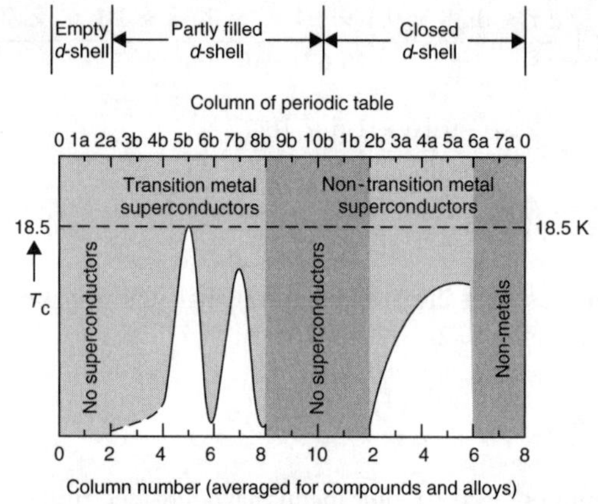

Figure 5.26 *The variation of* T_c *with position in the periodic table (from Rose, Shepard and Wulff, 1966; courtesy of John Wiley and Sons).*

$M^{-1/2}$, where M is the atomic mass of the isotope. Since both the frequency of atomic vibrations and the velocity of elastic waves also varies as $M^{-1/2}$, the interaction between electrons and lattice vibrations (i.e. electron–phonon interaction) must be at least one cause of superconductivity.

The theory of superconductivity indicates that the electron–electron attraction is strongest between electrons in pairs, such that the resultant momentum of each pair is exactly the same and the individual electrons of each pair have opposite spin. With this particular form of ordering the total electron energy (i.e. kinetic and interaction) is lowered and effectively introduces a finite energy gap between this organized state and the usual, more excited state of motion. The gap corresponds to a thin shell at the Fermi surface, but does not produce an insulator or semiconductor, because the application of an electric field causes the whole Fermi distribution, together with gap, to drift to an unsymmetrical position, so causing a current to flow. This current remains even when the electric field is removed, since the scattering which is necessary to alter the displaced Fermi distribution is suppressed.

At 0 K all the electrons are in paired states but, as the temperature is raised, pairs are broken by thermal activation, giving rise to a number of normal electrons in equilibrium with the superconducting pairs. With increasing temperature the number of broken pairs increases until at T_c they are finally eliminated together with the energy gap; the superconducting state then reverts to the normal conducting state. The superconductivity transition is a second-order transformation and a plot of C/T as a function of T^2 deviates from the linear behavior exhibited by normal conducting metals, the electronic contribution being zero at 0 K. The main theory of superconductivity, due to the attempts of Bardeen, Cooper and Schrieffer (BCS) to relate T_c to the strength of the interaction potential, the density of states at the Fermi surface and to the average frequency of lattice vibration involved in the scattering, provides some explanation for the variation of T_c with the e/a ratio for a wide range of alloys, as shown in Figure 5.26. The main effect is attributable to the change in density of states with e/a ratio. Superconductivity is thus favored in compounds of polyvalent atoms with crystal structures having a high density of states at the Fermi surface. Compounds with high T_c values, such as Nb_3Sn (18.1 K), Nb_3Al (17.5 K), V_3Si (17.0 K) and V_3Ga (16.8 K), all crystallize with the β-tungsten structure and have an e/a ratio close to 4.7; T_c is very sensitive to the degree of order and to deviation from the stoichiometric ratio, so values probably correspond to the non-stoichiometric condition.

The magnetic behavior of superconductivity is as remarkable as the corresponding electrical behavior, as shown in Figure 5.27 by the Meissner effect for an ideal (structurally perfect) superconductor. It is observed for a specimen placed in a magnetic field ($H < H_c$), which is then cooled down below

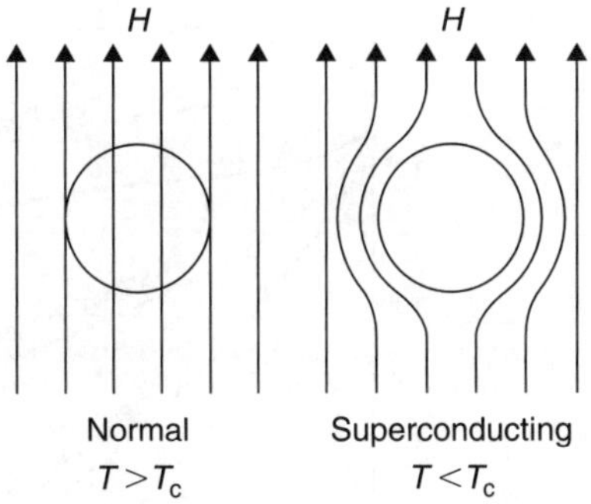

Figure 5.27 *The Meissner effect, shown by the expulsion of magnetic flux when the specimen becomes superconducting.*

T_c, that magnetic lines of force are pushed out. The specimen is a perfect diamagnetic material with zero inductance as well as zero resistance. Such a material is termed an ideal type I superconductor. An ideal type II superconductor behaves similarly at low field strengths, with $H < H_{c1} < H_c$, but then allows a gradual penetration of the field returning to the normal state when penetration is complete at $H > H_{c2} > H_c$. In detail, the field actually penetrates to a small extent in type I superconductors when it is below H_c and in type II superconductors when H is below H_{c1}, and decays away at a penetration depth ≈100–10 nm.

The observation of the Meissner effect in type I superconductors implies that the surface between the normal and superconducting phases has an effective positive energy. In the absence of this surface energy, the specimen would break up into separate fine regions of superconducting and normal material to reduce the work done in the expulsion of the magnetic flux. A negative surface energy exists between the normal and superconducting phases in a type II superconductor and hence the superconductor exists naturally in a state of finely separated superconducting and normal regions. By adopting a 'mixed state' of normal and superconducting regions the volume of interface is maximized, while at the same time keeping the volume of normal conduction as small as possible. The structure of the mixed state is believed to consist of lines of normal phases parallel to the applied field through which the field lines run, embedded in a superconducting matrix. The field falls off with distances from the center of each line over the characteristic distance λ, and vortices or whirlpools of supercurrents flow around each line; the flux line, together with its current vortex, is called a fluxoid. At H_{c1}, fluxoids appear in the specimen and increase in number as the magnetic field is raised. At H_{c2}, the fluxoids completely fill the cross-section of the sample and type II superconductivity disappears. Type II superconductors are of particular interest because of their high critical fields, which makes them potentially useful for the construction of high-field electromagnetics and solenoids. To produce a magnetic field of ≈10 T with a conventional solenoid would cost more than 10 times that of a superconducting solenoid wound with Nb_3Sn wire. By embedding Nb wire in a bronze matrix it is possible to form channels of Nb_3Sn by interdiffusion. The conventional installation would require considerable power, cooling water and space, whereas the superconducting solenoid occupies little space, has no steady-state power consumption and uses relatively little liquid helium. It is necessary, however, for the material to carry useful currents without resistance in such high fields, which is not usually the case in annealed homogeneous type II superconductors. Fortunately, the critical current density is extremely sensitive to microstructure and is markedly increased by precipitation hardening, cold work, radiation damage, etc., because the lattice defects introduced pin the fluxoids and tend to immobilize them. Figure 5.28 shows the influence of metallurgical treatment on the critical current density.

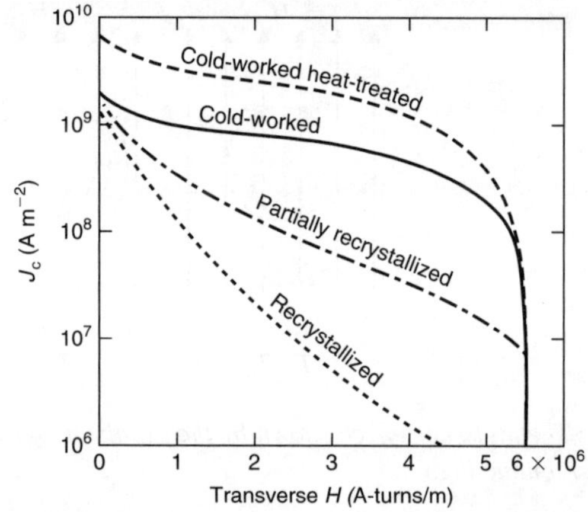

Figure 5.28 *The effect of processing on the* J_c–H *curve of an Nb–25% Zr alloy wire which produces a fine precipitate and raises* J_c *(from Rose, Shepard and Wulff, 1966; courtesy of John Wiley and Sons).*

5.7.5 Oxide superconductors

In 1986 a new class of 'warm' superconductors, based on mixed ceramic oxides, was discovered by J. G. Bednorz and K. A. Müller. These lanthanum–copper oxide superconductors had a T_c around 35 K, well above liquid hydrogen temperature. Since then, three mixed oxide families have been developed with much higher T_c values, all around 100 K. Such materials give rise to optimism for superconductor technology; first, in the use of liquid nitrogen rather than liquid hydrogen and, second, in the prospect of producing a room temperature superconductor.

The first oxide family was developed by mixing and heating the three oxides Y_2O_3, BaO and CuO. This gives rise to the mixed oxide $YBa_2Cu_3O_{7-x}$, sometimes referred to as 1–2–3 compound or YBCO. The structure is shown in Figure 5.29 and is basically made by stacking three perovskite-type unit cells one above the other; the top and bottom cells have barium ions at the center and copper ions at the corners, the middle cell has yttrium at the center. Oxygen ions sit halfway along the cell edges but planes, other than those containing barium, have some missing oxygen ions (i.e. vacancies denoted by x in the oxide formula). This structure therefore has planes of copper and oxygen ions containing vacancies, and copper–oxygen ion chains perpendicular to them. YBCO has a T_c value of about 90 K, which is virtually unchanged when yttrium is replaced by other rare earth elements. The second family of oxides are Bi–Ca–Sr–Cu–O_x materials with the metal ions in the ratio of 2:1:1:1, 2:1:2:2 or 2:2:2:3, respectively. The 2:1:1:1 oxide has only one copper–oxygen layer between the bismuth–oxygen layers, the 2:1:2:2 two and the 2:2:2:3 three, giving rise to an increasing T_c up to about 105 K. The third family is based on Tl–Ca–Ba–Cu–O with a 2:2:2:3 structure having three copper–oxygen layers and a T_c of about 125 K.

While these oxide superconductors have high T_c values and high critical magnetic field (H_c) values, they unfortunately have very low values of J_c, the critical current density. A high J_c is required if they are to be used for powerful superconducting magnets. Electrical applications are therefore unlikely until the J_c value can be raised by several orders of magnitude comparable to those of conventional superconductors, i.e. 10^6 A cm^{-2}. The reason for the low J_c is thought to be largely due to the grain boundaries in polycrystalline materials, together with dislocations, voids and impurity particles. Single crystals show J_c values around 10^5 A cm^{-2} and textured materials, produced by melt growth

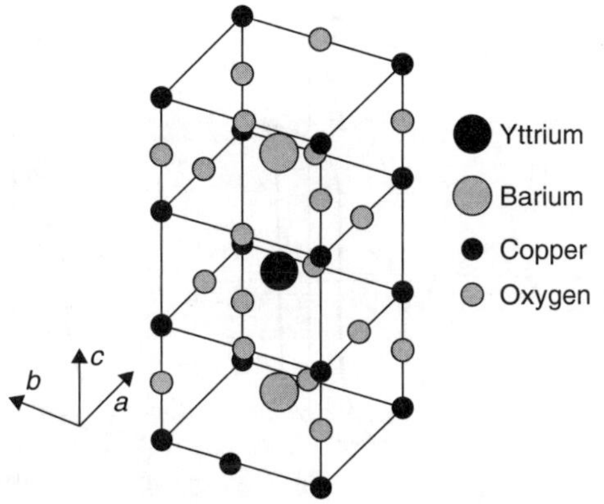

Figure 5.29 *Structure of 1–2–3 compound; the unit cell of the 90 K superconducting perovskite,* $YBa_2Cu_3O_{7-x}$*, where* $x \sim 0$ *(by courtesy of P. J. Hirst, Superconductivity Research Group, University of Birmingham, UK).*

techniques, about 10^4 A cm^{-2}, but both processes have limited commercial application. Electronic applications appear to be more promising, since it is in the area of thin (1 μm) films that high J_c values have been obtained. By careful deposition control, epitaxial and single-crystal films having $J_c \geq 10^6$ A cm^{-2} with low magnetic field dependence have been produced.

5.8 Magnetic properties

5.8.1 Magnetic susceptibility

When a metal is placed in a magnetic field of strength H, the field induced in the metal is given by

$$B = H + 4\pi I, \tag{5.16}$$

where I is the intensity of magnetization. The quantity I is a characteristic property of the metal, and is related to the susceptibility per unit volume of the metal, which is defined as

$$\kappa = I/H. \tag{5.17}$$

The susceptibility is usually measured by a method which depends upon the fact that when a metal specimen is suspended in a non-uniform transverse magnetic field, a force proportional to $\kappa V \cdot H \cdot \mathrm{d}H/\mathrm{d}x$, where V is the volume of the specimen and $\mathrm{d}H/\mathrm{d}x$ is the field gradient measured transversely to the lines of force, is exerted upon it. This force is easily measured by attaching the specimen to a sensitive balance, and one type commonly used is that designed by Sucksmith. In this balance the distortion of a copper–beryllium ring, caused by the force on the specimen, is measured by means of an optical or electromechanical system. Those metals for which κ is negative, such as copper, silver, gold and bismuth, are repelled by the field and are termed diamagnetic materials. Most metals, however, have positive κ values (i.e. they are attracted by the field) and are either paramagnetic (when κ is small) or ferromagnetic (when κ is very large). Only four pure metals – iron, cobalt and nickel from the transition series, and gadolinium from the rare earth series – are ferromagnetic ($\kappa \approx 1000$) at room temperature, but there are several ferromagnetic alloys and some contain no

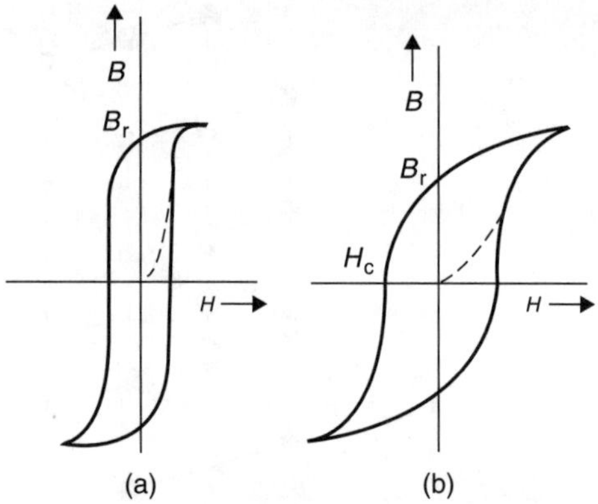

Figure 5.30 *B–H curves for soft (a) and hard (b) magnets.*

metals which are themselves ferromagnetic. The Heusler alloy, which contains manganese, copper and aluminum, is one example; ferromagnetism is due to the presence of one of the transition metals.

The ability of a ferromagnetic metal to concentrate the lines of force of the applied field is of great practical importance, and while all such materials can be both magnetized and demagnetized, the ease with which this can be achieved usually governs their application in the various branches of engineering. Materials may be generally classified either as magnetically soft (temporary magnets) or as magnetically hard (permanent magnets), and the difference between the two types of magnet may be inferred from Figure 5.30. Here, H is the magnetic field necessary to induce a field of strength B inside the material. Upon removal of the field H, a certain residual magnetism B_r, known as the remanence residual, is left in the specimen, and a field H_c, called the coercive force, must be applied in the opposite direction to remove it. A soft magnet is one which is easy both to magnetize and to demagnetize and, as shown in Figure 5.30a, a low value of H is sufficient to induce a large field B in the metal, while only a small field H_c is required to remove it; a hard magnet is a material that is magnetized and demagnetized with difficulty (Figure 5.30b).

5.8.2 Diamagnetism and paramagnetism

Diamagnetism is a universal property of the atom, since it arises from the motion of electrons in their orbits around the nucleus. Electrons moving in this way represent electrical circuits and it follows from Lenz's law that this motion is altered by an applied field in such a manner as to set up a repulsive force. The diamagnetic contribution from the valency electrons is small, but from a closed shell it is proportional to the number of electrons in it and to the square of the radius of the 'orbit'. In many metals this diamagnetic effect is outweighed by a paramagnetic contribution, the origin of which is to be found in the electron spin. Each electron behaves like a small magnet and in a magnetic field can take up one of two orientations, either along the field or in the other opposite direction, depending on the direction of the electron spin. Accordingly, the energy of the electron is either decreased or increased and may be represented conveniently by the band theory. Thus, if we regard the band of energy levels as split into two halves (Figure 5.31a), each half associated with electrons of opposite spin, it follows that, in the presence of the field, some of the electrons will transfer their allegiance from one band to the other until the Fermi energy level is the same in both. It is clear, therefore, that in this state there will be a larger number of electrons which have their energy lowered by the field than have their energy raised. This condition defines paramagnetism, since there will be an excess of unpaired spins which give rise to a resultant magnetic moment.

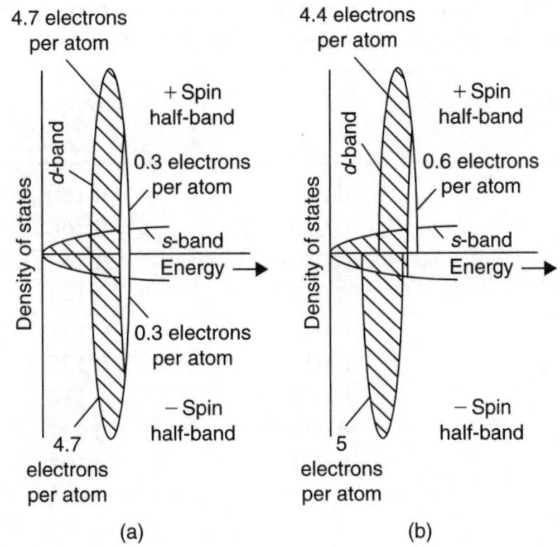

Figure 5.31 *Schematic representation of paramagnetic nickel (a) and ferromagnetic nickel (b) (after Raynor, 1988; by courtesy of Inst. of Materials).*

It is evident that an insulator will not be paramagnetic, since the bands are full and the lowered half-band cannot accommodate those electrons which wish to 'spill over' from the raised half-band. On the other hand, it is not true, as one might expect, that conductors are always paramagnetic. This follows because in some elements the natural diamagnetic contribution outweighs the paramagnetic contribution; in copper, for example, the newly filled *d*-shell gives rise to a larger diamagnetic contribution.

5.8.3 Ferromagnetism

The theory of ferromagnetism is difficult and at present not completely understood. Nevertheless, from the electron theory of metals it is possible to build up a band picture of ferromagnetic materials which goes a long way to explain not only their ferromagnetic properties, but also the associated high resistivity and electronic specific heat of these metals compared to copper. In recent years considerable experimental work has been done on the electronic behavior of the transition elements, and this suggests that the electronic structure of iron is somewhat different to that of cobalt and nickel.

Ferromagnetism, like paramagnetism, has its origin in the electron spin. In ferromagnetic materials, however, permanent magnetism is obtained and this indicates that there is a tendency for electron spins to remain aligned in one direction even when the field has been removed. In terms of the band structure this means that the half-band associated with one spin is automatically lowered when the vacant levels at its top are filled by electrons from the top of the other (Figure 5.31b); the change in potential energy associated with this transfer is known as the exchange energy. Thus, while it is energetically favorable for a condition in which all the spins are in the same direction, an opposing factor is the Pauli Exclusion Principle, because if the spins are aligned in a single direction many of the electrons will have to go into higher quantum states with a resultant increase in kinetic energy. In consequence, the conditions for ferromagnetism are stringent, and only electrons from partially filled *d* or *f* levels can take part. This condition arises because only these levels have (1) vacant levels available for occupation and (2) a high density of states, which is necessary if the increase in kinetic energy accompanying the alignment of spins is to be smaller than the decrease in exchange energy.

Table 5.1 *Radii (nm) of electronic orbits of atoms of transition metals of first long period (after Slater,* Quantum Theory of Matter*).*

Element	*3d*	*4s*	*Atomic radius in metal (nm)*
Sc	0.061	0.180	0.160
Ti	0.055	0.166	0.147
V	0.049	0.152	0.136
Cr	0.045	0.141	0.128
Mn	0.042	0.131	0.128
Fe	0.039	0.122	0.128
Co	0.036	0.114	0.125
Ni	0.034	0.107	0.125
Cu	0.032	0.103	0.128

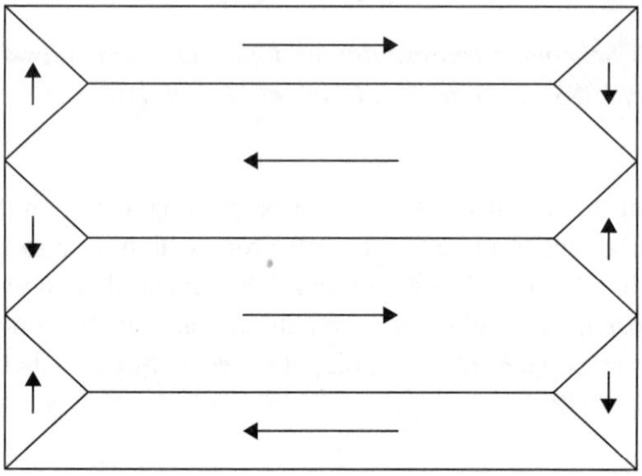

Figure 5.32 *Simple domain structure in a ferromagnetic material. The arrows indicate the direction of magnetization in the domains.*

Both of these conditions are fulfilled in the transition and rare-earth metals, but of all the metals in the long periods only the elements iron, cobalt and nickel are ferromagnetic at room temperature, gadolinium just above RT ($T_c \approx 16°C$) and the majority are, in fact, strongly paramagnetic. This observation has led to the conclusion that the exchange interactions are most favorable, at least for the iron group of metals, when the ratio of the atomic radius to the radius of the unfilled shell, i.e. the d-shell, is somewhat greater than 3 (see Table 5.1). As a result of this condition it is hardly surprising that there are a relatively large number of ferromagnetic alloys and compounds, even though the base elements themselves are not ferromagnetic.

In ferromagnetic metals the strong interaction results in the electron spins being spontaneously aligned, even in the absence of an applied field. However, a specimen of iron can exist in an unmagnetized condition because such an alignment is limited to small regions, or domains, which statistically oppose each other. These domains are distinct from the grains of a polycrystalline metal and in general there are many domains in a single grain, as shown in Figure 5.32. Under the application of a magnetic field the favorably oriented domains grow at the expense of the others by the migration of the domain boundaries until the whole specimen appears fully magnetized. At high field strengths it is also possible for unfavorably oriented domains to 'snap-over' into more favorable orientations quite suddenly, and this process, which can often be heard using sensitive equipment, is known as the Barkhausen effect.

The state in which all the electron spins are in complete alignment is possible only at low temperatures. As the temperature is raised the saturation magnetization is reduced, falling slowly at first and then increasingly rapidly, until a critical temperature, known as the Curie temperature, is reached. Above this temperature, T_c, the specimen is no longer ferromagnetic, but becomes paramagnetic, and for the metals iron, cobalt and nickel this transition occurs at 780°C, 1075°C and 365°C respectively. Such a cooperative process may be readily understood from thermodynamic reasoning, since the additional entropy associated with the disorder of the electron spins makes the disordered (paramagnetic) state thermodynamically more stable at high temperatures. This behavior is similar to that shown by materials which undergo the order–disorder transformation and, as a consequence, ferromagnetic metals exhibit a specific heat peak of the form previously shown (see Figure 5.3b).

A ferromagnetic crystal in its natural state has a domain structure. From Figure 5.32 it is clear that by dividing itself into domains the crystal is able to eliminate those magnetic poles which would otherwise occur at the surface. The width of the domain boundary or Bloch wall is not necessarily small, however, and in most materials is of the order of 100 atoms in thickness. By having a wide boundary the electron spins in neighboring atoms are more nearly parallel, which is a condition required to minimize the exchange energy. On the other hand, within any one domain the direction of magnetization is parallel to a direction of easy magnetization (i.e. $\langle 1\,0\,0\rangle$ in iron, $\langle 1\,1\,1\rangle$ in nickel and $\langle 0\,0\,1\rangle$ in cobalt) and as one passes across a boundary the direction of magnetization rotates away from one direction of easy magnetization to another. To minimize this magnetically disturbed region, the crystal will try to adopt a boundary which is as thin as possible. Consequently, the boundary width adopted is one of compromise between the two opposing effects, and the material may be considered to possess a magnetic interfacial or surface energy.

5.8.4 Magnetic alloys

The work done in moving a domain boundary depends on the energy of the boundary, which in turn depends on the magnetic anisotropy. The ease of magnetization also depends on the state of internal strain in the material and the presence of impurities. Both these latter factors affect the magnetic 'hardness' through the phenomenon of magnetostriction, i.e. the lattice constants are slightly altered by the magnetization so that a directive influence is put upon the orientation of magnetization of the domains. Materials with internal stresses are hard to magnetize or demagnetize, while materials free from stresses are magnetically soft. Hence, since internal stresses are also responsible for mechanical hardness, the principle which governs the design of magnetic alloys is to make permanent magnetic materials as mechanically hard and soft magnets as mechanically soft as possible.

Magnetically soft materials are used for transformer laminations and armature stampings, where a high permeability and a low hysteresis are desirable: iron–silicon or iron–nickel alloys are commonly used for this purpose. In the development of magnetically soft materials it is found that those elements which form interstitial solid solutions with iron are those which broaden the hysteresis loop most markedly. For this reason, it is common to remove such impurities from transformer iron by vacuum melting or hydrogen annealing. However, such processes are expensive and, consequently, alloys are frequently used as 'soft' magnets, particularly iron–silicon and iron–nickel alloys (because silicon and nickel both reduce the amount of carbon in solution). The role of Si is to form a γ-loop and hence remove transformation strains and also improve orientation control. In the production of iron–silicon alloys the factors which are controlled include the grain size, the orientation difference from one grain to the next, and the presence of non-magnetic inclusions, since all are major sources of coercive force. The coercive force increases with decreasing grain size because the domain pattern in the neighborhood of a grain boundary is complicated owing to the orientation difference between two adjacent grains. Complex domain patterns can also arise at the free surface of the metal unless these are parallel to a direction of easy magnetization. Accordingly, to minimize the coercive force, rolling and annealing schedules are adopted to produce a preferred oriented material with a strong 'cube texture', i.e. one with two ⟨1 0 0⟩ directions in the plane of the sheet (see Chapter 6). This procedure is extremely important, since transformer material is used in the form of thin sheets to minimize eddy-current losses. Fe–Si–B in the amorphous state is finding increasing application in transformers.

The iron–nickel series, *Permalloys*, present many interesting alloys and are used chiefly in communication engineering, where a high permeability is a necessary condition. The alloys in the range 40–55% nickel are characterized by a high permeability and at low field strengths this may be as high as 15 000 compared with 500 for annealed iron. The 50% alloy, *Hypernik*, may have a permeability which reaches a value of 70 000, but the highest initial and maximum permeability occurs in the composition range of the $FeNi_3$ superlattice, provided the ordering phenomenon is suppressed. An interesting development in this field is in the heat treatment of the alloys while in a strong magnetic field. By such a treatment, the permeability of *Permalloy* 65 has been increased to about 260 000. This effect is thought to be due to the fact that, during alignment of the domains, plastic deformation is possible and magnetostrictive strains may be relieved.

Magnetically hard materials are used for applications where a 'permanent' magnetic field is required, but where electromagnets cannot be used, such as in electric clocks, meters, etc. Materials commonly used for this purpose include *Alnico* (Al–Ni–Co) alloys, *Cunico* (Cu–Ni–Co) alloys, ferrites (barium and strontium), samarium–cobalt alloys ($SmCo_5$ and $Sm_2(Co, Fe, Cu, Zr)_{17}$) and *Neomax* ($Nd_2Fe_{14}B$). The *Alnico* alloys have high remanence but poor coercivities; the ferrites have rather low remanence but good coercivities, together with very cheap raw material costs. The rare-earth magnets have a high performance but are rather costly, although the Nd-based alloys are cheaper than the Sm-based ones.

In the development of magnetically hard materials, the principle is to obtain, by alloying and heat treatment, a matrix containing finely divided particles of a second phase. These fine precipitates, usually differing in lattice parameter from the matrix, set up coherency strains in the lattice which affect the domain boundary movement. Alloys of copper–nickel–iron, copper–nickel–cobalt and aluminum–nickel–cobalt are of this type. An important advance in this field is to make the particle size of the alloy so small, i.e. less than 100 nm diameter, that each grain contains only a single domain. Then magnetization can occur only by the rotation of the direction of magnetization *en bloc*. *Alnico* alloys containing 6–12% Al, 14–25% Ni, 0–35% Co, 0–8% Ti, 0–6% Cu in 40–70% Fe depend on this feature and are the most commercially important permanent magnet materials. They are precipitation-hardened alloys and are heat treated to produce rod-like precipitates (30 nm × 100 nm) lying along ⟨1 0 0⟩ in the bcc matrix. During magnetic annealing the rods form along the ⟨1 0 0⟩ axis nearest to the direction of the field, when the remanence and coercivity are markedly increased; $Sm_2(Co, Fe, Cu, Zr)_{17}$ alloys also rely on the pinning of magnetic domains by fine precipitates. Clear correlation exists between mechanical hardness and intrinsic coercivity. $SmCo_5$ magnets depend on the very high magnetocrystalline anisotropy of this compound and the individual grains are single-domain particles. The big advantage of these magnets over the *Alnico* alloys is their much higher coercivities.

The Heusler alloys, copper–manganese–aluminum, are of particular interest because they are made up from non-ferromagnetic metals and yet exhibit ferromagnetic properties. The magnetism in this group of alloys is associated with the compound Cu_2MnAl, evidently because of the presence of manganese atoms. The compound has the Fe_3Al-type superlattice when quenched from 800°C, and in this state is ferromagnetic, but when the alloy is slowly cooled it has a γ-brass structure and is non-magnetic, presumably because the correct exchange forces arise from the lattice rearrangement on ordering. A similar behavior is found in both the copper–manganese–gallium and the copper–manganese–indium systems.

The order–disorder phenomenon is also of magnetic importance in many other systems. As discussed previously, when ordering is accompanied by a structural change, i.e. cubic to tetragonal,

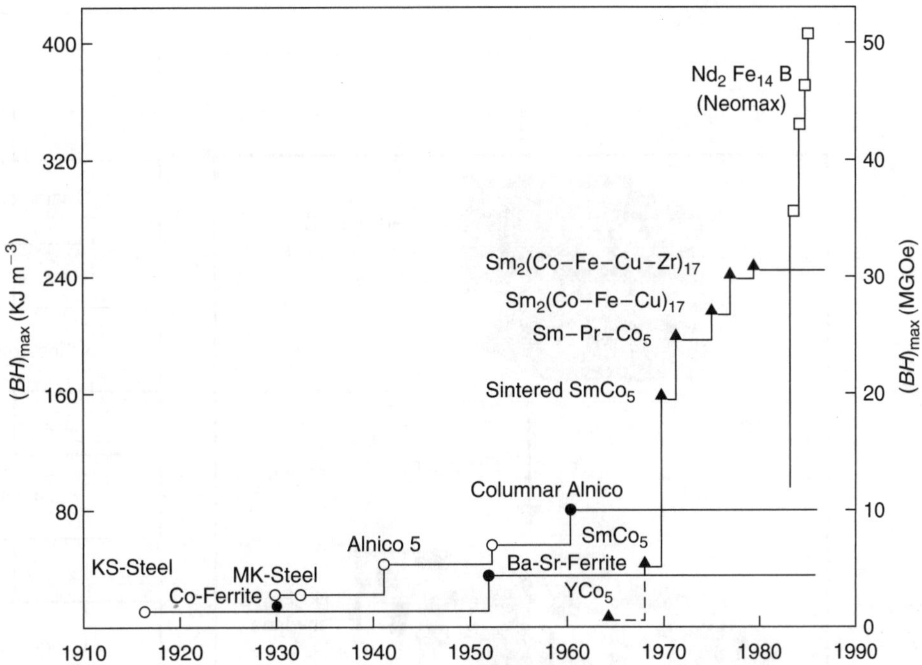

Figure 5.33 *The variation of $(BH)_{max}$ with time over this century (courtesy of I. R. Harris).*

coherency strains are set up which often lead to magnetic hardness. In FePt, for example, extremely high coercive forces are produced by rapid cooling. However, because the change in mechanical properties accompanying the transformation is found to be small, it has been suggested that the hard magnetic properties in this alloy are due to the small particle-size effect, which arises from the finely laminated state of the structure.

While the much cheaper but lower performance magnets such as ferrites have a significant market share of applications, the rare-earth (RE) magnets have revolutionized the properties and applications of permanent magnets. A parameter which illustrates the potential of these materials is the maximum energy product $(BH)_{max}$. The larger the value of $(BH)_{max}$ the smaller the volume of magnet required to produce a given magnetic flux. This is illustrated in Figure 5.33, where the neodymium–iron–boron materials have $(BH)_{max}$ in excess of 400 kJ m^{-3}, an order of magnitude stronger than the ferrites. In the drive for miniaturization the $Nd_2Fe_{14}B$ materials are unrivaled. They are also finding applications where a very strong permanent field is required, such as MRI scanners. The main process route for these magnets, shown in Figure 5.34, consists of powdering the coarse-grained cast ingot, aligning the fine powder along the easy magnetization axis, compacting and then sintering to produce a fully dense magnet. The alloy is very active with respect to hydrogen gas so that, on exposure at room temperature and around 1 bar pressure, the bulk alloy absorbs the hydrogen, particularly in the grain boundary region. The differential and overall volume expansion results in the bulk alloy ‘decrepitating’ (separating into parts with a crying sound) into very friable particulate matter which consists of fine, grain boundary debris and grains of $Nd_2Fe_{14}B$ which are of the order of $\sim$100 μm in size. The phenomenon of ‘hydrogen decrepitation’ (HD) has been incorporated into the process route. Apart from economically producing powder, the HD powder is extremely friable, which substantially aids the subsequent jet-milling process. With this modified processing route substantial savings of between 15% and 25% can be achieved in the cost of magnet production. The majority of NdFeB magnets are now made by this process.

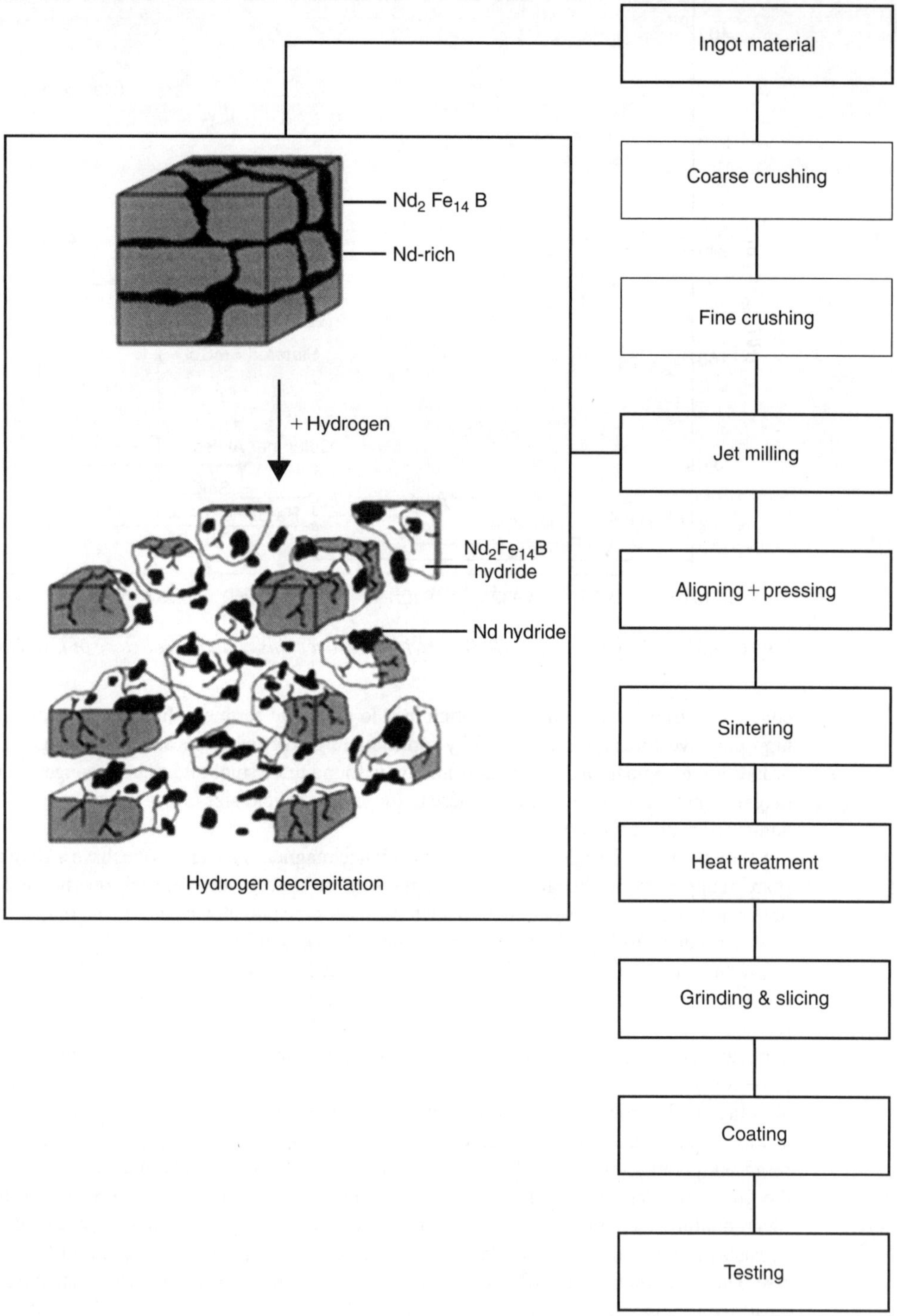

Figure 5.34 *A schematic representation of the HD process (courtesy of I. R. Harris).*

5.8.5 Anti-ferromagnetism and ferrimagnetism

Apart from the more usual dia-, para- and ferromagnetic materials, there are certain substances which are termed anti-ferromagnetic; in these, the net moments of neighboring atoms are aligned in opposite directions, i.e. anti-parallel. Many oxides and chlorides of the transition metals are examples, including both chromium and α-manganese, and also manganese–copper alloys. Some of the relevant features of anti-ferromagnetism are similar in many respects to ferromagnetism, and are summarized as follows:

1. In general, the magnetization directions are aligned parallel or anti-parallel to crystallographic axes, e.g. in MnI and CoO the moment of the Mn^{2+} and Co^{2+} ions are aligned along a cube edge of the unit cell. The common directions are termed directions of anti-ferromagnetism.
2. The degree of long-range anti-ferromagnetic ordering progressively decreases with increasing temperature and becomes zero at a critical temperature, T_n, known as the Néel temperature; this is the anti-ferromagnetic equivalent of the Curie temperature.
3. An anti-ferromagnetic domain is a region in which there is only one common direction of anti-ferromagnetism; this is probably affected by lattice defects and strain.

The most characteristic property of an anti-ferromagnetic material is that its susceptibility χ shows a maximum as a function of temperature, as shown in Figure 5.35a. As the temperature is raised from 0 K the interaction which leads to anti-parallel spin alignment becomes less effective until at T_n the spins are free.

Similar characteristic features are shown in the resistivity curves due to scattering as a result of spin disorder. However, the application of neutron diffraction techniques provides a more direct method of studying anti-ferromagnetic structures, as well as giving the magnetic moments associated with the ions of the metal. There is a magnetic scattering of neutrons in the case of certain magnetic atoms, and owing to the different scattering amplitudes of the parallel and anti-parallel atoms, the

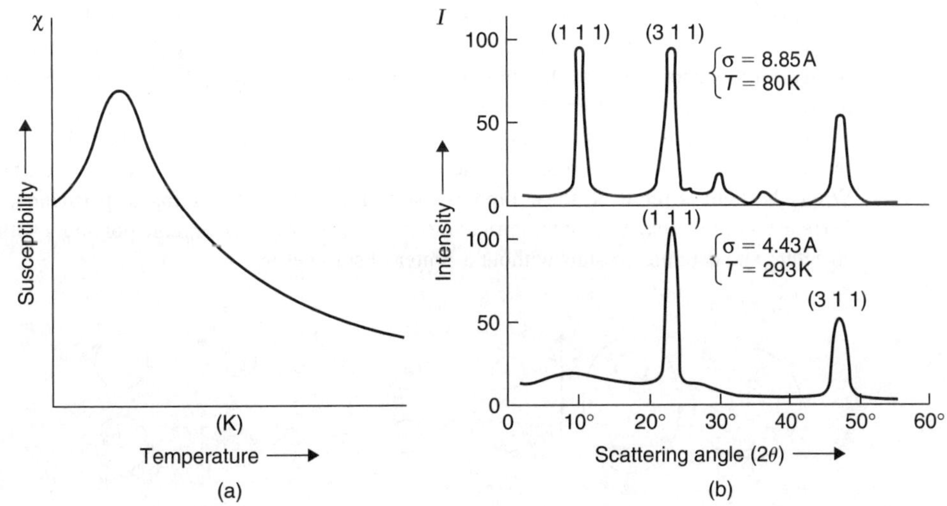

Figure 5.35 *(a) Variation of magnetic susceptibility with temperature for an anti-ferromagnetic material. (b) Neutron diffraction pattern from the anti-ferromagnetic powder MnO above and below the critical temperature for ordering (after Shull and Smart, 1949).*

possibility arises of the existence of superlattice lines in the anti-ferromagnetic state. In manganese oxide (MnO), for example, the parameter of the magnetic unit cell is 0.885 nm, whereas the chemical unit cell (NaCl structure) is half this value, 0.443 nm. This atomic arrangement is analogous to the structure of an ordered alloy and the existence of magnetic superlattice lines below the Néel point (122 K) has been observed, as shown in Figure 5.35b.

Some magnetic materials have properties which are intermediate between those of anti-ferromagnetic and ferromagnetic. This arises if the moments in one direction are unequal in magnitude to those in the other, as, for example, in magnetite (Fe_3O_4), where the ferrous and ferric ions of the $FeO \cdot Fe_2O_3$ compound occupy their own particular sites. Néel has called this state *ferrimagnetism* and the corresponding materials are termed ferrites. Such materials are of importance in the field of electrical engineering because they are ferromagnetic without being appreciably conducting; eddy-current troubles in transformers are therefore not so great. Strontium ferrite is extensively used in applications such as electric motors, because of these properties and low material costs.

5.9 Dielectric materials

5.9.1 Polarization

Dielectric materials, usually those which are covalent or ionic, possess a large energy gap between the valence band and the conduction band. These materials exhibit high electrical resistivity and have important applications as insulators, which prevent the transfer of electrical charge, and capacitors, which store electrical charge. Dielectric materials also exhibit piezoelectric and ferroelectric properties.

Application of an electric field to a material induces the formation of dipoles (i.e. atoms or groups of atoms with an unbalanced charge or moment), which become aligned with the direction of the applied field. The material is then polarized. This state can arise from several possible mechanisms – electronic, ionic or molecular, as shown in Figure 5.36a–c. With electronic polarization, the electron clouds of an atom are displaced with respect to the positively charged ion core setting up an electric dipole with moment μ_e. For ionic solids in an electric field, the bonds between the ions are elastically deformed and the anion–cation distances become closer or further apart, depending on the direction of the field. These induced dipoles produce polarization and may lead to dimensional changes. Molecular polarization occurs in molecular materials, some of which contain natural dipoles. Such materials are described as polar and for these the influence of an applied field will change the polarization by displacing the atoms and thus changing the dipole moment (i.e. atomic polarizability) or by causing the molecule as a whole to rotate to line up with the imposed field (i.e. orientation polarizability). When the field is removed these dipoles may remain aligned, leading to permanent polarization. Permanent dipoles exist in asymmetrical molecules such as H_2O, organic polymers with asymmetric structure and ceramic crystals without a center of symmetry.

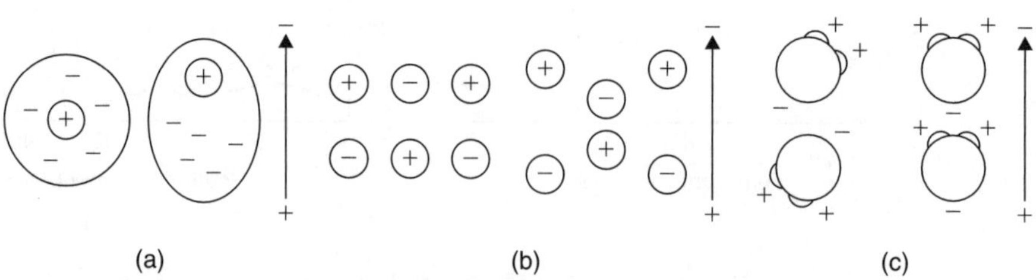

Figure 5.36 *Application of field to produce polarization by electronic (a), ionic (b) and molecular (c) mechanisms.*

5.9.2 Capacitors and insulators

In a capacitor the charge is stored in a dielectric material which is easily polarized and has a high electrical resistivity $\sim 10^{11}\,\mathrm{V\,A^{-1}\,m}$ to prevent the charge flowing between conductor plates. The ability of the material to polarize is expressed by the permittivity ε, and the relative permittivity or dielectric constant κ is the ratio of the permittivity of the material and the permittivity of a vacuum ε_0. While a high κ is important for a capacitor, a high dielectric strength or breakdown voltage is also essential. The dielectric constant κ values for vacuum, water, polyethylene, *Pyrex* glass, alumina and barium titanate are 1, 78.3, 2.3, 4, 6.5 and 3000 respectively.

Structure is an important feature of dielectric behavior. Glassy polymers and crystalline materials have a lower κ than their amorphous counterparts. Polymers with asymmetric chains have a high κ because of the strength of the associated molecular dipole; thus, polyvinyl chloride (PVC) and polystyrene (PS) have larger κ values than polyethylene (PE). $BaTiO_3$ has an extremely high κ value because of its asymmetrical structure. Frequency response is also important in dielectric applications and depends on the mechanism of polarization. Materials which rely on electronic and ionic dipoles respond rapidly to frequencies of 10^{13}–10^{16} Hz, but molecular polarization solids, which require groups of atoms to rearrange, respond less rapidly. Frequency is also important in governing dielectric loss due to heat and usually increases when one of the contributions to polarization is prevented. This behavior is common in microwave heating of polymer adhesives; preferential heating in the adhesive due to dielectric losses starts the thermosetting reaction. For moderate increases, raising the voltage and temperature increases the polarizability and leads to a higher dielectric constant. Nowadays, capacitor dielectrics combine materials with different temperature dependence to yield a final product with a small linear temperature variation. These materials are usually titanates of Ba, Ca, Mg, Sr and rare-earth metals.

For an insulator, the material must possess a high electrical resistivity, a high dielectric strength to prevent breakdown of the insulator at high voltages, a low dielectric loss to prevent heating, and small dielectric constant to hinder polarization and hence charge storage. Materials increasingly used are alumina, aluminum nitride, glass-ceramics, steatite porcelain and glasses.

5.9.3 Piezoelectric materials

When stress is applied to certain materials an electric polarization is produced proportional to the stress applied. This is the well-known piezoelectric effect. Conversely, dilatation occurs on application of an electric field. Common materials displaying this property are quartz, $BaTiO_3$, $Pb(Ti, Zr)O_3$ or PZT and Na or $LiNbO_3$. For quartz, the piezoelectric constant d relating strain ε to field strength $F(\varepsilon = \mathrm{d} \times F)$ is $2.3 \times 10^{-12}\,\mathrm{m\,V^{-1}}$, whereas for PZT it is $250 \times 10^{-12}\,\mathrm{m\,V^{-1}}$. The piezoelectric effect is used in transducers, which convert sound waves to electric fields or vice versa. Applications range from microphones, where a few millivolts are generated, to military devices creating several kilovolts and from small sub-nanometer displacements in piezoelectrically deformed mirrors to large deformations in power transducers.

5.9.4 Pyroelectric and ferroelectric materials

Some materials, associated with low crystal symmetry, are observed to acquire an electric charge when heated; this is known as pyroelectricity. Because of the low symmetry, the centers of gravity of the positive and negative charges in the unit cell are separated, producing a permanent dipole moment. Moreover, alignment of individual dipoles leads to an overall dipole moment which is non-zero for the crystal. Pyroelectric materials are used as detectors of electromagnetic radiation in a wide band from ultraviolet to microwave, in radiometers and in thermometers sensitive to changes of temperature as

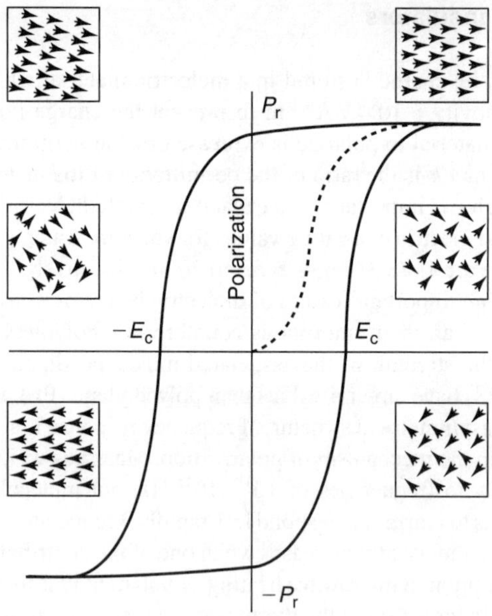

Figure 5.37 *Hysteresis loop for ferroelectric materials, showing the influence of electric field* E *on polarization* P.

small as 6×10^{-6}°C. Pyroelectric TV camera tubes have also been developed for long-wavelength infrared imaging and are useful in providing visibility through smoke. Typical materials are strontium barium niobate and PZT with Pb_2FeNbO_6 additions to broaden the temperature range of operation.

Ferroelectric materials are those which retain a net polarization when the field is removed and is explained in terms of the residual alignment of permanent dipoles. Not all materials that have permanent dipoles exhibit ferroelectric behavior because these dipoles become randomly arranged as the field is removed so that no net polarization remains. Ferroelectrics are related to the pyroelectrics; for the former materials the direction of spontaneous polarization can be reversed by an electric field (Figure 5.37), whereas for the latter this is not possible. This effect can be demonstrated by a polarization versus field hysteresis loop similar in form and explanation to the *B–H* magnetic hysteresis loop (see Figure 5.30). With increasing positive field all the dipoles align to produce a saturation polarization. As the field is removed a remanent polarization P_r remains due to a coupled interaction between dipoles. The material is permanently polarized, and a coercive field E_c has to be applied to randomize the dipoles and remove the polarization.

Like ferromagnetism, ferroelectricity depends on temperature and disappears above an equivalent Curie temperature. For $BaTiO_3$, ferroelectricity is lost at 120°C when the material changes crystal structure. By analogy with magnetism there is also a ferroelectric analog of anti-ferromagnetism and ferrimagnetism. $NaNbO_3$, for example, has a T_c, of 640°C and anti-parallel electric dipoles of unequal moments characteristic of a ferrielectric material.

5.10 Optical properties

5.10.1 Reflection, absorption and transmission effects

The optical properties of a material are related to the interaction of the material with electromagnetic radiation, particularly visible light. The electromagnetic spectrum is shown in Figure 5.1, from which it can be seen that the wavelength λ varies from 10^4 m for radio waves down to 10^{-14} m for γ-rays and the corresponding photon energies vary from 10^{-10} to 10^8 eV.

Photons incident on a material may be reflected, absorbed or transmitted. Whether a photon is absorbed or transmitted by a material depends on the energy gap between the valency and conduction bands and the energy of the photon. The band structure for metals has no gap and so photons of almost any energy are absorbed by exciting electrons from the valency band into a higher energy level in the conduction band. Metals are thus opaque to all electromagnetic radiation from radio waves, through the infrared, the visible to the ultraviolet, but are transparent to high-energy X-rays and γ-rays. Much of the absorbed radiation is re-emitted as radiation of the same wavelength (i.e. reflected). Metals are both opaque and reflective, and it is the wavelength distribution of the reflected light, which we see, that determines the color of the metal. Thus, copper and gold reflect only a certain range of wavelengths and absorb the remaining photons, i.e. copper reflects the longer-wavelength red light and absorbs the shorter-wavelength blue. Aluminum and silver are highly reflective over the complete range of the visible spectrum and appear silvery.

Because of the gaps in their band structure, non-metals may be transparent. The wavelength for visible light varies from about 0.4 to 0.7 μm so that the maximum band-gap energy for which absorption of visible light is possible is given by $E = hc/\lambda = (4.13 \times 10^{-15}) \times (3 \times 10^8)/(4 \times 10^{-7}) = 3.1$ eV. The minimum band-gap energy is 1.8 eV. Thus, if the photons have insufficient energy to excite electrons in the material to a higher energy level, they may be transmitted rather than absorbed and the material is transparent. In high-purity ceramics and polymers, the energy gap is large and these materials are transparent to visible light. In semiconductors, electrons can be excited into acceptor levels or out of donor levels and phonons having sufficient energy to produce these transitions will be absorbed. Semiconductors are therefore opaque to short wavelengths and transparent to long.[3] The band structure is influenced by crystallinity and hence materials such as glasses and polymers may be transparent in the amorphous state but opaque when crystalline.

High-purity non-metallics such as glasses, diamond or sapphire (Al_2O_3) are colorless but are changed by impurities. For example, small additions of Cr^{3+} ions (Cr_2O_3) to Al_2O_3 produce a ruby color by introducing impurity levels within the band gap of sapphire which give rise to absorption of specific wavelengths in the visible spectrum. Coloring of glasses and ceramics is produced by addition of transition metal impurities which have unfilled d-shells. The photons easily interact with these ions and are absorbed; Cr^{3+} gives green, Mn^{2+} yellow and Co^{2+} blue–violet coloring.

In photochromic sunglasses the energy of light quanta is used to produce changes in the ionic structure of the glass. The glass contains silver (Ag^+) ions as a dopant which are trapped in the disordered glass network of silicon and oxygen ions: these are excited by high-energy quanta (photons) and change to metallic silver, causing the glass to darken (i.e. light energy is absorbed). With a reduction in light intensity, the silver atoms re-ionize. These processes take a small period of time relying on absorption and non-absorption of light.

5.10.2 Optical fibers

Modern communication systems make use of the ability of optical fibers to transmit light signals over large distances. Optical guidance by a fiber is produced (see Figure 5.38) if a core fiber of refractive index n_1 is surrounded by a cladding of slightly lower index n_2 such that total internal reflection occurs confining the rays to the core: typically the core is about 100 μm and $n_1 - n_2 \approx 10^{-2}$. With such a simple optical fiber, interference occurs between different modes, leading to a smearing of the signals. Later designs use a core in which the refractive index is graded, parabolically, between

[3] Figure 4.37b shows a dislocation source in the interior of a silicon crystal observed using infrared light.

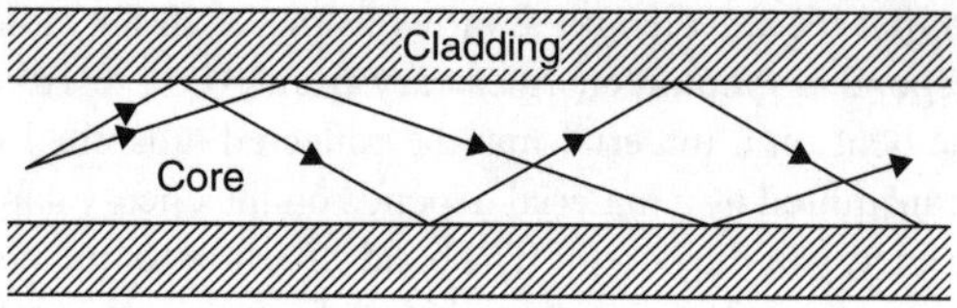

Figure 5.38 *Optical guidance in a multimode fiber.*

the core axis and the interface with the cladding. This design enables modulated signals to maintain their coherency. In vitreous silica, the refractive index can be modified by additions of dopants such as P_2O_5 and GeO_2 which raise it and B_2O_5 and F which lower it. Cables are sheathed to give strength and environmental protection; PE and PVC are commonly used for limited fire-hazard conditions.

5.10.3 Lasers

A laser (Light Amplification by Stimulated Emission of Radiation) is a powerful source of coherent light (i.e. monochromatic and all in phase). The original laser material, still used, is a single crystal rod of ruby, i.e. Al_2O_3 containing dopant Cr^{3+} ions in solid solution. Nowadays, lasers can be solid, liquid or gaseous materials and ceramics, glasses and semiconductors. In all cases, electrons of the laser material are excited into a higher energy state by some suitable stimulus (see Figure 5.39). In a device this is produced by the photons from a flash tube, to give an intense source of light surrounding the rod to concentrate the energy into the laser rod. Alternatively, an electrical discharge in a gas is used. The ends of the laser rod are polished flat and parallel, and then silvered such that one end is totally reflecting and the other partially transmitting.

In the ruby laser, a xenon flash lamp excites the electrons of the Cr^{3+} ions into higher energy states. Some of these excited electrons decay back to their ground state directly and are not involved in the laser process. Other electrons decay into a metastable intermediate state before further emission returns them to the ground state. Some of the electrons in the metastable state will be spontaneously emitted after a short (~ms) rest period. Some of the resultant photons are retained in the rod because of the reflectivity of the silvered ends and stimulate the release of other electrons from the metastable state. Thus, one photon releases another such that an avalanche of emissions is triggered, all in phase with the triggering photon. The intensity of the light increases as more emissions are stimulated until a very high intensity, coherent, collimated 'burst' of light is transmitted through the partially silvered end, lasting a few nanoseconds and with considerable intensity.

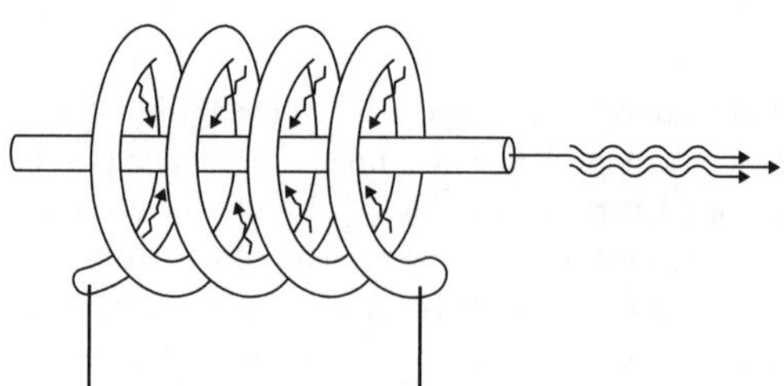

Figure 5.39 *Schematic diagram of a laser.*

5.10.4 Ceramic 'windows'

Many ceramics, usually oxides, have been prepared in optically transparent or translucent forms (translucent means that incident light is partly reflected and partly transmitted). Examples include aluminum oxide, magnesium oxide, their double oxide or spinel, and chalcogenides (zinc sulfide, zinc selenide). Very pure raw materials of fine particle size are carefully processed to eliminate voids and control grain sizes. Thus, translucent alumina is used for the arc tube of high-pressure sodium lamps; a grain size of 25 μm gives the best balance of translucency and mechanical strength.

Ceramics are also available to transmit electromagnetic radiation with wavelengths which lie below or above the visible range of 400–700 nm (e.g. infrared, microwave, radar, etc.). Typical candidate materials for development include vitreous silica, cordierite glass-ceramics and alumina.

5.10.5 Electro-optic ceramics

Certain special ceramics combine electrical and optical properties in a unique manner. Lead lanthanum zirconium titanate, known as PLZT, is a highly transparent ceramic which becomes optically birefringent when electrically charged. This phenomenon is utilized as a switching mechanism in arc-welding goggles, giving protection against flash blindness. The PLZT plate is located between two 'crossed' sheets of polarizing material. A small impressed d.c. voltage on the PLZT plate causes it to split the incident light into two rays vibrating in different planes. One of these rays can pass through the inner polar sheet and enter the eye. A sudden flash of light will activate photodiodes in the goggles, reduce the impressed voltage and cause rapid darkening of the goggles.

Problems

5.1 Assuming that the vacancy concentration of a close packed metal is 10^{-4} at its melting point and that $D_0 = 10^{-4}\ \text{m}^2\ \text{s}^{-1}$, where $D = D_0 \exp(-E_D/\mathbf{k}T)$ and D is the self-diffusion coefficient, answer questions (i)–(vi), which relate to diffusion by a vacancy mechanism in a close-packed metal:

(i) What are the vacancy concentrations at $\frac{1}{4}$, $\frac{1}{2}$ and $\frac{3}{4}$ T_m (in K)?
(ii) Estimate the diffusion coefficient of the vacancies at $\frac{1}{4}$, $\frac{1}{2}$ and $\frac{3}{4}$ T_m.
(iii) Estimate the self-diffusion coefficient for the metal at $\frac{1}{4}$, $\frac{1}{2}$ and $\frac{3}{4}$ T_m.
(iv) How far does a vacancy diffuse at $T_m/2$ in 1 hour?
(v) How far does an atom diffuse at $T_m/2$ in 1 hour?
(vi) If copper melts at 1065°C, estimate E_f. (Boltzmann's constant $\mathbf{k} = 8.6 \times 10^{-5}\ \text{eV}\ \text{K}^{-1}$.)

5.2 The diffusivity of lithium in silicon is $10^{-9}\ \text{m}^2\ \text{s}^{-1}$ at 1376 K and $10^{-10}\ \text{m}^2\ \text{s}^{-1}$ at 968 K. What are the values of E_D and D_0 for diffusion of lithium in silicon? (E_D is the activation energy for diffusion in $\text{J}\ \text{mol}^{-1}$ and $\mathbf{R} = 8.314\ \text{J}\ \text{mol}^{-1}\ \text{K}^{-1}$.)

5.3 The melting endotherm of a sample of an impure material has been analyzed to determine the fraction, f, of sample melted at each temperature T_s:

f	0.099	0.122	0.164	0.244	0.435
T_s (K)	426.0	426.5	427.0	427.5	428.0

The fraction, f, of the sample melted at temperature T_s is given by

$$T_s = T_0 - \Delta T/f,$$

where $\Delta T = (T_0 - T_m)$, T_0 is the melting point of the pure and T_m the impure sample.
The van't Hoff equation,

$$\Delta T = \left(\frac{\mathbf{R}T_0^2}{\Delta H_f}\right) x_2,$$

relates ΔT to mole fraction of impurity present, where **R** is the gas constant (8.31 J K^{-1} mol^{-1}), ΔH_f is the enthalpy of fusion and x_2 the mole fraction of impurity.

If the enthalpy of fusion of the pure material is 25.5 kJ mol^{-1}, use the above data to determine graphically the lowering of the melting point and hence determine the mole fraction of impurity present in the sample.

5.4 The resistivity of intrinsic germanium is 0.028 Ωm at 385 K and 2.74×10^{-4} Ωm at 714 K. Assume that the hole and electron mobilities both vary as $T^{-3/2}$.
(a) Determine the band-gap energy E_g.
(b) At what wavelength would you expect the onset of optical absorption?

5.5 The magnetic susceptibility (χ) of iron is temperature dependent according to $\chi \propto 1/(T - T_c)$, where T_c is the Curie temperature. At 900°C, χ has a value of 2.5×10^{-4}. T_c for iron is 770°C. Determine the susceptibility at 800°C.

5.6 The current flowing around a superconducting loop of wire decays according to

$$i(t) = i(0)e^{-\frac{R}{L}t},$$

where R = resistance and L = self-inductance. What is the largest resistance a 1 m diameter loop of superconducting wire, 1 mm^2 cross-sectional area, can sustain if it is to maintain a current flow of 1 A for one year without appreciable loss (<1%)? (Given: a loop with diameter $2a$ and wire thickness $2r$ has a self-inductance $L = \mu_0 a\left[\ln\left(\frac{8a}{r}\right) - \frac{7}{4}\right]$, where $\mu_0 = 4\pi \times 10^{-7}$.)

Further reading

Anderson, J. C., Leaver, K. D., Rawlins, R. D. and Alexander, J. M. (1990). *Materials Science*. Chapman & Hall, London.

Braithwaite, N. and Weaver, G. (eds) (1990). *Open University Materials in Action Series*. Butterworths, London.

Cullity, B. D. (1972). *Introduction to Magnetic Materials*. Addison-Wesley, Wokingham.

Hume-Rothery, W. and Coles, B. R. (1946, 1969). *Atomic Theory for Students of Metallurgy*. Institute of Metals, London.

Porter, D. A. and Easterling, K. E. (1992). *Phase Transformations in Metals and Alloys*, 2nd edn. Van Nostrand Reinhold, Wokingham.

Raynor, G. V. (1947, 1988). *Introduction to Electron Theory of Metals*. Institute of Metals, London.

Shewmon, P. G. (1989). *Diffusion in Solids*. Minerals, Metals and Materials Soc., Warrendale, USA.

Swalin, R. A. (1972). *Thermodynamics of Solids*. Wiley, Chichester.

Warn, J. R. W. (1969). *Concise Chemical Thermodynamics*. Van Nostrand, New York.

Chapter 6
Mechanical properties I

6.1 Mechanical testing procedures

6.1.1 Introduction

Real crystals, however carefully prepared, contain lattice imperfections which profoundly affect those properties sensitive to structure. Careful examination of the mechanical behavior of materials can give information on the nature of these atomic defects. In some branches of industry the common mechanical tests, such as tensile, hardness, impact, creep and fatigue tests, may be used not to study the 'defect state', but to check the quality of the product produced against a standard specification. Whatever its purpose, the mechanical test is of importance in the development of both materials science and engineering properties. It is inevitable that a large number of different machines for performing the tests are in general use. This is because it is often necessary to know the effect of temperature and strain rate at different levels of stress depending on the material being tested. Consequently, no attempt is made here to describe the details of the various testing machines. The elements of the various tests are outlined below.

6.1.2 The tensile test

In a tensile test the ends of a test piece are fixed into grips, one of which is attached to the load-measuring device on the tensile machine and the other to the straining device. The strain is usually applied by means of a motor-driven crosshead and the elongation of the specimen is indicated by its relative movement. The load necessary to cause this elongation may be obtained from the elastic deflection of either a beam or proving ring, which may be measured by using hydraulic, optical or electromechanical methods. The last method (where there is a change in the resistance of strain gauges attached to the beam) is, of course, easily adapted into a system for autographically recording the load–elongation curve.

The load–elongation curves for both polycrystalline mild steel and copper are shown in Figure 6.1a and b. The corresponding stress (load per unit area, P/A) versus strain (change in length per unit length, dl/l) curves may be obtained knowing the dimensions of the test piece. At low stresses the deformation is elastic, reversible and obeys Hooke's law with stress linearly proportional to strain. The proportionality constant connecting stress and strain is known as the elastic modulus and may be either (a) the elastic or Young's modulus, E, (b) the rigidity or shear modulus μ, or (c) the bulk modulus K, depending on whether the strain is tensile, shear or hydrostatic compressive, respectively. Young's modulus, bulk modulus, shear modulus and Poisson's ratio ν, the ratio of lateral contractions to longitudinal extension in uniaxial tension, are related according to

$$K = \frac{E}{3(1-2\nu)}, \quad \mu = \frac{E}{2(1+\nu)}, \quad E = \frac{9K\mu}{3K+\mu}. \tag{6.1}$$

In general, the elastic limit is an ill-defined stress, but for impure iron and low-carbon steels the onset of plastic deformation is denoted by a sudden drop in load, indicating both an upper and lower

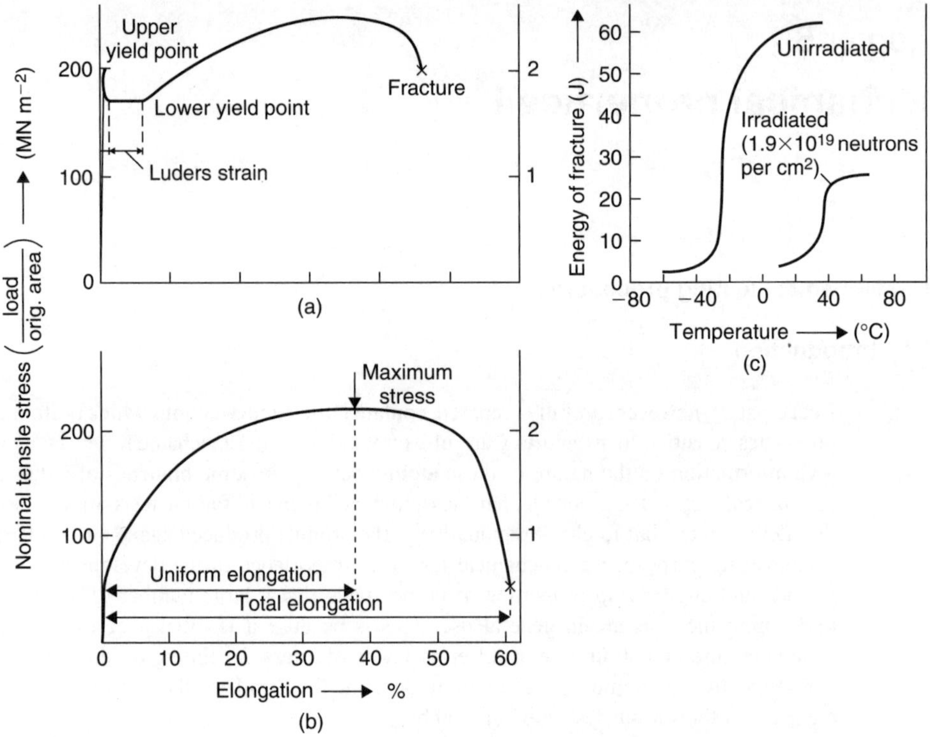

Figure 6.1 *Stress–elongation curves for: (a) impure iron, (b) copper, (c) ductile–brittle transition in mild steel (after Churchman, Mogford and Cottrell, 1957).*

yield point.[1] This yielding behavior is characteristic of many metals, particularly those with bcc structure containing small amounts of solute element (see Section 6.4.6). For materials not showing a sharp yield point, a conventional definition of the beginning of plastic flow is the 0.1% proof stress, in which a line is drawn parallel to the elastic portion of the stress–strain curve from the point of 0.1% strain.

For control purposes the tensile test gives valuable information on the tensile strength (TS = maximum load/original area) and ductility (percentage reduction in area or percentage elongation) of the material. When it is used as a research technique, however, the exact shape and fine details of the curve, in addition to the way in which the yield stress and fracture stress vary with temperature, alloying additions and grain size, are probably of greater significance.

The increase in stress from the initial yield up to the TS indicates that the specimen hardens during deformation (i.e. work hardens). On straining beyond the TS the metal still continues to work-harden, but at a rate too small to compensate for the reduction in cross-sectional area of the test piece. The deformation then becomes unstable, such that as a localized region of the gauge length strains more than the rest, it cannot harden sufficiently to raise the stress for further deformation in this region above that to cause further strain elsewhere. A neck then forms in the gauge length, and further deformation is confined to this region until fracture. Under these conditions, the reduction in area

[1] Load relaxations are obtained only on 'hard' beam Polanyi-type machines, where the beam deflection is small over the working load range. With 'soft' machines, those in which the load-measuring device is a soft spring, rapid load variations are not recorded because the extensions required are too large, while in dead-loading machines no load relaxations are possible. In these latter machines sudden yielding will show as merely an extension under constant load.

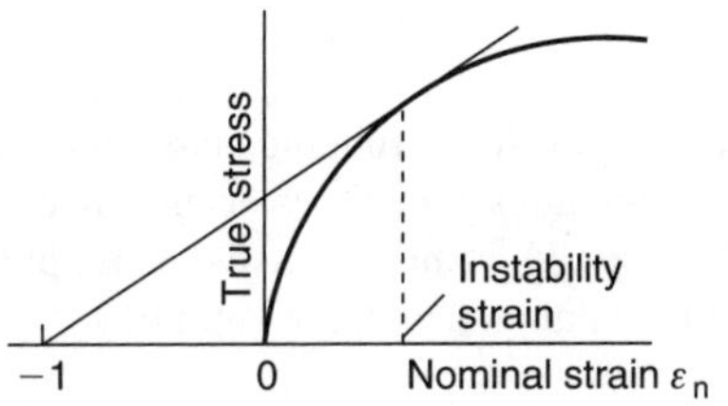

Figure 6.2 *Considère's construction.*

$(A_0 - A_1)/A_0$, where A_0 and A_1 are the initial and final areas of the neck, gives a measure of the localized strain, and is a better indication than the strain to fracture measured along the gauge length.

True stress–true strain curves are often plotted to show the work hardening and strain behavior at large strains. The true stress σ is the load P divided by the area A of the specimen at that particular stage of strain, and the total true strain in deforming from initial length l_0 to length l_1 is $\varepsilon = \int_{l_0}^{l_1} (\mathrm{d}l/l) = \ln(l_1/l_0)$. The true stress–strain curves often fit the Ludwig relation $\sigma = k\varepsilon^n$, where n is a work-hardening coefficient ≈ 0.1–0.5 and k the strength coefficient. Plastic instability, or necking, occurs when an increase in strain produces no increase in load supported by the specimen, i.e. $\mathrm{d}P = 0$, and hence since $P = \sigma A$, then

$$\mathrm{d}P = A\mathrm{d}\sigma + \sigma\mathrm{d}A = 0$$

defines the instability condition. During deformation, the specimen volume is essentially constant (i.e. $\mathrm{d}V = 0$) and from

$$\mathrm{d}V = \mathrm{d}(lA) = A\mathrm{d}l + l\mathrm{d}A = 0,$$

we obtain

$$\frac{\mathrm{d}\sigma}{\sigma} = -\frac{\mathrm{d}A}{A} = \frac{\mathrm{d}l}{l} = \mathrm{d}\varepsilon. \tag{6.2}$$

Thus, necking occurs at a strain at which the slope of the true stress–true strain curve equals the true stress at that strain, i.e. $\mathrm{d}\sigma/\mathrm{d}\varepsilon = \sigma$. Alternatively, since $k\varepsilon^n = \sigma = \mathrm{d}\sigma/\mathrm{d}\varepsilon = nk\varepsilon^{n-1}$, then $\varepsilon = n$ and necking occurs when the true strain equals the strain-hardening exponent. The instability condition may also be expressed in terms of the conventional (nominal strain),

$$\begin{aligned}\frac{\mathrm{d}\sigma}{\mathrm{d}\varepsilon} &= \frac{\mathrm{d}\sigma}{\mathrm{d}\varepsilon_n}\frac{\mathrm{d}\varepsilon_n}{\mathrm{d}\varepsilon} = \frac{\mathrm{d}\sigma}{\mathrm{d}\varepsilon_n}\left(\frac{\mathrm{d}l/l_0}{\mathrm{d}l/l}\right) = \frac{\mathrm{d}\sigma}{\mathrm{d}\varepsilon_n}\frac{l}{l_0} \\ &= \frac{\mathrm{d}\sigma}{\mathrm{d}\varepsilon_n}(1+\varepsilon_n) = \sigma,\end{aligned} \tag{6.3}$$

which allows the instability point to be located using Considère's construction (see Figure 6.2), by plotting the true stress against nominal strain and drawing the tangent to the curve from $\varepsilon_n = -1$ on the strain axis. The point of contact is the instability stress and the tensile strength is $\sigma/(1+\varepsilon_n)$.

Tensile specimens can also give information on the type of fracture exhibited. Usually in polycrystalline metals transgranular fractures occur (i.e. the fracture surface cuts through the grains) and the 'cup-and-cone' type of fracture is extremely common in really ductile metals such as copper. In this, the fracture starts at the center of the necked portion of the test piece and at first grows roughly perpendicular to the tensile axis, so forming the 'cup', but then, as it nears the outer surface, it turns into a 'cone' by fracturing along a surface at about 45° to the tensile axis. In detail the 'cup' itself consists of many irregular surfaces at about 45° to the tensile axis, which gives the fracture a fibrous appearance. Cleavage is also a fairly common type of transgranular fracture, particularly in materials of bcc structure when tested at low temperatures. The fracture surface follows certain crystal planes (e.g. {1 0 0} planes), as is shown by the grains revealing large bright facets, but the surface also appears granular, with 'river lines' running across the facets where cleavage planes have been torn apart. Intercrystalline fractures sometimes occur, often without appreciable deformation. This type of fracture is usually caused by a brittle second phase precipitating out around the grain boundaries, as shown by copper containing bismuth or antimony.

6.1.3 Indentation hardness testing

The hardness of a metal, defined as the resistance to penetration, gives a conveniently rapid indication of its deformation behavior. The hardness tester forces a small sphere, pyramid or cone into the surface of the metals by means of a known applied load, and the hardness number (Brinell or Vickers diamond pyramid) is then obtained from the diameter of the impression. The hardness may be related to the yield or tensile strength of the metal, since during the indentation, the material around the impression is plastically deformed to a certain percentage strain. The Vickers hardness number (VPN) is defined as the load divided by the pyramidal area of the indentation, in kgf mm^{-2}, and is about three times the yield stress for materials which do not work harden appreciably. The Brinell hardness number (BHN) is defined as the stress P/A, in kgf mm^{-2}, where P is the load and A the surface area of the spherical cap forming the indentation. Thus,

$$\text{BHN} = P/\left(\frac{\pi}{2}D^2\right)\{1 - [1 - (d/D)^2]^{1/2}\},$$

where d and D are the indentation and indentor diameters respectively. For consistent results the ratio d/D should be maintained constant and small. Under these conditions soft materials have similar values of BHN and VPN. Hardness testing is of importance in both control work and research, especially where information on brittle materials at elevated temperatures is required.

6.1.4 Impact testing

A material may have a high tensile strength and yet be unsuitable for shock loading conditions. To determine this the impact resistance is usually measured by means of the notched or unnotched Izod or Charpy impact test. In this test a load swings from a given height to strike the specimen, and the energy dissipated in the fracture is measured. The test is particularly useful in showing the decrease in ductility and impact strength of materials of bcc structure at moderately low temperatures. For example, carbon steels have a relatively high ductile–brittle transition temperature (Figure 6.1c) and, consequently, they may be used with safety at sub-zero temperatures only if the transition temperature is lowered by suitable alloying additions or by refining the grain size. Nowadays, increasing importance is given to defining a fracture toughness parameter K_c for an alloy, since many alloys contain small cracks which, when subjected to some critical stress, propagate; K_c defines the critical combination of stress and crack length. Brittle fracture is discussed more fully in Chapter 7.

6.1.5 Creep testing

Creep is defined as plastic flow under constant stress, and although the majority of tests are carried out under constant load conditions, equipment is available for reducing the loading during the test to compensate for the small reduction in cross-section of the specimen. At relatively high temperatures creep appears to occur at all stress levels, but the creep rate increases with increasing stress at a given temperature. For the accurate assessment of creep properties, it is clear that special attention must be given to the maintenance of the specimen at a constant temperature, and to the measurement of the

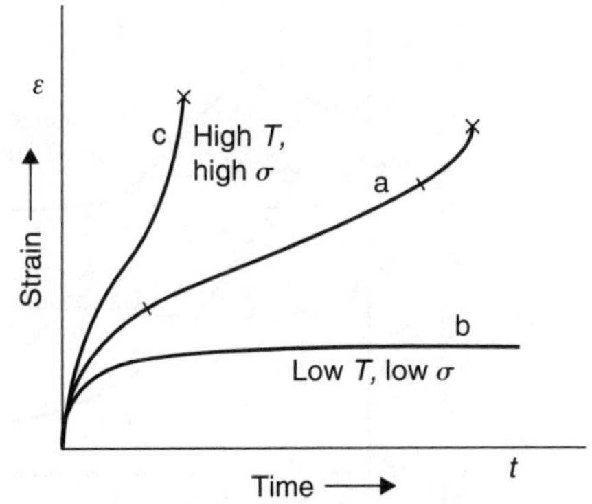

Figure 6.3 *Typical creep curves.*

small dimensional changes involved. This latter precaution is necessary, since in many materials a rise in temperature by a few tens of degrees is sufficient to double the creep rate. Figure 6.3a shows the characteristics of a typical creep curve and, following the instantaneous strain caused by the sudden application of the load, the creep process may be divided into three stages, usually termed primary or transient creep, secondary or steady-state creep and tertiary or accelerating creep. The characteristics of the creep curve often vary, however, and the tertiary stage of creep may be advanced or retarded if the temperature and stress at which the test is carried out is high or low respectively (see Figure 6.3b and c). Creep is discussed more fully in Section 6.9.

6.1.6 Fatigue testing

The fatigue phenomenon is concerned with the premature fracture of metals under repeatedly applied low stresses, and is of importance in many branches of engineering (e.g. aircraft structures). Several different types of testing machines have been constructed in which the stress is applied by bending, torsion, tension or compression, but all involve the same principle of subjecting the material to constant cycles of stress. To express the characteristics of the stress system, three properties are usually quoted: these include (1) the maximum range of stress, (2) the mean stress and (3) the time period for the stress cycle. Four different arrangements of the stress cycle are shown in Figure 6.4, but the reverse and the repeated cycle tests (e.g. 'push–pull') are the most common, since they are the easiest to achieve in the laboratory.

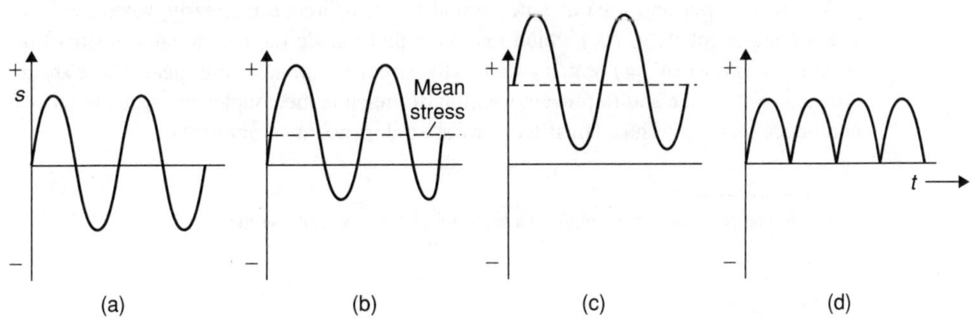

Figure 6.4 *Alternative forms of stress cycling: (a) reversed; (b) alternating (mean stress ≠ zero), (c) fluctuating and (d) repeated.*

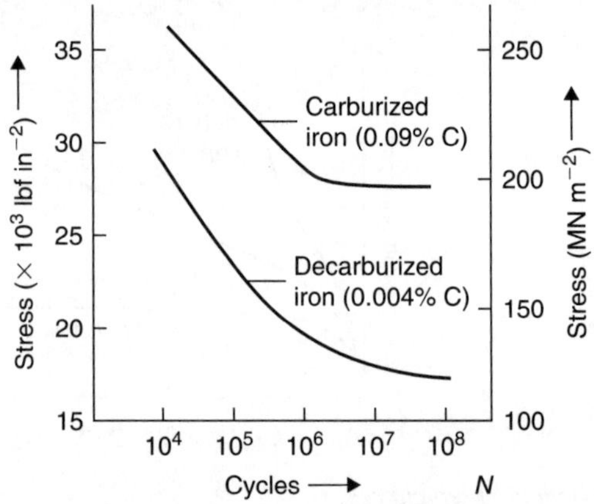

Figure 6.5 *S–N curve for carburized and decarburized iron.*

The standard method of studying fatigue is to prepare a large number of specimens free from flaws, and to subject them to tests using a different range of stress, S, on each group of specimens. The number of stress cycles, N, endured by each specimen at a given stress level is recorded and plotted, as shown in Figure 6.5. This S–N diagram indicates that some metals can withstand indefinitely the application of a large number of stress reversals, provided the applied stress is below a limiting stress known as the endurance limit. For certain ferrous materials when they are used in the absence of corrosive conditions the assumption of a safe working range of stress seems justified, but for non-ferrous materials and for steels when they are used in corrosive conditions a definite endurance limit cannot be defined. Fatigue is discussed in more detail in Section 6.11.

6.2 Elastic deformation

It is well known that metals deform both elastically and plastically. Elastic deformation takes place at low stresses and has three main characteristics, namely (1) it is reversible, (2) stress and strain are linearly proportional to each other according to Hooke's Law, and (3) it is usually small (i.e. $<1\%$ elastic strain).

The stress at a point in a body is usually defined by considering an infinitesimal cube surrounding that point and the forces applied to the faces of the cube by the surrounding material. These forces may be resolved into components parallel to the cube edges and when divided by the area of a face give the nine stress components shown in Figure 6.6. A given component σ_{ij} is the force acting in the j-direction per unit area of face normal to the i-direction. Clearly, when $i=j$ we have normal stress components (e.g. σ_{xx}) which may be either tensile (conventionally positive) or compressive (negative), and when $i \neq j$ (e.g. σ_{xy} or τ_{xy}) the stress components are shear. These shear stresses exert couples on the cube and to prevent rotation of the cube the couples on opposite faces must balance, and hence $\sigma_{ij}=\sigma_{ji}$.[2] Thus, stress has only six independent components.

[2] The nine components of stress σ_{ij} form a second-rank tensor usually written as:

$$\begin{matrix} \sigma_{xx} & \sigma_{xy} & \sigma_{xz} \\ \sigma_{yx} & \sigma_{yy} & \sigma_{yz} \\ \sigma_{zx} & \sigma_{zy} & \sigma_{zz} \end{matrix}$$

and is known as the stress tensor.

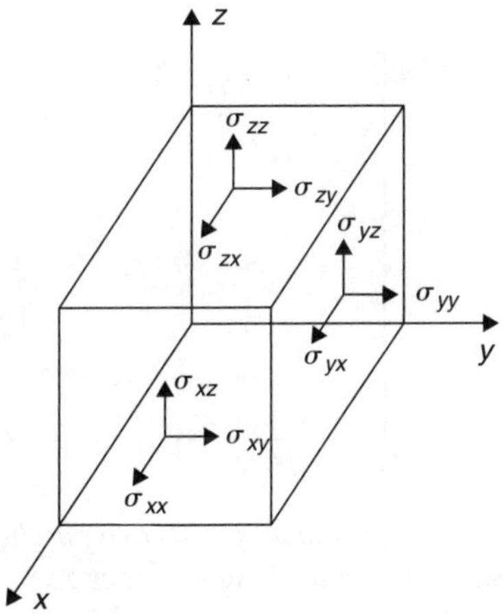

Figure 6.6 *Normal and shear stress components.*

When a body is strained, small elements in that body are displaced. If the initial position of an element is defined by its coordinates (x, y, z) and its final position by $(x + u, y + v, z + w)$, then the displacement is (u, v, w). If this displacement is constant for all elements in the body, no strain is involved, only a rigid translation. For a body to be under a condition of strain the displacements must vary from element to element. A uniform strain is produced when the displacements are linearly proportional to distance. In one dimension then $u = ex$, where $e = \mathrm{d}u/\mathrm{d}x$ is the coefficient of proportionality or nominal tensile strain. For a three-dimensional uniform strain, each of the three components u, v, w is made a linear function in terms of the initial elemental coordinates, i.e.

$$\begin{aligned} u &= e_{xx}x + e_{xy}y + e_{xz}z \\ v &= e_{yx}x + e_{yy}y + e_{yz}z \\ w &= e_{zx}x + e_{zy}y + e_{zz}z. \end{aligned}$$

The strains $e_{xx} = \mathrm{d}u/\mathrm{d}x, e_{yy} = \mathrm{d}v/\mathrm{d}y, e_{zz} = \mathrm{d}w/\mathrm{d}z$ are the tensile strains along the x-, y- and z-axes, respectively. The strains e_{xy}, e_{yz}, etc. produce shear strains and in some cases a rigid-body rotation. The rotation produces no strain and can be allowed for by rotating the reference axes (see Figure 6.7). In general, therefore, $e_{ij} = \varepsilon_{ij} + \omega_{ij}$, with ε_{ij} the strain components and ω_{ij} the rotation components. If, however, the shear strain is defined as the angle of shear, this is twice the corresponding shear strain component, i.e. $\gamma = 2\varepsilon_{ij}$. The strain tensor, like the stress tensor, has nine components, which are usually written as:

$$\begin{matrix} \varepsilon_{xx} & \varepsilon_{xy} & \varepsilon_{xz} \\ \varepsilon_{yx} & \varepsilon_{yy} & \varepsilon_{yz} \\ \varepsilon_{zx} & \varepsilon_{zy} & \varepsilon_{zz} \end{matrix} \quad \text{or} \quad \begin{matrix} \varepsilon_{xx} & \frac{1}{2}\gamma_{xy} & \frac{1}{2}\gamma_{xz} \\ \frac{1}{2}\gamma_{yx} & \varepsilon_{yy} & \frac{1}{2}\gamma_{yz} \\ \frac{1}{2}\gamma_{zx} & \frac{1}{2}\gamma_{zy} & \varepsilon_{zz}, \end{matrix}$$

where ε_{xx} etc. are tensile strains and γ_{xy}, etc. are shear strains. All the simple types of strain can be produced from the strain tensor by setting some of the components equal to zero. For example, a pure dilatation (i.e. change of volume without change of shape) is obtained when $\varepsilon_{xx} = \varepsilon_{yy} = \varepsilon_{zz}$ and all other components are zero. Another example is a uniaxial tensile test when the tensile strain along the

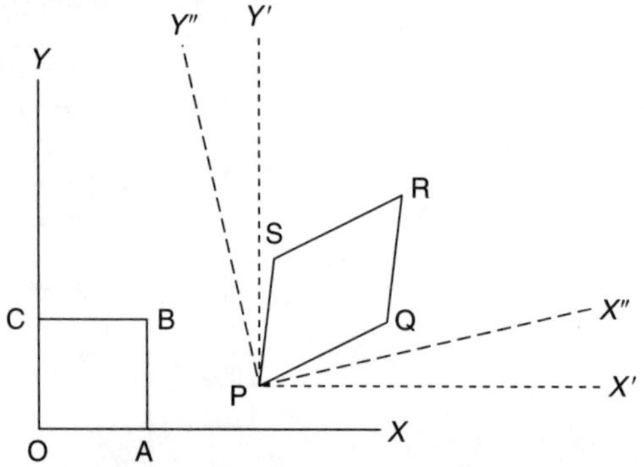

Figure 6.7 *Deformation of a square OABC to a parallelogram PQRS involving: (i) a rigid-body translation OP allowed for by redefining new axes X′Y′, (ii) a rigid-body rotation allowed for by rotating the axes to X″Y″, and (iii) a change of shape involving both tensile and shear strains.*

x-axis is simply $e = \varepsilon_{xx}$. However, because of the strains introduced by lateral contraction, $\varepsilon_{yy} = -\nu e$ and $\varepsilon_{zz} = -\nu e$, where ν is Poisson's ratio; all other components of the strain tensor are zero.

At small elastic deformations, the stress is linearly proportional to the strain. This is Hooke's law and in its simplest form relates the uniaxial stress to the uniaxial strain by means of the modulus of elasticity. For a general situation, it is necessary to write Hooke's law as a linear relationship between six stress components and the six strain components, i.e.

$$
\begin{aligned}
\sigma_{xx} &= c_{11}\varepsilon_{xx} + c_{12}\varepsilon_{yy} + c_{13}\varepsilon_{zz} + c_{14}\gamma_{yz} + c_{15}\gamma_{zx} + c_{16}\gamma_{xy} \\
\sigma_{yy} &= c_{21}\varepsilon_{xx} + c_{22}\varepsilon_{yy} + c_{23}\varepsilon_{zz} + c_{24}\gamma_{yz} + c_{25}\gamma_{zx} + c_{26}\gamma_{xy} \\
\sigma_{zz} &= c_{31}\varepsilon_{xx} + c_{32}\varepsilon_{yy} + c_{33}\varepsilon_{zz} + c_{34}\gamma_{yz} + c_{35}\gamma_{zx} + c_{36}\gamma_{xy} \\
\tau_{yz} &= c_{41}\varepsilon_{xx} + c_{42}\varepsilon_{yy} + c_{43}\varepsilon_{zz} + c_{44}\gamma_{yz} + c_{45}\gamma_{zx} + c_{46}\gamma_{xy} \\
\tau_{zx} &= c_{51}\varepsilon_{xx} + c_{52}\varepsilon_{yy} + c_{53}\varepsilon_{zz} + c_{54}\gamma_{yz} + c_{55}\gamma_{zx} + c_{56}\gamma_{xy} \\
\tau_{xy} &= c_{61}\varepsilon_{xx} + c_{62}\varepsilon_{yy} + c_{63}\varepsilon_{zz} + c_{64}\gamma_{yz} + c_{65}\gamma_{zx} + c_{66}\gamma_{xy}.
\end{aligned}
$$

The constants $c_{11}, c_{12}, \ldots, c_{ij}$ are called the elastic stiffness constants.[3]

Taking account of the symmetry of the crystal, many of these elastic constants are equal or become zero. Thus, in cubic crystals there are only three independent elastic constants c_{11}, c_{12} and c_{44} for the three independent modes of deformation. These include the application of (1) a hydrostatic stress p to produce a dilatation Θ given by

$$p = -\frac{1}{3}(c_{11} + 2c_{12})\Theta = -\kappa\Theta,$$

where κ is the bulk modulus, (2) a shear stress on a cube face in the direction of the cube axis defining the shear modulus $\mu = c_{44}$, and (3) a rotation about a cubic axis defining a shear modulus $\mu_1 = (c_{11} - c_{12})/2$. The ratio μ/μ_1 is the elastic anisotropy factor and in elastically isotropic crystals it is unity with $2c_{44} = c_{11} - c_{12}$; the constants are all interrelated, with $c_{11} = \kappa + 4\mu/3,\ c_{12} = \kappa - 2\mu/3$ and $c_{44} = \mu$.

[3] Alternatively, the strain may be related to the stress, e.g. $\varepsilon_x = s_{11}\sigma_{xx} + s_{12}\sigma_{yy} + s_{13}\sigma_{zz} + \ldots$, in which case the constants $s_{11}, s_{12}, \ldots, s_{ij}$ are called elastic compliances.

Table 6.1 *Elastic constants of cubic crystals (GN m^{-2}).*

Metal	c_{11}	c_{12}	c_{44}	$2c_{44}/(c_{11} - c_{12})$
Na	006.0	004.6	005.9	8.5
K	004.6	003.7	002.6	5.8
Fe	237.0	141.0	116.0	2.4
W	501.0	198.0	151.0	1.0
Mo	460.0	179.0	109.0	0.77
Al	108.0	62.0	28.0	1.2
Cu	170.0	121.0	75.0	3.3
Ag	120.0	90.0	43.0	2.9
Au	186.0	157.0	42.0	3.9
Ni	250.0	160.0	118.0	2.6
β-Brass	129.1	109.7	82.4	8.5

Table 6.1 shows that most metals are far from isotropic and, in fact, only tungsten is isotropic; the alkali metals and β-compounds are mostly anisotropic. Generally, $2c_{44} > (c_{11} - c_{12})$ and, hence, for most elastically anisotropic metals E is maximum in the $\langle 1\ 1\ 1 \rangle$ and minimum in the $\langle 1\ 0\ 0 \rangle$ directions. Molybdenum and niobium are unusual in having the reverse anisotropy when E is greatest along $\langle 1\ 0\ 0 \rangle$ directions. Most commercial materials are polycrystalline, and consequently they have approximately isotropic properties. For such materials the modulus value is usually independent of the direction of measurement because the value observed is an average for all directions, in the various crystals of the specimen. However, if during manufacture a preferred orientation of the grains in the polycrystalline specimen occurs, the material will behave, to some extent, like a single crystal and some 'directionality' will take place.

6.3 Plastic deformation

6.3.1 Slip and twinning

The limit of the elastic range cannot be defined exactly but may be considered to be that value of the stress below which the amount of plasticity (irreversible deformation) is negligible, and above which the amount of plastic deformation is far greater than the elastic deformation. If we consider the deformation of a metal in a tensile test, one or other of two types of curve may be obtained. Figure 6.1a shows the stress–strain curve characteristic of iron, from which it can be seen that plastic deformation begins abruptly and continues initially with no increase in stress. Figure 6.1b shows a stress–strain curve characteristic of copper, from which it will be noted that the transition to the plastic range is gradual. No abrupt yielding takes place and in this case the stress required to start macroscopic plastic flow is known as the flow stress.

Once the yield or flow stress has been exceeded plastic or permanent deformation occurs, and this is found to take place by one of two simple processes, slip (or glide) and twinning. During slip, shown in Figure 6.8a, the top half of the crystal moves over the bottom half along certain crystallographic planes, known as slip planes, in such a way that the atoms move forward by a whole number of lattice vectors; as a result the continuity of the lattice is maintained. During twinning (Figure 6.8b) the atomic movements are not whole lattice vectors, and the lattice generated in the deformed region, although the same as the parent lattice, is oriented in a twin relationship to it. It will also be observed

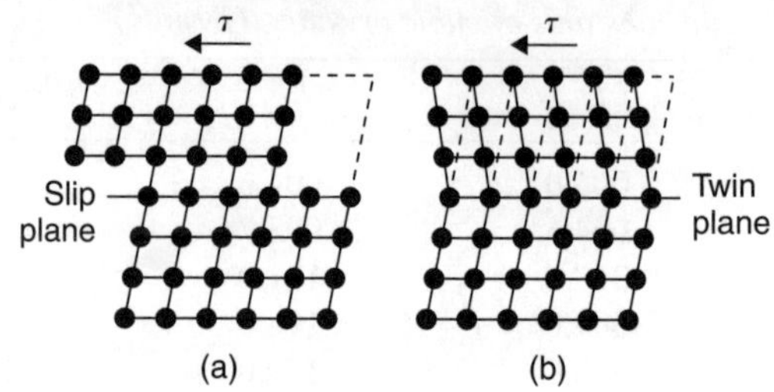

Figure 6.8 *Slip and twinning in a crystal.*

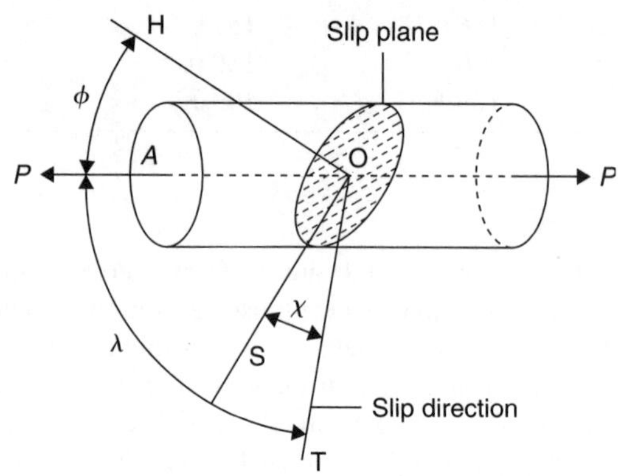

Figure 6.9 *Relation between the slip plane, slip direction and the axis of tension for a cylindrical crystal.*

that, in contrast to slip, the sheared region in twinning occurs over many atom planes, the atoms in each plane being moved forward by the same amount relative to those of the plane below them.

6.3.2 Resolved shear stress

All working processes such as rolling, extrusion, forging, etc. cause plastic deformation and, consequently, these operations will involve the processes of slip or twinning outlined above. The stress system applied during these working operations is often quite complex, but for plastic deformation to occur the presence of a shear stress is essential. The importance of shear stresses becomes clear when it is realized that these stresses arise in most processes and tests even when the applied stress itself is not a pure shear stress. This may be illustrated by examining a cylindrical crystal of area A in a conventional tensile test under a uniaxial load P. In such a test, slip occurs on the slip plane, shown shaded in Figure 6.9, the area of which is $A/\cos\phi$, where ϕ is the angle between the normal to the plane OH and the axis of tension. The applied force P is spread over this plane and may be resolved into a force normal to the plane along OH, $P\cos\phi$, and a force along OS, $P\sin\phi$. Here, OS is the line of greatest slope in the slip plane and the force $P\sin\phi$ is a shear force. It follows that the applied stress (force/area) is made up of two stresses, a normal stress $(P/A)\cos^2\phi$ tending to pull the atoms apart, and a shear stress $(P/A)\cos\phi\sin\phi$ trying to slide the atoms over each other.

In general, slip does not take place down the line of greatest slope unless this happens to coincide with the crystallographic slip of direction. It is necessary, therefore, to know the resolved shear stress on the slip plane and in the slip direction. Now, if OT is taken to represent the slip direction, the resolved shear stress will be given by

$$\sigma = P\cos\phi \sin\phi \cos\chi/A,$$

where χ is the angle between OS and OT. Usually this formula is written more simply as

$$\sigma = P\cos\phi \cos\lambda/A, \tag{6.4}$$

where λ is the angle between the slip direction OT and the axis of tension. It can be seen that the resolved shear stress has a maximum value when the slip plane is inclined at 45° to the tensile axis, and becomes smaller for angles either greater than or less than 45°. When the slip plane becomes more nearly perpendicular to the tensile axis ($\phi > 45°$), it is easy to imagine that the applied stress has a greater tendency to pull the atoms apart than to slide them. When the slip plane becomes more nearly parallel to the tensile axis ($\phi < 45°$), the shear stress is again small but in this case it is because the area of the slip plane, $A/\cos\phi$, is correspondingly large.

A consideration of the tensile test in this way shows that it is shear stresses which lead to plastic deformation, and for this reason the mechanical behavior exhibited by a material will depend, to some extent, on the type of test applied. For example, a ductile material can be fractured without displaying its plastic properties if tested in a state of hydrostatic or triaxial tension, since under these conditions the resolved shear stress on any plane is zero. Conversely, materials which normally exhibit a tendency to brittle behavior in a tensile test will show ductility if tested under conditions of high shear stresses and low tension stresses. In commercial practice, extrusion approximates closely to a system of high shear stress, and it is common for normally brittle materials to exhibit some ductility when deformed in this way (e.g. when extruded).

Worked example

A single crystal of iron is pulled along [1 2 3]. Which is the first slip system to operate?

Solution

Slip plane	*Cos ϕ*	*Burgers vector*	*Cos λ*	*Schmid factor $\times\sqrt{28}\times\sqrt{42}$*
(1 1 0)	$3/\sqrt{28}$	$1/2[1\ \bar{1}\ 1]$	$2/\sqrt{42}$	6
		$1/2[1\ \bar{1}\ \bar{1}]$	$-4/\sqrt{42}$	−12
$(1\ \bar{1}\ 0)$	$-1/\sqrt{28}$	$1/2[1\ 1\ 1]$	$6/\sqrt{42}$	−6
		$1/2[1\ 1\ \bar{1}]$	0	0
(1 0 1)	$4/\sqrt{28}$	$1/2[1\ 1\ \bar{1}]$	0	0
		$1/2[1\ \bar{1}\ \bar{1}]$	$-4/\sqrt{42}$	−16 ←
$(1\ 0\ \bar{1})$	$-2/\sqrt{28}$	$1/2[1\ 1\ 1]$	$6/\sqrt{42}$	−12
		$1/2[1\ \bar{1}\ 1]$	$2/\sqrt{42}$	−4
(0 1 1)	$5/\sqrt{28}$	$1/2[1\ 1\ \bar{1}]$	0	0
		$1/2[\bar{1}\ 1\ \bar{1}]$	$-2/\sqrt{42}$	−10
$(0\ 1\ \bar{1})$	$-1/\sqrt{28}$	$1/2[1\ 1\ 1]$	$6/\sqrt{42}$	−6
		$1/2[\bar{1}\ 1\ 1]$	$4/\sqrt{42}$	−4

Slip will occur on $(1\ 0\ 1)[1\ \bar{1}\ \bar{1}]$ first.

6.3.3 Relation of slip to crystal structure

An understanding of the fundamental nature of plastic deformation processes is provided by experiments on single crystals only, because if a polycrystalline sample is used the result obtained is the average behavior of all the differently oriented grains in the material. Such experiments with single crystals show that, although the resolved shear stress is a maximum along lines of greatest slope in planes at 45° to the tensile axis, slip occurs preferentially along certain crystal planes and directions. Three well-established laws governing the slip behavior exist, namely: (1) the direction of slip is almost always that along which the atoms are most closely packed, (2) slip usually occurs on the most closely packed plane, and (3) from a given set of slip planes and directions, the crystal operates on that system (plane and direction) for which the resolved shear stress is largest. The slip behavior observed in fcc metals shows the general applicability of these laws, since slip occurs along $\langle 1\ 1\ 0\rangle$ directions in $\{1\ 1\ 1\}$ planes. In cph metals slip occurs along $\langle 1\ 1\ \bar{2}\ 0\rangle$ directions, since these are invariably the closest packed, but the active slip plane depends on the value of the axial ratio. Thus, for the metals cadmium and zinc, c/a is 1.886 and 1.856 respectively, the planes of greatest atomic density are the $\{0\ 0\ 0\ 1\}$ basal planes and slip takes place on these planes. When the axial ratio is appreciably smaller than the ideal value of $c/a = 1.633$ the basal plane is not so closely packed, nor so widely spaced, as in cadmium and zinc, and other slip planes operate. In zirconium ($c/a = 1.589$) and titanium ($c/a = 1.587$), for example, slip takes place on the $\{1\ 0\ \bar{1}\ 0\}$ prism planes at room temperature and on the $\{1\ 0\ \bar{1}\ 1\}$ pyramidal planes at higher temperatures. In magnesium the axial ratio ($c/a = 1.624$) approximates to the ideal value, and although only basal slip occurs at room temperature, at temperatures above 225°C slip on the $\{1\ 0\ \bar{1}\ 1\}$ planes has also been observed. Bcc metals have a single well-defined close-packed $\langle 1\ 1\ 1\rangle$ direction, but several planes of equally high density of packing, i.e. $\{1\ 1\ 2\}$, $\{1\ 1\ 0\}$ and $\{1\ 2\ 3\}$. The choice of slip plane in these metals is often influenced by temperature and a preference is shown for $\{1\ 1\ 2\}$ below $T_m/4$, $\{1\ 1\ 0\}$ from $T_m/4$ to $T_m/2$ and $\{1\ 2\ 3\}$ at high temperatures, where T_m is the melting point. Iron often slips on all the slip planes at once in a common $\langle 1\ 1\ 1\rangle$ slip direction, so that a slip line (i.e. the line of intersection of a slip plane with the outer surface of a crystal) takes on a wavy appearance.

6.3.4 Law of critical resolved shear stress

This law states that slip takes place along a given slip plane and direction when the shear stress reaches a critical value. In most crystals the high symmetry of atomic arrangement provides several crystallographic equivalent planes and directions for slip (i.e. cph crystals have three systems made up of one plane containing three directions, fcc crystals have 12 systems made up of four planes each with three directions, while bcc crystals have many systems) and in such cases slip occurs first on that plane and along that direction for which the maximum stress acts (law 3 above). This is most easily demonstrated by testing in tension a series of zinc single crystals. Then, because zinc is cph in structure only one plane is available for the slip process and the resultant stress–strain curve will depend on the inclination of this plane to the tensile axis. The value of the angle ϕ is determined by chance during the process of single-crystal growth, and consequently all crystals will have different values of ϕ, and the corresponding stress–strain curves will have different values of the flow stress, as shown in Figure 6.10a. However, because of the criterion of a critical resolved shear stress, a plot of resolved shear stress (i.e. the stress on the glide plane in the glide direction) versus strain should be a common curve, within experimental error, for all the specimens. This plot is shown in Figure 6.10b.

The importance of a critical shear stress may be demonstrated further by taking the crystal which has its basal plane oriented perpendicular to the tensile axis, i.e. $\phi = 0°$, and subjecting it to a bend test. In contrast to its tensile behavior, where it is brittle it will now appear ductile, since the shear stress on the slip plane is only zero for a tensile test and not for a bend test. On the other hand, if we

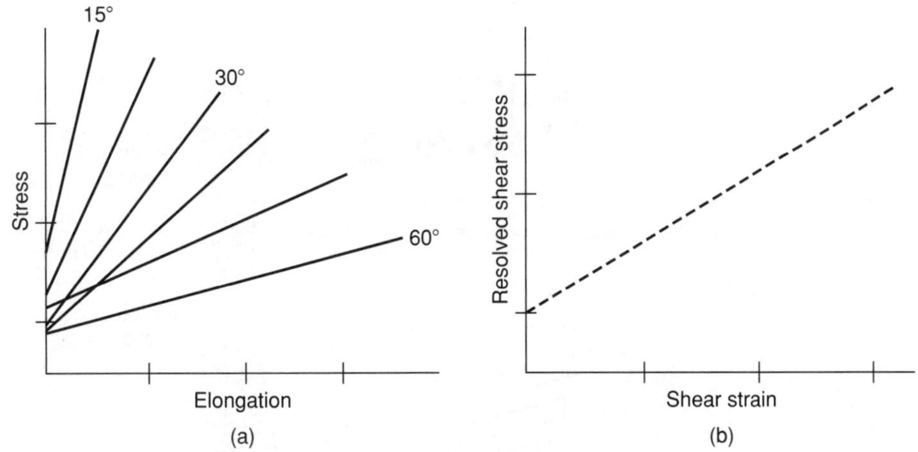

Figure 6.10 *Schematic representation of variation of stress versus elongation with orientation of basal plane (a) and constancy of revolved shear stress (b).*

take the crystal with its basal plane oriented parallel to the tensile axis (i.e. $\phi = 90°$) this specimen will appear brittle whatever stress system is applied to it. For this crystal, although the shear force is large, owing to the large area of the slip plane, $A/\cos\phi$, the resolved shear stress is always very small and insufficient to cause deformation by slipping.

6.3.5 Multiple slip

The fact that slip bands, each consisting of many slip lines, are observed on the surface of deformed crystals shows that deformation is inhomogeneous, with extensive slip occurring on certain planes, while the crystal planes lying between them remain practically undeformed. Figure 6.11a and b shows such a crystal in which the set of planes shear over each other in the slip direction. In a tensile test, however, the ends of a crystal are not free to move 'sideways' relative to each other, since they are constrained by the grips of the tensile machine. In this case, the central portion of the crystal is altered in orientation, and rotation of both the slip plane and slip direction into the axis of tension occurs, as shown in Figure 6.11c. This behavior is more conveniently demonstrated on a stereographic projection of the crystal by considering the rotation of the tensile axis relative to the crystal rather than vice versa. This is illustrated in Figure 6.12a for the deformation of a crystal with fcc structure. The tensile axis, P, is shown in the unit triangle and the angles between P and $[\bar{1}\ 0\ 1]$, and P and $(1\ 1\ 1)$ are equal to λ and ϕ respectively. The active slip system is the $(1\ 1\ 1)$ plane and the $[\bar{1}\ 0\ 1]$ direction, and as deformation proceeds the change in orientation is represented by the point, P, moving along the zone, shown broken in Figure 6.12a, towards $[\bar{1}\ 0\ 1]$, i.e. λ decreasing and ϕ increasing.

As slip occurs on the one system, the primary system, the slip plane rotates away from its position of maximum resolved shear stress until the orientation of the crystal reaches the $[0\ 0\ 1]$–$[\bar{1}\ 1\ 1]$ symmetry line. Beyond this point, slip should occur equally on both the primary system and a second system (the conjugate system) $(\bar{1}\ \bar{1}\ 1)$–$[0\ 1\ 1]$, since these two systems receive equal components of shear stress. Subsequently, during the process of multiple or duplex slip the lattice will rotate so as to keep equal stresses on the two active systems, and the tensile axis moves along the symmetry line towards $[\bar{1}\ 1\ 2]$. This behavior agrees with early observations on virgin crystals of aluminum and

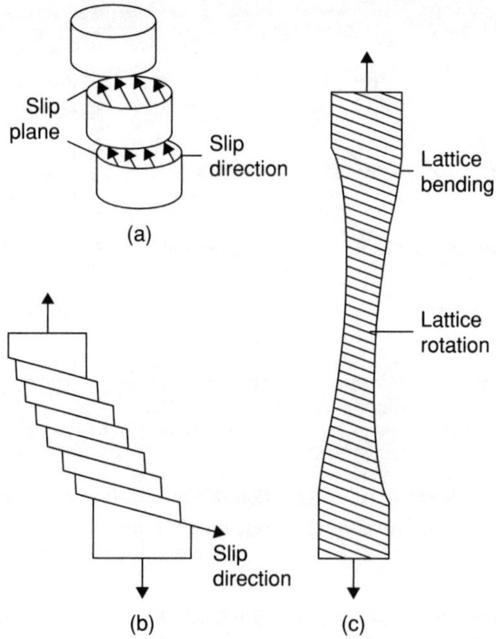

Figure 6.11 *(a) and (b) show the slip process in an unconstrained single crystal; (c) illustrates the plastic bending in a crystal gripped at its ends.*

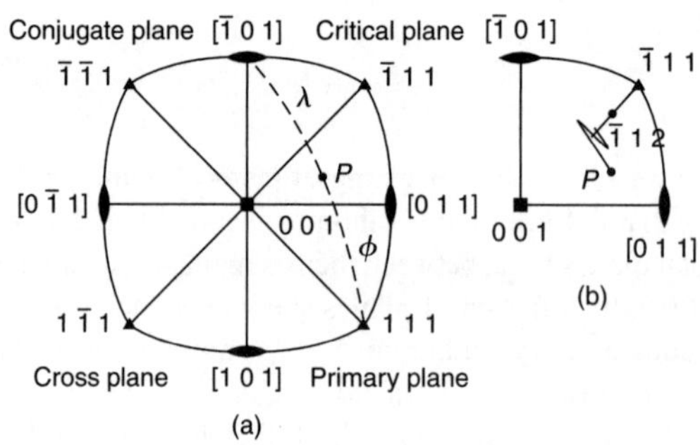

Figure 6.12 *Stereographic representation of slip systems in fcc crystals (a) and overshooting of the primary slip system (b).*

copper, but not with those made on certain alloys, or pure metal crystals given special treatments (e.g. quenched from a high temperature or irradiated with neutrons). Results from the latter show that the crystal continues to slip on the primary system after the orientation has reached the symmetry line, causing the orientation to overshoot this line, i.e. to continue moving towards $[\bar{1}\ 0\ 1]$, in the direction of primary slip. After a certain amount of this additional primary slip the conjugate system suddenly operates, and further slip concentrates itself on this system, followed by overshooting in the opposite direction. This behavior, shown in Figure 6.12b, is understandable when it is remembered that slip on the conjugate system must intersect that on the primary system, and to do this is presumably more difficult than to 'fit' a new slip plane in the relatively undeformed region between those planes on which slip has already taken place. This intersection process is more difficult in materials which have a low stacking-fault energy (e.g. α-brass).

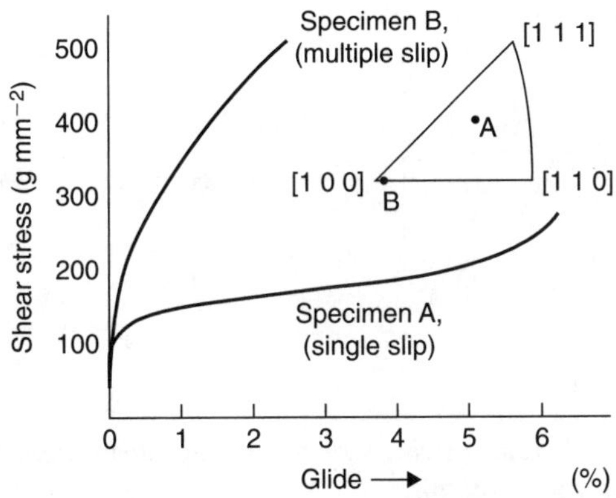

Figure 6.13 *Stress–strain curves for aluminum deformed by single and multiple slip (after Lücke and Lange, 1950).*

6.3.6 Relation between work hardening and slip

The curves of Figure 6.1 show that following the yield phenomenon a continual rise in stress is required to continue deformation, i.e. the flow stress of a deformed metal increases with the amount of strain. This resistance of the metal to further plastic flow as the deformation proceeds is known as work hardening. The degree of work hardening varies for metals of different crystal structure, and is low in hexagonal metal crystals such as zinc or cadmium, which usually slip on one family of planes only. The cubic crystals harden rapidly on working but even in this case when slip is restricted to one slip system (see the curve for specimen A, Figure 6.13), the coefficient of hardening, defined as the slope of the plastic portion of the stress–strain curve, is small. Thus, this type of hardening, like overshoot, must be associated with the interaction which results from slip on intersecting families of planes. This interaction will be dealt with more fully in Section 6.6.2.

6.4 Dislocation behavior during plastic deformation

6.4.1 Dislocation mobility

The ease with which crystals can be plastically deformed at stresses many orders of magnitude less than the theoretical strength ($\tau_t = \mu b/2\pi a$) is quite remarkable, and due to the mobility of dislocations. Figure 6.14a shows that as a dislocation glides through the lattice it moves from one symmetrical lattice position to another and at each position the dislocation is in neutral equilibrium, because the atomic forces acting on it from each side are balanced. As the dislocation moves from these symmetrical lattice positions some imbalance of atomic forces does exist, and an applied stress is required to overcome this lattice friction. As shown in Figure 6.14b, an intermediate displacement of the dislocation also leads to an approximately balanced force system.

The lattice friction depends rather sensitively on the dislocation width w and has been shown by Peierls and Nabarro to be given by

$$\tau \approx \mu \exp[-2\pi w/b] \tag{6.5}$$

for the shear of a rectangular lattice of interplanar spacing a with $w = \mu b/2\pi(1 - \nu)\tau_t = a/(1 - \nu)$. The friction stress is therefore often referred to as the Peierls–Nabarro stress. The two opposing

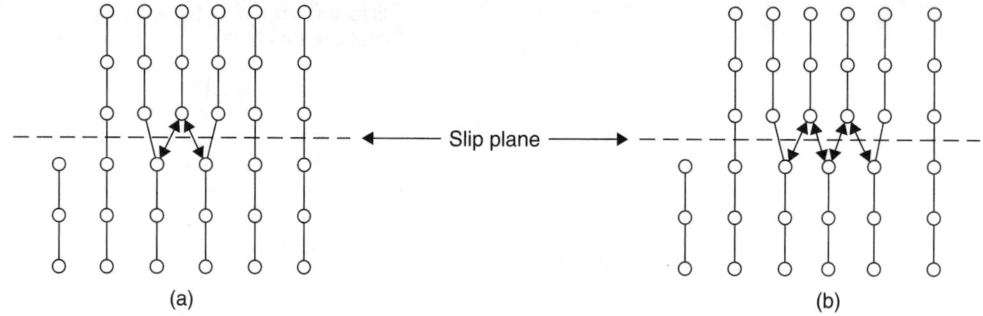

Figure 6.14 *Diagram showing structure of edge dislocation during gliding from equilibrium (a) to metastable position (b).*

factors affecting w are (1) the elastic energy of the crystal, which is reduced by spreading out the elastic strains, and (2) the misfit energy, which depends on the number of misaligned atoms across the slip plane. Metals with close-packed structures have extended dislocations and hence w is large. Moreover, the close-packed planes are widely spaced, with weak alignment forces between them (i.e. have a small b/a factor). These metals have highly mobile dislocations and are intrinsically soft. In contrast, directional bonding in crystals tends to produce narrow dislocations, which leads to intrinsic hardness and brittleness. Extreme examples are ionic and ceramic crystals and the covalent materials such as diamond and silicon. The bcc transition metals display intermediate behavior (i.e. intrinsically ductile above room temperatures but brittle below).

Direct measurements of dislocation velocity v have now been made in some crystals by means of the etch-pitting technique; the results of such an experiment are shown in Figure 6.15. Edge dislocations move faster than screws, because of the frictional drag of jogs on screws, and the velocity of both varies rapidly with applied stress τ according to an empirical relation of the form $v = (\tau/\tau_0)^n$, where τ_0 is the stress for unit speed and n is an index which varies for different materials. At high stresses the velocity may approach the speed of elastic waves $\approx 10^3$ m s^{-1}. The index n is usually low (<10) for intrinsically hard, covalent crystals such as Ge, ≈ 40 for bcc crystals and high (≈ 200) for intrinsically soft fcc crystals. It is observed that a critical applied stress is required to start the dislocations moving and denotes the onset of microplasticity. A macroscopic tensile test is a relatively insensitive measure of the onset of plastic deformation and the yield or flow stress measured in such a test is related not to the initial motion of an individual dislocation, but to the motion of a number of dislocations at some finite velocity, e.g. ~ 10 nm s^{-1}, as shown in Figure 6.16a. Decreasing the temperature of the test or increasing the strain rate increases the stress level required to produce the same finite velocity (see Figure 6.16b), i.e. displacing the velocity–stress curve to the right. Indeed, hardening the material by any mechanism has the same effect on the dislocation dynamics. This observation is consistent with the increase in yield stress with decreasing temperature or increasing strain rate. Most metals and alloys are hardened by cold working or by placing obstacles (e.g. precipitates) in the path of moving dislocations to hinder their motion. Such strengthening mechanisms increase the stress necessary to produce a given finite dislocation velocity in a similar way to that found by lowering the temperature.

6.4.2 Variation of yield stress with temperature and strain rate

The high Peierls–Nabarro stress, which is associated with materials with narrow dislocations, gives rise to a short-range barrier to dislocation motion. Such barriers are effective only over an atomic spacing or so, hence thermal activation is able to aid the applied stress in overcoming them. Thermal activation helps a portion of the dislocation to cross the barrier after which glide then proceeds by the

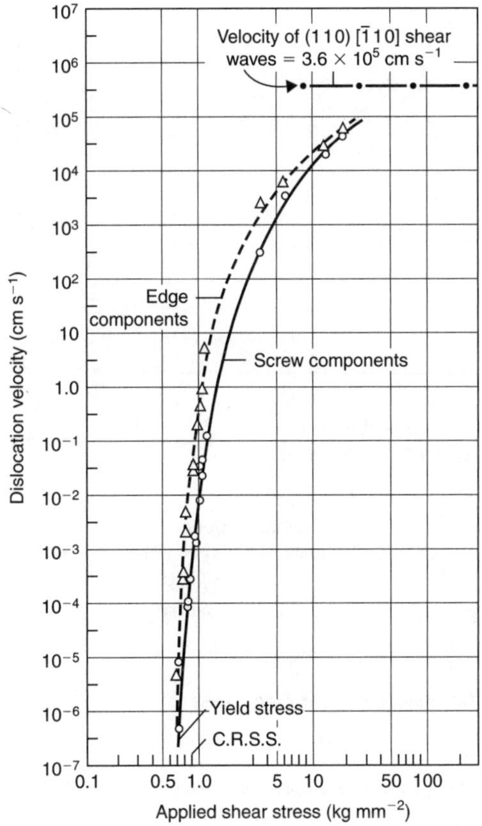

Figure 6.15 *Stress dependence of the velocity of edge and screw dislocations in lithium fluoride (from Johnston and Gilman, 1959; courtesy of the American Physical Society).*

sideways movement of kinks. (This process is shown in Figure 6.28, Section 6.4.8.) Materials with narrow dislocations therefore exhibit a significant temperature sensitivity; intrinsically hard materials rapidly lose their strength with increasing temperature, as shown schematically in Figure 6.17a. In this diagram the (yield stress/modulus) ratio is plotted against T/T_m to remove the effect of modulus which decreases with temperature. Figure 6.17b shows that materials which exhibit a strong temperature-dependent yield stress also exhibit a high strain-rate sensitivity, i.e. the higher the imposed strain rate, the higher the yield stress. This arises because thermal activation is less effective at the faster rate of deformation.

In bcc metals a high lattice friction to the movement of a dislocation may arise from the dissociation of a dislocation on several planes. As discussed in Chapter 3, when a screw dislocation with Burgers vector $a/2[1\ 1\ 1]$ lies along a symmetry direction it can dissociate on three crystallographically equivalent planes. If such a dissociation occurs, it will be necessary to constrict the dislocation before it can glide in any one of the slip planes. This constriction will be more difficult to make as the temperature is lowered so that the large temperature dependence of the yield stress in bcc metals, shown in Figure 6.17a and also Figure 6.29, may be due partly to this effect. In fcc metals the dislocations lie on $\{1\ 1\ 1\}$ planes, and although a dislocation will dissociate in any given $(1\ 1\ 1)$ plane, there is no direction in the slip plane along which the dislocation could also dissociate on other planes; the temperature dependence of the yield stress is small, as shown in Figure 6.17a. In cph metals the dissociated dislocations moving in the basal plane will also have a small Peierls force

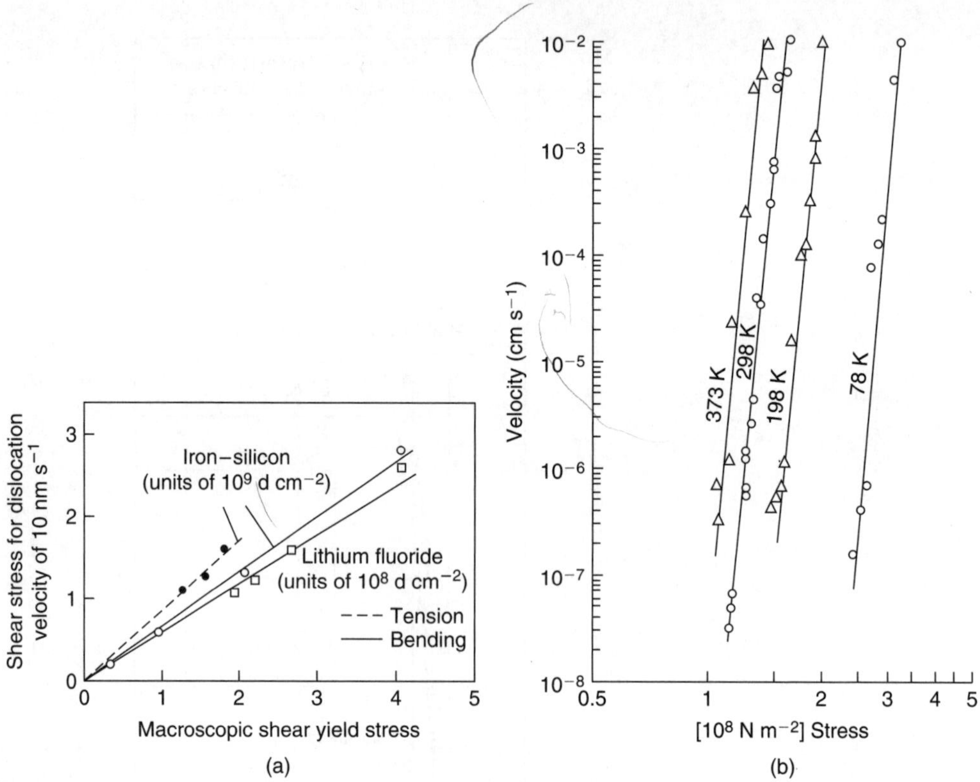

Figure 6.16 *(a) Correlation between stress to cause dislocation motion and the macro-yield stresses of crystals. (b) Edge dislocation motions in Fe–3% Si crystals (after Stein and Low, 1960; courtesy of the American Physical Society).*

and be glissile with low temperature dependence. However, screw dislocations moving on non-basal planes (i.e. prismatic and pyramidal planes) may have a high Peierls force because they are able to extend in the basal plane, as shown in Figure 6.18. Hence, constrictions will once again have to be made before the screw dislocations can advance on non-basal planes. This effect contributes to the high critical shear stress and strong temperature dependence of non-basal glide observed in this crystal system, as mentioned in Chapter 3.

6.4.3 Dislocation source operation

When a stress is applied to a material the specimen plastically deforms at a rate governed by the strain rate of the deformation process (e.g. tensile testing, rolling, etc.) and the strain rate imposes a particular velocity on the mobile dislocation population. In a crystal of dimensions $L_1 \times L_2 \times 1$ cm, shown in Figure 6.19, a dislocation with velocity v moves through the crystal in time $t = L_1/v$ and produces a shear strain b/L_2, i.e. the strain rate is bv/L_1L_2. If the density of glissible dislocations is ρ, the total number of dislocations which become mobile in the crystal is $\rho L_1 L_2$ and the overall strain rate is thus given by

$$\gamma = \frac{b}{L_2}\frac{v}{L_1}\rho L_1 L_2 = \rho b v. \tag{6.6}$$

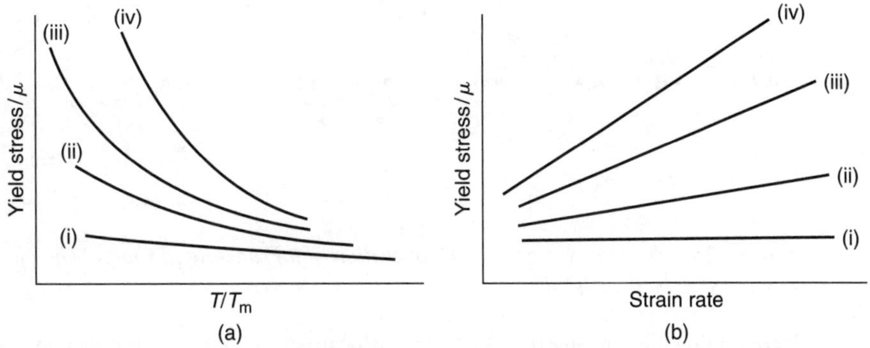

Figure 6.17 *Variation of yield stress with: (a) temperature and (b) strain rate for crystals with (i) fcc, (ii) bcc, (iii) ionic-bonded and (iv) covalent-bonded structure.*

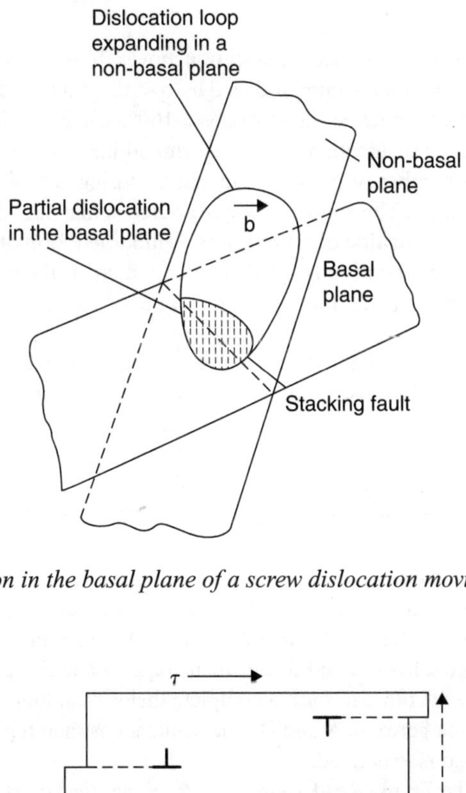

Figure 6.18 *Dissociation in the basal plane of a screw dislocation moving on a non-basal glide plane.*

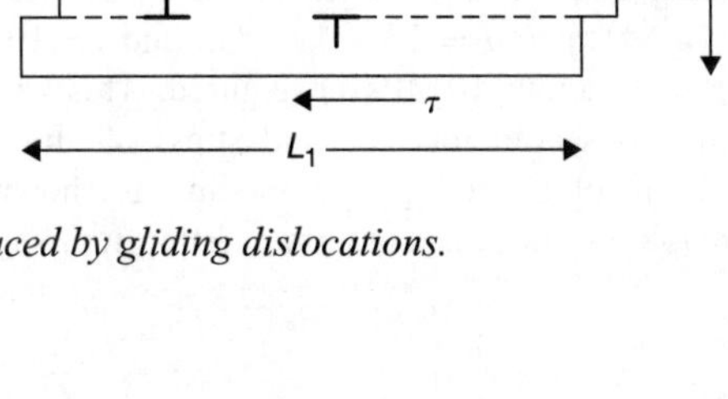

Figure 6.19 *Shear produced by gliding dislocations.*

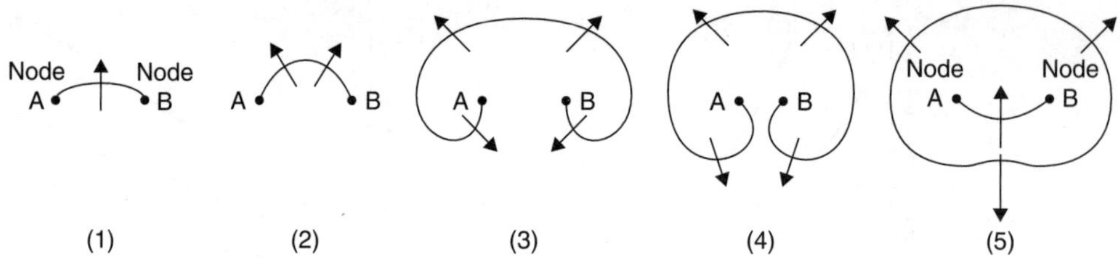

Figure 6.20 *Successive stages in the operation of a Frank–Read source. The plane of the paper is assumed to be the slip plane.*

At conventional strain rates (e.g. 1 s^{-1}) the dislocations would be moving at quite moderate speeds of a few cm s^{-1} if the mobile density $\approx 10^7$ cm^{-2}. During high-speed deformation the velocity approaches the limiting velocity. The shear strain produced by these dislocations is given by

$$\gamma = \rho b\bar{x}, \tag{6.7}$$

where $\bar{x}$ is the average distance a dislocation moves. If the distance $x \simeq 10^{-4}$ cm (the size of an average sub-grain) the maximum strain produced by $\rho \approx 10^7$ is about $(10^7 \times 3 \times 10^{-8} \times 10^{-4})$, which is only a fraction of 1%. In practice, shear strains >100% can be achieved, and hence to produce these large strains many more dislocations than the original ingrown dislocations are required. To account for the increase in number of mobile dislocations during straining the concept of a dislocation source has been introduced. The simplest type of source is that due to Frank and Read, and accounts for the regenerative multiplication of dislocations. A modified form of the Frank–Read source is the multiple cross-glide source, first proposed by Koehler, which, as the name implies, depends on the cross-slip of screw dislocations and is therefore more common in metals of intermediate and high stacking-fault energy.

Figure 6.20 shows a Frank–Read source consisting of a dislocation line fixed at the nodes A and B (fixed, for example, because the other dislocations that join the nodes do not lie in slip planes). Because of its high elastic energy (≈4 eV per atom plane threaded by a dislocation) the dislocation possesses a line tension tending to make it shorten its length as much as possible (position 1, Figure 6.20). This line tension T is roughly equal to $\alpha\mu b^2$, where μ is the shear modulus, b the Burgers vector and α a constant usually taken to be about $\frac{1}{2}$. Under an applied stress the dislocation line will bow out, decreasing its radius of curvature until it reaches an equilibrium position in which the line tension balances the force due to the applied stress. Increasing the applied stress causes the line to decrease its radius of curvature further until it becomes semicircular (position 2). Beyond this point it has no equilibrium position so it will expand rapidly, rotating about the nodes and taking up the succession of forms indicated by 3, 4 and 5. Between stages 4 and 5 the two parts of the loop below AB meet and annihilate each other to form a complete dislocation loop, which expands into the slip plane and a new line source between A and B. The sequence is then repeated and one unit of slip is produced by each loop that is generated.

To operate the Frank–Read source the force applied must be sufficient to overcome the restoring force on the dislocation line due to its line tension. Referring to Figure 6.21, this would be $2T\mathrm{d}\theta/2 > \tau bl\mathrm{d}\theta/2$, and if $T \sim \mu b^2/2$ the stress to do this is about $\mu b/l$, where μ and b have their usual meaning and l is the length of the Frank–Read source; the substitution of typical values ($\mu = 4 \times 10^{10}$ N m^{-2}, $b = 2.5 \times 10^{-10}$ m and $l = 10^{-6}$ m) into this estimate shows that a critical shear stress of about 10 MPa is required. This value is somewhat less than, but of the same order as, that observed for the yield stress of virgin pure metal single crystals. Another source mechanism involves multiple cross-slip, as shown in Figure 6.22. It depends on the Frank–Read principle but does not require a dislocation segment to be anchored by nodes. Thus,

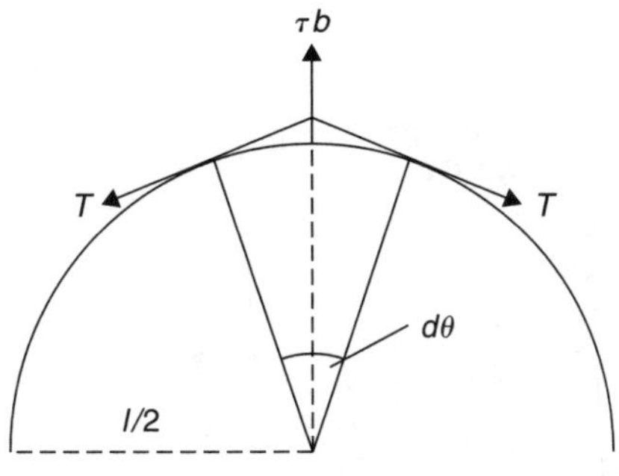

Figure 6.21 *Geometry of Frank–Read source used to calculate the stress to operate.*

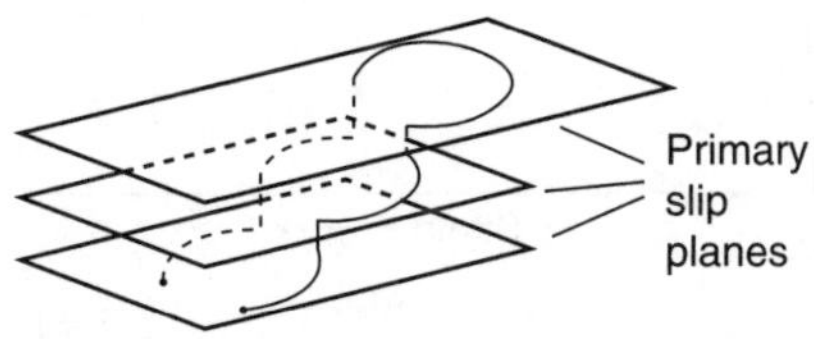

Figure 6.22 *Cross-slip multiplication source.*

if part of a moving screw dislocation undergoes double cross-slip the two pieces of edge dislocation on the cross-slip plane effectively act as anchoring points for a new source. The loop expanding on the slip plane parallel to the original plane may operate as a Frank–Read source, and any loops produced may in turn cross slip and become a source. This process therefore not only increases the number of dislocations on the original slip plane, but also causes the slip band to widen.

The concept of the dislocation source accounts for the observation of slip bands on the surface of deformed metals. The amount of slip produced by the passage of a single dislocation is too small to be observable as a slip line or band under the light microscope. To be resolved it must be at least 300 nm in height and hence $\approx$1000 dislocations must have operated in a given slip band. Moreover, in general, the slip band has considerable width, which tends to support the operation of the cross-glide source as the predominant mechanism of dislocation multiplication during straining.

Worked example

A Frank–Read source is operated by an applied stress of magnitude $10^{-4}\mu$, where μ is the shear modulus. If the limiting speed of a dislocation is 10^3 m s^{-1}, show that the source could nucleate a slip band which is observed in the light microscope to form in about 10^{-6} s.

Solution

To observe a slip band in the light microscope, which has a resolution of approximately the wavelength of light, requires about 1000 dislocations to have emanated from the source.

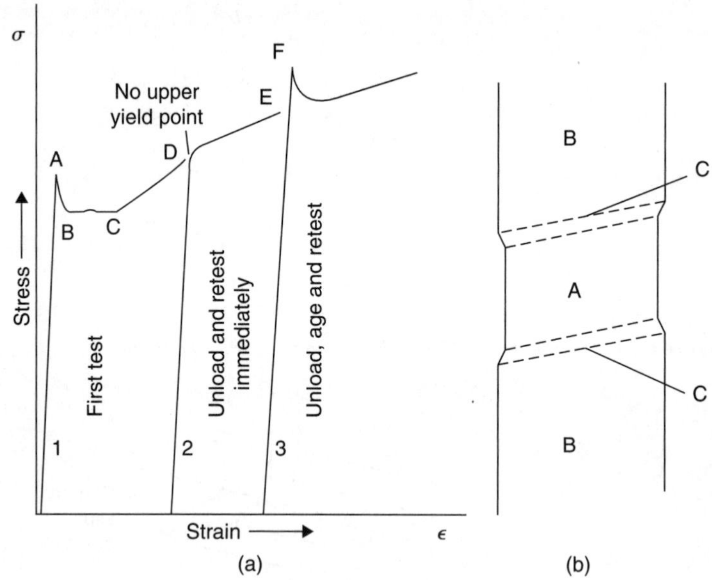

Figure 6.23 *Schematic representation of strain ageing (a) and Lüders band formation (b).*

The stress to operate the source is $\tau = \mu b/\ell$, where μ is the shear modulus, b the Burgers vector and ℓ the source length. Time to produce one dislocation loop takes $t \sim \ell/v$, where v is the dislocation velocity. Thus, to nucleate a slip band of 1000 dislocations requires total time

$$t = 10^3\ell/v = \frac{10^3\mu b}{\tau v} = \frac{10^3\mu \times 2.5 \times 10^{-10}}{10^{-4}\mu \times 10^3} = 2.5 \times 10^{-6}\ \text{s}.$$

6.4.4 Discontinuous yielding

In some materials the onset of macroscopic plastic flow begins in an abrupt manner with a yield drop in which the applied stress falls, during yielding, from an upper to a lower yield point. Such yield behavior is commonly found in iron containing small amounts of carbon or nitrogen as impurity. The main characteristics of the yield phenomenon in iron may be summarized as follows.

6.4.4.1 Yield point

A specimen of iron during tensile deformation (Figure 6.23a, curve 1) behaves elastically up to a certain high load A, known as the upper yield point, and then it suddenly yields plastically. The important feature to note from this curve is that the stress required to maintain plastic flow immediately after yielding has started is lower than that required to start it, as shown by the fall in load from A to B (the lower yield point). A yield point elongation to C then occurs, after which the specimen work hardens and the curve rises steadily and smoothly.

6.4.4.2 Overstraining

The yield point can be removed temporarily by applying a small preliminary plastic strain to the specimen. Thus, if after reaching the point D, for example, the specimen is unloaded and a second test is made fairly soon afterwards, a stress–strain curve of type 2 will be obtained. The specimen deforms elastically up to the unloading point D, and the absence of a yield point at the beginning of plastic flow is characteristic of a specimen in an overstrained condition.

6.4.4.3 Strain-age hardening

If a specimen which has been overstrained to remove the yield point is allowed to rest, or age, before retesting, the yield point returns, as shown in Figure 6.23a, curve 3. This process, which is accompanied by hardening (as shown by the increased stress, EF, to initiate yielding) is known as strain ageing or, more specifically, strain-age hardening. In iron, strain ageing is slow at room temperature but is greatly speeded up by annealing at a higher temperature. Thus, a strong yield point returns after an ageing treatment of only a few seconds at 200°C, but the same yield point will take many hours to develop if ageing is carried out at room temperature.

6.4.4.4 Lüders band formation

Closely related to the yield point is the formation of Lüders bands. These bands are markings on the surface of the specimen which distinguish those parts of the specimen that have yielded, A, from those which have not, B. Arrival at the upper yield point is indicated by the formation of one or more of these bands and, as the specimen passes through the stage of the yield point elongation, these bands spread along the specimen and coalesce until the entire gauge length has been covered. At this stage the whole of the material within the gauge length has been overstrained, and the yield point elongation is complete. The growth of a Lüders band is shown diagrammatically in Figure 6.23b. It should be noted that the band is a macroscopic band crossing all the grains in the cross-section of a polycrystalline specimen, and thus the edges of the band are not necessarily the traces of individual slip planes. A second point to observe is that the rate of plastic flow in the edges of a band can be very high, even in an apparently slow test; this is because the zones, marked C in Figure 6.23b, are very narrow compared with the gauge length.

These Lüders bands frequently occur in drawing and stamping operations, when the surface markings in relief are called stretcher strains. These markings are unsightly in appearance and have to be avoided on many finished products. The remedy consists of overstraining the sheet prior to pressing operations, by means of a temper roll, or roller leveling, pass so that the yield phenomenon is eliminated. It is essential, once this operation has been performed, to carry out pressing before the sheet has time to strain-age; the use of a 'non-ageing' steel is an alternative remedy.

These yielding effects are influenced by the presence of small amounts of carbon or nitrogen atoms interacting with dislocations. The yield point can be removed by annealing at 700°C in a wet hydrogen atmosphere, and cannot subsequently be restored by any strain-ageing treatment. Conversely, exposing the decarburized specimen to an atmosphere of dry hydrogen containing a trace of hydrocarbon at 700°C for as little as one minute restores the yield point. The carbon and nitrogen atoms can also be removed from solution in other ways – for example, by adding to the iron such elements as molybdenum, manganese, chromium, vanadium, niobium or titanium, which have a strong affinity for forming carbides or nitrides in steels. For this reason, these elements are particularly effective in removing the yield point and producing a non-strain-ageing steel.

The carbon/nitrogen atoms are important in the yielding process because they interact with the dislocations and immobilize them. This locking of the dislocations is brought about because the strain energy due to the distortion of a solute atom can be relieved if it fits into a structural region where the local lattice parameter approximates to that of the natural lattice parameter of the solute. Such a condition will be brought about by the segregation of solute atoms to the dislocations, with large substitutional atoms taking up lattice positions in the expanded region and small ones in the compressed region; small interstitial atoms will tend to segregate to interstitial sites below the half-plane. Thus, where both dislocations and solute atoms are present in the lattice, interactions of the stress field can occur, resulting in a lowering of the strain energy of the system. This provides a driving force tending to attract solute atoms to dislocations and if the necessary time for diffusion is allowed, a solute atom 'atmosphere' will form around each dislocation.

When a stress is applied to a specimen in which the dislocations are locked by carbon atoms the dislocations are not able to move at the stress level at which free dislocations are normally mobile. With increasing stress, yielding occurs when dislocations suddenly become mobile, either by breaking away from the carbon atmosphere or by nucleating fresh dislocations at stress concentrations. At this high stress level the mobile dislocation density increases rapidly. The lower yield stress is then the stress at which free dislocations continue to move and produce plastic flow. The overstrained condition corresponds to the situation where the mobile dislocations, brought to rest by unloading the specimen, are set in motion again by reloading before the carbon atmospheres have time to develop by diffusion. If, however, time is allowed for diffusion to take place, new atmospheres can re-form and immobilize the dislocations again. This is the strain-aged condition when the original yield characteristics reappear.

The upper yield point in conventional experiments on polycrystalline materials is the stress at which initially yielded zones trigger yield in adjacent grains. As more and more grains are triggered, the yield zones spread across the specimen and form a Lüders band.

The propagation of yield is thought to occur when a dislocation source operates and releases an avalanche of dislocations into its slip plane, which eventually pile up at a grain boundary or other obstacle. The stress concentration at the head of the pile-up acts with the applied stress on the dislocations of the next grain and operates the nearest source, so that the process is repeated in the next grain. The applied shear stress σ_y at which yielding propagates is given by

$$\sigma_y = \sigma_i + (\sigma_c r^{1/2})d^{-1/2}, \tag{6.8}$$

where r is the distance from the pile-up to the nearest source, $2d$ is the grain diameter and σ_c is the stress required to operate a source which involves unpinning a dislocation τ_c at that temperature. Equation (6.8) reduces to the Hall–Petch equation $\sigma_y = \sigma_i + k_y d^{-1/2}$, where σ_i is the 'friction' stress term and k_y the grain size dependence parameter ($=m^2 \tau_c r^{1/2}$) discussed in Section 6.4.11.

6.4.5 Yield points and crystal structure

The characteristic feature of discontinuous yielding is that at the yield point the specimen goes from a condition where the availability of mobile dislocations is limited to one where they are in abundance, the increase in mobile density largely arising from dislocation multiplication at the high stress level. A further feature is that not all the dislocations have to be immobilized to observe a yield drop. Indeed, this is not usually possible because specimen handling, non-axial loading, scratches, etc. give rise to stress concentrations that provide a small local density of mobile dislocations (i.e. pre-yield microstrain).

For materials with a high Peierls–Nabarro (P–N) stress, yield drops may be observed even when they possess a significant mobile dislocation density. A common example is that observed in silicon; this is an extremely pure material with no impurities to lock dislocations, but usually the dislocation density is quite modest (10^7 m m^{-3}) and possesses a high P–N stress.

When these materials are pulled in a tensile test, the overall strain rate $\dot{\gamma}$ imposed on the specimen by the machine has to be matched by the motion of dislocations according to the relation $\dot{\gamma} = \rho b v$.

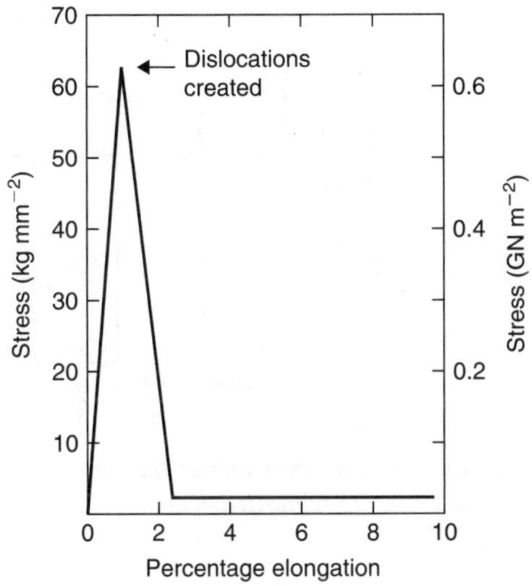

Figure 6.24 *Yield point in a copper whisker.*

However, because ρ is small the individual dislocations are forced to move at a high speed v, which is only attained at a high stress level (the upper yield stress) because of the large P–N stress. As the dislocations glide at these high speeds, rapid multiplication occurs and the mobile dislocation density increases rapidly. Because of the increased value of the term ρ, a lower average velocity of dislocations is then required to maintain a constant strain rate, which means a lower glide stress. The stress that can be supported by the specimen thus drops during initial yielding to the lower yield point, and does not rise again until the dislocation–dislocation interactions caused by the increased ρ produce a significant work hardening.

In the fcc metals, the P–N stress is quite small and the stress to move a dislocation is almost independent of velocity up to high speeds. If such metals are to show a yield point, the density of mobile dislocations must be reduced virtually to zero. This can be achieved as shown in Figure 6.24 by the tensile testing of whisker crystals which are very perfect. Yielding begins at the stress required to create dislocations in the perfect lattice, and the upper yield stress approaches the theoretical yield strength. Following multiplication, the stress for glide of these dislocations is several orders of magnitude lower.

Bcc transition metals such as iron are intermediate in their plastic behavior between the fcc metals and diamond cubic Si and Ge. Because of the significant P–N stress these bcc metals are capable of exhibiting a sharp yield point, even when the initial mobile dislocation density is not zero, as shown by the calculated curves of Figure 6.25. However, in practice, the dislocation density of well-annealed pure metals is about 10^{10} m m^{-3} and too high for any significant yield drop without an element of dislocation locking by carbon atoms.

It is evident that discontinuous yielding can be produced in all the common metal structures provided the appropriate solute elements are present and correct testing procedure adopted. The effect is particularly strong in the bcc metals and has been observed in α-iron, molybdenum, niobium, vanadium and β-brass, each containing a strongly interacting interstitial solute element. The hexagonal metals (e.g. cadmium and zinc) can also show the phenomenon provided interstitial nitrogen atoms are added. The copper- and aluminum-based fcc alloys also exhibit yielding behavior, but often to a lesser degree. In this case it is substitutional atoms (e.g. zinc in α-brass and copper in aluminum alloys) which are responsible for the phenomenon (see Section 6.4.7).

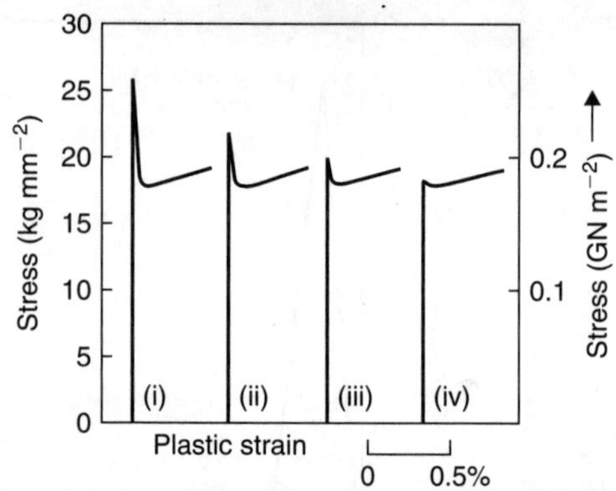

Figure 6.25 *Calculated stress–strain curves showing influence of initial dislocation density on the yield drop in iron for* n = 35 *with: (i)* 10^1 cm^{-2}*, (ii)* 10^3 cm^{-2}*, (iii)* 10^5 cm^{-2} *and (iv)* 10^7 cm^{-2} *(after Hahn, 1962).*

Worked example

A low-carbon steel exhibits a yield point when tensile tested at a strain rate of $1\,s^{-1}$. If the density of mobile dislocations before and after the yield phenomenon is 10^{11} and $10^{14}\,m^{-2}$ respectively, estimate:

(i) The dislocation velocity at the upper yield point, and
(ii) The magnitude of the yield drop.

(Take the lattice parameter of the alloy to be 0.28 nm and the stress dependence of the dislocation velocity to have an exponent $n = 35$.)

Solution

(i) Plastic strain rate is $\dot{\varepsilon} = \phi b \rho_m \bar{v}$. Thus, at the upper yield point (uyp), $1 = 0.5 \times \frac{\sqrt{3}}{2} \times 2.8 \times 10^{-10} \times 10^{11} \times \bar{v}$, since $b = (\sqrt{3}/2)a$ and $\phi \sim 0.5$.

$$\therefore \bar{v} = 8.06 \times 10^{-2}\,\text{m s}^{-1}.$$

(ii) The average velocity $\bar{v} = (\sigma/\sigma_o)^n$. Since $\dot{\varepsilon}$ is the same at both upper and lower yield points, then

$$0.5 \times 10^{11} \times (\sigma_{uyp}/\sigma_o)^{35} = 0.5 \times 10^{14} \times (\sigma_{lyp}/\sigma_o)^{35}$$

$$(\sigma_{uyp}/\sigma_{lyp})^{35} = 10^3.$$

The ratio of upper to lower yield points, $\sigma_{uyp}/\sigma_{lyp} = 10^{3/35} = 1.2$.

6.4.6 Discontinuous yielding in ordered alloys

Discontinuous yield points have been observed in a wide variety of A_3B-type alloys. Figure 6.26 shows the development of the yield point in Ni_3Fe on ageing. The addition of Al speeds up the kinetics of

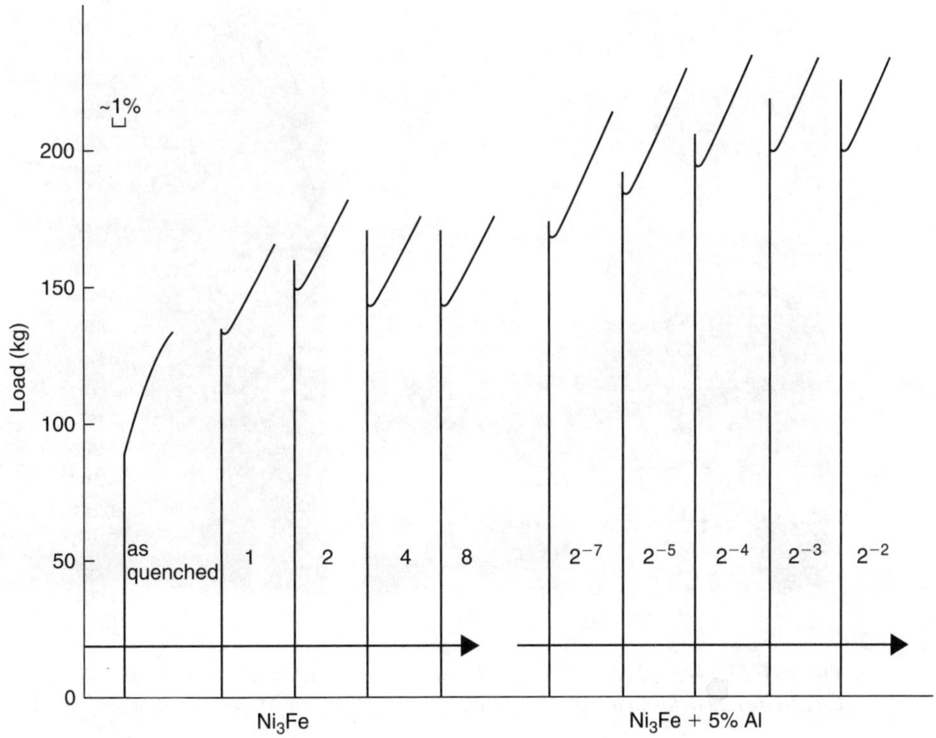

Figure 6.26 *Development of a yield point with ageing at 490°C for the times in days indicated. (a) Ni_3Fe. (b) $Ni_3Fe + 5\%$ Al. The tests are at room temperature.*

ordering and therefore the onset of the yield point. Ordered materials deform by superdislocation motion and the link between yield points and superdislocations is confirmed by the observation that, in Cu_3Au, for example, a transition from groups of single dislocations to more randomly arranged superdislocation pairs takes place at $\sim S = 0.7$ (see Chapter 3), and this coincides with the onset of a large yield drop and rapid rise in work hardening.

Sharp yielding may be explained by at least two mechanisms, namely (1) cross-slip of the superdislocation onto the cube plane to lower the APB energy effectively pinning it and (2) dislocation locking by rearrangement of the APB on ageing. The shear APB between a pair of superdislocations is likely to be energetically unstable, since there are many like bonds across the interface and thermal activation will modify this sharp interface by atomic rearrangement. This APB-locking model will give rise to sharp yielding because the energy required by the lead dislocation in creating sharp APB is greater than that released by the trailing dislocation initially moving across diffuse APB. Experimental evidence favors the APB model and weak-beam electron microscopy (see Figure 6.27) shows that the superdislocation separation for a shear APB corresponds to an energy of $48 \pm 5\,mJ\,m^{-2}$, whereas a larger dislocation separation corresponding to an APB energy of $25 \pm 3\,mJ\,m^{-2}$ was observed for a strained and aged Cu_3Au.

6.4.7 Solute–dislocation interaction

Iron containing carbon or nitrogen shows very marked yield point effects and there is a strong elastic interaction between these solute atoms and the dislocations. The solute atoms occupy interstitial

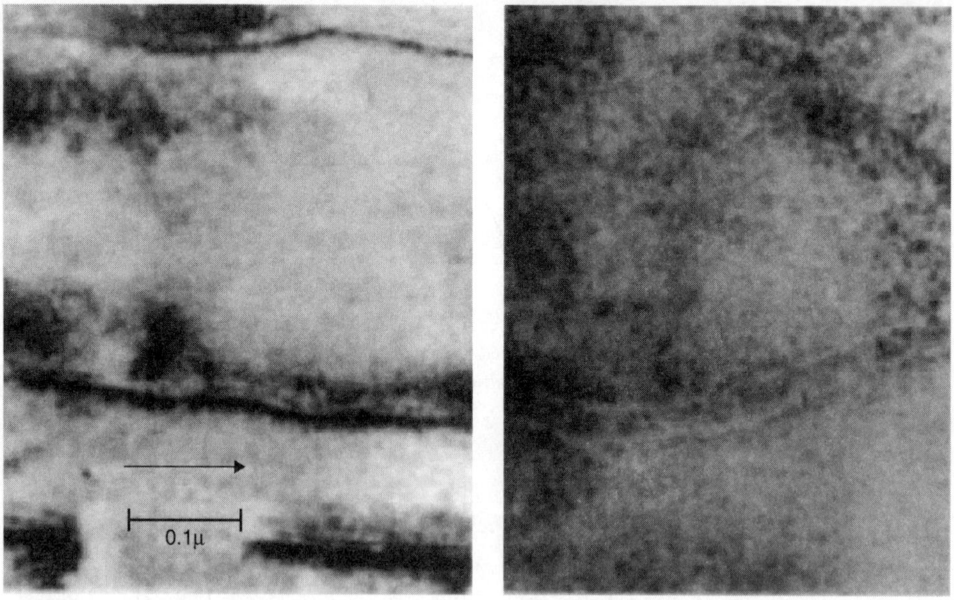

Figure 6.27 *Weak-beam micrographs showing separation of superdislocation partials in Cu_3Au. (a) As deformed. (b) After ageing at 225°C (after Morris and Smallman, 1975).*

sites in the lattice and produce large tetragonal distortions, as well as large-volume expansions. Consequently, they can interact with both shear and hydrostatic stresses and can lock screw as well as edge dislocations. Strong yielding behavior is also expected in other bcc metals, provided they contain interstitial solute elements. On the other hand, in the case of fcc metals the arrangement of lattice positions around either interstitial or substitutional sites is too symmetrical to allow a solute atom to produce an asymmetrical distortion, and the atmosphere locking of screw dislocations, which requires a shear stress interaction, would appear to be impossible. Then by this argument, since the screw dislocations are not locked, a drop in stress at the yield point should not be observed. Nevertheless, yield points are observed in fcc materials and one reason for this is that unit dislocations in fcc metals dissociate into pairs of partial dislocations which are elastically coupled by a stacking fault. Moreover, since their Burgers vectors intersect at 120° there is no orientation of the line of the pair for which both can be pure screws. At least one of them must have a substantial edge component, and a locking of this edge component by hydrostatic interactions should cause a locking of the pair, although it will undoubtedly be weaker.

In its quantitative form the theory of solute atom locking has been applied to the formation of an atmosphere around an edge dislocation due to hydrostatic interaction. Since hydrostatic stresses are scalar quantities, no knowledge is required in this case of the orientation of the dislocation with respect to the interacting solute atom, but it is necessary in calculating shear stresses interactions.[4] Cottrell and Bilby have shown that if the introduction of a solute atom causes a volume

[4] To a first approximation a solute atom does not interact with a screw dislocation, since there is no dilatation around the screw; a second-order dilatation exists, however, which gives rise to a non-zero interaction falling off with distance from the dislocation according to $1/r^2$. In real crystals, anisotropic elasticity will lead to first-order size effects, even with screw dislocations, and hence a substantial interaction is to be expected.

change Δv at some point in the lattice where the hydrostatic pressure of the stress field is p, the interaction energy is

$$V = p\Delta v = K\Theta\Delta v, \tag{6.9}$$

where K is the bulk modulus and Θ is the local dilatation strain. The dilatation strain at a point (R, θ) from a positive edge dislocation is $b(1-2\nu) \times \sin\theta/2\pi R(1-\nu)$, and substituting $K = 2\mu(1+\nu)/3(1-2\nu)$, where μ is the shear modulus and ν Poisson's ratio, we get the expression:

$$\begin{aligned} V_{(R,\theta)} &= b(1+\nu)\mu\Delta v \sin\theta/3\pi R(1-\nu) \\ &= A\sin\theta/R. \end{aligned} \tag{6.10}$$

This is the interaction energy at a point whose polar coordinates with respect to the center of the dislocation are R and θ. We note that V is positive on the upper side $(0 < \theta < \pi)$ of the dislocation for a large atom $(\Delta v > 0)$ and negative on the lower side, which agrees with the qualitative picture of a large atom being repelled from the compressed region and attracted into the expanded one.

It is expected that the site for the strongest binding energy V_{max} will be at a point $\theta = 3\pi/2$, $R = r_0 \approx b$; and using known values of μ, ν and Δv in equation (6.10), we obtain $A \approx 3 \times 10^{-29}\,\mathrm{N\,m^2}$ and $V_{max} \approx 1\,\mathrm{eV}$ for carbon or nitrogen in α-iron. This value is almost certainly too high because of the limitations of the interaction energy equation in describing conditions near the center of a dislocation, and a more realistic value obtained from experiment (e.g. internal friction experiments) is $V_{max} \approx \frac{1}{2}$ to $\frac{3}{4}$ eV. For a substitutional solute atom such as zinc in copper, Δv is not only smaller but also easier to calculate from lattice parameter measurements. Thus, if r and $r(1+\varepsilon)$ are the atomic radii of the solvent and solute respectively, where ε is the misfit value, the volume change Δv is $4\pi r^3\,\varepsilon$ and equation (6.10) becomes

$$\begin{aligned} V &= 4(1+\nu)\mu b\varepsilon r^3 \sin\theta/3(1-\nu)R \\ &= A\sin\theta/R. \end{aligned} \tag{6.11}$$

Taking the known values $\mu = 40\,\mathrm{GN\,m^{-2}}$, $\nu = 0.36$, $b = 2.55 \times 10^{-10}\,\mathrm{m}$, r_0 and $\varepsilon = 0.06$, we find $A \approx 5 \times 10^{-30}\,\mathrm{N\,m^2}$, which gives a much lower binding energy, $V_{max} = \frac{1}{8}$ eV.

The yield phenomenon is particularly strong in iron because an additional effect is important; this concerns the type of atmosphere a dislocation gathers round itself, which can be either condensed or dilute. During the strain-ageing process, migration of the solute atoms to the dislocation occurs and two important cases arise. First, if all the sites at the center of the dislocation become occupied the atmosphere is then said to be condensed; each atom plane threaded by the dislocation contains one solute atom at the position of maximum binding, together with a diffuse cloud of other solute atoms further out. If, on the other hand, equilibrium is established before all the sites at the center are saturated, a steady state must be reached in which the probability of solute atoms leaving the center can equal the probability of their entering it. The steady-state distribution of solute atoms around the dislocations is then given by the relation

$$C_{(R,\theta)} = c_0 \exp\left[V_{(R,\theta)}/\mathbf{k}T\right],$$

where c_0 is the concentration far from a dislocation, $\mathbf{k}$ is Boltzmann's constant, T is the absolute temperature and C the local impurity concentration at a point near the dislocation where the binding energy is V. This is known as the dilute or Maxwellian atmosphere. Clearly, the form of an atmosphere will be governed by the concentration of solute atoms at the sites of maximum binding energy V_{max}, and for a given alloy (i.e. c_0 and V_{max} fixed) this concentration will be

$$C_{V_{max}} = c_0 \exp(V_{max}/\mathbf{k}T) \tag{6.12}$$

as long as $C_{V_{max}}$ is less than unity. The value of $C_{V_{max}}$ depends only on the temperature, and as the temperature is lowered $C_{V_{max}}$ will eventually rise to unity. By definition the atmosphere will then have passed from a dilute to a condensed state. The temperature at which this occurs is known as the condensation temperature T_c, and can be obtained by substituting the value $C_{V_{max}} = 1$ in equation (6.12) when

$$T_c = V_{max}/\mathbf{k}\ln(1/c_0). \tag{6.13}$$

Substituting the value of V_{max} for iron, i.e. $\frac{1}{2}$ eV, in this equation we find that only a very small concentration of carbon or nitrogen is necessary to give a condensed atmosphere at room temperature, and with the usual concentration strong yielding behavior is expected up to temperatures of about 400°C.

In the fcc structure, although the locking between a solute atom and a dislocation is likely to be weaker, condensed atmospheres are still possible if this weakness can be compensated for by sufficiently increasing the concentration of the solution. This may be why examples of yielding in fcc materials have been mainly obtained from alloys. Solid solution alloys of aluminum usually contain less than 0.1 at.% of solute element, and these show yielding in single crystals only at low temperature (e.g. liquid nitrogen temperature, −196°C), whereas supersaturated alloys show evidence of strong yielding even in polycrystals at room temperature; copper dissolved in aluminum has a misfit value $\varepsilon \approx 0.12$, which corresponds to $V_{max} = \frac{1}{4}$ eV, and from equation (6.13) it can be shown that a 0.1 at.% alloy has a condensation temperature $T_c = 250$ K. Copper-based alloys, on the other hand, usually form extensive solid solutions and, consequently, concentrated alloys may exhibit strong yielding phenomena.

The best-known example is α-brass and, because $V_{max} \approx \frac{1}{8}$ eV, a dilute alloy containing 1 at.% zinc has a condensation temperature $T_c \approx 300$ K. At low zinc concentrations (1–10%) the yield point in brass is probably solely due to the segregation of zinc atoms to dislocations. At higher concentrations, however, it may also be due to short-range order.

6.4.8 Dislocation locking and temperature

The binding of a solute atom to a dislocation is short range in nature, and is effective only over an atomic distance or so (Figure 6.28). Moreover, the dislocation line is flexible and this enables yielding to begin by throwing forward a small length of dislocation line, only a few atomic spacings long, beyond the position marked x_2. The applied stress then separates the rest of the dislocation line from its anchorage by pulling the sides of this loop outward along the dislocation line, i.e. by double kink movement. Such a breakaway process would lead to a yield stress which depends sensitively on temperature, as shown in Figure 6.29a. It is observed, however, that k_y, the grain-size dependence parameter in the Hall–Petch equation, in most annealed bcc metals is almost independent of temperature down to the range (<100 K) where twinning occurs, and that practically all the large temperature dependence is due to σ_i (see Figure 6.29b). To explain this observation it is argued that when locked dislocations exist initially in the material, yielding starts by unpinning them if they are weakly locked (this corresponds to the condition envisaged by Cottrell–Bilby), but if they are strongly locked it starts instead by the creation of new dislocations at points of stress concentration. This is an athermal process and thus k_y is almost independent of temperature. Because of the rapid diffusion

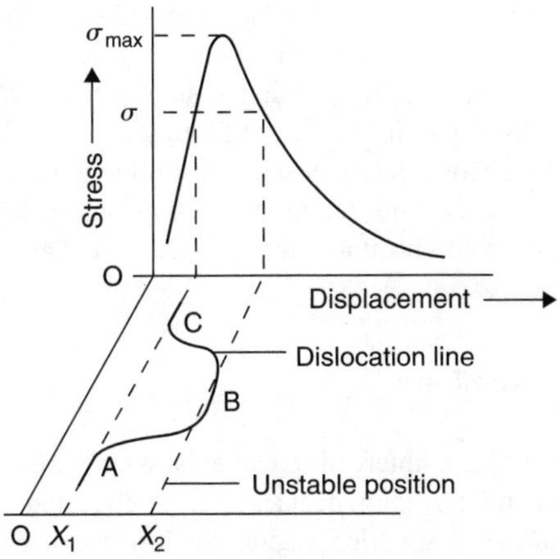

Figure 6.28 *Stress–displacement curve for the breakaway of a dislocation from its atmosphere (after Cottrell, 1957; courtesy of the Institution of Mechanical Engineers).*

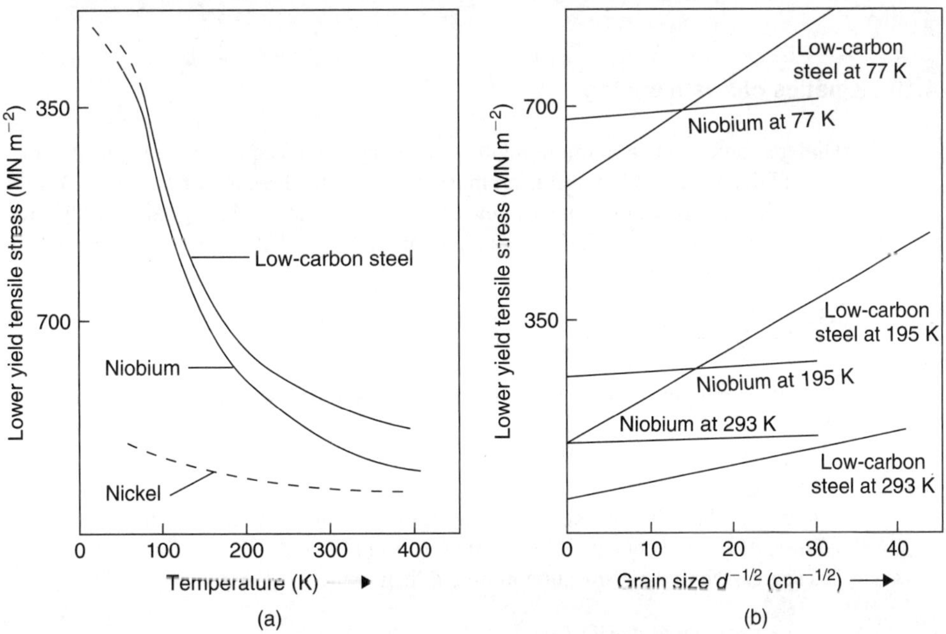

Figure 6.29 *Variation of lower yield stress with temperature (a) and grain size (b), for low-carbon steel and niobium; the curve for nickel is shown in (a) for comparison (after Adams, Roberts and Smallman, 1960; Hull and Mogford, 1958).*

of interstitial elements the conventional annealing and normalizing treatments should commonly produce strong locking. In support of this theory, it is observed that k_y is dependent on temperature in the very early stages of ageing following either straining or quenching, but on subsequent ageing k_y becomes temperature independent. The interpretation of k_y therefore depends on the degree of ageing.

Direct observations of crystals that have yielded show that the majority of the strongly anchored dislocations remain locked and do not participate in the yielding phenomenon. Thus, large numbers of dislocations are generated during yielding by some other mechanism than breaking away from Cottrell atmospheres, and the rapid dislocation multiplication, which can take place at the high stress levels, is now considered the most likely possibility. Prolonged ageing tends to produce coarse precipitates along the dislocation line and unpinning by bowing out between them should easily occur before grain boundary creation. This unpinning process would also give k_y independent of temperature.

6.4.9 Inhomogeneity interaction

A different type of elastic interaction can exist which arises from the different elastic properties of the solvent matrix and the region near a solute. Such an inhomogeneity interaction has been analyzed for both a rigid and a soft spherical region; the former corresponds to a relatively hard impurity atom and the latter to a vacant lattice site. The results indicate that the interaction energy is of the form B/r^2, where B is a constant involving elastic constants and atomic size. It is generally believed that the inhomogeneity effect is small for solute–dislocation interactions but dominates the size effect for vacancy–dislocation interaction. The kinetics of ageing support this conclusion.

6.4.10 Kinetics of strain ageing

Under a force F an atom migrating by thermal agitation acquires a steady drift velocity $v = DF/\mathbf{k}T$ (in addition to its random diffusion movements) in the direction of F, where D is the coefficient of diffusion. The force attracting a solute atom to a dislocation is the gradient of the interaction energy $\mathrm{d}V/\mathrm{d}r$ and hence $v = (D/\mathbf{k}T)(A/r^2)$. Thus, atoms originally at a distance r from the dislocation reach it in a time given approximately by

$$t = r/v = r^3\mathbf{k}T/AD.$$

After this time t the number of atoms to reach unit length of dislocation is

$$n(t) = \pi r^2 c_0 = \pi c_0[(AD/\mathbf{k}T)t]^{2/3},$$

where c_0 is the solute concentration in uniform solution in terms of the number of atoms per unit volume. If ρ is the density of dislocations (cm cm^{-3}) and f the fraction of the original solute which has segregated to the dislocation in time t, then

$$f = \pi\rho[(AD/\mathbf{k}T)t]^{2/3}. \tag{6.14}$$

This expression is valid for the early stages of ageing, and may be modified to fit the later stages by allowing for the reduction in the matrix concentration as ageing proceeds, such that the rate of flow is proportional to the amount left in the matrix,

$$\mathrm{d}f/\mathrm{d}t = \pi\rho(AD/\mathbf{k}T)^{2/3}(2/3)t^{-1/3}(1-f),$$

which when integrated gives

$$f = 1 - \exp\{-\pi\rho[(AD/\mathbf{k}T)t]^{2/3}\}. \tag{6.15}$$

This reduces to the simpler equation (6.14) when the exponent is small, and is found to be in good agreement with the process of segregation and precipitation on dislocations in several bcc metals. For carbon in α-Fe, Harper determined the fraction of solute atom still in solution using an internal friction technique and showed that log $(1-f)$ is proportional to $t^{2/3}$; the slope of the line is $\pi\rho(AD/\mathbf{k}T)$ and evaluation of this slope at a series of temperatures allows the activation energy for the process to be determined from an Arrhenius plot. The value obtained for α-iron is 84 kJ mol^{-1}, which is close to that for the diffusion of carbon in ferrite.

The inhomogeneity interaction is considered to be the dominant effect in vacancy–dislocation interactions, with $V=-B/r^2$, where B is a constant; this compares with the size effect for which $V=-A/r$ would be appropriate for the interstitial–dislocation interaction. It is convenient, however, to write the interaction energy in the general form $V=-A/r^n$ and, hence, following the treatment previously used for the kinetics of strain ageing, the radial velocity of a point defect towards the dislocation is

$$V=(D/\mathbf{k}T)(nA/r^{n+1}). \tag{6.16}$$

The number of a particular point defect species that reaches the dislocation in time t is

$$\begin{aligned} n(t) &= \pi r^2 c_0 \\ &= \pi c_0[ADn(n+2)/\mathbf{k}T]^{2/(n+2)}t^{2/(n+2)} \end{aligned} \tag{6.17}$$

and when $n=2$ then $n(t)\propto t^{1/2}$, and when $n=1$, $n(t)\propto t^{2/3}$. Since the kinetics of ageing in quenched copper follow $t^{1/2}$ initially, the observations confirm the importance of the inhomogeneity interaction for vacancies.

6.4.11 Influence of grain boundaries on plasticity

It might be thought that when a stress is applied to a polycrystalline metal, every grain in the sample deforms as if it were an unconstrained single crystal. This is not the case, however, and the fact that the aggregate does not deform in this manner is indicated by the high yield stress of polycrystals compared with that of single crystals. This increased strength of polycrystals immediately poses the question – is the hardness of a grain caused by the presence of the grain boundary or by the orientation difference of the neighboring grains? It is now believed that the latter is the case but that the structure of the grain boundary itself may be of importance in special circumstances, such as when brittle films, due to bismuth in copper or cementite in steel, form around the grains or when the grains slip past each other along their boundaries during high-temperature creep. The importance of the orientation change across a grain boundary to the process of slip has been demonstrated by experiments on 'bamboo'-type specimens, i.e. where the grain boundaries are parallel to each other and all perpendicular to the axis of tension. Initially, deformation occurs by slip only in those grains most favorably oriented, but later spreads to all the other grains as those grains which are deformed first work harden. It is then found that each grain contains wedge-shaped areas near the grain boundary, as shown in Figure 6.30a, where slip does not operate, which indicates that the continuance of slip from one grain to the next is difficult. From these observations it is natural to enquire what happens in a completely polycrystalline metal where the slip planes must in all cases make contact with a grain boundary. It will be clear that the polycrystalline aggregate must be stronger because, unlike the deformation of bamboo-type samples, where it is not necessary to raise the stress sufficiently high to operate those slip planes which made contact with a grain boundary, all the slip planes within any grain of a polycrystalline aggregate make contact with a grain boundary, but nevertheless have to be operated. The importance

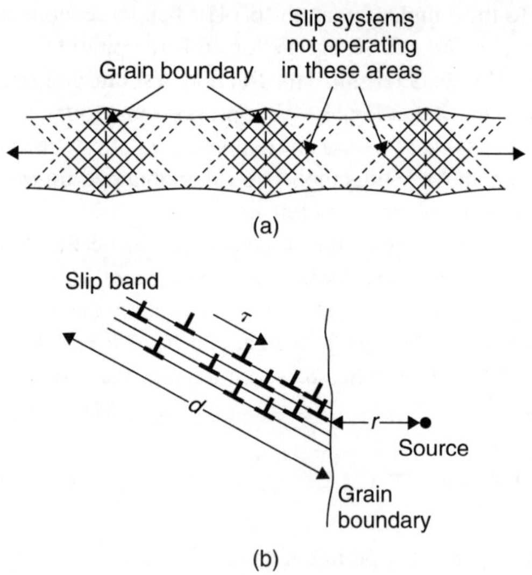

Figure 6.30 *(a) Grain-boundary blocking of slip. (b) Blocking of a slip band by a grain boundary.*

of the grain size on a strength is emphasized by Figure 6.29b, which shows the variation in lower yield stress, σ_y, with grain diameter, $2d$, for low-carbon steel. The smaller the grain size, the higher the yield strength according to a relation of the form

$$\sigma_y = \sigma_i + kd^{-1/2}, \tag{6.18}$$

where σ_i is a lattice friction stress and k a constant usually denoted k_y to indicate yielding. Because of the difficulties experienced by a dislocation in moving from one grain to another, the process of slip in a polycrystalline aggregate does not spread to each grain by forcing a dislocation through the boundary. Instead, the slip band which is held up at the boundary gives rise to a stress concentration at the head of the pile-up group of dislocations which acts with the applied stress and is sufficient to trigger off sources in neighboring grains. If τ_i is the stress a slip band could sustain if there were no resistance to slip across the grain boundary, i.e. the friction stress, and τ the higher stress sustained by a slip band in a polycrystal, then $(\tau - \tau_i)$ represents the resistance offered by the boundary, which reaches a limiting value when slip is induced in the next grain. The influence of grain size can be explained if the length of the slip band is proportional to d, as shown in Figure 6.30(b). Thus, since the stress concentration a short distance r from the end of the slip band is proportional to $(d/4r)^{1/2}$, the maximum shear stress at a distance r ahead of a slip band carrying an applied stress τ in a polycrystal is given by $(\tau - \tau_i)[d/4r]^{1/2}$ and lies in the plane of the slip band. If this maximum stress has to reach a value τ_{max} to operate a new source at a distance r, then

$$(\tau - \tau_i)[d/4r]^{1/2} = \tau_{max}$$

or, rearranging,

$$\tau = \tau_i + (\tau_{max} 2r^{1/2})d^{-1/2},$$

which may be written as

$$\tau = \tau_i + k_s d^{-1/2}.$$

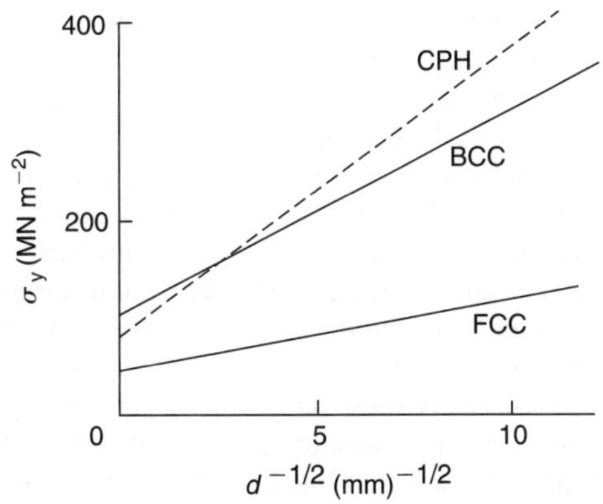

Figure 6.31 *Schematic diagram showing the grain-size dependence of the yield stress for crystals of different crystal structure.*

It then follows that the tensile flow curve of a polycrystal is given by

$$\sigma = m(\tau_i + k_s d^{-1/2}), \tag{6.19}$$

where m is the orientation factor relating the applied tensile stress σ to the shear stress, i.e. $\sigma = m\tau$. For a single crystal the m-factor has a minimum value of 2 as discussed, but in polycrystals deformation occurs in less favorably oriented grains and sometimes (e.g. hexagonal, intermetallics, etc.) on 'hard' systems, and so the m-factor is significantly higher. From equation (6.18) it can be seen that $\sigma_i = m\tau_i$ and $k = mk_s$.

While there is an orientation factor on a macroscopic scale in developing the critical shear stress within the various grains of a polycrystal, so there is a local orientation factor in operating a dislocation source ahead of a blocked slip band. The slip plane of the sources will not, in general, lie in the plane of maximum shear stress, and hence τ_{max} will need to be such that the shear stress, τ_c, required to operate the new source must be generated in the slip plane of the source. In general, the local orientation factor dealing with the orientation relationship of adjacent grains will differ from the macroscopic factor of slip plane orientation relative to the axis of stress, so that $\tau_{max} = \frac{1}{2}m'\tau_c$. For simplicity, however, it will be assumed $m' = m$ and hence the parameter k in the Petch equation is given by $k = m^2\tau_c r^{1/2}$.

It is clear from the above treatment that the parameter k depends essentially on two main factors. The first is the stress to operate a source dislocation, and this depends on the extent to which the dislocations are anchored or locked by impurity atoms. Strong locking implies a large τ_c and hence a large k; the converse is true for weak locking. The second factor is contained in the parameter m, which depends on the number of available slip systems. A multiplicity of slip systems enhances the possibility for plastic deformation and so implies a small k. A limited number of slip systems available would imply a large value of k. It then follows, as shown in Figure 6.31, that for (1) fcc metals, which have weakly locked dislocations and a multiplicity of slip systems, k will generally be small, i.e. there is only a small grain size dependence of the flow stress, for (2) cph metals, k will be large because of the limited slip systems, and for (3) bcc metals, because of the strong locking, k will be large.

Each grain does not deform as a single crystal in simple slip, since, if this were so, different grains would then deform in different directions with the result that voids would be created at the grain boundaries. Except in high-temperature creep, where the grains slide past each other along their boundaries, this does not happen and each grain deforms in coherence with its neighboring grains. However, the fact that the continuity of the metal is maintained during plastic deformation must mean that each grain is deformed into a shape that is dictated by the deformation of its neighbors. Such behavior will, of course, require the operation of several slip systems, and von Mises has shown that to allow this unrestricted change of shape of a grain requires at least five independent shear modes. The deformation of metal crystals with cubic structure easily satisfies this condition so that the polycrystals of these metals usually exhibit considerable ductility, and the stress–strain curve generally lies close to that of single crystals of extreme orientations deforming under multiple slip conditions. The hexagonal metals do, however, show striking differences between their single-crystal and polycrystalline behavior. This is because single crystals of these metals deform by a process of basal plane slip, but the three shear systems (two independent) which operate do not provide enough independent shear mechanisms to allow unrestricted changes of shape in polycrystals. Consequently, to prevent gaps opening up at grain boundaries during the deformation of polycrystals, some additional shear mechanisms, such as non-basal slip and mechanical twinning, must operate. Hence, because the resolved stress for non-basal slip and twinning is greater than that for basal-plane slip, yielding in a polycrystal is prevented until the applied stress is high enough to deform by these mechanisms.

6.4.12 Superplasticity

A number of materials, particularly two-phase eutectic or eutectoid alloys, have been observed to exhibit large elongations ($\approx$1000%) without fracture, and such behavior has been termed superplasticity. Several metallurgical factors have been put forward to explain superplastic behavior and it is now generally recognized that the effect can be produced in materials either (1) with a particular structural condition or (2) tested under special test conditions. The particular structural condition is that the material has a very fine grain size and the presence of a two-phase structure is usually of importance in maintaining this fine grain size during testing. Materials which exhibit superplastic behavior under special test conditions are those for which a phase boundary moves through the strained material during the test (e.g. during temperature cycling).

In general, the superplastic material exhibits a high strain-rate sensitivity. Thus, the plastic flow of a solid may be represented by the relation

$$\sigma = K\dot{\varepsilon}^m, \tag{6.20}$$

where σ is the stress, $\dot{\varepsilon}$ the strain rate and m an exponent generally known as the strain-rate sensitivity. When $m = 1$ the flow stress is directly proportional to strain rate and the material behaves as a Newtonian viscous fluid, such as hot glass. Superplastic materials are therefore characterized by high m-values, since this leads to increased stability against necking in a tensile test. Thus, for a tensile specimen length l with cross-sectional area A under an applied load P, then $\mathrm{d}l/l = -\mathrm{d}A/A$ and, introducing the time factor, we obtain

$$\dot{\varepsilon} = -(1/A)\mathrm{d}A/\mathrm{d}t$$

and if, during deformation, the equation $\sigma = K\dot{\varepsilon}^m$ is obeyed, then

$$\mathrm{d}A/\mathrm{d}t = (P/K)^{1/m} A^{\{(1-(1/m)\}}. \tag{6.21}$$

For most metals and alloys $m \approx 0.1$–0.2 and the rate at which A changes is sensitively dependent on A, and hence once necking starts the process rapidly leads to failure. When $m = 1$, the rate of change of area is independent of A and, as a consequence, any irregularities in specimen geometry are not accentuated during deformation. The resistance to necking therefore depends sensitively on m, and increases markedly when $m \gtrsim 0.5$. Considering, in addition, the dependence of the flow stress on strain, then

$$\sigma = K^1 \varepsilon^n \dot{\varepsilon}^m \tag{6.22}$$

and, in this case, the stability against necking depends on a factor $(1 - n - m)/m$, but n-values are not normally very high. Superplastic materials such as Zn–Al eutectoid, Pb–Sn eutectic, Al–Cu eutectic, etc. have m values approaching unity at elevated temperatures.

The total elongation increases as m increases and, with increasing microstructural fineness of the material (grain size or lamella spacing), the tendency for superplastic behavior is increased. Two-phase structures are advantageous in maintaining a fine grain size during testing, but exceptionally high ductilities have been produced in several commercially pure metals (e.g. Ni, Zn and Mg), for which the fine grain size was maintained during testing at a particular strain rate and temperature.

It follows that there must be several possible conditions leading to superplasticity. Generally, it is observed metallographically that the grain structure remains remarkably equiaxed during extensive deformation and that grain boundary sliding is a common deformation mode in several superplastic alloys. While grain boundary sliding can contribute to the overall deformation by relaxing the five independent mechanisms of slip, it cannot give rise to large elongations without bulk flow of material (e.g. grain boundary migration). In polycrystals, triple junctions obstruct the sliding process and give rise to a low m-value. Thus, to increase the rate sensitivity of the boundary shear it is necessary to lower the resistance to sliding from barriers, relative to the viscous drag of the boundary; this can be achieved by grain boundary migration. Indeed, it is observed that superplasticity is controlled by grain boundary diffusion.

The complete explanation of superplasticity is still being developed, but it is already clear that, during deformation, individual grains or groups of grains with suitably aligned boundaries will tend to slide. Sliding continues until obstructed by a protrusion in a grain boundary, when the local stress generates dislocations which slip across the blocked grain and pile up at the opposite boundary until the back stress prevents further generation of dislocations and thus further sliding. At the temperature of the test, dislocations at the head of the pile-up can climb into and move along grain boundaries to annihilation sites. The continual replacement of these dislocations would permit grain boundary sliding at a rate governed by the rate of dislocation climb, which in turn is governed by grain boundary diffusion. It is important that any dislocations created by local stresses are able to traverse yielded grains and this is possible only if the 'dislocation cell size' is larger than, or at least of the order of, the grain size, i.e. a few microns. At high strain rates and low temperatures the dislocations begin to tangle and form cell structures, and superplasticity then ceases.

The above conditions imply that any metal in which the grain size remains fine during deformation could behave superplastically; this conclusion is borne out in practice. The stability of grain size can, however, be achieved more readily with a fine microduplex structure, as observed in some Fe–20Cr–6Ni alloys when hot-worked to produce a fine dispersion of austenite and ferrite. Such stainless steels have an attractive combination of properties (strength, toughness, fatigue strength, weldability and corrosion resistance) and, unlike the usual range of two-phase stainless steels, have good hot workability if 0.5Ti is added to produce a random distribution of TiC rather than $Cr_{23}C_6$ at ferrite–austenite boundaries.

Superplastic forming is now an established and growing industry, largely using vacuum forming to produce intricate shapes with high draw ratios. Two alloys which have achieved engineering importance are *Supral* (containing Al–6Cu–0.5Zr) and *IMI 318* (containing Ti–6Al–4V). *Supral* is deformed at 460°C and *IMI 318* at 900°C under argon. Although the process is slow, the loads required are also low and the process can be advantageous in the press-forming field to replace some of the present expensive and complex forming technology.

Table 6.2 *Twinning elements for some common metals.*

Structure	*Plane*	*Direction*	*Metals*
Cph	$\{1\ 0\ \bar{1}\ 2\}$	$\langle 1\ 0\ \bar{1}\ \bar{1}\rangle$	Zn, Cd, Be, Mg
Bcc	$\{1\ 1\ 2\}$	$\langle 1\ 1\ 1\rangle$	Fe, β-brass, W, Ta, Nb, V, Cr, Mo
Fcc	$\{1\ 1\ 1\}$	$\langle 1\ 1\ 2\rangle$	Cu, Ag, Au, Ag–Au, Cu–Al
Tetragonal	$\{3\ 3\ 1\}$	–	Sn
Rhombohedral	$\{0\ 0\ 1\}$	–	Bi, As, Sb

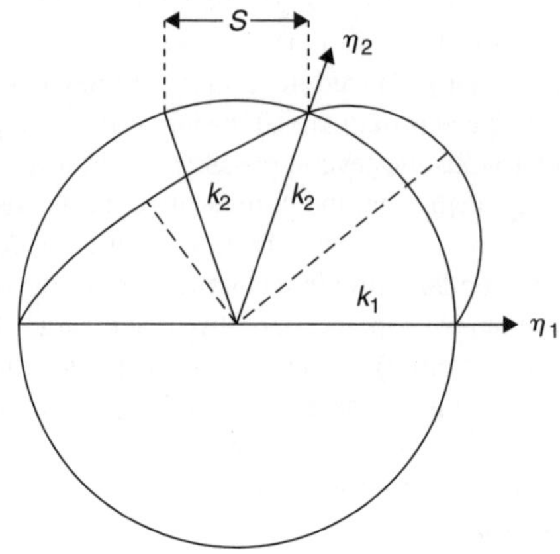

Figure 6.32 *Crystallography of twinning.*

6.5 Mechanical twinning

6.5.1 Crystallography of twinning

Mechanical twinning plays only a minor part in the deformation of the common metals such as copper or aluminum, and its study has consequently been neglected. Nevertheless, twinning does occur in all the common crystal structures under some conditions of deformation. Table 6.2 shows the appropriate twinning elements for the common structures.

The geometrical aspects of twinning can be represented with the aid of a unit sphere, shown in Figure 6.32. The twinning plane k_1 intersects the plane of the drawing in the shear direction η_1. On twinning, the unit sphere is distorted to an ellipsoid of equal volume, and the shear plane k_1 remains unchanged during twinning, while all other planes become tilted. Distortion of planes occurs in all cases except k_1 and k_2. The shear strain, s, at unit distance from the twinning plane is related to the angle between k_1 and k_2. Thus, the amount of shear is fixed by the crystallographic nature of the two undistorted planes. In the bcc lattice, the two undistorted planes are the $(1\ 1\ 2)$ and $(1\ 1\ \bar{2})$ planes, displacement occurring in a $[1\ 1\ 1]$ direction a distance of 0.707 lattice vectors. The twinning elements are thus:

k_1	k_2	η_1	η_2	*Shear*
$(1\ 1\ 2)$	$(1\ 1\ \bar{2})$	$[1\ 1\ \bar{1}]$	$[1\ 1\ 1]$	0.707

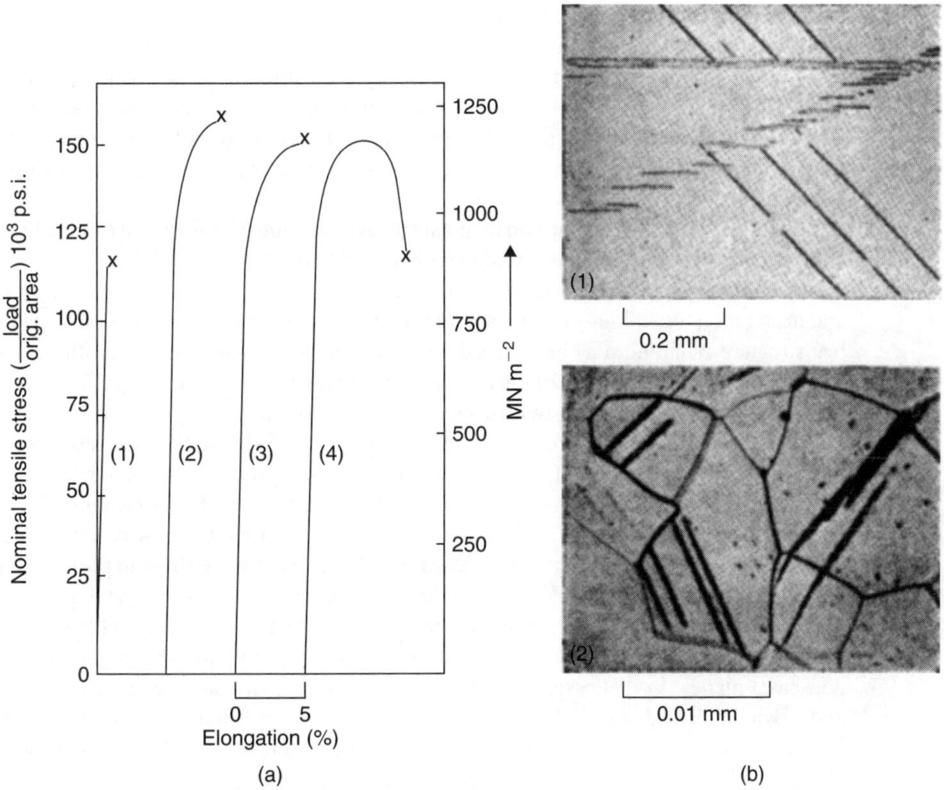

Figure 6.33 *(a) Effect of grain size on the stress–strain curves of specimens of niobium extended at a rate of* $2.02 \times 10^{-4}\,s^{-1}$ *at 20 K: (1) grain size* $2d = 1.414$ *mm, (2) grain size* $2d = 0.312$ *mm, (3) grain size* $2d = 0.0951$ *mm, (4) grain size* $2d = 0.0476$ *mm. (b) Deformation twins in specimen 1 and specimen 3 extended to fracture. Etched in 95%* HNO_3 *+ 5% HF (after Adams, Roberts and Smallman, 1960).*

where k_1 and k_2 denote the first and second undistorted planes respectively, and η_1 and η_2 denote directions lying in k_1 and k_2, respectively, perpendicular to the line of intersection of these planes. k_1 is also called the composition or twinning plane, while η_1 is called the shear direction. The twins consist of regions of crystal in which a particular set of $\{1\,1\,2\}$ planes (the k_1 set of planes) is homogeneously sheared by 0.707 in a $\langle 1\,1\,1 \rangle$ direction (the η_1 direction). The same atomic arrangement may be visualized by a shear of 1.414 in the *reverse* $\langle 1\,1\,1 \rangle$ direction, but this larger shear has never been observed.

6.5.2 Nucleation and growth of twins

During the development of mechanical twins, thin lamellae appear very quickly ($\approx$ speed of sound) and these thicken with increasing stress by the steady movement of the twin interface. New twins are usually formed in bursts and are sometimes accompanied by a sharp audible click, which coincides with the appearance of irregularities in the stress–strain curve, as shown in Figure 6.33. The rapid production of clicks is responsible for the so-called twinning cry (e.g. in tin).

Although most metals show a general reluctance to twin, when tested under suitable conditions they can usually be made to do so. As mentioned in Section 6.3.1, the shear process involved in twinning must occur by the movement of partial dislocations and, consequently, the stress to cause twinning will depend not only on the line tension of the source dislocation, as in the case of slip, but also on the surface tension of the twin boundary. The stress to cause twinning is therefore usually greater than that required for slip, and at room temperature deformation will nearly always occur by slip in preference to twinning. As the deformation temperature is lowered the critical shear stress for slip increases and then, because the general stress level will be high, the process of deformation twinning is more likely.

Twinning is most easily achieved in metals of cph structure where, because of the limited number of slip systems, twinning is an essential and unavoidable mechanism of deformation in polycrystalline specimens (see Section 6.4.11), but in single crystals the orientation of the specimen, the stress level and the temperature of deformation are all important factors in the twinning process. In metals of the bcc structure twinning may be induced by impact at room temperature or with more normal strain rates at low temperature, where the critical shear stress for slip is very high. In contrast, only a few fcc metals have been made to twin, even at low temperatures.

In zinc single crystals it is observed that there is no well-defined critical resolved shear stress for twinning such as exists for slip, and that a very high stress indeed is necessary to nucleate twins. In most crystals, slip usually occurs first and twin nuclei are then created by means of the very high stress concentration which exists at dislocation pile-ups. Once formed, the twins can propagate provided the resolved shear stress is higher than a critical value, because the stress to propagate a twin is much lower than that to nucleate it. This can be demonstrated by deforming a crystal oriented in such a way that basal slip is excluded, i.e. when the basal planes are nearly parallel to the specimen axis. Even in such an oriented crystal it is found that the stress to cause twinning is higher than that for slip on non-basal planes. In this case, non-basal slip occurs first, so that when a dislocation pile-up arises and a twin is formed, the applied stress is so high that an avalanche or burst of twins results.

It is also believed that in the bcc metals twin nucleation is more difficult than twin propagation. One possible mechanism is that nucleation is brought about by the stress concentration at the head of a piled-up array of dislocations produced by a burst of slip as a Frank–Read source operates. Such behavior is favored by impact loading, and it is well known that twin lamellae known as Neumann bands are produced this way in α-iron at room temperature. At normal strain rates, however, it should be easier to produce a slip burst suitable for twin nucleation in a material with strongly locked dislocations, i.e. one with a large k-value (as defined by equation (6.19)), than one in which the dislocation locking is relatively slight (small k-values). In this context it is interesting to note that both niobium and tantalum have a small k-value and, although they can be made to twin, do so with reluctance compared, for example, with α-iron.

In all the bcc metals the flow stress increases so rapidly with decreasing temperature (see Figure 6.29), that even with moderate strain rates ($10^{-4}\,s^{-1}$) α-iron will twin at 77 K, while niobium with its smaller value of k twins at 20 K. The type of stress–strain behavior for niobium is shown in Figure 6.33a. The pattern of behavior is characterized by small amounts of slip interspersed between extensive bursts of twinning in the early stages of deformation. Twins, once formed, may themselves act as barriers, allowing further dislocation pile-up and further twin nucleation. The action of twins as barriers to slip dislocations could presumably account for the rapid work hardening observed at 20 K.

Fcc metals do not readily deform by twinning, but it can occur at low temperatures, and even at 0°C, in favorably oriented crystals. The apparent restriction of twinning to certain orientations and low temperatures may be ascribed to the high shear stress attained in tests on crystals with these orientations, since the stress necessary to produce twinning is high. Twinning has been confirmed in heavily rolled copper. The exact mechanism for this twinning is not known, except that it must occur by the propagation of a half-dislocation and its associated stacking fault across each plane of a set of parallel (1 1 1) planes. For this process the half-dislocation must climb onto successive twin planes, as below for bcc iron.

6.5.3 Effect of impurities on twinning

It is well established that solid solution alloying favors twinning in fcc metals. For example, silver–gold alloys twin far more readily than the pure metals. Attempts have been made to correlate this effect with stacking-fault energy and it has been shown that the twinning stress of copper-based alloys increases with increasing stacking-fault energy. Twinning is also favored by solid solution alloying in bcc metals, and alloys of Mo–Re, W–Re and Nb–V readily twin at room temperature. In this case it has been suggested that the lattice frictional stress is increased and the ability to cross-slip reduced by alloying, thereby confining slip dislocations to bands where stress multiplication conducive to twin nucleation occurs.

6.5.4 Effect of prestrain on twinning

Twinning can be suppressed in most metals by a certain amount of prestrain; the ability to twin may be restored by an ageing treatment. It has been suggested that the effect may be due to the differing dislocation distribution produced under different conditions. For example, niobium will normally twin at $-196°C$, when a heterogeneous arrangement of elongated screw dislocations capable of creating the necessary stress concentrations is formed. Room temperature prestrain, however, inhibits twin formation as the regular network of dislocations produced provides more mobile dislocations and homogenizes the deformation.

6.5.5 Dislocation mechanism of twinning

In contrast to slip, the shear involved in the twinning process is homogeneous throughout the entire twinning region, and each atom plane parallel to the twinning plane moves over the one below it by only a fraction of a lattice spacing in the twinning direction. Nevertheless, mechanical twinning is thought to take place by a dislocation mechanism for the same reasons as slip, but the dislocations that cause twinning are partial and not unit dislocations. From the crystallography of the process it can be shown that twinning in the cph lattice, in addition to a simple shear on the twinning plane, must be accompanied by a localized rearrangement of the atoms, and furthermore, only in the bcc lattice does the process of twinning consist of a simple shear on the twinning plane (e.g. a twinned structure in this lattice can be produced by a shear of $1/\sqrt{2}$ in a $\langle 1\ 1\ 1 \rangle$ direction on a $\{1\ 1\ 2\}$ plane).

An examination of Figure 6.8 shows that the main problem facing any theory of twinning is to explain how twinning develops homogeneously through successive planes of the lattice. This could be accomplished by the movement of a single twinning (partial) dislocation successively from plane to plane. One suggestion, similar in principle to the crystal growth mechanism, is the pole mechanism proposed by Cottrell and Bilby, illustrated in Figure 6.34a. Here, OA, OB and OC are dislocation lines. The twinning dislocation is OC, which produces the correct shear as it sweeps through the twin plane about its point of emergence O, and OA and OB form the pole dislocation, being partly or wholly of screw character with a pitch equal to the spacing of the twinning layers. The twinning dislocation rotates round the pole dislocation and, in doing so, not only produces a monolayer sheet of twinned crystal but also climbs up the 'pole' to the next layer. The process is repeated and a thick layer of twin is built up.

The dislocation reaction involved is as follows. The line AOB represents a unit dislocation with a Burgers vector $a/2[1\ 1\ 1]$ and that part OB of the line lies in the $(1\ 1\ 2)$ plane. Then, under the action of stress, dissociation of this dislocation can occur according to the reaction:

$$a/2[1\ 1\ 1] \rightarrow a/3[1\ 1\ 2] + a/6[1\ 1\ \bar{1}].$$

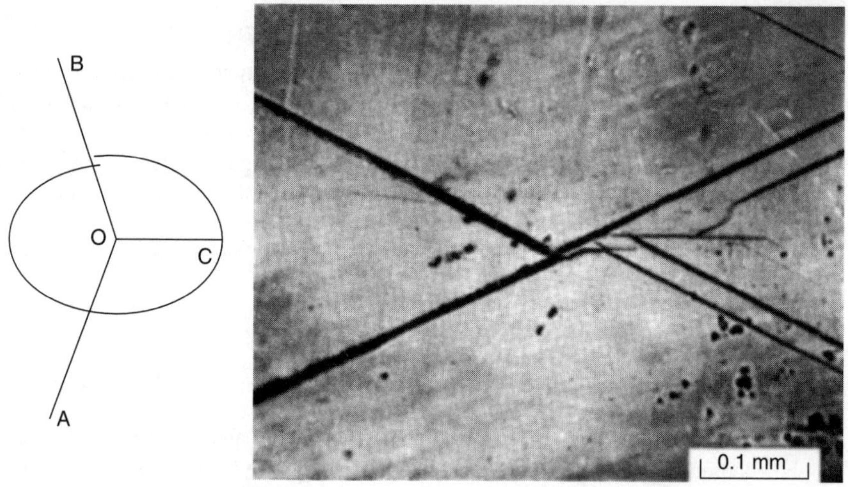

(a) (b)

Figure 6.34 *(a) Diagram illustrating the pole mechanism of twinning. (b) The formation of a crack at a twin intersection in silicon–iron (after Hull, 1960).*

The dislocation with vector $a/6[1\ 1\ \bar{1}]$ forms a line OC lying in one of the other $\{1\ 1\ 2\}$ twin planes (e.g. the $(\bar{1}\ 2\ 1)$ plane) and produces the correct twinning shear. The line OB is left with a Burgers vector $a/3[1\ 1\ 2]$, which is of pure edge type and sessile in the $(1\ 1\ 2)$ plane.

6.5.6 Twinning and fracture

It has been suggested that a twin, like a grain boundary, may present a strong barrier to slip and that a crack can be initiated by the pile-up of slip dislocations at the twin interface (see Figure 7.33). In addition, cracks may be initiated by the intersection of twins, and examples are common in molybdenum, silicon–iron (bcc) and zinc (cph). Figure 6.34b shows a very good example of crack nucleation in 3% silicon–iron; the crack has formed along an $\{0\ 0\ 1\}$ cleavage plane at the intersection of two $\{1\ 1\ 2\}$ twins, and part of the crack has developed along one of the twins in a zigzag manner while still retaining $\{0\ 0\ 1\}$ cleavage facets.

In tests at low temperature on bcc and cph metals both twinning and fracture readily occur, and this has led to two conflicting views. First, that twins are nucleated by the high stress concentrations associated with fracture and, second, that the formation of twins actually initiates the fracture. It is probable that both effects occur.

6.6 Strengthening and hardening mechanisms

6.6.1 Point defect hardening

The introduction of point defects into materials to produce an excess concentration of either vacancies or interstitials often gives rise to a significant change in mechanical properties (Figures 6.35 and 6.36). For aluminum the shape of the stress–strain curve is very dependent on the rate of cooling and a large increase in the yield stress may occur after quenching. We have already seen in Chapter 3 that quenched-in vacancies result in clustered vacancy defects and these may harden the material.

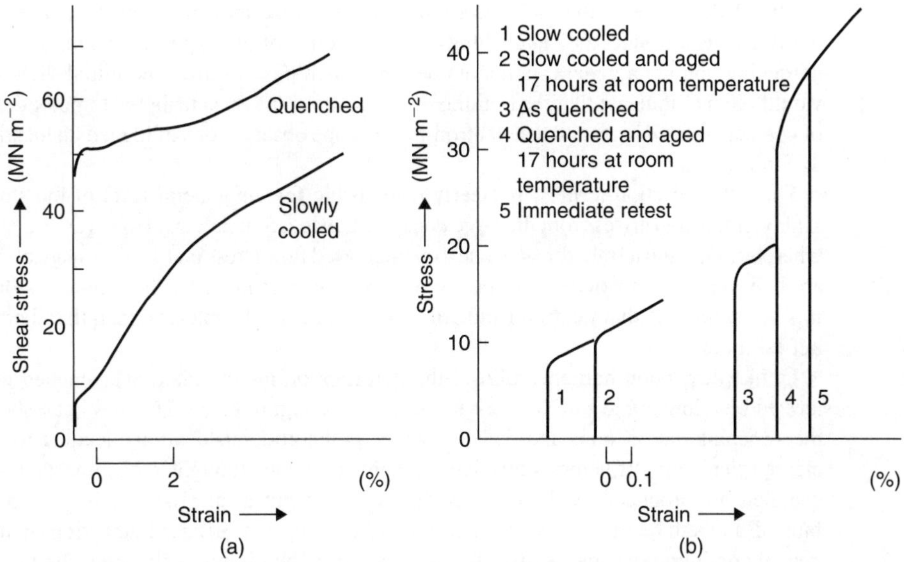

Figure 6.35 *Effect of quenching on the stress–strain curves from: (a) aluminum (after Maddin and Cottrell, 1955) and (b) gold (after Adams and Smallman, unpublished).*

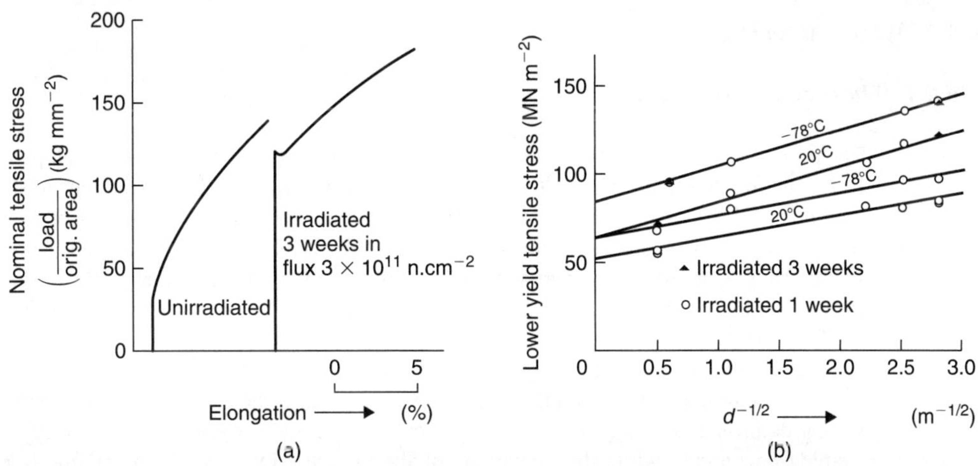

Figure 6.36 *(a) Stress–strain curves for unirradiated and irradiated fine-grained polycrystalline copper, tested at 20°C. (b) Variation of yield stress with grain size and neutron dose (after Adams and Higgins, 1959).*

Similarly, irradiation by high-energy particles may produce irradiation hardening (see Figure 6.36). Information on the mechanisms of hardening can be obtained from observation of the dependence of the lower yield stress on grain size. The results, reproduced in Figure 6.36b, show that the relation $\sigma_y = \sigma_i + k_y d^{-1/2}$, which is a general relation describing the propagation of yielding in materials, is obeyed.

This dependence of the yield stress, σ_y, on grain size indicates that the hardening produced by point defects introduced by quenching or irradiation is of two types: (1) an initial dislocation source hardening and (2) a general lattice hardening which persists after the initial yielding. The k_y term would seem to indicate that the pinning of dislocations may be attributed to point defects in the form of coarsely spaced jogs, and the electron-microscope observations of jogged dislocations would seem to confirm this.

The lattice friction term σ_i is clearly responsible for the general level of the stress–strain curve after yielding and arises from the large density of dislocation defects. However, the exact mechanisms whereby loops and tetrahedra give rise to an increased flow stress is still controversial. Vacancy clusters are believed to be formed *in situ* by the disturbance introduced by the primary collision, and hence it is not surprising that neutron irradiation at 4 K hardens the material, and that thermal activation is not essential.

Unlike dispersion-hardened alloys, the deformation of irradiated or quenched metals is characterized by a low initial rate of work hardening (see Figure 6.35). This has been shown to be due to the sweeping out of loops and defect clusters by the glide dislocations, leading to the formation of cleared channels. Diffusion-controlled mechanisms are not thought to be important, since defect-free channels are produced by deformation at 4 K. The removal of prismatic loops, both unfaulted and faulted, and tetrahedra can occur as a result of the strong coalescence interactions with screws to form helical configurations and jogged dislocations when the gliding dislocations and defects make contact. Clearly, the sweeping-up process occurs only if the helical and jogged configurations can glide easily. Resistance to glide will arise from jogs not lying in slip planes and also from the formation of sessile jogs (e.g. Lomer–Cottrell dislocations in fcc crystals).

6.6.2 Work hardening

6.6.2.1 Theoretical treatment

The properties of a material are altered by cold working, i.e. deformation at a low temperature relative to its melting point, but not all the properties are improved, for although the tensile strength, yield strength and hardness are increased, the plasticity and general ability to deform decreases. Moreover, the physical properties such as electrical conductivity, density and others are all lowered. Of these many changes in properties, perhaps the most outstanding are those that occur in the mechanical properties; the yield stress of mild steel, for example, may be raised by cold work from 170 up to $1050\,MN\,m^{-2}$.

Such changes in mechanical properties are, of course, of interest theoretically, but they are also of great importance in industrial practice. This is because the rate at which the material hardens during deformation influences both the power required and the method of working in the various shaping operations, while the magnitude of the hardness introduced governs the frequency with which the component must be annealed (always an expensive operation) to enable further working to be continued.

Since plastic flow occurs by a dislocation mechanism, the fact that work hardening occurs means that it becomes difficult for dislocations to move as the strain increases. All theories of work hardening depend on this assumption, and the basic idea of hardening, put forward by Taylor in 1934, is that some dislocations become 'stuck' inside the crystal and act as sources of internal stress, which oppose the motion of other gliding dislocations.

One simple way in which two dislocations could become stuck is by elastic interaction. Thus, two parallel-edge dislocations of opposite sign moving on parallel slip planes in any sub-grain may become stuck, as a result of the interaction discussed in Chapter 3. G. I. Taylor assumed that dislocations become stuck after traveling an average distance, L, while the density of dislocations reaches ρ, i.e.

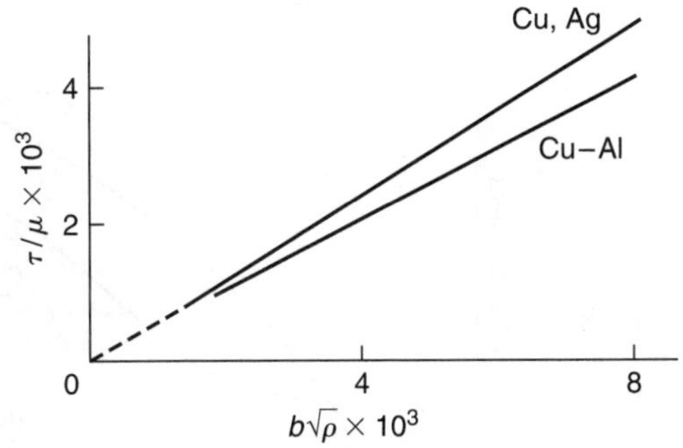

Figure 6.37 *Dependence of flow stress on (dislocation density)$^{1/2}$ for Cu, Ag and Cu–Al.*

work hardening is due to the dislocations getting in each other's way. The flow stress is then the stress necessary to move a dislocation in the stress field of those dislocations surrounding it. This stress τ is quite generally given by

$$\tau = \alpha\mu b/l, \tag{6.23}$$

where μ is the shear modulus, b the Burgers vector, l the mean distance between dislocations, which is $\approx\rho^{-1/2}$, and α a constant; in the Taylor model $\alpha = 1/8\pi(1-\nu)$ where ν is Poisson's ratio. Figure 6.37 shows such a relationship for Cu–Al single crystals and polycrystalline Ag and Cu.

In his theory Taylor considered only a two-dimensional model of a cold-worked metal. However, because plastic deformation arises from the movement of dislocation loops from a source, it is more appropriate to assume that, when the plastic strain is γ, N dislocation loops of side L (if we assume for convenience that square loops are emitted) have been given off per unit volume. The resultant plastic strain is then given by

$$\gamma = NL^2b \tag{6.24}$$

and l by

$$l \approx [1/\rho^{1/2}] = [1/4LN]^{1/2}. \tag{6.25}$$

Combining these equations, the stress–strain relation

$$\tau = \text{const} \cdot (b/L)^{1/2}\gamma^{1/2} \tag{6.26}$$

is obtained. Taylor assumed L to be a constant, i.e. the slip lines are of constant length, which results in a parabolic relationship between τ and γ.

Taylor's assumption that during cold work the density of dislocations increases has been amply verified, and indeed the parabolic relationship between stress and strain is obeyed, to a first approximation, in many polycrystalline aggregates where deformation in all grains takes place by multiple slip. Experimental work on single crystals shows, however, that the work or strain hardening curve may deviate considerably from parabolic behavior, and depends not only on crystal structure but also on other variables such as crystal orientation, purity and surface conditions (see Figures 6.38 and 6.39).

The crystal structure is important (see Figure 6.38) in that single crystals of some hexagonal metals slip only on one family of slip planes, those parallel to the basal plane, and these metals show a low rate of work hardening. The plastic part of the stress–strain curve is also more nearly linear than parabolic,

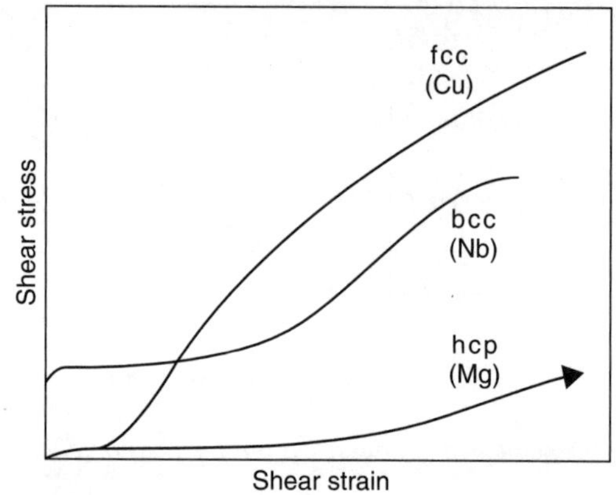

Figure 6.38 *Stress–strain curves of single crystals (after Hirsch and Mitchell, 1967; courtesy of the National Research Council of Canada).*

with a slope which is extremely small: this slope ($d\tau/d\gamma$) becomes even smaller with increasing temperature of deformation. Cubic crystals, on the other hand, are capable of deforming in a complex manner on more than one slip system, and these metals normally show a strong work-hardening behavior. The influence of temperature depends on the stress level reached during deformation and on other factors which must be considered in greater detail. However, even in cubic crystals the rate of work hardening may be extremely small if the crystal is restricted to slip on a single slip system. Such behavior points to the conclusion that strong work hardening is caused by the mutual interference of dislocations gliding on intersecting slip planes.

Many theories of work hardening similar to that of Taylor exist but all are oversimplified, since work hardening depends not so much on individual dislocations as on the group behavior of large numbers of them. It is clear, therefore, that a theoretical treatment which would describe the complete stress–strain relationship is difficult, and consequently the present-day approach is to examine the various stages of hardening and then attempt to explain the mechanisms likely to give rise to the different stages. The work-hardening behavior in metals with a cubic structure is more complex than in most other structures because of the variety of slip systems available, and it is for this reason that much of the experimental evidence is related to these metals, particularly those with fcc structures.

6.6.2.2 Three-stage hardening

The stress–strain curve of a fcc single crystal is shown in Figure 6.39 and three regions of hardening are experimentally distinguishable. Stage I, or the easy glide region, immediately follows the yield point and is characterized by a low rate of work hardening θ_I up to several per cent glide; the length of this region depends on orientation, purity and size of the crystals. The hardening rate $(\theta_I/\mu) \sim 10^{-4}$ and is of the same order as for hexagonal metals. Stage II, or the linear hardening region, shows a rapid increase in work-hardening rate. The ratio $(\theta_{II}/\mu) = (d\tau/d\gamma)/\mu$ is of the same order of magnitude for all fcc metals, i.e. 1/300, although this is $\approx$1/150 for orientations at the corners of the stereographic triangle. In this stage short slip lines are formed during straining quite suddenly, i.e. in a short increment of stress $\Delta\tau$, and thereafter do not grow either in length or intensity. The mean length of the slip lines, $L \approx 25\ \mu m$, decreases with increasing strain. Stage III, or the parabolic hardening region, the onset of which is markedly dependent on temperature, exhibits a low rate of work hardening, θ_{III}, and the appearance of coarse slip bands. This stage sets in at a strain which

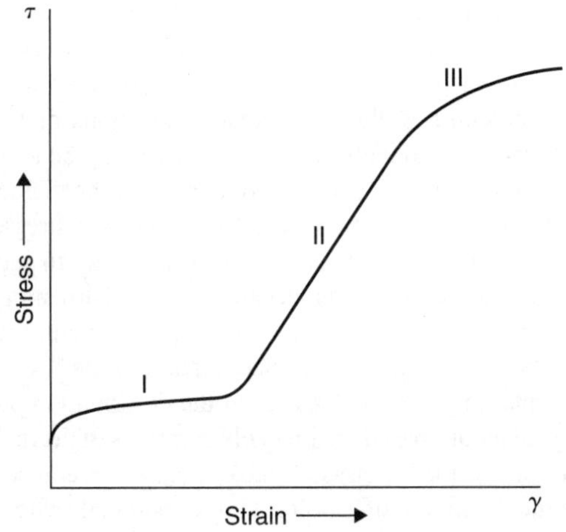

Figure 6.39 *Stress–strain curve showing the three stages of work hardening.*

increases with decreasing temperature and is probably associated with the annihilation of dislocations as a consequence of cross-slip.

The low stacking-fault energy metals exhibit all three work-hardening stages at room temperature, but metals with a high stacking-fault energy often show only two stages of hardening. It is found, for example, that at 78 K aluminum behaves like copper at room temperature and exhibits all three stages, but at room temperature and above, Stage II is not clearly developed and Stage III starts before Stage II becomes at all predominant. This difference between aluminum and the noble metals is due not only to the difference in melting point, but also to the difference in stacking-fault energies, which affects the width of extended dislocations. The main effect of a change of temperature of deformation is, however, a change in the onset of Stage III; the lower the temperature of deformation, the higher is the stress τ_{111} corresponding to the onset of Stage III.

The Stage I easy glide region in cubic crystals, with its small linear hardening, corresponds closely to the hardening of cph crystals, where only one glide plane operates. It occurs in crystals oriented to allow only one glide system to operate, i.e. for orientations near the [0 1 1] pole of the unit triangle (Figure 6.12). In this case the slip distance is large, of the order of the specimen diameter, with the probability of dislocations slipping out of the crystal. Electron microscope observations have shown that the slip lines on the surface are very long ($\approx$1 mm) and closely spaced, and that the slip steps are small corresponding to the passage of only a few dislocations. This behavior obviously depends on such variables as sample size and oxide films, since these influence the probability of dislocations passing out of the crystal. It is also to be expected that the flow stress in easy glide will be governed by the ease with which sources begin to operate, since there is no slip on a secondary slip system to interfere with the movement of primary glide dislocations.

As soon as another glide system becomes activated there is a strong interaction between dislocations on the primary and secondary slip systems, which gives rise to a steep increase in work hardening. It is reasonable to expect that easy glide should end, and turbulent flow begin, when the crystal reaches an orientation for which two or more slip systems are equally stressed, i.e. for orientations on the symmetry line between [0 0 1] and [1 1 1]. However, easy glide generally ends before symmetrical orientations are reached and this is principally due to the formation of deformation bands to accommodate the rotation of the glide plane in fixed grips during tensile tests. This rotation leads to a high resolved stress on the secondary slip system, and its operation gives rise to those lattice irregularities which cause some dislocations to become 'stopped' in the crystal. The transformation to Stage II then occurs.

The characteristic feature of deformation in Stage II is that slip takes place on both the primary and secondary slip systems. As a result, several new lattice irregularities may be formed, which will include (1) forest dislocations, (2) Lomer–Cottrell barriers and (3) jogs produced either by moving dislocations cutting through forest dislocations or by forest dislocations cutting through source dislocations. Consequently, the flow stress τ may be identified, in general terms, with a stress which is sufficient to operate a source and then move the dislocations against (1) the internal elastic stresses from the forest dislocations, (2) the long-range stresses from groups of dislocations piled up behind barriers and (3) the frictional resistance due to jogs. In a cold-worked metal all these factors may exist to some extent, but because a linear hardening law can be derived by using any one of the various contributory factors, there have been several theories of Stage II hardening, namely (1) the pile-up theory, (2) the forest theory and (3) the jog theory. All have been shown to have limitations in explaining various features of the deformation process, and have given way to a more phenomenological theory based on direct observations of the dislocation distribution during straining.

Observations on fcc and bcc crystals have revealed several common features of the microstructure, which include the formation of dipoles, tangles and cell structures with increasing strain. The most detailed observations have been made for copper crystals, and these are summarized below to illustrate the general pattern of behavior. In Stage I, bands of dipoles are formed (see Figure 6.40a), elongated normal to the primary Burgers vector direction. Their formation is associated with isolated forest dislocations and individual dipoles are about 1 μm in length and $\lesssim$10 nm wide. Different patches are arranged at spacings of about 10 μm along the line of intersection of a secondary slip plane. With increasing strain in Stage I, the size of the gaps between the dipole clusters decreases and therefore the stress required to push dislocations through these gaps increases. Stage II begins (see Figure 6.40b) when the applied stress plus internal stress resolved on the secondary systems is sufficient to activate secondary sources near the dipole clusters. The resulting local secondary slip leads to local interactions between primary and secondary dislocations both in the gaps and in the clusters of dipoles, the gaps being filled with secondary dislocations and short lengths of other dislocations formed by interactions (e.g. Lomer–Cottrell dislocations in fcc crystals and $a\langle 1\ 0\ 0\rangle$-type dislocations in bcc crystals). Dislocation barriers are thus formed surrounding the original sources.

In Stage II (see Figure 6.40c) it is proposed that dislocations are stopped by elastic interaction when they pass too close to an existing tangled region with high dislocation density. The long-range internal stresses due to the dislocations piling up behind are partially relieved by secondary slip, which transforms the discrete pile-up into a region of high dislocation density containing secondary dislocation networks and dipoles. These regions of high dislocation density act as new obstacles to dislocation glide, and since every new obstacle is formed near one produced at a lower strain, two-dimensional dislocation structures are built up forming the walls of an irregular cell structure. With increasing strain the number of obstacles increases, the distance a dislocation glides decreases and therefore the slip line length decreases in Stage II. The structure remains similar throughout Stage II but is reduced in scale. The obstacles are in the form of ribbons of high densities of dislocations which, like pile-ups, tend to form sheets. The work-hardening rate depends mainly on the effective radius of the obstacles, and this has been considered in detail by Hirsch and co-workers and shown to be a constant fraction k of the discrete pile-up length on the primary slip system. In general, the work-hardening rate is given by $\theta_{11} = k\mu/3\pi$ and, for an fcc crystal, the small variation in k with orientation and alloying element is able to account for the variation of θ_{11} with those parameters.

The dislocation arrangement in metals with other structures is somewhat similar to that of copper, with differences arising from stacking-fault energy. In Cu–Al alloys the dislocations tend to be confined more to the active slip planes, the confinement increasing with decreasing γ_{SF}. In Stage I dislocation multipoles are formed as a result of dislocations of opposite sign on parallel nearby slip planes 'pairing up' with one another. Most of these dislocations are primaries. In Stage II the density of secondary dislocations is much less $\left(\approx \frac{1}{3}\right)$ than that of the primary dislocations. The secondary slip occurs in bands and in each band slip on one particular secondary plane predominates. In niobium,

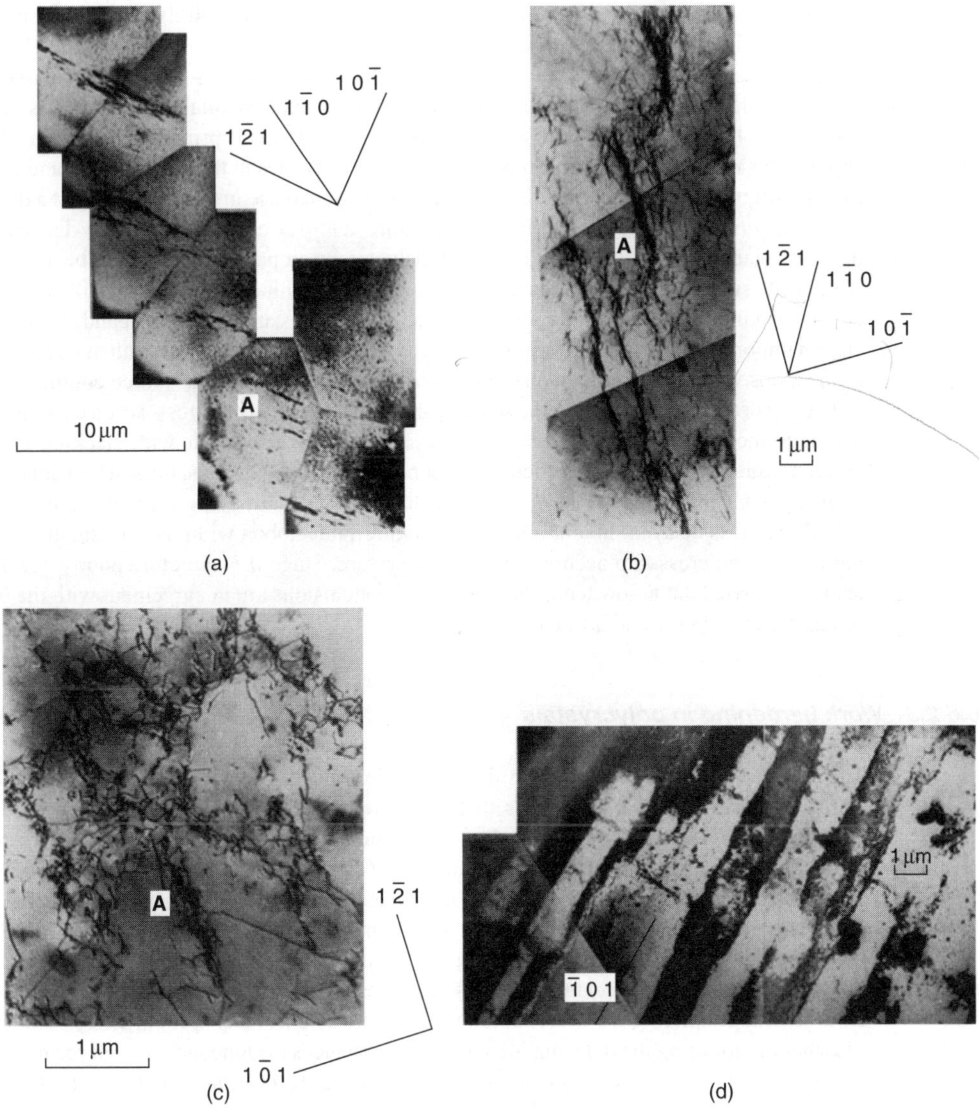

Figure 6.40 *Dislocation structure observed in copper single crystals deformed in tension to: (a) Stage I, (b) end of easy glide and beginning of Stage II, (c) top of Stage II and (d) Stage III (after Steeds, 1963; Crown copyright; reproduced by permission of the Controller, HM Stationery Office).*

a metal with high γ_{SF}, the dislocation distribution is rather similar to copper. In Mg, typical of cph metals, Stage I is extensive and the dislocations are mainly in the form of primary edge multipoles, but forest dislocations threading the primary slip plane do not appear to be generated.

From the curve shown in Figure 6.39 it is evident that the rate of work hardening decreases in the later stages of the test. This observation indicates that at a sufficiently high stress or temperature the dislocations held up in Stage II are able to move by a process which at lower stresses and temperature had been suppressed. The onset of Stage III is accompanied by cross-slip, and the slip lines are broad, deep and consist of segments joined by cross-slip traces. Electron metallographic observations on sections of deformed crystal inclined to the slip plane (see Figure 6.40d) show the formation of a cell structure in the form of boundaries, approximately parallel to the primary slip plane of spacing about 1–3 μm, plus other boundaries extending normal to the slip plane as a result of cross-slip.

The simplest process which is in agreement with the experimental observations is that the screw dislocations held up in Stage II cross-slip and possibly return to the primary slip plane by double cross-slip. By this mechanism, dislocations can bypass the obstacles in their glide plane and do not have to interact strongly with them. Such behavior leads to an increase in slip distance and a decrease in the accompanying rate of work hardening. Furthermore, it is to be expected that screw dislocations leaving the glide plane by cross-slip may also meet dislocations on parallel planes and be attracted by those of opposite sign. Annihilation then takes place and the annihilated dislocation will be replaced, partly at least, from the original source. This process, if repeated, can lead to slip-band formation, which is also an important experimental feature of Stage III. Hardening in Stage III is then due to the edge parts of the loops which remain in the crystal and increase in density as the source continues to operate.

The importance of the value of the stacking-fault energy, γ, on the stress–strain curve is evident from its importance to the process of cross-slip. Low values of γ give rise to wide stacking-fault 'ribbons', and consequently cross-slip is difficult at reasonable stress levels. Thus, the screws cannot escape from their slip plane, the slip distance is small, the dislocation density is high and the transition from Stage II to Stage III is delayed. In aluminum the stacking-fault ribbon width is very small because γ has a high value, and cross-slip occurs at room temperature. Stage II is therefore poorly developed unless testing is carried out at low temperatures. These conclusions are in agreement with the observations of dislocation density and arrangement.

6.6.2.3 *Work hardening in polycrystals*

The dislocation structure developed during the deformation of fcc and bcc polycrystalline metals follows the same general pattern as that in single crystals: primary dislocations produce dipoles and loops by interaction with secondary dislocations, which give rise to local dislocation tangles gradually developing into three-dimensional networks of sub-boundaries. The cell size decreases with increasing strain, and the structural differences that are observed between various metals and alloys are mainly in the sharpness of the sub-boundaries. In bcc metals, and fcc metals with high stacking-fault energy, the tangles rearrange into sharp boundaries but in metals of low stacking-fault energy the dislocations are extended, cross-slip is restricted and sharp boundaries are not formed even at large strains. Altering the deformation temperature also has the effect of changing the dislocation distribution; lowering the deformation temperature reduces the tendency for cell formation, as shown in Figure 6.41. For a given dislocation distribution the dislocation density is simply related to the flow stress τ by an equation of the form:

$$\tau = \tau_0 + \alpha\mu b\rho^{1/2}, \tag{6.27}$$

where α is a constant at a given temperature ≈ 0.5; τ_0 for fcc metals is zero (see Figure 6.37). The work-hardening rate is determined by the ease with which tangled dislocations rearrange themselves and is high in materials with low γ, i.e. brasses, bronzes and austenitic steels, compared to Al and bcc metals. In some austenitic steels, work hardening may be increased and better sustained by a strain-induced phase transformation (see Chapter 7).

Grain boundaries affect work hardening by acting as barriers to slip from one grain to the next. In addition, the continuity criterion of polycrystals enforces complex slip in the neighborhood of the boundaries, which spreads across the grains with increasing deformation. This introduces a dependence of work-hardening rate on grain size which extends to several percent elongation. After this stage, however, the work-hardening rate is independent of grain size and for fcc polycrystals is

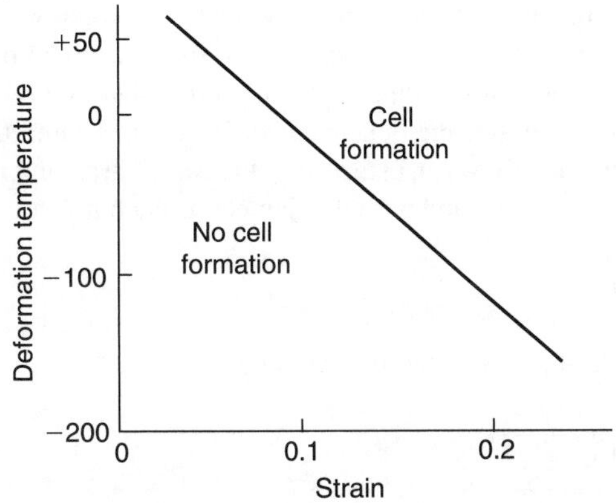

Figure 6.41 *Influence of deformation strain and temperature on the formation of a cell structure in α-iron.*

about $\mu/40$, which, allowing for the orientation factors, is roughly comparable with that found in single crystals deforming in multiple slip. Thus, from the relations $\sigma = m\tau$ and $\varepsilon = \gamma/m$, the average resolved shear stress on a slip plane is rather less than half the applied tensile stress, and the average shear strain parallel to the slip plane is rather more than twice the tensile elongation. The polycrystal work-hardening rate is thus related to the single-crystal work-hardening rate by the relation

$$d\sigma/d\varepsilon = m^2 d\tau/d\gamma. \tag{6.28}$$

For bcc metals with a multiplicity of slip systems the ease of cross-slip m is more nearly 2, so that the work-hardening rate is low. In polycrystalline cph metals the deformation is complicated by twinning, but in the absence of twinning $m \approx 6.5$, and hence the work-hardening rate is expected to be more than an order of magnitude greater than for single crystals, and also higher than the rate observed in fcc polycrystals, for which $m \approx 3$.

6.6.2.4 Dispersion-hardened alloys

On deforming an alloy containing incoherent, non-deformable particles, the rate of work hardening is much greater than that shown by the matrix alone (see Figure 7.10). The dislocation density increases very rapidly with strain because the particles produce a turbulent and complex deformation pattern around them. The dislocations gliding in the matrix leave loops around particles either by bowing between the particles or by cross-slipping around them; both these mechanisms are discussed in Chapter 7. The stresses in and around particles may also be relieved by activating secondary slip systems in the matrix. All these dislocations spread out from the particle as strain proceeds and, by intersecting the primary glide plane, hinder primary dislocation motion and lead to intense work hardening. A dense tangle of dislocations is built up at the particle and a cell structure is formed with the particles predominantly in the cell walls.

At small strains ($\lesssim 1\%$), work hardening probably arises from the back-stress exerted by the few Orowan loops around the particles, as described by Fisher, Hart and Pry. The stress–strain curve is reasonably linear with strain ε according to

$$\sigma = \sigma_i + \alpha\mu f^{3/2}\varepsilon,$$

with the work hardening depending only on f, the volume fraction of particles. At larger strains the 'geometrically necessary' dislocations stored to accommodate the strain gradient which arises because one component deforms plastically more than the other determine the work hardening. A determination of the average density of dislocations around the particles with which the primary dislocations interact allows an estimate of the work-hardening rate, as initially considered by Ashby. Thus, for a given strain ε and particle diameter d the number of loops per particle is

$$n \sim \varepsilon d/b$$

and the number of particles per unit volume

$$N_{\mathrm{v}} = 3f/4\pi r^2, \quad \text{or} \quad 6f/\pi d^3.$$

The total number of loops per unit volume is nN_{v} and hence the dislocation density $\rho = nN_{\mathrm{v}}\pi d = 6f\varepsilon/db$. The stress–strain relationship from equation (6.27) is then

$$\sigma = \sigma_i + \alpha\mu(fb/d)^{1/2}\varepsilon^{1/2} \tag{6.29}$$

and the work-hardening rate

$$\mathrm{d}\sigma/\mathrm{d}\varepsilon = \alpha'\mu(f/d)^{1/2}(b/\varepsilon)^{1/2}. \tag{6.30}$$

Alternative models taking account of the detailed structure of the dislocation arrays (e.g. Orowan, prismatic and secondary loops) have been produced to explain some of the finer details of dispersion-hardened materials. However, this simple approach provides a useful working basis for real materials. Some additional features of dispersion-strengthened alloys are discussed in Chapter 7.

6.6.2.5 *Work hardening in ordered alloys*

A characteristic feature of alloys with long-range order is that they work-harden more rapidly than in the disordered state. θ_{11} for Fe–Al with a B2 ordered structure is $\approx\mu/50$ at room temperature, several times greater than a typical fcc or bcc metal. However, the density of secondary dislocations in Stage II is relatively low and only about $1/100$ of that of the primary dislocations. One mechanism for the increase in work-hardening rate is thought to arise from the generation of anti-phase domain boundary (APB) tubes. A possible geometry is shown in Figure 6.42a; the superdislocation partials shown each contain a jog produced, for example, by intersection with a forest dislocation, which are non-aligned along the direction of the Burgers vector. When the dislocation glides and the jogs move non-conservatively a tube of APBs is generated. Direct evidence for the existence of tubes from weak-beam electron microscope studies was first reported for Fe–30 at.% Al. The micrographs show faint lines along $\langle 1\ 1\ 1\rangle$, the Burgers vector direction, and are about 3 nm in width. The images are expected to be weak, since the contrast arises from two closely spaced overlapping faults, the second effectively canceling the displacement caused by the first, and are visible only when superlattice reflections are excited. APB tubes have since been observed in other compounds.

Theory suggests that jogs in superdislocations in screw orientations provide a potent hardening mechanism, estimated to be about eight times as strong as that resulting from pulling out of APB tubes on non-aligned jogs on edge dislocations. The major contributions to the stress to move a dislocation are (1) τ_{s}, the stress to generate point defects or tubes, and (2) the interaction stress τ_{i} with dislocations on neighboring slip planes, and $\tau_{\mathrm{s}} + \tau_{\mathrm{i}} = \frac{3}{4}\alpha_{\mathrm{s}}\mu(\rho_{\mathrm{f}}/\rho_{\mathrm{p}})\varepsilon$. Thus, with $\alpha_{\mathrm{s}} = 1.3$ and provided $\rho_{\mathrm{f}}/\rho_{\mathrm{p}}$ is constant and small, linear hardening with the observed rate is obtained.

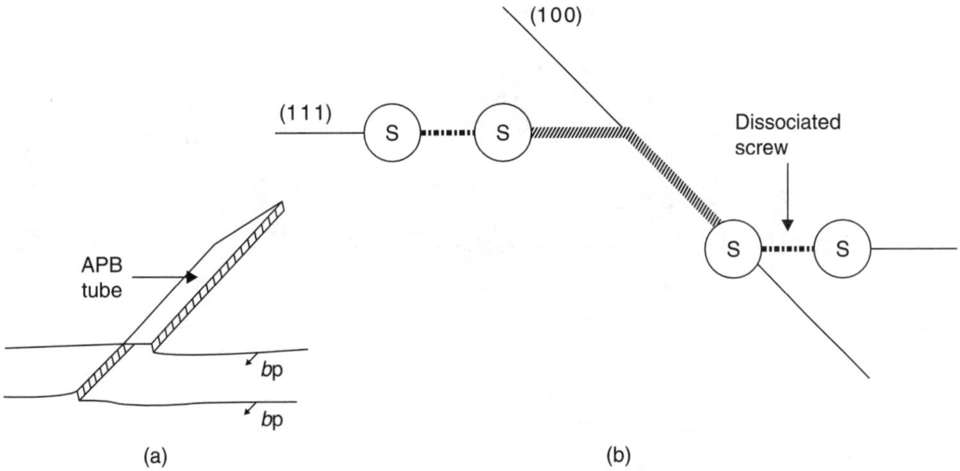

Figure 6.42 *Schematic diagram of superdislocation: (a) with non-aligned jogs which, after glide, produce an APB tube and (b) cross-slipped onto the cube plane to form a Kear–Wilsdorf (K–W) lock.*

In crystals with A_3B order only one rapid stage of hardening is observed compared with the normal three-stage hardening of fcc metals. Moreover, the temperature dependence of θ_{11}/μ increases with temperature and peaks at $\sim 0.4T_m$. It has been argued that the APB tube model is unable to explain why anomalously high work-hardening rates are observed for those single-crystal orientations favorable for single slip on $\{1\ 1\ 1\}$ planes alone. An alternative model to APB tubes has been proposed based on cross-slip of the leading unit dislocation of the superdislocation. If the second unit dislocation cannot follow exactly in the wake of the first, both will be pinned.

For alloys with $L1_2$ structure the cross-slip of a screw superpartial with $b = \frac{1}{2}[\bar{1}\ 0\ 1]$ from the primary (1 1 1) plane to the (0 1 0) plane was first proposed by Kear and Wilsdorf. The two $\frac{1}{2}[\bar{1}\ 0\ 1]$ superpartials, one on the (1 1 1) plane and the other on the (0 1 0) plane, are of course dissociated into $\langle 1\ 1\ 2\rangle$-type partials and the whole configuration is sessile. This dislocation arrangement is known as a Kear–Wilsdorf (K–W) lock and is shown in Figure 6.42b. Since cross-slip is thermally activated, the number of locks and therefore the resistance to (1 1 1) glide will increase with increasing temperature. This could account for the increase in yield stress with temperature, while the onset of cube slip at elevated temperatures could account for the peak in the flow stress.

Cube cross-slip and cube slip has now been observed in a number of $L1_2$ compounds by TEM. There is some TEM evidence that the APB energy on the cube plane is lower than that on the (1 1 1) plane (see Chapter 8) to favor cross-slip, which would be aided by the torque, arising from elastic anisotropy, exerted between the components of the screw dislocation pair.

6.6.3 Development of preferred orientation

6.6.3.1 Crystallographic aspects

When a polycrystalline metal is plastically deformed the individual grains tend to rotate into a common orientation. This preferred orientation is developed gradually with increasing deformation, and although it becomes extensive above about 90% reduction in area, it is still inferior to that of a good single crystal. The degree of texture produced by a given deformation is readily shown on a monochromatic X-ray transmission photograph, since the grains no longer reflect uniformly into the

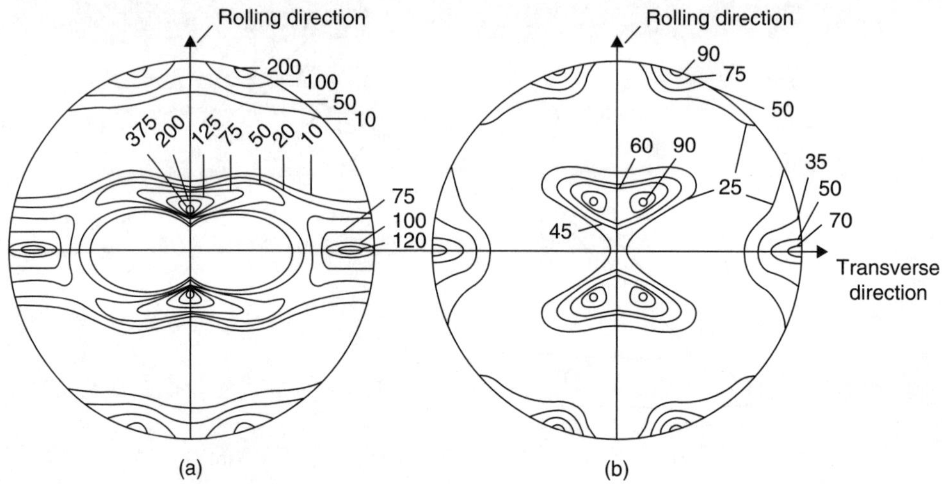

Figure 6.43 *(1 1 1) pole figures from copper (a) and α-brass (b) after 95% deformation (intensities in arbitrary units).*

Table 6.3 *Deformation textures in metals with common crystal structures.*

Structure	*Wire (fiber texture)*	*Sheet (rolling texture)*
Bcc	[1 1 0]	{1 1 2}⟨1 $\bar{1}$ 0⟩ to {1 0 0} ⟨0 1 1⟩
Fcc	[1 1 1], [1 0 0] double fiber	{1 1 0} ⟨1 1 2⟩ to {3 $\bar{5}$ 1}⟨1 1 2⟩
Cph	[2 1 0]	{0 0 0 1} ⟨1 0 0 0⟩

diffraction rings but only into certain segments of them. The results are usually described in terms of an ideal orientation, such as $[u, v, w]$ for the fibre texture developed by drawing or swaging, and $\{h\,k\,l\}$ $\langle u\,v\,w\rangle$ for a rolling texture for which a plane of the form $(h\,k\,l)$ lies parallel to the rolling plane and a direction of the type $\langle u\,v\,w\rangle$ is parallel to the rolling direction. However, the scatter about the ideal orientation can only be represented by means of a pole figure, which describes the spread of orientation about the ideal orientation for a particular set of $(h\,k\,l)$ poles (see Figure 6.43).

In tension, the grains rotate in such a way that the movement of the applied stress axis is towards the operative slip direction as discussed in Section 6.3.5 and for compression the applied stress moves towards the slip plane normal. By considering the deformation process in terms of the particular stresses operating and applying the appropriate grain rotations it is possible to predict the stable end-grain orientation and hence the texture developed by extensive deformation. Table 6.3 shows the predominant textures found in different metal structures for both wires and sheet.

For fcc metals a marked transition in deformation texture can be effected either by lowering the deformation temperature or by adding solid solution alloying elements which lower the stacking-fault energy. The transition relates to the effect on deformation modes of reducing stacking-fault energy or thermal energy, deformation banding and twinning becoming more prevalent and cross-slip less important at lower temperatures and stacking-fault energies. This texture transition can be achieved in most fcc metals by alloying additions and by altering the rolling temperature. Al, however, has a high fault energy and because of the limited solid solubility it is difficult to lower by alloying. The extreme types of rolling texture, shown by copper and 70/30 brass, are given in Figure 6.43a and b.

In bcc metals there are no striking examples of solid solution alloying effects on deformation texture, the preferred orientation developed being remarkably insensitive to material variables. However, material variables can affect cph textures markedly. Variations in c/a ratio alone cause alterations in the orientation developed, as may be appreciated by consideration of the twinning modes, and it is also possible that solid solution elements alter the relative values of critical resolved shear stress for different deformation modes. Processing variables are also capable of giving a degree of control in hexagonal metals. No texture, stable to further deformation, is found in hexagonal metals and the angle of inclination of the basal planes to the sheet plane varies continuously with deformation. In general, the basal plane lies at a small angle ($<45°$) to the rolling plane, tilted either towards the rolling direction (Zn, Mg) or towards the transverse direction (Ti, Zr, Be, Hf).

The deformation texture cannot, in general, be eliminated by an annealing operation even when such a treatment causes recrystallization. Instead, the formation of a new annealing texture usually results, which is related to the deformation texture by standard lattice rotations.

6.6.3.2 Texture hardening

The flow stress in single crystals varies with orientation according to Schmid's law and hence materials with a preferred orientation will also show similar plastic anisotropy, depending on the perfection of the texture. The significance of this relationship is well illustrated by a crystal of beryllium, which is cph and capable of slip only on the basal plane, a compressive stress approaching $\approx 2000\ \text{MN m}^{-2}$ applied normal to the basal plane produces negligible plastic deformation. Polycrystalline beryllium sheet, with a texture such that the basal planes lie in the plane of the sheet, shows a correspondingly high strength in biaxial tension. When stretched uniaxially the flow stress is also quite high, when additional (prismatic) slip planes are forced into action even though the shear stress for their operation is five times greater than for basal slip. During deformation there is little thinning of the sheet, because the $\langle 1\,1\,\bar{2}\,0\rangle$ directions are aligned in the plane of the sheet. Other hexagonal metals, such as titanium and zirconium, show less marked strengthening in uniaxial tension because prismatic slip occurs more readily, but resistance to biaxial tension can still be achieved. Applications of texture hardening lie in the use of suitably textured sheet for high biaxial strength, e.g. pressure vessels, dent resistance, etc. Because of the multiplicity of slip systems, cubic metals offer much less scope for texture hardening. Again, a consideration of single-crystal deformation gives the clue; for, whereas in a hexagonal crystal m can vary from 2 (basal planes at 45° to the stress axis) to infinity (when the basal planes are normal), in an fcc crystal m can vary only by a factor of 2 with orientation, and in bcc crystals the variation is rather less. In extending this approach to polycrystalline material certain assumptions have to be made about the mutual constraints between grains. One approach gives $m = 3.1$ for a random aggregate of fcc crystals and the calculated orientation dependence of σ/τ for fiber texture shows that a rod with $\langle 1\,1\,1\rangle$ or $\langle 1\,1\,0\rangle$ texture ($\sigma/\tau = 3.664$) is 20% stronger than a random structure; the cube texture ($\sigma/\tau = 2.449$) is 20% weaker.

If conventional mechanical properties were the sole criterion for texture-hardened materials, then it seems unlikely that they would challenge strong precipitation-hardened alloys. However, texture hardening has more subtle benefits in sheet metal forming in optimizing fabrication performance. The variation of strength in the plane of the sheet is readily assessed by tensile tests carried out in various directions relative to the rolling direction. In many sheet applications, however, the requirement is for through-thickness strength (e.g. to resist thinning during pressing operations). This is more difficult to measure and is often assessed from uniaxial tensile tests by measuring the ratio of the strain in the width direction to that in the thickness direction of a test piece. The strain ratio R is given by

$$\begin{aligned} R &= \varepsilon_w/\varepsilon_t = \ln(w_0/w)/\ln(t_0/t) \\ &= \ln(w_0/w)/\ln(wL/w_0L_0), \end{aligned} \tag{6.31}$$

where w_0, L_0 and t_0 are the original dimensions of width, length and thickness, and w, L and t are the corresponding dimensions after straining, which is derived assuming no change in volume occurs.

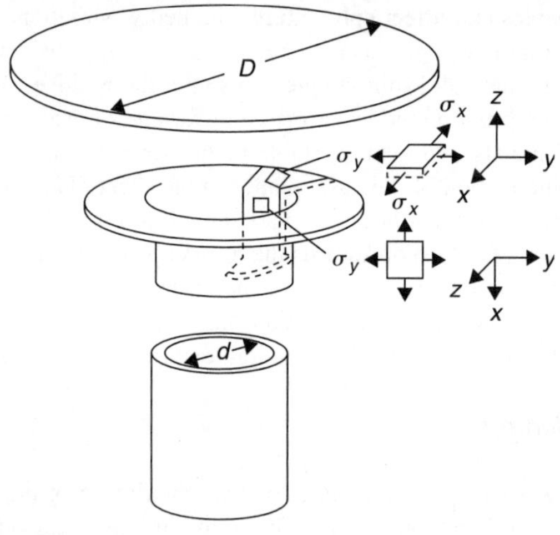

Figure 6.44 *Schematic diagram of the deep-drawing operations, indicating the stress systems operating in the flange and the cup wall. Limiting drawing ratio is defined as the ratio of the diameter of the largest blank which can satisfactorily complete the draw* (D_{max}) *to the punch diameter (d) (after Dillamore, Smallman and Wilson, 1969; courtesy of the Institute of Materials, Minerals and Mining).*

The average strain ratio $\overline{R}$, for tests at various angles in the plane of the sheet, is a measure of the normal anisotropy, i.e. the difference between the average properties in the plane of the sheet and that property in the direction normal to the sheet surface. A large value of R means that there is a lack of deformation modes oriented to provide strain in the through-thickness direction, indicating a high through-thickness strength.

In deep drawing, schematically illustrated in Figure 6.44, the dominant stress system is radial tension combined with circumferential compression in the drawing zone, while that in the base and lower cup wall (i.e. central stretch-forming zone) is biaxial tension. The latter stress is equivalent to a through-thickness compression, plus a hydrostatic tension which does not affect the state of yielding. Drawing failure occurs when the central stretch-forming zone is insufficiently strong to support the load needed to draw the outer part of the blank through the die. Clearly, differential strength levels in these two regions, leading to greater ease of deformation in the drawing zone compared with the stretching zone, would enable deeper draws to be made: this is the effect of increasing the $\overline{R}$ value, i.e. high through-thickness strength relative to strength in the plane of the sheet will favor drawability. This is confirmed in Figure 6.45, where deep drawability as determined by limiting drawing ratio (i.e. ratio of maximum drawable blank diameter to final cup diameter) is remarkably insensitive to ductility and, by inference from the wide range of materials represented in the figure, to absolute strength level. Here it is noted that for hexagonal metals slip occurs readily along $\langle 1\ 1\ \bar{2}\ 0\rangle$, thus contributing no strain in the c-direction, and twinning only occurs on $\{1\ 0\ \bar{1}\ 2\}$ when the applied stress nearly parallel to the c-axis is compressive for $c/a > \sqrt{3}$ and tensile for $c/a < \sqrt{3}$. Thus, titanium, $c/a < \sqrt{3}$, has a high strength in through-thickness compression, whereas Zn with $c/a < \sqrt{3}$ has low through-thickness strength when the basal plane is oriented parallel to the plane of the sheet. In contrast, hexagonal metals with $c/a > \sqrt{3}$ would have a high R for $\{1\ 0\ \bar{1}\ 0\}$ parallel to the plane of the sheet.

Texture hardening is much less in the cubic metals, but fcc materials with $\{1\ 1\ 1\}\ \langle 1\ 1\ 0\rangle$ slip system and bcc with $\{1\ 1\ 0\}\ \langle 1\ 1\ 1\rangle$ are expected to increase R when the texture has component with $\{1\ 1\ 1\}$ and $\{1\ 1\ 0\}$ parallel to the plane of the sheet. The range of values of $\overline{R}$ encountered

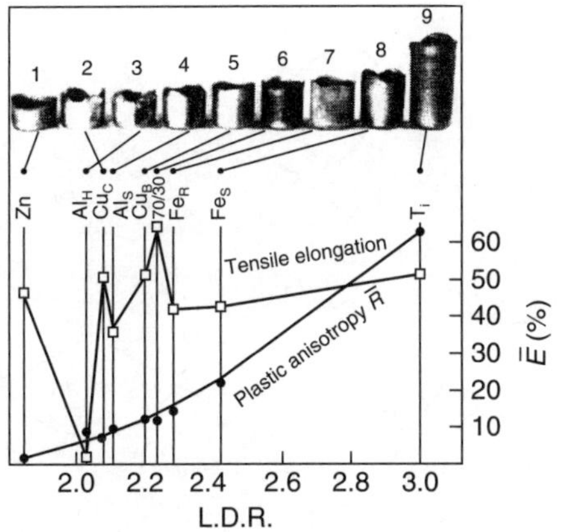

Figure 6.45 *Limiting draw ratios (LDR) as a function of average values of* R *and of elongation to fracture measured in tensile tests at 0°, 45° and 90° to the rolling direction (after Wilson, 1966; courtesy of the Institute of Materials, Minerals and Mining).*

in cubic metals is much less. Face-centered cubic metals have $\overline{R}$ ranging from about 0.3 for cube texture, $\{1\ 0\ 0\}\ \langle 0\ 0\ 1\rangle$, to a maximum, in textures so far attained, of just over 1.0. Higher values are sometimes obtained in body-centered cubic metals. Values of $\overline{R}$ in the range 1.4–1.8 obtained in aluminum-killed low-carbon steel are associated with significant improvements in deep-drawing performance compared with rimming steel, which has $\overline{R}$-values between 1.0 and 1.4. The highest values of R in steels are associated with texture components with $\{1\ 1\ 1\}$ parallel to the surface, while crystals with $\{1\ 0\ 0\}$ parallel to the surface have a strongly depressing effect on $\overline{R}$.

In most cases it is found that the R values vary with testing direction and this has relevance in relation to the strain distribution in sheet metal forming. In particular, ear formation on pressings generally develops under a predominant uniaxial compressive stress at the edge of the pressing. The ear is a direct consequence of the variation in strain ratio for different directions of uniaxial stressing, and a large variation in R value, where $\Delta R = (R_{max} - R_{min})$ generally correlates with a tendency to form pronounced ears. On this basis we could write a simple recipe for good deep-drawing properties in terms of strain ratio measurements made in a uniaxial tensile test as high R and low ΔR. Much research is aimed at improving forming properties through texture control.

6.7 Macroscopic plasticity

6.7.1 Tresca and von Mises criteria

In dislocation theory it is usual to consider the flow stress or yield stress of ductile metals under simple conditions of stressing. In practice, the engineer deals with metals under more complex conditions of stressing (e.g. during forming operations) and hence needs to correlate yielding under combined stresses with that in uniaxial testing. To achieve such a yield stress criterion it is usually assumed that the metal is mechanically isotropic and deforms plastically at constant volume, i.e. a hydrostatic state of stress does not affect yielding. In assuming plastic isotropy, macroscopic shear is allowed to take place along lines of maximum shear stress and crystallographic slip is ignored, and the yield stress in tension is equal to that in compression, i.e. there is no Bauschinger effect.

A given applied stress state in terms of the principal stresses σ_1, σ_2, σ_3 which act along three principal axes, X_1, X_2 and X_3, may be separated into the hydrostatic part (which produces changes

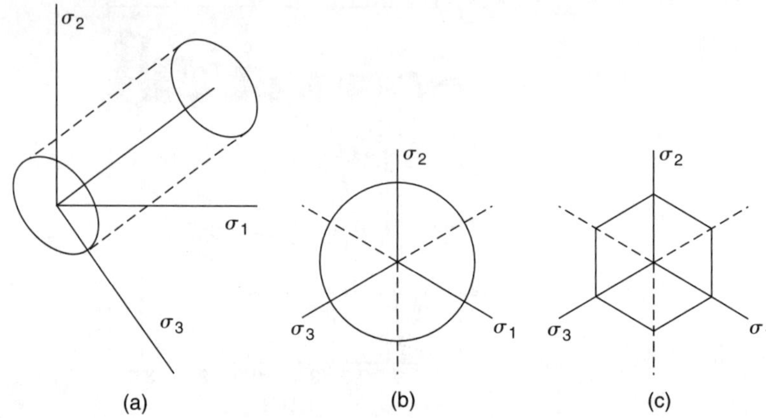

Figure 6.46 *Schematic representation of the yield surface with: (a) principal stresses σ_1, σ_2 and σ_3, (b) von Mises yield criterion and (c) Tresca yield criterion.*

in volume) and the deviatoric components (which produce changes in shape). It is assumed that the hydrostatic component has no effect on yielding and hence the more the stress state deviates from pure hydrostatic, the greater the tendency to produce yield. The stresses may be represented on a stress–space plot (see Figure 6.46a), in which a line equidistant from the three stress axes represents a pure hydrostatic stress state. Deviation from this line will cause yielding if the deviation is sufficiently large, and define a yield surface which has sixfold symmetry about the hydrostatic line. This arises because the conditions of isotropy imply equal yield stresses along all three axes, and the absence of the Bauschinger effect implies equal yield stresses along σ_1 and $-\sigma_1$. Taking a section through stress space, perpendicular to the hydrostatic line, gives the two simplest yield criteria satisfying the symmetry requirements corresponding to a regular hexagon and a circle.

The hexagonal form represents the Tresca criterion (see Figure 6.46c), which assumes that plastic shear takes place when the maximum shear stress attains a critical value k equal to shear yield stress in uniaxial tension. This is expressed by

$$\tau_{\max} = \frac{\sigma_1 - \sigma_3}{2} = k, \tag{6.32}$$

where the principal stresses $\sigma_1 > \sigma_2 > \sigma_3$. This criterion is the isotropic equivalent of the law of resolved shear stress in single crystals. The tensile yield stress $Y = 2k$ is obtained by putting $\sigma_1 = Y$, $\sigma_2 = \sigma_3 = 0$.

The circular cylinder is described by the equation:

$$(\sigma_1 - \sigma_2)^2 + (\sigma_2 - \sigma_3)^2 + (\sigma_3 - \sigma_1)^2 = \text{constant} \tag{6.33}$$

and is the basis of the von Mises yield criterion (see Figure 6.46b). This criterion implies that yielding will occur when the shear energy per unit volume reaches a critical value given by the constant. This constant is equal to $6k^2$ or $2Y^2$, where k is the yield stress in simple shear, as shown by putting $\sigma_2 = 0$, $\sigma_1 = \sigma_3$, and Y is the yield stress in uniaxial tension when $\sigma_2 = \sigma_3 = 0$. Clearly, $Y = 3k$ compared to $Y = 2k$ for the Tresca criterion and, in general, this is found to agree somewhat closer with experiment.

In many practical working processes (e.g. rolling), the deformation occurs under approximately plane strain conditions with displacements confined to the X_1X_2 plane. It does not follow that the

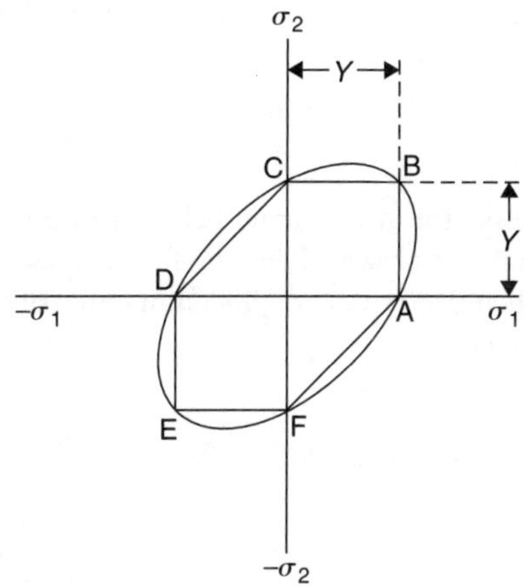

Figure 6.47 *The von Mises yield ellipse and Tresca yield hexagon.*

stress in this direction is zero and, in fact, the deformation conditions are satisfied if $\sigma_3 = \frac{1}{2}(\sigma_1 + \sigma_2)$, so that the tendency for one pair of principal stresses to extend the metal along the X_3-axis is balanced by that of the other pair to contract it along this axis. Eliminating σ_3 from the von Mises criterion, the yield criterion becomes

$$(\sigma_1 - \sigma_2) = 2k$$

and the plane strain yield stress, i.e. when $\sigma_2 = 0$, given when

$$\sigma_1 = 2k = 2Y/\sqrt{3} = 1.15Y.$$

For plane strain conditions, the Tresca and von Mises criteria are equivalent and two-dimensional flow occurs when the shear stress reaches a critical value. The above condition is thus equally valid when written in terms of the deviatoric stresses $\sigma_1', \sigma_2', \sigma_3'$ defined by equations of the type:

$$\sigma_1' = \sigma_1 - \tfrac{1}{3}(\sigma_1 + \sigma_2 + \sigma_3).$$

Under plane stress conditions, $\sigma_3 = 0$ and the yield surface becomes two-dimensional and the von Mises criterion becomes

$$\sigma_1^2 + \sigma_1\sigma_2 + \sigma_2^2 = 3k^2 = Y^2, \tag{6.34}$$

which describes an ellipse in the stress plane. For the Tresca criterion the yield surface reduces to a hexagon inscribed in the ellipse, as shown in Figure 6.47. Thus, when σ_1 and σ_2 have opposite signs, the Tresca criterion becomes $\sigma_1 - \sigma_2 = 2k - Y$ and is represented by the edges of the hexagon CD and FA. When they have the same sign, then $\sigma_1 = 2k = Y$ or $\sigma_2 = 2k = Y$ and defines the hexagon edges AB, BC, DE and EF.

6.7.2 Effective stress and strain

For an isotropic material, a knowledge of the uniaxial tensile test behavior, together with the yield function, should enable the stress–strain behavior to be predicted for any stress system. This is achieved by defining an effective stress–effective strain relationship such that if $\sigma = K\varepsilon^n$ is the uniaxial stress–strain relationship then we may write

$$\overline{\sigma} = K\overline{\varepsilon}^n \tag{6.35}$$

for any state of stress. The stress–strain behavior of a thin-walled tube with internal pressure is a typical example, and it is observed that the flow curves obtained in uniaxial tension and in biaxial torsion coincide when the curves are plotted in terms of effective stress and effective strain. These quantities are defined by:

$$\overline{\sigma} = \frac{\sqrt{2}}{2}[(\sigma_1 - \sigma_2)^2 + (\sigma_2 - \sigma_3)^2 + (\sigma_3 - \sigma_1)^2]^{1/2} \tag{6.36}$$

and

$$\overline{\varepsilon} = \frac{\sqrt{2}}{3}[(\varepsilon_1 - \varepsilon_2)^2 + (\varepsilon_2 - \varepsilon_3)^2 + (\varepsilon_3 - \varepsilon_1)^2]^{1/2}, \tag{6.37}$$

where $\varepsilon_1, \varepsilon_2$ and ε_3 are the principal strains, all of which reduce to the axial normal components of stress and strain for a tensile test. It should be emphasized, however, that this generalization holds only for isotropic media and for constant loading paths, i.e. $\sigma_1 = \alpha\sigma_2 = \beta\sigma_3$, where α and β are constants independent of the value of σ_1.

6.8 Annealing

6.8.1 General effects of annealing

When a metal is cold worked, by any of the many industrial shaping operations, changes occur in both its physical and mechanical properties. While the increased hardness and strength which result from the working treatment may be of importance in certain applications, it is frequently necessary to return the metal to its original condition to allow further forming operations (e.g. deep drawing) to be carried out for applications where optimum physical properties, such as electrical conductivity, are essential. The treatment given to the metal to bring about a decrease of the hardness and an increase in the ductility is known as annealing. This usually means keeping the deformed metal for a certain time at a temperature higher than about one-third the absolute melting point.

Cold working produces an increase in dislocation density; for most metals ρ increases from the value of 10^{10}–10^{12} lines m^{-2} typical of the annealed state, to 10^{12}–10^{13} after a few percent deformation, and up to 10^{15}–10^{16} lines m^{-2} in the heavily deformed state. Such an array of dislocations gives rise to a substantial strain energy stored in the lattice, so that the cold-worked condition is thermodynamically unstable relative to the undeformed one. Consequently, the deformed metal will try to return to a state of lower free energy, i.e. a more perfect state. In general, this return to a more equilibrium structure cannot occur spontaneously but only at elevated temperatures, where thermally activated processes such as diffusion, cross-slip and climb take place. Like all non-equilibrium processes the rate of approach to equilibrium will be governed by an Arrhenius equation of the form:

$$\text{Rate} = A\exp[-Q/\mathbf{k}T],$$

where the activation energy Q depends on impurity content, strain, etc.

The formation of atmospheres by strain ageing is one method whereby the metal reduces its excess lattice energy, but this process is unique in that it usually leads to a further increase in the

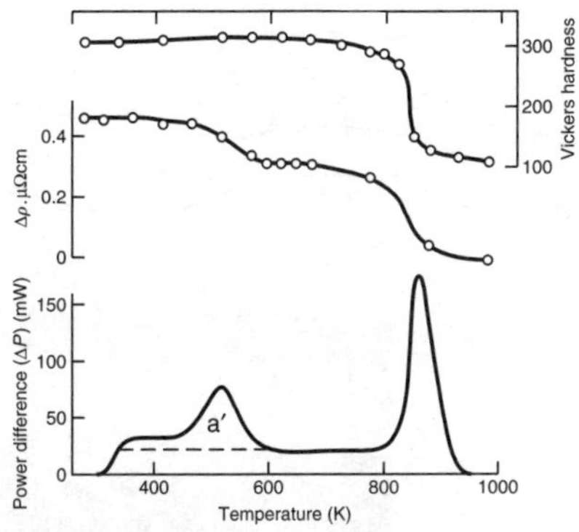

Figure 6.48 *Rate of release of stored energy (ΔP), increment in electrical resistivity (Δρ) and hardness (VPN) for specimens of nickel deformed in torsion and heated at 6 K min*$^{-1}$ *(Clareborough, Hargreaves and West, 1955).*

structure-sensitive properties rather than a reduction to the value characteristic of the annealed condition. It is necessary, therefore, to increase the temperature of the deformed metal above the strain-ageing temperature before it recovers its original softness and other properties.

The removal of the cold-worked condition occurs by a combination of three processes, namely: (1) recovery, (2) recrystallization and (3) grain growth. These stages have been successfully studied using light microscopy, transmission electron microscopy or X-ray diffraction; mechanical property measurements (e.g. hardness); and physical property measurements (e.g. density, electrical resistivity and stored energy). Figure 6.48 shows the change in some of these properties on annealing. During the recovery stage the decrease in stored energy and electrical resistivity is accompanied by only a slight lowering of hardness, and the greatest simultaneous change in properties occurs during the primary recrystallization stage. However, while these measurements are no doubt striking and extremely useful, it is necessary to understand them to correlate such studies with the structural changes by which they are accompanied.

6.8.2 Recovery

This process describes the changes in the distribution and density of defects with associated changes in physical and mechanical properties which take place in worked crystals before recrystallization or alteration of orientation occurs. It will be remembered that the structure of a cold-worked metal consists of dense dislocation networks, formed by the glide and interaction of dislocations, and, consequently, the recovery stage of annealing is chiefly concerned with the rearrangement of these dislocations to reduce the lattice energy and does not involve the migration of large-angle boundaries. This rearrangement of the dislocations is assisted by thermal activation. Mutual annihilation of dislocations is one process.

When the two dislocations are on the same slip plane, it is possible that as they run together and annihilate they will have to cut through intersecting dislocations on other planes, i.e. 'forest' dislocations. This recovery process will therefore be aided by thermal fluctuations, since the activation energy for such a cutting process is small. When the two dislocations of opposite sign are not on the same slip plane, climb or cross-slip must first occur, and both processes require thermal activation.

One of the most important recovery processes which leads to a resultant lowering of the lattice strain energy is rearrangement of the dislocations into cell walls. This process in its simplest form was originally termed polygonization and is illustrated schematically in Figure 6.49, whereby dislocations

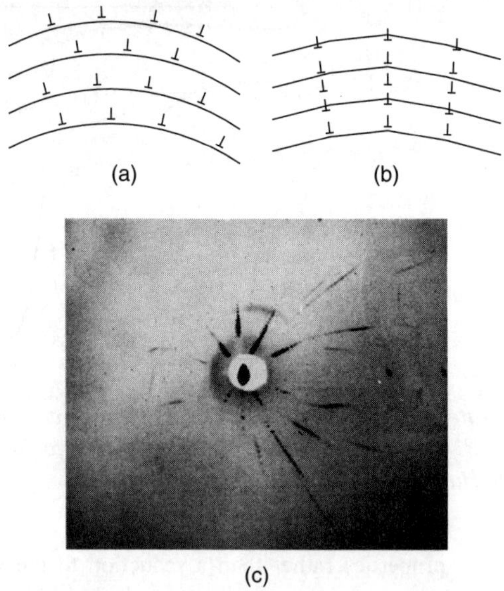

Figure 6.49 *(a) Random arrangement of excess parallel edge dislocations. (b) Alignment into dislocation walls. (c) Laue photograph of polygonized zinc (after Cahn, 1949; courtesy of Institute of Materials, Minerals and Mining).*

all of one sign align themselves into walls to form small-angle or sub-grain boundaries. During deformation a region of the lattice is curved, as shown in Figure 6.49a, and the observed curvature can be attributed to the formation of excess edge dislocations parallel to the axis of bending. On heating, the dislocations form a sub-boundary by a process of annihilation and rearrangement. This is shown in Figure 6.49b, from which it can be seen that it is the excess dislocations of one sign which remain after the annihilation process that align themselves into walls.

Polygonization is a simple form of sub-boundary formation and the basic movement is climb, whereby the edge dislocations change their arrangement from a horizontal to a vertical grouping. This process involves the migration of vacancies to or from the edge of the half-planes of the dislocations (see Section 3.3.4). The removal of vacancies from the lattice, together with the reduced strain energy of dislocations which results, can account for the large change in both electrical resistivity and stored energy observed during this stage, while the change in hardness can be attributed to the rearrangement of dislocations and to the reduction in the density of dislocations.

The process of polygonization can be demonstrated using the Laue method of X-ray diffraction. Diffraction from a bent single crystal of zinc takes the form of continuous radial streaks. On annealing, these asterisms break up into spots, as shown in Figure 6.49c, where each diffraction spot originates from a perfect polygonized sub-grain, and the distance between the spots represents the angular misorientation across the sub-grain boundary. Direct evidence for this process is observed in the electron microscope, where, in heavily deformed polycrystalline aggregates at least, recovery is associated with the formation of sub-grains out of complex dislocation networks by a process of dislocation annihilation and rearrangement. In some deformed metals and alloys the dislocations are already partially arranged in sub-boundaries, forming diffuse cell structures by dynamical recovery (see Figure 6.40). The conventional recovery process is then one in which these cells sharpen and grow. In other metals, dislocations are more uniformly distributed after deformation, with hardly any cell

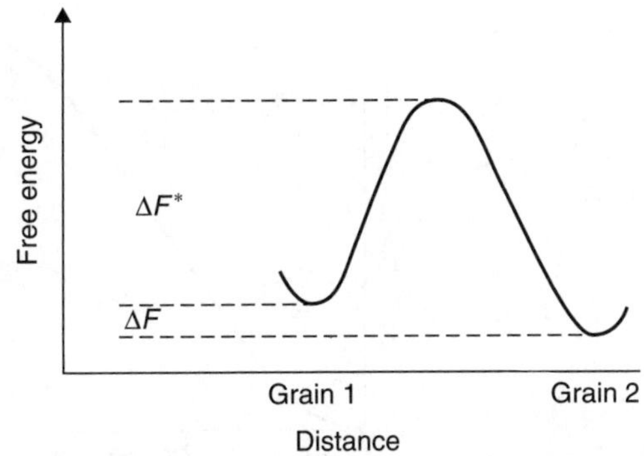

Figure 6.50 *Variation in free energy during grain growth.*

structure discernible, and the recovery process then involves formation, sharpening and growth of sub-boundaries. The sharpness of the cell structure formed by deformation depends on the stacking-fault energy of the metal, the deformation temperature and the extent of deformation (see Figure 6.41).

6.8.3 Recrystallization

The most significant changes in the structure-sensitive properties occur during the primary recrystallization stage. In this stage the deformed lattice is completely replaced by a new unstrained one by means of a nucleation and growth process, in which practically stress-free grains grow from nuclei formed in the deformed matrix. The orientation of the new grains differs considerably from that of the crystals they consume, so that the growth process must be regarded as incoherent, i.e. it takes place by the advance of large-angle boundaries separating the new crystals from the strained matrix.

During the growth of grains, atoms get transferred from one grain to another across the boundary. Such a process is thermally activated, as shown in Figure 6.50, and by the usual reaction-rate theory the frequency of atomic transfer one way is

$$\nu \exp\left(-\frac{\Delta F}{\mathbf{k}T}\right) \mathrm{s}^{-1} \tag{6.38}$$

and in the reverse direction

$$\nu \exp\left(-\frac{\Delta F^* + \Delta F}{\mathbf{k}T}\right) \mathrm{s}^{-1}, \tag{6.39}$$

where ΔF is the difference in free energy per atom between the two grains, i.e. supplying the driving force for migration, and ΔF^* is an activation energy. For each net transfer the boundary moves forward a distance b and the velocity v is given by

$$v = M\Delta F, \tag{6.40}$$

where M is the mobility of the boundary, i.e. the velocity for unit driving force, and is thus

$$M = \frac{b\nu}{\mathbf{k}T} \exp\left(\frac{\Delta S^*}{\mathbf{k}}\right) \exp\left(-\frac{\Delta E^*}{\mathbf{k}T}\right). \tag{6.41}$$

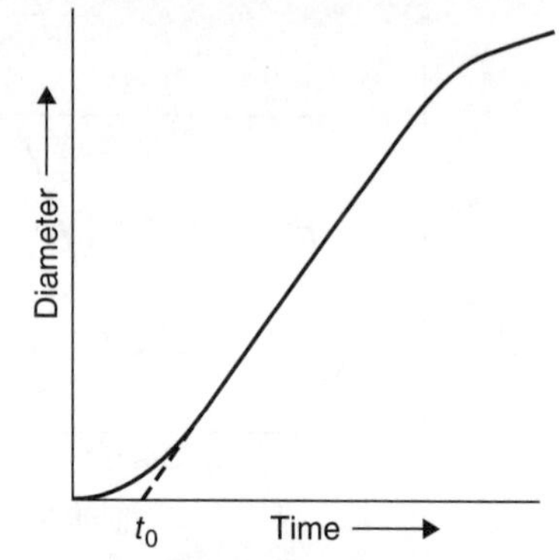

Figure 6.51 *Variation of grain diameter with time at a constant temperature.*

Generally, the open structure of high-angle boundaries should lead to a high mobility. However, they are susceptible to the segregation of impurities, low concentrations of which can reduce the boundary mobility by orders of magnitude. In contrast, special boundaries which are close to a CSL are much less affected by impurity segregation and hence can lead to higher relative mobility.

It is well known that the rate of recrystallization depends on several important factors, namely: (1) the amount of prior deformation (the greater the degree of cold work, the lower the recrystallization temperature and the smaller the grain size), (2) the temperature of the anneal (as the temperature is lowered, the time to attain a constant grain size increases exponentially[5]) and (3) the purity of the sample (e.g. zone-refined aluminum recrystallizes below room temperature, whereas aluminum of commercial purity must be heated several hundred degrees). The role these variables play in recrystallization will be evident once the mechanism of recrystallization is known. This mechanism will now be outlined.

Measurements, using the light microscope, of the increase in diameter of a new grain as a function of time at any given temperature can be expressed as shown in Figure 6.51. The diameter increases linearly with time until the growing grains begin to impinge on one another, after which the rate necessarily decreases. The classical interpretation of these observations is that nuclei form spontaneously in the matrix after a so-called nucleation time, t_0, and these nuclei then proceed to grow steadily as shown by the linear relationship. The driving force for the process is provided by the stored energy of cold work contained in the strained grain on one side of the boundary relative to that on the other side. Such an interpretation would suggest that the recrystallization process occurs in two distinct stages, i.e. first nucleation and then growth.

During the linear growth period the radius of a nucleus is $R = G\,(t - t_0)$, where G, the growth rate, is $\mathrm{d}R/\mathrm{d}t$ and, assuming the nucleus is spherical, the volume of the recrystallized nucleus is $4\pi/3\,G^3(t - t_0)^3$. If the number of nuclei that form in a time increment $\mathrm{d}t$ is $N\mathrm{d}t$ per unit volume of unrecrystallized matrix, and if the nuclei do not impinge on one another, then for unit total volume

$$f = \frac{4}{3}\pi NG^3 \int_0^t (t - t_0)^3 \mathrm{d}t$$

or

$$f = \frac{\pi}{3} G^3 t^4. \tag{6.42}$$

[5] The velocity of linear growth of new crystals usually obeys an exponential relationship of the form $\nu = \nu_0 \exp[-Q/\mathbf{R}T]$.

This equation is valid in the initial stages when $f << 1$. When the nuclei impinge on one another the rate of recrystallization decreases and is related to the amount untransformed $(1-f)$ by

$$f = 1 - \exp\left(-\frac{\pi}{3}NG^3t^4\right), \tag{6.43}$$

where, for short times, equation (6.42) reduces to equation (6.41). This Johnson–Mehl equation is expected to apply to any phase transformation where there is random nucleation, constant N and G, and small t_0. In practice, nucleation is not random and the rate not constant, so that equation (6.43) will not strictly apply. For the case where the nucleation rate decreases exponentially, Avrami developed the equation:

$$f = 1 - \exp(-kt^n), \tag{6.44}$$

where k and n are constants, with $n \approx 3$ for a fast and $n \approx 4$ for a slow decrease of nucleation rate. Provided there is no change in the nucleation mechanism, n is independent of temperature but k is very sensitive to temperature T; clearly, from equation (6.43), $k = \pi NG^3/3$, and both N and G depend on T.

An alternative interpretation is that the so-called incubation time t_0 represents a period during which small nuclei, of a size too small to be observed in the light microscope, are growing very slowly. This latter interpretation follows from the recovery stage of annealing. Thus, the structure of a recovered metal consists of sub-grain regions of practically perfect crystal and, thus, one might expect the 'active' recrystallization nuclei to be formed by the growth of certain sub-grains at the expense of others.

The process of recrystallization may be pictured as follows. After deformation, polygonization of the bent lattice regions on a fine scale occurs and this results in the formation of several regions in the lattice where the strain energy is lower than in the surrounding matrix; this is a necessary primary condition for nucleation. During this initial period when the angles between the sub-grains are small and less than one degree, the sub-grains form and grow quite rapidly. However, as the sub-grains grow to such a size that the angles between them become of the order of a few degrees, the growth of any given sub-grain at the expense of the others is very slow. Eventually one of the sub-grains will grow to such a size that the boundary mobility begins to increase with increasing angle. A large-angle boundary, $\theta \approx 30$–$40°$, has a high mobility because of the large lattice irregularities or 'gaps' which exist in the boundary transition layer. The atoms on such a boundary can easily transfer their allegiance from one crystal to the other. This sub-grain is then able to grow at a much faster rate than the other sub-grains which surround it and so acts as the nucleus of a recrystallized grain. The further it grows, the greater will be the difference in orientation between the nucleus and the matrix it meets and consumes, until it finally becomes recognizable as a new strain-free crystal separated from its surroundings by a large-angle boundary.

The recrystallization nucleus therefore has its origin as a sub-grain in the deformed microstructure. Whether it grows to become a strain-free grain depends on three factors: (1) the stored energy of cold work must be sufficiently high to provide the required driving force, (2) the potential nucleus should have a size advantage over its neighbors and (3) it must be capable of continued growth by existing in a region of high lattice curvature (e.g. transition band), so that the growing nucleus can quickly achieve a high-angle boundary. *In situ* experiments in the HVEM have confirmed these factors. Figure 6.52a shows the as-deformed substructure in the transverse section of rolled copper, together with the orientations of some selected areas. The sub-grains are observed to vary in width from 50 to 500 nm, and exist between regions 1 and 8 as a transition band across which the orientation changes sharply. On heating to 200°C, the sub-grain region 2 grows into the transition region and the orientation of the new grain well developed at 300°C is identical to the original sub-grain (Figure 6.52b).

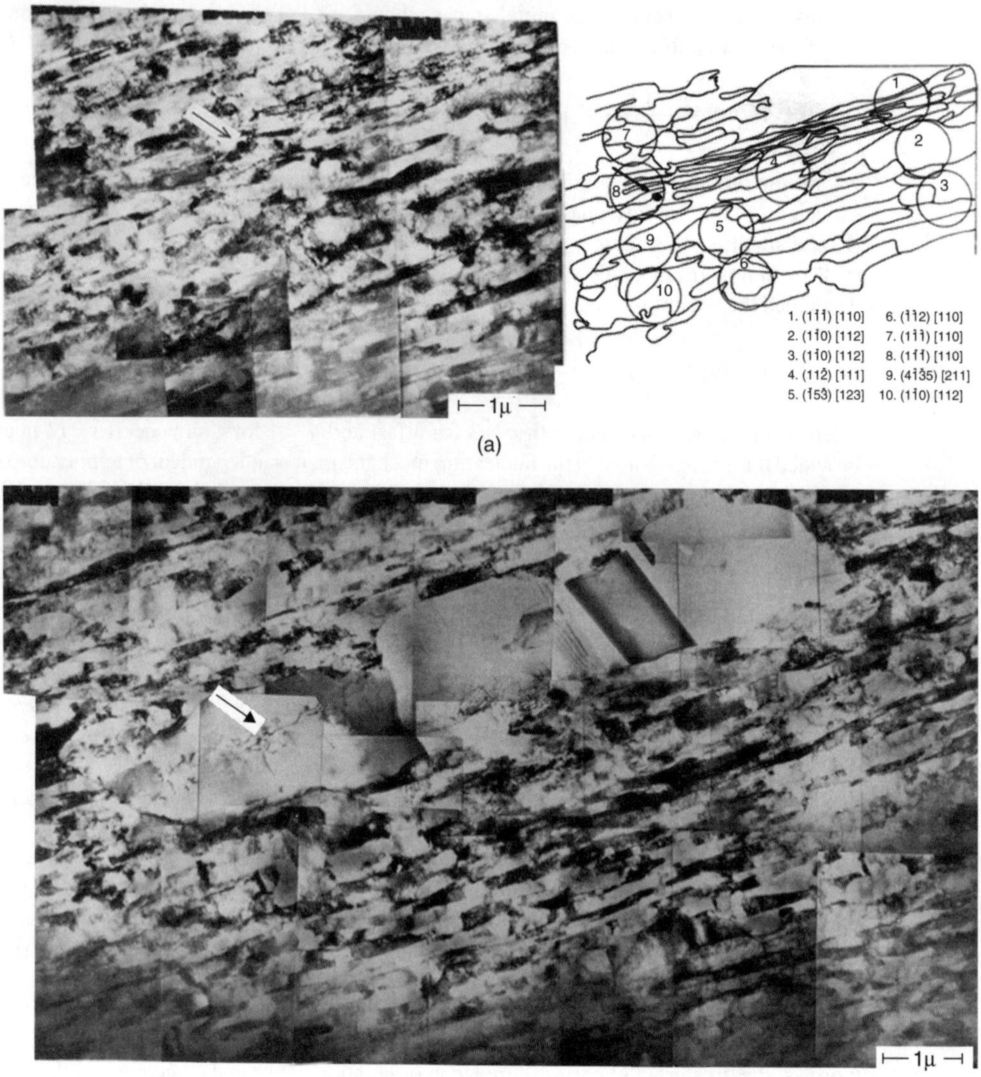

Figure 6.52 *Electron micrographs of copper. (a) Cold-rolled 95% at room temperature, transverse section. (b) Heated to 300°C in the HVEM.*

With this knowledge of recrystallization the influence of several variables known to affect the recrystallization behavior of a metal can now be understood. Prior deformation, for example, will control the extent to which a region of the lattice is curved. The larger the deformation, the more severely will the lattice be curved and, consequently, the smaller will be the size of a growing sub-grain when it acquires a large-angle boundary. This must mean that a shorter time is necessary at any given temperature for the sub-grain to become an 'active' nucleus or, conversely, that the higher the annealing temperature, the quicker will this stage be reached. In some instances, heavily cold-worked metals recrystallize without any significant recovery owing to the formation of strain-free cells during deformation. The importance of impurity content on recrystallization temperature is also evident from the effect impurities have on obstructing sub-boundary dislocation and grain boundary mobility.

The intragranular nucleation of strain-free grains, as discussed above, is considered as abnormal sub-grain growth, in which it is necessary to specify that some sub-grains acquire a size advantage and are able to grow at the expense of the normal sub-grains. It has been suggested that nuclei may also be formed by a process involving the rotation of individual cells so that they coalesce with neighboring cells to produce larger cells by volume diffusion and dislocation rearrangement.

In some circumstances, intergranular nucleation is observed in which an existing grain boundary bows out under an initial driving force equal to the difference in free energy across the grain boundary. This strain-induced boundary migration is irregular and is from a grain with low strain (i.e. large cell size) to one of larger strain and smaller cell size. For a boundary to grow in this way the strain energy difference per unit volume across the boundary must be sufficient to supply the energy increase to bow out a length of boundary ≈ 1 μm.

Segregation of solute atoms to, and precipitation on, the grain boundary tends to inhibit intergranular nucleation and gives an advantage to intragranular nucleation, provided the dispersion is not too fine. In general, the recrystallization behavior of two-phase alloys is extremely sensitive to the dispersion of the second phase. Small, finely dispersed particles retard recrystallization by reducing both the nucleation rate and the grain boundary mobility, whereas large, coarsely dispersed particles enhance recrystallization by increasing the nucleation rate. During deformation, zones of high dislocation density and large misorientations are formed around non-deformable particles, and, on annealing, recrystallization nuclei are created within these zones by a process of polygonization by sub-boundary migration. Particle-stimulated nucleation occurs above a critical particle size, which decreases with increasing deformation. The finer dispersions tend to homogenize the microstructure (i.e. dislocation distribution), thereby minimizing local lattice curvature and reducing nucleation.

The formation of nuclei becomes very difficult when the spacing of second-phase particles is so small that each developing sub-grain interacts with a particle before it becomes a viable nucleus. The extreme case of this is SAP (sintered aluminum powder), which contains very stable, close-spaced oxide particles. These particles prevent the rearrangement of dislocations into cell walls and their movement to form high-angle boundaries, and hence SAP must be heated to a temperature very close to the melting point before it recrystallizes.

6.8.4 Grain growth

When primary recrystallization is complete (i.e. when the growing crystals have consumed all the strained material), the material can lower its energy further by reducing its total area of grain surface. With extensive annealing it is often found that grain boundaries straighten, small grains shrink and larger ones grow. The general phenomenon is known as grain growth, and the most important factor governing the process is the surface tension of the grain boundaries. A grain boundary has a surface tension, T (= surface free energy per unit area), because its atoms have a higher free energy than those within the grains. Consequently, to reduce this energy a polycrystal will tend to minimize the area of its grain boundaries and, when this occurs, the configuration taken up by any set of grain boundaries (see Figure 6.53) will be governed by the condition that

$$T_A/\sin A = T_B/\sin B = T_C/\sin C. \tag{6.45}$$

Most grain boundaries are of the large-angle type with their energies approximately independent of orientation, so that for a random aggregate of grains $T_A = T_B = T_C$ and the equilibrium grain boundary angles are each equal to 120°. Figure 6.53b shows an idealized grain in two dimensions surrounded by others of uniform size, and it can be seen that the equilibrium grain shape takes the form of a polygon of six sides with 120° inclusive angles. All polygons with either more or less than this number of sides cannot be in equilibrium. At high temperatures where the atoms are mobile,

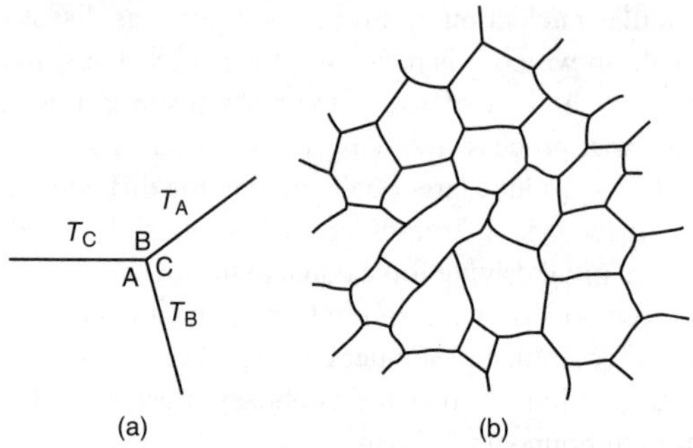

Figure 6.53 *(a) Relation between angles and surface tensions at a grain boundary triple point. (b) Idealized polygonal grain structure.*

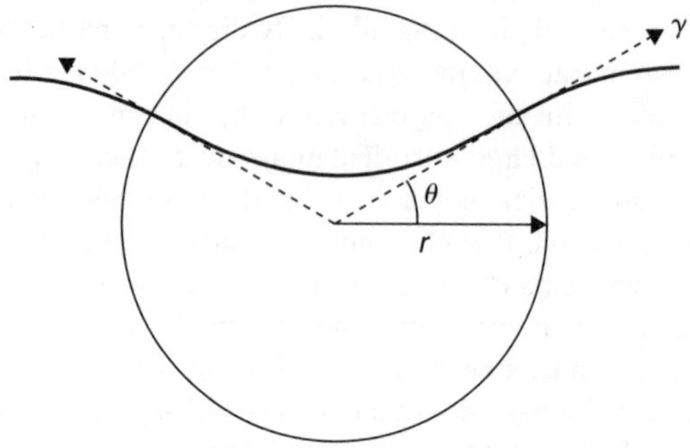

Figure 6.54 *Diagram showing the drag exerted on a boundary by a particle.*

a grain with fewer sides will tend to become smaller, under the action of the grain boundary surface tension forces, while one with more sides will tend to grow.

Second-phase particles have a major inhibiting effect on boundary migration and are particularly effective in the control of grain size. The pinning process arises from surface tension forces exerted by the particle–matrix interface on the grain boundary as it migrates past the particle. Figure 6.54 shows that the drag exerted by the particle on the boundary, resolved in the forward direction, is

$$F = \pi r\gamma \sin 2\theta,$$

where γ is the specific interfacial energy of the boundary; $F = F_{\max} = \pi r\gamma$ when $\theta = 45°$. Now if there are N particles per unit volume, the volume fraction is $4\pi r^3 N/3$ and the number n intersecting unit area of boundary is given by

$$n = 3f/2\pi r^2. \tag{6.46}$$

For a grain boundary migrating under the influence of its own surface tension the driving force is $2\gamma/R$, where R is the minimum radius of curvature and, as the grains grow, R increases and the driving force decreases until it is balanced by the particle drag, when growth stops. If $R \sim d$, the mean grain diameter, then the critical grain diameter is given by the condition

$$nF \approx 2\gamma/d_{\text{crit}}$$

or

$$d_{\text{crit}} \approx 2\gamma(2\pi r^2/3f\pi r\gamma) = 4r/3f. \tag{6.47}$$

This Zener drag equation overestimates the driving force for grain growth by considering an isolated spherical grain. A heterogeneity in grain size is necessary for grain growth, and taking this into account gives a revised equation:

$$d_{\text{crit}} \approx \frac{\pi r}{3f}\left[\frac{3}{2} - \frac{2}{Z}\right], \tag{6.48}$$

where Z is the ratio of the diameters of growing grains to the surrounding grains. This treatment explains the successful use of small particles in refining the grain size of commercial alloys.

During the above process growth is continuous and a uniform coarsening of the polycrystalline aggregate usually occurs. Nevertheless, even after growth has finished the grain size in a specimen which was previously severely cold-worked remains relatively small, because of the large number of nuclei produced by the working treatment. Exaggerated grain growth can often be induced, however, in one of two ways, namely: (1) by subjecting the specimen to a critical strain-anneal treatment or (2) by a process of secondary recrystallization. By applying a critical deformation (usually a few percent strain) to the specimen the number of nuclei will be kept to a minimum, and if this strain is followed by a high-temperature anneal in a thermal gradient some of these nuclei will be made more favorable for rapid growth than others. With this technique, if the conditions are carefully controlled, the whole of the specimen may be turned into one crystal, i.e. a single crystal. The term secondary recrystallization describes the process whereby a specimen which has been given a primary recrystallization treatment at a low temperature is taken to a higher temperature to enable the abnormally rapid growth of a few grains to occur. The only driving force for secondary recrystallization is the reduction of grain boundary free energy, as in normal grain growth, and, consequently, certain special conditions are necessary for its occurrence. One condition for this 'abnormal' growth is that normal continuous growth is impeded by the presence of inclusions, as is indicated by the exaggerated grain growth of tungsten wire containing thoria, or the sudden coarsening of deoxidized steel at about 1000°C. A possible explanation for the phenomenon is that in some regions the grain boundaries become free (e.g. if the inclusions slowly dissolve or the boundary tears away) and as a result the grain size in such regions becomes appreciably larger than the average (Figure 6.55a). It then follows that the grain boundary junction angles between the large grain and the small ones that surround it will not satisfy the condition of equilibrium discussed above. As a consequence, further grain boundary movement to achieve 120° angles will occur, and the accompanying movement of a triple junction point will be as shown in Figure 6.55b. However, when the dihedral angles at each junction are approximately 120° a severe curvature in the grain boundary segments between the junctions will arise, and this leads to an increase in grain boundary area. Movement of these curved boundary segments towards their centers of curvature must then take place and this will give rise to the configuration shown in Figure 6.55c. Clearly, this sequence of events can be repeated and continued growth of the large grains will result.

The behavior of the dispersed phase is extremely important in secondary recrystallization and there are many examples in metallurgical practice where the control of secondary recrystallization with dispersed particles has been used to advantage. One example is in the use of Fe–3% Si in the production of strip for transformer laminations. This material is required with (1 1 0) [0 0 1] 'Goss' texture because of the [0 0 1] easy direction of magnetization, and it is found that the presence of MnS particles favors the growth of secondary grains with the appropriate Goss texture. Another example is in the removal of the pores during the sintering of metal and ceramic powders, such as alumina and metallic carbides. The sintering process is essentially one of vacancy creep involving the diffusion of vacancies from the pore of radius r to a neighboring grain boundary, under a

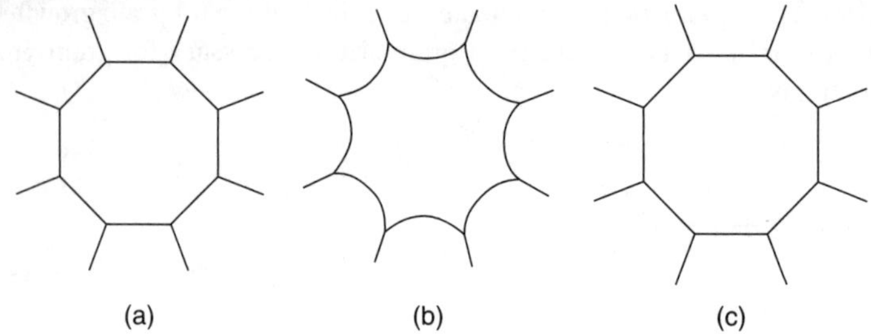

Figure 6.55 *Grain growth during secondary recrystallization.*

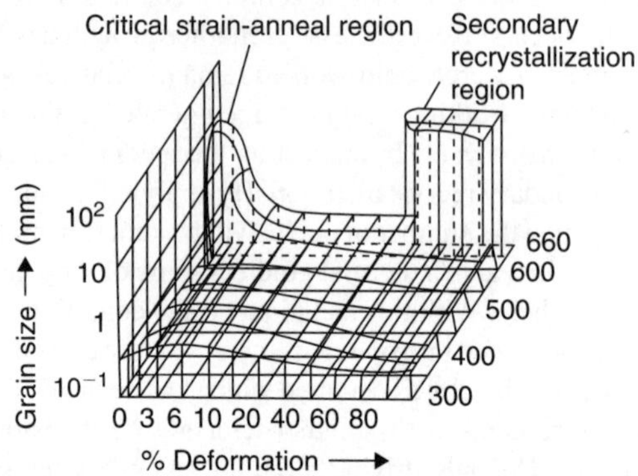

Figure 6.56 *Relation between grain size, deformation and temperature for aluminum (after Burgers, courtesy of Akademie-Verlags-Gesellschaft).*

driving force $2\gamma_S/r$, where γ_S is the surface energy. In practice, sintering occurs fairly rapidly up to about 95% full density because there is a plentiful association of boundaries and pores. When the pores become very small, however, they are no longer able to anchor the grain boundaries against the grain growth forces, and hence the pores sinter very slowly, since they are stranded within the grains some distance from any boundary. To promote total sintering, an effective dispersion is added. The dispersion is critical, however, since it must produce sufficient drag to slow down grain growth, during which a particular pore is crossed by several migrating boundaries, but not sufficiently large to give rise to secondary recrystallization when a given pore would be stranded far from any boundary.

The relation between grain size, temperature and strain is shown in Figure 6.56 for commercially pure aluminum. From this diagram it is clear that either a critical strain-anneal treatment or a secondary recrystallization process may be used for the preparation of perfect strain-free single crystals.

6.8.5 Annealing twins

A prominent feature of the microstructures of most annealed fcc metals and alloys is the presence of many straight-sided bands that run across grains. These bands have a twinned orientation relative to their neighboring grain and are referred to as annealing twins (see Chapter 3). The parallel boundaries

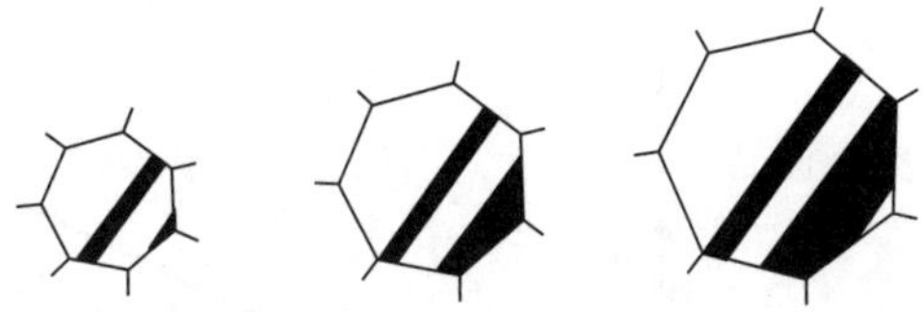

Figure 6.57 *Formation and growth of annealing twins (from Burke and Turnbull, 1952).*

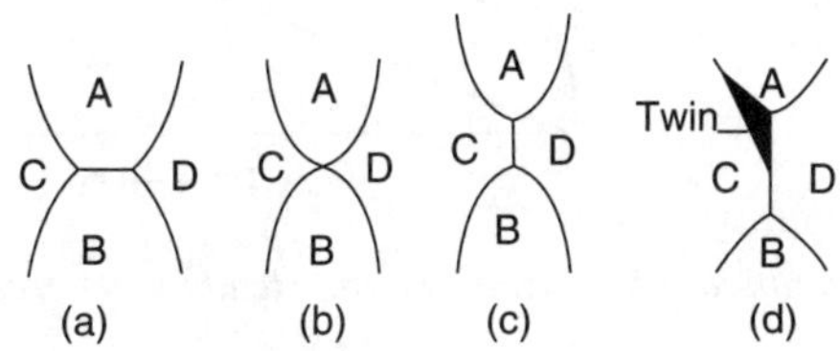

Figure 6.58 *Nucleation of an annealing twin during grain growth.*

usually coincide with a (1 1 1) twinning plane with the structure coherent across it, i.e. both parts of the twin hold a single (1 1 1) plane in common.

As with formation of deformation twins, it is believed that a change in stacking sequence is all that is necessary to form an annealing twin. Such a change in stacking sequence may occur whenever a properly oriented grain boundary migrates. For example, if the boundary interface corresponds to a (1 1 1) plane, growth will proceed by the deposition of additional (1 1 1) planes in the usual stacking sequence **ABCABC**.... If, however, the next newly deposited layer falls into the wrong position, the sequence **ABCABCB** is produced, which constitutes the first layer of a twin. Once a twin interface is formed, further growth may continue with the sequence in reverse order, **ABCABC|BACB**... until a second accident in the stacking sequence completes the twin band, **ABCABCBACBACBABC**. When a stacking error, such as that described above, occurs the number of nearest neighbors is unchanged, so that the ease of formation of a twin interface depends on the relative value of the interface energy. If this interface energy is low, as in copper, where $\gamma_{\text{twin}} < 20\,\text{mJ}\,\text{m}^{-2}$ twinning occurs frequently while, if it is high, as in aluminum, the process is rare.

Annealing twins are rarely (if ever) found in cast metals because grain boundary migration is negligible during casting. Worked and annealed metals show considerable twin band formation; after extensive grain growth a coarse-grained metal often contains twins which are many times wider than any grain that was present shortly after recrystallization. This indicates that twin bands grow in width, during grain growth, by migration in a direction perpendicular to the (1 1 1) composition plane, and one mechanism whereby this can occur is illustrated schematically in Figure 6.57. This shows that a twin may form at the corner of a grain, since the grain boundary configuration will then have a lower interfacial energy. If this happens the twin will then be able to grow in width because one of its sides forms part of the boundary of the growing grain. Such a twin will continue to grow in width until a second mistake in the positioning of the atomic layers terminates it; a complete twin band is then formed. In copper and its alloys, $\gamma_{\text{twin}}/\gamma_{\text{gb}}$ is low and hence twins occur frequently, whereas in aluminum the corresponding ratio is very much higher and so twins are rare.

Twins may develop according to the model shown in Figure 6.58 where, during grain growth, a grain contact is established between grains C and D. Then if the orientation of grain D is close to the twin orientation of grain C, the nucleation of an annealing twin at the grain boundary, as shown in Figure 6.58d, will lower the total boundary energy. This follows because the twin/D interfaces will be reduced to about 5% of the normal grain boundary energy, the energies of the C/A and twin/A interfaces will be approximately the same, and the extra area of interface C/twin has only a very low

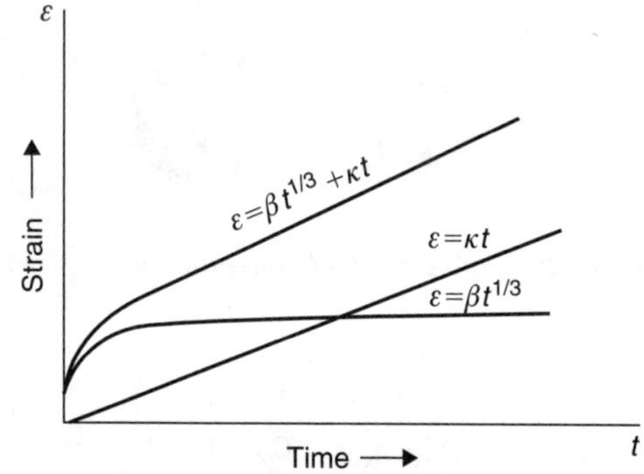

Figure 6.59 *Combination of transient and steady-state creep.*

energy. This model indicates that the number of twins per unit grain boundary area depends only on the number of new grain contacts made during grain growth, irrespective of grain size and annealing temperature.

6.8.6 Recrystallization textures

The preferred orientation developed by cold work often changes on recrystallization to a totally different preferred orientation. To explain this observation, Barrett and (later) Beck have put forward the 'oriented growth' theory of recrystallization textures, in which it is proposed that nuclei of many orientations initially form but, because the rate of growth of any given nucleus depends on the orientation difference between the matrix and growing crystal, the recrystallized texture will arise from those nuclei which have the fastest growth rate in the cold-worked matrix, i.e. those bounded by large-angle boundaries. It then follows that, because the matrix has a texture, all the nuclei which grow will have orientations that differ by 30–40° from the cold-worked texture. This explains why the new texture in fcc metals is often related to the old texture, by a rotation of approximately 30–40° around ⟨1 1 1⟩ axes, in bcc metals by 30° about ⟨1 1 0⟩ and in hcp by 30° about ⟨0 0 0 1⟩. However, while it is undoubtedly true that oriented growth provides a selection between favorably and unfavorably oriented nuclei, there are many observations to indicate that the initial nucleation is not entirely random. For instance, because of the crystallographic symmetry one would expect grains appearing in an fcc texture to be related to rotations about all four ⟨1 1 1⟩ axes, i.e. eight orientations arising from two possible rotations about each of the four ⟨1 1 1⟩ axes. All these possible orientations are rarely (if ever) observed.

To account for such observations, and for those cases where the deformation texture and the annealing texture show strong similarities, oriented nucleation is considered to be important. The oriented nucleation theory assumes that the selection of orientations is determined in the nucleation stage. It is generally accepted that all recrystallization nuclei pre-exist in the deformed matrix, as sub-grains, which become more perfect through recovery processes prior to recrystallization. It is thus most probable that there is some selection of nuclei determined by the representation of the orientations in the deformation texture, and that the oriented nucleation theory should apply in some cases. In many cases the orientations which are strongly represented in the annealing texture are very weakly represented in the deformed material. The most striking example is the 'cube' texture, (1 0 0) [0 0 1], found in most fcc pure metals which have been annealed following heavy rolling reductions. In this texture, the cube axes are extremely well aligned along the sheet axes, and its behavior resembles that of a single crystal. It is thus clear that cube-oriented grains or sub-grains must have a very high initial growth rate in order to form the remarkably strong quasi-single-crystal cube texture. The percentage of cubically aligned grains increases with increased deformation, but the sharpness of the textures is profoundly affected by alloying additions. The amount of alloying addition required to suppress the texture depends on those factors which affect the stacking-fault energy, such as the lattice misfit of the solute atom in the solvent lattice, valency, etc., in much the same way as that described for the transition of a pure metal deformation texture.

In general, however, if the texture is to be altered a distribution of second phase must either be present before cold rolling or be precipitated during annealing. In aluminum, for example, the amount of cube texture can be limited in favor of retained rolling texture by limiting the amount of grain growth with a precipitate dispersion of Si and Fe. By balancing the components, earing can be minimized in drawn aluminum cups. In aluminum-killed steels A1N precipitation prior to recrystallization produces a higher proportion of grains with $\{1\ 1\ 1\}$ planes parallel to the rolling plane and a high $\overline{R}$-value suitable for deep drawing. The A1N dispersion affects sub-grain growth, limiting the available nuclei and increasing the orientation selectivity, thereby favoring the high-energy $\{1\ 1\ 1\}$ grains. Improved $\overline{R}$-values in steels in general are probably due to the combined effect of particles in homogenizing the deformed microstructure and in controlling the subsequent sub-grain growth. The overall effect is to limit the availability of nuclei with orientations other than $\{1\ 1\ 1\}$.

6.9 Metallic creep

6.9.1 Transient and steady-state creep

Creep is the process by which plastic flow occurs when a constant stress is applied to a metal for a prolonged period of time. After the initial strain ε_0 which follows the application of the load, creep usually exhibits a rapid transient period of flow (stage 1) before it settles down to the linear steady-stage stage 2, which eventually gives way to tertiary creep and fracture. Transient creep, sometimes referred to as β-creep, obeys a $t^{1/3}$ law. The linear stage of creep is often termed steady-state creep and obeys the relation

$$\varepsilon = \kappa t. \tag{6.49}$$

Consequently, because both transient and steady-state creep usually occur together during creep at high temperatures, the complete curve (Figure 6.59) during the primary and secondary stages of creep fits the equation

$$\varepsilon = \beta t^{1/3} + \kappa t \tag{6.50}$$

extremely well. In contrast to transient creep, steady-state creep increases markedly with both temperature and stress. At constant stress the dependence on temperature is given by

$$\dot{\varepsilon}_{ss} = d\varepsilon/dt = \text{const.} \exp[-Q/\mathbf{k}T], \tag{6.51}$$

where Q is the activation energy for steady-state creep, while at constant temperature the dependence on stress σ (compensated for modulus E) is

$$\dot{\varepsilon}_{ss} = \text{const.}(\sigma/E)^n. \tag{6.52}$$

Steady-state creep is therefore described by the equation:

$$\dot{\varepsilon}_{ss} = A(\sigma/E)^n \exp(-Q/\mathbf{k}T). \tag{6.53}$$

The basic assumption of the mechanism of steady-state creep is that during the creep process the rate of recovery r (i.e. decrease in strength, $\mathrm{d}\sigma/\mathrm{d}t$) is sufficiently fast to balance the rate of work hardening $h = (\mathrm{d}\sigma/\mathrm{d}\varepsilon)$. The creep rate $(\mathrm{d}\varepsilon/\mathrm{d}t)$ is then given by

$$\mathrm{d}\varepsilon/\mathrm{d}t = (\mathrm{d}\sigma/\mathrm{d}t)/(\mathrm{d}\sigma/\mathrm{d}\varepsilon) = r/h. \tag{6.54}$$

To prevent work hardening, both the screw and edge parts of a glissile dislocation loop must be able to escape from tangled or piled-up regions. The edge dislocations will, of course, escape by climb, and since this process requires a higher activation energy than cross-slip, it will be the rate-controlling process in steady-state creep. The rate of recovery is governed by the rate of climb, which depends on diffusion and stress such that

$$r = A(\sigma/E)^p D = A(\sigma/E)^p D_0 \exp[-Q/\mathbf{k}T],$$

where D is a diffusion coefficient and the stress term arises because recovery is faster the higher the stress level and the closer dislocations are together. The work-hardening rate decreases from the initial rate h_0 with increasing stress, i.e. $h = h_0(E/\sigma)^q$, thus

$$\dot{\varepsilon}_{ss} = r/h = B(\sigma/E)^n D, \tag{6.55}$$

where B $(=A/h_0)$ is a constant and n $(=p+q)$ is the stress exponent.

The structure developed in creep arises from the simultaneous work hardening and recovery. The dislocation density ρ increases with ε and the dislocation network gets finer, since dislocation spacing is proportional to $\rho^{-1/2}$. At the same time, the dislocations tend to reduce their strain energy by mutual annihilation and rearrange to form low-angle boundaries, and this increases the network spacing. Straining then proceeds at a rate at which the refining action just balances the growth of the network by recovery; the equilibrium network size being determined by the stress. Although dynamical recovery can occur by cross-slip, the rate-controlling process in steady-state creep is climb, whereby edge dislocations climb out of their glide planes by absorbing or emitting vacancies; the activation energy is therefore that of self-diffusion. Structural observations confirm the importance of the recovery process to steady-state creep. These show that sub-grains form within the original grains and, with increasing deformation, the sub-grain angle increases while the dislocation density within them remains constant.[6] The climb process may, of course, be important in several different ways. Thus, climb may help a glissile dislocation to circumvent different barriers in the structure, such as a sessile dislocation, or it may lead to the annihilation of dislocations of opposite sign on different glide planes. Moreover, because creep-resistant materials are rarely pure metals, the climb process may also be important in allowing a glissile dislocation to get round a precipitate or move along a grain boundary. A comprehensive analysis of steady-state creep, based on the climb of dislocations, has been given by Weertman.

The activation energy for creep Q may be obtained experimentally by plotting $\ln \dot{\varepsilon}_{ss}$ versus $1/T$, as shown in Figure 6.60. Usually above $0.5T_m$, Q corresponds to the activation energy for self-diffusion E_{SD}, in agreement with the climb theory, but below $0.5T_m$, $Q < E_{SD}$, possibly corresponding to pipe diffusion. Figure 6.61 shows that three creep regimes may be identified and the temperature range where $Q = E_{SD}$ can be moved to higher temperatures by increasing the strain rate. Equation

[6] Sub-grains do not always form during creep and in some metallic solid solutions where the glide of dislocations is restrained due to the dragging of solute atoms, the steady-state substructure is essentially a uniform distribution of dislocations.

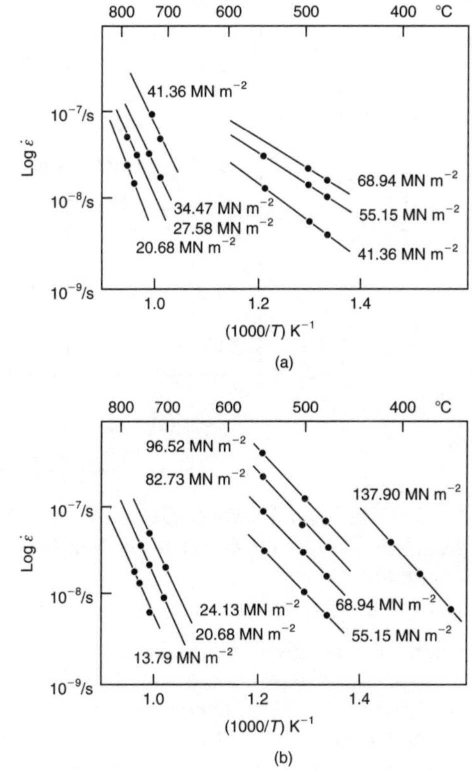

Figure 6.60 *Log $\dot{\varepsilon}$ versus $1/T$ for Ni–Al_2O_3 (a) and Ni–67Co–Al_2O_3 (b), showing the variation in activation energy above and below $0.5T_m$ (after Hancock, Dillamore and Smallman, 1972; courtesy of Institute of Materials, Minerals and Mining).*

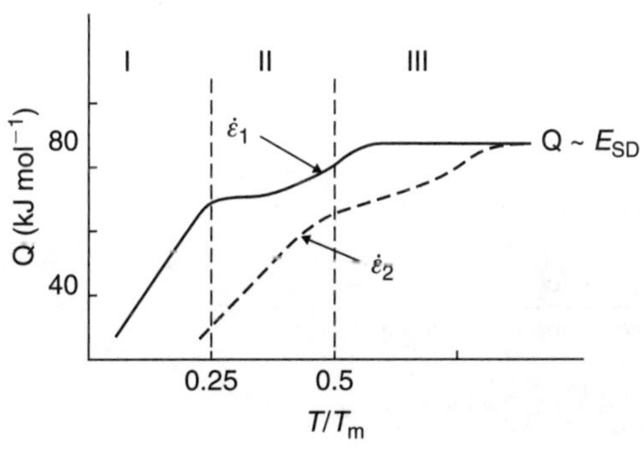

Figure 6.61 *Variation in activation energy Q with temperature for aluminum.*

(6.55) shows that the stress exponent n can be obtained experimentally by plotting $\ln \dot{\varepsilon}_{ss}$ versus $\ln \sigma$, as shown in Figure 6.62, where $n \approx 4$. While n is generally about 4 for dislocation creep, Figure 6.63 shows that n may vary considerably from this value depending on the stress regime; at low stresses (i.e. regime I) creep occurs not by dislocation glide and climb, but by stress-directed flow of vacancies.

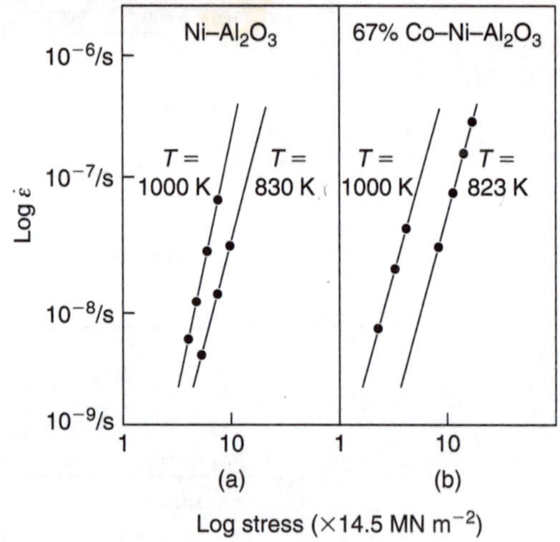

Figure 6.62 *Log $\dot{\varepsilon}$ versus log σ for Ni–Al_2O_3 (a) and Ni–67Co–Al_2O_3 (b) (after Hancock, Dillamore and Smallman, 1972; courtesy of Institute of Materials, Minerals and Mining).*

Worked example

Creep data for a light alloy are given in the table

Stress (*N mm^{-2}*)	*Temperature* (*K*)	*Minimum creep rate* (*s^{-1}*)
8.9	600	1×10^{-5}
5.0	600	1×10^{-6}
5.0	640	5×10^{-6}

Calculate the expected steady-state creep rate at a constant stress of 2.8 N mm^{-2} at (i) 600 K and (ii) 640 K.

Solution

The creep equation: $\dot{\varepsilon}_{ss} = A\sigma^n \exp(-Q/\mathbf{R}T)$.

At constant $T = 600$ K,

$$n = \frac{\log \dot{\varepsilon}_1 - \log \dot{\varepsilon}_2}{\log \sigma_1 - \log \sigma_2} = \frac{-5 - (-6)}{\log 8.9 - \log 5} = 4$$

Therefore, at 2.8 MPa, 600 K,

$$-5 - \log \dot{\varepsilon} = 4 \times (\log 8.9 - \log 2.8)$$

$$\therefore \dot{\varepsilon} = 0.98 \times 10^{-7} \text{s}^{-1}.$$

At constant $\sigma = 5.0$ MPa,

$$\ln 10^{-6} = \ln(A\sigma^n) - \frac{Q}{\mathbf{R}T_1}$$

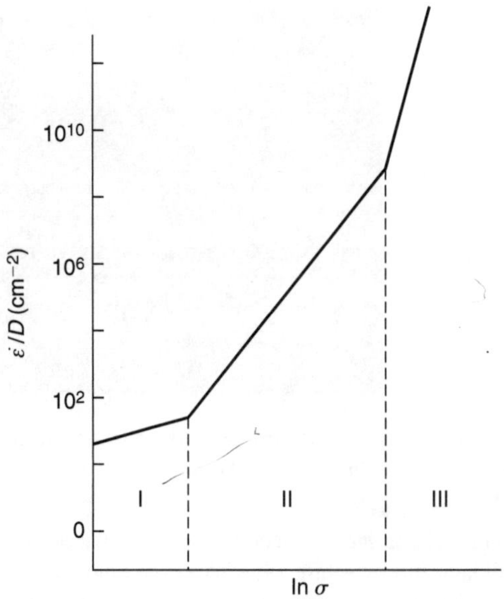

Figure 6.63 *Schematic diagram showing influence of stress on diffusion-compensated steady-state creep.*

$$\ln 5 \times 10^{-6} = \ln(A\sigma^n) - \frac{Q}{\mathbf{R}T_2}$$

$$\therefore Q = (\ln 5 \times 10^{-6} - \ln 10^{-6}) \times \frac{8.314}{1/600 - 1/640} = 128.4 \times 10^3\,\mathrm{J\,mol^{-1}} = 128.4\,\mathrm{kJ\,mol^{-1}}.$$

At constant $\sigma = 2.8$ MPa,

$$\ln \dot{\varepsilon} - \ln(0.98 \times 10^{-7}) = \frac{Q}{\mathbf{R}}\left(\frac{1}{600} - \frac{1}{640}\right) = \frac{128.4 \times 10^3}{8.314} \times \left(\frac{1}{600} - \frac{1}{640}\right)$$

$$\therefore \dot{\varepsilon} = 4.9 \times 10^{-7}\mathrm{s}^{-1}.$$

6.9.2 Grain boundary contribution to creep

In the creep of polycrystals at high temperatures, the grain boundaries themselves are able to play an important part in the deformation process due to the fact that they may (1) slide past each other or (2) create vacancies. Both processes involve an activation energy for diffusion and therefore may contribute to steady-state creep.

Grain boundary sliding during creep was inferred initially from the observation of steps at the boundaries, but the mechanism of sliding can be demonstrated on bi-crystals. Figure 6.64 shows a good example of grain boundary movement in a bi-crystal of tin, where the displacement of the straight grain boundary across its middle is indicated by marker scratches. Grain boundaries, even when specially produced for bi-crystal experiments, are not perfectly straight, and after a small amount of sliding at the boundary interface, movement will be arrested by protuberances. The grains are then locked, and the rate of slip will be determined by the rate of plastic flow in the protuberances. As a result, the rate

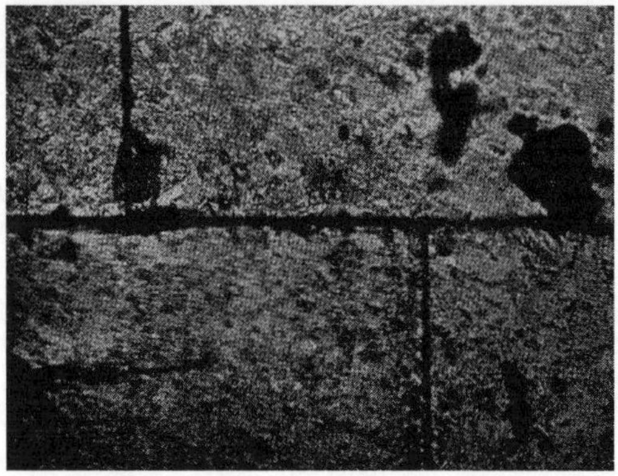

Figure 6.64 *Grain boundary sliding on a bi-crystal tin (after Puttick and King, 1952; courtesy of Institute of Materials, Minerals and Mining).*

of slip along a grain boundary is not constant with time, because the dislocations first form into piled-up groups, and later these become relaxed. Local relaxation may be envisaged as a process in which the dislocations in the pile-up climb towards the boundary. In consequence, the activation energy for grain boundary slip may be identified with that for steady-state creep. After climb, the dislocations are spread more evenly along the boundary, and are thus able to give rise to grain boundary migration, when sliding has temporarily ceased, which is proportional to the overall deformation.

A second creep process which also involves the grain boundaries is one in which the boundary acts as a source and sink for vacancies. The mechanism depends on the migration of vacancies from one side of a grain to another, as shown in Figure 6.65, and is often termed Herring–Nabarro creep, after the two workers who originally considered this process. If, in a grain of sides d under a stress σ, the atoms are transported from faces BC and AD to the faces AB and DC the grain creeps in the direction of the stress. To transport atoms in this way involves creating vacancies on the tensile faces AB and DC and destroying them on the other compressive faces by diffusion along the paths shown.

On a tensile face AB the stress exerts a force σb^2 (or $\sigma\Omega^{2/3}$) on each surface atom and so does work $\sigma b^2 \times b$ each time an atom moves forward one atomic spacing b (or $\Omega^{1/3}$) to create a vacancy. The energy of vacancy formation at such a face is thus reduced to $(E_f - \sigma b^3)$ and the concentration of vacancies in equilibrium correspondingly increased to $c_\tau = \exp[(-E_f + \sigma b^3)/\mathbf{k}T] = c_0 \exp(\sigma b^3/\mathbf{k}T)$. The vacancy concentration on the compressive faces will be reduced to $c_c = c_0 \exp(-\sigma b^3/\mathbf{k}T)$. Vacancies will therefore flow down the concentration gradient, and the number crossing a face under tension to one under compression will be given by Fick's law as

$$\phi = -D_v d^2(c_T - c_c)/\alpha d,$$

where D_v is the vacancy diffusivity and α relates to the diffusion length. Substituting for c_T, c_c and $D = (D_v c_0 b^3)$ leads to

$$\phi = 2dD \sinh(\sigma b^3/\mathbf{k}T)/\alpha b^3.$$

Each vacancy created on one face and annihilated on the other produces a strain $\varepsilon = b^3/d^3$, so that the creep strain rate $\dot{\varepsilon} = \phi(b^3/d^3)$. At high temperatures and low stresses this reduces to

$$\dot{\varepsilon}_{H-N} = 2D\sigma b^3/\alpha d^2 bf\mathbf{k}T = B_{H-N}D\sigma\omega/d^2\mathbf{k}T, \tag{6.56}$$

where the constant $B_{H-N} \sim 10$.

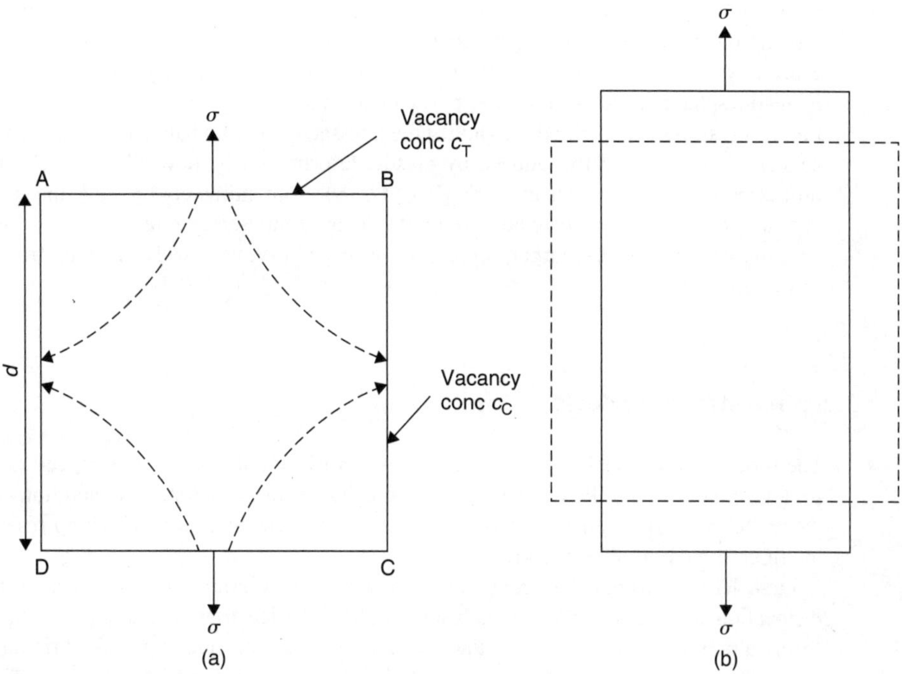

Figure 6.65 *Schematic representation of Herring–Nabarro creep; with $c_T > c_c$ vacancies flow from the tensile faces to the longitudinal faces (a) to produce creep, as shown in (b).*

In contrast to dislocation creep, Herring–Nabarro creep varies linearly with stress and occurs at $T \approx 0.8T_m$ with $\sigma \approx 10^6\,\text{N m}^{-2}$. The temperature range over which vacancy–diffusion creep is significant can be extended to much lower temperatures (i.e. $T \approx 0.5T_m$) if the vacancies flow down the grain boundaries rather than through the grains. Equation (6.56) is then modified for Coble or grain boundary diffusion creep, and is given by

$$\dot{\varepsilon}_{\text{Coble}} = B_c D_{gb} \sigma \Omega \omega / \mathbf{k} T d^3, \tag{6.57}$$

where ω is the width of the grain boundary. Under such conditions (i.e. $T \approx 0.5$–$0.6T_m$ and low stresses), diffusion creep becomes an important creep mechanism in a number of high-technology situations, and has been clearly identified in magnesium-based canning materials used in gas-cooled reactors.

6.9.3 Tertiary creep and fracture

Tertiary creep and fracture are logically considered together, since the accelerating stage represents the initiation of conditions which lead to fracture. In many cases the onset of accelerating creep is an indication that voids or cracks are slowly but continuously forming in the material, and this has been confirmed by metallography and density measurements. The type of fracture resulting from tertiary creep is not transcrystalline but grain boundary fracture. Two types of grain boundary fracture have been observed. The first occurs principally at the triple point formed where three grain boundaries meet, and sliding along boundaries on which there is a shear stress produces stress concentrations at the point of conjunction sufficiently high to start cracks. However, under conditions of slow strain rate for long times, which would be expected to favor recovery, small holes form on grain boundaries, especially those perpendicular to the tensile axis, and these gradually grow and coalesce.

Second-phase particles play an important part in the nucleation of cracks and cavities by concentrating stress in sliding boundaries and at the intersection of slip bands with particles, but these stress concentrations are greatly reduced by plastic deformation by power-law creep and by diffusional processes. Cavity formation and early growth is therefore intimately linked to the creep process itself and the time-to-fracture correlates well with the minimum creep rate for many structural materials. Fracture occurs when the larger, more closely spaced cavities coalesce. Creep fracture is discussed further in Chapter 7.

6.9.4 Creep-resistant alloy design

The problem of the design of engineering creep-resistant alloys is complex, and the optimum alloy for a given service usually contains several constituents in various states of solution and precipitation. Nevertheless, it is worth considering some of the principles underlying creep-resistant behavior in the light of the preceding theories.

First, let us consider the strengthening of the solid solution by those mechanisms which cause dislocation locking and those which contribute to lattice friction hardening. The former include solute atoms interacting with (1) the dislocation or (2) the stacking fault. Friction hardening can arise from (1) the stress fields around individual atoms (i.e. the Mott–Nabarro effect), (2) clusters of solute atoms in solid solutions, (3) by increasing the separation of partial dislocations and so making climb, cross-slip and intersection more difficult, (4) by the solute atoms becoming attached to jogs and thereby impeding climb, and (5) by influencing the energies of formation and migration of vacancies. The alloy can also be hardened by precipitation, and it is significant that many of the successful industrial creep-resistant alloys are of this type (e.g. the nickel alloys, and both ferritic and austenitic steels).

The effectiveness of these various methods of conferring strength on the alloy will depend on the conditions of temperature and stress during creep. All the effects should play some part during fast primary creep, but during the slow secondary creep stage the impeding of dislocation movement by solute locking effects will probably be small. This is because modern creep-resistant alloys are in service up to temperatures of about two-thirds the absolute melting point ($T/T_m \approx \frac{2}{3}$) of the parent metal, whereas above about $T/T_m \approx \frac{1}{2}$ solute atoms will migrate as fast as dislocations. Hardening which relies on clusters will be more difficult to remove than that which relies upon single atoms and should be effective up to higher temperatures. However, for any hardening mechanism to be really effective, whether it is due to solute atom clusters or actual precipitation, the rate of climb and cross-slip past the barriers must be slow. Accordingly, the most probable role of solute alloying elements in modern creep-resistant alloys is in reducing the rate of climb and cross-slip processes. The three hardening mechanisms listed as 3, 4 and 5 above are all effective in this way. From this point of view, it is clear that the best parent metals on which to base creep-resistant alloys will be those in which climb and cross-slip are difficult; these include the fcc and cph metals of low stacking-fault energy, for which the slip dislocations readily dissociate. Generally, the creep rate is described by the empirical relation

$$\dot{\varepsilon} = A(\sigma/E)^n(\gamma)^m D, \tag{6.58}$$

where A is a constant, n and m stress and fault energy exponents respectively, and D the diffusivity; for fcc materials $m \approx 3$ and $n \approx 4$. The reason for the good creep strength of austenitic and Ni-based materials containing Co, Cr, etc. arises from their low fault energy and also because of their relatively high melting point when D is small.

From the above discussion it appears that a successful creep-resistant material would be an alloy, the composition of which gives a structure with a hardened solid–solution matrix containing a sufficient number of precipitated particles to force glissile partial dislocations either to climb or to cross-slip to circumvent them. The constitution of the *Nimonic* alloys, which consist of a nickel matrix containing dissolved chromium, titanium, aluminum and cobalt, is in accordance with these principles, and since no large atomic size factors are involved it appears that one of the functions of these additions is to lower the stacking-fault energy and thus widen the separation of the partial dislocations. A second object of the titanium and aluminum alloy additions[7] is to produce precipitation, and in the *Nimonic* alloys much of the precipitate is Ni_3Al. This precipitate is isomorphous with the matrix, and while it has a parameter difference ($\approx\frac{1}{2}\%$) small enough to give a low interfacial energy, it is nevertheless sufficiently large to give a source of hardening. Thus, since the energy of the interface provides the driving force for particle growth, this low-energy interface between particle and matrix ensures a low rate of particle growth and hence a high service temperature.

Grain boundary precipitation is advantageous in reducing grain boundary sliding. Alternatively, the weakness of the grain boundaries may be eliminated altogether by using single-crystal material. *Nimonic* alloys used for turbine blades have been manufactured in single-crystal form by directional solidification (see Chapters 2 and 8).

Dispersions are effective in conferring creep strength by two mechanisms. First, the particle will hinder a dislocation and force it to climb and cross-slip. Second, and more important, is the retarding effect on recovery as shown by some dispersions, Cu–Al_2O_3 (extruded), SAP (sintered alumina powder) and Ni–ThO_2, which retain their hardness almost to the melting point. A comparison of SAP with a 'conventional' complex aluminum alloy shows that at 250°C there is little to choose between them but at 400°C SAP is several times stronger. Generally, the dislocation network formed by strain hardening interconnects the particles and is thereby anchored by them. To do this effectively, the particle must be stable at the service temperature and remain finely dispersed. This depends on the solubility C, diffusion coefficient D and interfacial energy γ_1, since the time to dissolve the particle is $t = r^4\mathbf{k}T/DC\gamma_1 R^2$. In precipitation-hardening alloys, C is appreciable and D offers little scope for adjustment; great importance is therefore placed on γ_1 as for the Ni_3 (TiAl) phase in *Nimonics*, where it is very low.

Figure 6.62 shows that $n \approx 4$ both above and below $0.5T_m$ for the Ni–Al_2O_3 and Ni–Co–Al_2O_3 alloys that were completely recrystallized, which contrasts with values very much greater than 4 for extruded TD nickel and other dispersion-strengthened alloys[8] containing a dislocation substructure. This demonstrates the importance of substructure and probably indicates that in completely recrystallized alloys containing a dispersoid, the particles control the creep behavior, whereas in alloys containing a substructure the dislocation content is more important. Since $n \approx 4$ for the Ni– and Ni–Co–Al_2O_3 alloys in both temperature regimes, the operative deformation mechanism is likely to be the same, but it is clear from the activation energies, listed in Table 6.4, that the rate-controlling thermally activated process changes with temperature. The activation energy is greater at the higher temperature when it is also, surprisingly, composition (or stacking-fault energy) independent.

Such behavior may be explained, if it is assumed that the particles are bypassed by cross-slip (see Chapter 7) and this process is easy at all temperatures, but it is the climb of the edge segments of the cross-slipped dislocations that is rate controlling. At low temperatures, climb would proceed by pipe diffusion so that the composition dependence relates to the variation in the ease of pipe diffusion along dislocations of different widths. At high temperatures, climb occurs by bulk diffusion and the

[7] The chromium forms a spinel with NiO and hence improves the oxidation resistance.

[8] To analyze these it is generally necessary to introduce a threshold (or friction) stress σ_0, so that the effective stress is $(\sigma - \sigma_0)$.

Table 6.4 *Experimentally determined parameters from creep of Ni–Al_2O_3 and Ni–Co–Al_2O_3 alloys.*

Alloy	*Test temperature*				
	773 K		*1000 K*		
	Q (kJ mol^{-1})	*A* (s^{-1})	*Q* (kJ mol^{-1})	*A* (s^{-1})	A/D_0
Ni	85	1.67×10^{16}	276	1.1×10^{28}	5.5×10^{28}
Ni–67% Co	121	9.95×10^{19}	276	2.2×10^{28}	5.8×10^{28}

absence of any composition dependence is due to the fact that in these alloys the jog distribution is determined mainly by dislocation/particle interactions and not, as in single-phase alloys and in dispersion-strengthened alloys containing a substructure, by the matrix stacking-fault energy. The optimum creep resistance of dispersion-strengthened alloys is produced when a uniform dislocation network in a fibrous grain structure is anchored by the particles and recovery is minimized. Such a structure can reduce the creep rate by several orders of magnitude from that given in Figure 6.62, but it depends critically upon the working and heat treatment used in fabricating the alloy.

Second-phase particles can also inhibit diffusion creep. Figure 6.66 shows the distribution of particles before and after diffusion creep, and indicates that the longitudinal boundaries tend to collect precipitates as vacancies are absorbed and the boundaries migrate inwards, while the tensile boundaries acquire a PFZ. Such a structural change has been observed in Mg–0.5% Zr (*Magnox ZR55*) at 400°C and is accompanied by a reduced creep rate. It is not anticipated that diffusion is significantly affected by the presence of particles and hence the effect is thought to be due to the particles affecting the vacancy-absorbing capabilities of the grain boundaries. Whatever mechanism is envisaged for the annihilation of vacancies at a grain boundary, the climb-glide of grain boundary dislocations is likely to be involved and such a process will be hindered by the presence of particles.

6.10 Deformation mechanism maps

The discussion in this chapter has emphasized that, over a range of stress and temperature, an alloy is capable of deforming by several alternative and independent mechanisms, e.g. dislocation creep with either pipe diffusion at low temperatures and lattice diffusion at high temperatures being the rate-controlling mechanism, and diffusional creep with either grain-boundary diffusion or lattice diffusion being important. In a particular range of temperature, one of these mechanisms is dominant and it is therefore useful in engineering applications to identify the operative mechanism for a given stress–temperature condition, since it is ineffective to change the metallurgical factors to influence, for example, a component deforming by power-law creep controlled by pipe diffusion if the operative mechanism is one of Herring–Nabarro creep.

The various alternative mechanisms are displayed conveniently on a deformation mechanism map in which the appropriate stress, i.e. shear stress or equivalent stress, compensated by modulus on a log scale, is plotted against homologous temperature T/T_m as shown in Figure 6.67 for nickel and a nickel-based superalloy with a grain size of 100 μm. By comparing the diagrams it is evident that solid solution strengthening and precipitation hardening have raised the yield stress and reduced the dislocation creep field. The shaded boxes shown in Figure 6.67 indicate the typical stresses and temperatures to which a turbine blade would be subjected; it is evident that the mechanism of creep during operation has changed and, indeed, the creep rate is reduced by several orders of magnitude.

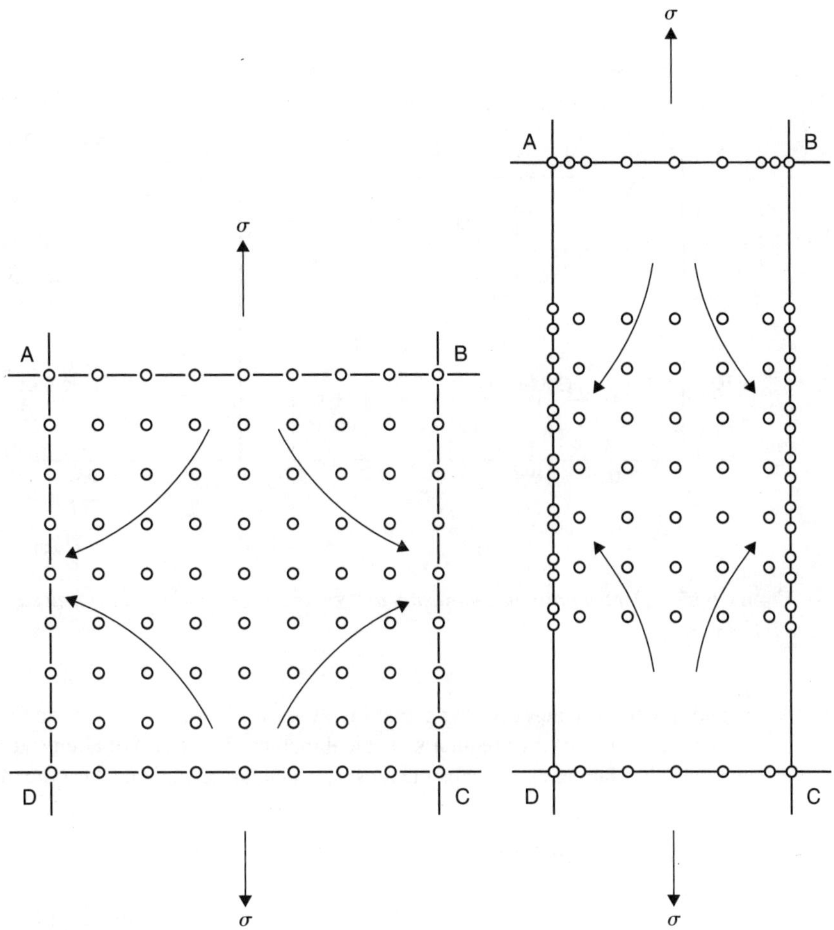

Figure 6.66 *Schematic diagram showing the distribution of second-phase particles before and after diffusion creep.*

6.11 Metallic fatigue

6.11.1 Nature of fatigue failure

The term fatigue applies to the behavior of a metal which, when subjected to a cyclically variable stress of sufficient magnitude (often below the yield stress), produces a detectable change in mechanical properties. In practice, a large number of service failures are due to fatigue, and so engineers are concerned mainly with fatigue failure, where the specimen is actually separated into two parts. Some of these failures can be attributed to poor design of the component, but in some cases can be ascribed to the condition of the material. Consequently, the treatment of fatigue may be conveniently divided into three aspects: (1) engineering considerations, (2) gross metallurgical aspects, and (3) fine-scale structural and atomic changes.

The fatigue conditions which occur in service are usually extremely complex. Common failures are found in axles where the eccentric load at a wheel or pulley produces a varying stress which is a maximum in the skin of the axle. Other examples, such as the flexure stresses produced in aircraft

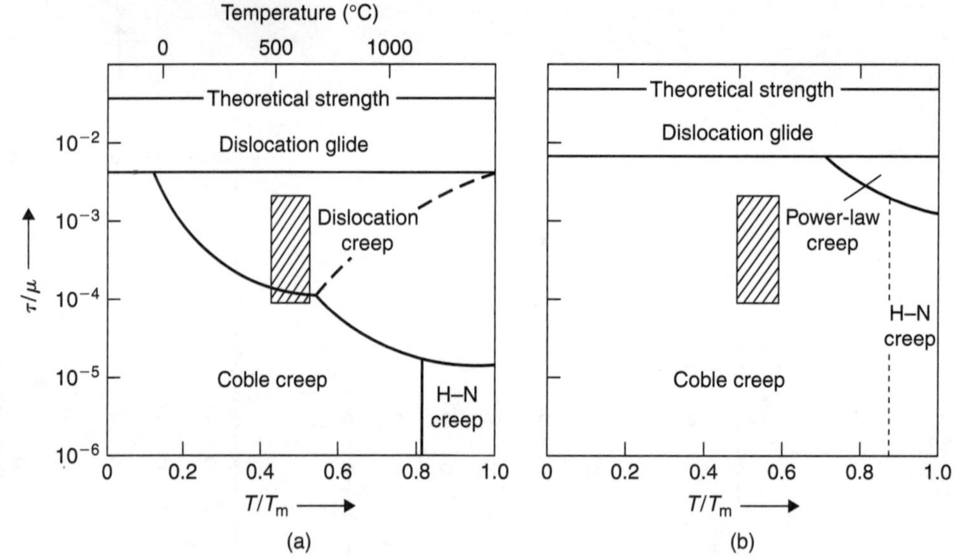

Figure 6.67 *Deformation mechanism maps for nickel (a) and nickel-based superalloy (b) (after M. F. Ashby).*

wings and in undercarriages during ground taxiing, do, however, emphasize that the stress system does not necessarily vary in a regular sinusoidal manner. The series of aircraft disasters attributed to pressurized cabin failures is perhaps the most spectacular example of this type of fatigue failure.

6.11.2 Engineering aspects of fatigue

In laboratory testing of materials the stress system is usually simplified, and both the Woehler and push–pull types of test are in common use. The results are usually plotted on the familiar *S–N* curve (i.e. stress versus the number of cycles to failure, usually plotted on a logarithmic scale). Ferritic steels may be considered to exhibit a genuine fatigue limit with a fatigue ratio $S/\mathrm{TS} \approx 0.5$. However, other materials, such as aluminum- or copper-based alloys, certainly those of the age-hardening variety, definitely do not show a sharp discontinuity in the *S–N* curve. For these materials no fatigue limit exists and all that can be specified is the endurance limit at *N* cycles. The importance of the effect is illustrated by the behavior of commercial aluminum-based alloys containing zinc, magnesium and copper. Such an alloy may have a TS of 617 MN m^{-2} but the fatigue stress for a life of 10^8 cycles is only 154 MN m^{-2} (i.e. a fatigue ratio at 10^8 cycles of 0.25).

The amplitude of the stress cycle to which the specimen is subjected is the most important single variable in determining its life under fatigue conditions, but the performance of a material is also greatly affected by various other conditions, which may be summarized as follows:

1. *Surface preparation.* Since fatigue cracks frequently start at or near the surface of the component, the surface condition is an important consideration in fatigue life. The removal of machining marks and other surface irregularities invariably improves the fatigue properties. Putting the surface layers under compression by shot peening or surface treatment improves the fatigue life.

2. *Effect of temperature*. Temperature affects the fatigue properties in much the same way as it does the tensile strength (TS); the fatigue strength is highest at low temperatures and decreases gradually with rising temperature. For mild steel the ratio of fatigue limit to TS remains fairly constant at about 0.5, while the ratio of fatigue limit to yield stress varies over much wider limits. However, if the temperature is increased above about 100°C, both the tensile strength and the fatigue strength of mild steel show an increase, reaching a maximum value between 200 and 400°C. This increase, which is not commonly found in other materials, has been attributed to strain ageing.
3. *Frequency of stress cycle*. In most metals the frequency of the stress cycle has little effect on the fatigue life, although lowering the frequency usually results in a slightly reduced fatigue life. The effect becomes greater if the temperature of the fatigue test is raised, when the fatigue life tends to depend on the total time of testing rather than on the number of cycles. With mild steel, however, experiments show that the normal speed effect is reversed in a certain temperature range and the number of cycles to failure increases with decrease in the frequency of the stress cycle. This effect may be correlated with the influence of temperature and strain rate on the TS. The temperature at which the tensile strength reaches a maximum depends on the rate of strain, and it is therefore not surprising that the temperature at which the fatigue strength reaches a maximum depends on the cyclic frequency.
4. *Mean stress*. For conditions of fatigue where the mean stress, i.e.

$$\Delta\sigma N_{\mathrm{f}}^{a} = (\sigma_{\max} + \sigma_{\min})/2,$$

does not exceed the yield stress σ_{y}, then the relationship

$$\Delta\sigma N_{\mathrm{f}}^{a} = \text{const.}, \tag{6.59}$$

known as Basquin's law, holds over the range 10^2 to $\approx 10^5$ cycles, i.e. N less than the knee of the S–N curve, where $a \approx \frac{1}{10}$ and N_{f} is the number of cycles to failure. For low cycle fatigue with $\Delta\sigma > \sigma_{\mathrm{y}}$, then Basquin's law no longer holds, but a reasonable relationship

$$\Delta\varepsilon_{\mathrm{p}} N_{\mathrm{f}}^{b} = D^{b} = \text{const.}, \tag{6.60}$$

known as the Coffin–Manson law, is found, where $\Delta\varepsilon_{\mathrm{p}}$ is the plastic strain range, $b \approx 0.6$ and D is the ductility of the material. If the mean stress becomes tensile a lowering of the fatigue limit results. Several relationships between fatigue limit and mean stress have been suggested, as illustrated in Figure 6.68a. However, there is no theoretical reason why a material should follow any given relationship and the only safe rule on which to base design is to carry out prior tests on the material concerned to determine its behavior under conditions similar to those it will meet in service. Another common engineering relationship frequently used, known as Miner's concept of cumulative damage, is illustrated in Figure 6.68b. This hypothesis states that damage can be expressed in terms of the number of cycles applied divided by the number to produce failure at a given stress level. Thus, if a maximum stress of value S_1 is applied to a specimen for n_1 cycles which is less than the fatigue life N_1, and then the maximum stress is reduced to a value equal to S_2, the specimen is expected to fail after n_2 cycles, since according to Miner the following relationship will hold:

$$n_1/N_1 + n_2/N_2 + \ldots = \Sigma n/N = 1. \tag{6.61}$$

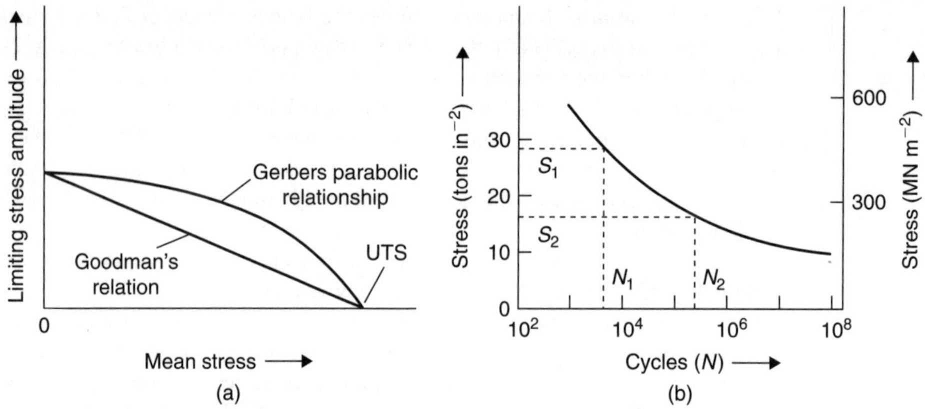

Figure 6.68 *Fatigue relationships.*

5. *Environment.* Fatigue occurring in a corrosive environment is usually referred to as corrosion fatigue. It is well known that corrosive attack by a liquid medium can produce etch pits which may act as notches, but when the corrosive attack is simultaneous with fatigue stressing, the detrimental effect is far greater than just a notch effect. Moreover, from microscopic observations the environment appears to have a greater effect on crack propagation than on crack initiation. For most materials even atmospheric oxygen decreases the fatigue life by influencing the speed of crack propagation, and it is possible to obtain a relationship between fatigue life and the degree of vacuum in which the specimen has been held.

It is now well established that fatigue starts at the surface of the specimen. This is easy to understand in the Woehler test because, in this test, it is there that the stress is highest. However, even in push–pull fatigue, the surface is important for several reasons: (1) slip is easier at the surface than in the interior of the grains, (2) the environment is in contact with the surface and (3) any specimen misalignment will always give higher stresses at the surface. Accordingly, any alteration in surface properties must bring about a change in the fatigue properties. The best fatigue resistance occurs in materials with a worked surface layer produced by polishing with emery, shot-peening or skin-rolling the surface. This beneficial effect of a worked surface layer is principally due to the fact that the surface is put into compression, but the increased TS as a result of work hardening also plays a part. Electropolishing the specimen by removing the surface layers usually has a detrimental effect on the fatigue properties, but other common surface preparations such as nitriding and carburizing, both of which produce a surface layer which is in compression, may be beneficial. Conversely, such surface treatments as the decarburizing of steels and the cladding of aluminum alloys with pure aluminum increase their susceptibility to fatigue.

The alloy composition and thermal and mechanical history of the specimen are also of importance in the fatigue process. Any treatment which increases the hardness or yield strength of the material will increase the level of the stress needed to produce slip and, as we shall see later, since the fundamental processes of fatigue are largely associated with slip, this leads directly to an increase in fatigue strength. It is also clear that grain size is a relevant factor: the smaller the grain size, the higher is the fatigue strength at a given temperature.

The fatigue processes in stable alloys are essentially the same as those of pure metals but there is, of course, an increase in fatigue strength. However, the processes in unstable alloys and in materials exhibiting a yield point are somewhat different. In fatigue, as in creep, structural instability frequently leads to enhancement of the fundamental processes. In all cases the approach to equilibrium is more complete, so that in age-hardening materials, solution-treated specimens become harder and fully aged specimens become softer. The changes which occur are local rather than general, and are associated with the enhanced diffusion brought about by the production of vacancies during the fatigue test. Clearly, since vacancy mobility is a thermally activated process, such effects can be suppressed at sufficiently low temperatures.

In general, non-ferrous alloys do not exhibit the type of fatigue limit shown by mild steel. One exception to this generalization is the alloy aluminum/2–7% magnesium/0.5% manganese, and it is interesting to note that this alloy also has a sharp yield point and shows Lüders markings in an ordinary tensile test. Accordingly, it has been suggested that the fatigue limit occupies a similar place in the field of alternating stresses to that filled by the yield point in unidirectional stressing. Stresses above the fatigue limit readily unlock the dislocations from their solute atom atmospheres, while below the fatigue limit most dislocations remain locked. In support of this view, it is found that when the carbon and nitrogen content of mild steel is reduced, by annealing in wet hydrogen, striking changes take place in the fatigue limit (Figure 6.5) as well as in the sharp yield point.

6.11.3 Structural changes accompanying fatigue

Observations of the structural details underlying fatigue hardening show that in polycrystals large variations in slip-band distributions and the amount of lattice misorientation exist from one grain to another. Because of such variations it is difficult to typify structural changes, so that in recent years this structural work has been carried out more and more on single crystals; in particular, copper has received considerable attention as being representative of a typical metal. Such studies have now established that fatigue occurs as a result of slip, the direction of which changes with the stress cycle, and that the process continues throughout the whole of the test (shown, for example, by interrupting a test and removing the slip bands by polishing; the bands reappear on subsequent testing).

Moreover, four stages in the fatigue life of a specimen are distinguishable; these may be summarized as follows. In the early stages of the test, the whole of the specimen hardens. After about 5% of the life, slip becomes localized and persistent slip bands appear; they are termed persistent because they reappear and are not permanently removed by electropolishing. Thus, reverse slip does not continue throughout the whole test in the bulk of the metal (the matrix). Electron microscope observations show that metal is extruded from the slip bands and that fine crevices called intrusions are formed within the band. During the third stage of the fatigue life the slip bands grow laterally and become wider, and at the same time cracks develop in them. These cracks spread initially along slip bands, but in the later stages of fracture the propagation of the crack is often not confined to certain crystallographic directions and catastrophic rupture occurs. These two important crack growth stages, i.e. stage I in the slip band and stage II roughly perpendicular to the principal stress, are shown in Figure 6.69 and are influenced by the formation of localized (persistent) slip bands (i.e. PSBs). However, PSBs are not clearly defined in low stacking-fault energy, solid solution alloys.

Cyclic stressing therefore produces plastic deformation which is not fully reversible and the build-up of dislocation density within grains gives rise to fatigue hardening with an associated structure which is characteristic of the strain amplitude and the ability of the dislocations to cross-slip, i.e. temperature and SFE. The non-reversible flow at the surface leads to intrusions, extrusions and crack formation in PSBs. These two aspects will now be considered separately and in greater detail.

6.11.3.1 Fatigue hardening

If a single or polycrystalline specimen is subjected to many cycles of alternating stress, it becomes harder than a similar specimen extended unidirectionally by the same stress applied only once.

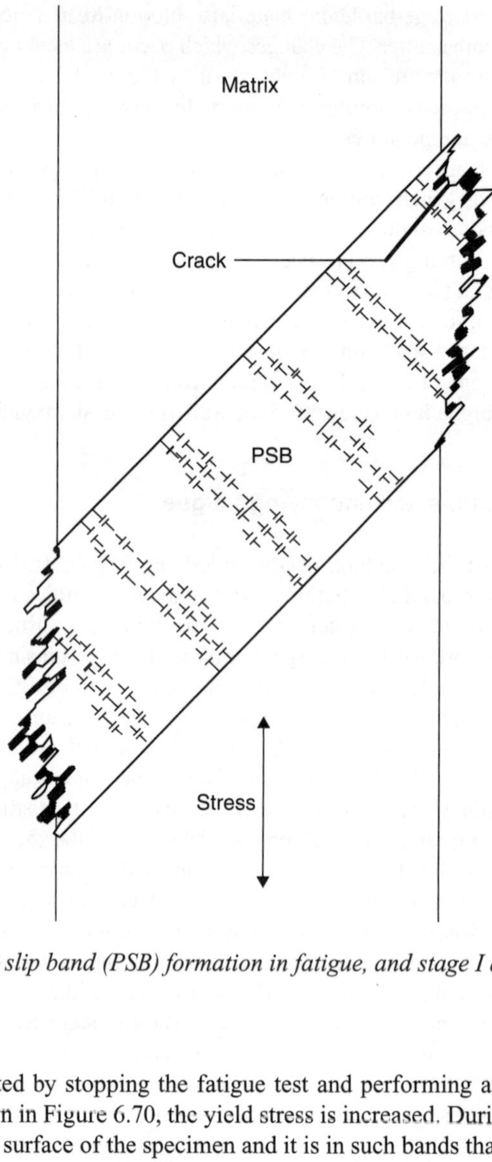

Figure 6.69 *Persistent slip band (PSB) formation in fatigue, and stage I and stage II crack growth.*

This may be demonstrated by stopping the fatigue test and performing a static tensile test on the specimen when, as shown in Figure 6.70, the yield stress is increased. During the process, persistent slip bands appear on the surface of the specimen and it is in such bands that cracks eventually form. The behavior of a fatigue-hardened specimen has two unusual features when compared with an ordinary work-hardened material. The fatigue-hardened material, having been stressed symmetrically, has the same yield stress in compression as in tension, whereas the work-hardened specimen (e.g. prestrained in tension) exhibits a Bauschinger effect, i.e. weaker in compression than tension. It arises from the fact that the obstacles behind the dislocation are weaker than those resisting further dislocation motion, and the pile-up stress causes it to slip back under a reduced load in the reverse direction. The other important feature is that the temperature dependence of the hardening produced by fatigue is significantly greater than that of work hardening and, because of the similarity with the behavior of metals hardened by quenching and by irradiation, it has been attributed to the effect of vacancies and dislocation loops created during fatigue.

At the start of cyclic deformation the initial slip bands (Figure 6.71a) consist largely of primary dislocations in the form of dipole and multipole arrays; the number of loops is relatively small because

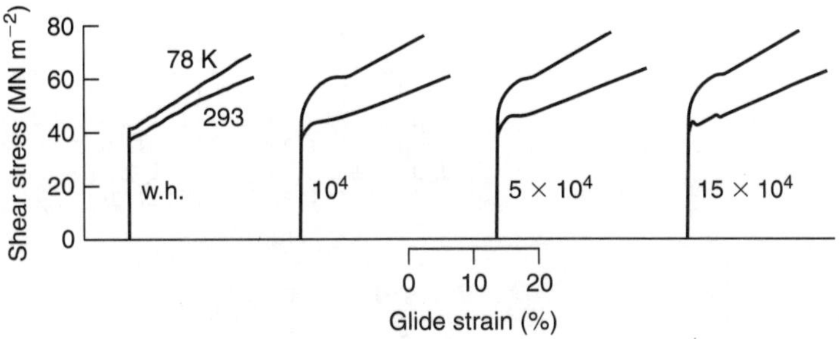

Figure 6.70 *Stress–strain curves for copper after increasing amounts of fatigue testing (after Broom and Ham, 1959).*

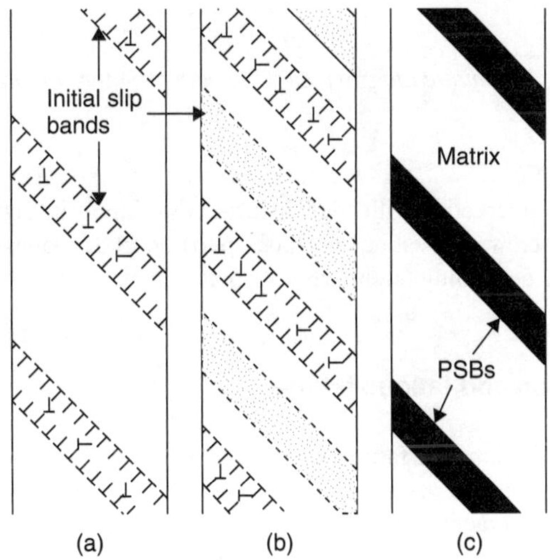

Figure 6.71 *Formation of persistent slip bands (PSBs) during fatigue.*

the frequency of cross-slip is low. As the specimen work-hardens slip takes place between the initial slip bands, and the new slip bands contain successively more secondary dislocations because of the internal stress arising from nearby slip bands (Figure 6.71b). When the specimen is completely filled with slip bands, the specimen has work hardened and the softest regions are now those where slip occurred originally, since these bands contain the lowest density of secondary dislocations. Further slip and the development of PSBs takes place within these original slip bands, as shown schematically in Figure 6.71c.

As illustrated schematically in Figure 6.72, TEM of copper crystals shows that the main difference between the matrix and the PSBs is that in the matrix the dense arrays of edge dislocation (di- and multipoles) are in the form of large veins occupying about 50% of the volume, whereas they form a 'ladder'-type structure within walls occupying about 10% of the volume in PSBs. The PSBs are the active regions in the fatigue process while the matrix is associated with the inactive parts of the specimen between the PSBs. Steady-state deformation then takes place by the to-and-fro glide of the same dislocations in the matrix, whereas an equilibrium between dislocation multiplication and annihilation exists in the PSBs. Multiplication occurs by bowing-out of the walls and annihilation

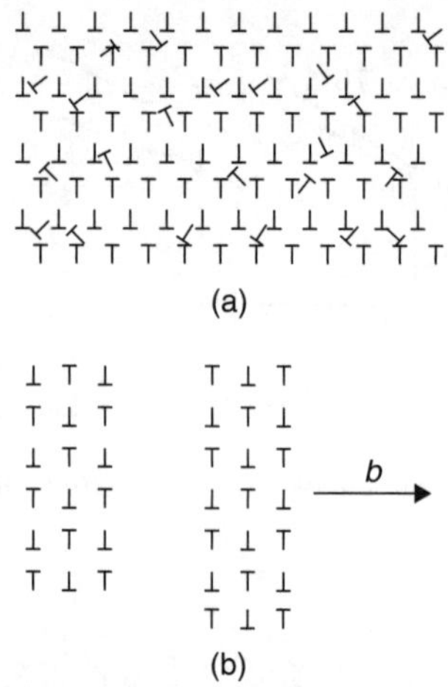

Figure 6.72 *Schematic diagram showing vein structure of matrix (a) and ladder structure of PSBs (b).*

takes place by interaction with edge dislocations of opposite sign ($\approx 75b$ apart) on glide planes in the walls and of screw dislocations ($\approx 200b$ apart) on glide planes in the low-dislocation channels, the exact distance depending on the ease of cross-slip.

6.11.4 Crack formation and fatigue failure

Extrusions, intrusions and fatigue cracks can be formed at temperatures as low as 4 K where thermally activated movement of vacancies does not take place. Such observations indicate that the formation of intrusions and cracks cannot depend on either chemical or thermal action and the mechanism must be a purely geometrical process which depends on cyclic stressing.

Two general mechanisms have been suggested. The first, the Cottrell 'ratchet' mechanism, involves the use of two different slip systems with different directions and planes of slip, as is shown schematically in Figure 6.73. The most favored source (e.g. S_1 in Figure 6.73a) produces a slip step on the surface at P during a tensile half-cycle. At a slightly greater stress in the same half-cycle, the second source S_2 produces a second step at Q (Figure 6.73b). During the compression half-cycle, the source S_1 produces a surface step of opposite sign at P′ (Figure 6.73c), but, owing to the displacing action of S_2, this is not in the same plane as the first and thus an intrusion is formed. The subsequent operation of S_2 produces an extrusion at QQ′ (Figure 6.73d) in a similar manner. Such a mechanism requires the operation of two slip systems and, in general, predicts the occurrence of intrusions and extrusions with comparable frequency, but not in the same slip band.

The second mechanism, proposed by Mott, involves cross-slip resulting in a column of metal extruded from the surface and a cavity is left behind in the interior of the crystal. One way in which this could happen is by the cyclic movement of a screw dislocation along a closed circuit of crystallographic planes, as shown in Figure 6.74. During the first half-cycle the screw dislocation glides along two faces ABCD and BB′ C′ C of the band, and during the second half-cycle returns along the faces B′ C′ A′ D and A′ D′ DA. Unlike the Cottrell mechanism this process can be operated with a single slip direction, provided cross-slip can occur.

Neither mechanism can fully explain all the experimental observations. The interacting slip mechanism predicts the occurrence of intrusions and extrusions with comparable frequency but not, as

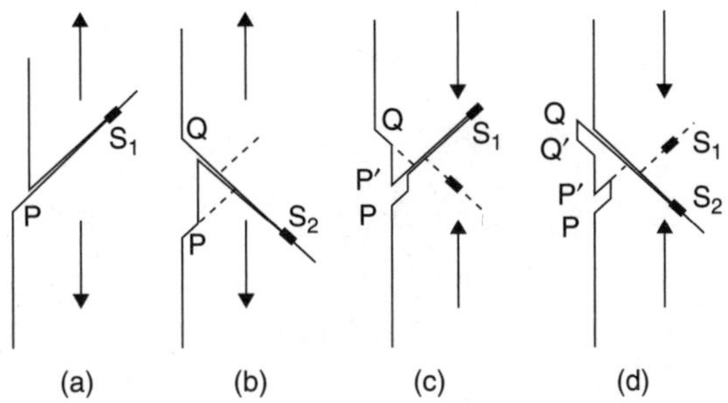

Figure 6.73 *Formation of intrusions and extrusions (after Cottrell, 1959; courtesy of John Wiley and Sons).*

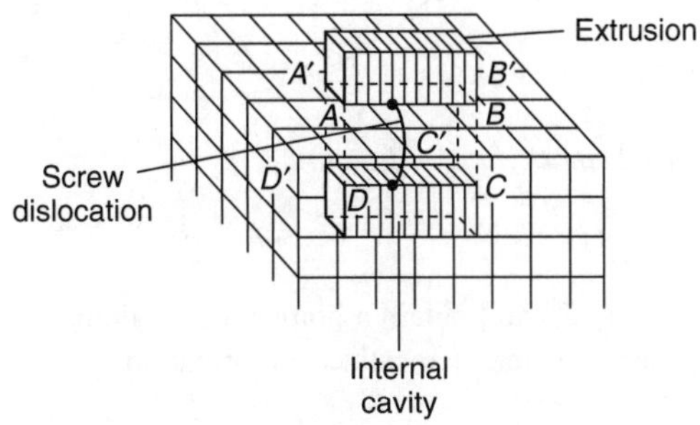

Figure 6.74 *Formation of an extrusion and associated cavity by the Mott mechanism.*

is often found, in the same slip band. With the cross-slip mechanism, there is no experimental evidence to show that cavities exist beneath the material being extruded. It may well be that different mechanisms operate under different conditions.

In a polycrystalline aggregate the operation of several slip modes is necessary and intersecting slip unavoidable. Accordingly, the widely differing fatigue behavior of metals may be accounted for by the relative ease with which cross-slip occurs. Thus, those factors which affect the onset of stage III in the work-hardening curve will also be important in fatigue, and conditions suppressing cross-slip would, in general, increase the resistance to fatigue failure, i.e. low stacking-fault energy and low temperatures. Aluminum would be expected to have poor fatigue properties on this basis, but the unfavorable fatigue characteristics of the high-strength aluminum alloys is probably also due to the unstable nature of the alloy and to the influence of vacancies.

In pure metals and alloys, transgranular cracks initiate at intrusions in PSBs or at sites of surface roughness associated with emerging planar slip bands in low SFE alloys. Often the microcrack forms at the PSB/matrix interface where the stress concentration is high. In commercial alloys containing inclusions or second-phase particles, the fatigue behavior depends on the particle size. Small particles $\approx 0.1\ \mu m$ can have beneficial effects by homogenizing the slip pattern and delaying fatigue-crack nucleation. Larger particles reduce the fatigue life by both facilitating crack nucleation by slip band/particle interaction and increasing crack growth rates by interface decohesion and voiding within the plastic zone at the crack tip. The formation of voids at particles on grain boundaries can lead to intergranular separation and crack growth. The preferential deformation of 'soft' precipitate-free zones (PFZs) associated with grain boundaries in age-hardened alloys also provides a mechanism of intergranular fatigue-crack initiation and growth. To improve the fatigue behavior it is therefore

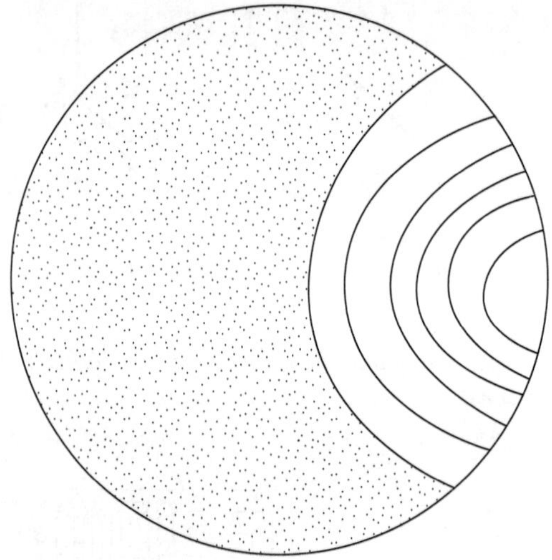

Figure 6.75 *A schematic fatigue fracture.*

necessary to avoid PFZs and obtain a homogeneous deformation structure and uniform precipitate distribution by heat treatment; localized deformation in PFZs can be restricted by a reduction in grain size.

From the general appearance of a typical fatigue fracture, shown in Figure 6.75, one can distinguish two distinct regions. The first is a relatively smooth area, through which the fatigue crack has spread slowly. This area usually has concentric marks about the point of origin of the crack which correspond to the positions at which the crack was stationary for some period. The remainder of the fracture surface shows a typically rough transcrystalline fracture where the failure has been catastrophic. Electron micrographs of the relatively smooth area show that this surface is covered with more or less regular contours perpendicular to the direction of the propagation front. These fatigue striations represent the successive positions of the propagation front and are spaced further apart the higher the velocity of propagation. They are rather uninfluenced by grain boundaries and in metals where cross-slip is easy (e.g. mild steel or aluminum) may be wavy in appearance. Generally, the lower the ductility of the material, the less well defined are the striations.

Stage II growth is rate controlling in the fatigue failure of most engineering components, and is governed by the stress intensity at the tip of the advancing crack. The striations seen on the fracture surface may form by a process of plastic blunting at the tip of the crack, as shown in Figure 6.76. In (a) the crack under the tensile loading part of the cycle generates shear stresses at the tip. With increasing tensile load the crack opens up and a new surface is created (b), separation occurs in the slip band and 'ears' are formed at the end of the crack. The plastic deformation causes the crack to be both extended and blunted (c). On the compressive part of the cycle the crack begins to close (d). The shear stresses are reversed and with increasing load the crack almost closes (e). In this part of the cycle the new surface folds and the ears correspond to the new striations on the final fracture surface. A one-to-one correlation therefore exists between the striations and the opening and closing with ear formation. Crack growth continues in this manner until it is long enough to cause the final instability when either brittle or ductile (due to the reduced cross-section not being able to carry the load) failure occurs. In engineering alloys, rather than pure metals, which contain inclusions or second-phase particles, cracking or voiding occurs ahead of the original crack tip rather than in the ears when the tensile stress or strain reaches a critical value. This macroscopic stage of fracture is clearly of importance

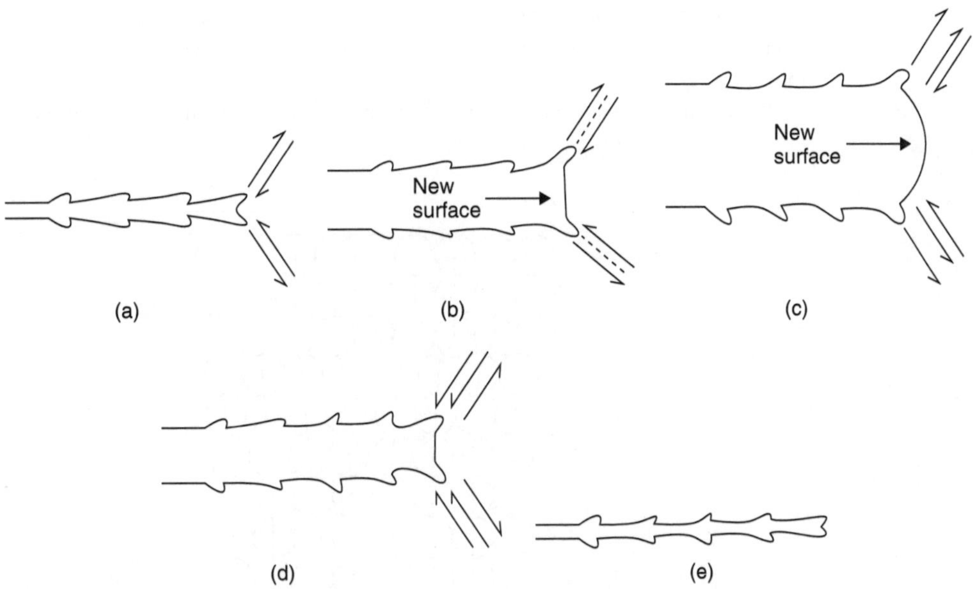

Figure 6.76 *Schematic illustration of the formation of fatigue striations.*

to engineers in predicting the working life of a component and has been successfully treated by the application of fracture mechanics, as discussed in Chapter 7.

6.11.5 Fatigue at elevated temperatures

At ambient temperature the fatigue process involves intracrystalline slip and surface initiation of cracks, followed by transcrystalline propagation. When fatigued at elevated temperatures $\gtrsim 0.5\,T_m$, pure metals and solid solutions show the formation of discrete cavities on grain boundaries, which grow, link up and finally produce failure. It is probable that vacancies produced by intracrystalline slip give rise to a supersaturation which causes the vacancies to condense on those grain boundaries that are under a high shear stress where the cavities can be nucleated by a sliding or ratchet mechanism. It is considered unlikely that grain boundary sliding contributes to cavity growth, increasing the grain size decreases the cavity growth because of the change in boundary area. *Magnox* (Mg) and alloys used in nuclear reactors up to $0.75T_m$ readily form cavities, but the high-temperature nickel-base alloys do not show intergranular cavity formation during fatigue at temperatures within their normal service range, because intracrystalline slip is inhibited by γ' precipitates. Above about $0.7T_m$, however, the γ' precipitates coarsen or dissolve and fatigue then produce cavities and eventually cavity failure.

Problems

6.1 (a) In the diagram the dislocation line is taken to point into the paper. Mark on the diagram the Burgers vector. Under the action of the shear stress shown, which way would the dislocation move?

(b) If $\underline{u}=[1\ 1\ \bar{2}]$, $\underline{b}=1/2[\bar{1}\ 1\ 0]$ and $n=(1\ 1\ 1)$, and the shear stress shown is replaced by a uniaxial compressive stress σ along [2 0 1], deduce in which direction the dislocation would move.

What would be the magnitude of the resolved shear stress on the dislocation?

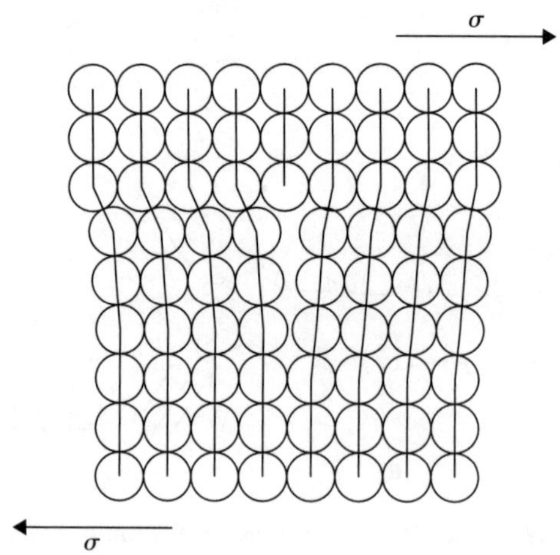

6.2 A single crystal of aluminum is pulled along [1 1 0]. Which slip system or slip systems operate first?

6.3 Estimate the shear stress at the upper yield point and the yield drop (shear stress) that occurs when the mobile dislocation density increases by two orders of magnitude from the initial density of $10^5\ \text{cm}^{-2}$. (Take the strain rate to be $10^{-3}\ \text{s}^{-1}$, τ_o the stress for unit dislocation velocity of $1\ \text{cm s}^{-1}$ to be $2.8\times10^4\ \text{N cm}^{-2}$, n to be 20 and b, the Burgers vector, to be 2×10^{-8} cm.)

6.4 The strengthening of a polycrystalline metal is provided by grain refinement and dispersion of particles. The tensile yield stress of the metal is 400 MPa when the grain size is 0.32 mm and 300 MPa when $d=1$ mm. Calculate the average distance between the particles. Assume the shear modulus of the metal $\mu=80$ GPa and $b=0.25$ nm.

6.5 A steel with a grain size of 25 μm has a yield stress of 200 MPa and with a grain size of 9 μm a yield stress of 300 MPa. A dispersion of non-deformable particles is required to raise the strength to 500 MPa in a steel with grain size 100 μm. What would be the required dispersion spacing? (Assume the shear modulus $\mu=80$ GPa and the Burgers vector $b=0.2$ nm.)

6.6 The deformation mechanism map given in the figure below shows three fields of creep for each of which the creep rate $\dot{\varepsilon}$ (s^{-1}) is represented by an expression of the form $\dot{\varepsilon}=A\sigma^n\exp(-Q/\mathbf{R}T)$. The constant A is 1.5×10^5, $5.8\times10^5/d^2$ and $10^{-9}/d^3$ for dislocation creep, Herring–Nabarro creep and Coble creep respectively ($d=$ grain size in m), while the stress exponent n is 5, 1 and 1 and the activation energy Q (kJ mol^{-1}) 550, 550 and 400. The stress σ is in MPa. Assuming that the grain size of the material is 1 mm and given the gas constant $\mathbf{R}=8.3\ \text{J mol}^{-1}\ \text{K}^{-1}$:

(i) Label the three creep fields
(ii) Calculate the stress level σ in MPa of the boundary AB
(iii) Calculate the temperature (K) of the boundary AC.

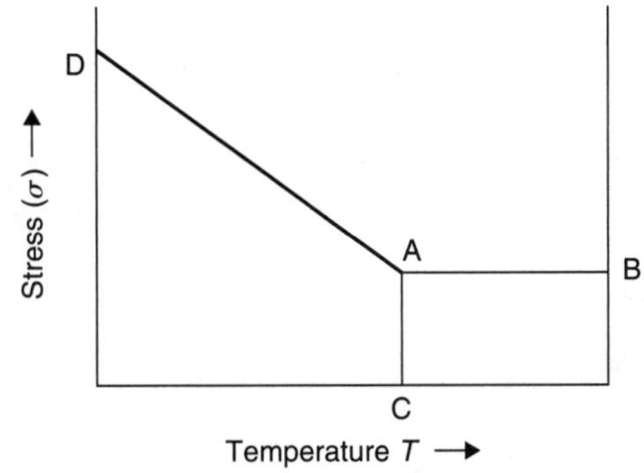

6.7 During the strain ageing of a mild steel specimen, the yield point returned after 1302, 420, 90 and 27 seconds when aged at 50, 65, 85 and 100°C respectively. Determine the activation energy for the diffusion of carbon in α-iron.

6.8 In a high-temperature application an alloy is observed to creep at an acceptable steady-state rate under a stress of 70 MPa at a temperature of 1250 K. If metallurgical improvements would allow the alloy to operate at the same creep rate but at a higher stress level of 77 MPa, estimate the new temperature at which the alloy would operate under the original stress conditions. (Take stress exponent n to be 5, and activation energy for creep Q to be 200 kJ mol^{-1}.)

6.9 Cyclic fatigue of an aluminum alloy showed it failed under a stress range $\Delta\sigma = 280$ MPa after 10^5 cycles, and for a range 200 MPa after 10^7 cycles. Using Basquin's law, estimate the life of the component subjected to a stress range of 150 MPa.

Further reading

Argon, A. (1969). *The Physics of Strength and Plasticity*. MIT Press, Cambridge, MA.
Cottrell, A. H. (1964). *Mechanical Properties of Matter*. John Wiley, Chichester.
Cottrell, A. H. (1964). *The Theory of Crystal Dislocations*. Blackie, Glasgow.
Dislocations and Properties of Real Metals (1984). Conf. Metals Society.
Evans, R. W. and Wilshire, B. (1993). *Introduction to Creep*. Institute of Materials, London.
Freidel, J. (1964). *Dislocations*. Pergamon Press, London.
Hirsch, P. B. (ed.) (1975). *The Physics of Metals. 2. Defects*. Cambridge University Press, Cambridge.
Hirth, J. P. and Lothe, J. (1984). *Theory of Dislocations*. McGraw-Hill, New York.

Chapter 7

Mechanical properties II – Strengthening and toughening

7.1 Introduction

The manufacture of metals and materials which possess considerable strength at both room and elevated temperatures is of great practical importance. We have already seen how alloying, solute–dislocation interaction, grain size control and cold working can give rise to an increased yield stress. Of these methods, refining the grain size is of universal application to materials in which the yield stress has a significant dependence upon grain size. In certain alloy systems, it is possible to produce an additional increase in strength and hardness by heat treatment alone. Such a method has many advantages, since the required strength can be induced at the most convenient stage of production or fabrication; moreover, the component is not sent into service in a highly stressed, plastically deformed state. The basic requirement for such a special alloy is that it should undergo a phase transformation in the solid state. One type of alloy satisfying this requirement, already considered, is that which can undergo an order–disorder reaction; the hardening accompanying this process (similar in many ways to precipitation hardening) is termed order hardening. However, conditions for this form of hardening are quite stringent, so that the two principal hardening methods, commonly used for alloys, are based upon (1) precipitation from a supersaturated solid solution and (2) eutectoid decomposition.

In engineering applications, strength is, without doubt, an important parameter. However, it is by no means the only important one and usually a material must provide a combination of properties. Some ductility is generally essential, enabling the material to relieve stress concentrations by plastic deformation and to resist fracture. The ability of materials to resist crack propagation and fracture, known generally as toughness, will be discussed in this chapter. Fracture can take many forms; some special forms, such as brittle fracture by cleavage, ductile fracture by microvoid coalescence, creep fracture by triple-point cracking and fatigue cracking, will be examined.

This chapter primarily concerns alloy behavior, partly because of the inherent versatility of alloy systems and partly because the research background to much of the current understanding of strength, toughness and fracture is essentially metallurgical. However, it is often possible to extend the basic principles to non-metallic materials, particularly in the case of fracture processes. This will be apparent later, in Chapter 10, when we describe how the unique transformation characteristics of zirconia can be used to inhibit crack propagation in a brittle ceramic such as alumina. Methods for toughening glasses are described in the same chapter. In Chapter 11 we consider the strengthening and toughening effects produced when plastics, metals and ceramics are reinforced with filaments to form composite materials.

7.2 Strengthening of non-ferrous alloys by heat treatment

7.2.1 Precipitation hardening of Al–Cu alloys

7.2.1.1 Precipitation from supersaturated solid solution

The basic requirements of a precipitation-hardening alloy system is that the solid solubility limit should decrease with decreasing temperature, as shown in Figure 7.1 for the Al–Cu system. During

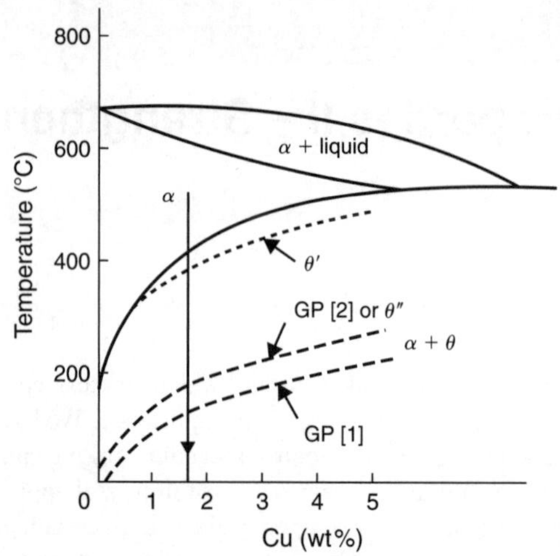

Figure 7.1 *Al-rich Al–Cu binary diagram showing GP[1], θ″ and θ′ solvus lines (dotted).*

the precipitation-hardening heat treatment procedure the alloy is first solution heat-treated at the high temperature and then rapidly cooled by quenching into water or some other cooling medium. The rapid cooling suppresses the separation of the θ-phase so that the alloy exists at the low temperature in an unstable supersaturated state. If, however, after quenching, the alloy is allowed to 'age' for a sufficient length of time, the second phase precipitates out. This precipitation occurs by a nucleation and growth process, fluctuations in solute concentration providing small clusters of atoms in the lattice which act as nuclei for the precipitate. However, the size of the precipitate becomes finer as the temperature at which precipitation occurs is lowered, and extensive hardening of the alloy is associated with a critical dispersion of the precipitate. If, at any given temperature, ageing is allowed to proceed too far, coarsening of the particles occurs (i.e. the small ones tend to redissolve and the large ones to grow still larger, as discussed in Section 7.2.6) and the numerous finely dispersed, small particles are gradually replaced by a smaller number of more widely dispersed, coarser particles. In this state the alloy becomes softer, and it is then said to be in the overaged condition (see Figure 7.2).

7.2.1.2 Changes in properties accompanying precipitation

The actual quenching treatment gives rise to small changes in many of the mechanical and physical properties of alloys, because both solute atoms and point defects in excess of the equilibrium concentration are retained during the process, and because the quench itself often produces lattice strains. Perhaps the property most markedly affected is the electrical resistance and this is usually considerably increased. In contrast, the mechanical properties are affected relatively much less.

On ageing, the change in properties in a quenched material is more marked and, in particular, the mechanical properties often show striking modifications. For example, the tensile strength of *Duralumin* (i.e. an aluminum–4% copper alloy containing magnesium, silicon and manganese) may be raised from 0.21 to 0.41 GN m^{-2}, while that of a Cu–2Be alloy may be increased from 0.46 to 1.23 GN m^{-2}. The structure-sensitive properties such as hardness, yield stress, etc. are, of course, extremely dependent on the structural distribution of the phases and, consequently, such alloys usually exhibit softening as the finely dispersed precipitates coarsen.

A simple theory of precipitation, involving the nucleation and growth of particles of the expected new equilibrium phase, leads one to anticipate that the alloy would show a single hardening peak,

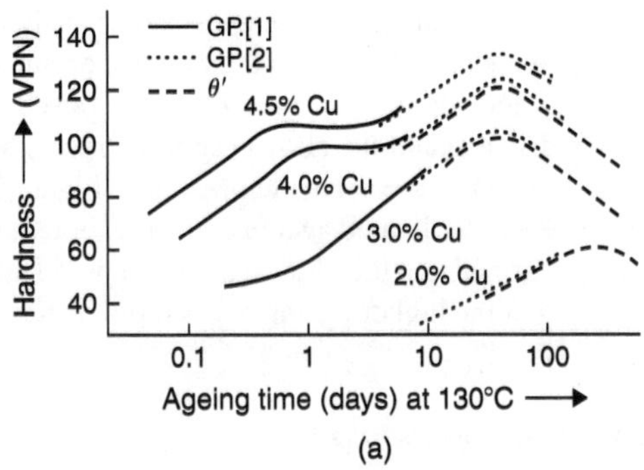

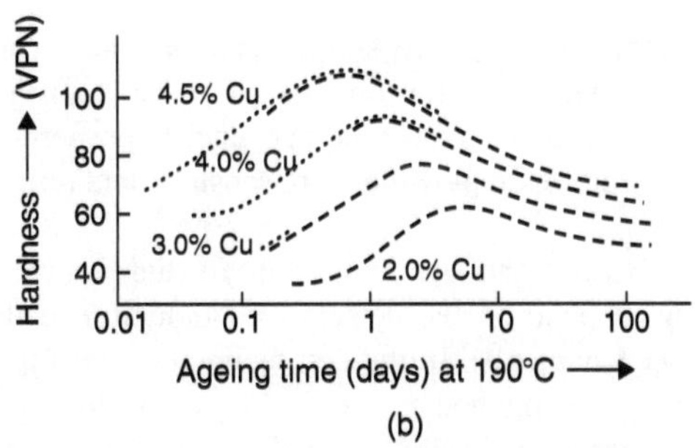

Figure 7.2 *The ageing of aluminum–copper alloys at 130°C (a) and 190°C (b) (after Silcock, Heal and Hardy, 1953–4).*

the electrical resistivity a decrease and the lattice parameter an increase (assuming the solute atom is smaller than the solvent atom) as the solute is removed from solution. Such property changes are found in practice, but only at low supersaturations and high ageing temperatures. At higher supersaturations and lower ageing temperatures the various property changes are not consistent with such a simple picture of precipitation; the alloy may show two or more age-hardening peaks, and the electrical resistivity and lattice parameter may not change in the anticipated manner. A hardening process which takes place in two stages is shown in aluminum–copper alloys (Figure 7.2a), where the initial hardening occurs without any attendant precipitation being visible in the light microscope and, moreover, is accompanied by a decrease in conductivity and no change in lattice parameter. Such behavior may be accounted for if precipitation is a process involving more than one stage. The initial stage of precipitation, at the lower ageing temperatures, involves a clustering of solute atoms on the solvent lattice sites to form zones or clusters, coherent with the matrix; the zones cannot be seen in the light microscope and for this reason this stage was at one time termed pre-precipitation. At a later stage of the ageing process these clusters break away from the matrix lattice to form distinct particles with their own crystal structure and a definite interface. These hypotheses were confirmed originally by structural studies using X-ray diffraction techniques, but nowadays the so-called pre-precipitation effects can be observed directly in the electron microscope.

Even though clustering occurs, the general kinetic behavior of the precipitation process is in agreement with that expected on thermodynamic grounds. From Figure 7.2 it is evident that the rate of ageing increases markedly with increasing temperature, while the peak hardness decreases. Two-stage hardening takes place at low ageing temperatures and is associated with high maximum hardness, while single-stage hardening occurs at higher ageing temperatures, or at lower ageing temperatures for lower solute contents.

Another phenomenon commonly observed in precipitation-hardening alloys is reversion or retrogression. If an alloy hardened by ageing at low temperature is subsequently heated to a higher ageing temperature it softens temporarily, but becomes harder again on more prolonged heating. This temporary softening, or reversion of the hardening process, occurs because the very small nuclei or zones precipitated at the low temperature are unstable when raised to the higher ageing temperature, and consequently they redissolve and the alloy becomes softer. The temperature above which the nuclei or zones dissolve is known as the solvus temperature; Figure 7.1 shows the solvus temperatures for GP zones, θ'', θ' and θ. On prolonged ageing at the higher temperature larger nuclei, characteristic of that temperature, are formed and the alloy again hardens. Clearly, the reversion process is reversible, provided re-hardening at the higher ageing temperature is not allowed to occur.

7.2.1.3 Structural changes during precipitation

Early metallographic investigations showed that the microstructural changes which occur during the initial stages of ageing are on too fine a scale to be resolved by the light microscope, yet it is in these early stages that the most profound changes in properties are found. Accordingly, to study the process, it is necessary to employ the more sensitive and refined techniques of X-ray diffraction and electron microscopy.

The two basic X-ray techniques, important in studying the regrouping of atoms during the early stages of ageing, depend on the detection of radiation scattered away from the main diffraction lines or spots (see Chapter 4). In the first technique, developed independently by Guinier and Preston in 1938, the Laue method is used. They found that the single-crystal diffraction pattern of an aluminum–copper alloy developed streaks extending from an aluminum lattice reflection along $\langle 1\ 0\ 0\rangle_{\mathrm{Al}}$ directions. This was attributed to the formation of copper-rich regions of plate-like shape on $\{1\ 0\ 0\}$ planes of the aluminum matrix (now called Guinier–Preston zones or GP zones). The net effect of the regrouping is to modify the scattering power of, and spacing between, very small groups of $\{1\ 0\ 0\}$ planes throughout the crystal. However, being only a few atomic planes thick, the zones produce the diffraction effect typical of a two-dimensional lattice, i.e. the diffraction spot becomes a diffraction streak. In recent years the Laue method has been replaced by a single-crystal oscillation technique employing monochromatic radiation, since interpretation is made easier if the wavelength of the X-rays used is known. The second technique makes use of the phenomenon of scattering of X-rays at small angles (see Chapter 4). Intense small-angle scattering can often be observed from age-hardening alloys (as shown in Figures 7.3 and 7.5) because there is usually a difference in electron density between the precipitated zone and the surrounding matrix. However, in alloys such as aluminum–magnesium or aluminum–silicon the technique is of no value because in these alloys the small difference in scattering power between the aluminum and silicon or magnesium atoms, respectively, is insufficient to give rise to appreciable scattering at small angles.

With the advent of the electron microscope the ageing of aluminum alloys was one of the first subjects to be investigated with the thin-foil transmission method. Not only can the detailed structural changes which occur during the ageing process be followed, but electron diffraction pictures taken from selected areas of the specimen while it is still in the microscope enable further important information on the structure of the precipitated phase to be obtained. Moreover, under some conditions the interaction of moving dislocations and precipitates can be observed. This naturally leads to a more complete understanding of the hardening mechanism.

Both the X-ray and electron-microscope techniques show that in virtually all age-hardening systems the initial precipitate is not the same structure as the equilibrium phase. Instead, the ageing sequence zones → intermediate precipitates → equilibrium precipitate is followed. This sequence occurs because the equilibrium precipitate is incoherent with the matrix, whereas the transition structures are either fully coherent, as in the case of zones, or at least partially coherent. Then, because of the importance of the surface energy and strain energy of the precipitate to the precipitation process, the system follows such a sequence in order to have the lowest free energy in all stages of precipitation.

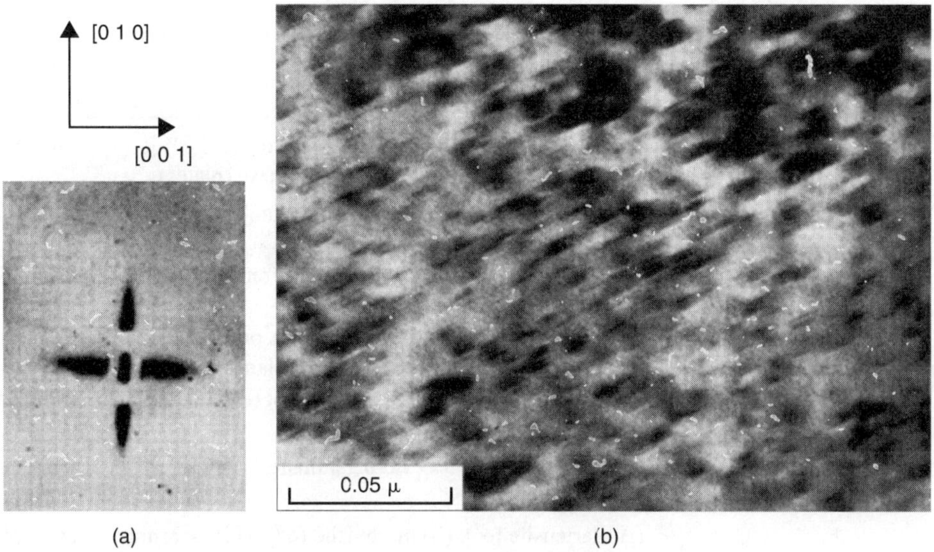

Figure 7.3 *(a) Small-angle X-ray pattern from aluminum–4% copper single crystal taken with molybdenum* Kα *radiation at a sample to film distance of 4 cm (after Guinier and Fournet, 1955; courtesy of John Wiley and Sons). (b) Electron micrograph of aluminum–4% copper aged 16 hours at 130°C, showing GP[1] zones (after Nicholson, Thomas and Nutting, 1958–9).*

The surface energy of the precipitates dominates the process of nucleation when the interfacial energy is large (i.e. when there is a discontinuity in atomic structure, somewhat like a grain boundary, at the interface between the nucleus and the matrix), so that for the incoherent type of precipitate the nuclei must exceed a certain minimum size before they can nucleate a new phase. To avoid such a slow mode of precipitation a coherent type of precipitate is formed instead, for which the size effect is relatively unimportant. The condition for coherence usually requires the precipitate to strain its equilibrium lattice to fit that of the matrix, or to adopt a metastable lattice. However, in spite of both a higher volume free energy and a higher strain energy, the transition structure is more stable in the early stages of precipitation because of its lower interfacial energy.

When the precipitate does become incoherent the alloy will, nevertheless, tend to reduce its surface energy as much as possible, by arranging the orientation relationship between the matrix and the precipitate so that the crystal planes which are parallel to, and separated by, the bounding surface have similar atomic spacings. Clearly, for these habit planes, as they are called, the better the crystallographic match, the less will be the distortion at the interface and the lower the surface energy. This principle governs the precipitation of many alloy phases, as shown by the frequent occurrence of the Widmanstätten structure, i.e. plate-shaped precipitates lying along prominent crystallographic planes of the matrix. Most precipitates are plate shaped because the strain energy factor is least for this form.

The existence of a precipitation sequence is reflected in the ageing curves and, as we have seen in Figure 7.2, often leads to two stages of hardening. The zones, by definition, are coherent with the matrix, and as they form the alloy becomes harder. The intermediate precipitate may be coherent with the matrix, in which case a further increase of hardness occurs, or only partially coherent, when either hardening or softening may result. The equilibrium precipitate is incoherent and its formation always leads to softening. These features are best illustrated by a consideration of some actual age-hardening systems.

Precipitation reactions occur in a wide variety of alloy systems, as shown in Table 7.1. The aluminum–copper alloy system exhibits the greatest number of intermediate stages in its precipitation

Table 7.1 *Some common precipitation-hardening systems.*

Base metal	*Solute*	*Transition structure*	*Equilibrium precipitate*
Al	Cu	(i) Plate-like solute-rich GP[1] zones on $\{1\ 0\ 0\}_{Al}$; (ii) ordered zones of GP[2]; (iii) θ'-phase (plates).	θ-$CuAl_2$
	Ag	(i) Spherical solute-rich zones; (ii) platelets of hexagonal γ' on $\{1\ 1\ 1\}_{Al}$.	γ-Ag_2Al
	Mg, Si	(i) GP zones rich in Mg and Si atoms on $\{1\ 0\ 0\}_{Al}$ planes; (ii) ordered zones of β'.	β-Mg_2Si (plates)
	Mg, Cu	(i) GP zones rich in Mg and Cu atoms on $\{1\ 0\ 0\}_{Al}$ planes; (ii) S′ platelets on $\{0\ 2\ 1\}_{Al}$ planes.	S-Al_2CuMg (laths)
	Mg, Zn	(i) Spherical zones rich in Mg and Zn; (ii) platelets of η'-phase on $\{1\ 1\ 1\}_{Al}$.	η-$MgZn_2$ (plates)
Cu	Be	(i) Be-rich regions on $\{1\ 0\ 0\}_{Cu}$ planes; (ii) γ'.	γ-CuBe
	Co	Spherical GP zones.	β-Co plates
Fe	C	(i) Martensite (α'); (ii) martensite (α''); (iii) ε-carbide.	Fe_3C plates cementite
	N	(i) Nitrogen martensite (α'); (ii) martensite (α'') disks.	Fe_4N
Ni	Al, Ti	γ' cubes	γ-$Ni_3(AlTi)$

process, and consequently is probably the most widely studied. When the copper content is high and the ageing temperature low, the sequence of stages followed is GP[1], GP[2], θ' and θ ($CuAl_2$). On ageing at higher temperatures, however, one or more of these intermediate stages may be omitted and, as shown in Figure 7.2, corresponding differences in the hardness curves can be detected. The early stages of ageing are due to GP[1] zones, which are interpreted as plate-like clusters of copper atoms segregated onto $\{1\ 0\ 0\}$ planes of the aluminum matrix. A typical small-angle X-ray scattering pattern and thin-foil transmission electron micrograph from GP[1] zones are shown in Figure 7.3. The plates are only a few atomic planes thick (giving rise to the $\langle 1\ 0\ 0\rangle$ streaks in the X-ray pattern), but are about 10 nm long, and hence appear as bright or dark lines on the electron micrograph.

GP[2] is best described as a coherent intermediate precipitate rather than a zone, since it has a definite crystal structure; for this reason the symbol θ'' is often preferred. These precipitates, usually of maximum thickness 10 nm and up to 150 nm in diameter, have a tetragonal structure which fits perfectly with the aluminum unit cell in the *a* and *b* directions but not in the *c*. The structure postulated has a central plane which consists of 100% copper atoms, the next two planes a mixture of copper and aluminum and the other two basal planes of pure aluminum, giving an overall composition of $CuAl_2$. Because of their size, θ'' precipitates are easily observed in the electron microscope, and because of the ordered arrangements of copper and aluminum atoms within the structure, their presence gives rise to intensity maxima on the diffraction streaks in an X-ray photograph. Since the *c* parameter (0.78 nm) differs from that of aluminum (0.404 nm) the aluminum planes parallel to the plate are distorted by elastic coherency strains. Moreover, the precipitate grows with the *c*-direction normal to the plane of the plate, so that the strain fields become larger as it grows and at peak hardness extend from one precipitate particle to the next (see Figure 7.4a). The direct observation of coherency strains confirms the theories of hardening based on the development of an elastically strained matrix (see next section).

The transition structure θ' is tetragonal; the true unit cell dimensions are $a = 0.404$ and $c = 0.58$ nm and the axes are parallel to $\langle 1\ 0\ 0\rangle_{Al}$ directions. The strains around the θ' plates can be relieved, however, by the formation of a stable dislocation loop around the precipitate and such a loop has been observed around small θ' plates in the electron microscope, as shown in Figure 7.4b. The long-range

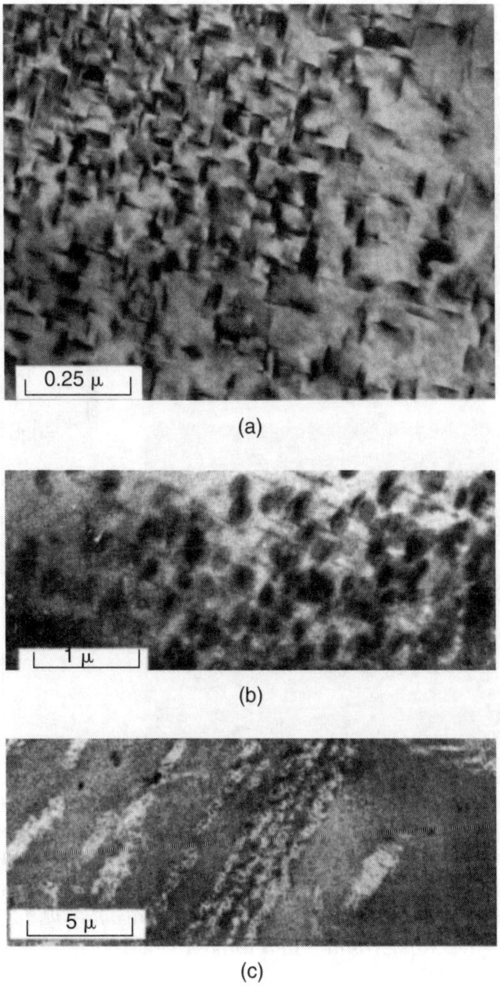

Figure 7.4 *Electron micrographs from Al–4Cu: (a) aged 5 hours at 160°C showing θ″ plates; (b) aged 12 hours at 200°C showing a dislocation ring round θ″ plates; (c) aged 3 days at 160°C showing θ″ precipitated on helical dislocations (after Nicholson, Thomas and Nutting, 1958–9).*

strain fields of the precipitate and its dislocation largely cancel. Consequently, it is easier for glide dislocations to move through the lattice of the alloy containing an incoherent precipitate such as θ' than a coherent precipitate such as θ'', and the hardness falls.

The θ structure is also tetragonal, with $a = 0.606$ and $c = 0.487$ nm. This equilibrium precipitate is incoherent with the matrix and its formation always leads to softening, since coherency strains disappear.

7.2.2 Precipitation hardening of Al–Ag alloys

Investigations using X-ray diffraction and electron microscopy have shown the existence of three distinct stages in the age-hardening process, which may be summarized: silver-rich clusters →

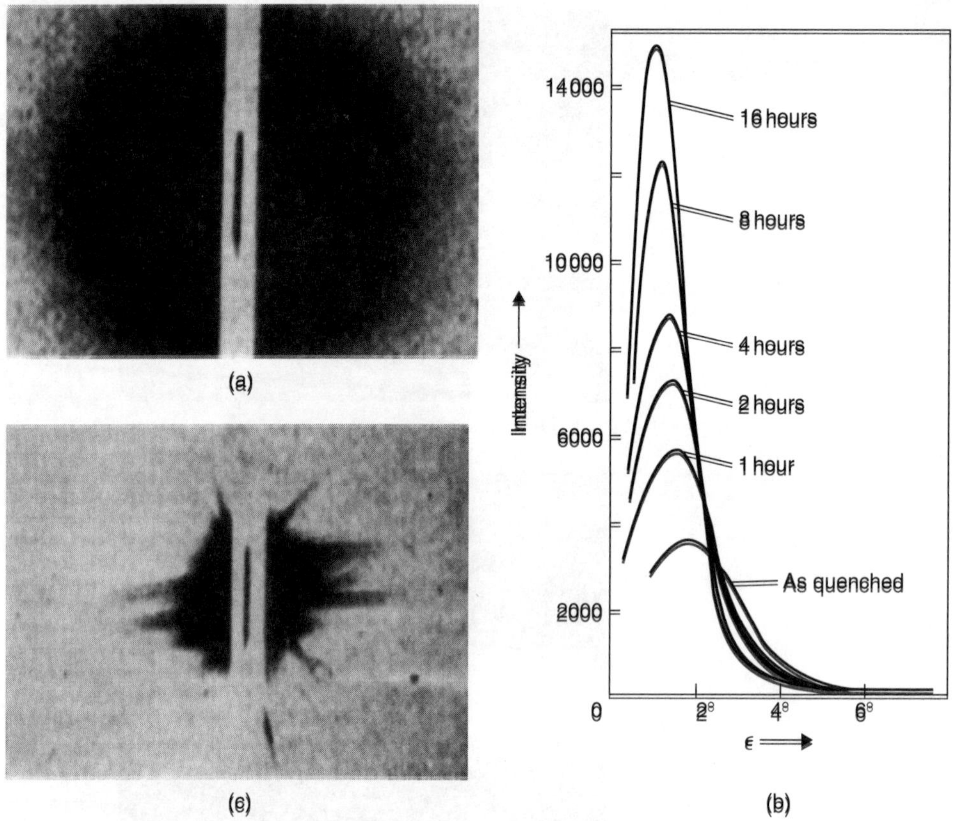

Figure 7.5 *Small-angle scattering of CuKα radiation by polycrystalline Al–Ag. (a) After quenching from 520°C (after Guinier and Walker, 1953). (b) The change in ring intensity and ring radius on ageing at 120°C (after Smallman and Westmacott, unpublished). (c) After ageing at 140°C for 10 days (after Guinier and Walker, 1953).*

intermediate hexagonal $\gamma' \Rightarrow$ equilibrium hexagonal γ. The hardening is associated with the first two stages in which the precipitate is coherent and partially coherent with the matrix, respectively.

During the quench and in the early stages of ageing, silver atoms cluster into small spherical aggregates and a typical small-angle X-ray picture of this stage, shown in Figure 7.5a, has a diffuse ring surrounding the trace of the direct beam. The absence of intensity in the center of the ring (i.e. at (0 0 0)) is attributed to the fact that clustering takes place so rapidly that there is left a shell-like region surrounding each cluster which is low in silver content. On ageing, the clusters grow in size and decrease in number, and this is characterized by the X-ray pattern showing a gradual decrease in ring diameter. The concentration and size of clusters can be followed very accurately by measuring the intensity distribution across the ring as a function of ageing time. This intensity may be represented (see Chapter 4) by an equation of the form:

$$I(\varepsilon) = Mn^2[\exp(-2\pi^2R^2\varepsilon^2/3\lambda^2) - \exp(-2\pi^2R_1^2\varepsilon^2/3\lambda^2)]^2 \tag{7.1}$$

and for values of ε greater than that corresponding to the maximum intensity, the contribution of the second term, which represents the denuded region surrounding the cluster, can be neglected.

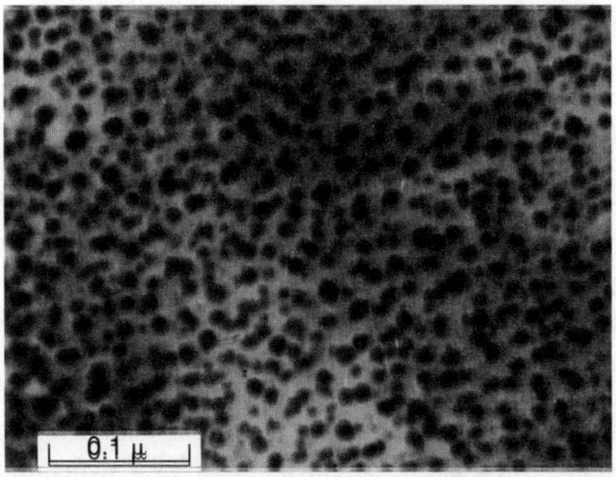

(a)

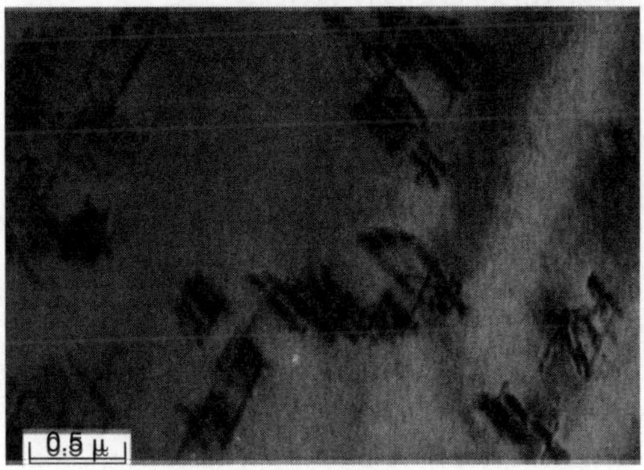

(b)

Figure 7.6 *Electron micrographs from Al–Ag alloy: (a) aged 5 hours at 160°C showing spherical zones and (b) aged 5 days at 160°C showing γ' precipitate (after Nicholson, Thomas and Nutting, 1958–9).*

Figure 7.5b shows the variation in the X-ray intensity, scattered at small angles (SAS) with cluster growth, on ageing an aluminum–silver alloy at 120°C. An analysis of this intensity distribution, using equation (7.1), indicates that the size of the zones increases from 2 to 5 nm in just a few hours at 120°C. These zones may, of course, be seen in the electron microscope and Figure 7.6a is an electron micrograph showing spherical zones in an aluminum–silver alloy aged for 5 hours at 160°C; the diameter of the zones is about 10 nm, in good agreement with that deduced by X-ray analysis. The zone shape is dependent upon the relative diameters of solute and solvent atoms. Thus, solute atoms such as silver and zinc which have atomic sizes similar to aluminum give rise to spherical zones, whereas solute atoms such as copper which have a high misfit in the solvent lattice form plate-like zones.

With prolonged annealing, the formation and growth of platelets of a new phase, γ', occur. This is characterized by the appearance in the X-ray pattern of short streaks passing through the trace of the direct beam (Figure 7.5c). The γ' platelet lies parallel to the $\{1\ 1\ 1\}$ planes of the matrix and its structure has lattice parameters very close to that of aluminum. However, the structure is hexagonal and, consequently, the precipitates are easily recognizable in the electron microscope by the stacking-fault contrast within them, as shown in Figure 7.6b. Clearly, these precipitates are never fully coherent with the matrix, but nevertheless, in this alloy system, where the zones are spherical and have little or no coherency strain associated with them, and where no coherent intermediate precipitate is formed, the partially coherent γ' precipitates do provide a greater resistance to dislocation movement than zones and a second stage of hardening results.

The same principles apply to the constitutionally more complex ternary and quaternary alloys as to the binary alloys. Spherical zones are found in aluminum–magnesium–zinc alloys as in aluminum–zinc, although the magnesium atom is some 12% larger than the aluminum atom. The intermediate precipitate forms on the $\{1\ 1\ 1\}_{Al}$ planes and is partially coherent with the matrix, with little or no strain field associated with it. Hence, the strength of the alloy is due purely to dispersion hardening and the alloy softens as the precipitate becomes coarser. In nickel-based alloys the hardening phase is the ordered γ'-Ni_3Al; this γ' is an equilibrium phase in the Ni–Al and Ni–Cr–Al systems and a metastable phase in Ni–Ti and Ni–Cr–Ti. These systems form the basis of the 'superalloys' (see Chapter 8), which owe their properties to the close matching of the γ' and the fcc matrix. The two phases have very similar lattice parameters ($\lesssim$0.25%, depending on composition) and the coherency (interfacial energy $\gamma_I \approx$ 10–20 mJ m^{-2}) confers a very low coarsening rate on the precipitate, so that the alloy overages extremely slowly, even at $0.7T_m$.

7.2.3 Mechanisms of precipitation hardening

7.2.3.1 *The significance of particle deformability*

The strength of an age-hardening alloy is governed by the interaction of moving dislocations and precipitates. The obstacles in precipitation-hardening alloys which hinder the motion of dislocations may be either (1) the strains around GP zones, (2) the zones or precipitates themselves, or both. Clearly, if it is the zones themselves which are important, it will be necessary for the moving dislocations either to cut through them or go round them. Thus, merely from elementary reasoning, it would appear that there are at least three causes of hardening, namely: (1) coherency strain hardening, (2) chemical hardening, i.e. when the dislocation cuts through the precipitate, or (3) dispersion hardening, i.e. when the dislocation goes round or over the precipitate.

The relative contributions will depend on the particular alloy system but, generally, there is a critical dispersion at which the strengthening is a maximum, as shown in Figure 7.7. In the small-particle regime the precipitates, or particles, are coherent and deformable as the dislocations cut through them, while in the larger-particle regime the particles are incoherent and non-deformable as the dislocations bypass them. For deformable particles, when the dislocations pass through the particle, the intrinsic properties of the particle are of importance and alloy strength varies only weakly with particle size. For non-deformable particles, when the dislocations bypass the particles, the alloy strength is independent of the particle properties but is strongly dependent on particle size and dispersion strength, decreasing as particle size or dispersion increases. The transition from deformable to non-deformable particle-controlled deformation is readily recognized by the change in microstructure, since the 'laminar' undisturbed dislocation flow for the former contrasts with the turbulent plastic flow for non-deformable particles. The latter leads to the production of a high density of dislocation loops, dipoles and other debris, which results in a high rate of work hardening. This high rate of work hardening is a distinguishing feature of all dispersion-hardened systems.

7.2.3.2 *Coherency strain hardening*

The precipitation of particles having a slight misfit in the matrix gives rise to stress fields which hinder the movement of gliding dislocations. For the dislocations to pass through the regions of

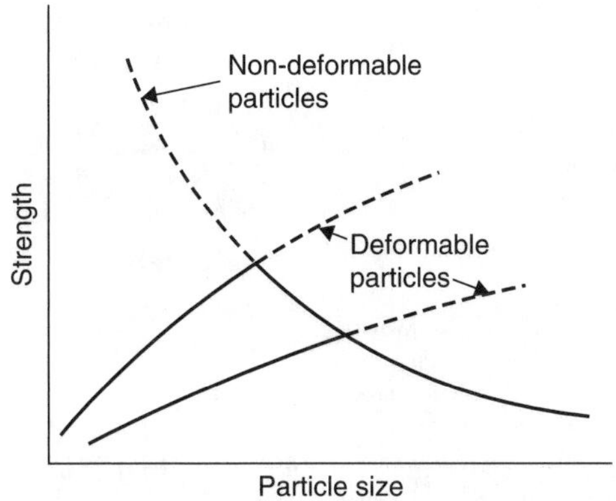

Figure 7.7 *Variation of strength with particle size, defining the deformable and non-deformable particle regimes.*

internal stress, the applied stress must be at least equal to the average internal stress, and for spherical particles this is given by

$$\tau = 2\mu\varepsilon f, \tag{7.2}$$

where μ is the shear modulus, ε is the misfit of the particle and f is the volume fraction of precipitate. This suggestion alone, however, cannot account for the critical size of dispersion of a precipitate at which the hardening is a maximum, since equation (7.2) is independent of L, the distance between particles. To explain this, Mott and Nabarro consider the extent to which a dislocation can bow round a particle under the action of a stress τ. Like the bowing stress of a Frank–Read source, this is given by

$$r = \alpha\mu b/\tau, \tag{7.3}$$

where r is the radius of curvature to which the dislocation is bent, which is related to the particle spacing. Hence, in the hardest age-hardened alloys, where the yield strength is about $\mu/100$, the dislocation can bend to a radius of curvature of about 100 atomic spacings, and since the distance between particles is of the same order it would appear that the dislocation can avoid the obstacles and take a form like that shown in Figure 7.8a. With a dislocation line taking up such a configuration, in order to produce glide, each section of the dislocation line has to be taken over the adverse region of internal stress without any help from other sections of the line – the alloy is then hard. If the precipitate is dispersed on too fine a scale (e.g. when the alloy has been freshly quenched or lightly aged), the dislocation is unable to bend sufficiently to lie entirely in the regions of low internal stress. As a result, the internal stresses acting on the dislocation line largely cancel and the force resisting its movement is small – the alloy then appears soft. When the dispersion is on a coarse scale, the dislocation line is able to move between the particles, as shown in Figure 7.8b, and the hardening is again small.

For coherency strain hardening, the flow stress depends on the ability of the dislocation to bend and thus experience more regions of adverse stress than of aiding stress. The flow stress therefore depends on the treatment of averaging the stress, and recent attempts separate the behavior of small and large coherent particles. For small coherent particles the flow stress is given by

$$\tau = 4.1\mu\varepsilon^{3/2}f^{1/2}(r/b)^{1/2}, \tag{7.4}$$

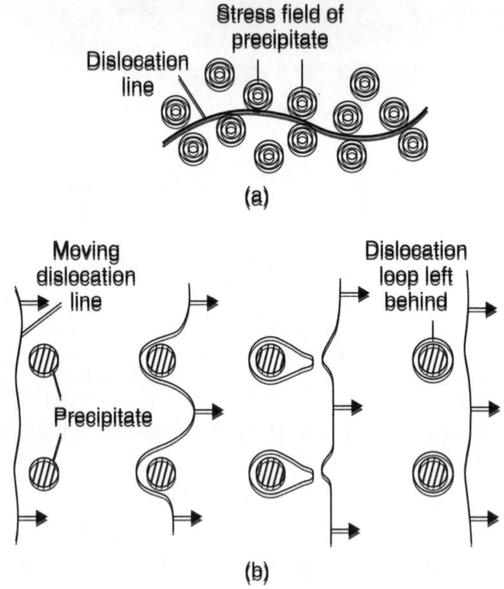

Figure 7.8 *Schematic representation of a dislocation: (a) curling round the stress fields from precipitates and (b) passing between widely spaced precipitates (Orowan looping).*

which predicts a greater strengthening than the simple arithmetic average of equation (7.2). For large coherent particles

$$\tau = 0.7\mu f^{1/2}(\varepsilon b^3/r^3)^{1/4}. \tag{7.5}$$

7.2.3.3 Chemical hardening

When a dislocation actually passes through a zone, as shown in Figure 7.9, a change in the number of solvent–solute near-neighbors occurs across the slip plane. This tends to reverse the process of clustering and, hence, additional work must be done by the applied stress to bring this about. This process, known as chemical hardening, provides a short-range interaction between dislocations and precipitates, and arises from three possible causes: (1) the energy required to create an additional particle/matrix interface with energy γ_I per unit area which is provided by a stress

$$\tau \approx \alpha\gamma_I^{3/2}(fr)^{1/2}/\mu b^2, \tag{7.6}$$

where α is a numerical constant, (2) the additional work required to create an anti-phase boundary inside the particle with ordered structure, given by

$$\tau \approx \beta\gamma_{APB}^{3/2}(fr)^{1/2}/\mu b^2, \tag{7.7}$$

where β is a numerical constant, and (3) the change in width of a dissociated dislocation as it passes through the particle where the stacking-fault energy differs from the matrix (e.g. Al–Ag, where $\Delta\gamma_{SF} \approx 100\,\text{mJ}\,\text{m}^{-2}$ between Ag zones and Al matrix), so that

$$\tau \approx \Delta\gamma_{SF}/b. \tag{7.8}$$

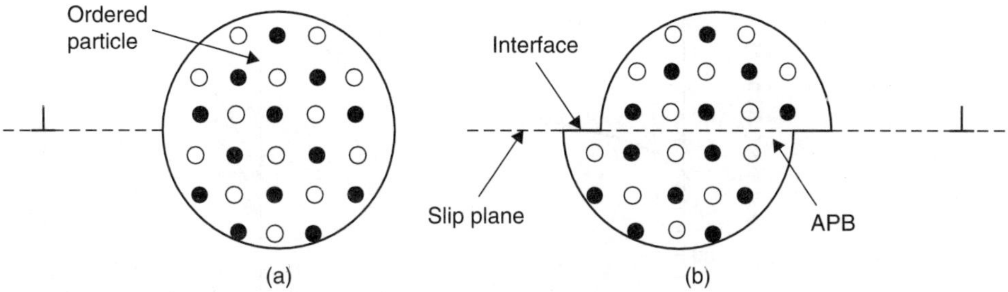

Figure 7.9 *Ordered particle (a) cut by dislocations (b) to produce new interface and APB.*

Usually $\gamma_I < \gamma_{APB}$ and so γ_I can be neglected, but the ordering within the particle requires the dislocations to glide in pairs. This leads to a strengthening given by

$$\tau = (\gamma_{APB}/2b)[(4\gamma_{APB}rf/\pi T)^{1/2} - f], \tag{7.9}$$

where T is the dislocation line tension.

7.2.3.4 Dispersion hardening

In dispersion hardening it is assumed that the precipitates do not deform with the matrix and that the yield stress is the stress necessary to expand a loop of dislocation between the precipitates. This will be given by the Orowan stress

$$\tau = \alpha\mu b/L, \tag{7.10}$$

where L is the separation of the precipitates. As discussed above, this process will be important in the later stages of precipitation when the precipitate becomes incoherent and the misfit strains disappear. A moving dislocation is then able to bypass the obstacles, as shown in Figure 7.8b, by moving in the clean pieces of crystal between the precipitated particles. The yield stress decreases as the distance between the obstacles increases in the overaged condition. However, even when the dispersion of the precipitate is coarse a greater applied stress is necessary to force a dislocation past the obstacles than would be the case if the obstruction were not there. Some particle or precipitate strengthening remains, but the majority of the strengthening arises from the dislocation debris left around the particles, giving rise to high work hardening.

7.2.3.5 Hardening mechanisms in Al–Cu alloys

The actual hardening mechanism which operates in a given alloy will depend on several factors, such as the type of particle precipitated (e.g. whether zone, intermediate precipitate or stable phase), the magnitude of the strain and the testing temperature. In the earlier stages of ageing (i.e. before overageing) the coherent zones are cut by dislocations moving through the matrix, and hence both coherency strain hardening and chemical hardening will be important, e.g. in such alloys as aluminum–copper, copper–beryllium and iron–vanadium–carbon. In alloys such as aluminum–silver and aluminum–zinc, however, the zones possess no strain field, so that chemical hardening will be the most important contribution. In the important high-temperature creep-resistant nickel alloys, the precipitate is of the

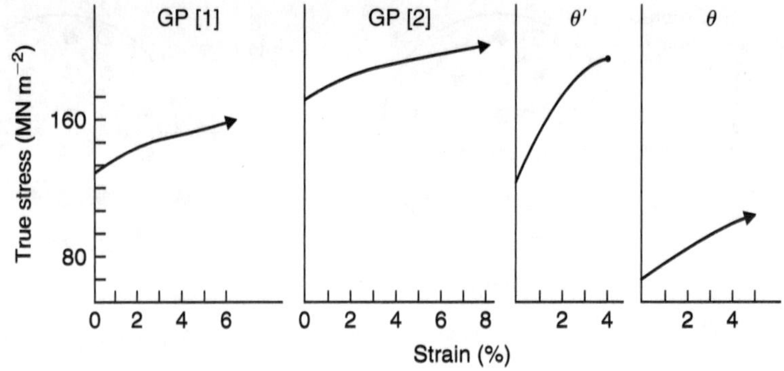

Figure 7.10 *Stress–strain curves from single crystals of aluminum–4% copper containing GP[1] zones, GP[2], zones, θ′ precipitates and θ precipitates respectively (after Fine, Bryne and Kelly, 1961; courtesy of Taylor & Francis).*

Ni_3Al form, which has a low particle/matrix misfit and hence chemical hardening due to dislocations cutting the particles is again predominant. To illustrate that more than one mechanism of hardening is in operation in a given alloy system, let us examine the mechanical behavior of an aluminum–copper alloy in more detail.

Figure 7.10 shows the deformation characteristics of single crystals of an aluminum–copper (nominally 4%) alloy in various structural states. The curves were obtained by testing crystals of approximately the same orientation, but the stress–strain curves from crystals containing GP[1] and GP[2] zones are quite different from those for crystals containing θ' or θ precipitates. When the crystals contain either GP[1] or GP[2] zones, the stress–strain curves are very similar to those of pure aluminum crystals, except that there is a two- or threefold increase in the yield stress. In contrast, when the crystals contain either θ' or θ precipitates the yield stress is less than for crystals containing zones, but the initial rate of work hardening is extremely rapid. In fact, the stress–strain curves bear no similarity to those of a pure aluminum crystal. It is also observed that when θ' or θ is present as a precipitate, deformation does not take place on a single slip system but on several systems; the crystal then deforms, more nearly as a polycrystal does and the X-ray pattern develops extensive asterism. These factors are consistent with the high rate of work hardening observed in crystals containing θ' or θ precipitates.

The separation of the precipitates cutting any slip plane can be deduced from both X-ray and electron-microscope observations. For the crystals, relating to Figure 7.10, containing GP[1] zones this value is 15 nm and for GP[2] zones it is 25 nm. It then follows from equation (7.3) that to avoid these precipitates the dislocations would have to bow to a radius of curvature of about 10 nm. To do this requires a stress several times greater than the observed flow stress and, in consequence, it must be assumed that the dislocations are forced through the zones. Furthermore, if we substitute the observed values of the flow stress in the relation $\mu b/\tau = L$, it will be evident that the bowing mechanism is unlikely to operate unless the particles are about 60 nm apart. This is confirmed by electron-microscope observations which show that dislocations pass through GP zones and coherent precipitates, but bypass non-coherent particles. Once a dislocation has cut through a zone, however, the path for subsequent dislocations on the same slip plane will be easier, so that the work-hardening rate of crystals containing zones should be low, as shown in Figure 7.10. The straight, well-defined slip bands observed on the surfaces of crystals containing GP[1] zones also support this interpretation.

If the zones possess no strain field, as in aluminum–silver or aluminum–zinc alloys, the flow stress would be entirely governed by the chemical hardening effect. However, the zones in aluminum–copper alloys do possess strain fields, as shown in Figure 7.4, and consequently the stresses around a

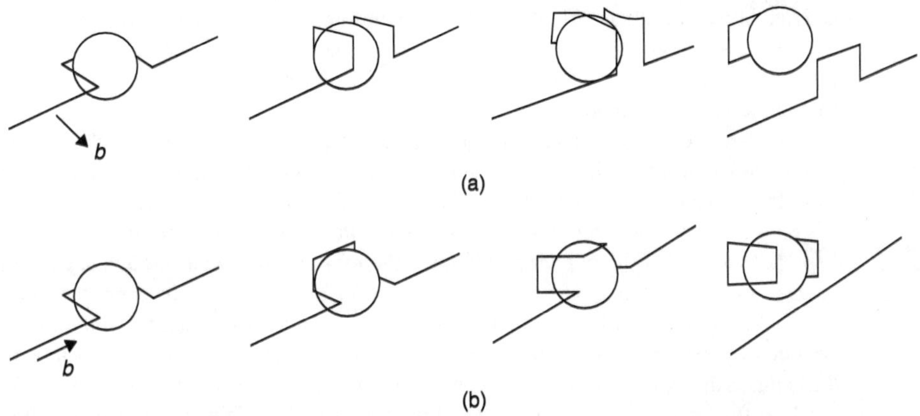

Figure 7.11 *Cross-slip of edge (a) and screw (b) dislocation over a particle producing prismatic loops in the process.*

zone will also affect the flow stress. Each dislocation will be subjected to the stresses due to a zone at a small distance from the zone.

It will be remembered from Chapter 6 that temperature profoundly affects the flow stress if the barrier which the dislocations have to overcome is of a short-range nature. For this reason, the flow stress of crystals containing GP[1] zones will have a larger dependence on temperature than that of those containing GP[2] zones. Thus, while it is generally supposed that the strengthening effect of GP[2] zones is greater than that of GP[1], and this is true at normal temperatures (see Figure 7.10), at very low temperatures it is probable that GP[1] zones will have the greater strengthening effect due to the short-range interactions between zones and dislocations.

The θ' and θ precipitates are incoherent and do not deform with the matrix, so that the critical resolved shear stress is the stress necessary to expand a loop of dislocation between them. This corresponds to the overaged condition and the hardening to dispersion hardening. The separation of the θ particles is greater than that of the θ', being somewhat greater than 1 μm, and the initial flow stress is very low. In both cases, however, the subsequent rate of hardening is high because, as suggested by Fisher, Hart and Pry, the gliding dislocation interacts with the dislocation loops in the vicinity of the particles (see Figure 7.8b). The stress–strain curves show, however, that the rate of work hardening falls to a low value after a few percent strain, and these authors attribute the maximum in the strain-hardening curve to the shearing of the particles. This process is not observed in crystals containing θ precipitates at room temperature and, consequently, it seems more likely that the particles will be avoided by cross-slip. If this is so, prismatic loops of dislocation will be formed at the particles, by the mechanism shown in Figure 7.11, and these will give approximately the same mean internal stress as that calculated by Fisher, Hart and Pry, but a reduced stress on the particle. The maximum in the work-hardening curve would then correspond to the stress necessary to expand these loops; this stress will be of the order of $\mu b/r$, where r is the radius of the loop, which is somewhat greater than the particle size. At low temperatures cross-slip is difficult and the stress may be relieved either by initiating secondary slip or by fracture.

7.2.4 Vacancies and precipitation

It is clear that because precipitation is controlled by the rate of atomic migration in the alloy, temperature will have a pronounced effect on the process. Moreover, since precipitation is a thermally activated process, other variables such as time of annealing, composition, grain size and prior cold work are also important. However, the basic treatment of age-hardening alloys is solution treatment followed by quenching, and the introduction of vacancies by the latter process must play an important role in the kinetic behavior.

It has been recognized that, near room temperature, zone formation in alloys such as aluminum–copper and aluminum–silver occurs at a rate many orders of magnitude greater than that calculated from the diffusion coefficient of the solute atoms. In aluminum–copper, for example, the formation of zones is already apparent after only a few minutes at room temperature, and is complete after an hour or two, so that the copper atoms must therefore have moved through several atomic spacings in that time. This corresponds to an apparent diffusion coefficient of copper in aluminum of about 10^{-20}–10^{-22} m^2 s^{-1}, which is many orders of magnitude faster than the value of 5×10^{-29} m^2 s^{-1} obtained by extrapolation of high-temperature data. Many workers have attributed this enhanced diffusion to the excess vacancies retained during the quenching treatment. Thus, since the expression for the diffusion coefficient at a given temperature contains a factor proportional to the concentration of vacancies at that temperature, if the sample contains an abnormally large vacancy concentration then the diffusion coefficient should be increased by the ratio c_Q/c_o, where c_Q is the quenched-in vacancy concentration and c_o is the equilibrium concentration. The observed clustering rate can be accounted for if the concentration of vacancies retained is about 10^{-3}–10^{-4}.

The observation of loops by transmission electron microscopy allows an estimate of the number of excess vacancies to be made, and in all cases of rapid quenching the vacancy concentration in these alloys is somewhat greater than 10^{-4}, in agreement with the predictions outlined above. Clearly, as the excess vacancies are removed, the amount of enhanced diffusion diminishes, which agrees with the observations that the isothermal rate of clustering decreases continuously with increasing time. In fact, it is observed that D decreases rapidly at first and then remains at a value well above the equilibrium value for months at room temperature; the process is therefore separated into what is called the fast and slow reactions. A mechanism proposed to explain the slow reaction is that some of the vacancies quenched-in are trapped temporarily and then released slowly. Measurements show that the activation energy in the fast reaction ($\approx$0.5 eV) is smaller than in the slow reaction ($\approx$1 eV) by an amount which can be attributed to the binding energy between vacancies and trapping sites. These traps are very likely small dislocation loops or voids formed by the clustering of vacancies. The equilibrium matrix vacancy concentration would then be greater than that for a well-annealed crystal by a factor $\exp[\gamma\Omega/r\mathbf{k}T]$, where γ is the surface energy, Ω the atomic volume and r the radius of the defect (see Chapter 3). The experimental diffusion rate can be accounted for if $r \approx 2$ nm, which is much smaller than the loops and voids usually seen, but they do exist. The activation energy for the slow reaction would then be $E_D - (\gamma\Omega/r)$ or approximately 1 eV for $r \approx 2$ nm.

Other factors known to affect the kinetics of the early stages of ageing (e.g. altering the quenching rate, interrupted quenching and cold work) may also be rationalized on the basis that these processes lead to different concentrations of excess vacancies. In general, cold working the alloy prior to ageing causes a decrease in the rate of formation of zones, which must mean that the dislocations introduced by cold work are more effective as vacancy sinks than as vacancy sources. Cold working or rapid quenching therefore have opposing effects on the formation of zones. Vacancies are also important in other aspects of precipitation hardening. For example, the excess vacancies, by condensing to form a high density of dislocation loops, can provide nucleation sites for intermediate precipitates. This leads to the interesting observation in aluminum–copper alloys that cold working or rapid quenching, by producing dislocations for nucleation sites, have the same effect on the formation of the θ' phase but, as we have seen above, the opposite effect on zone formation. It is also interesting to note that screw dislocations, which are not normally favorable sites for nucleation, can also become sites for preferential precipitation when they have climbed into helical dislocations by absorbing vacancies, and have thus become mainly of edge character. The long arrays of θ' phase observed in aluminum–copper alloys, shown in Figure 7.4c, have probably formed on helices in this way. In some of these alloys, defects containing stacking faults are observed, in addition to the dislocation loops and helices, and examples have been found where such defects nucleate an intermediate precipitate having a hexagonal structure. In aluminum–silver alloys it is also found that the helical dislocations introduced by quenching absorb silver and degenerate into long narrow stacking faults on {1 1 1} planes; these stacking-fault defects then act as nuclei for the hexagonal γ' precipitate.

Many commercial alloys depend critically on the interrelation between vacancies, dislocations and solute atoms, and it is found that trace impurities significantly modify the precipitation process. Thus, trace elements which interact strongly with vacancies inhibit zone formation, e.g. Cd, In and Sn prevent zone formation in slowly quenched Al–Cu alloys for up to 200 days at 30°C. This delays the age-hardening process at room temperature, which gives more time for mechanically fabricating the quenched alloy before it gets too hard, thus avoiding the need for refrigeration. On the other hand, Cd increases the density of θ' precipitate by increasing the density of vacancy loops and helices which act as nuclei for precipitation and by segregating to the matrix/θ' interfaces, thereby reducing the interfacial energy.

Since grain boundaries absorb vacancies, in many alloys there is a grain boundary zone relatively free from precipitation. The Al–Zn–Mg alloy is one commercial alloy which suffers grain boundary weakness, but it is found that trace additions of Ag have a beneficial effect in refining the precipitate structure and removing the precipitate free grain boundary zone. Here it appears that Ag atoms stabilize vacancy clusters near the grain boundary and also increase the stability of the GP zone, thereby raising the GP zone solvus temperature. Similarly, in the 'Concorde' alloy, *RR58* (basically Al–2.5Cu–1.2Mg with additions), Si addition (0.25Si) modifies the as-quenched dislocation distribution, inhibiting the nucleation and growth of dislocation loops and reducing the diameter of helices. The S-precipitate (Al_2CuMg) is homogeneously nucleated in the presence of Si rather than heterogeneously nucleated at dislocations, and the precipitate grows directly from zones, giving rise to improved and more uniform properties.

Apart from speeding up the kinetics of ageing, and providing dislocations nucleation sites, vacancies may play a structural role when they precipitate cooperatively with solute atoms to facilitate the basic atomic arrangements required for transforming the parent crystal structure to that of the product phase. In essence, the process involves the systematic incorporation of excess vacancies, produced by the initial quench or during subsequent dislocation loop annealing, in a precipitate zone or plate to change the atomic stacking. A simple example of θ' formation in Al–Cu is shown schematically in Figure 7.12. Ideally, the structure of the θ'' phase in Al–Cu consists of layers of copper on {1 0 0} separated by three layers of aluminum atoms. If a next-nearest neighbor layer of aluminum atoms from the copper layer is removed by condensing a vacancy loop, an embryonic θ' unit cell with Al in the correct **AAA** ... stacking sequence is formed (Figure 7.12b). Formation of the final $CuAl_2$ θ' fluorite structure requires shuffling only half of the copper atoms into the newly created next-nearest neighbor space and concurrent relaxation of the Al atoms to the correct θ' interplanar distances (Figure 7.12c).

The structural incorporation of vacancies in a precipitate is a non-conservative process, since atomic sites are eliminated. There exist equivalent conservative processes in which the new precipitate structure is created from the old by the nucleation and expansion of partial dislocation loops with predominantly shear character. Thus, for example, the **BABAB** {1 0 0} plane stacking sequence of the fcc structure can be changed to **BAABA** by the propagation of an $a/2\langle 1\ 0\ 0\rangle$ shear loop in the {1 0 0} plane, or to **BAAAB** by the propagation of a pair of $a/2\langle 1\ 0\ 0\rangle$ partials of opposite sign on adjacent planes. Again, the **AAA** stacking resulting from the double shear is precisely that required for the embryonic formation of the fluorite structure from the fcc lattice.

In visualizing the role of lattice defects in the nucleation and growth of plate-shaped precipitates, a simple analogy with Frank and Shockley partial dislocation loops is useful. In the formation of a Frank loop, a layer of hcp material is created from the fcc lattice by the (non-conservative) condensation of a layer of vacancies in {1 1 1}. Exactly the same structure is formed by the (conservative) expansion of a Shockley partial loop on a {1 1 1} plane. In the former case a semi-coherent 'precipitate'

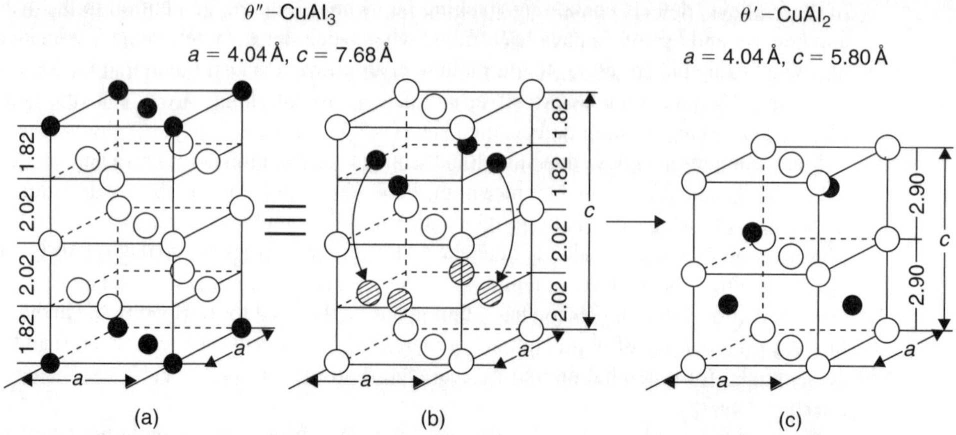

Figure 7.12 *Schematic diagram showing the transition of θ'' to θ' in Al–Cu by the vacancy mechanism. Vacancies from annealing loops are condensed on a next-nearest Al plane from the copper layer in θ'' to form the required* ***AAA*** *Al stacking. Formation of the θ' fluorite structure then requires only slight redistribution of the copper atom layer and relaxation of the Al layer spacings (courtesy of K. H. Westmacott).*

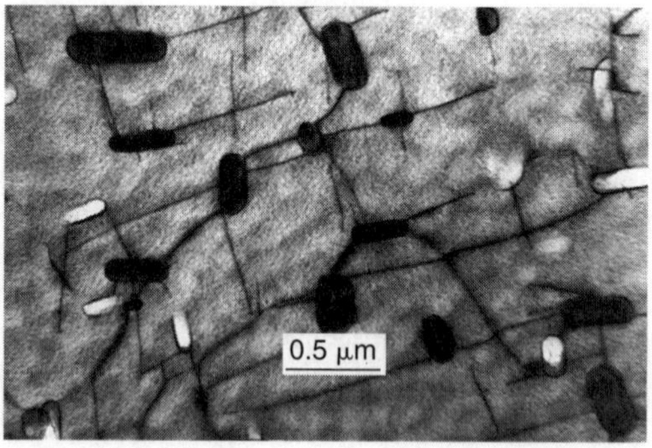

Figure 7.13 *The formation of semi-coherent Cu needles in Fe–1% Cu (courtesy of K. H. Westmacott).*

is produced bounded by an $a/3\langle 1\ 1\ 1\rangle$ dislocation, and in the latter a coherent one bounded by an $a/6\langle 1\ 1\ 2\rangle$. Continued growth of precipitate plates occurs by either process or a combination of processes. Of course, formation of the final precipitate structure requires, in addition to these structural rearrangements, the long-range diffusion of the correct solute atom concentration to the growing interface.

The growth of a second-phase particle with a disparate size or crystal structure relative to the matrix is controlled by two overriding principles – the accommodation of the volume and shape change, and the optimized use of the available deformation mechanisms. In general, volumetric transformation strains are accommodated by vacancy or interstitial condensation, or prismatic dislocation loop punching, while deviatoric strains are relieved by shear loop propagation. An example is shown in Figure 7.13. The formation of semi-coherent Cu needles in Fe–1% Cu is accomplished by the generation of shear loops in the precipitate/matrix interface. Expansion of the loops into the matrix and incorporation into nearby precipitate interfaces leads to a complete network of dislocations interconnecting the precipitates.

7.2.5 Duplex ageing

In non-ferrous heat treatment there is considerable interest in double (or duplex) ageing treatments to obtain the best microstructure consistent with optimum properties. It is now realized that it is unlikely that the optimum properties will be produced in alloys of the precipitation-hardening type by a single quench and ageing treatment. For example, while the interior of grains may develop an acceptable precipitate size and density, in the neighborhood of efficient vacancy sinks, such as grain boundaries, a precipitate-free zone (PFZ) is formed which is often associated with overageing in the boundary itself. This heterogeneous structure gives rise to poor properties, particularly under stress corrosion conditions.

Duplex ageing treatments have been used to overcome this difficulty. In Al–Zn–Mg, for example, it was found that storage at room temperature before heating to the ageing temperature leads to the formation of finer precipitate structure and better properties. This is just one special example of two-step or multiple ageing treatments which have commercial advantages and have been found to be applicable to several alloys. Duplex ageing gives better competitive mechanical properties in Al alloys (e.g. Al–Zn–Mg alloys) with much enhanced corrosion resistance, since the grain boundary zone is removed. It is possible to obtain strengths of 267–308 MN/m^{-2} in Mg–Zn–Mn alloys, which have very good strength/weight ratio applications, and nickel alloys also develop better properties with multiple ageing treatments.

The basic idea of all heat treatments is to 'seed' a uniform distribution of stable nuclei at the low temperature which can then be grown to optimum size at the higher temperature. In most alloys, there is a critical temperature T_c above which homogeneous nucleation of precipitate does not take place, and in some instances has been identified with the GP zone solvus. On ageing above T_c there is a certain critical zone size above which the zones are able to act as nuclei for precipitates and below which the zones dissolve.

In general, the ageing behavior of Al–Zn–Mg alloys can be divided into three classes, which can be defined by the temperature ranges involved:

1. Alloys quenched and aged above the GP zone solvus (i.e. the temperature above which the zones dissolve, which is above ~155°C in a typical Al–Zn–Mg alloy). Then, since no GP zones are ever formed during heat treatment, there are no easy nuclei for subsequent precipitation and a very coarse dispersion of precipitates results, with nucleation principally on dislocations.
2. Alloys quenched and aged below the GP zone solvus. GP zones form continuously and grow to a size at which they are able to transform to precipitates. The transformation will occur rather more slowly in the grain boundary regions due to the lower vacancy concentration there, but since ageing will always be below the GP zone solvus, no PFZ is formed other than a very small (~30 nm) solute-denuded zone due to precipitation in the grain boundary.
3. Alloys quenched below the GP zone solvus and aged above it (e.g. quenched to room temperature and aged at 180°C for Al–Zn–Mg). This is the most common practical situation. The final dispersion of precipitates and the PFZ width are controlled by the nucleation treatment below 155°C, where GP zone size distribution is determined. A long nucleation treatment gives a fine dispersion of precipitates and a narrow PFZ.

It is possible to stabilize GP zones by addition of trace elements. These have the same effect as raising T_c, so that alloys are effectively aged below T_c. One example is Ag to Al–Zn–Mg, which raises T_c from 155 to 185°C, another is Si to Al–Cu–Mg, another Cu to Al–Mg–Si and yet another Cd or Sn to Al–Cu alloys. It is then possible to get uniform distribution and optimum properties by single ageing, and is an example of achieving by chemistry what can similarly be done with physics during multiple ageing. Whether it is best to alter the chemistry or to change the physics for a given alloy usually depends on other factors (e.g. economics).

7.2.6 Particle coarsening

With continued ageing at a given temperature, there is a tendency for the small particles to dissolve and the resultant solute to precipitate on larger particles, causing them to grow, thereby lowering the total interfacial energy. This process is termed particle coarsening, or sometimes Ostwald ripening. The driving force for particle growth is the difference between the concentration of solute (S_{F}) in equilibrium with small particles of radius r and that in equilibrium with larger particles. The variation of solubility with surface curvature is given by the Gibbs–Thomson or Thomson–Freundlich equation:

$$\ln (S_{\mathrm{F}}/S) = 2\gamma\Omega/\mathbf{k}Tr, \tag{7.11}$$

where S is the equilibrium concentration, γ the particle/matrix interfacial energy and Ω the atomic volume; since $2\gamma\Omega \ll \mathbf{k}Tr$, then $S_{\mathrm{F}} = S[1 + 2\gamma\Omega/\mathbf{k}Tr]$.

To estimate the coarsening rate of a particle it is necessary to consider the rate-controlling process for material transfer. Generally, the rate-limiting factor is considered to be diffusion through the matrix, and the rate of change of particle radius is then derived from the equation:

$$4\pi r^2(\mathrm{d}r/\mathrm{d}t) = D4\pi R^2(\mathrm{d}S/\mathrm{d}R),$$

where $\mathrm{d}S/\mathrm{d}R$ is the concentration gradient across an annulus at a distance R from the particle center. Rewriting the equation after integration gives

$$\mathrm{d}r/\mathrm{d}t = -D(S_{\mathrm{F}} - S_{\mathrm{a}})/r, \tag{7.12}$$

where S_{a} is the average solute concentration a large distance from the particle and D is the solute diffusion coefficient. When the particle solubility is small, the total number of atoms contained in particles may be assumed constant, independent of particle size distribution. Further consideration shows that

$$(S_{\mathrm{a}} - S_{\mathrm{F}}) = \{2\gamma\Omega S/\mathbf{k}T\}[(1/\bar{r}) - (1/r)]$$

and combining with equation (7.11) gives the variation of particle growth rate with radius according to

$$\mathrm{d}r/\mathrm{d}t = \{2DS\gamma\Omega/\mathbf{k}Tr\}[(1/\bar{r}) - (1/r)]. \tag{7.13}$$

This function is plotted in Figure 7.14, from which it is evident that particles of radius less than $\bar{r}$ are dissolving at increasing rates with decreasing values of r. All particles of radius greater than $\bar{r}$ are growing but the graph shows a maximum for particles twice the mean radius. Over a period of time the number of particles decreases discontinuously when particles dissolve, and ultimately the system would tend to form one large particle. However, before this state is reached the mean radius $\bar{r}$ increases and the growth rate of the whole system slows down.

A more detailed theory than that, due to Greenwood, outlined above has been derived by Lifshitz and Slyozov, and by Wagner, taking into consideration the initial particle size distribution. They show that the mean particle radius varies with time according to

$$\bar{r}^3 - \bar{r}_0^3 = Kt, \tag{7.14}$$

where $\bar{r}_0$ is the mean particle radius at the onset of coarsening and K is a constant given by

$$K = 8DS\gamma\Omega/9\mathbf{k}T.$$

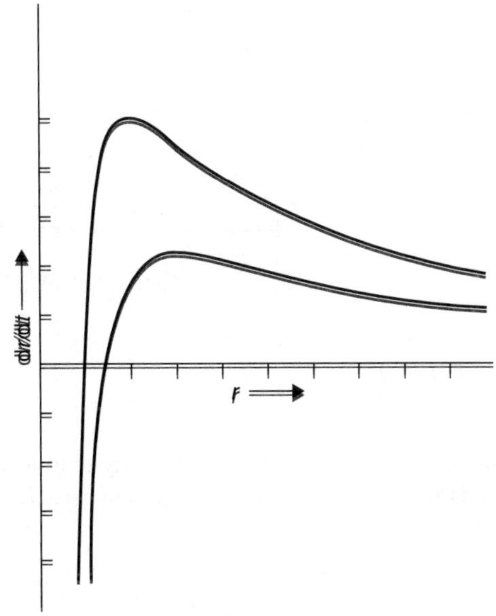

Figure 7.14 *The variation of growth rate* $d\bar{r}/dt$ *with particle radius* $\bar{r}$ *for diffusion-controlled growth, for two values of* γ. *The value of* γ *for the lower curve is 1.5 times that for the upper curve. Particles of radius equal to the mean radius of all particles in the system at any instant are neither growing nor dissolving. Particles of twice this radius are growing at the fastest rate. The smallest particles are dissolving at a rate approximately proportional to* $\bar{r}^2$ *(after Greenwood, 1968; courtesy of Institute of Materials, Minerals and Mining).*

This result is almost identical to that obtained by integrating equation (7.13) in the elementary theory and assuming that the mean radius is increasing at half the rate of that of the fastest growing particle.

Coarsening rate equations have also been derived assuming that the most difficult step in the process is for the atom to enter into solution across the precipitate/matrix interface; the growth is then termed interface controlled. The appropriate rate equation is

$$dr/dt = -C(S_r - S_a)$$

and leads to a coarsening equation of the form

$$\bar{r}^2 - \bar{r}_0^2 = (64CS\gamma\Omega t/8kT), \tag{7.15}$$

where C is some interface constant.

Measurements of coarsening rates carried out so far support the analysis basis on diffusion control of the particle growth. The most detailed results have been obtained for nickel-based systems, particularly the coarsening of γ' (Ni_3Al–Ti or Si), which show a good $\bar{r}^3$ versus t relationship over a wide range of temperatures. Strains due to coherency and the fact that γ' precipitates are cube shaped do not seriously affect the analysis in these systems. Concurrent measurements of $\bar{r}$ and the solute concentration in the matrix during coarsening have enabled values for the interfacial energy $\approx$13 mJ m^{-2} to be determined. In other systems the agreement between theory and experiment is generally less precise, although generally the cube of the mean particle radius varies linearly with time, as shown in Figure 7.15 for the growth of Mn precipitates in an Mg–Mn alloy.

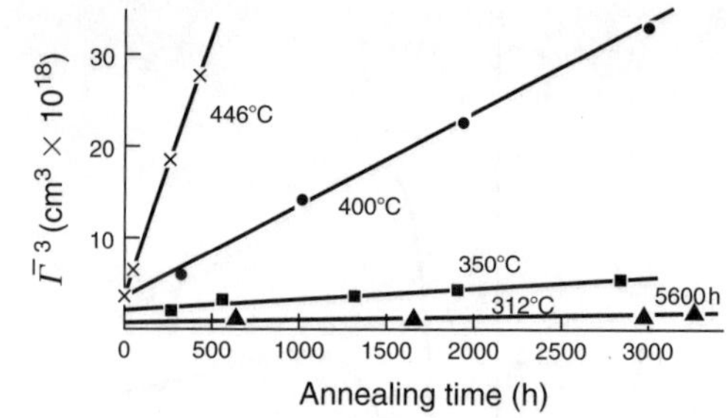

Figure 7.15 *The variation of $\bar{r}^3$ with time of annealing for manganese precipitates in a magnesium matrix (after Smith, 1967).*

Because of the ease of nucleation, particles may tend to concentrate on grain boundaries, and hence grain boundaries may play an important part in particle growth. For such a case, the Thomson–Freundlich equation becomes

$$\ln (S_r/S) = (2\gamma - \gamma_g)\Omega/\mathbf{k}Tx,$$

where γ_g is the grain boundary energy per unit area and 2 × the particle thickness, and their growth follows a law of the form:

$$r_f^4 - r_0^4 = Kt, \tag{7.16}$$

where the constant K includes the solute diffusion coefficient in the grain boundary and the boundary width. The activation energy for diffusion is lower in the grain boundary than in the matrix and this leads to a less strong dependence on temperature for the growth of grain boundary precipitates. For this reason their preferential growth is likely to be predominant only at relatively low temperature.

Worked example

During the ageing of a Cu–Co alloy, the spherical Co precipitates coarsen such that they double their initial size after 14 hours at 527°C and treble their initial size after 8 hours at 577°C. Calculate the activation energy for the process, which may be assumed to be volume diffusion controlled ($\mathbf{R} = 8.31\,\mathrm{J\,mol^{-1}\,K^{-1}}$).

Solution

The Ostwald ripening equation for coarsening is $\bar{r}_t^3 - \bar{r}_o^3 = 8\gamma c v_m Dt/9\mathbf{R}T$, where $\bar{r}_t$ and $\bar{r}_o$ are the mean particle radius at time t and 0 respectively, γ is the surface free energy of the particle/matrix interface, v_m is the molar volume of precipitate, c the equilibrium solute concentration of the matrix and D the diffusivity of the diffusing species. Over the temperature range of the experiment γ, c and v_m can be assumed independent of temperature.

Thus, at $T_1 = 800\,\mathrm{K},\ \bar{r}_t = 2\bar{r}_o,\ t_1 = 5 \times 10^4\,\mathrm{s}$

$T_2 = 850\,\mathrm{K},\ \bar{r}_t = 3\bar{r}_o,\ t_2 = 2.88 \times 10^4\,\mathrm{s}.$

Hence,

$$\bar{r}_o^3(8-1) = 7\bar{r}_o^3 = \text{constant} \times D_1 \times 5 \times 10^4/800 \text{ for } T_1 = 800\,\mathrm{K} \tag{i}$$

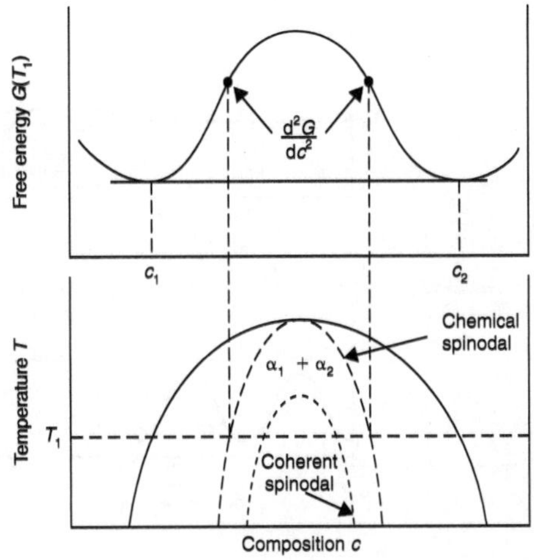

Figure 7.16 *Variation of chemical and coherent spinodal with composition.*

and

$$\bar{r}_0^3(27-1) = 26\bar{r}_0^3 = \text{constant} \times D_2 \times 2.88 \times 10^4/850 \text{ for } T_2 = 850\,\text{K}. \quad \text{(ii)}$$

From (i), $D_1 = 56\bar{r}_0^3/\text{constant} \times 500$.
From (ii), $D_2 = 221\bar{r}_0^3/\text{constant} \times 288$.
Thus, $D_1/D_2 = 0.146$, and from Arrhenius rate theory

$$D_1/D_2 = \exp\left[-(Q/\mathbf{R})(1/T_1 - 1/T_2)\right]$$

$$\ln(D_1/D_2) = -(Q/\mathbf{R})(1/T_1 - 1/T_2)$$

$$\ln(0.146) = -(Q/8.31)(1/800 - 1/850)$$

$Q = 2.2 \times 10^5\,\text{J}\,\text{mol}^{-1}$.

7.2.7 Spinodal decomposition

For any alloy composition where the free energy curve has a negative curvature, i.e. $(\mathrm{d}^2G/\mathrm{d}c^2) < 0$, small fluctuations in composition that produce A-rich and B-rich regions will bring about a lowering of the total free energy. At a given temperature the alloy must lie between two points of inflection (where $\mathrm{d}^2G/\mathrm{d}c^2 = 0$) and the locus of these points at different temperatures is depicted on the phase diagram by the chemical spinodal line (see Figure 7.16).

For an alloy c_0 quenched inside this spinodal, composition fluctuations increase very rapidly with time and have a time constant $\tau = -\lambda/4\pi^2 D$, where λ is the wavelength of composition modulations in one dimension and D is the interdiffusion coefficient. For such a kinetic process, shown in Figure 7.17, 'uphill' diffusion takes place, i.e. regions richer in solute than the average become richer, and poorer become poorer until the equilibrium compositions c_1 and c_2 of the A-rich and B-rich regions are formed. As for normal precipitation, interfacial energy and strain energy influency the decomposition. During the early stages of decomposition the interface between A-rich and B-rich regions is diffuse

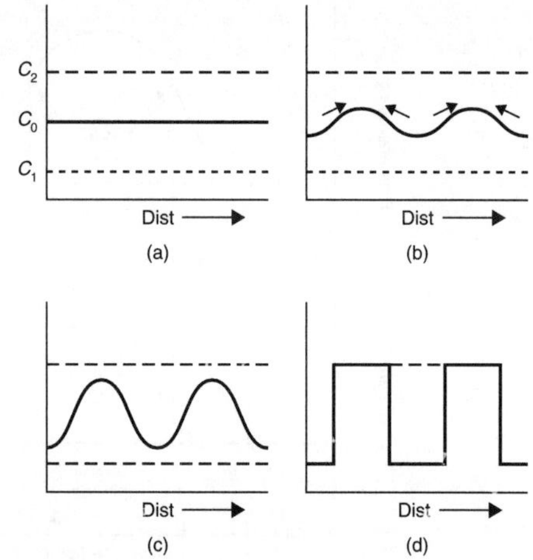

Figure 7.17 *Composition fluctuations in a spinodal system.*

and the interfacial energy becomes a gradient energy, which depends on the composition gradient across the interface according to

$$\Delta G_{\text{int}} = K(\Delta c/\lambda)^2, \tag{7.17}$$

where λ is the wavelength and Δc the amplitude of the sinusoidal composition modulation, and K depends on the difference in bond energies between like and unlike atom pairs. The coherency strain energy term is related to the misfit ε between regions A and B, where $\varepsilon = (1/a)\mathrm{d}a/\mathrm{d}c$, the fractional change in lattice parameter a per unit composition change, and is given for an elastically isotropic solid by

$$\Delta G_{\text{strain}} = \varepsilon^2 \Delta c^2 EV/(1-\nu), \tag{7.18}$$

with E Young's modulus, ν Poisson's ratio and V the molar volume. The total free energy change arising from a composition fluctuation is therefore

$$\Delta G = \left[\frac{\mathrm{d}^2G}{\mathrm{d}c^2} + \frac{2K}{\lambda^2} + (2\varepsilon^2 EV/(1-\nu))\right]\Delta c^2/2 \tag{7.19}$$

and a homogeneous solid solution will decompose spinodally provided

$$-(\mathrm{d}^2G/\mathrm{d}c^2) > (2K/\lambda^2) + (2\varepsilon^2 EV/(1-\nu)). \tag{7.20}$$

For $\lambda = \infty$, the condition $[(\mathrm{d}^2G/\mathrm{d}c^2) + (2\varepsilon^2 EV/1-\nu)] = 0$ is known as the coherent spinodal, as shown in Figure 7.16. The λ of the composition modulations has to satisfy the condition

$$\lambda^2 > 2K/[\mathrm{d}^2G/\mathrm{d}c^2 + (2\varepsilon^2 EV/(1-\nu))] \tag{7.21}$$

and decreases with increasing degree of undercooling below the coherent spinodal line. A λ-value of 5–10 nm is favored, since shorter λ-values have too sharp a concentration gradient and longer values have too large a diffusion distance. For large misfit values, a large undercooling is required to overcome the strain energy effect. In cubic crystals, E is usually smaller along ⟨1 0 0⟩ directions and the high strain energy is accommodated more easily in the elastically soft directions, with composition modulations localized along this direction.

Spinodal decompositions have now been studied in a number of systems such as Cu–Ni–Fe, Cu–Ni–Si, Ni–12Ti and Cu–5Ti exhibiting 'side-bands' in X-ray small-angle scattering, satellite spots in electron diffraction patterns and characteristic modulation of structure along ⟨1 0 0⟩ in electron micrographs. Many of the alloys produced by splat cooling might be expected to exhibit spinodal decomposition, and it has been suggested that in some alloy systems GP zones form in this way at high supersaturations, because the GP zone solvus (see Figure 7.1) gives rise to a metastable coherent miscibility gap.

The spinodally decomposed microstructure is believed to have unusually good mechanical stability under fatigue conditions.

7.3 Strengthening of steels by heat treatment

7.3.1 Time–temperature–transformation diagrams

Eutectoid decomposition occurs in both ferrous (e.g. iron–carbon) and non-ferrous (e.g. copper–aluminum, copper–tin) alloy systems, but it is of particular importance industrially in governing the hardening of steels. In the iron–carbon system (see Figure 2.20) the γ-phase, austenite, which is a solid solution of carbon in fcc iron, decomposes on cooling to give a structure known as pearlite, composed of alternate lamellae of cementite (Fe_3C) and ferrite. However, when the cooling conditions are such that the alloy structure is far removed from equilibrium, an alternative transformation may occur. Thus, on very rapid cooling, a metastable phase called martensite, which is a supersaturated solid solution of carbon in ferrite, is produced. The microstructure of such a transformed steel is not homogeneous but consists of plate-like needles of martensite embedded in a matrix of the parent austenite. Apart from martensite, another structure known as bainite may also be formed if the formation of pearlite is avoided by cooling the austenite rapidly through the temperature range above 550°C, and then holding the steel at some temperature between 250 and 550°C. A bainitic structure consists of plate-like grains of ferrite, somewhat like the plates of martensite, inside which carbide particles can be seen.

The structure produced when austenite is allowed to transform isothermally at a given temperature can be conveniently represented by a diagram of the type shown in Figure 7.18, which plots the time necessary at a given temperature to transform austenite of eutectoid composition to one of the three structures: pearlite, bainite or martensite. Such a diagram, made up from the results of a series of isothermal decomposition experiments, is called a TTT curve, since it relates the transformation product to the time at a given temperature. It will be evident from such a diagram that a wide variety of structures can be obtained from the austenite decomposition of a particular steel; the structure may range from 100% coarse pearlite, when the steel will be soft and ductile, to fully martensitic, when the steel will be hard and brittle. It is because this wide range of properties can be produced by the transformation of a steel that it remains a major constructional material for engineering purposes.

From the TTT curve it can be seen that, just below the critical temperature, A_1, the rate of transformation is slow even though the atomic mobility must be high in this temperature range. This is because any phase change involving nucleation and growth (e.g. the pearlite transformation) is faced with nucleation difficulties, which arise from the necessary surface and strain energy contributions to the nucleus. Of course, as the transformation temperature approaches the temperature corresponding to the knee of the curve, the transformation rate increases. The slowness of the transformation below the knee of the TTT curve, when bainite is formed, is also readily understood, since atomic migration is slow at these lower temperatures and the bainite transformation depends on diffusion. The lower part of the TTT curve below about 250–300°C indicates, however, that the transformation speeds up

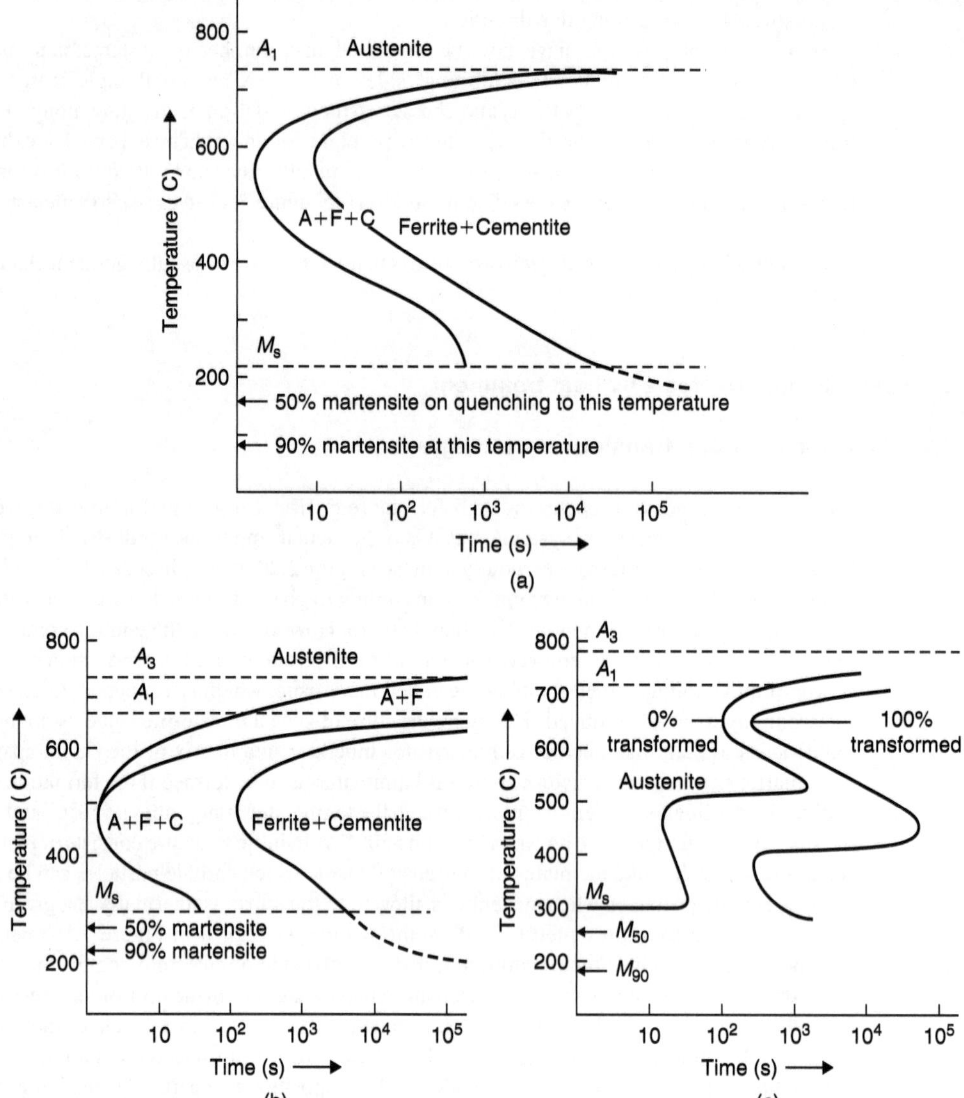

Figure 7.18 *TTT curves for: (a) eutectoid, (b) hypo-eutectoid and (c) low alloy (e.g. Ni–Cr–Mo) steels (after* ASM Metals Handbook*).*

again and takes place exceedingly fast, even though atomic mobility in this temperature range must be very low. For this reason, it is concluded that the martensite transformation does not depend on the speed of migration of carbon atoms and, consequently, it is often referred to as a diffusionless transformation. The austenite only starts transforming to martensite when the temperature falls below a critical temperature, usually denoted by M_s. Below M_s the percentage of austenite transformed to martensite is indicated on the diagram by a series of horizontal lines.

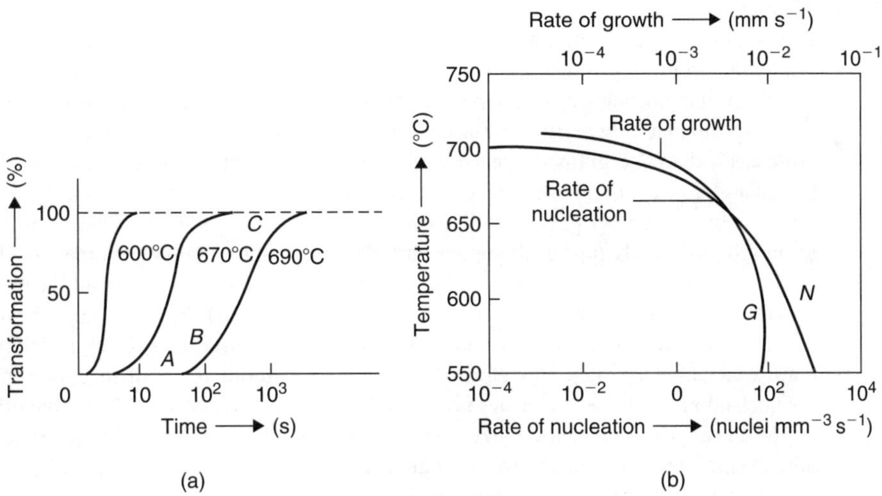

Figure 7.19 *Effect of temperature on: (a) amount of pearlite formed with time and (b) rate of nucleation and rate of growth of pearlite (after Mehl and Hagel, 1956).*

The M_s temperature may be predicted for steels containing various alloying elements in weight percent by the formula, due to Steven and Haynes, given by M_s (°C) = 561–474C–33Mn–17Ni–17Cr–21Mo.

7.3.2 Austenite–pearlite transformation

7.3.2.1 Nucleation and growth of pearlite

If a homogeneous austenitic specimen of eutectoid composition were to be transferred quickly to a bath held at some temperature between 720 and 550°C, decomposition curves of the form shown in Figure 7.19a would be obtained. These curves, typical of a nucleation and growth process, indicate that the transformation undergoes an incubation period, an accelerating stage and a decelerating stage; the volume transformed into pearlite has the time dependence described by the Avrami equation (6.44). When the transformation is in its initial stage the austenite contains a few small pearlite nodules, each of which grow during the period A to B (see curve obtained at 690°C) and, at the same time, further nuclei form. The percentage of austenite transformed is quite small, since the nuclei are small and their total volume represents only a fraction of the original austenite. During the B to C stage the transformation rate accelerates, since as each nodule increases in size the area of contact between austenite and pearlite regions also increases: the larger the pearlite volumes, the greater is the surface area upon which to deposit further transformation products. At C, the growing nodules begin to impinge on each other, so that the area of contact between pearlite and austenite decreases and, from this stage onwards, the larger the nodules the lower is the rate of transformation. Clearly, the rate of transformation depends on (1) the rate of nucleation of pearlite nodules, N (i.e. the number of nuclei formed in unit volume in unit time), and (2) the rate of growth of these nodules, G (i.e. the rate that the radius of the nodule increases with time). The variation of N and G with temperature for a eutectoid steel is shown in Figure 7.19b.

The rate of nucleation increases with decreasing temperature down to the knee of the curve and in this respect is analogous to other processes of phase precipitation where hysteresis occurs (see

Chapter 2). In addition, the nucleation rate is very structure sensitive, so that nucleation occurs readily in regions of high energy where the structure is distorted. In homogeneous austenite the nucleation of pearlite occurs almost exclusively at grain boundaries and, for this reason, the size of the austenite grains, prior to quenching, has an important effect on hardenability (a term which denotes the depth in a steel to which a fully martensitic structure can be obtained). Coarse-grained steels can be hardened more easily than fine-grained steels because, to obtain maximum hardening in a steel, the decomposition of austenite to pearlite should be avoided, and this is more easily accomplished if the grain boundary area, or the number of potential pearlite nucleation sites, is small. Thus, an increase in austenite grain size effectively pushes the upper part of the TTT curve to longer times, so that, with a given cooling rate, the knee can be avoided more easily. The structure sensitivity of the rate of nucleation is also reflected in other ways. For example, if the austenite grain is heterogeneous, pearlite nucleation is observed at inclusions as well as at grain boundaries. Moreover, plastic deformation during transformation increases the rate of transformation, since the introduction of dislocations provides extra sites for nucleation, while the vacancies produced by plastic deformation enhance the diffusion process.

The rate of growth of pearlite, like the rate of nucleation, also increases with decreasing temperature down to the knee of the curve, even though it is governed by the diffusion of carbon, which, of course, decreases with decreasing temperature. The reason for this is that the interlamellar spacing of the pearlite also decreases rapidly with decreasing temperature, and because the carbon atoms do not have to travel so far, the carbon supply is easily maintained. In contrast to the rate of nucleation, however, the rate of growth of pearlite is quite structure insensitive and therefore is indifferent to the presence of grain boundaries or inclusions. These two factors are important in governing the size of the pearlite nodules produced. If, for instance, the steel is transformed just below A_1, where the rate of nucleation is very low in comparison with the rate of growth (i.e. the ratio N/G is small), very large nodules are developed. Then, owing to the structure insensitivity of the growth process, the few nodules formed are able to grow across grain boundaries, with the result that pearlite nodules larger than the original austenite grain size are often observed. By comparison, if the steel is transformed at a lower temperature, just above the knee of the TTT curve where N/G is large, the rate of nucleation is high and the pearlite nodule size is correspondingly small.

7.3.2.2 *Mechanism and morphology of pearlite formation*

The growth of pearlite from austenite clearly involves two distinct processes: (1) a redistribution of carbon (since the carbon concentrates in the cementite and avoids the ferrite) and (2) a crystallographic change (since the structure of both ferrite and cementite differs from that of austenite). Of these two processes it is generally agreed that the rate of growth is governed by the diffusion of carbon atoms, and the crystallographic change occurs as readily as the redistribution of carbon will allow. The active nucleus of the pearlite nodule may be either a ferrite or cementite platelet, depending on the conditions of temperature and composition which prevail during the transformation, but usually it is assumed to be cementite. The nucleus may form at a grain boundary, as shown in Figure 7.20a, and after its formation the surrounding matrix is depleted of carbon, so that conditions favor the nucleation of ferrite plates adjacent to the cementite nucleus (Figure 7.20b). The ferrite plates in turn reject carbon atoms into the surrounding austenite and this favors the formation of cementite nuclei, which then continue to grow. At the same time as the pearlite nodule grows sideways, the ferrite and cementite lamellae advance into the austenite, since the carbon atoms rejected ahead of the advancing ferrite diffuse into the path of the growing cementite (Figure 7.20c). Eventually, a cementite plate of different orientation forms and this acts as a new nucleus, as shown in Figure 7.20d and e.

Homogeneous austenite, when held at a constant temperature, produces pearlite at a constant rate and with a constant interlamellar spacing. However, the interlamellar spacing decreases with decreasing temperature, and becomes irresolvable in the optical microscope as the temperature approaches

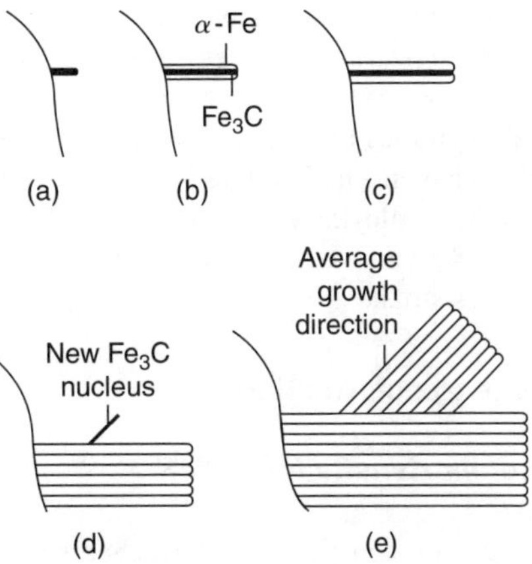

Figure 7.20 *Nucleation and growth of pearlite nodules. (a) Initial* Fe_3C *nucleus. (b)* Fe_3C *plate full grown, α-Fe now nucleated. (c) α-Fe plates now full grown, new* Fe_3C *plates nucleated. (d)* Fe_3C *nucleus of different orientation forms and original nodule grows. (e) New nodule at advanced stage of growth (after Mehl and Hagel, 1956).*

that corresponding to the knee of the curve. An increase in hardness occurs as the spacing decreases. Zener explains the dependence of interlamellar spacing on temperature in the following way. If the interlamellar spacing is large, the diffusion distance of the carbon atoms in order to concentrate in the cementite is also large, and the rate of carbon redistribution is correspondingly slow. Conversely, if the spacing is small the area, and hence energy, of the ferrite/cementite interfaces become large. In consequence, such a high proportion of the free energy released in the austenite to pearlite transformation is needed to provide the interfacial energy that little will remain to provide the 'driving force' for the change. Thus, a balance between these two opposing conditions is necessary to allow the formation of pearlite to proceed, and at a constant temperature the interlamellar spacing remains constant. However, because the free energy change, ΔG, accompanying the transformation increases with increasing degree of undercooling, larger interfacial areas can be tolerated as the temperature of transformation is lowered, with the result that the interlamellar spacing decreases with decreasing temperature.

The majority of commercial steels are not usually of the euctectoid composition (0.8% carbon), but hypo-eutectoid (i.e. $<0.8\%$ carbon). In such steels, pro-eutectoid ferrite is first formed before the pearlite reaction begins and this is shown in the TTT curve by a third decomposition line. From Figure 7.18b it can be seen that the amount of pro-eutectoid ferrite decreases as the isothermal transformation temperature is lowered. The morphology of the precipitated ferrite depends on the usual precipitation variables (i.e. temperature, time, carbon content and grain size), and growth occurs preferentially at grain boundaries and on certain crystallographic planes. The Widmanstätten pattern with ferrite growing along $\{1\ 1\ 1\}$ planes of the parent austenite is a familiar structure of these steels.

7.3.2.3 *Influence of alloying elements on pearlite formation*

With the exception of cobalt, all alloying elements in small amounts retard the transformation of austenite to pearlite. These elements decrease both the rate of nucleation, N, and the rate of growth, G, so that the top part of the TTT curve is displaced towards longer times. This has considerable technological importance, since in the absence of such alloying elements, a steel can only transform into the harder constituents of bainite or martensite if it is in the form of very thin sections, so that the cooling rate will be fast enough to avoid crossing the knee of the TTT curve during the cooling process and hence avoid pearlite transformation. For this reason, most commercially heat-treatable steels contain one or more of the elements chromium, nickel, manganese, vanadium, molybdenum or tungsten. Cobalt increases both N and G, and its effect on the pearlite interlamellar spacing is contrary to the other elements in that it decreases the spacing.

With large additions of alloying elements, the simple form of TTT curve often becomes complex, as shown in Figure 7.18c. Thus, to obtain any desired structure by heat treatment a detailed knowledge of the TTT curve is essential.

7.3.3 Austenite–martensite transformation

7.3.3.1 Crystallography of martensite formation

Martensite, the hardening constituent in quenched steels, is formed at temperatures below about 200°C. The regions of the austenite which have transformed to martensite are lenticular in shape and may easily be recognized by etching, or from the distortion they produce on the polished surface of the alloy. These relief effects, shown schematically in Figure 7.21, indicate that the martensite needles have been formed not with the aid of atomic diffusion but by a shear process, since if atomic mobility were allowed the large strain energy associated with the transformed volume would then be largely avoided. The lenticular shape of a martensite needle is a direct consequence of the stresses produced in the surrounding matrix by the shear mechanism of the transformation and is exactly analogous to the similar effect found in mechanical twinning. The strain energy associated with martensite is tolerated because the growth of such sheared regions does not depend on diffusion, and since the regions are coherent with the matrix they are able to spread at great speed through the crystal. The large free energy change associated with the rapid formation of the new phase outweighs the strain energy, so that there is a net lowering of free energy.

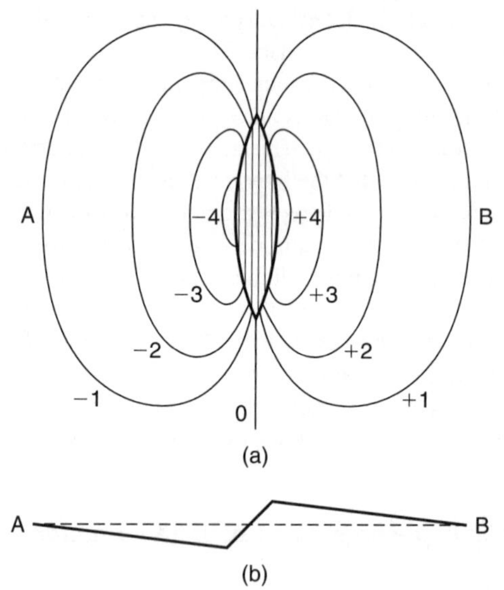

Figure 7.21 *Schematic diagram of the observed shape deformation produced by a martensite plate. (a) Contour lines on an originally flat surface. (b) Section of the surface through AB. Vertical scale much exaggerated (after Bilby and Christian, 1956; courtesy of the Institute of Materials, Minerals and Mining).*

Direct TEM observations of martensite plates have shown that there are two main types of martensite, one with a twinned structure (see Figure 7.22), known as acicular martensite, and the other with a high density of dislocations but few or no twins, called massive martensite.

In contrast to the pearlite transformation, which involves both a redistribution of carbon atoms and a structural change, the martensite transformation involves only a change in crystal structure.

The structure cell of martensite is body-centered tetragonal, which is a distorted form of a body-centered cubic structure, and hence may be regarded as a supersaturated solution of carbon in α-iron. X-ray examination shows that while the c/a ratio of the bct structure of martensite increases with increasing carbon content, the curve of c/a ratio against composition extrapolates back to $c/a = 1$ for zero carbon content, and the lattice parameter is equal to that of pure α-iron (Figure 7.23).

From the crystallographic point of view, the most important experimental data in any martensite transformation are the orientation relations of the two phases and the habit plane. In steel, there are three groups of orientations often quoted; those due to Kurdjumov and Sachs, Nishiyama, and Greninger and Troiano, respectively. According to the Kurdjumov–Sachs relation, in iron–carbon alloys with 0.5–1.4% carbon, a $\{1\ 1\ 1\}_\gamma$ plane of the austenite lattice is parallel to the $\{1\ 1\ 0\}_\alpha$ plane of the martensite, with a $\langle 1\ 1\ 0\rangle_\gamma$ axis of the former parallel to a $\langle 1\ 1\ 1\rangle_\alpha$ axis of the latter; the associated habit plane is $\{2\ 2\ 5\}_\gamma$. In any one crystal there are 24 possible variants of the Kurdjumov–Sachs relationship, consisting of 12 twin pairs, both orientations of a pair having the same habit plane. However, for general discussion it is usual to choose one relation which may be written

$$(1\ 1\ 1)_\gamma \| (1\ 0\ 1)_\alpha \text{ with } [1\ \bar{1}\ 0]_\gamma \| [1\ 1\ \bar{1}]_\alpha.$$

In the composition range 1.5–1.8% carbon, the habit plane changes to $\approx \{2\ 5\ 9\}_\gamma$ with the orientation relationship unspecified. This latter type of habit plane has also been reported by Nishiyama for iron–nickel alloys (27–34% nickel), for which the orientation relationship is of the form

$$(1\ 1\ 1)_\gamma \| (1\ 0\ 1)_\alpha \text{ with } [1\ \bar{2}\ 1]_\gamma \| [1\ 0\ \bar{1}]_\alpha.$$

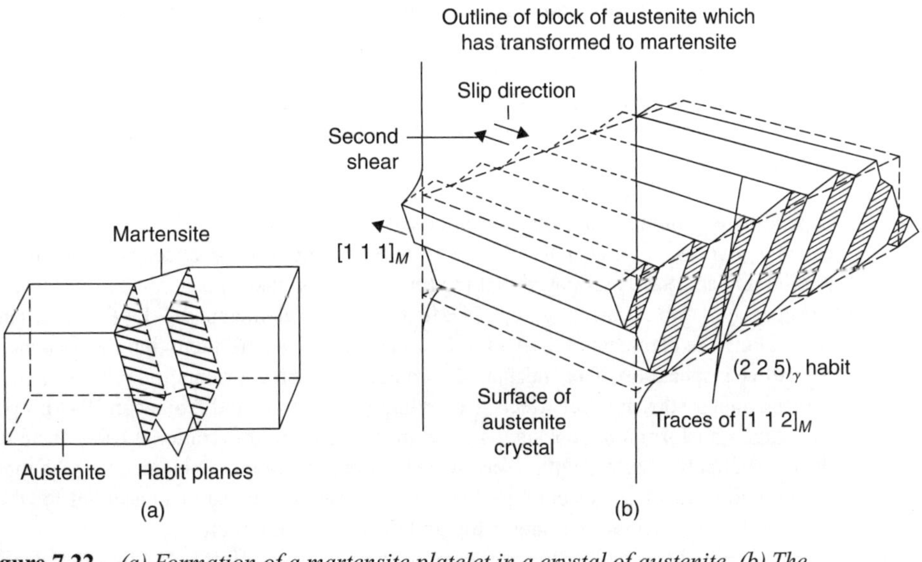

Figure 7.22 *(a) Formation of a martensite platelet in a crystal of austenite. (b) The inhomogeneous twinning shear within the platelet (after Kelly and Nutting, 1960; courtesy of the Royal Society).*

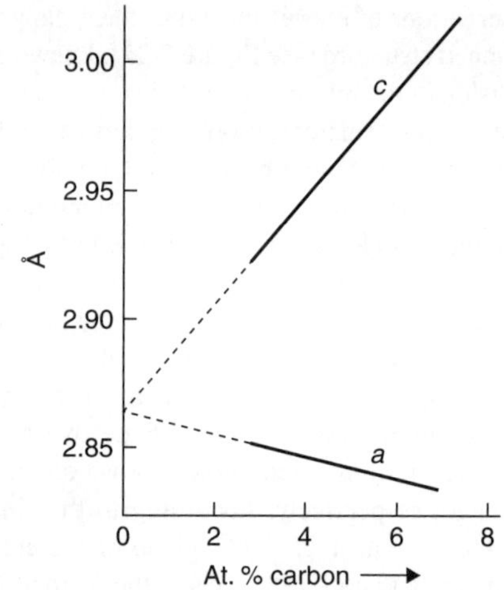

Figure 7.23 *Variation of* c *and* a *parameters with carbon content in martensite (after Kurdjumov, 1948).*

However, Greninger and Troiano have shown by precision orientation determinations that irrational relationships are very probable, and that in a ternary iron–nickel–carbon alloy (0.8% carbon, 22% nickel), $(1\ 1\ 1)_\gamma$ is approximately 1° from $(1\ 0\ 1)_\alpha$ with $[1\ \bar{2}\ 1]_\gamma$ approximately 2° from $[1\ 0\ 1]_\alpha$, and is associated with a habit plane about 5° from (2 5 9).

7.3.3.2 Mechanism of martensite formation

The martensite transformation is diffusionless, and therefore martensite forms without any interchange in the position of neighboring atoms. Accordingly, the observed orientation relationships are a direct consequence of the atom movements that occur during the transformation. The first suggestion of a possible transformation mechanism was made by Bain in 1934. He suggested that, since austenite may be regarded as a body-centered tetragonal structure of axial ratio $\sqrt{2}$, the transformation merely involves a compression of the c-axis of the austenite unit cell and expansion of the α-axis. The interstitially dissolved carbon atoms prevent the axial ratio from going completely to unity and, depending on composition, the c/a ratio will be between 1.08 and 1.0. Clearly, such a mechanism can only give rise to three martensite orientations whereas, in practice, 24 result. To account for this, Kurdjumov and Sachs proposed that the transformation takes place not by one shear process but by a sequence of two shears (Figure 7.24), first along the elements $(1\ 1\ 1)_\gamma\langle 1\ 1\ 2\rangle_\gamma$, and then a minor shear along the elements $(1\ 1\ 2)_\alpha\langle 1\ 1\ 1\rangle_\alpha$; these elements are the twinning elements of the fcc and bcc lattice respectively. This mechanism predicts the correct orientation relations, but not the correct habit characteristics or relief effects. Accordingly, Greninger and Troiano in 1941 proposed a different two-stage transformation, consisting of an initial shear on the irrational habit plane which produces the relief effects, together with a second shear along the twinning elements of the martensite lattice. If slight adjustments in spacing are then allowed, the mechanism can account for the relief effects, habit plane, the orientation relationship and the change of structure.

Further additions to these theories have been made in an effort to produce the ideal general theory of the crystallography of martensite transformation. Bowles, for example, replaces the first shear of the Greninger–Troiano mechanism by the general type of homogeneous deformation in which

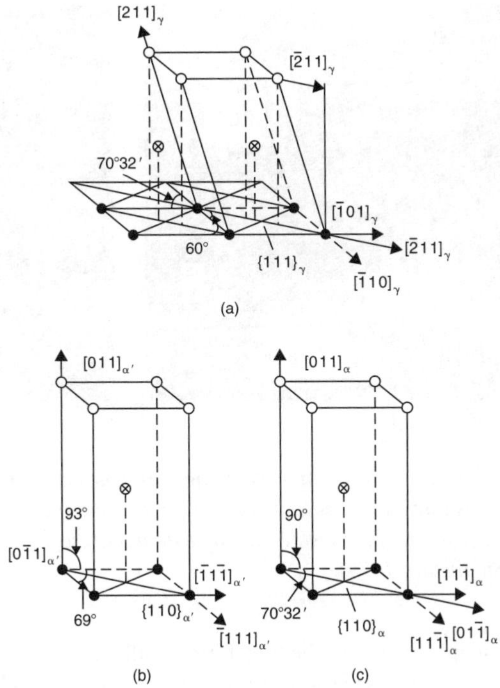

Figure 7.24 *Shear mechanisms of Kurdjumov and Sachs. (a) Face-centered austenite with $\{1\ 1\ 1\}_\gamma$ in horizontal plane. (b) Body-centered tetragonal martensite (α). (c) Cubic ferrite (α) (after Bowles and Barrett, 1952).*

the habit plane remains invariant, i.e. all directions in this plane are unrotated and unchanged in length. However, in all such cases the problem resolves itself into one of determining whether a homogeneous strain can transform the γ-lattice into the α-lattice, while preserving coherency at the boundary between them. The homogeneous strain does not do this, so that some reasonable additional type of strain has to be added.

This shear can occur either by twinning or by slip, the mode prevailing depending on the composition and cooling rate. Between carbon contents of 0.2% and 0.5% the martensite changes from dislocated martensite arranged in thin lathes or needles to twinned acicular martensite arranged in plates. In the martensite formed at low C contents (e.g. Fe–Ni alloys) the thin lathes lie parallel to each other, with a $\{1\ 1\ 1\}_\gamma$ habit, to form pockets of massive martensite with jagged boundaries due to the impingement of other nearby pockets of lathes. The inhomogeneous shear produced by deformation twinning occurs on $\{1\ 1\ 2\}$ planes in the martensite, so that each martensite plate is made up of parallel twin plates of thickness 2–50 nm. By operation of such a complex transformation mode with a high index habit plan the system maintains an invariant interfacial plane.

Because of the shears involved and the speed of the transformation it is attractive to consider that dislocations play an important role in martensite formation. Some insight into the basic dislocation mechanisms has been obtained by *in situ* observations during either cooling below M_s or by straining, but unfortunately only for Ni–Cr austenitic steels with low stacking-fault energy (i.e. $\gamma \approx 20\ \text{mJ m}^{-2}$). For these alloys it has been found that stacking faults are formed either by emitting partial dislocations with $b = a/6\langle 1\ 1\ 2\rangle$ from grain boundaries or by the dissociation of unit dislocations with $b = a/2\langle 1\ 1\ 0\rangle$. In regions of the grain where on cooling or deformation a high density of stacking faults developed, the corresponding diffraction pattern revealed cph ε-martensite. On subsequent deformation

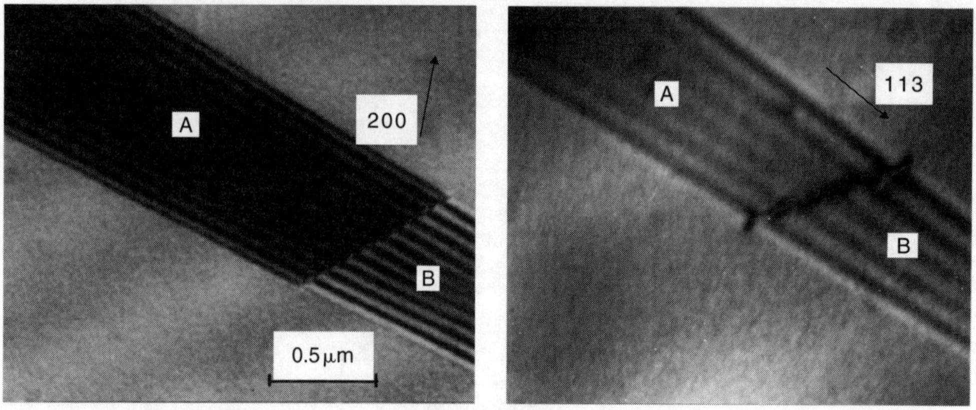

(a) (b)

Figure 7.25 *Electron micrographs showing: (a) contrast from overlapping faults on (1 1 1); A is extrinsic and B is intrinsic in nature; (b) residual contrast arising from a supplementary displacement across the faults which is intrinsic in nature for both faults A and B (after Brooks, Loretto and Smallman, 1979).*

or cooling, regions of ε-martensite transform rapidly into bcc α-martensite and, indeed, the only way in which α-martensite was observed to form was from an ε nucleus.

Because straining or cooling can be interrupted during the *in situ* experiments it was possible to carry out a detailed analysis of the defect structure formed prior to a region becoming recognizably (from diffraction patterns) martensitic. In this way it has been shown that the interplanar spacing across the individual stacking faults in the austenite decreased to the (0 0 0 1) spacing appropriate to ε-martensite. Figure 7.25 shows micrographs which reveal this change of spacing; no contrast is expected in Figure 7.25b if the faulted {1 1 1} planes remained at the fcc spacing, since the condition of invisibility $g \cdot R = n$ is obeyed. The residual contrast observed arises from the supplementary displacement ΔR across the faults which, from the white outer fringe, is positive (intrinsic) in nature for both faults and $\approx 2\%$ of the {1 1 1} spacing. The formation of regions of α from ε could also be followed, although in this case the speed of the transformation precluded detailed analysis. Figure 7.26 shows a micrograph taken after the formation of α-martensite and this, together with continuous observations, show that the martensite/matrix interface changes from {1 1 1} to the well-known {2 2 5} as it propagates. Clearly, one of the important roles that the formation of ε-martensite plays in acting as a precursor for the formation of α-martensite is in the generation of close-packed planes with **ABAB** stacking, so that atomic shuffles can subsequently transform these planes to {1 1 0} bcc which are, of course, stacked **ABAB** (see Figure 7.27). The α-martensite forms in dislocation pile-ups where the $a/6\langle 1\,1\,2\rangle$ partials are forced closer together by the applied stress. The volume of effective bcc material increases as more dislocations join the pile-up, until the nucleus formed by this process reaches a critical size and rapid growth takes place. The martensite initially grows perpendicular to, and principally on, one side of the $\{1\,1\,1\}_\gamma$ slip plane associated with the nucleus, very likely corresponding to the side of the dislocations with missing half-planes, since α-martensite is less dense than austenite.

7.3.3.3 Kinetics of martensite formation

One of the most distinctive features of the martensite transformation is that in most systems martensite is formed only when the specimen is cooling, and that the rate of martensite formation is negligible if

Figure 7.26 *Electron micrograph showing an α-martensite plate, the austenite–martensite interface and the faults in the austenite matrix (after Brooks, Loretto and Smallman, 1979).*

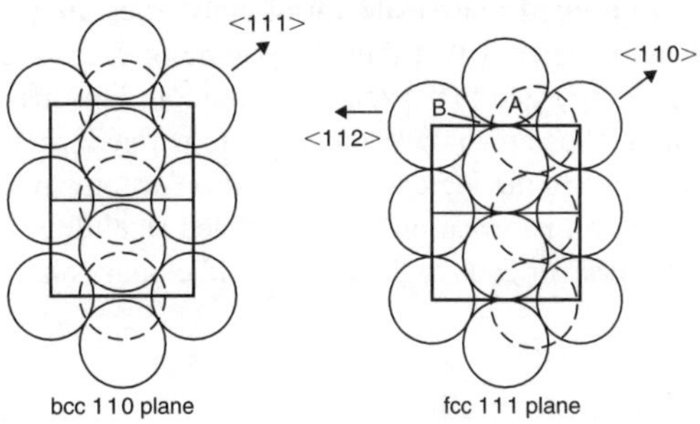

Figure 7.27 *A shear of* a/6 *⟨1 1 2⟩ moves atoms in the fcc structure from A sites to B sites, and after half this shear the structure has pseudo-bcc packing (after Brooks, Loretto and Smallman, 1979).*

cooling is stopped. For this reason, the reaction is often referred to as an athermal[1] transformation, and the percentage of austenite transformed to martensite is indicated on the TTT curve by a series of horizontal lines. The transformation begins at a temperature M_S, which is not dependent on cooling rate, but is dependent on prior thermal and mechanical history, and on composition. For example, it is well established that the M_S temperature decreases approximately linearly with increasing concentration of solutes such as carbon, nickel or manganese.

Speed of formation

The observation that martensite plates form rapidly and at a rate which is temperature independent shows that thermal activation is not required for the growth process. Electronic methods show that the martensite needles form, in iron–nickel–carbon alloys, for example, in about 10^{-7} s and, moreover, that the linear growth velocity is about 10^3 m s^{-1} even at very low temperatures. Such observations show that the activation energy for the growth of a martensite plate is virtually zero, and that the velocity of growth approaches the speed of sound in the matrix. Sometimes a 'burst phenomenon' is exhibited, as, for example, in iron–nickel alloys, when the stresses produced by one martensite plate assist in the nucleation of others. The whole process is autocatalytic and about 25% of the transformation can occur in the time interval of an audible click.

[1] In some alloys, such as iron–manganese–carbon and iron–manganese–nickel, the martensitic transformation occurs isothermally. For these systems, growth is still very rapid but the nuclei are formed by thermal activation.

The effect of applied stress

Since the formation of martensite involves a homogeneous distortion of the parent structure, it is expected that externally applied stresses will be of importance. Plastic deformation is effective in forming martensite above the M_s temperature, provided the temperature does not exceed a critical value usually denoted by M_d. However, cold work above M_d may either accelerate or retard the transformation on subsequent cooling. Even elastic stresses, when applied above the M_s temperature and maintained during cooling, can affect the transformation; uniaxial compression or tensile stresses raise the M_s temperature while hydrostatic stresses lower the M_s temperature.

Stabilization

When cooling is interrupted below M_s, stabilization of the remaining austenite often occurs. Thus, when cooling is resumed martensite forms only after an appreciable drop in temperature. Such thermal stabilization has been attributed by some workers to an accumulation of carbon atoms on those dislocations important to martensite formation. This may be regarded as a direct analog of the yield phenomenon. The temperature interval before transformation is resumed increases with holding time and is analogous to the increase in yield drop accompanying carbon build-up on strain ageing. Furthermore, when transformation in a stabilized steel does resume, it often starts with a 'burst', which in this case is analogous to the lower yield elongation.

7.3.4 Austenite–bainite transformation

The bainite reaction has many features common to both the pearlite and martensite reactions. The pearlite transformation involves the redistribution of carbon followed by a structure change, the martensite transformation involves the structure change alone and, in contrast, the bainite transformation involves a structure change followed by the redistribution of carbon, which precipitates as a carbide. Consequently, the austenite–bainite decomposition may be regarded as a martensite transformation involving the diffusion of carbon atoms, so that, in this case, the rate of coherent growth is necessarily slow compared with that of martensite. Lower bainite is hardly distinguishable from martensite tempered at the same temperature, while upper bainite exhibits an acicular structure. The metallographic appearance of the transformed steel is found to alter continuously between these two extremes, the actual structure exhibited being governed by the diffusion rate of the carbon, which in turn depends on the temperature of the transformation. The hardness of the reaction product also increases continuously with decreasing temperature, lower bainite being harder than upper bainite, which is harder than most fine pearlite.

The ferrite in bainite has a martensite-like appearance and is, in most cases, clearly distinguished from both ferrite and pro-eutectoid ferrite formed in the pearlite range. The bainitic ferrite exhibits the same surface relief effects as martensite, while pro-eutectoid ferrite and pearlite do not. Such surface tilting is further evidence for a shear-like transformation, but the orientation relationship between austenite and bainite is not necessarily the same as that between austenite and martensite. In fact, the bainitic ferrite has the same orientation with respect to the parent austenite as does pro-eutectoid ferrite, which suggests that ferrite may nucleate bainite.

7.3.5 Tempering of martensite

The presence of martensite in a quenched steel, while greatly increasing its hardness and TS, causes the material to be brittle. Such behavior is hardly surprising, since the formation of martensite is accompanied by severe matrix distortions. The hardness and strength of martensite increase sharply with increase in C content. Contributions to the strength arise from the carbon in solution, carbides precipitated during the quench, dislocations introduced during the transformation and the grain size.

Although the martensite structure is thermodynamically unstable, the steel will remain in this condition more or less indefinitely at room temperature because, for a change to take place, bulk diffusion of carbon, with an activation energy Q of approximately 83 kJ mol^{-1} atom is necessary. However, because there is an exponential variation of the reaction rate with temperature, the steel will be able slowly to approach the equilibrium structure at a slightly elevated temperature, i.e. rate of reaction $= A\exp[-Q/\mathbf{k}T]$. Thus, by a carefully controlled tempering treatment, the quenching stresses can be relieved and some of the carbon can precipitate from the supersaturated solid solution to form a finely dispersed carbide phase. In this way, the toughness of the steel can be vastly improved with very little detriment to its hardness and tensile properties.

The structural changes which occur on tempering may be considered to take place in three stages. In the primary stage, fine particles of a cph carbide phase (ε-carbide) of composition about $Fe_{2,4}C$, precipitates, with the corresponding formation of low-carbon martensite. This low-carbon martensite grows at the expense of the high-carbon martensite until, at the end of this stage, the structure consists of retained austenite, ε-carbide and martensite of reduced tetragonality. During the second stage any retained austenite in the steel begins to transform isothermally to bainite, while the third stage is marked by the formation of cementite platelets. The precipitation of cementite is accompanied by a dissolution of the ε-carbide phase so that the martensite loses its remaining tetragonality and becomes bcc ferrite. The degree to which these three stages overlap will depend on the temperature of the anneal and the carbon content. In consequence, the final structure produced will be governed by the initial choice of steel and the properties, and hence thermal treatment, required. Alloying elements, with the exception of Cr, affect the tempering of martensite. Plain carbon steels soften above 100°C owing to the early formation of ε-carbide, whereas in Si-bearing steels the softening is delayed to above 250°C, since Si stabilizes ε-carbide and delays its transformation to cementite. Alloying additions (see Table 7.2) thus enable the improvement in ductility to be achieved at higher tempering temperatures.

When a steel specimen is quenched prior to tempering, quenching cracks often occur. These are caused by the stresses which arise from both the transformation and the differential expansion produced when different parts of the specimen cool at different rates. To minimize such cracking, the desired properties of toughness and strength are often produced in the steel by alternative heat-treatment schedules; examples of these schedules are summarized in Figure 7.28, from which it will become evident that advantage is taken of the shape of the TTT curve to economize on the time the specimen is in the furnace, and also to minimize quenching stresses. During conventional

Table 7.2 *Influence of alloying additions on tempering.*

Element	*Retardation in tempering per 1% addition*	*Ratio of retardation of tempering to depression of* M_s
C	−40	Negative
Co	8	>8
Cr	0	0
Mn	8	0.24
Mo	17	0.8
Ni	8	0.5
Si	20	1.8
V	30	>1.0
W	10	0.9

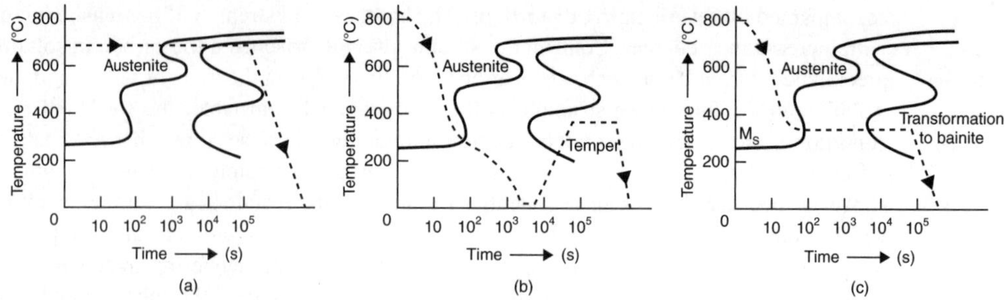

Figure 7.28 *Diagrams showing the heat-treatment procedure during isothermal annealing (a), martempering (b) and austempering (c).*

annealing, for example, the steel is heated above the upper critical temperature and allowed to cool slowly in the furnace. In isothermal annealing the steel is allowed to transform in the furnace, but when it has completely transformed, the specimen is removed from the furnace and allowed to air-cool, thereby saving furnace time. In martempering, the knee of the TTT curve is avoided by rapid cooling, but the quench is interrupted above M_S and the steel allowed to cool relatively slowly through the martensite range. With this treatment the thermal stresses set up by very rapid cooling are reduced. Such a procedure is possible because at the holding temperature there is ample time for the temperature to become equal throughout the sample before the transformation begins, and as a result the transformation occurs much more uniformly. After the transformation is complete, tempering is carried out in the usual way. In austempering, quenching is again arrested above M_S and a bainite product, having similar properties to tempered martensite, is allowed to form.

Alloying elements also lower the M_S temperatures and, consequently, greater stresses and distortion are introduced during quenching. This can be minimized by austempering and martempering as discussed above, but such treatments are expensive. Alloying elements should therefore be chosen to produce the maximum retardation of tempering for minimum depression of M_S; Table 7.2 shows that (1) C should be as low as possible, (2) Si and Co are particularly effective, and (3) Mo is the preferred element of the Mo, W, V group, since it is easier to take into solution than V and is cheaper than W.

Some elements, particularly Mo and V, produce quite high tempering temperatures. In quantities above about 1% for Mo and 0.5% for V, a precipitation reaction is also introduced, which has its maximum hardening effect at 550°C. This phenomenon of increased hardness by precipitation at higher temperatures is known as secondary hardening and may be classified as a fourth stage of tempering. $2–2\frac{1}{2}$Mo addition produces adequate temper resistance and changes the precipitate to Mo_2C, which is more resistant to overageing than Cr_7C_3, which is present in most alloy steels. High V additions lead to undissolved V_4C_3 at the quenching temperature, but 0.5V in conjunction with 2Mo does not form a separate carbide during tempering but dissolves in the Mo_2C. Cr also dissolves in Mo_2C but lowers the lattice parameters of the carbide and hence lowers the temper resistance by decreasing the matrix/carbide mismatch. However, 1Cr may be tolerated without serious reduction in temper resistance and reduces the tendency to quench crack. Si decreases the lattice parameter of matrix ferrite and hence increases temper resistance. A typical secondary hardening steel usually contains 0.4C, 2Mo, 0.5V, 0.5Si and 1.5Cr, with 1.8 GN m^{-2} TS and 15% elongation.

7.3.6 Thermomechanical treatments

To produce steels with an improved strength/ductility ratio the heat treatment may be modified to include a deformation operation in the process. The combined use of mechanical working and heat treatment is generally called thermomechanical treatment (THT). Three types of treatment have proved successful with martensitic and bainitic steels. These may be classified as follows:

1. Deformation in the stable austenite range above A_3 before transformation, i.e. (HTHT).
2. Deformation below A_1 before transformation; this (LTHT) low-temperature thermomechanical treatment is called ausforming.
3. Deformation during isothermal transformation to pearlite, i.e. below A_3, known as isoforming.

The main advantage of HTHT is in grain refinement, and steels such as silicon steels that recrystallize slowly are particularly suitable. It can, however, be applied to low-alloy high-carbon tool steels which are not suitable for ausforming, with significant increases in strength and toughness. The fatigue limit is also improved in many steels provided the deformation is limited to 25–30%. In ausforming, the deformation is usually carried out in the range 450–550°C and hence the steel must have a large bay in the TTT diagram to enable the deformation to be carried out. A suitable steel is Fe–0.35C–0.5Mn–1.5Ni–1.25Cr–0.25Mo, for which the strength increases by about 4.6–7.7 MN m^{-2} for each percent of deformation. The properties are improved as the deformation temperature is lowered, provided it is not below M_s, and with high deformation treatments (>70%) strengths up to about $E/70$ with good ductility have been achieved. A very fine carbide dispersion is produced in the austenite during deformation, together with a high density of dislocations. The removal of carbon from solution in the austenite means that during transformation the martensite formed is less supersaturated in C, and thus has lower tetragonality and is more ductile. The carbides also pin the dislocations in the austenite, helping to retain some of them together with those formed during the transformation. The martensite formed is therefore heavily dislocated with relatively stable dislocations (compared to those which would be formed by deforming martensite at room temperature), and has superior strength and toughness. Such steels are, of course, somewhat difficult to machine.

Isoforming has potential in improving the toughness of low-alloy steels. During isoforming to pearlite the normal ferrite/pearlite structure is modified, by the polygonization of sub-grains in the ferrite and the spheroidizing of cementite particles. Isoforming to bainite is also possible.

7.4 Fracture and toughness

7.4.1 Griffith microcrack criterion

Most materials break at a stress well below the theoretical fracture stress, which is that stress, σ_t, required to pull apart two adjoining layers of atoms. This stress varies with the distance between the atom planes and, as shown in Figure 7.29, may be approximated to a sine curve of wavelength such that

$$\sigma = \sigma_t \sin\left(\frac{2\pi u}{\lambda}\right) \approx \sigma_t \left(\frac{2\pi u}{\lambda}\right),$$

where u is the displacement from the equilibrium spacing b. From Hooke's law $\sigma = (Eu/b)$ and hence $\sigma_t = \lambda E/2\pi b$. Now in pulling apart the two atomic planes it is necessary to supply the surface energy γ and hence

$$\gamma = \int_0^{\lambda/2} \sigma_t \sin\left(\frac{2\pi u}{\lambda}\right) du = \frac{\lambda \sigma_t}{2\pi},$$

so that the theoretical tensile strength is given by

$$\sigma_t = \sqrt{(E\gamma/b)}. \tag{7.22}$$

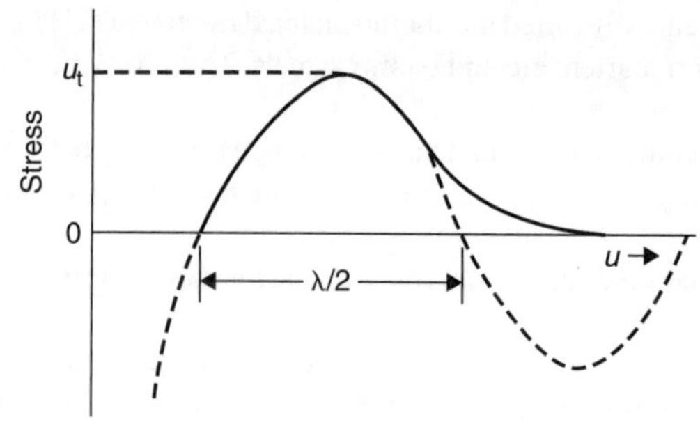

Figure 7.29 *Model for estimating the theoretical fracture strength* σ_t.

Glass fibers and both metallic and non-metallic whiskers have strengths approaching this ideal value of about $E/10$, but bulk metals, even when tested under favorable conditions (e.g. at 4 K), can rarely withstand stresses above $E/100$. Griffith, in 1920, was the first to suggest that this discrepancy was due to the presence of small cracks which propagate through the crystal and cause fracture. Griffith's theory deals with elastic cracks, where at the tip of the crack atomic bonds exist in every stage of elongation and fracture. As the crack grows, each of the bonds in its path take up the strain, and the work done in stretching and breaking these bonds becomes the surface energy γ of the fractured faces. If separation of the specimen between two atomic layers occurs in this way, the theoretical strength only needs to be exceeded one point at a time, and the applied stress will be much lower than the theoretical fracture stress. The work done in breaking the bonds has to be supplied by the applied force, or by the elastic energy of the system. Assuming for a crack of length $2c$ that an approximately circular region of radius c is relieved of stress σ and hence strain energy $(\sigma^2/2E)\pi c^2$ by the presence of the crack, the condition when the elastic strain of energy balances the increase of surface energy is given by

$$\frac{d}{dc}\left(\frac{\pi c^2 \sigma^2}{2E}\right) = \frac{d}{dc}(2c\gamma)$$

and leads to the well-known Griffith formula,

$$\sigma = \sqrt{\frac{2\gamma E}{\pi c}} \approx \sqrt{\frac{\gamma E}{2c}}, \tag{7.23}$$

for the smallest tensile stress σ able to propagate a crack of length $2c$. The Griffith criterion therefore depends on the assumption that the crack will spread if the decrease in elastic strain energy resulting from an increase in $2c$ is greater than the increase in surface energy due to the increase in the surface area of the crack.

Griffith's theory has been verified by experiments on glasses and polymers at low temperatures, where a simple process of fracture by the propagation of elastic cracks occurs. In such 'weak' brittle fractures there is little or no plastic deformation and γ is mainly the surface energy ($\approx$1–10 J m^{-2}) and the fracture strength $\sigma_f \approx 10^{-5}E$. In crystalline solids, however, the cracks are not of the elastic type and a plastic zone exists around the crack tip, as shown in Figure 7.30. In such specimens, fracture cannot occur unless the applied tensile stress σ is magnified to the theoretical strength σ_t. For an atomically sharp crack (where the radius of the root of the crack r is of the order of b) of length $2c$ it can be shown that the magnified stress σ_m will be given by $\sigma_m = \sigma\sqrt{(c/r)}$, which, if the crack is to propagate, must be equal to the theoretical fracture stress of the material at the end of the crack. It follows that substituting this value of σ_t in equation (7.22) leads to the Griffith formula of equation (7.23).

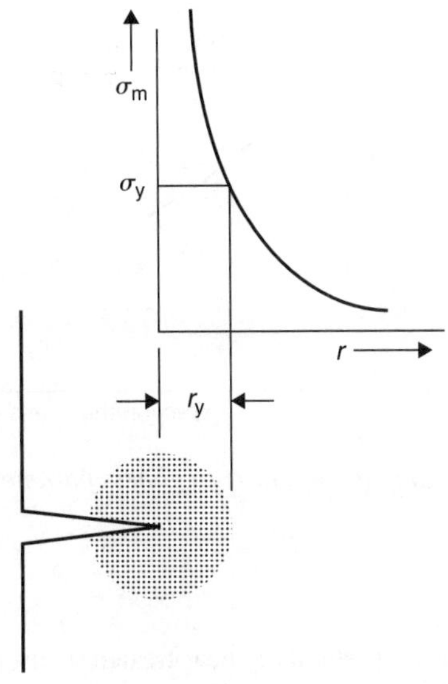

Figure 7.30 *Variation of stress from the tip of a crack and the extent of the plastic zone, radius* r_y.

Figure 7.30 shows the way the magnified stress drops off with distance from the tip of the crack. Clearly, at some distance r_y the stress reaches the yield stress and plastic flow occurs. There is thus a zone of plastic flow around the tip of radius r_y. The larger the plastic zone, as in ductile metals, the more energy is absorbed in fracture. In ceramics this zone is usually small.

In 'strong' fractures γ is greatly increased by the contribution of the plastic work around the crack tip, which increases the work required for crack propagation. The term γ must now be replaced by $(\gamma + \gamma_p)$, where γ_p is the plastic work term; generally $(\gamma + \gamma_p)$ is replaced by G, the strain energy release rate, so that equation (7.23) becomes the Orowan–Irwin relationship:

$$\sigma = \sqrt{(EG/\pi c)}. \qquad (7.24)$$

Here, G might be $\sim 10^4\ \mathrm{J\,m^{-2}}$ and $\sigma_f \approx 10^{-2}$–$10^{-3}\,E$.

Worked example

Steel with a lattice friction stress of $6.9 \times 10^8\ \mathrm{N\,m^{-2}}$, dislocation density of $10^{16}\ \mathrm{m^{-2}}$, surface energy $\gamma_s = 1\ \mathrm{J\,m^{-2}}$ contains microcracks 10^{-4} m in length. Will the component yield or fracture in service?

Solution

Assuming direct stress is two times the shear stress, to yield,

$$\begin{aligned}\sigma_y &= \sigma_i + 4\mu b\sqrt{\rho} = 2 \times 6.9 \times 10^8 + 4 \times 8 \times 10^9 \times 0.2 \times 10^{-9} \times \sqrt{10^{16}} \\ &= 13.8 \times 10^8 + 6.4 \times 10^8 = 20.2 \times 10^8\ \mathrm{Pa} = 2.02\ \mathrm{GPa}.\end{aligned}$$

To fracture, $\sigma_f = \sqrt{2E\gamma_s/\pi a} = \sqrt{2 \times 210 \times 10^9 \times 1/\pi \times 10^{-4}} = 0.36 \times 10^8$ Pa. $\sigma_f \ll \sigma_y$, so fracture will take place before yield.

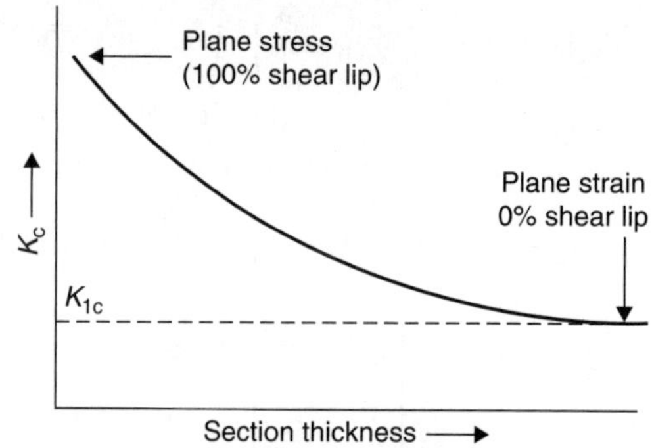

Figure 7.31 *Variation in the fracture toughness parameter with section thickness.*

7.4.2 Fracture toughness

In engineering structures, particularly heat-treated steels, cracks are likely to arise from weld defects, inclusions, surface damage, etc. and it is necessary to design structures with the knowledge that cracks are already present and capable of propagation at stresses below the macroscopic yield stress as measured in a tensile test. Since different materials show different crack propagation characteristics (e.g. hard steel and glass) it is necessary for the design engineers to find the limiting design stress in terms of some property or parameter of the material. For this reason, a fracture toughness parameter is now being employed to measure the tendency of cracks of given dimensions to propagate under particular stress conditions.

In Section 7.4.1 it was shown that $\sigma\sqrt{(\pi c)}=\sqrt{(EG)}$, which indicates that fast fracture will occur when a material is stressed to a value σ and the crack reaches some critical size, or alternatively when a material containing a crack is subjected to some critical stress σ, i.e. the critical combination of stress and crack length for fast fracture is a constant, $\sqrt{(EG)}$ for the material, where E is Young's modulus and G is the strain energy release rate. The term $\sigma\sqrt{(\pi c)}$ is given the symbol K and is called the stress intensity factor, with units MN $m^{-3/2}$. Fast fracture will then occur when $K=K_c$, where K_c $[=\sqrt{(EG_c)}]$ is the critical stress intensity factor, or more commonly the fracture toughness parameter.

The fracture toughness of a material can alter markedly depending on whether the elastic–plastic field ahead of the crack approximates to plane strain or plane stress conditions, much larger values being obtained under plane stress conditions as operate in thin sheets. The important and critical factor is the size of the plastic zone in relation to the thickness of the section containing the crack. When this is small, as in thick plates and forgings, plane strain conditions prevail and the hydrostatic tension ahead of the crack results in a semi-brittle 'flat' fracture. When the value is large, as in thin sheets of ductile metals, plane stress conditions operate and the tension at the crack front is smaller, giving rise to a more ductile mode of failure. At intermediate values a mixed fracture, with a flat center bordered by shear lips, is obtained. Thus, without changing the structure or properties of the materials in any way it is possible to produce a large difference in fracture toughness by changing the section thickness. For thick sections, when a state of complete constraint is more nearly approached, the values of K_c and G_c approach minimum limiting values. These values are denoted by K_{1c} and G_{1c} and are considered to be material constants; the subscript '1' denotes the first mode of crack extension, i.e. the opening mode (see Figure 7.31).

The general procedure in measuring the fracture toughness parameter is to introduce a crack of suitable size into a specimen of suitable dimension and geometry. The specimen is then loaded at a slow rate and the crack extension measured up to the critical condition. The measurement of K_{1c}

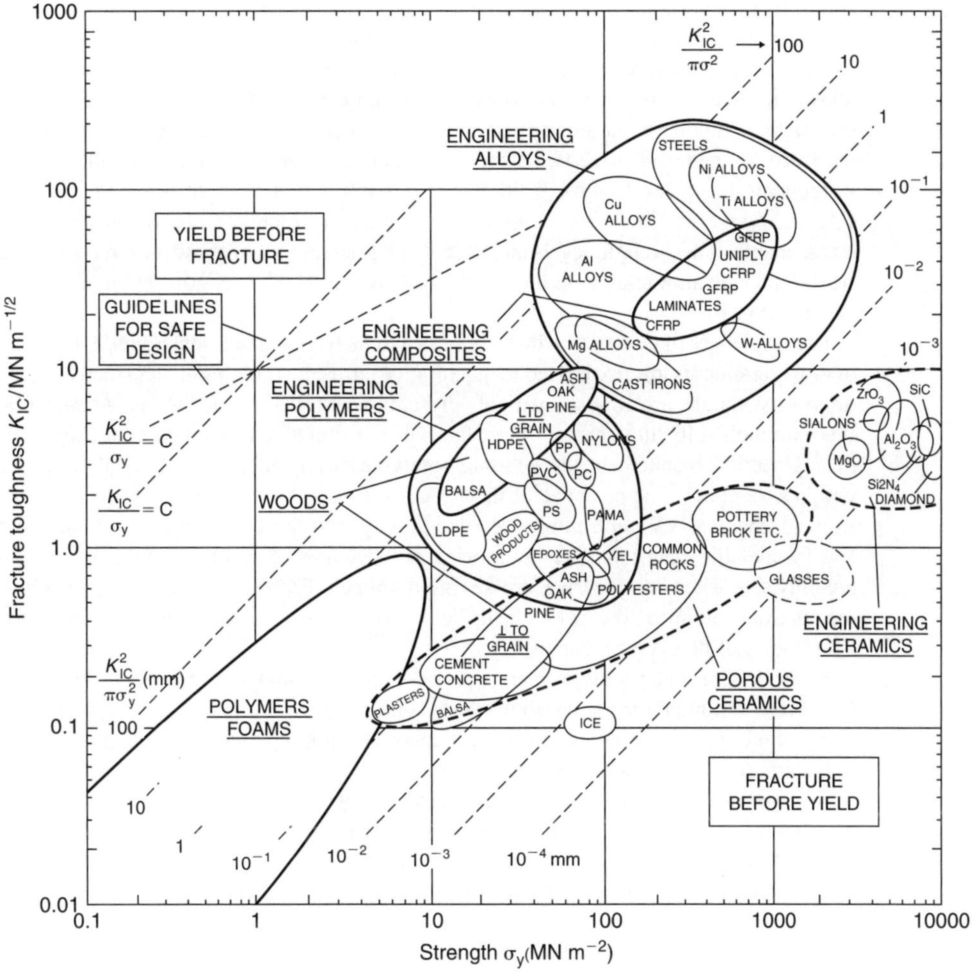

Fracture-toughness, K_{IC}, plotted against strength, σ_y. The contours show the value of K^2_{IC}, $\pi\,\sigma^2_y$—roughly, the diameter of the process zone at a crack tip (units: mm). The guidelines of constant K_{IC}/σ_y and $K^2{}_{IC}/\sigma_y$ are used in yield-before-break and leak-before-break design.

Figure 7.32 *Ashby property chart of fracture toughness versus strength (Ashby, 1992).*

will be valid if the plastic zone size is small (by a factor 10) in relation to the cross-section of the specimen. The zone size r_y may be obtained by equating the stress field of the crack at $r = r_y$ to the strength σ_y of the material and is given by

$$r_y = K^2_{1c}/2\pi\sigma^2_y.$$

An Ashby property chart of fracture toughness versus strength, given in Figure 7.32, shows that the size of the zone (broken lines) varies from atomic dimensions for brittle ceramics to tens of centimeters for ductile metals.

In designing safe structures for a given load, the structure is required to yield before it breaks. For a minimum detectable crack size of $2c$ this condition is given by $(K_{1c}/\sigma_y) \geq \sqrt{\pi c}$. The safest material is the one with the greatest value of K_{1c}/σ_y. Clearly, a high yield stress σ_y is also required. The chart shows that steel satisfies both these requirements and indicates why it is still the best material for highly stressed structures where weight is not important.

Fracture toughness requirements are now written into the general specification of high-technology alloys and hence it is necessary to determine the effect of heat treatment and alloying additions on fracture toughness parameters. Processes such as ausforming and controlled rolling improve the fracture toughness of certain steels. Carbon has a considerable effect and there are advantages in reducing the C-level below 0.1% where possible. High-strength low-alloy (HSLA) steels have $C \lesssim 0.1\%$, and the Nb, V and Ti additions form fine carbides which, together with the small grain sizes, enable good strength levels and acceptable fracture toughness values to be achieved. Maraging steels with high alloy and low carbon ($<0.04\%$) give very high strength combined with high toughness (see Chapter 8).

The brittleness of ceramics is directly linked to their high notch sensitivity. The fracture toughness of most ceramics is low: expressed in quantitative terms, it is commonly less than 8 MN $m^{-3/2}$. Flaws, often very minute, are almost invariably present in ceramics and act as stress-concentrating notches. It is extremely difficult to prevent these flaws from forming during manufacture and service. In terms of engineering practice, brittle ceramic components are intolerant of misalignment and poor fits within assemblies. The presence of flaws is also responsible for the variability or 'spread' of results from mechanical tests and introduces uncertainty into the design process. (The 'spread' is much less for metallic materials.) Design procedures have moved well beyond the principle that ceramics are only safe when compressive stresses are dominant. Probabilistic assessments of mechanical test results from ceramics now tackle the difficult task of allowing for randomness in the size, shape and distribution of flaws (see Chapter 10).

Despite these underlying problems, progress has been made in producing tough 'ductile ceramics'. Fracture toughness values above 20 MN $m^{-3/2}$ have been achieved. Typical measures include elimination of flaws (microcracks, pores), incorporation of crack-retarding phases (in zirconia) and reduction of average grain size (below 1 mm).

At ambient temperature, fracture surfaces of ceramics are conchoidal (glassy), intergranular or cleavage in character, depending on the material. The strain at fracture is very small, being about 0.001. At elevated temperatures, under creep conditions, fracture strain is greater. Conventional ceramics have little or no capacity for slip and crack tips therefore tend to remain sharp. As the crack propagates, the load-supporting cross-section gets smaller and the general level of stress increases so that failure can be sudden. These remarks refer to tensile stresses, since compressive stresses will, of course, tend to close the crack.

In some polycrystalline ceramics, such as magnesia, the von Mises criterion for maintenance of cohesions is not satisfied. Slip is limited and cracks are not effectively blunted. However, raising the temperature often enables the necessary minimum of five independent deformation modes to operate, leading to ductility. Any treatments which eliminate surface flaws will naturally enhance this ductility. (Against this background, one can readily appreciate why fabrication methods for polycrystalline ceramic components, in contrast to those for metals, are not based upon bulk deformation.) At these higher temperatures the environment becomes increasingly more likely to react with surfaces of the ceramic: it may even penetrate an open texture and cause crack blunting. The mechanisms of flow in polycrystalline ceramics at elevated temperatures are similar to those encountered in metallic systems (e.g. glide and climb of dislocations, vacancy diffusion, grain boundary sliding). The similarity between deformation processes in polycrystalline ceramics and metals is evident if one compares the layout of the corresponding deformation mechanism maps. A minor phase is usually present at grain boundary surfaces in ceramics, functioning as a ceramic bond, and if it softens with rise in temperature, then the grains may be able to slide past each other. These regions tend to liquefy before the actual grains so that it is advisable not to exceed the solidus temperature. Grain boundary surfaces are particularly susceptible to the nucleation of cracks. As in the final accelerating stage of metallic creep, these cracks form in planes normal to the direction of applied stress.

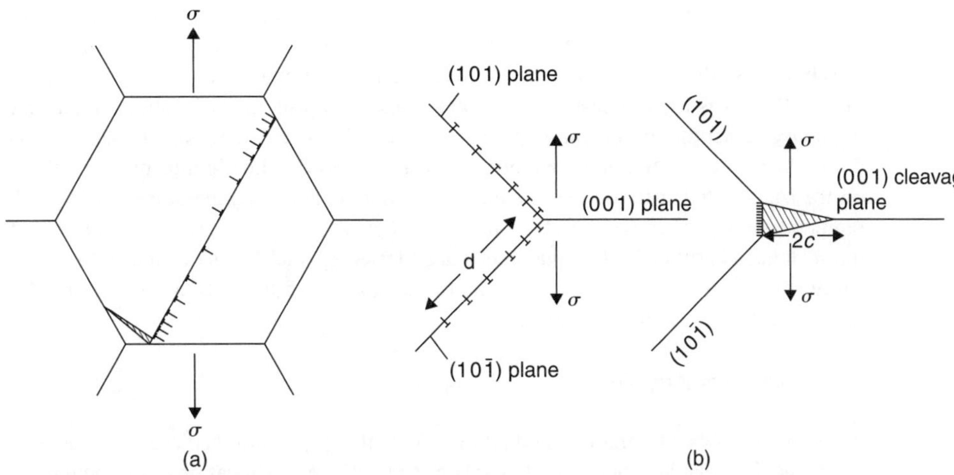

Figure 7.33 *Formation of a crack: (a) by the piling up of dislocations against a grain boundary after strain and (b) by the dislocations on (1 0 1) and (1 0 $\bar{1}$) planes running together (after Cottrell, 1958, courtesy of American Institute of Mechanical Engineers).*

7.4.3 Cleavage and the ductile–brittle transition

The fracture toughness versus strength chart, shown in Figure 7.32, indicates the wide spread of values for the different classes of material. Metals dissipate much energy in the plastic zone, which accounts for the difference between the fracture energy G_{1c} and the true surface energy γ. The larger the zone, the more energy is absorbed; G_{1c} is high and so is K_{1c} ($\approx 100\,\mathrm{MN\,m^{-3/2}}$). Ceramics and glasses fracture without much plastic flow to blunt the cracks by simply breaking atomic bonds, leading to cleavage; for these materials K_{1c} is less than $10\,\mathrm{MN\,m^{-3/2}}$.

At low temperatures some metals, notably steels, also become brittle and fracture by cleavage. Since they are ductile at room temperature this transition to brittle cleavage behavior is quite spectacular and has led to several engineering catastrophes. In general, brittle cleavage can occur in metals with bcc and cph under the appropriate conditions, while in fcc materials it does not. The most important factor linking these three different structures is the Peierls stress and the way the yield stress varies with temperature. In steel, for example, the yield stress increases rapidly with lowering of temperature below room temperature such that plastic deformation at the crack tip is minimized and the fracture mechanism changes from plastic tearing to cleavage. Even in these materials some plastic deformation does occur.

Several models have been suggested for the process whereby glide dislocations are converted into microcracks. The simplest mechanism, postulated by Stroh, is that involving a pile-up of dislocations against a barrier, such as a grain boundary. The applied stress pushes the dislocations together and a crack forms beneath their coalesced half-planes, as shown in Figure 7.33a. A second mechanism of crack formation, suggested by Cottrell, is that arising from the junction of two intersecting slip planes. A dislocation gliding in the (1 0 1) plane coalesces with one gliding in the (1 0 $\bar{1}$) plane to form a new dislocation which lies in the (0 0 1) plane according to the reaction:

$$a/2[\bar{1}\ \bar{1}\ 1] + a/2[1\ 1\ 1] \rightarrow a[0\ 0\ 1].$$

The new dislocation, which has a Burgers vector $a[0\ 0\ 1]$, is a pure edge dislocation and, as shown in Figure 7.33b, may be considered as a wedge, one lattice constant thick, inserted between the faces of the (0 0 1) planes. It is considered that the crack can then grow by means of other dislocations in the

(1 0 1) and (1 0 $\bar{1}$) planes running into it. Although the mechanism readily accounts for the observed (1 0 0) cleavage plane of bcc metals, examples have not been directly observed.

While these dislocation coalescence mechanisms may operate in single-phase materials, in two-phase alloys it is usually easier to nucleate cracks by piling up dislocations at particles (e.g. grain boundary cementite or cementite lamellae in pearlite). The pile-up stress then leads to cracking of either the particle or the particle/matrix interface. A brittle–ductile transition can then be explained on the basis of the criterion that the material is ductile at any temperature, if the yield stress at that temperature is smaller than the stress necessary for the growth of these microcracks, but if it is larger the material is brittle. If cleavage cracks are formed by such a dislocation mechanism, the Griffith formula may be rewritten to take account of the number of dislocations, n, forming the crack. Thus, rearranging Griffith's formula we have

$$\sigma(2c\sigma/E) = \gamma = 2c\sigma^2/E,$$

where the product of length $2c$ and strain σ/E is the maximum displacement between the faces of the crack. This displacement will depend on the number of dislocations forming the cleavage wedge and may be interpreted as a pile-up of n edge dislocations, each of Burgers vector a, so that equation (7.23) becomes

$$\sigma(na) = \gamma \tag{7.25}$$

and gives a general criterion for fracture. Physically, this means that a number of glide dislocations, n, run together and in doing so cause the applied stress acting on them to do some work, which for fracture to occur must be at least sufficient to supply the energy to create the new cracked faces, i.e. $(\gamma + \gamma_p)$.

Qualitatively, we would expect those factors which influence the yield stress also to have an effect on the ductile–brittle fracture transition. The lattice 'friction' term σ_i, dislocation locking term k_y, and grain size $2d$ should also all be important because any increase in σ_i and k_y, or decrease in the grain size, will raise the yield stress with a corresponding tendency to promote brittle failure.

These conclusions have been put on a quantitative basis by Cottrell, who considered the stress needed to grow a crack at or near the tensile yield stress, σ_y, in specimens of grain diameter $2d$. Let us consider first the formation of a microcrack. If τ_y is the actual shear stress operating, the effective shear stress acting on a glide band is only $(\tau_y - \tau_i)$, where, it will be remembered, τ_i is the 'friction' stress resisting the motion of unlocked dislocations arising from the Peierls–Nabarro lattice stress, intersecting dislocations or groups of impurities. The displacement na is then given by

$$na = [(\tau_y - \tau_i)/\mu]d, \tag{7.26}$$

where μ is the shear modulus and d is the length of the slip band containing the dislocations (assumed here to be half the grain diameter). Once a microcrack is formed, however, the whole applied tensile stress normal to the crack acts on it, so that σ can be written as ($\tau_y \times$ constant), where the constant is included to account for the ratio of normal stress to shear stress. Substituting for na and σ in the Griffith formula (equation (7.25)), then fracture should be able to occur at the yield point when $\sigma = \sigma_y$ and

$$\tau_y(\tau_y - \tau_i)d = C\mu\gamma, \tag{7.27}$$

where C is a constant. The importance of the avalanche of dislocations produced at the yield drop can be seen if we replace τ_y by (constant $\times\ \sigma_y$), τ_i by (constant $\times\ \sigma_i$) and use the Petch relationship $\sigma_y = \sigma_i + k_y d^{-1/2}$, when equation (7.27) becomes

$$(\sigma_i d^{1/2} + k_y)k_y = \beta\mu\gamma, \tag{7.28}$$

where β is a constant which depends on the stress system; $\beta \approx 1$ for tensile deformation and $\beta \approx \frac{1}{3}$ for a notched test.

This is a general equation for the propagation of a crack at the lower yield point and shows what factors are likely to influence the fracture process. Alternative models for growth-controlled cleavage fracture have been developed to incorporate the possibility of carbide particles nucleating cracks. Such models emphasize the importance of yield parameters and grain size as well as carbide thickness. Coarse carbides give rise to low fracture stresses, thin carbides to high fracture stresses and ductile behavior.

7.4.4 Factors affecting brittleness of steels

Many of the effects of alloying, heat treatment and condition of testing on brittle fracture can be rationalized on the basis of the above 'transition' equation.

7.4.4.1 Ductile–brittle transition

Under conditions where the value of the left-hand side of equation (7.28) is less than the value of the right-hand side, ductile behavior should be observed; when the left-hand side exceeds the right-hand side the behavior should be brittle. Since the right-hand side of equation (7.28) varies only slowly with temperature, it is the way in which changes occur in values of the terms on the left of the equation which are important in determining the ductile–brittle transition. Thus, in a given material brittleness should be favored by low temperatures and high strain rate, because these give rise to large values of σ_i and k_y, and by large grain sizes. On the right-hand side, the typical effect of a sharp notch is to raise the transition temperature of structural steel from around 100 K for a normal tensile test into the range of 200–300 K, because the value of β is lowered.

7.4.4.2 Effects of composition and grain size

At a constant temperature, because the values of σ_i and k_y remain fixed, the transition point will occur at a critical grain size, above which the metal is brittle and below which it is ductile.

The inclusion in equation (7.28) of the grain-size term, $d^{1/2}$, in combination with the σ_i, term, enables many previous metallurgical misunderstandings to be cleared up. It shows that there is no simple connection between hardness and brittleness, since hardening produced by refining the grain size reduces the brittleness, whereas hardening due to an increase in σ_i increases the brittleness.

Heat treatment is generally used to control the grain size of the sample and refine the structure. 'Killed' steel has very good notch toughness, because aluminum additions refine the grain size. Manganese reduces the grain size and by combining with carbon also reduces the k_y value, so that this addition is especially beneficial in improving low-temperature ductility. It is fairly evident that an improved notch toughness steel, compared with that used for welded ships in World War II, is given by increasing the manganese content and decreasing the carbon content, i.e. a high manganese-to-carbon ratio. Other additions, particularly nickel and chromium, have a similar effect on low-temperature ductility.

The Group 6A metals (Cr, Mo and W) are more susceptible to brittle fracture than the Group 5A metals (V, Nb and Ta). A comparison of these metals in terms of cleavage fracture is difficult, however, since Cr, Mo and W are susceptible to grain boundary fracture because segregation of impurities to such regions reduces the effective surface energy γ. However, even if this effect is eliminated by lowering the impurity level, it appears that Ta, Nb and V are more ductile than Fe, Mo, Cr and W, presumably because they have a lower k_y/μ ratio and a higher γ value.

7.4.4.3 Work hardening and irradiation hardening

Small amounts of plastic deformation at room temperature, which overcome the yield point and unlock some of the dislocations, improve the ductility at low temperatures. The room-temperature ductility of chromium is similarly affected by small amounts of plastic deformation at 400°C. In general, however, plastic deformation which leads to work hardening embrittles the metal because it raises the σ_i contribution, due to the formation of intersecting dislocations, vacancy aggregates and other lattice defects.

The importance of twins in fracture is not clear as there are several mechanisms other than twinning for the formation of a crack which can initiate fracture, and there is good evidence that microcracks form in steel in the absence of twins and that cracks start at inclusions. Nevertheless, twinning and cleavage are generally found under similar conditions of temperature and strain rate in bcc transition metals, probably because both phenomena occur at high stress levels. The nucleation of a twin requires a higher stress than the propagation of the twin interface.

Irradiation hardening also embrittles the metal. According to the theory of this type of hardening outlined in Chapter 6, radiation damage can produce an increase in both k_y (migration to dislocations of vacancies or interstitials) and σ_i (formation of dislocation loops and other aggregated defects). However, for steel, radiation hardening is principally due to an increased σ_i contribution, presumably because the dislocations in mild steel are already too heavily locked with carbon atoms for any change in the structure of the dislocation to make any appreciable difference to k_y. Nevertheless, a neutron dose of 1.9×10^{23} n m^{-2} will render a typical fine-grained, unnotched mild steel, which is normally ductile at −196°C, quite brittle. Moreover, experiments on notched fine-grained steel samples (see Figure 7.1c) show that this dose increases the ductile–brittle transition temperature by 65°C.

7.4.4.4 Microstructure

The change in orientation at individual grain boundaries impedes the propagation of the cleavage crack by (1) creating cleavage steps, (2) causing localized deformation and (3) tearing near the grain boundary. It is the extra work done (γ_p) in such processes which increases the apparent surface energy γ to ($\gamma_s + \gamma_p$). It follows, therefore, that the smaller the distance a crack is able to propagate without being deviated by a change of orientation of the cleavage plane, the greater is the resistance to brittle fracture. In this respect, the coarser high-temperature products of steel, such as pearlite and upper bainite, have inferior fracture characteristics compared with the finer lower bainite and martensite products. The fact that coarse carbides promote cleavage while fine carbides lead to ductile behavior has already been discussed.

7.4.5 Hydrogen embrittlement of steels

It is well known that ferritic and martensitic steels are severely embrittled by small amounts of hydrogen. The hydrogen may be introduced during melting and retained during the solidification of massive steel castings. Plating operations (e.g. Cd plating of steel for aircraft parts) may also lead to hydrogen embrittlement. Hydrogen can also be introduced during acid pickling or welding, or by exposure to H_2S atmospheres.

The chief characteristics of hydrogen embrittlement are its (1) strain-rate sensitivity, (2) temperature dependence and (3) susceptibility to produce delayed fracture (see Figure 7.34). Unlike normal brittle fracture, hydrogen embrittlement is enhanced by slow strain rates and, consequently, notched impact tests have little significance in detecting this type of embrittlement. Moreover, the phenomenon is not more common at low temperatures, but is most severe in some intermediate temperature range around

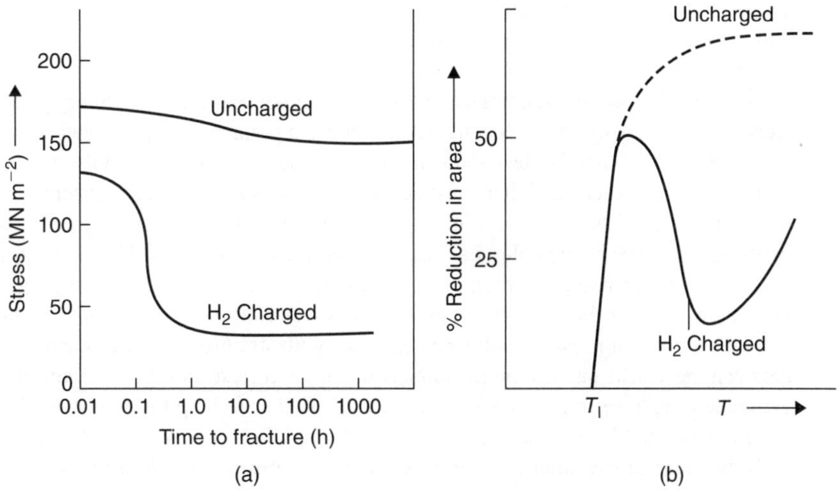

Figure 7.34 *Influence of hydrogen on fracture behavior, showing time dependence (a) and temperature dependence (b).*

room temperature (i.e. −100 to 100°C). These effects have been taken to indicate that hydrogen must be present in the material and must have a high mobility in order to cause embrittlement in polycrystalline aggregates.

A commonly held concept of hydrogen embrittlement is that monatomic hydrogen precipitates at internal voids or cracks as molecular hydrogen, so that as the pressure builds up it produces fracture. Alternatively, it has been proposed that the critical factor is the segregation of hydrogen, under applied stress, to regions of triaxial stress just ahead of the tip of the crack, and when a critical hydrogen concentration is obtained, a small crack grows and links up with the main crack. Hydrogen may also exist in the void or crack but it is considered that this has little effect on the fracture behavior, and it is only the hydrogen in the stressed region that causes embrittlement. Neither model considers the Griffith criterion, which must be satisfied if cracks are to continue spreading.

An application of the fracture theory may be made to this problem. Thus, if hydrogen collects in microcracks and exerts internal pressure P, the pressure may be directly added to the external stress to produce a total stress $(P+p)$ for propagation. Thus, the crack will spread when

$$(P+p)\pi a = (\gamma_s + \gamma_p), \tag{7.29}$$

where the surface energy is made up from a true surface energy γ_s and a plastic work term γ_p. The possibility that hydrogen causes embrittlement by becoming adsorbed on the crack surfaces thereby lowering γ is thought to be small, since the plastic work term γ_p is the major term controlling γ, whereas adsorption would mainly affect γ_s.

Supersaturated hydrogen atoms precipitate as molecular hydrogen gas at a crack nucleus, or the interface between non-metallic inclusions and the matrix. The stresses from the build-up of hydrogen pressure are then relieved by the formation of small cleavage cracks. Clearly, while the crack is propagating, an insignificant amount of hydrogen will diffuse to the crack and, as a consequence, the pressure inside the crack will drop. However, because the length of the crack has increased, if a sufficiently large and constant stress is applied, the Griffith criterion will still be satisfied and completely brittle fracture can, in theory, occur. Thus, in iron single crystals, the presence or absence of hydrogen appears to have little effect during crack propagation because the crack has little difficulty spreading through the crystal. In polycrystalline material, however, the hydrogen must be both present and mobile, since propagation occurs during tensile straining.

When a sufficiently large tensile stress is applied such that $(p+P)$ is greater than that required by the Griffith criterion, the largest and sharpest crack will start to propagate, but will eventually be stopped at a microstructural feature, such as a grain boundary, as previously discussed. The pressure in the crack will then be less than in adjacent cracks which have not been able to propagate. A concentration gradient will then exist between such cracks (since the concentration is proportional to the square root of the pressure of hydrogen), which provides a driving force for diffusion, so that the hydrogen pressure in the enlarged crack begins to increase again. The stress to propagate the crack decreases with increase in length of crack, and since p is increased by straining, a smaller increment ΔP of pressure may be sufficient to get the crack restarted. The process of crack propagation followed by a delay time for pressure build-up continues with straining until the specimen fails when the area between the cracks can no longer support the applied load. In higher strain-rate tests, the hydrogen is unable to diffuse from one stopped crack to another to help the larger crack get started before it becomes blunted by plastic deformation at the tip. The decrease in the susceptibility to hydrogen embrittlement in specimens tested at low temperatures results from the lower pressure build-up at these temperatures since $PV = 3n\mathbf{R}T$, and also because hydrogen has a lower mobility.

7.4.6 Intergranular fracture

Intergranular brittle failures are often regarded as a special class of fracture. In many alloys, however, there is a delicate balance between the stress required to cause a crack to propagate by cleavage and that needed to cause brittle separation along grain boundaries. Although the energy absorbed in crack propagation may be low compared to cleavage fractures, much of the analysis of cleavage is still applicable if it is considered that chemical segregation to grain boundaries or crack faces lowers the surface energy γ of the material. Fractures at low stresses are observed in austenitic chromium–nickel steels, due to the embrittling effect of intergranular carbide precipitation at grain boundaries. High transition temperatures and low fracture stresses are also common in tungsten and molybdenum as a result of the formation of thin second-phase films due to small amounts of oxygen, nitrogen or carbon. Similar behavior is observed in the embrittlement of copper by antimony and iron by oxygen, although in some cases the second-phase films cannot be detected.

A special intergranular failure, known as temper embrittlement, occurs in some alloy steels when tempered in the range 500–600°C. This phenomenon is associated with the segregation of certain elements or combinations of elements to the grain boundaries. The amount segregated is very small ($\sim$ a monolayer), but the species and amount has been identified by AES on specimens fractured intergranularly within the ultra-high vacuum of the Auger system. Group VIB elements are known to be the most surface active in iron but, fortunately, they combine readily with Mn and Cr, thereby effectively reducing their solubility. Elements in Groups IVB and VB are less surface active but often co-segregate in the boundaries with Ni and Mn. In Ni–Cr steels, the co-segregation of Ni–P and Ni–Sb occurs, but Mo additions can reduce the tendency for temper embrittlement. Since carbides are often present in the grain boundaries, these can provide the crack nucleus under the stress concentration from dislocation pile-ups either by cracking or by decohesion of the ferrite/carbide interface, particularly if the interfacial energy has been lowered by segregation.

7.4.7 Ductile failure

Ductile failure was introduced in Chapter 3 because of the role played by voids in the failure processes, which occurs by void nucleation, growth and coalescence. The nucleation of voids often takes place at

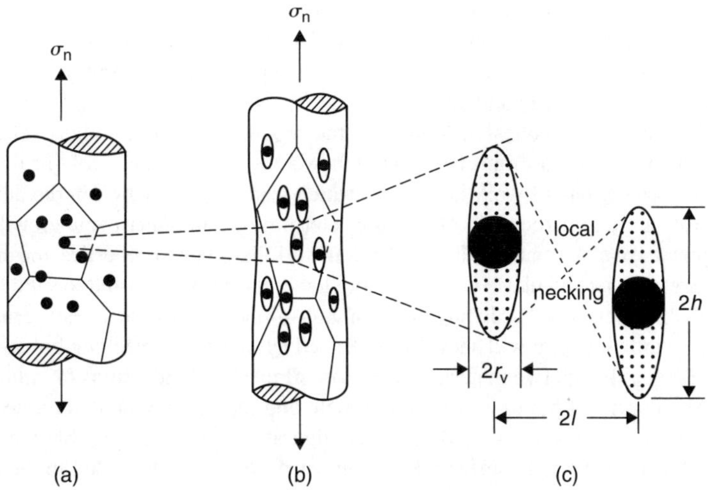

Figure 7.35 *Schematic representation of ductile fracture. (a) Voids nucleate at inclusions. (b) Voids elongate as the specimen extends. (c) Voids coalesce to cause fracture when their length 2h is about equal to their separation (after Ashby* et al.*, 1979).*

inclusions. The dislocation structure around particle inclusions leads to a local rate of work hardening higher than the average, and the local stress on reaching some critical value σ_c will cause fracture of the inclusion or decohesion of the particle/matrix interface, thereby nucleating a void. The critical nucleation strain ε_n can be estimated and lies between 0.1 and 1.0 depending on the model. For dispersion-hardening materials where dislocation loops are generated the stress on the interface due to the nearest prismatic loop, at distance r, is $\mu b/r$, and this will cause separation of the interface when it reaches the theoretical strength of the interface, of order γ_w/b. The parameter r is given in terms of the applied shear strain ε, the particle diameter d and the length k equal to half the mean particle spacing as $r = 4kb/\varepsilon d$. Hence, void nucleation occurs on a particle of diameter d after a strain ε, given by $\varepsilon = 4k\gamma_w/\mu db$. Any stress concentration effect from other loops will increase with particle size, thus enhancing the particle size dependence of strain to voiding.

Once nucleated, the voids grow until they coalesce to provide an easy fracture path. A spherical-shaped void concentrates stress under tensile conditions and, as a result, elongates initially at about C ($\approx$2) times the rate of the specimen, but as it becomes ellipsoidal the growth rate slows until finally the elongated void grows at about the same rate as the specimen. At some critical strain, the plasticity becomes localized, the voids rapidly coalesce and fracture occurs. The localization of the plasticity is thought to take place when the voids reach a critical distance of approach, given when the void length $2h$ is approximately equal to the separation, as shown in Figure 7.35. The true strain for coalescence is then

$$\begin{aligned}\varepsilon &= (1/C)\ln[\alpha(2l - 2r_v)/2r_v] \\ &\approx (1/C)\ln[\alpha(1/f_v^{1/2} - 1)],\end{aligned} \tag{7.30}$$

where $\alpha \approx 1$ and f_v is the volume fraction of inclusions.

Void growth leading to failure will be much more rapid in the necked portion of a tensile sample following instability than during stable deformation, since the stress system changes in the neck from uniaxial tension to approximately plane strain tension. Thus, the overall ductility of a specimen will depend strongly on the macroscopic features of the stress–strain curves which (from Considère's criterion) determine the extent of stable deformation, as well as on the ductile rupture process of void nucleation and growth. Nevertheless, an equation of the form of (7.30) reasonably describes the fracture strain for cup and cone failures.

The work of decohesion influences the progress of voiding and is effective in determining the overall ductility in a simple tension test in two ways. The onset of voiding during uniform deformation depresses the rate of work hardening, which leads to a reduction in the uniform strain, and the void density and size at the onset of necking determine the amount of void growth required to cause ductile rupture. Thus, for matrices having similar work-hardening properties, the one with the least tendency to 'wet' the second phase will show both lower uniform strain and lower necking strain. For matrices with different work-hardening potential but similar work of decohesion, the matrix having the lower work-hardening rate will show the lower reduction prior to necking but the greater reduction during necking, although two materials will show similar total reductions to failure.

The degree of bonding between particle and matrix may be determined from voids on particles annealed to produce an equilibrium configuration by measuring the contact angle θ of the matrix surface to the particle surface. Resolving surface forces tangential to the particle, then the specific interface energy γ_I is given approximately in terms of the matrix surface energy γ_m and the particle surface energy γ_P as $\gamma_I = \gamma_P - \gamma_m \cos\theta$. The work of separation of the interface γ_w is then given by

$$\gamma_w = \gamma_P + \gamma_m - \gamma_I = \gamma_m(1 + \cos\theta). \quad (7.31)$$

Measurements show that the interfacial energy of TD nickel is low and hence exhibits excellent ductility at room temperature. Specific additions (e.g. Zr to TD nickel, and Co to Ni–Al_2O_3 alloys) are also effective in lowering the interfacial energy, thereby causing the matrix to 'wet' the particle and increase the ductility. Because of their low γ_I, dispersion-hardened materials have superior mechanical properties at high temperatures compared with conventional hardened alloys.

7.4.8 Rupture

If the ductile failure mechanisms outlined above are inhibited then ductile rupture occurs (see Figure 7.36). Specimens deformed in tension ultimately reach a stage of mechanical instability when the deformation is localized either in a neck or in a shear band. With continued straining the cross-section reduces to zero and the specimen ruptures, the strain-to-rupture depending on the amount of strain before and after localization. These strains are influenced by the work-hardening behavior and strain-rate sensitivity. Clearly, rupture is favored when void nucleation and/or growth is inhibited. This will occur if (1) second-phase particles are removed by zone refining or dissolution at high temperatures, (2) the matrix/particle interface is strong and ε_n is high, (3) the stress state minimizes plastic constraint and plane strain conditions (e.g. single crystals and thin sheets), (4) the work-hardening rate and strain-rate sensitivity is high as for superplastic materials (in some superplastic materials voids do not form, but in many others they do, and it is the growth and coalescence processes which are suppressed), and (5) there is stress relief at particles by recovery or dynamic recrystallization. Rupture is observed in most fcc materials, usually associated with dynamic recrystallization.

7.4.9 Voiding and fracture at elevated temperatures

Creep usually takes place above $0.3T_m$ with a rate given by $\dot{\varepsilon} = B\sigma^n$, where B and n are material parameters, as discussed in Chapter 6. Under such conditions ductile failure of a transgranular nature, similar to the ductile failure found commonly at low temperatures, may occur, when voids nucleated at inclusions within the grains grow during creep deformation and coalesce to produce fracture.

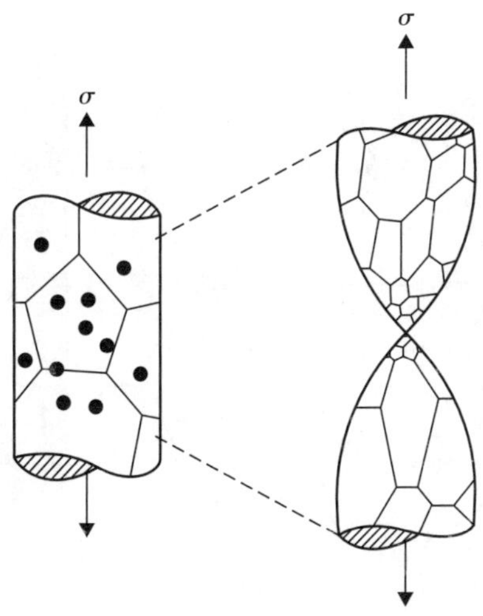

Figure 7.36 *Schematic representation of rupture with dynamic recrystallization (after Ashby* et al., *1979).*

However, because these three processes are occurring at $T \approx 0.3T_m$, local recovery is taking place, and this delays both the onset of void nucleation and void coalescence. More commonly at lower stresses and longer times to fracture, intergranular rather than transgranular fracture is observed. In this situation, grain boundary sliding leads to the formation of either wedge cracks or voids on those boundaries normal to the tensile axis, as shown schematically in Figure 7.37b. This arises because grain boundary sliding produces a higher local strain rate on an inclusion in the boundary than in the body of the grain, i.e. $\dot{\varepsilon}_{local} \approx \dot{\varepsilon}(fd/2r)$, where $f \approx 0.3$ is the fraction of the overall strain due to sliding. The local strain therefore reaches the critical nucleation strain ε_n much earlier than inside the grain.

The time to fracture t_f is observed to be $\propto (1/\dot{\varepsilon}_{ss})$, which confirms that fracture is controlled by power-law creep even though the rounded shape of grain boundary voids indicates that local diffusion must contribute to the growth of the voids. One possibility is that the void nucleation, even in the boundary, occupies a major fraction of the lifetime t_f, but a more likely general explanation is that the nucleated voids or cracks grow by local diffusion controlled by creep in the surrounding grains. Figure 7.37c shows the voids growing by diffusion, but between the voids the material is deforming by power-law creep, since the diffusion fields of neighboring voids do not overlap. Void growth therefore depends on coupled diffusion and power-law creep, with the creep deformation controlling the rate of cavity growth. It is now believed that most intergranular creep fractures are governed by this type of mechanism.

At very low stresses and high temperatures where diffusion is rapid and power-law creep negligible, the diffusion fields of the growing voids overlap. Under these conditions, the grain boundary voids are able to grow entirely by boundary diffusion; void coalescence then leads to fracture by a process of creep cavitation (Figure 7.38). In uniaxial tension the driving force arises from the process of taking atoms from the void surface and depositing them on the face of the grain that is almost perpendicular to the tensile axis, so that the specimen elongates in the direction of the stress and work is done. The vacancy concentration near the tensile boundary is $c_0\exp(\sigma\Omega/\mathbf{k}T)$ and near the void of radius r is $c_0 \exp(2\gamma\Omega/r\mathbf{k}T)$, as discussed previously in Chapter 6, where Ω is the atomic volume and γ the surface energy per unit area of the void. Thus, vacancies flow usually by grain boundary diffusion from the boundaries to the voids when $\sigma \geq 2\gamma/r$, i.e. when the chemical potential difference $(\sigma\Omega - 2\gamma\Omega/r)$ between the two sites is negative. For a void $r \approx 10^{-6}$ m and $\gamma \approx 1\,\mathrm{J\,m^{-2}}$ the minimum stress for

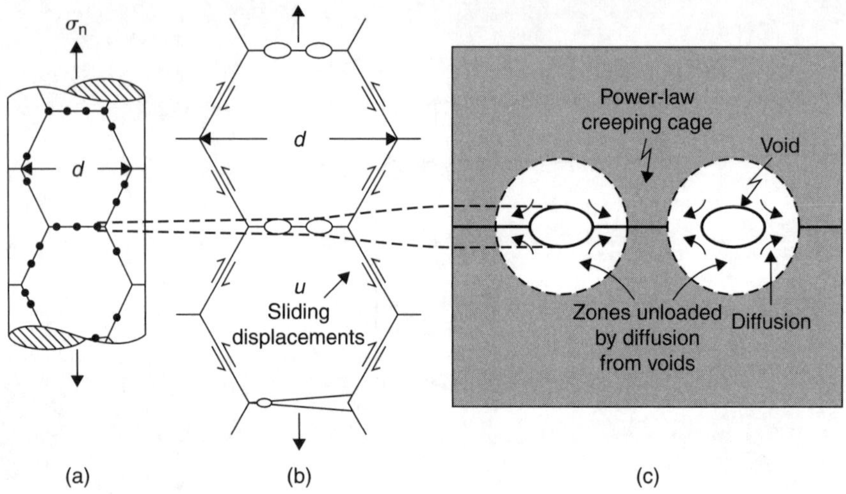

Figure 7.37 *Intergranular, creep-controlled, fracture. Voids nucleated by grain boundary sliding (a and b) and growth by diffusion (c) (after Ashby* et al.*, 1979).*

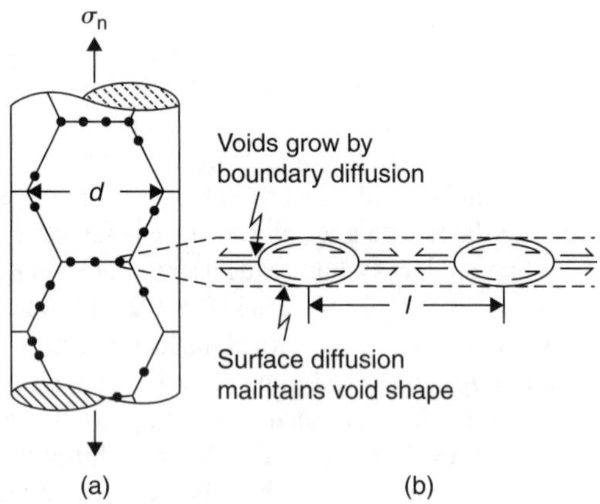

Figure 7.38 *Voids lying on 'tensile' grain boundaries (a) grown by grain boundary diffusion (b) (after Ashby* et al.*, 1979).*

hole growth is $\approx$2 MN m^{-2}. In spite of being pure diffusional controlled growth, the voids may not always maintain their equilibrium near-spherical shape. Rapid surface diffusion is required to keep the balance between growth rate and surface redistribution, and with increasing stress the voids become somewhat flattened.

7.4.10 Fracture mechanism maps

The fracture behavior of a metal or alloy in different stress and temperature regimes can be summarized conveniently by displaying the dominant mechanisms on a fracture mechanism map. Seven mechanisms have been identified, three for brittle behavior including cleavage and intergranular brittle fracture, and four ductile processes. Figure 7.39 shows schematic maps for fcc and bcc materials respectively. Not all the fracture regimes are exhibited by fcc materials, and even some of the ductile processes can be inhibited by altering the metallurgical variables. For example, intergranular creep

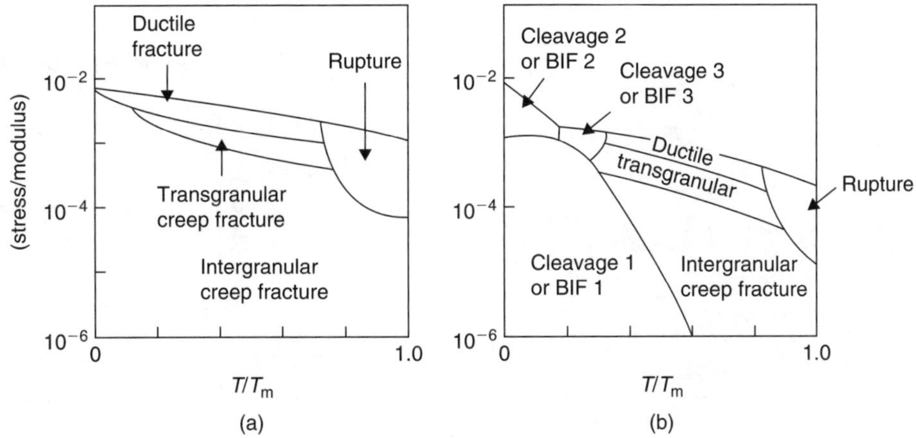

Figure 7.39 *Schematic fracture mechanism maps for fcc (a) and bcc (b) materials (Gandhi and Ashby, 1979).*

fracture is absent in high-purity aluminum but occurs in commercial-purity material and, because the dispersoid suppresses dynamic recrystallization in TD nickel, rupture does not take place, whereas it does in *Nimonic* alloys at temperatures where the γ' and carbides dissolve.

In the bcc metals, brittle behavior is separated into three fields; a brittle failure from a pre-existing crack, well below general yield, is called either cleavage 1 or brittle intergranular fracture BIF1, depending on the fracture path. An almost totally brittle failure from a crack nucleated by slip or twinning, below general yield, is called either cleavage 2 or BIF2, and a cleavage or brittle boundary failure after general yield and with measurable strain to failure is called either cleavage 3 or BIF3. In many cases, mixed transgranular and intergranular fractures are observed, as a result of small changes in impurity content, texture or temperature, which cause the crack to deviate from one path to another; no distinction is then made in the regime between cleavage and BIF. While maps for only two structures are shown in Figure 7.39, it is evident that as the bonding changes from metallic to ionic and covalent the fracture mechanism fields will move from left to right: refractory oxides and silicates, for example, exhibit only the three brittle regimes and intergranular creep fracture.

7.4.11 Crack growth under fatigue conditions

Engineering structures such as bridges, pressure vessels and oil rigs all contain cracks, and it is necessary to assess the safe life of the structure, i.e. the number of stress cycles the structure will sustain before a crack grows to a critical length and propagates catastrophically. The most effective approach to this problem is by the use of fracture mechanics. Under static stress conditions, the state of stress near a crack tip is described by K, the stress intensity factor, but in cyclic loading K varies over a range ΔK ($=K_{\max} - K_{\min}$). The cyclic stress intensity ΔK increases with time at constant load, as shown in Figure 7.40a, because the crack grows. Moreover, for a crack of length a the rate of crack growth ($\mathrm{d}a/\mathrm{d}N$) in μm per cycle varies with ΔK according to the Paris–Erdogan equation,

$$\mathrm{d}a/\mathrm{d}N = C(\Delta K)^m, \tag{7.32}$$

where C and m are constants, with m between 2 and 4. A typical crack growth rate curve is shown in Figure 7.40b and exhibits the expected linear relationship over part of the range. The upper limit corresponds to K_{Ic}, the fracture toughness of the material, and the lower limit of ΔK is called the

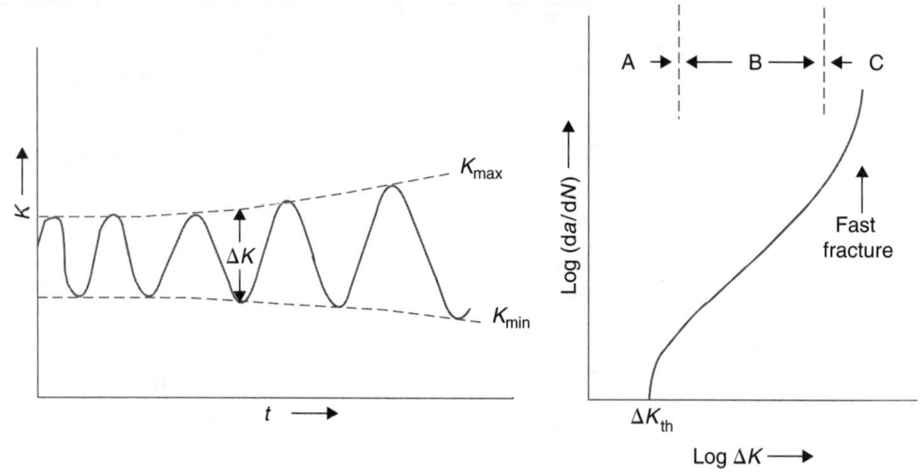

Figure 7.40 *(a) Increase in stress intensity ΔK during fatigue. (b) Variation of crack growth rate with increasing ΔK.*

threshold for crack growth (ΔK_{th}). Clearly, when the stress intensity factor is less than ΔK_{th} the crack will not propagate at that particular stress and temperature, and hence ΔK_{th} is of significance in design criteria. If the initial crack length is a_0 and the critical length a_c, then the number of cycles to catastrophic failure will be given by

$$N_f = \int_{a_0}^{a_c} \mathrm{d}a/C(\Delta K)^m = \int_{a_0}^{a_c} \mathrm{d}a/C[\Delta\sigma\sqrt{(\pi a)}]^m. \tag{7.33}$$

The mean stress level is known to affect the fatigue life and therefore $\mathrm{d}a/\mathrm{d}N$. If the mean stress level is increased for a constant value of ΔK, K_{max} will increase and thus as K_{max} approaches K_{1c} the value of $\mathrm{d}a/\mathrm{d}N$ increases rapidly in practice, despite the constant value of ΔK.

A survey of fatigue fractures indicates there are four general crack growth mechanisms: (1) striation formation, (2) cleavage, (3) void coalescence and (4) intergranular separation; some of these mechanisms have been discussed in Chapter 6. The crack growth behavior shown in Figure 7.40b can be divided into three regimes, which exhibit different fracture mechanisms. In regime A, there is a considerable influence of microstructure, mean stress and environment on the crack growth rate. In regime B, failure generally occurs, particularly in steels, by a transgranular ductile striation mechanism and there is often little influence of microstructure, mean stress or mild environments on crack growth. The degree of plastic constraint which varies with specimen thickness also appears to have little effect. At higher growth rate, exhibited in regime C, the growth rates become extremely sensitive to both microstructure and mean stress, with a change from striation formation to fracture modes normally associated with non-cyclic deformation, including cleavage and intergranular fracture.

7.5 Atomistic modeling of mechanical behavior

One key aim of the physical metallurgist is to understand how properties of materials depend on their microstructures. While the various characterization and microscopy techniques introduced in

Chapter 4 enable one to 'see' material microstructures at different length scales, to link up to material properties, one still needs accurate descriptions of the physics of the microstructures and their effects. For simpler problems such as the long-range elastic properties of straight dislocations (Section 3.3.5), one can resort to analytical solution of the corresponding linear elasticity problem, but more complicated problems, such as the effects of the (non-linear) cores of dislocations or the behavior of groups of curved dislocations, cannot be solved analytically and are therefore best studied by computer simulation techniques. In fact, a long-standing dream of materials scientists has been to develop computational capabilities to predict material behavior directly from 'first principles', i.e. from the quantum mechanical description of electrons, namely, the Schrödinger equation. The corresponding field, termed 'computational materials science' (CMS), has been one of the fastest growing fields in the last two decades, and so this is briefly introduced here, as an epilog to the current discussion on mechanical behavior.

7.5.1 Multiscale modeling

A wide range of techniques have been developed for doing CMS (Table 7.3), including (i) first principles or *ab initio* methods, (ii) molecular statics and dynamics (MS and MD), (iii) Monte Carlo (MC) methods, (iv) phase-field kinetic methods, (v) dislocation dynamics (DD), in addition to the more traditional finite element method (FEM) for continuum simulations. These methods fall into a 'multiscale' perspective, meaning that each method best handles problems within a certain length scale, from nano to micro (first principles, MS, MD), then micro to meso (DD, phase field), and meso to macro (FEA), with the anticipation that the outcome of one scale is fed into the next one up as input, so that one can hope to calculate macroscopic behavior starting from quantum mechanics. Thus, *ab initio* methods are for solving the Schrödinger equation directly, and because of the computational efforts involved, are applicable to nano-sized clusters of atoms of the order of 10^3. MS and MD rely on established 'interatomic' potentials as input to compute, in the case of MD, the trajectories of the atoms by integrating Newton's second law or, in the case of MS, the equilibrium positions of the atoms at zero temperature. MC is a class of methods for computing integrals by random number generation, and is particularly useful in computing thermodynamic functions which are ensemble averages in integral forms. DD is for calculating group dislocation behavior and works by treating the dislocations as 'elastic' curves which interact with one another through some simplified laws. Phase-field methods are based on kinetic laws such as the Langevin equation and are used to predict phase evolution, often taking into account fluctuations. Finally, FEM is for solving continuum problems governed by known constitutive laws, by dividing the domain in question into 'finite elements', so that the problem is converted into one of inverting algebraic matrices, which can be effectively handled by high-speed computers. Thus, in moving up each scale along the multiscale route described above, one loses rigor and accuracy, in exchange for the capability to handle larger systems.

In reality, although tremendous progress concerning the 'multiscale' approach has been made, parallel to the rapid growth of the power of computers during the past few decades, successful examples of accurate prediction of macroscopic, engineering behavior of materials remain extremely rare to date. The problem is twofold. First, while the capabilities at both the nano/micro and the macro ends are better established, those at the meso scale are still not accurate or robust enough. Thus, while FEM procedures for macroplasticity are well established, and the same is true for *ab initio* and MS/MD methods for accurate computation of the stability of phases, properties of clusters of atoms, monolithic defects such as interfaces, dislocations and so on, it remains very difficult to link up these two ends, especially for real materials with usually complicated microstructures. The corresponding methods such as DD or phase-field models are usually rather *ad hoc*, with a significant amount of unknowns hidden in the actual implementation. Secondly, while the spatial scales can be bridged as described above, problems with long temporal scales are much more difficult to

Table 7.3 *Multiscale approach of CMS.*

Length scale	*Size/complexity involved*	*Methods*
Nano	10^2–10^3 atoms	*Ab initio*, molecular statics (MS), molecular dynamics (MD), Monte Carlo (MC)
Micro	$\sim 10^6$ atoms	MS, MD, MC
Meso	Semi-continuum, bridging micro and macro scales	Dislocation dynamics (DD), phase-field kinetic models, etc.
Macro	Continuum	Finite element method (FEM), etc.

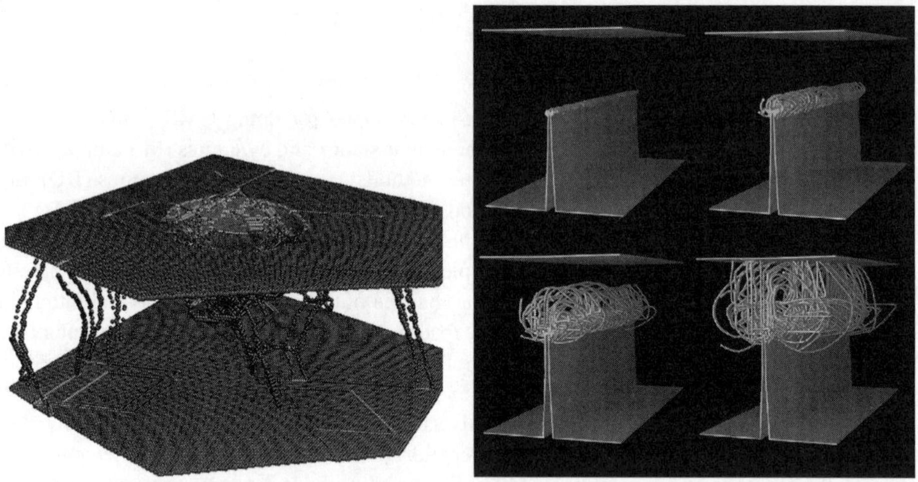

(a) (b)

Figure 7.41 *MD simulations involving millions of atoms. (a) Nanoindentation in aluminum thin film (by J. Li). (b) Crack propagation in an fcc crystal (by F. Abraham). To reveal the generated dislocations, only atoms with high energies are shown.*

simulate. Current supercomputers can at best handle MD problems with a real time scale of only up to ~1 ms, and this drops rapidly to the nanosecond regime as the size of the problem increases beyond $\sim 10^6$ atoms. Figure 7.41 shows two examples of large-scale MD simulations involving $\sim 10^6$ atoms, concerning the evolution of dislocation structures during nanoindentation (Figure 7.41a) and crack propagation (Figure 7.41b) in fcc prototypic crystals. Although impressive to look at, both examples involve time scales in the nanosecond regime and so their relevance to real experimental conditions is unclear. Many important material integrity problems such as creep, corrosion, time-dependent fracture, etc. involve time scales of months and years and, at present, they are simply out of the reach of the multiscale CMS approach.

7.5.2 Atomistic simulations of defects

While further breakthrough in the meso length scale and in bridging time scales is needed in order for the multiscale approach in Table 7.3 to work, CMS has been making important contributions in

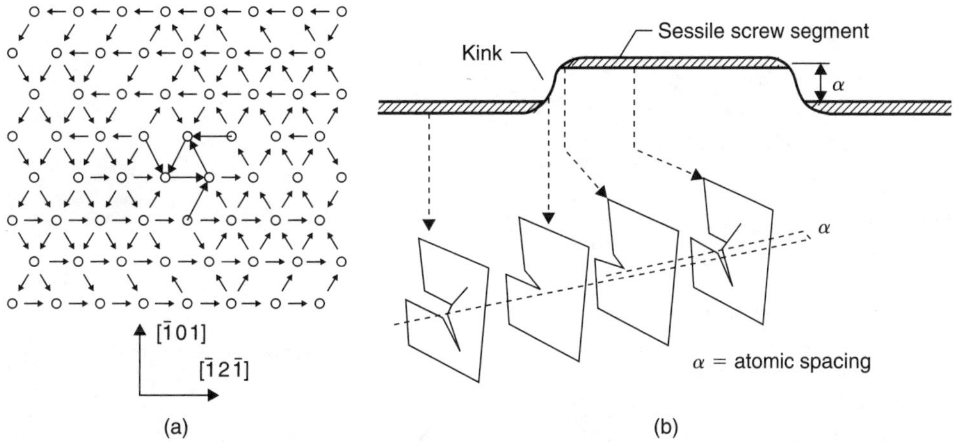

Figure 7.42 *(a) Differential displacement map of 1/2<1 1 1> screw dislocation in iron-like bcc metal, predicted by MS (Wen and Ngan, 2000). (b) Schematic showing the movement of the screw dislocation by the kink-pair mechanism (Li, Ngan and Gumbsch, 2003).*

the understanding of material behaviors within the more traditional models or frameworks described earlier in this book, by providing reliable prediction of defect properties, such as stacking-fault energies or activation energies for mobility, which are crucial input in these models. One area much investigated is dislocation core effects. In crystalline materials the yield properties are principally determined by the mobility of individual dislocations, rather than by their group behavior. While the elastic properties of a dislocation discussed in Section 3.3.5 determine its long-range interaction effects, the mobility of the dislocation is controlled by its 'non-linear' core (see equation (3.10)). The geometry of the core governs whether the dislocation is mobile or not, and is determined by crystal symmetry as well as the energy of the planar fault(s) into which it is dissociated. Predicting the geometry of dislocation cores can be done by MS or MD methods. As an example, Figure 7.42a shows the core of screw dislocations in iron-like bcc materials computed by MS. In the differential displacement map shown in Figure 7.42a, the circles represent atoms viewed along the dislocation axis, and the length of each arrow drawn between a pair of atoms represents the magnitude of the relative displacement between the two atoms, due to the strain field of the dislocation. The Burgers vector content of the dislocation is evidently stored along three $\{1\ 0\ 1\}$ planes separated by 120°, as schematized in Figure 7.42b, and so the ground-state configuration would be rather immobile or sessile. The dislocation can only move forward by forming a pair of kinks and propagating them in opposite directions along the dislocation line, as shown in Figure 7.42b. Since the enantiomer, or mirror-image form, of the configuration shown in Figure 7.41a can also exist, a variety of structurally different kinks can exist in the bcc structure, as shown in Figure 7.43a. However, MS calculations predict that all the other kinks have considerably higher energy than the 'APB' kink, and so only the APB kink is favorable in practice. The only way to involve only the favorable APB kink in the dislocation's motion is to have the dislocation changing path after each atomic step to an intersecting glide plane, as shown in Figure 7.43b. The zigzag motion of screw dislocations, known as 'pencil glide', is long thought to be responsible for the observation of wavy slip traces in deformed bcc metals. Such a peculiar way of dislocation motion can only be interpreted by results from atomistic simulations.

A closely related application is to use atomistic simulations to predict energies of competing stacking faults in complicated structures. For example, dislocations in ordered structures (Section 3.6.5) may in principle dissociate into different faults, such as anti-phase domain boundaries, superlattice

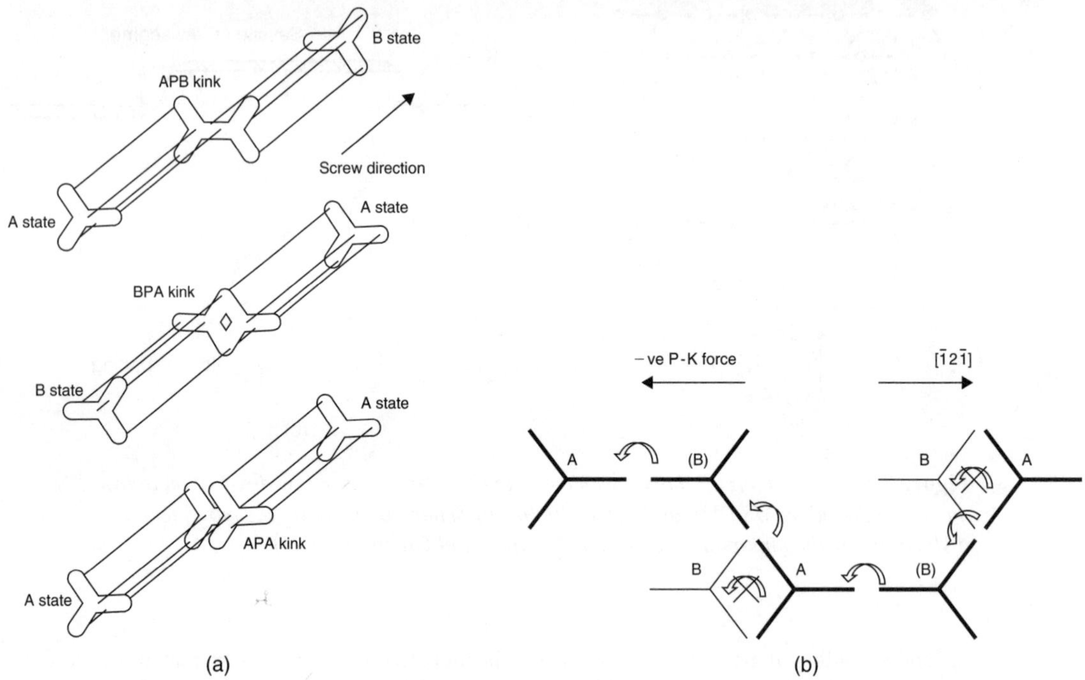

Figure 7.43 *(a) Possible kink configurations in iron-like bcc (Li, Ngan and Gumbsch, 2003). (b) Zigzag motion of screw dislocation to involve only the lowest energy kink (Wen and Ngan, 2000).*

intrinsic stacking faults or complex stacking faults. The understanding of the properties of structural intermetallics has benefitted enormously from accurate *ab initio* calculations of the various fault energies.

Problems

7.1 During age hardening of an aluminum alloy the maximum hardness could be achieved by ageing at 327°C for 10 hours or 280 hours at 227°C. How long would it take at 257°C? If the alloy then contains precipitates 10^{-7} m in diameter separated by 10^{-6} m, estimate the tensile yield stress.

7.2 In certain Al–Cu alloys, enhanced diffusion occurs following quenching from an elevated temperature. From the data given below for an Al–4 wt% Cu alloy, calculate the energies for formation, E_f, and motion, E_m, of vacancies by assuming that the activation energy for diffusion in an annealed alloy equals $E_f + E_m$. For an alloy at 25°C, show graphically how the factor $\Re$, by which quenching increases diffusivity, varies with quenching temperature, T_q, between 25 and 550°C.

Diffusion coefficients of Cu in Al–4 wt% Cu:

T(°C)	D (m^2s^{-1})	Heat treatment
500	3.63×10^{-14}	Annealed
25	8.61×10^{-30}	Annealed
25	2.76×10^{-21}	Quenched from 500°C

7.3 A steel is strengthened by a dispersion of chromium carbides of initial mean radius 0.25 μm. The operating temperature is to be 650°C and material strength is insufficient for particles in excess of 0.75 μm radius. Estimate the component lifetime if the interfacial energy is 500 mJ m^{-2} and coarsening is controlled by carbon diffusion. Solubility of carbon is 0.05 wt% in the alloy steel, $D_o = 2 \times 10^{-5} m^2 s^{-1}$, $Q = 130$ kJ mol^{-1} for carbon diffusion, $\mathbf{R} = 8.31$ J mol^{-1} K^{-1}.

How would the lifetime be affected if Cr diffusion was limiting? Solubility of Cr is 20 wt%, $D_o = 1.6 \times 10^{-4} m^2 s^{-1}$ and $Q = 240$ kJ mol^{-1} for Cr diffusion, Boltzmann's constant $\mathbf{k} = 1.38 \times 10^{-23}$ J K^{-1}.

7.4 The kinetics of the austenite-to-pearlite transformations obey the Avrami relationship $f = 1 - \exp(Kt^n)$ where f is the fraction transformed in time t and K, n are constants. Using the fraction transformed–time data given below, determine the total time required for 95% of the austenite to transform to pearlite:

Fraction transformed (f)	Time (t)
0.2	280 s
0.6	425 s

7.5 In order to strengthen a ferrous alloy it is possible to refine the grain size or the second-phase dispersion. Which route would you choose and why?

7.6 A thick steel plate had a microcrack of 5 mm and a fracture toughness $K_c \sim 40$ MN $m^{-3/2}$. Determine the stress at which fast fracture takes place.

7.7 The crack growth rate equation in a steel component $da/dN = c(\Delta K)^m$ has values of $a = 0.1$ mm, $c = 2 \times 10^{-13}(\text{MN m}^{-2})^{-4} \text{m}^{-1}$, $m = 4$ and $K_c = 54$ MN $m^{-3/2}$. Calculate the lifetime to failure if the component is subjected to an alternating stress from 0 to 180 MN m^{-2} about the mean $\Delta\sigma/2$.

7.8 Explain what is meant by the terms *elastic stress concentration factor*, as relating to a circular hole in a plate subjected to uniform tensile stress, and *stress intensity factor*, as relating to a sharp crack similarly subjected to uniform stress. State the dimensions associated with these terms.

(a) A cylindrical pressure vessel contains large circular openings into which nozzles are welded. If the welding process induces cracks of length 10 mm, lying normal to the hoop stress, indicate how the critical pressure at which the vessel would fracture could be estimated.

(b) A thin steel sheet, of dimensions 1 m × 1 m, contains a central hole of diameter 20 mm and is coated with a brittle lacquer, which fractures at a tensile strain of 0.1%. If Young's modulus for the steel is 210 GPa, explain carefully what you would observe as a stress applied to the top and bottom edges of the plate is increased to 75 MPa. How would the observations differ if the hole were 80 mm in diameter? A second, similar plate contains not a hole, but a central crack, of total length 80 mm oriented normal to the applied stress. If the applied stress is 75 MPa, estimate the distance ahead of the crack tip over which the lacquer would be observed to fracture.

Further reading

Ashby, M. F. and Jones, D. R. H. (1980). *Engineering Materials – An introduction to their properties and applications*. Pergamon.

Bilby, B. A. and Christian, J. W. (1956). *The Mechanism of Phase Transformations in Metals*. Institute of Metals.

Bowles and Barrett, *Progress in Metal Physics*, **3**, 195. Pergamon Press.
Charles, J. A., Greenwood, G. W. and Smith, G. C. (1992). *Future Developments of Metals and Ceramics*. Institute of Materials, London.
Honeycombe, R. W. K. (1981). *Steels, Microstructure and Properties*. Edward Arnold, London.
Kelly, A. and MacMillan, N. H. (1986). *Strong Solids*. Oxford Science Publications, Oxford.
Kelly, A. and Nicholson, R. B. (eds) (1971). *Strengthening Methods in Crystals*. Elsevier, New York.
Knott, J. (1973). *Fundamentals of Fracture Mechanics*. Butterworths, London.
Knott, J. F. and Withey, P. (1993). *Fracture Mechanics, Worked Examples*. Institute of Materials, London.
Pickering, F. B. (1978). *Physical Metallurgy and the Design of Steels*. Applied Science Publishers, London.
Porter, D. A. and Easterling, K. E. (1992). *Phase Transformations in Metals and Alloys*, 2nd edn. Chapman & Hall, London.
Raabe, D. (1998). *Computational Materials Science*. Wiley-VCH, Weinheim.

Chapter 8
Advanced alloys

8.1 Introduction

In this chapter we will outline some of the developments and properties of modern metallic alloys. Crucial to these materials have been the significant developments that have taken place in manufacturing, made possible by a more detailed understanding of the manufacturing process itself and of the behavior of the material during both processing and in-service performance. Casting techniques in particular have advanced much over the past decade and now provide reliable clean material with precision. Process modeling is developing to the extent that the process designer is able to take the microstructural specification for a given composition, which controls the properties of the material, and define an optimum manufacturing route to provide the desired material and performance. Modern alloys therefore depend on the proper integration of alloy composition and structure with processing to produce the desired properties and performance.

8.2 Commercial steels

8.2.1 Plain carbon steels

Carbon is an effective, cheap, hardening element for iron and hence a large tonnage of commercial steels contains very little alloying element. They may be divided conveniently into low-carbon (<0.3% C), medium-carbon (0.3–0.7% C) and high-carbon (0.7–1.7% C) steels. Figure 8.1 shows the effect of carbon on the strength and ductility. The low-carbon steels combine moderate strength

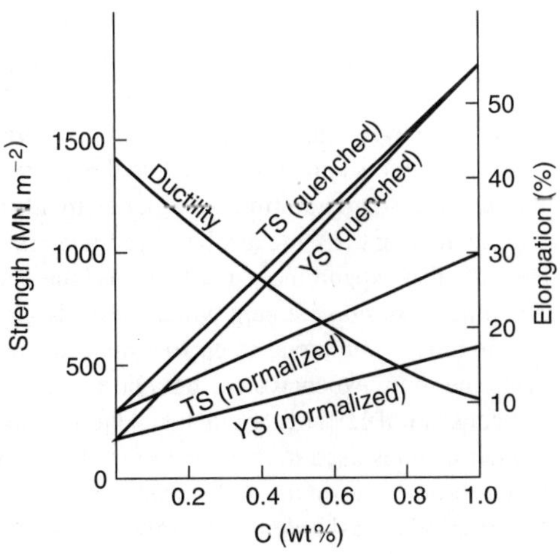

Figure 8.1 *Influence of carbon content on the strength and ductility of steel.*

with excellent ductility and are used extensively for their fabrication properties in the annealed or normalized condition for structural purposes, i.e. bridges, buildings, cars and ships. Even above about 0.2% C, however, the ductility is limiting for deep-drawing operations, and brittle fracture becomes a problem, particularly for welded thick sections. Improved low-carbon steels (<0.2% C) are produced by deoxidizing or 'killing' the steel with Al or Si, or by adding Mn to refine the grain size. It is now more common, however, to add small amounts (<0.1%) of Nb, which reduces the carbon content by forming NbC particles. These particles not only restrict grain growth, but also give rise to strengthening by precipitation hardening within the ferrite grains. Other carbide formers, such as Ti, may be used but because Nb does not deoxidize, it is possible to produce a semi-killed steel ingot which, because of its reduced ingot pipe, gives increased tonnage yield per ingot cast.

Medium-carbon steels are capable of being quenched to form martensite and tempered to develop toughness with good strength. Tempering in higher-temperature regions (i.e. 350–550°C) produces a spheroidized carbide which toughens the steel sufficiently for use as axles, shafts, gears and rails. The process of ausforming can be applied to steels with this carbon content to produce even higher strengths without significantly reducing the ductility. The high-carbon steels are usually quench hardened and lightly tempered at 250°C to develop considerable strength with sufficient ductility for springs, dies and cutting tools. Their limitations stem from their poor hardenability and their rapid softening properties at moderate tempering temperatures.

8.2.2 Alloy steels

In low/medium alloy steels, with total alloying content up to about 5%, the alloy content is governed largely by the hardenability and tempering requirements, although solid solution hardening and carbide formation may also be important. Some of these aspects have already been discussed, the main conclusions being that Mn and Cr increase hardenability and generally retard softening and tempering; Ni strengthens the ferrite and improves hardenability and toughness; Cu behaves similarly but also retards tempering; Co strengthens ferrite and retards softening on tempering; Si retards and reduces the volume change to martensite, and both Mo and V retard tempering and provide secondary hardening.

In larger amounts, alloying elements either open up the austenite phase field, as shown in Figure 8.2a, or close the γ-field (Figure 8.2b). 'Full' metals with atoms like hard spheres (e.g. Mn, Co, Ni) favor close-packed structures and open the γ-field, whereas the stable bcc transition metals (e.g. Ti, V, Cr, Mo) close the field and form what is called a γ-loop. The development of austenitic steels, an important class of ferrous alloys, is dependent on the opening of the γ-phase field. The most common element added to iron to achieve this effect is Ni, as shown in Figure 8.2a. From this diagram the equilibrium phases at lower temperatures for alloys containing 4–40% Ni are ferrite and austenite. In practice, it turns out that it is unnecessary to add the quantity of Ni to reach the γ-phase boundary at room temperature, since small additions of other elements tend to depress the γ/α transformation temperature range, so making the γ metastable at room temperature. Interstitial C and N, which most ferrous alloys contain, also expand the γ-field because there are larger interstices in the fcc than the bcc structure. The other common element which expands the γ-field is Mn. Small amounts (<1%) are usually present in most commercial steels to reduce the harmful effect of FeS. Up to 2% Mn may be added to replace the more expensive Ni, but additions in excess of this concentration have little commercial significance until 12% Mn is reached. Hadfield's steel contains 12–14% Mn and 1% C, is noted for its toughness, and is used in railway points, drilling machines and rock crushers. The steel is water-quenched to produce austenite. The fcc structure has good fracture resistance and, having a low stacking-fault energy, work-hardens very rapidly. During the abrasion and work hardening the hardening is further intensified by a partial strain transformation of the austenite to martensite; this principle is also used in the sheet-forming of stainless steels (see below).

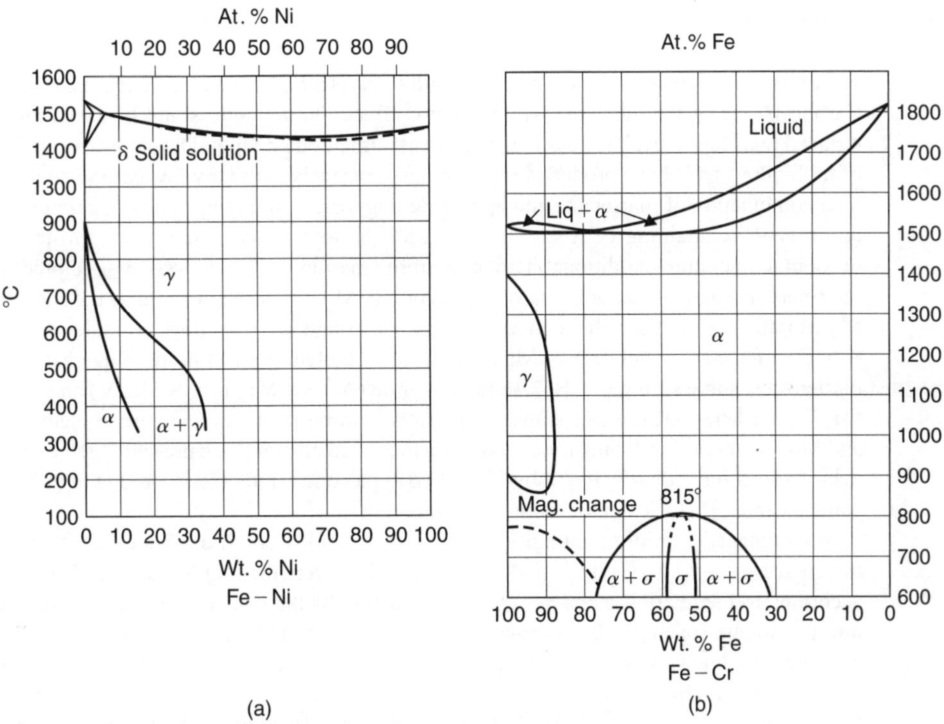

Figure 8.2 *Effect of Ni (a) and Cr (b) on γ-field (from Smithells, 1967).*

To make the austenitic steels resistant to oxidation and corrosion (see Chapter 9) the element Cr is usually added in concentrations greater than 12%. Chromium closes the γ-field, however, and with very low carbon contents single-phase austenite cannot be produced with the stainless (>12%) composition. These alloys form the stainless (ferritic) irons and are easily fabricated for use as furnace components. Increasing the carbon content expands the γ-loop and in the medium-carbon range Cr contents with good stainless qualities (≈15–18%) can be quench-hardened for cutlery purposes, where martensite is required to give a hard, sharp cutting edge. The combination of both Cr and Ni (i.e. 18/8) produces the metastable austenitic stainless steel which is used in chemical plant construction, kitchenware and surgical instruments because of its ductility, toughness and cold-working properties. Metastable austenitic steels have good press-forming properties because the strain-induced transformation to martensite provides an additional strengthening mechanism to work hardening, and moreover counteracts any drawing instability by forming martensite in the locally thinned, heavily deformed regions.

High-strength transformable stainless steels with good weldability to allow fabrication of aircraft and engine components have been developed from the 0.05–0.1% C/12% Cr/stainless steels by secondary hardening addition (1.5–2% Mo; 0.3–0.5% V). Small additions of Ni or Mn (2%) are also added to counteract the ferrite-forming elements Mo and V to make the steel fully austenitic at the high temperatures. Air quenching to give α followed by tempering at 650°C to precipitate Mo_2C produces a steel with high yield strength (0.75 GN m^{-2}), high TS (1.03 GN m^{-2}), and good elongation and impact properties. Even higher strengths can be achieved with stainless (12–16% Cr; 0.05% C) steels which, although austenitic at room temperature (5% Ni, 2% Mn), transform on cooling to −78°C. The steel is easily fabricated at room temperature, cooled to control the transformation and finally tempered at 650–700°C to precipitate Mo_2C.

8.2.3 Maraging steels

A serious limitation in producing high-strength steels is the associated reduction in fracture toughness. Carbon is one of the elements which mostly affects the toughness and hence in alloy steels it is reduced to as low a level as possible, consistent with good strength. Developments in the technology of high-alloy steels have produced high strengths in steels with very low carbon contents (<0.03%) by a combination of martensite and age hardening, called maraging. The maraging steels are based on an Fe–Ni containing between 18% and 25% Ni to produce massive martensite on air cooling to room temperature. Additional hardening of the martensite is achieved by precipitation of various intermetallic compounds, principally Ni_3Mo or $Ni_3(Mo, Ti)$, brought about by the addition of roughly 5% Mo and 8% Co as well as small amounts of Ti and Al; the alloys are solution heat-treated at 815°C and aged at about 485°C. Many substitutional elements can produce age hardening in Fe–Ni martensites, some strong (Ti, Be), some moderate (Al, Nb, Mn, Mo, Si, Ta, V) and other weak (Co, Cu, Zr) hardeners. There can, however, be rather strong interactions between elements such as Co and Mo, in that the hardening produced when these two elements are present together is much greater than if added individually. It is found that A_3B-type compounds are favored at high Ni or (Ni + Co) contents and A_2B Laves phases at lower contents.

In the unaged condition maraging steels have a yield strength of about $0.7\,\mathrm{GN\,m^{-2}}$. On ageing this increases up to $2.0\,\mathrm{GN\,m^{-2}}$ and the precipitation strengthening is due to an Orowan mechanism according to the relation $\sigma = \sigma_0 + (\alpha\mu b/L)$, where σ_0 is the matrix strength, α a constant and L the interprecipitate spacing. The primary precipitation-strengthening effect arises from the (Co + Mo) combination, but Ti plays a double role as a supplementary hardener and a refining agent to tie up residual carbon. The alloys generally have good weldability, resistance to hydrogen embrittlement and stress corrosion, but are used mainly (particularly the 18% Ni alloy) for their excellent combination of high strength and toughness.

8.2.4 High-strength low-alloy (HSLA) steels

The requirement for structural steels to be welded satisfactorily has led to steels with lower C (<0.1%) content. Unfortunately, lowering the C content reduces the strength and this has to be compensated for by refining the grain size. This is difficult to achieve with plain C-steels rolled in the austenite range, but the addition of small amounts of strong carbide-forming elements (e.g. <0.1% Nb) causes the austenite boundaries to be pinned by second-phase particles and fine grain sizes (<10 μm) to be produced by controlled rolling. Nitrides and carbonitrides as well as carbides, predominantly fcc and mutually soluble in each other, may feature as suitable grain refiners in HSLA steels; examples include AlN, Nb(CN), V(CN), (NbV)CN, TiC and Ti(CN). The solubility of these particles in the austenite decreases in the order VC, TiC, NbC while the nitrides, with generally lower solubility, decrease in solubility in the order VN, AlN, TiN and NbN. Because of the low solubility of NbC, Nb is perhaps the most effective grain size controller. However, Al, V and Ti are effective in high-nitrogen steels, Al because it forms only a nitride, V and Ti by forming V(CN) and Ti(CN), which are less soluble in austenite than either VC or TiC.

The major strengthening mechanism in HSLA steels is grain refinement, but the required strength level is usually obtained by additional precipitation strengthening in the ferrite. VC, for example, is more soluble in austenite than NbC, so if V and Nb are used in combination, then on transformation of the austenite to ferrite, NbC provides the grain refinement and VC precipitation strengthening; Figure 8.3 shows a stress–strain curve from a typical HSLA steel.

Solid-solution strengthening of the ferrite is also possible. Phosphorus is normally regarded as deleterious due to grain boundary segregation, but it is a powerful strengthener, second only to carbon. In car construction, where the design pressure is for lighter bodies and energy saving, HSLA

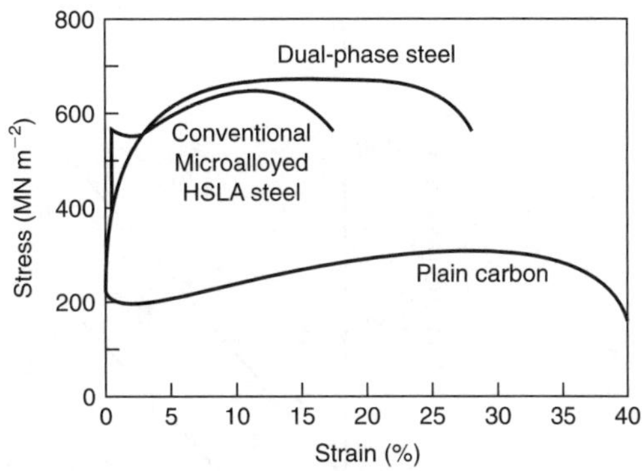

Figure 8.3 *Stress–strain curves for plain carbon, HSLA and dual-phase steels.*

steels, rephosphorized and bake-hardened to increase the strength further, have allowed sheet gauges to be reduced by 10–15% while maintaining dent resistance. The bake hardening arises from the locking of dislocations with interstitials, as discussed in Chapter 6, during the time at the temperature of the paint-baking stage of manufacture.

8.2.5 Dual-phase (DP) steels

Much research into the deformation behavior of speciality steels has been aimed at producing improved strength while maintaining good ductility. The conventional means of strengthening by grain refinement, solid-solution additions (Si, P, Mn) and precipitation hardening by V, Nb or Ti carbides (or carbonitrides) have been extensively explored and a conventionally treated HSLA steel would have a lower yield stress of 550 MN m^{-2}, a TS of 620 MN m^{-2} and a total elongation of about 18%. In recent years an improved strength–ductility relationship has been found for low-carbon, low-alloy steels rapidly cooled from an annealing temperature at which the steel consisted of a mixture of ferrite and austenite. Such steels have a microstructure containing principally low-carbon, fine-grained ferrite intermixed with islands of fine martensite and are known as dual-phase steels. Typical properties of this group of steels would be a TS of 620 MN m^{-2}, a 0.2% offset flow stress of 380 MN m^{-2} and a 3% offset flow stress of 480 MN m^{-2}, with a total elongation $\approx$28%.

The implications of the improvement in mechanical properties are evident from an examination of the nominal stress–strain curves, shown in Figure 8.3. The dual-phase steel exhibits no yield discontinuity but work-hardens rapidly so as to be just as strong as the conventional HSLA steel when both have been deformed by about 5%. In contrast to ferrite–pearlite steels, the work-hardening rate of dual-phase steel increases as the strength increases. The absence of discontinuous yielding in dual-phase steels is an advantage during cold-pressing operations and this feature, combined with the way in which they sustain work hardening to high strains makes them attractive materials for sheet-forming operations. The flow stress and tensile strength of dual-phase steels increase as the volume fraction of hard phase increases with a corresponding decrease in ductility; about 20% volume fraction of martensite produces the optimum properties.

The dual phase is produced by annealing in the $(\alpha + \gamma)$ region, followed by cooling at a rate which ensures that the γ-phase transforms to martensite, although some retained austenite is also usually present, leading to a mixed martensite–austenite (M–A) constituent. To allow air cooling after annealing, microalloying elements are added to low-carbon–manganese–silicon steel, particularly vanadium or molybdenum and chromium. Vanadium in solid solution in the austenite increases the hardenability but the enhanced hardenability is due mainly to the presence of fine carbonitride

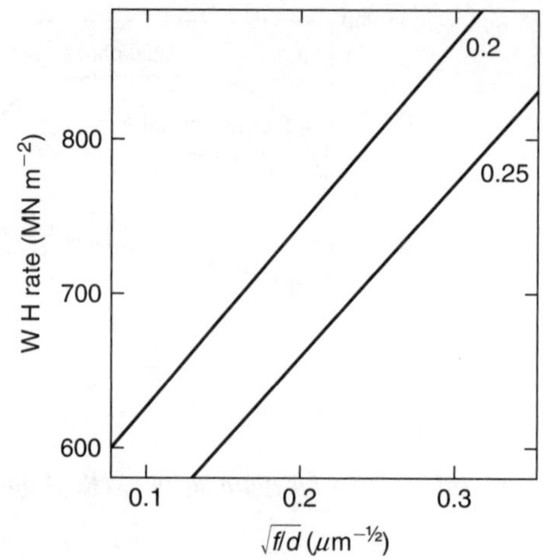

Figure 8.4 *Dependence of work-hardening rate on (volume fraction f/particle size)$^{1/2}$ for a dual-phase steel at strain values of 0.2 and 0.25 (after Balliger and Gladman, 1981).*

precipitates which are unlikely to dissolve in either the austenite or the ferrite at the temperatures employed and thus inhibit the movement of the austenite/ferrite interface during the post-anneal cooling.

The martensite structure found in dual-phase steels is characteristic of plate martensite having internal microtwins. The retained austenite can transform to martensite during straining, thereby contributing to the increased strength and work hardening. Interruption of the cooling, following intercritical annealing, can lead to stabilization of the austenite with an increased strength on subsequent deformation. The ferrite grains ($\approx 5\,\mu m$) adjacent to the martensite islands are generally observed to have a high dislocation density resulting from the volume and shape change associated with the austenite to martensite transformation. Dislocations are also usually evident around retained austenitic islands due to differential contraction of the ferrite and austenite during cooling.

Some deformation models of DP steels assume both phases are ductile and obey the Ludwig relationship, with equal strain in both phases. Measurements by several workers have, however, clearly shown a partitioning of strain between the martensite and ferrite, with the mixed (M–A) constituent exhibiting no strain until deformations well in excess of the maximum uniform strain. Models based on the partitioning of strain predict a linear relationship between yield stress, TS and volume fraction of martensite, but a linear relationship is not sensitive to the model. An alternative approach is to consider the microstructure as approximating to that of a dispersion-strengthened alloy. This would be appropriate when the martensite does not deform and still be a good approximation when the strain difference between the two phases is large. Such a model affords an explanation of the high work-hardening rate, as outlined in Chapter 6, arising from the interaction of the primary dislocations with the dense 'tangle' of dislocations generated in the matrix around the hard phase islands.

Several workers have examined DP steels to determine the effect of size and volume fraction of the hard phase. Figure 8.4 shows the results at two different strain values and confirms the linear relationship between work-hardening rate ($d\sigma/d\varepsilon$) and $(f/d)^{1/2}$ predicted by the dispersion-hardening theory (see Chapter 6). Increasing the hard-phase volume fraction while keeping the island diameter constant increases the work-hardening rate, increases the TS but decreases the elongation. At constant volume fraction of hard phase, decreasing the mean island diameter produces no effect on the tensile strength but increases the work-hardening rate and the maximum uniform elongation (Figure 8.5).

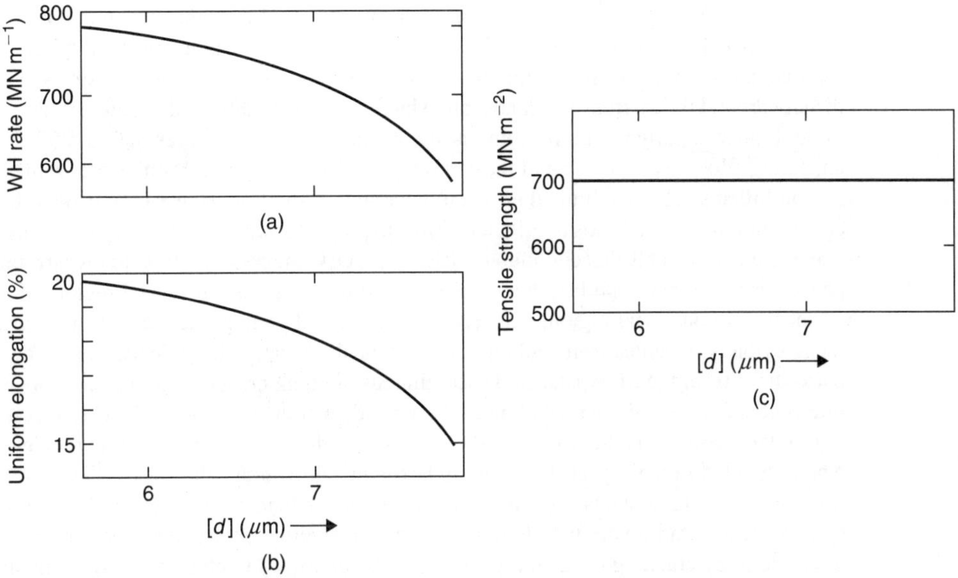

Figure 8.5 *Effect of second-phase particles size* d *at constant volume fraction* f *on: (a) work-hardening rate, (b) elongation and (c) tensile strength (after Balliger and Gladman, 1981).*

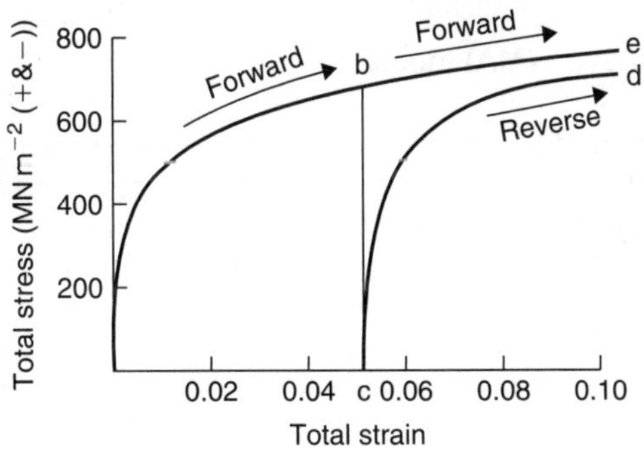

Figure 8.6 *Bauschinger tests for a 0.06% C–1.5% Mn–0.85% Si dual-phase steel (courtesy of D. V. Wilson).*

Thus, the strength is improved by increasing the volume fraction of hard phase, while the work hardening and ductility are improved by reducing the hard-phase island size. Although dual-phase steels contain a complex microstructure it appears from their mechanical behavior that they can be considered as agglomerates of non-deformable hard particles, made up of martensite and/or bainite and/or retained austenite, in a ductile matrix of ferrite. Consistent with the dispersion-strengthened model, the Bauschinger effect, where the flow stress in compression is less than that in tension, is rather large in dual-phase steels, as shown in Figure 8.6, and increases with increase in martensite content up to about 25%. The Bauschinger effect arises from the long-range back-stress exerted by the martensite islands, which adds to the applied stress in reversed straining.

The ferrite grain size can give significant strengthening at small strains, but an increasing proportion of the strength arises from work hardening and this is independent of grain-size changes from about

3 to 30 μm. Solid-solution strengthening of the ferrite (e.g. by silicon) enhances the work-hardening rate; P, Mn and V are also beneficial. The absence of a sharp yield point must imply that the dual-phase steel contains a high density of mobile dislocations. The microstructure exhibits such a dislocation density around the martensite islands, but why these remain unpinned at ambient temperature is still in doubt, particularly as strain ageing is significant on ageing between 423 and 573 K. Intercritical annealing allows a partitioning of the carbon to produce very low carbon ferrite, while aluminum- or silicon-killed steels have limited nitrogen remaining in solution. However, it is doubtful whether the concentration of interstitials is sufficiently low to prevent strain ageing at low temperature; hence it is considered more likely that continuous yielding is due to the residual stress fields surrounding second-phase islands. Two possibilities then arise: (1) yielding can start in several regions at the same time rather than in one local region, which initiates a general yield process catastrophically, and (2) any local region is prevented from yielding catastrophically because the glide band has to overcome a high back-stress from the M–A islands. Discontinuous yielding on ageing at higher temperatures is then interpreted in terms of the relaxation of these residual stresses, followed by classical strain-ageing.

In dual-phase steels the n value ≈ 0.2 gives the high and sustained work-hardening rate required when stretch formability is the limiting factor in fabrication. However, when fracture *per se* is limiting, dual-phase steels probably perform no better than other steels with controlled inclusion content. Tensile failure of dual-phase steels is initiated either by decohesion of the martensite–ferrite interface or by cracking of the martensite islands. Improved fracture behavior is obtained when the martensite islands are unconnected, when the martensite–ferrite interface is free from precipitates to act as stress raisers, and when the hard phase is relatively tough. The optimum martensite content is considered to be 20%, because above this level void formation at hard islands increases markedly.

8.2.6 Mechanically alloyed (MA) steels

For strengthening at high temperatures, dispersion strengthening with oxide, nitride or carbide particles is an attractive possibility. Such dispersion-strengthened materials are usually produced by powder processing, a special form of which is known as mechanical alloying (MA).

Mechanical alloying is a dry powder, high-energy ball-milling process in which the particles of elemental or pre-alloyed powder are continuously welded together and broken apart until a homogeneous mixture of the matrix material and dispersoid is produced. Mechanical alloying is not simply mixing on a fine scale but one in which true alloying occurs. The final product is then consolidated by a combination of high temperature and pressure (i.e. extrusion of canned powder) or hot isostatic pressing (i.e. HIPing). Further processing is by thermomechanical processing (TMP) to produce either (1) fine equiaxed grains for good room-temperature strength and good fatigue strength or (2) coarser, elongated grains to give good high-temperature stress–rupture strength and thermal-fatigue resistance.

Various types of ferrous alloy have been made by mechanical alloying, including 17% Cr–7% Ni–1.2% Al precipitation-hardened austenitic martensitic steel and Fe–25Cr–6Al–2Y. However, the most highly developed material is the 20% Cr–4.5% Al ferritic stainless steel, dispersion strengthened with 0.5% Y_2O_3 (*MA 956*). *MA 956*, which has been made into various fabricated forms, has extremely good high-temperature strength (0.2% proof strength is 200 MN m^{-2} at 600°C, 100 MN m^{-2} at 1000°C and 75 MN m^{-2} at 1200°C).

The high-strength capability is combined with exceptional high-temperature oxidation and corrosion resistance, associated with the formation of an aluminum oxide scale which is an excellent barrier to carbon. No carburization occurs in hydrogen–methane mixtures at 1000°C. Sulfidation resistance is also good.

MA 956 was originally developed for use in sheet form in gas-turbine combustors but, with its combination of high strength up to 1300°C, corrosion resistance and formability, the alloy has found

Table 8.1 *Compositions of some carbon and alloy steels.*

BS designation	*AISI-SAE number**	*% C*	*% Si*	*% Mn*	*% Ni*	*% Cr*	*% others*
040A20	1020	0.18–0.23		0.30–0.50			
080A62	1060	0.60–0.65		0.70–0.90			
070A78	1080	0.75–0.82		0.60–0.80			
150M36	1340	0.32–0.40		1.30–1.70			
212M44	1140	0.40–0.48	0.25 max	1.00–1.40			0.12–0.2 S
527A19	5120	0.17–0.22		0.70–0.90		0.70–0.90	
665H20	4620	0.17–0.23	0.10–0.35	0.35–0.75	1.50–2.00		0.20–0.30 Mo
708M40	4140	0.36–0.44	0.10–0.35	0.70–1.00		0.90–1.20	0.15–0.25 Mo

*Approximately equivalent composition.

many other applications in power stations, including oil and coal burners and swirlers, and fabricated tube assemblies for fluid-bed combustion.

8.2.7 Designation of steels

The original system for labeling wrought steels was devised in 1941 and used En numbers. This system was replaced in 1976 by the British Standard (BS) designation of steels, which uses a six-unit system. Essentially, it enables the code to express composition, steel type and supply requirements. The latter is shown by three letters: M means supply to specified mechanical properties, H supply to hardenability requirements and A supply to chemical analysis requirements. For convenience, steels are divided into types: namely, carbon and carbon–manganese steels, free-cutting steels, high-alloy steels and alloy steels. For example, carbon and carbon–manganese steels are designated by mean of Mn/letter/mean of C. Thus, 080H41 signifies 0.6–1.0 Mn/hardenability requirement/ 0.38–0.45 C. Free-cutting steels are designated by 200–240/letter/mean of C. Thus, 225M44 signifies free-cutting 0.2–0.3 S/mechanical properties requirement/0.4–0.48 C with 1.3–1.7 Mn. High-alloy steels include stainless and valve steels. The designation is similar to the AISI system and is given by 300–499/letters/variants 11–19. Thus, 304S15 (previously known as Type 304 as used by the AISI) signifies 0.06 max. C/8–11 Ni/17.5–19 Cr. Alloy steels are designated by 500–999/letter/mean of C. Thus, 500–519 are Ni steels, 520–539 Cr steels, 630–659 Ni–Cr steels, 700–729 Cr–Mo steels and 800–839 Ni–Cr–Mo steels. Typically, 530M40 signifies 0.36–0.44 C/0.9–1.2 Cr, supplied to mechanical properties.

Tables 8.1 and 8.2 give the compositions of typical carbon, alloy and stainless steels.

8.3 Cast irons

In the iron–carbon system (Chapter 2), carbon is thermodynamically more stable as graphite than cementite. At the low carbon contents of typical steels, graphite is not formed, however, because of the sluggishness of the reaction to graphite. But when the carbon content is increased to that typical of cast irons (2–4% C), either graphite or cementite may separate depending on the cooling rate, chemical (alloy) composition and heat treatment (see Figure 8.7). When the carbon exists as cementite, the cast irons are referred to as white because of the bright fracture produced by this brittle constituent. In gray cast irons the carbon exists as flakes of graphite embedded in the ferrite–pearlite

Table 8.2 *Compositions and properties of some stainless steels.*

Steel	*BS designation*	*% C*	*% Cr*	*% Ni*	*Others*	*Tensile strength ($MN\ m^{-2}$)*	*Yield strength ($MN\ m^{-2}$)*	*% elongation*	*Condition*
Austenitic									
304	304S15	0.08	18–20	8.0–10.5		517	207	30	Annealed
						1276	966	9	Cold worked
316	316S16	0.08	16–18	10–14	2–3% Mo	517	207	30	Annealed
321	321S12	0.08	17–19	9–12	Ti (5 × % C)	586	241	55	Annealed
347	347S17	0.08	17–19	9–13	Nb (10 × % C)	621	241	50	Annealed
Ferritic									
430	430S15	0.12	16–18			448	207	22	Annealed
Martensitic									
416	416S21	0.15	12–14		0.60% Mo	1241	966	18	Quenched and tempered
431	431S29	0.20	15–17	1.25–2.50		1379	1035	16	
Precipitation hardening									
17–4		0.07	16–18	3–5	0.15–0.45% Nb	1310	1172	10	Age hardened
17–7		0.09	16–18	6.5–7.8	0.75–1.25% Al	1655	1586	6	

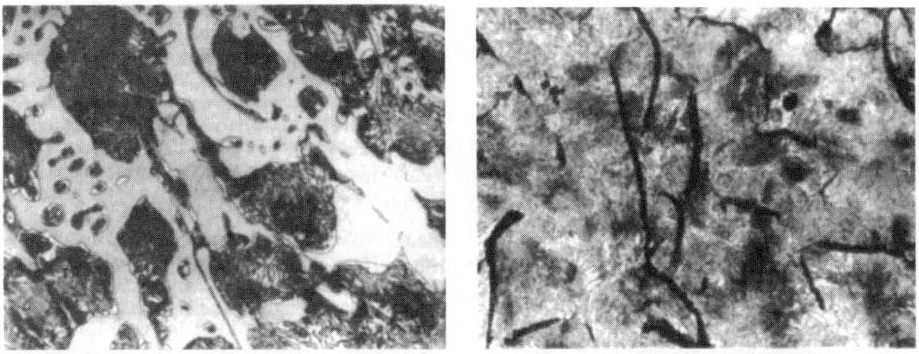

(a) (b)

Figure 8.7 *Microstructure of cast irons: (a) white iron and (b) gray iron (×400). (a) shows cementite (white) and pearlite; (b) shows graphite flakes, some ferrite (white) and a matrix of pearlite.*

matrix and these impart a dull gray appearance to the fracture. When both cementite and graphite are present a 'mottled' iron is produced.

High cooling rates, which tend to stabilize the cementite, and the presence of carbide formers give rise to white irons. The addition of graphite-forming elements (Si, Ni) produces gray irons, even when rapidly cooled if the Si is above 3%. These elements, particularly Si, alter the eutectic composition which may be taken into account by using the carbon equivalent of the cast iron, given by [total % C + (% Si + %P)/3], rather than the true carbon content. Phosphorus is present in most cast irons as a low melting point phosphide eutectic, which improves the fluidity of the iron by lengthening the solidification period; this favors the decomposition of cementite. Gray cast iron is used for a wide variety of applications because of its good strength/cost ratio. It is easily cast into intricate shapes and has good machinability, since the chips break off easily at the graphite flakes. It also has a high damping capacity and hence is used for lathe and other machine frames where vibrations need to be damped out. The limited strength and ductility of gray cast iron may be improved by small additions of the carbide formers (Cr, Mo), which reduce the flake size and refine the pearlite. The main use of white irons is as a starting material for malleable cast iron, in which the cementite in the casting is decomposed by annealing. Such irons contain sufficient Si (<1.3%) to promote the decomposition process during the heat treatment, but not enough to produce graphite flakes during casting. White-heart malleable iron is made by heating the casting in an oxidizing environment (e.g. hematite iron ore at 900°C for 3–5 days). In thin sections the carbon is oxidized to ferrite and, in thick sections, ferrite at the outside gradually changes to graphite clusters in a ferrite–pearlite matrix near the inside. Black-heart malleable iron is made by annealing the white iron in a neutral packing (i.e. iron silicate slag) when the cementite is changed to rosette-shaped graphite nodules in a ferrite matrix. The deleterious cracking effect of the graphite flakes is removed by this process and a cast iron which combines the casting and machinability of gray iron with good strength and ductility, i.e. TS 350 MN m^{-2} and 5–15% elongation is produced. It is therefore used widely in engineering and agriculture, where intricate-shaped articles with good strength are required.

Even better mechanical properties (550 MN m^{-2}) can be achieved in cast irons, without destroying the excellent casting and machining properties, by the production of a spherulitic graphite. The spherulitic nodules are roughly spherical in shape and are composed of a number of graphite crystals, which grow radially from a common nucleus with their basal planes normal to the radial growth axis. This form of growth habit is promoted in an as-cast gray iron by the addition of small amounts of Mg or Ce to the molten metal in the ladle, which changes the interfacial energy between the graphite and the liquid. Good strength, toughness and ductility can thus be obtained in castings that are too thick in section for malleabilizing and can replace steel castings and forgings in certain applications.

Heat treating the ductile cast iron produces austempered ductile iron (ADI), with an excellent combination of strength, fracture toughness and wear resistance for a wide variety of applications in automotive, rail and heavy engineering industries. A typical composition is 3.5–4.0% C, 2–2.5% Si, 0.03–0.06% Mg, 0.015% maximum S and 0.06% maximum P. Alloying elements such as Cu and Ni may be added to enhance the heat treatability. Heat treatment of the cast ductile iron (graphite nodules in a ferrite matrix) consists of austenization at 950°C for 1–3 hours during which the matrix becomes fully austenitic, saturated with carbon as the nodules dissolve. The fully austenized casting is then quenched to around 350°C and austempered at this temperature for 1–3 hours. The austempering temperature is the most important parameter in determining the mechanical properties of ADI; high austempering temperatures (i.e. 350–400°C) result in high ductility and toughness and lower yield and tensile strengths, whereas lower austempering temperatures (250–300°C) result in high yield and tensile strengths, high wear resistance, and lower ductility and toughness. After austempering the casting is cooled to room temperature.

The desired microstructure of ADI is acicular ferrite plus stable, high-carbon austenite, where the presence of Si strongly retards the precipitation of carbides. When the casting is austempered for longer times than that to produce the desired structure, carbides are precipitated in the ferrite to produce bainite. Low austempering temperatures (~250°C) lead to cementite precipitation, but at the higher austempering temperatures (300–400°C) transition carbides are formed, ε-carbides at the lower temperatures and η-carbides at the higher. With long austempering times the high-carbon austenite precipitates χ-carbide at the ferrite–austenite boundaries. The formation of bainite does not result in any catastrophic change in properties but produces a gradual deterioration with increasing time of austempering. Typically, ADI will have a tensile strength of 1200–1500 MN m^{-2}, an elongation of 6–10% and $K_{Ic} \approx 80$ MN m$^{-3/2}$. With longer austempering the elongation drops to a few percent and the K_{Ic} reduces to 40 MN m$^{-3/2}$. The formation of χ-carbide at the ferrite–austenite boundaries must be avoided, since this leads to more brittle fracture. Generally, the strength is related to the volume fraction of austenite and the ferrite spacing. Figure 8.8 shows the microstructure of Si spheroidal graphite (SG) iron and the corresponding fracture mode.

8.4 Superalloys

8.4.1 Basic alloying features

These alloys have been developed for high-temperature service and include iron-, cobalt- and nickel-based materials, although nowadays they are principally nickel based. The production of these alloys over several decades (see Figure 8.9) illustrates the transition in the development of engineering materials from basic alloy composition achievements to a more process-dominated control.

In these alloys γ' (Ni_3Al) and γ^* (Ni_3Nb) are the principal strengtheners by chemical and coherency strain hardening. The ordered γ'-Ni_3Al phase is an equilibrium second phase in both the binary Ni–Al and Ni–Cr–Al systems and a metastable phase in the Ni–Ti and Ni–Cr–Ti systems, with close matching of the γ' and the fcc matrix. The two phases have very similar lattice parameters ($\lesssim$0.25%, depending on composition) and the coherency (interfacial energy $\gamma_I \approx 10$–20 mJ m^2) confers a very low coarsening rate on the precipitate so that the alloy overages extremely slowly even at $0.7T_m$. In alloys containing Nb, a metastable Ni_3Nb phase occurs but, although ordered and coherent, it is less stable than γ' at high temperatures.

Another source of strengthening is due to solid-solution hardening; Cr is a major element, Co may be added up to 20% and Mo, W and Ta up to a total of 15%. These elements also dissolve in γ' so

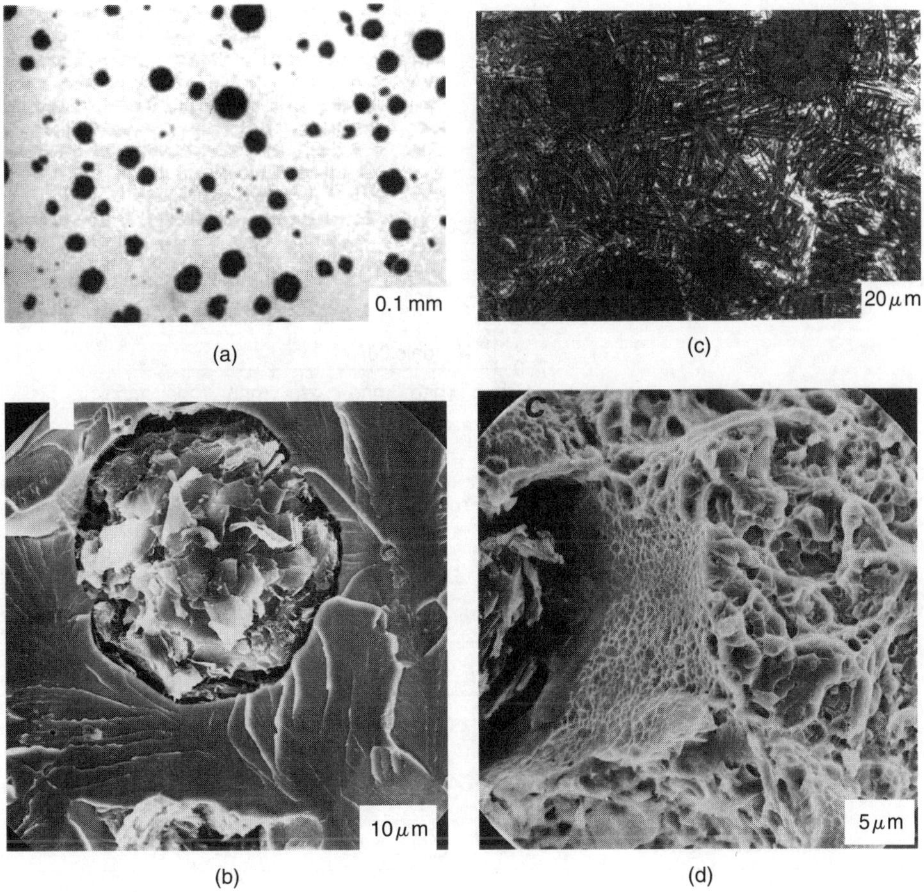

Figure 8.8 *Microstructure and fracture mode of silicon spheroidal graphite (SG) iron: (a, b) as-cast and (c, d) austempered at 350°C for 1 h (L. Sidjanin and R. E. Smallman, 1992; courtesy of Institute of Materials, Minerals and Mining).*

that the hardening effect may be twofold. Additions of Cr, like Co, also increase the γ' solvus and lower the stacking-fault energy.

In high-temperature service, the properties of the grain boundaries are as important as the strengthening by γ' within the grains. Grain boundary strengthening is produced mainly by precipitation of chromium and refractory metal carbides; small additions of Zr and B improve the morphology and stability of these carbides. Optimum properties are developed by multistage heat treatment; the intermediate stages produce the desired grain boundary microstructure of carbide particles enveloped in a film of γ' and the other stages produce two size ranges of γ' for the best combination of strength at both intermediate and high temperatures. Table 8.3 indicates the effect of the different alloying elements.

Some of the nickel-based alloys have a tendency to form an embrittling σ-phase (based on the composition FeCr) after long-term in-service applications, when composition changes occur removing σ-resisting elements such as Ni and enhancing σ-promoting elements such as Cr, Mo or W. This tendency is predicted in alloy design by a technique known as Phacomp (phase computation), based on Pauling's model of hybridization of 3*d* electrons in transition metals. While a fraction of the 3*d* orbitals hybridize with *p* and *s* orbitals to create the metallic bond, the remainder forms non-bonding

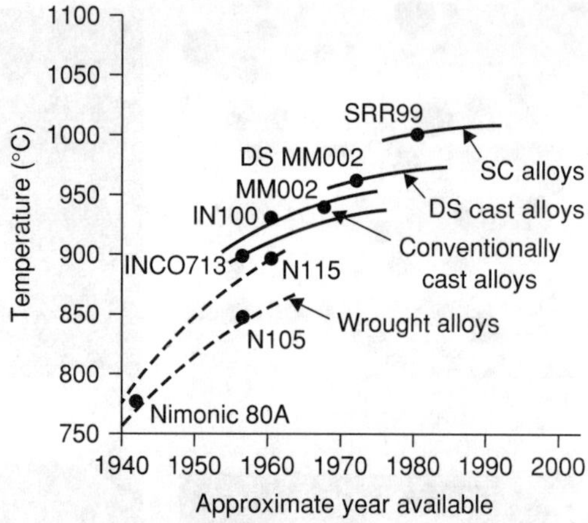

Figure 8.9 *Increases in temperature capability for turbine blade alloys, based on creep rupture in 1000 h at 150 MN m^{-2} (from Driver, 1985, by permission of Institute of Materials, Minerals and Mining).*

Table 8.3 *Influence of various alloying additions in superalloys.*

Influence	*Cr*	*Al*	*Ti*	*Co*	*Mo*	*W*	*B*	*Zr*	C	*Nb*	*Hf*	*Ta*
Matrix strengthening	√			√	√	√						
γ' formers		√	√							√		√
Carbide formers	√		√		√	√				√	√	√
Grain boundary strengthening							√	√	√		√	
Oxide scale formers	√	√										

orbitals which partly fill the electron holes in the d-shell, increasing through the transition series to give electron hole numbers N_v for Cr (4.66), Mn (3.66), Fe (3.66), Co (1.71) and Ni (0.66). Computation shows that the γ/σ phase relation depends on the average hole number $\overline{N}_V$, given by

$$\overline{N}_V = \sum_{i=1}^{n} m_i (N_v)_i,$$

where m_i is the atomic fraction of the ith element of electron hole number N_v and n is the number of elements in the alloy. The limit of γ-phase stability is reached at $\overline{N}_v \approx 2.5$.

8.4.2 Nickel-based superalloy development

A major application of superalloys is in turbine materials, jet engines, both disk and blades. Initial disk alloys were *Inco 718* and *Inco 901* produced by conventional casting ingot, forged billet and forged disk route. These alloys were developed from austenitic steels, which are still used in industrial turbines, but were later replaced by *Waspaloy* and *Astroloy* as stress and temperature requirements increased. These alloys (Table 8.4) were turbine blade alloys with a suitably modified heat treatment for disks. However, blade material is designed for creep, whereas disk material requires tensile strength

Table 8.4 *Composition of some Ni-based Superalloys*

	Composition (wt%)									
	Ni-Alloy	*Cr*	*Co*	*Ti*	*Al*	*Mo*	*C*	*Zr*	*B*	*Others*
Waspaloy	Balance	19.5	13.5	3.0	1.3	–	0.08	0.06	0.006	
Astroloy	Balance	15.0	17.0	3.5	4.0	5.25	0.06	–	0.030	
Inco 718	Balance	19.0	3.0	0.9	0.6	3.0	0.04	–	–	Fe 20 Nb 5.2
Inco 901	42	13.0	–	3.0	0.3	5.7	0.04	–	–	Bal. Fe

coupled with low cycle fatigue life to cope with the stress changes in the flight cycle. To meet these requirements *Waspaloy* (Table 8.4) was thermomechanically processed (TMP) to give a fine grain size and a 40% increase in tensile strength over the corresponding blade material, but at the expense of creep life. Similar improvements for disks have been produced in *Inco 901* by TMP. More highly alloyed nickel-based disks suffer from excessive ingot segregation, which makes grain size control difficult. Further development led to alloys produced by powder processing by gas atomization of a molten stream of metal in an inert argon atmosphere and consolidating the resultant powder by HIPing to near-net shape. Such products are limited in stress application because of inclusions in the powder and, hence, to realize the maximum advantage of this process it is necessary to produce 'superclean' material by electron beam or plasma melting.

Improvements in turbine materials were initially developed by increasing the volume fraction of γ' in changing *Nimonic 80A* up to *Nimonic 115*. Unfortunately, increasing the (Ti + Al) content lowers the melting point, thereby narrowing the forging range, which makes processing more difficult. Improved high-temperature oxidation and hot corrosion performance has led to the introduction of aluminide and overlay coatings, and subsequently the development of *IN 738* and *IN 939* with much improved hot-corrosion resistance.

Further improvements in superalloys have depended on alternative manufacturing routes, particularly using modern casting technology. Vacuum casting was first used to retain high Ti and Al contents without oxidation loss. With 9–11% (Ti + Al), a 70% volume fraction of γ' has been produced in *IN 100 (Nimocast PK 24)* which does not require supplementary solid solution strengtheners and therefore gives a saving in density.[1]

Additions of high melting point elements such as W extend the high-temperature capabilities at the expense of density. *M200* contains 12% W and 1% Nb but has limited ductility around 760°C, which can be improved by additions of hafnium. The significant improvement in ductility and reduced porosity produced by Hf has led to its addition to other alloys (e.g. *Mar 001* (*IN 100* + Hf) and *Mar 004* (*IN 713* + Hf) and *M002*, which contains 10% W, 2.5% Ta and 1.5% Hf).

Creep failures are known to initiate at transverse grain boundaries and, hence, it is reasonable to aim to eliminate them in the turbine blade to gain further improvement in performance. Technologically this was achieved by directional solidified (DS) castings with columnar grains aligned along the growth direction with no grain boundaries normal to that direction. By incorporating a geometric constriction into the mold or by the use of a seed crystal it has been possible to eliminate grain boundaries entirely and grow the blade as one single crystal (see Chapter 2).

The elimination of grain boundaries immediately removes the necessity for adding grain-boundary strengthening elements, such as C, B, Zr and Hf, to the superalloy. The removal of such elements raises the melting point and allows a higher solution heat-treatment temperature with consequent improvement in chemical homogeneity and more uniform distribution of γ' precipitates. Particularly

[1] A Pratt and Whitney version of *IN 100* (*B1900*) replaced Ti with Ta to improve the castability.

important, however, is the control of the growth direction along the [1 0 0] direction. The [1 0 0] alignment along the axis of the blade gives rise to an intrinsic high creep which enables thermal stresses caused by temperature gradients across the blade to be minimized.

Single-crystal blades have now been used successfully for both civil and military engines, *SSR99* replacing *Mar M002* but with improved tensile, creep and fatigue properties and a lower-density alloy *RR2000* replacing *IN 100*.

8.4.3 Dispersion-hardened superalloys

All γ'-hardened alloys experience a reduction in strength at elevated temperatures because of the solution of γ' precipitate. To produce improved high-temperature strength, alloys hardened by oxides (ODS or oxide dispersion strengthened), particularly thoria, have been developed. TD nickel (i.e. thoria dispersion-strengthened nickel or nickel with 2% ThO_2) and *TD-Nichrome* (i.e. nickel–20% Cr–2% ThO_2) are produced by mixing thoria sols with nickel-containing solutions to ensure a good dispersion. Drying the oxide powder mixture, followed by a hydrogen reduction process, produces a fine composite nickel–thoria powder. Compaction and controlled TMP during extrusion and rolling develops the structure and strength. These materials have excellent high-temperature creep resistance but poor low-temperature properties, which precludes aero-engine applications, although some other applications have been found. It is not possible to add the γ'-forming elements via the above process to produce lower-temperature strengthening because aluminum and titanium oxide are not reduced by hydrogen.

Mechanical alloying (MA) is a dry powder process and overcomes this problem. MA, a high-energy ball-milling process, produces a homogeneous mixture of the matrix material and dispersoid. The final product is then produced by TMP with the grain structure on recrystallization elongated in the direction of working.

These materials are generally anisotropic in their properties, but a range of MA materials is emerging with more acceptable dispersoids. *MA754* is a *Nichrome* matrix with Y_2O_3 dispersoids and *MA6000* is a more complex nickel superalloy-based yttria material (Ni–12Cr–2Mo–4W–2Ta–4.5Al–2.5Ti and 1.1Y_2O_3) suitable for turbine blades, having the creep characteristics of the γ'-strengthened materials at low temperatures combined with the advantages of dispersion strengthening at high temperature (see Figure 8.10). TMP limits the amount of γ' strengthening that can be introduced and, while *MA6000* can be run at higher temperatures than conventional nickel-based superalloys, application is then limited to areas where the loadings can be kept low.

8.5 Titanium alloys

8.5.1 Basic alloying and heat-treatment features

Since the emergence of titanium as a 'wonder metal' in the 1950s, the titanium industry has developed a wide range of alloys with different compositions (see Figure 8.11).[2] These alloys rely on the high strength/weight ratio, good resistance to corrosion, combined low thermal conductivity and thermal expansion of titanium, properties which make it attractive for aerospace applications in both engine and airframe components.

Titanium exists in the cph α form up to 882°C and then as bcc β to its melting point. Alloying additions change the temperature at which the α to β transition takes place, solutes that raise the transus are termed α-stabilizers and those that lower the β-transus temperature are termed β-stabilizers

[2] The Larson–Miller parameter ϕ is given by $\phi = T(A + \log_{10}t)$, where T is the temperature in degrees Kelvin, t the time in hours and A a constant, and defines the conditions to produce a given amount of plastic strain (e.g. 0.2%).

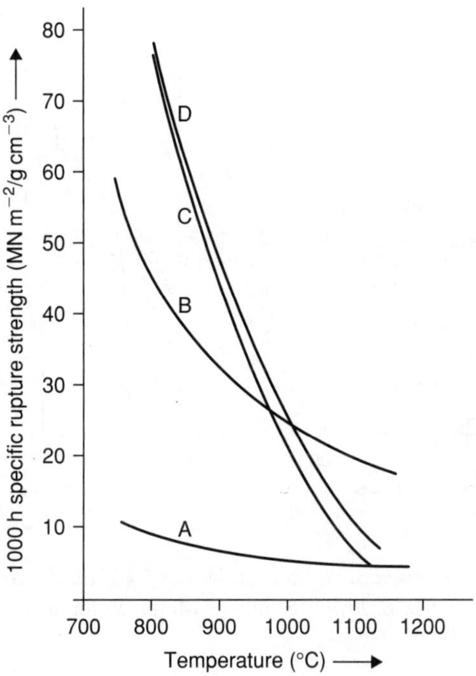

Figure 8.10 *Comparison of* MA6000 *with other high-strength nickel alloys: (A) TD-Nickel, (B)* MA6000*, (C) directionally solidified* Mar M200 *and (D) single-crystal* PWA 454.

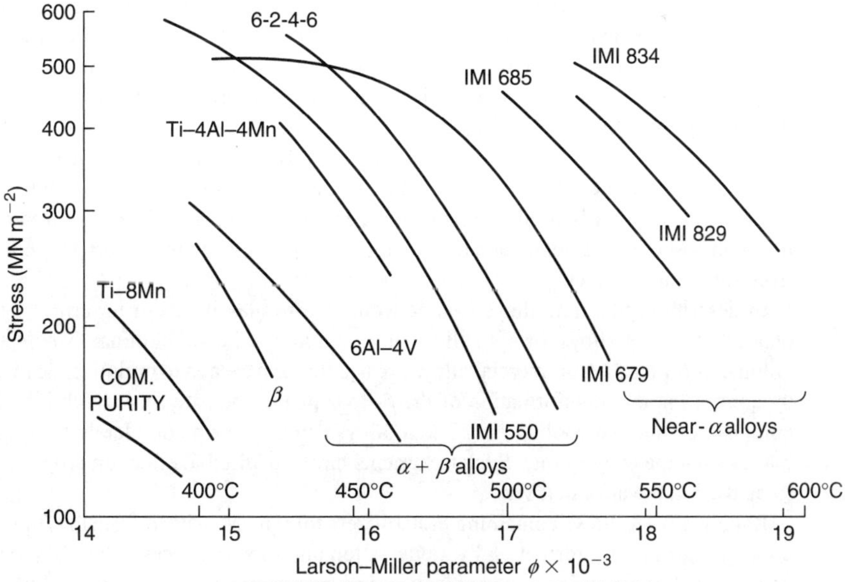

Figure 8.11 *Plot of stress versus Larson–Miller parameter ϕ for a range of titanium alloys.*

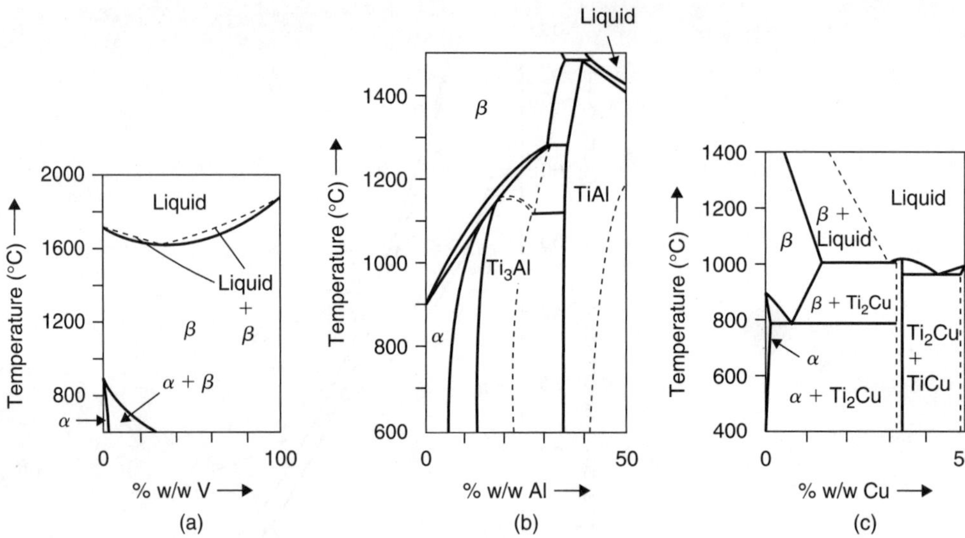

Figure 8.12 *Representative phase diagrams for Ti alloys. (a) Ti–V. (b) Ti–Al. (c) Ti–Cu (after E. A. Brandes and G. B. Brook,* Smithells Metals Reference Book, *1998).*

(Figure 8.12). The predominant α-stabilizer is aluminum. It is also an effective α-strengthening element at ambient and elevated temperatures up to 550°C, and thus a major constituent of most commercial alloys. The low density of aluminum is an important additional advantage. α-Phase strengthening is also achieved by additions of tin and zirconium. These metals exhibit extensive solubility in both α and β titanium but have little influence on the β-transus and are thus regarded as neutral additions. β-Stabilizers may be either β-isomorphous (i.e. have the bcc structure like β–Ti) or β-eutectoid elements. β-isomorphous elements have a limited α-solubility and are completely soluble in β-titanium, typical additions being molybdenum, vanadium and niobium. In contrast, β-eutectoid elements have a restricted solubility in β-titanium and form intermetallic compounds by eutectoid decomposition of the β-phase. In some alloy systems containing β-eutectoid elements, such as silicon or copper, the compound formation (i.e. respectively Ti_5Si_3 and Ti_2Cu) leads to an improvement in mechanical properties. Titanium will also take interstitial solutes in solid solution, hydrogen being a β-stabilizer while carbon, nitrogen and oxygen are strong α-stabilizers. To minimize gas in Ti leads to a high cost of manufacture and heat treatment requires vacuum or inert gas conditions and freedom from refractory contact.

In describing titanium alloys it is conventional to classify them in terms of the microstructural phase (α alloys, β alloys, $(\alpha + \beta)$ alloys or near-α alloys, i.e. predominantly α-phase but with a small volume of β-phase). Commercial alloys are usually heat treated to optimize the mechanical properties by controlling the transformation of the β- to α phase, the extent of which is governed by the alloy composition and the cooling rate. The α alloys can transform completely from the β- to α phase no matter what the cooling rate. Such treatments have a negligible effect on properties and α alloys tend to be used in the annealed state.

Rapidly cooled alloys containing β-stabilizers form martensitic α from the β-phase, whereas slower cooling rates favor α formation by a nucleation and growth process. Several morphologies of α can be produced by controlling the nucleation and growth mechanism; slow cooling, for example, tends to produce similarly aligned α platelets in colonies, combined with primary α at the grain boundaries. Faster cooling and higher α-stabilizer contents result in a basket-weave microstructure. Metastable-β,

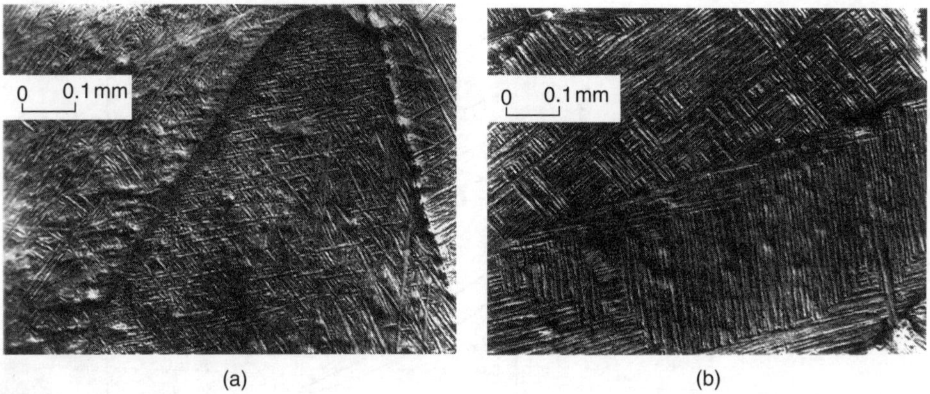

Figure 8.13 *Microstructure of near-α titanium alloy* (IMI 829) *initially β heat-treated at 1050°C for 1 h. (a) Oil quenched. (b) Air cooled (after Woodfield, Loretto and Smallman, 1988).*

when aged, precipitates fine α, giving increased strength. The less stable the β, the more α can be precipitated and hence the higher the strength attained.

β-Stabilizing elements improve strength by strengthening the β-phase. The microstructure consists of primary α combined with the β-phase, which can be strengthened by an ageing treatment to precipitate acicular-α. Further strengthening is achieved by the limited solubility of the β-stabilizing element. Generally, these alloys have poor ductility properties.

The most important alloys contain both α- and β-stabilizers which, after working and annealing, give good strength and fabrication properties. For good creep strength an α-titanium base, strengthened as much as possible by solute elements, is required. To meet this requirement the near-α alloys have been developed. These alloys combine the high α stability with sufficient β-stabilizer to give adequate strength. By β heat treating, the $(\alpha + \beta)$ microstructure changes to a totally transformed β structure containing basket-weave α. These alloys have good creep resistance and reasonable room-temperature properties. The basket-weave morphology is effective in inhibiting crack growth, and near-α alloys exhibit lower crack propagation rates than the $\alpha + \beta$ microstructures. Most of the commercial alloys which have been developed recently are of this type. The major factor influencing the post-forging microstructure is the cooling rate; an oil quench results in a basket-weave structure (see Figure 8.13a) and an air cool gives a typically aligned microstructure (see Figure 8.13b). The alloy is usually stress-relieved by annealing for 2 hours at 625°C or above.

8.5.2 Commercial titanium alloys

α alloys transform entirely to α on cooling from the β-phase, regardless of cooling rate. Commercially pure titanium with a nominal oxygen content of 500, 1000, 1550 and 2700 ppm respectively gives tensile properties which range from 450 to 640 MN m^{-2} (see Figure 8.14). These are used mainly in sheet form. Solid-solution strengthening by aluminum, tin or zirconium increases tensile and creep strength and *IMI 317* (Ti–5Al–2.5Sn) is typical.

Increasing the α-stabilizing composition increases creep strength but makes fabrication more difficult and can lead to embrittlement during prolonged exposure at temperature in service due to the formation of the coherent ordered phase α_2 (Ti_3Al). To avoid this it was empirically established that the aluminum equivalent Al* must be no greater than 9, where Al* (in wt%) is given by:

$$\text{Al}^* = \text{Al} + \frac{\text{Sn}}{3} + \frac{\text{Zr}}{6} + (10 \times \text{O}_2).$$

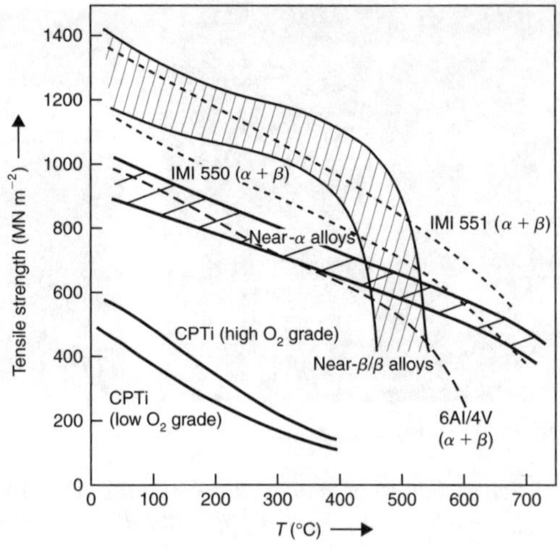

Figure 8.14 *Variation of tensile strength with temperature for a range of commercial titanium alloys.*

IMI 317 is difficult to fabricate and is often replaced by an *IMI 230*, which is an α-phase alloy containing the precipitation-hardening phase Ti_2Cu; it can be fabricated and welded, and has good strength and ductility up to 400°C.

β alloys contain enough β-stabilizing elements to maintain the bcc β-phase to room temperature. Unfortunately, bcc β–Ti alloys are prone to embrittlement. The binary alloys, titanium with Fe, Cr, Mn, Nb, Mo, Cr or V, all precipitate the embrittling ω-phase. More complex β–Ti alloys containing Cr also suffer embrittlement from $TiCr_2$. More stable alloys have been developed (e.g. (Ti–11.5Mo–6Zr–4.5Sn)) but are little used. However, β-phase alloys, such as Ti–10V–2Fe–3Al, have potential as airframe construction materials offering high strength (1250 MN m^{-2}) and toughness (45 MN $m^{-3/2}$) in relatively thick cross-sections (90 mm) and superior hot-working characteristics which are attractive in expensive forging operations.

Highly stabilized β alloys have been developed for burn resistance. Ti alloys can burn when they rub together in gas turbine engines and steel stators are commonly used to separate the blades in compressors. Replacement of steel components by Ti alloys is a considerable weight saving. Pratt and Whitney developed a composition (wt%) Ti–35V–15Cr where the burn resistance is associated with the high V-content which produces a volatile V_2O_5 to shield the component from oxygen. A cheaper version with lower V-content has been developed by Rolls-Royce with composition Ti–25V–15Cr–2Al–0.2C. This is made from a low-cost V/Cr/Al master alloy readily available in the steel industry. Unfortunately, the introduction of Al results in a B2 structure which embrittles the alloy at the operating temperature of 550°C when α is precipitated at the grain boundaries. Alpha is very much influenced by the presence of oxygen which is an α-stabilizer. The addition of carbon forms Ti(CO) carbides, which increases the oxygen concentration in the β-matrix and prevents oxygen segregation at the grain boundaries, with a corresponding reduction in grain boundary precipitation of α.

(α + β) alloys are probably the most widely used titanium alloys and contain alloying additions which strengthen both phases. These alloys are thermomechanically processed to control the size, shape and distribution of both α and β. The most versatile (α + β) alloy is *IMI 318*, which contains 6% Al and 4% V; it can be used at temperatures up to 350°C and has good forging and fabrication properties. It initially replaced steel as a disk material in jet engines, leading to 20% weight saving. Another important (α + β) alloy is *IMI 550* (Ti–6Al–2Sn–4Mo–0.5Si), which has higher strength and good creep resistance up to 400°C. (α + β) alloys remain the principal materials for fan disks and blades, and for low- and intermediate-pressure compressor disks and blades of current gas turbine engines. ($\alpha + \beta$) alloys with extra low interstitial (ELI) content are attractive as 'damage-tolerant' materials for critical airframe components. Ti–6Al–4V with low oxygen has a tensile strength 8% lower than the standard alloy but, more importantly, the minimum fracture toughness is 60 MN m$^{-3/2}$.

Near-α alloys have increased the strength and the volume of the more creep-resistant α-phase at the expense of the bcc β-phase, which imports good low-temperature strength and forgeability. *IMI 685* was the first titanium alloy to operate above 500°C. It contains Ti–6Al–5Zr–0.5Mo–0.2Si with Al and Zr instead of Sn as α-stabilizers, reduced Mo, the β-stabilizer, to minimize β at the α-needles and Si to improve creep resistance. These alloys are worked and heat treated in the β-range, have a tensile strength of about 1000 MN m^{-2} and give less than 0.1% creep strain in 100 h under a stress of 310 MN m^{-2} at 520°C. *IMI 829* (Ti–5.6Al–3.5Sn–3Zr–1Nb–0.25Mo–0.3Si) has been derived from *IMI 685* by replacing some of the Zr with the more potent strengthener Sn. It is β heat treated and has sufficient higher-temperature capability to be used in the hotter regions of engines. *IMI 834* (Ti–5.8Al–4Sn–3.5Zr–0.7Nb–0.5Mo–0.35Si–0.06C) has been developed for use up to 600°C and combines the high fracture toughness and crack propagation resistance of a transformed β-structure with the typical equiaxed structure of the $\alpha + \beta$ alloys, providing good fatigue resistance and ductility. The small addition of carbon allows a controlled high α/β heat treatment. Hot working is carried out in the $\alpha + \beta$ field and heat treatment involves solution treatment for 2 hours at 1025°C, consistent with about 15% primary α, followed by oil quenching prior to ageing for 2 hours at 700°C, then air cooling. With such good high-temperature properties the alloy is being specified for engine compressor applications.

8.5.3 Processing of titanium alloys

Some of the titanium alloys have excellent superplastic-forming characteristics and fabricating manufacturers have taken advantage of these properties in developing new processing technologies. Fine-grained thin sheets of *IMI 318* have been superplastically deformed at 900°C under slow strain-rate conditions to produce a variety of complex parts. In conjunction with diffusion bonding, a weight saving of about one-third has been achieved in, for example, the wing access panels of the A320 Airbus. Near-α alloys have also been shown to exhibit superplastic behavior.

Casting technology has also been developed. Investment casting of *IMI 318* is widely used but there is now increasing interest in the use of high-temperature alloy castings as weight-saving alternatives to steel and nickel alloys above 500°C. The introduction of HIPing for titanium castings has widened their potential application. For *IMI 829*, the best properties are obtained in the α/β HIP and solution-treated and aged condition, generally meeting the specification for the alloy in the wrought condition. Rapid solidification processing may also offer possibilities by incorporating Er_2O_3 and other rare-earth oxide dispersions into the titanium matrix. At present, the control of these dispersoids remains a problem.

To develop higher-performance materials than the near-α alloys into the 700–800°C range, attention is being given to strengthening high-temperature alloy materials, including titanium aluminides (see Section 8.6) with ceramic fibers such as SiC and B_4C.

8.6 Structural intermetallic compounds

8.6.1 General properties of intermetallic compounds

In terms of their properties, intermetallic compounds are generally regarded as a class of materials between metals and ceramics, which arises from the bonding being a mixture of metallic and covalent.

Table 8.5 *Comparison of physical properties of some intermetallic compounds.*

Compound	*Crystal structure*	*Melting temp. (°C)*	*Density (kg m^{-3})*	*Young's modulus/density*
Ni_3Al	$L1_2$ (ordered fcc)	1400	7500	45
Ni_3Si	$L1_2$ (ordered fcc)	1140	7300	
NiAl	B2 (ordered bcc)	1640	5860	35
Ti_3Si	$D0_{19}$ (ordered cph)	1600	4200	50
TiAl	$L1_0$ (ordered tetragonal)	1460	3910	24
FeAl	B2 (ordered bcc)	1300	5560	47

Intermetallics are intrinsically strong (and in the $L1_2$-ordered fcc compounds strength increases with temperature up to about 600°C) with high elastic modulus. The strong bonding and ordered structure also gives rise to lower self-diffusion coefficients and hence greater stability of diffusion-controlled properties. Some of the compounds of current interest are shown in Table 8.5. Intermetallics containing aluminum or silicon exhibit a resistance to oxidation and corrosion because of their adherent surface oxides. Those based on light elements have attractive low density giving rise to high specific properties particularly important in weight-saving applications.

Like ceramics, however, the greatest disadvantage of intermetallics is their low ductility, particularly at low and intermediate temperatures. The reasons for the lack of ductility vary from compound to compound, but include (1) a limited number of easy deformation modes to satisfy the von Mises criterion, (2) operation of dislocations with large slip vectors, (3) restricted cross-slip, (4) difficulty of transmitting slip across grain boundaries, (5) intrinsic grain boundary weakness, (6) segregation of deleterious solutes to grain boundaries, (7) covalent bonding and high Peierls–Nabarro stress, and (8) environmental susceptibility. It has been demonstrated, however, that some intermetallics can be ductilized by small alloying additions: Ni_3Al with boron, TiAl with Mn, Ti_3Al with Nb. This observation has encouraged recent research and development of intermetallics, and the possibility of application of those materials.

8.6.2 Nickel aluminides

Ni_3Al (nickel aluminide) is the ordered fcc γ'-phase and is a major strengthening component in superalloys. Ni_3Al single crystals are reasonably ductile but in polycrystalline form are quite brittle and fail by intergranular fracture at ambient temperatures. The basic slip system is $\{1\,1\,1\}\,\langle 1\,1\,0\rangle$ and has more than five independent slip modes, but still exhibits grain boundary brittleness. Remarkably, small additions of ~0.1 at.% boron produce elongations up to 50%. General explanations for this effect are that B segregates to grain boundaries, and (1) increases the cohesive strength of the boundary and (2) disorders the grain boundary region so that dislocation pile-up stresses can be relieved by slip across the boundary rather than by cracking. This general explanation is no doubt of significance but, additionally, there are distinct microstructural changes within the grains which must lead to a reduced friction stress and ease the operation of polyslip. For example, the addition of B reduces the occurrence of stacking-fault defects. Addition of solutes, such as B, are not expected to raise the stacking-fault energy and hence this effect possibly arises from the segregation of B to dislocations, preventing the superdislocation dissociation reactions (see Chapter 3, Section 3.6.5).

Microhardness measurements inside grains and away from grain boundaries indeed show that boron softens the grains. The ductilization effect is limited to nickel-rich aluminides and cannot be produced by carbon or other elements, although some substitutional solutes such as Pd, which

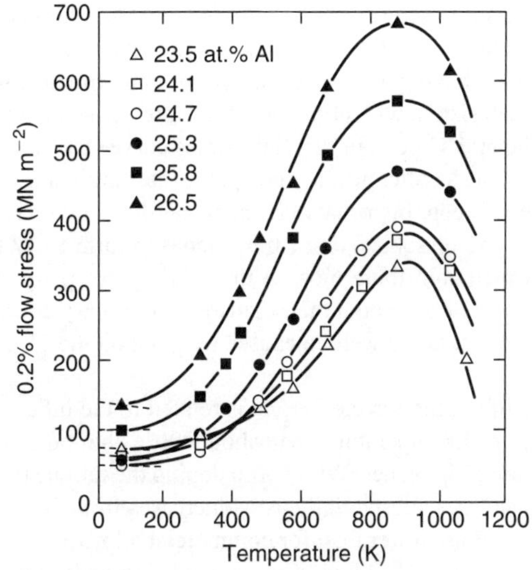

Figure 8.15 *Effect of aluminum content on the temperature dependence of the flow stress in Ni_3Al (after Noguchi, Oya and Suzuki, 1981).*

Table 8.6 *Anti-phase boundary energies in Ni_3Al.*

Alloy	γ_{111} *(mJ m^{-2})*	γ_{100} *(mJ m^{-2})*	$\gamma_{111}/\gamma_{100}$
Ni–23.5Al	183 ± 12	157 ± 8	1.17
Ni–24.5Al	179 ± 15	143 ± 7	1.25
Ni–25.5Al	175 ± 13	134 ± 8	1.31
Ni–26.5Al	178 ± 12	113 ± 10	1.51
Ni–23.5Al + 0.25B	170 ± 13	124 ± 8	1.37

substitutes for Ni, and Cu produce a small improvement in elongation. Small additions of Fe, Mn and Hf have also been claimed to improve fabricability. Grain size has been shown to influence the yield stress according to the Hall–Petch equation and B appears to lower the slope k_y and facilitate slip across grain boundaries. These alloys are also known to be environmentally sensitive. Hf, for example, which does not segregate to grain boundaries but still improves ductility, has a large misfit (11%) and possibly traps H from environmental reactions, such as $Al + H_2O \rightarrow Al_2O_3 + H$. Ti, which has a small misfit, does not improve the ductility.

The most striking property of Ni_3Al is the increasing yield stress with increasing temperature up to the peak temperature of 600°C (see Figure 8.15). This behavior is also observed in other $L1_2$ intermetallics, particularly Ni_3Si and Zr_3Al. This effect results from the thermally activated cross-slip of screw dislocations from the {1 1 1} planes to the {1 0 0} cube planes, where the APB energy is somewhat lower. The glide of superdislocations is made more difficult by the formation of Kear–Wilsdorf (K–W) locks (see Chapter 6) and their frequency increases with temperature. Electron-microscopy measurements of APB energies given in Table 8.6 shows that the APB energy on {1 0 0} decreases with aluminum content and this influences the composition dependence of the strength, shown in Figure 8.15. The cross-slip of screw dislocations from the {1 1 1} planes to cube planes also gives rise to a high work-hardening rate.

Although the study of creep in γ'-based materials is limited, it does appear to be inferior to that of superalloys. Above $0.6T_m$ creep displays the characteristic primary and secondary stages, with steady-state creep having a stress exponent of approximately 4 and an activation energy of around 400 kJ mol^{-1}, consistent with climb being the rate-controlling process. At intermediate temperatures (i.e. around the 600°C peak in the yield stress curve) the creep behavior does not display the three typical stages. Instead, after primary creep, the rate continuously increases with creep strain, a feature known as inverse creep. In primary creep, planar dissociation leads to an initial high creep rate which slows as the screws dissociate on $\{1\ 0\ 0\}$ planes to form K–W locks. However, it is the mobile edge dislocations which contribute most to the primary creep strain and their immobilization by climb dissociation which brings about the exhaustion of primary creep. The inverse creep regime is still not fully researched, but could well be caused by glide on the $\{1\ 0\ 0\}$ planes of the cross-slipped screw components.

The fatigue life in high-cycle fatigue is related to the influence of temperature on the yield stress and is invariant with temperature up to about 800°C, but falls off for higher temperatures with cracks propagating along slip planes. With boron doping the fatigue resistance is very sensitive to aluminum content and decreases substantially as Al increases from 24 to 26 at.%. Nevertheless, crack growth rates of $Ni_3Al + B$ are lower than for commercial alloys.

Hyperstoichiometric Ni_3Al with boron can be prepared by either vacuum melting and casting or from powders by HIPing. Fabrication into sheets is possible with intermediate anneals at 1000°C. At present, however, the application of Ni_3Al is not significant; Ni_3Al powders are used as bond coats to improve adherence of thermal spray coatings. Nevertheless, Ni_3Al alloys have been tested as heating elements, diesel engine components, glass-making molds and hot-forging dies, slurry-feed pumps in coal-fired boilers, hot-cutting wires and rubber extruders in the chemical industry. Ni_3Al-based alloys as matrix materials for composites are also being investigated.

Nickel aluminide (NiAl) has a cesium chloride or ordered β-brass structure and exists over a very wide range of composition either side of the stoichiometric 50/50 composition. It has a high melting point of 1600°C and exhibits a good resistance to oxidation. Even with such favorable properties it has not been commercially exploited because of its unfavorable mechanical properties. Because it is strongly ordered, low-temperature deformation occurs by an $a\langle 1\ 0\ 0\rangle$ dislocation vector and not by $a/2\langle 1\ 1\ 1\rangle$ superdislocations. $\{1\,1\,0\}\ \langle 1\,0\,0\rangle$ slip therefore leads to insufficient slip modes to satisfy the general plasticity criterion and in the polycrystalline condition β-NiAl is extremely brittle. The ductility does improve with increasing temperature but above 500°C the strength drops off considerably as a result of extensive glide and climb. Improvements in properties are potentially possible by refinement of the grain size and by using alloying additions to promote $\langle 1\ 1\ 1\rangle$ slip, as in FeAl, which has the same structure. In this respect, additions of Fe, Cr or Mn appear to be of interest. For high-temperature applications, ternary additions of Nb and Ta have been shown to improve creep strength through the precipitation of second phases and mechanical alloying with yttria or alumina is also beneficial.

A further commercial problem of this material is that conventional production by casting and fabrication is difficult, but production through a powder route followed by either HIPing or hot extrusion is more promising.

8.6.3 Titanium aluminides

Because of the limited scope for improvements in the properties of conventional titanium alloys above 650°C, either by alloy development or by TMP, increased attention is being given to the titanium intermetallics, Ti_3Al (α_2-phase) and TiAl (γ-phase). With low density, high modulus and good creep and oxidation resistance up to 900°C, they have considerable potential if the poor ductility at ambient temperatures could be improved. A comparison of Ti_3Al- and TiAl-based materials with conventional Ti alloys is given in Table 8.7.

Table 8.7 *Comparison of α_2 and γ superalloys with conventional titanium alloys.*

Property	*Titanium alloys*	*$(\alpha_2+\beta)$*	*$(\gamma+\alpha_2)$*
Density ($g\,cm^{-3}$)	4.54	4.84	4.04
E, stiffness ($GN\,m^{-2}$)	110	145	176
RT tensile strength ($MN\,m^{-2}$)		1100	620
HT (760°C) tensile strength ($MN\,m^{-2}$)		620	550
Max. creep temp. (°C)	540	730	900
RT ductility (%)	20	4–6	3
Service temp. ductility (%)	High	5–12	5–12

Electron microscopy studies of Ti_3Al or α_2 have shown that deformation by slip occurs at room temperature by coupled pairs of dislocations with $b=1/6\langle 1\ 1\ \bar{2}\ 0\rangle$, which glide only on $\{1\ 0\ \bar{1}\ 0\}$ planes and by very limited glide on $\{1\ 1\ \bar{2}\ \bar{1}\}$ with pairs of dislocation with $b=1/6\langle 1\ 1\ \bar{2}\ 6\rangle$. The ductility increases at higher temperatures due to climb of the $\langle 1\ 1\ \bar{2}\ 0\rangle$ dislocations and to the increased glide mobility of $1/6\langle 1\ 1\ \bar{2}\ 0\rangle$ and $1/6\langle 1\ 1\ \bar{2}\ 6\rangle$ dislocations through thermal activation. Only limited activity of the $\{0\ 0\ 0\ 1\}\ \langle 1\ 1\ \bar{2}\ 0\rangle$ slip systems is observed, even at high temperatures.

The most successful improvements in the ductility of Ti_3Al have been produced by the addition of β-stabilizing elements, particularly niobium, to produce α_2 alloys. An addition of 4 at.% Nb produces significant slip on $\{1\ 0\ \bar{1}\ 0\}\langle 1\ 1\ \bar{2}\ 0\rangle$, $\{0\ 0\ 0\ 1\}\langle 1\ 1\ \bar{2}\ 0\rangle$ and $\{1\ 1\ \bar{2}\ 1\}\langle 1\ 1\ \bar{2}\ 6\rangle$ as well as some slip on $\{1\ 0\ \bar{1}\ 1\}\ \langle 1\ 1\ \bar{2}\ 0\rangle$. This improvement is attributed to the decrease in covalency as Nb substitutes for Ti with a consequent reduction in the Peierls–Nabarro friction stress. Alloys based on α_2 are Ti–(23–25)Al–(8–18)Nb, of which Ti–24Al–11Nb has excellent spalling resistance. Most Ti_3Al+Nb alloys, such as super α_2, also contain other β-stabilizers including Mo and V, i.e. Ti–25Al–10Nb–3V–1Mo, which exhibits about 7% room temperature elongation. Alloying Ti_3Al with β-stabilizing elements to produce two-phase alloys significantly increases the fracture strength. These Ti_3Al-based alloys can be plasma-melted and cast followed by TMP in the $(\alpha_2+\beta)$ or β-range. The improved ductility of Ti_3Al alloys has led to aerospace applications in after-burners in jet engines, where it compares favorably in performance with superalloys and gives a 40% weight saving.

Developments are taking place in rapid solidification processing to include a second phase (e.g. rare-earth precipitates) and to provide powders, which may be consolidated by HIPing, to produce fully dense components with properties comparable to wrought products. There are also developments in intermetallic matrix composites by the addition of SiC or Al_2O_3 fibers (~10 μm). These have some attractive properties, but the fiber/intermetallic interface is still a problem.

The γ-phase Ti–(50–56)Al has an ordered fc tetragonal ($L1_0$) structure up to the m.p. 1460°C, with $c/a=1.02$ (Figure 8.16). Deformation by slip occurs on $\{1\ 1\ 1\}$ planes and, because of the tetragonality, there are two types of dislocations, namely ordinary dislocations $1/2\langle 1\ 1\ 0\rangle$ and superdislocations $\langle 0\ 1\ 1\rangle=1/2\langle 0\ 1\ 1\rangle+1/2\langle 0\ 1\ 1\rangle$. Another superdislocation $1/2\langle 1\ 1\ 2\rangle$ has also been reported.

At room temperature, deformation occurs by both ordinary and superdislocations. However, [0 1 1] and [1 0 1] superdislocations are largely immobile because segments of the trailing superpartials $1/6\langle 1\ 1\ 2\rangle$ type form faulted dipoles. The dissociated $1/2\langle 1\ 1\ 0\rangle$ dislocations bounding complex stacking faults are largely sessile because of the Peierls–Nabarro stress. Some limited twinning also occurs. The flow stress increases with increasing temperature up to 600°C as the superpartials become mobile and cross-slip from $\{1\ 1\ 1\}$ to $\{1\ 0\ 0\}$ to form K–W-type locks, the $1/2\langle 1\ 1\ 0\rangle$ slip activity increases and twinning is promoted.

The two-phase $(\gamma+\alpha_2)$ Ti–Al alloys have better ductility than single-phase γ with a maximum at 48 at.% Al. This improvement has been attributed to the reduced c/a with decreased Al, further promotion of twinning and the scavenging of O_2 and N_2 interstitials by α_2. The combination of high

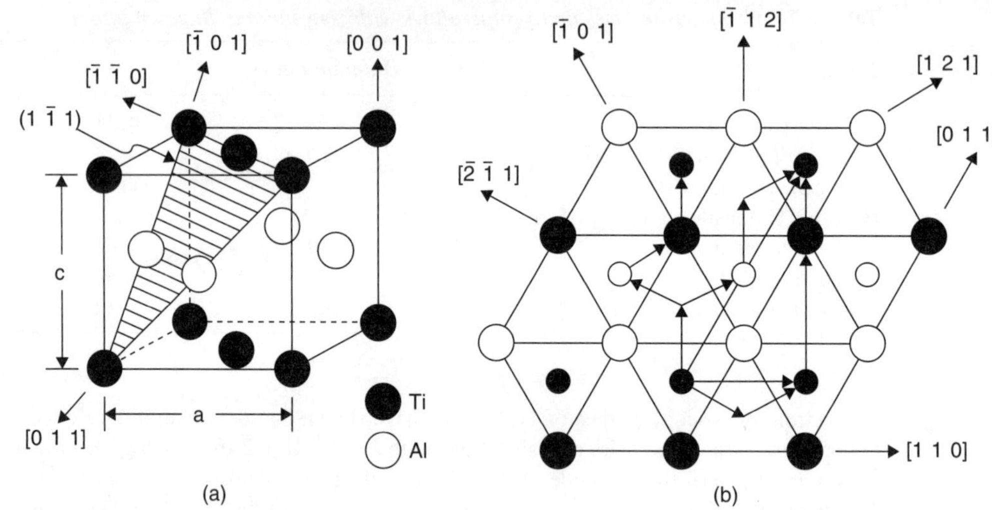

Figure 8.16 *Structure of TiAl ($L1_0$) (a) and (1 1 1) plane (b) showing slip vectors for possible dissociation reactions, e.g. ordinary dislocations 1/2[1 1 0], superdislocations [0 1 1] and 1/2[1 1 2], and twin dislocations 1/6[1 1 2] (after Kim and Froes, 1990).*

stiffness ($E = 175$ GPa at 20°C to 150 GPa at 700°C), density-normalized strength similar to cast Ni-based alloys, high temperature strength and reasonable oxidation resistance to 750°C, low thermal expansion coefficient and high thermal conductivity have led to a high level of interest in TiAl-based alloys. The major limitations to their application are the intrinsic low room temperature ductility (no better than 2–3%), the low fracture toughness (between 10 and 20 MPa $m^{1/2}$ at 20°C) and the high growth rate of fatigue cracks.

Alloy and process development have resulted in some successful applications of these alloys over the last 10 or so years. Cast turbochargers are now manufactured in Japan for cars and wrought exhaust valves were used in Formula 1 cars for some years. Major applications for these alloys are still awaited despite the success of these two applications; the high cost of processing is holding commercial developments back. In the case of thermomechanical processing the costs are high because the alloys are strong at normal hot-working temperatures and because some sort of protection (such as canning with steel) from oxidation must be used during working. In the case of castings the efficiency of material usage is very low, both because casting technology is not efficient and because melting and casting are difficult because of the reactivity of molten TiAl-based alloys.

The compositions of TiAl-based alloys which are of commercial interest lie within the range of about Ti/45–48 Al (at.%) but all alloys contain other elements in attempts to improve the properties of the binary alloy. Additions of Nb between about 5 and 8 at.% are important in improving oxidation resistance and also imparting some solid solution strengthening.

An understanding of the microstructures which can be obtained in cast or in wrought products of TiAl-based alloys requires knowledge of the phase changes that occur over the temperature range from the melting point to room temperature. The relevant part of the binary phase diagram between Ti and Al is shown in Figure 8.17, which also indicates schematically the influence of some alloying additions on phase boundaries.

The various phase transformations in the Ti–Al system offer the possibility of microstructural control both for the wrought route and for the casting route. Thus, cooling samples containing less than about 44 at.% Al the solidification will take place through the formation of β, which may or may not be removed via the peritectic reaction. Subsequent cooling of the α-phase results in precipitation of

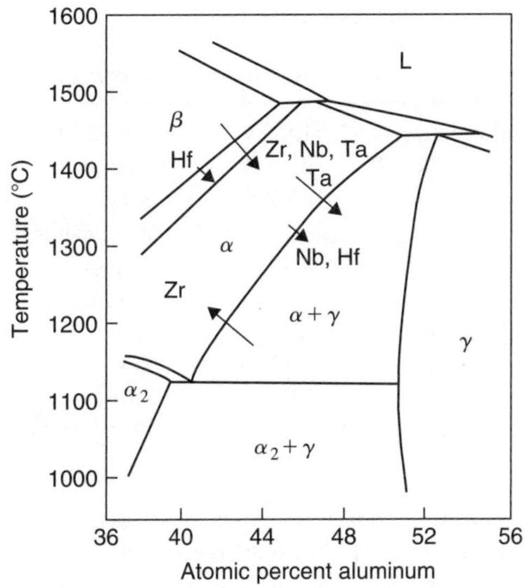

Figure 8.17 *Partial Ti–Al phase diagram showing the influence of ternary additions on the position of the various boundaries (courtesy M. H. Loretto).*

γ- which, under typical cooling rates encountered with castings, results in the formation of a lamellar structure. This 'fully lamellar' structure consists of parallel lamellae of γ and α and of twinned γ. These lamellae are formed on the (0 0 0 1) plane of the α-phase and thus their length is defined by the pre-existing α grain size. Somewhat slower cooling results in some of the γ lamellae coarsening at colony boundaries to form γ grains through local growth of the lamellae, to form a 'near fully lamellar' structure. Hot working in the two-phase region results in the formation of equiaxed γ and α grains; the ratio of the amounts being defined by the average alloy composition and the hot-working temperature. Subsequent cooling results in the equiaxed α grains forming lamellae to yield a duplex microstructure, or if extensive hot working is used, a structure consisting of γ and α grains is formed, termed 'near γ'.

If the cooling rate is increased, as in oil or water quenching, the α-phase transforms massively to γ if the Al content is above about 44 at.% (below this Al-content, α is retained) and this transformation offers a further opportunity for microstructural control in cast samples by heat treating in the two-phase field so that α can precipitate on all four $\{1\,1\,1\}$ planes, throughout the γ grains. The tendency to transform massively is strongly dependent upon the composition, which has important consequences upon the choice of alloy composition in cast samples.

8.6.4 Other intermetallic compounds

A number of intermetallic compounds are already used in areas which do not rely on stringent mechanical properties. Fe_3Al, for example, is used in fossil-fuel plants, where resistance to both sulfur attack and oxidation is important. Ni_3Si is used where resistance to hot sulfuric acid is required. There are several compounds with rare-earth elements used in magnet technology (see Chapter 5). PdIn is gold colored and a possible dental material. Zr_3Al has a low neutron capture cross-section and is a possible reactor material.

The β-compound NiTi (*Nitinol*) is an important shape-memory alloy. The shape-memory effect (SME) manifests itself when the alloy is deformed into a shape while in a low-temperature martensitic

Table 8.8 *Aluminum alloy designation systems.*

Wrought alloys	*Designation*	*Casting alloys*	*Designation*
99.00% (min.) aluminum	1XXX	99.00% (min.) aluminum	1XX.X
Copper	2XXX	Copper	2XX.X
Manganese	3XXX	Silicon with added copper and/or magnesium	3XX.X
Silicon	4XXX	Silicon	4XX.X
Magnesium	5XXX	Magnesium	5XX.X
Magnesium and silicon	6XXX	Zinc	6XX.X
Zinc	7XXX	Tin	7XX.X
Others	8XXX	Others	8XX.X

condition but regains its original shape when the stress is removed and it is heated above the martensitic regime. Strains of the order of 8% can be completely recovered by the reverse transformation of the deformed martensitic phase to the higher-temperature parent phase. The martensite transformation in these alloys is a thermoelastic martensitic transformation in which the martensite plates form and grow continuously as the temperature is lowered and are removed reversibly as the temperature is raised. NiTi was one of the original SME alloys, but there are many copper-based alloys which undergo a martensitic transformation, e.g. Cu–17Zr–7Al. Application of SME alloys relies on the characteristic that they can change shape repeatedly as a result of heating and cooling and exert a force as the shape changes. By composition control (increasing the Ni content or substitution of Cu lowers the M_S temperature of TiNi), the shape memory can be triggered by normal body temperature or any other convenient temperature to operate a device. Several biomedical applications have been developed in orthopedic devices (e.g. pulling fractures together), in orthodontics, in intrauterine contraceptives and in artificial hearts. Industrial applications include pipe couplings for ships which shrink during heating, electrical connectors, servo-mechanisms for driving recording pens, switches, actuators and thermostats.

8.7 Aluminum alloys

8.7.1 Designation of aluminum alloys

Aluminum has an attractive combination of properties (i.e. low density, strong, easy to fabricate) which can be developed and modified by alloying and processing. The basic physical metallurgy has been outlined in Chapter 7 and hence in this section some of the alloys developed for particular industries such as the transportation, construction, electrical and packaging industries will be considered.

Aluminum alloys are identified by a four-digit system[3] based on the main alloying element. This is summarized in Table 8.8. For wrought alloys the first digit identifies the alloy group and the second digit any modification to the original alloy, which is identified by the last two digits. The system is slightly different for casting alloys. The first digit again identifies the group, the second two digits identify the alloy and the last digit, preceded by a decimal point, indicates the product form (i.e. 0 for casting and 1 for ingot).

[3] The International Alloy Designation System (IADS) was first introduced by the Aluminum Association of the USA and is now standard for wrought alloys.

8.7.2 Applications of aluminum alloys

With the need for fuel economy and weight saving, aluminum alloys are increasingly used in cars, and its two most important properties are density and thermal conductivity. Over the past 15 years the aluminum content of cars has increased from around 5% to 13% by both volume and weight. In engines they are used for pistons, cylinder heads and sumps. Al–Si casting alloys of the 3XX.X series are being used for engine blocks and Al–Si pistons with cast iron cylinder liners for wear resistance. The superior thermal conductivity reduces the volume of coolant in the system. Aluminum wheels, vacuum cast or forged, are replacing conventional steel wheels in sports cars. Heat-treatable[4] 2XXX and 6XXX as well as 5XXX series can be used for body sheet. However, the modulus of aluminum is only one-third that of steel and hence significant design changes are necessary to maintain rigidity and stiffness. A straightforward increase in gauge thickness would lead to a doubling cost, which limits the replacement to 'quality' cars. For bulk market cars, gauge-for-gauge substitution for steel is a future objective with structural reinforcement to enhance body torsional characteristics. This is possible with the use of adhesives in a weld-bonding approach which can reduce the weight by half, and fuel consumption and CO_2 emission by almost 15%. Aluminum-structured vehicle technology (ASVT) is likely to be essential when emission control to reduce global warming effects is tightened.

The Honda NSX all-aluminum car[5] is manufactured with conventional design and assembly, and has three different 6000 series alloys for external panels and a structural subframe of 5182 alloy.

In aircraft construction use is made of the high strength–density ratio of the Al–Cu (2000 series) and Al–Zn–Mg (7000 series) alloys in extruded form for wing spars, fuselage and landing gear, and for the skin in plate or sheet form; typically 7075 (5.6Zn–2.5Mg–1.6Cu) is used in the T6 condition and 2024 (4.4Cu–1.5Mg–0.6Mn) in the T3 or T8 conditions. The alloys of the 7000 series have higher strength than the 2000 alloys but lower resistance to fracture. However, higher purity levels (e.g. 2124 alloy) give enhanced toughness. Alloys of both series lose strength above 100°C and are thus not suitable for supersonic aircraft. The *RR58* alloy used for the Concorde at temperatures up to 175°C was originally an early engine material, for compressor blades and impellers.

Aluminum is used extensively in the construction industry because of its light weight, resistance to atmospheric attack and surface finish. For decorative applications, dyed anodic films produce a permanent durable finish. Generally, the Al–Mg–Si 6000 series is used, i.e. 6063 medium-strength and 6082 higher-strength alloys in the T6 condition for extrusions or T4 where forming is required during fabrication. The Al–Cu (2014A) alloy is also used for heavily loaded primary structures.

The packaging industry also provides a large market for aluminum alloys. The main requirement is for low-cost, simple alloys capable of being formed, and the Al–Mg (5000) series is often used. Impurity control is essential and liquid metal filtering is necessary in the production of thin sheet.

In the electrical industry, electrical conductivity grade aluminum and higher-strength 5000 and 6000 series alloys are used for transmission lines, replacing the more expensive copper. Dispersion-strengthened alloys containing a fine dislocation substructure stabilized by small precipitates are used for electrical wiring.

8.7.3 Aluminum–lithium alloys

The advantages of aluminum–lithium alloys have been known for a long time, but lower density and increased elastic stiffness were offset by poor ductility and fracture performance. Basic Al–Li alloys

[4] Alloy treatment is usually described by a suffix letter and digit system (e.g. F – as fabricated, O – annealed, H – work-hardening and T – heat treated). Digits following H specify the work-hardened condition, and that following T the type of ageing treatment (e.g. T6 is solution heat treated and artificially aged, T4 solution heat treated and naturally aged, T3 solution heat treated and cold worked).

[5] Launched in the USA in 1992.

Table 8.9 *Composition of commercial aluminum–lithium alloys.*

Alloy	*Li*	*Cu*	*Mg*	*Fe*	*Si*	*Zr*
(8090) *Lital A*	2.5	1.3	0.7	≤0.2	≤0.1	0.12
(9091) *Lital B*	2.6	1.9	0.9	≤0.2	≤0.1	0.12

precipitate the $(Al_3Li)\delta'$, a spherical ordered precipitate. Precipitation hardening leads, however, to localized deformation with limited cross-slip and poor fracture behavior. Additions of copper to the alloy so that the Li/Cu ratio is high leads to the formation of both δ' and a T_1-phase (Al_2CuLi). This gives some improvement in fracture toughness by independent control of the two precipitates. In the quaternary system Al–Li–Cu–Mg the S-phase precipitates in addition to the δ' and T_1. The S-phase is better at dispersing slip than T_1 and with adjustment of composition can be made to dominate the structure. Both S and T_1 are nucleated heterogeneously on dislocations and the best results are obtained by cold working the alloy after solution treatment.

Commercial alloys based on this background are *Lital A, B* and *C*, which have been developed to match the (1) conventional medium-strength 2014-T6, (2) high-strength 7075-T6 and (3) damage-tolerant 2024-T3 alloys, with a 10% reduction in density and 10% improvement in stiffness (see Table 8.9).

Lital A in T6 sheet form typically has 365 MN m^{-2} 0.2% proof stress, 465 MN m^{-2} TS, 6% elongation, 66 MN m$^{-3/2}$ fracture toughness, an elastic modulus of 80 GN m^{-2} and density of 2550 kg m^{-3}. *Lital B* has roughly 10% improvement in strength. *Lital C* is a variant of the 8090 alloy and is heat treated to increase toughness (~76 MN m$^{-3/2}$) at the expense of strength (TS ~440 MN m^{-2}).

Lithium additions are also being made to conventional aluminum alloys. The addition of lithium has a major influence since Li possesses a significant vacancy binding energy of about 0.25 eV. Lithium atoms therefore trap vacancies and form Li–V aggregates. This decreases the concentration of mobile vacancies available for the transport of zone-forming atoms, and therefore inhibits the diffusion of Zn and Mg in 7075, and Si and Mg in 6061, into zones. Second, the Li–V aggregates, very probably present during quenching and immediately after ageing, act as heterogeneous sites for subsequent clustering of zone-forming atoms during ageing.

Additions of Li into either Al–2Mg–0.6Si–0.3Cu–0.3Cr (6061) or Al–5.9Zn–2.4Mg–1.7Cu (7075) modify the precipitation scheme and age-hardening behavior of the original alloys. The precipitates which form in the base alloys are inhibited or even suppressed. For the 6061 the addition of 0.7% Li retards the precipitation of needle-shaped GP zones and produces a ternary compound AlLiSi, whereas the addition of 2.0% Li results in the dominant precipitation of δ' and extremely delayed and limited formation of needle-shaped GP zones and AlLiSi. For 7075 the addition of 0.7% Li alters the conventional precipitation scheme from solute-rich GP zone $\rightarrow \eta' \rightarrow \eta MgZn_2$ into vacancy-rich GP zone $\rightarrow T' \rightarrow T(AlZn)_{49}Mg_{32}$, whereas the addition of 2.0% Li produces the dominant δ' precipitate and limited and delayed formation of T-phase. As a result, the age-hardening response relating to these major hardening phases in both base alloys is delayed or decreased. Such additions can produce narrower PFZs and give improved fracture properties.

A further commercial alloy is *UL*40, which is essentially a binary alloy containing 4% Li. The alloy is cast using a spray-deposition process resulting in a fine-grained microstructure, with uniform distribution of second phase, free from oxide. The high Li content alloy has a very low density (2400 kg m^{-3}) and is almost a third lighter than conventional aluminum and magnesium alloys. It extrudes well and can be welded with Al–Mg–Zr filler, producing components for aircraft and helicopters, such as pump housings and valves, and for yachting with good corrosion resistance.

8.7.4 Processing developments

8.7.4.1 *Superplastic aluminum alloys*

Superplastic forming is a cost-effective manufacturing process for producing both simple and complex shapes from aluminum alloy sheet, because of its low-cost tooling and short lead times for production. A range of alloys is available, including 2004 (Al–6Cu–0.4Zr) or *Supral*, 5083 SPF, 7475 SPF and *Lital* 8090 SPF (Al–Li–Mg–Cu). *Supral* and *Lital* dynamically recrystallize to a fine grain size during the early stages of deformation (~500°C), which is stabilized with $ZrAl_3$ particles. The grain size in 7475 is stabilized by TMP and submicron Cr particles and, in 5083, by an Mn-dispersoid. Components are formed by clamping the alloy sheet in a pressure chamber and then applying gas or air pressure to force the sheet slowly into contact with a tool surface; both the tool and sheet are maintained at the forming temperature throughout the process. During normal superplastic forming, the alloys tend to develop voiding. This void formation is minimized by forming in hydrostatic conditions by introducing a gas pressure on both the front and back surfaces of the sheet being formed. The sheet is forced against the tool surface with a small pressure differential.

8.7.4.2 *Rapid solidification processing of aluminum alloys*

Rapid solidification processing (RSP) has been applied to aluminum alloys to produce a fine grain size and extend solid solubility, particularly for the transition metals iron, molybdenum, chromium and zirconium, which usually have low solid solubility and low diffusion rates in aluminum. Interestingly, RSP alloys containing Fe and Cr, on annealing, precipitate metastable spherical quasi-crystals of icosahedral phase with fivefold symmetry. These are extremely stable and hardly coarsen after extensive heat treatment, which indicates a potential for alloy development.

A series of commercially available high-temperature Al–Fe–V–Si alloys has been developed and consist of very fine, spherical $Al_{13}(FeV)_3Si$ silicides uniformly dispersed throughout the matrix, which display much slower coarsening rates than other dispersoids. A typical alloy with 27 vol.% silicides is 8009 (Al–8.5Fe–1.3V–1.7Si) and, without any needle or platelet precipitates in the microstructure, has a $K_{Ic} \sim 29\,\mathrm{MN\,m^{-3/2}}$. The tensile properties as a function of temperature are shown in Figure 8.18 in comparison with a conventional 7075-T6 alloy. At all temperatures up to 480°C, the 8009 alloy has a higher specific stiffness than a Ti–6A1–4V alloy. The fatigue and creep rupture properties are better than conventional aluminum alloys with excellent corrosion resistance. These RSP silicide alloys can be readily fabricated into sheet, extruded or forged, and the combination of attractive properties makes them serious candidates for aerospace applications. Other alloys developed include Al–Cr–Zn–Mn, Al–8Fe–2Mo and Al–Li.

8.7.4.3 *Mechanical alloys of aluminum*

Mechanical alloys of aluminum contain dispersions of carbides or oxides, which not only produce dispersion strengthening but also stabilize a fine-grained structure. An advantage of these alloys arises because the strength is derived from the dispersoids and thus the composition of the alloy matrix can be designed principally for corrosion resistance and toughness rather than strength. Thus, the alloying elements which are usually added to conventional aluminum alloys for precipitation strengthening and grain-size control may be unnecessary.

Mechanical alloying is carried out with elemental powders and an organic process control agent, such as stearic acid, to balance the cold-welding and powder-fracture processes. No dispersoid is added because the oxide on the surface of the powders and process control agent are consolidated

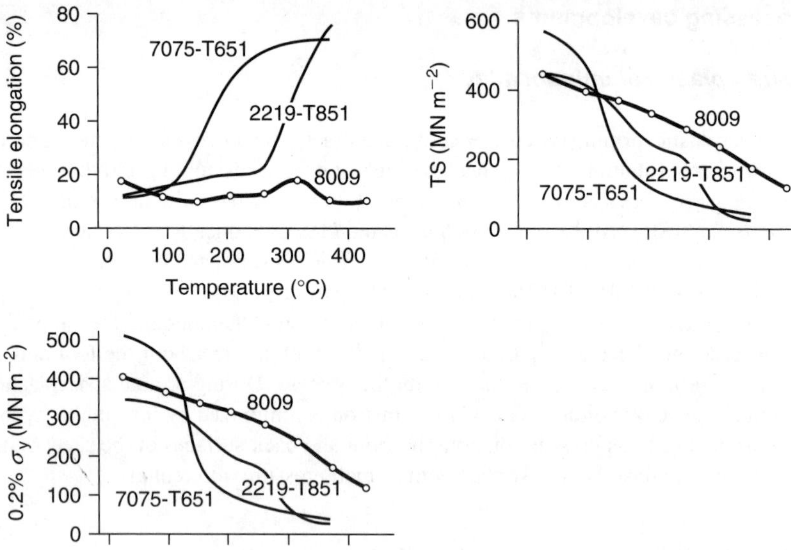

Figure 8.18 *Tensile properties of RSP Al alloy 8009 as a function of temperature compared with conventional aluminum alloys (from Gilman, 1990; courtesy of Institute of Materials, Minerals and Mining).*

during mechanical alloying as hydrated oxides and carbonates. The process produces a fine dispersion of ~20 nm particles in a dynamically recrystallized structure with grains as fine as 0.05 μm. Subsequent vacuum degassing at elevated temperature removes the H_2 and N_2 liberated, improves the homogenization of the matrix and reduces carbonates to Al_4C_3, which forms most of the dispersoid. The final grain size is around 0.1 μm. The powder is then compacted by HIPing or vacuum hot-pressing and conventionally extruded to produce a material with a stable grain size of 0.3 μm, with grain boundaries pinned by the dispersoid.

Mechanical alloys have been developed corresponding to the 2000, 5000 and 7000 aluminum series alloys. *IN 9021* is heat treatable by solution treatment and natural or elevated temperature ageing to give 500–560 MN m^{-2} proof stress, 570–600 MN m^{-2} TS, 12% elongation and 40 MN $m^{-3/2}$ K_{1c}. *IN 9052* is the equivalent of a 5000 series alloy, requiring no heat treatment and offering good strength in thick sections, 390 MN m^{-2} proof stress, 470 MN m^{-2} TS, 13% elongation and 46 MN $m^{-3/2}$ K_{1c}. Mechanically alloyed Al–Mg–Li offers inherent high strength in thick section, 430 MN m^{-2} proof stress, 500 MN m^{-2} TS, 10% elongation and 30 MN $m^{-3/2}$ K_{1c}.

Problems

8.1 What properties are required of steels for cold-forming applications?

8.2 Use of dual-phase steels is now more widespread in automobile applications. Describe the heat-treatment used for this type of steel and the microstructure developed. How are the mechanical properties optimized? What are the advantages gained by their use? Write down a typical dual-phase steel composition.

8.3 The composition of Cr steel for aircraft landing gear components is:

$$Fe - 0.40C - 0.70Mn - 0.8Cr - 1.8Ni - 0.25Mo - 1.6Si - 0.05V \text{ wt.\%}.$$

(a) What role does (i) vanadium, (ii) molybdenum and (iii) silicon play in the development of the required mechanical properties?

(b) Given that the yield stress is approximately 1650 MPa and the K_{1c} is approximately 60 Mpa $m^{1/2}$, estimate the critical crack size for catastrophic failure. Comment on the practical significance of this value.

8.4 The surfaces of steel specimens can be hardened by enrichment in their nitrogen content. One route is to maintain a nitrogen-rich atmosphere around a heated steel specimen. If this atmosphere gives a constant N content of 1.53 wt.% at 1000°C and the minimum hardness requires a nitrogen content of 0.25 wt.%, calculate the time required to achieve a hardened depth of (i) 1 μm and (ii) 1.75 μm under these conditions.

Diffusion of N in γ-Fe: $D_o = 9.1 \times 10^{-5}\,m^2\,s^{-1}$, $Q = 170\,kJ\,mol^{-1}$, $\mathbf{R} = 8.314\,J\,mol^{-1}\,K^{-1}$.

The error function is:

z	0.85	0.90	0.95	1.0	1.1	1.2	1.3	1.4	1.5	1.6	1.7
Erf(z)	0.770	0.797	0.8209	0.8427	0.8802	0.910	0.934	0.9523	0.9661	0.9763	0.9838

8.5 Recently, steel producers have produced automobile steels which are designed to bake-harden during the paint-drying process, which involves time at about 200°C. What is the magnitude of the increase in strength and outline the strengthening mechanism involved.

8.6 Steels for structural applications rarely have yield strengths in excess of 500 MPa, whereas steels having yield strengths up to 1500 MPa are used in aeroplane landing gear. Explain the reasons for this difference.

8.7 Outline the factors in the development of gas-turbine blades over the last 50 years that have required changes in the methods of processing nickel-based superalloys, including forging, investment casting, directional solidification and single-crystal production.

8.8 What are the principal advantages of directionally solidified and single-crystal nickel-based superalloys over conventionally cast material in relation to the mechanical performance of gas-turbine blades?

8.9 The intermetallic compounds NiAl and Ni_3Al are both fully ordered up to their melting point. NiAl deforms by the motion of unit dislocations, whereas Ni_3Al deforms by the operation of superdislocations.

(a) Describe the processes involved and explain the reasons for the different deformation behavior.

(b) Sketch how the yield stress of these two compounds varies with increase in temperature and explain the characteristics.

8.10 Over the last few decades the composition of line-pipe steels has changed significantly. Outline the changes and give the reasons for them.

8.11 Cast iron contains approximately 2.5% Si. What is its role and what is the desired microstructure in austempered ductile iron (ADI)? What are the two important heat treatments to achieve this structure?

8.12 A commonly used versatile titanium alloy contains 6% Al and 4% V. What is the role of these additions and what is the microstructure produced?

Further reading

Baker, C. (ed.) (1986). *Proc. 3rd International Aluminium–Lithium Conference*. Institute of Metals, London.

Bhadeshia, H. K. D. H. (1992). *Bainite in Steels*. Institute of Materials, London.
Honeycombe, R. W. K. (1981). *Steels, Micro-structure and Properties*. Edward Arnold, London.
Janowak, J. R. *et al.* (1984). A review of austempered ductile iron metallurgy. *First International Conference on ADI.*
Materials in defence (1988). *Metals and Materials*, **4**(7). Institute of Materials, London.
Meetham, G. W. (1981). *The Development of Gas Turbine Materials*. Applied Science, London.
Peters, H. (ed.) (1991). *Proc. 6th International Aluminium–Lithium Conference*. Deutsche Gesellschaft für Material-kunde.
Polmear, I. J. (1989). *Light Alloys*. Edward Arnold, London.
Sims, C. T. and Hagel, W. C. (eds) (1972). *Superalloys*. John Wiley, Chichester.
Sims, C. T., Hagel, W. C. and Stoloff, N. S. (1987). *Superalloys II*. John Wiley, Chichester.
Stoloff, N. S. (ed.) (1984). Ordered alloys. *International Metal Reviews*, **29**(3).
Stoloff, N. S. (1989). Physical and mechanical metallurgy of Ni_3Al and its alloys. *International Metal Reviews*, **34**(4).
Yoo, M. H. *et al.* (1993). Deformation and fracture of intermetallics. Overview No. 15. *Acta Metall. Mater.*, No. 4.
Kim, Y-W. (1989). Physical metallurgy of titanium aluminides. *TMS/ASM Symposium on High Temperature Aluminides and Intermetallics.*

Chapter 9

Oxidation, corrosion and surface treatment

9.1 The engineering importance of surfaces

The general truth of the engineering maxim that 'most problems are surface problems' is immediately apparent when one considers the nature of metallic corrosion and wear, the fatigue cracking of metals and the effect of catalysts on chemical reactions. For instance, with regard to corrosion, metal surfaces commonly oxidize in air at ambient temperatures to form a very thin oxide film (tarnish). This 'dry' corrosion is limited, destroys little of the metallic substrate and is not normally a serious problem. However, at elevated temperatures, nearly all metals and alloys react with their environment at an appreciable rate to form a thick, non-protective oxide layer (scale). Molten phases may form in the scale layer, being particularly dangerous because they allow rapid two-way diffusion of reacting species between the gas phase and the metallic substrate. 'Wet' or aqueous corrosion, in which electrochemical attack proceeds in the presence of water, can also destroy metallic surfaces and is responsible for a wide variety of difficult problems throughout all branches of industry. The principles and some examples of 'dry' and 'wet' corrosion will be discussed in Section 9.2.

Conventionally, the surface properties of steels are improved by machining to produce a smooth surface texture (superfinishing), mechanically working (shot-peening), introducing small atoms of carbon and/or nitrogen by thermochemical means (carburizing, nitriding, carbo-nitriding), applying protective coatings (galvanizing, electroplating), chemically converting (anodizing), etc. Many of these traditional methods employ a liquid phase (melt, electrolyte). In contrast, many of the latest generation of advanced methods for either coating or modifying material surfaces use vapors or high-energy beams of atoms/ions as the active media. Their successful application on a commercial scale has revealed the merits of developing a new philosophy of surface design and engineering. In Section 9.3, we examine some typical modern methods for improving surface behavior.

9.2 Metallic corrosion

9.2.1 Oxidation at high temperatures

9.2.1.1 Thermodynamics of oxidation

The tendency for a metal to oxidize, like any other spontaneous reaction, is indicated by the free energy change ΔG accompanying the formation of the oxide. Most metals readily oxidize because ΔG is negative for oxide formation. The free energy released by the combination of a fixed amount (1 mol) of the oxidizing agent with the metal is given by ΔG° and is usually termed the standard free energy of the reaction. ΔG° is, of course, related to ΔH°, the standard heat of reaction, and ΔS°, the standard change in entropy, by the Gibbs equation. The variation of the standard free energy change with absolute temperature for a number of metal oxides is shown in Figure 9.1. The noble metals, which are easily reduced, occur at the top of the diagram and the more reactive metals at the bottom. Some of these metals at the bottom (Al, Ti, Zr), however, resist oxidation at room temperature owing to the impermeability of the thin coherent oxide film which first forms.

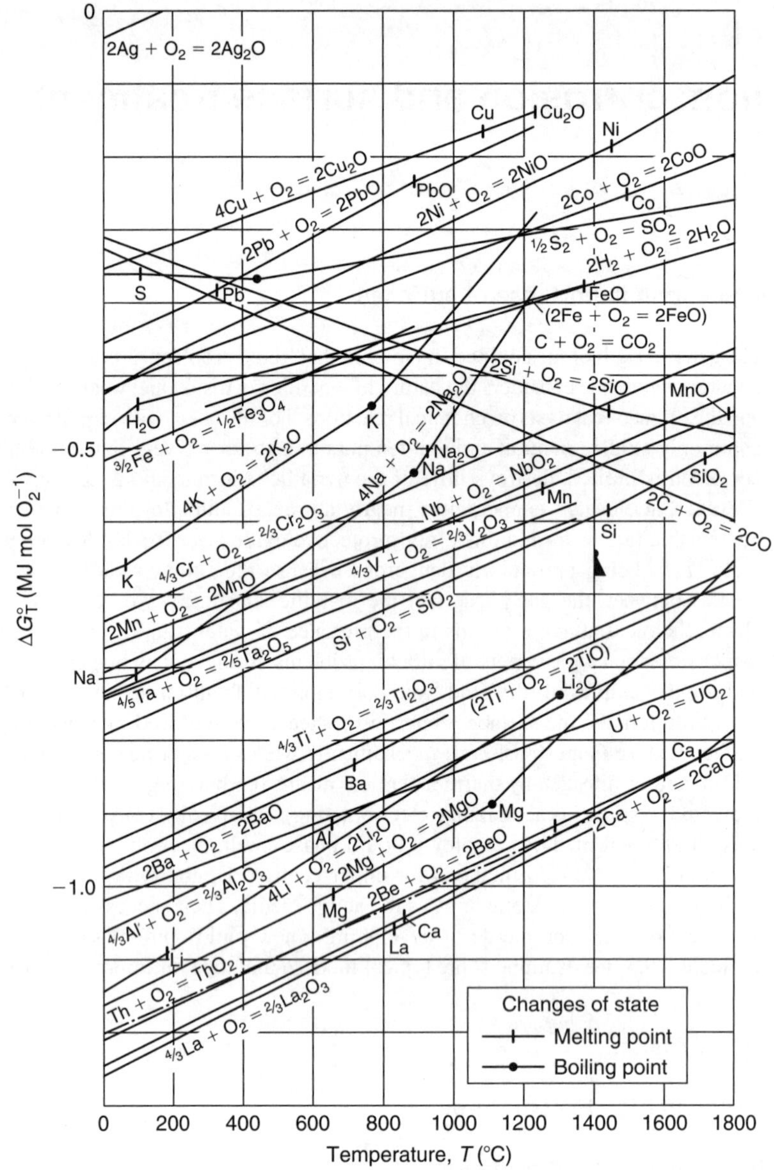

Figure 9.1 *Standard free energies of formation of oxides.*

The numerical value of ΔG° for oxidation reactions decreases with increase in temperature, i.e. the oxides become less stable. This arises from the decreased entropy accompanying the reaction, solid (metal) + gas (oxygen) → solid (oxide). The metal and oxide, being solids, have roughly the same entropy values and $d(\Delta G)/dT$ is thus almost equivalent to the entropy of the oxygen, i.e. 209.3 J deg^{-1} mol^{-1}. Most of the ΔG versus T lines therefore slope upwards at about this value, and any change in slope is due to a change in state. As expected, melting has a small effect on ΔS and hence ΔG, but transitions through the boiling point (e.g. ZnO at 970°C) and sublimation (e.g. Li_2O at 1330°C) have large effects. Exceptions to the positive slope of the ΔG versus T line occur for carbon oxidation to

CO or CO_2. In both cases the oxide product is gaseous and thus also has a high free energy. In the reaction $2C + O_2 \rightarrow 2CO$, two moles of gas are produced from one of oxygen so that

$$\begin{aligned}\Delta S &= (S_{\text{oxide}} - S_{\text{carbon}} - S_{\text{oxygen}}) \\ &= (2S^{\circ}_{\text{oxide}} - S^{\circ}_{\text{carbon}} - S^{\circ}_{\text{oxygen}}) \simeq S^{\circ}_{\text{oxide}}.\end{aligned} \tag{9.1}$$

The CO free energy versus temperature line has a downward slope of approximately this value. For the $C + O_2 \rightarrow CO_2$ reaction, one mole of CO_2 is produced from one mole of oxygen and hence $\Delta S \approx 0$; the CO_2 free energy line is thus almost horizontal. The carbon monoxide reaction is favored at high temperatures and consequently carbon is a very effective reducing agent, having a greater affinity for oxygen than most oxides.

Because of the positive slope to the ΔG° versus T line for most oxides in Figure 9.1, ΔG° tends to zero at some elevated temperature. This is known as the standard dissociation temperature when the oxide is in equilibrium with the pure element and oxygen at 1 atm pressure. In the case of gold, the oxide is not stable at room temperature, for silver Ag_2O dissociates when gently heated to about 200°C, and the oxides of the Pt group of metals around 1000°C. The other oxides dissociate at much higher temperatures. However, the temperature is affected by pressure since the free energy per mole of any gaseous phase varies with pressure P (atm) according to $G(P) = G^{\circ} + \mathbf{R}T \ln P$, whereas that for the solid phase is relatively unaffected. Thus, for the metal + oxygen → metal oxide reaction under standard conditions, $\Delta G^{\circ} = G^{\circ}_{\text{oxide}} - G^{\circ}_{\text{metal}} - G^{\circ}_{\text{oxygen}}$, and at P atm oxygen, $\Delta G = \Delta G^{\circ} - \mathbf{R}T \ln P_{O_2}$. The reaction is in equilibrium when $\Delta G = 0$ and hence

$$P_{O_2} = \exp[\Delta G^{\circ}/\mathbf{R}T] \tag{9.2}$$

is the equilibrium dissociation pressure of the oxide at the temperature T. If the pressure is lowered below this value the oxide will dissociate, if raised above the oxide is stable. The common metal oxides have very low dissociation pressures $\approx 10^{-10}\,\text{N m}^{-2}$ ($\sim 10^{-15}$ atm) at ordinary annealing temperatures and thus readily oxidize in the absence of reducing atmospheres.

The standard free energy change ΔG° is also related to the equilibrium constant K of the reaction. For the reaction discussed above, i.e.

$$Me + O_2 \rightarrow MeO_2,$$

the equilibrium constant $K = [MeO_2]/[Me][O_2]$ derived from the law of mass action. The active masses of the solid metal and oxide are taken equal to unity and that of the oxygen as its partial pressure under equilibrium conditions. The equilibrium constant at constant pressure, measured in atmospheres, is thus $K_P = 1/P_{O_2}$. It then follows that $\Delta G^{\circ} = -\mathbf{R}T \ln K_P$. To illustrate the use of these concepts, let us consider the reduction of an oxide to metal with the aid of a reducing agent (e.g. Cu_2O by steam). For the oxidation reaction,

$$4Cu + O_2 \rightarrow 2Cu_2O$$

and from Figure 9.1 at 1000 K, $\Delta G^{\circ} = -0.195\,\text{MJ mol}^{-1} = 1/P_{O_2}$, giving $P_{O_2} = 6.078 \times 10^{-6}\,\text{N m}^{-2}$. At 1000 K the equilibrium constant for the steam reaction $2H_2O \rightarrow 2H_2 + O_2$ is

$$K = P_{O_2}P^2_{H_2}/P^2_{H_2O} = 1.013 \times 10^{-15}\,\text{N m}^{-2}.$$

Thus, to reduce Cu_2O the term $P_{O_2} < 6 \times 10^{-11}$ in the steam reaction gives $P_{H_2}/P_{H_2O} > 10^{-5}$, so that an atmosphere of steam containing 1 in 10^5 parts of hydrogen is adequate to bright-anneal copper.

In any chemical reaction, the masses of the reactants and products are decreasing and increasing respectively during the reaction. The term chemical potential μ (=dG/dn) is used to denote the change of free energy of a substance in a reaction with change in the number of moles n, while the temperature, pressure and number of moles of the other substances are kept constant. Thus,

$$\mu_i = \mu_i^0 + \mathbf{R}T \ln P_i \tag{9.3}$$

and the free energy change of any reaction is equal to the arithmetical difference of the chemical potentials of all the phases present. So far, however, we have been dealing with ideal gaseous components and pure metals in our reaction. Generally, oxidation of alloys is of interest and we are then dealing with the solution of solute atoms in solvent metals. These are usually non-ideal solutions which behave as if they contain either more or less solute than they actually do. It is then convenient to use the activity of that component, a_i, rather than the partial pressure, P_i, or concentration, c_i. For an ideal solution $P_i = P_i^0 c_i$, whereas for non-ideal solutions $P_i = P_i^0 a_i$, such that a_i is an effective concentration equal to the ratio of the partial, or vapor, pressure of the ith component above the solution to its pressure in the standard state. The chemical potential may then be rewritten

$$\mu_i = \mu_i^0 + \mathbf{R}T \ln a_i, \tag{9.4}$$

where for an ideal gas mixture $a_i = P_i/P_i^0$ and by definition $P_i^0 = 1$. For the copper oxide reaction, the law of mass action becomes

$$K = \frac{a_{Cu_2O}^2}{a_{Cu}^4 a_{O_2}} = \frac{1}{P_{O_2}} = \exp[-\Delta G°/\mathbf{R}T], \tag{9.5}$$

where a_i^n is replaced by unity for any component present in equilibrium as a pure solid or liquid. Some solutions do behave ideally (e.g. Mn in Fe), obeying Raoult's law with $a_i = c_i$. Others tend to in dilute solution (e.g. Fe in Cu), but others deviate widely with a_i approximately proportional to c_i (Henry's law).

9.2.1.2 *Kinetics of oxidation*

Free energy changes indicate the probable stable reaction product but make no prediction of the rate at which this product is formed. During oxidation the first oxygen molecules to be absorbed on the metal surface dissociate into their component atoms before bonding chemically to the surface atoms of the metal. This process, involving dissociation and ionization, is known as chemisorption. After the build-up of a few adsorbed layers the oxide is nucleated epitaxially on the grains of the base metal at favorable sites, such as dislocations and impurity atoms. Each nucleated region grows, impinging on one another until the oxide film forms over the whole surface. Oxides are therefore usually composed of an aggregate of individual grains or crystals, and exhibit phenomena such as recrystallization, grain growth and creep involving lattice defects, just as in a metal.

If the oxide film initially produced is porous the oxygen is able to pass through and continue to react at the oxide/metal interface. Usually, however, the film is not porous and continued oxidation involves diffusion through the oxide layer. If oxidation takes place at the oxygen/oxide surface, then metal ions and electrons have to diffuse through from the underlying metal. When the oxidation reaction occurs at the metal/oxide interface, oxygen ions have to diffuse through the oxide and electrons migrate in the opposite direction to complete the reaction.

The growth of the oxide film may be followed by means of a thermobalance in conjunction with metallographic techniques. With the thermobalance it is possible to measure to a sensitivity of 10^{-7} g

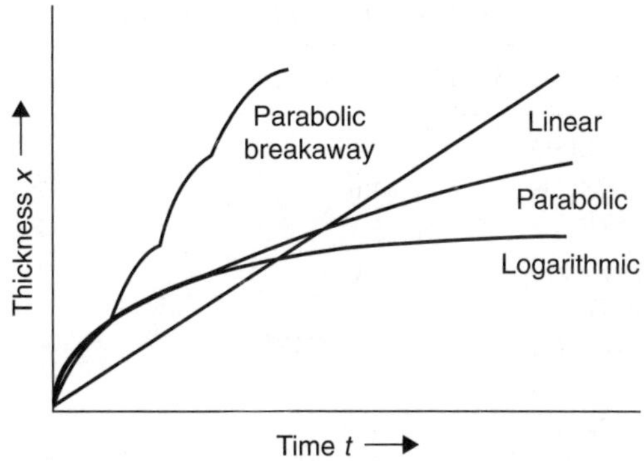

Figure 9.2 *Different forms of oxidation behavior in metals.*

in an accurately controlled atmosphere and temperature. The most common metallographic technique is ellipsometry, which depends on the change in the plane of polarization of a beam of polarized light on reflection from an oxide surface; the angle of rotation depends on the thickness of the oxide. Interferometry is also used, but more use is being made of replicas and thin films in the transmission electron microscope and the scanning electron microscope.

The rate at which the oxide film thickens depends on the temperature and the material, as shown in Figure 9.2. During the initial stages of growth at low temperatures, because the oxygen atoms acquire electrons from the surface metal atoms, a strong electric field is set up across the thin oxide film, pulling the metal atoms through the oxide. In this low-temperature range (e.g. Fe below 200°C) the thickness increases logarithmically with time ($x \propto \ln t$), the rate of oxidation falling off as the field strength diminishes.

At intermediate temperatures (e.g. 250–1000°C in Fe) the oxidation develops with time according to a parabolic law ($x^2 \propto t$) in nearly all metals. In this region the growth is a thermally activated process and ions pass through the oxide film by thermal movement, their speed of migration depending on the nature of the defect structure of the oxide lattice. Large stresses, either compressive or tensile, may often build up in oxide films and lead to breakaway effects when the protective oxide film cracks and spalls. Repeated breakaway on a fine scale can prevent the development of extensive parabolic growth and the oxidation assumes an approximately linear rate or even faster. The stresses in oxide film are related to the Pilling–Bedworth (P–B) ratio, defined as the ratio of the molecular volume of the oxide to the atomic volume of the metal from which the oxide is formed (Table 9.1). If the ratio is less than unity, as for Mg, Na and K, the oxide formed may be unable to give adequate protection against further oxidation right from the initial stages and, under these conditions, commonly found in alkali metals, linear oxidation ($x \propto t$) is obeyed. If, however, the P–B ratio is very much greater than unity, as for many of the transition metals, the oxide is too bulky and may also tend to spall.

At high temperatures oxide films thicken according to the parabolic rate law, $x^2 \propto t$, and the mechanism by which thickening proceeds has been explained by Wagner. As shown in Figure 9.3, point defects (see Chapter 3) diffuse through the oxide under the influence of a constant-concentration gradient. The defects are annihilated at one of the interfaces, causing a new lattice site to be formed. Specifically, zinc oxide thickens by the diffusion of zinc interstitials created at the metal/oxide interface through the oxide to the oxide/oxygen interface, where they are removed by the reaction

$$2Zn_i^{2+} + 4e + O_2 \rightarrow 2ZnO.$$

The concentration of zinc interstitials at the metal/oxide interface is maintained by the reaction

$$Zn_{(metal)} \rightarrow Zn_i^{2+} + 2e,$$

Table 9.1 *Some Pilling–Bedworth ratios.*

Metal and oxide	*Density of oxide (Mg m^{-3})*	*Pilling–Bedworth ratio*
Mg–MgO	3.6	0.8
Al–Al_2O_3	4.0	1.3
Ti–TiO_2	5.1	1.5
Zr–ZrO_2	5.6	1.5
Fe–Fe_2O_3	5.3	2.1
Cr–Cr_2O_3	5.1	2.1
Cu–Cu_2O	6.2	1.6
Ni–NiO	6.9	1.6
Si–SiO_2	2.7	1.9
U–UO_2	11.1	1.9
W–WO_3	7.3	3.3

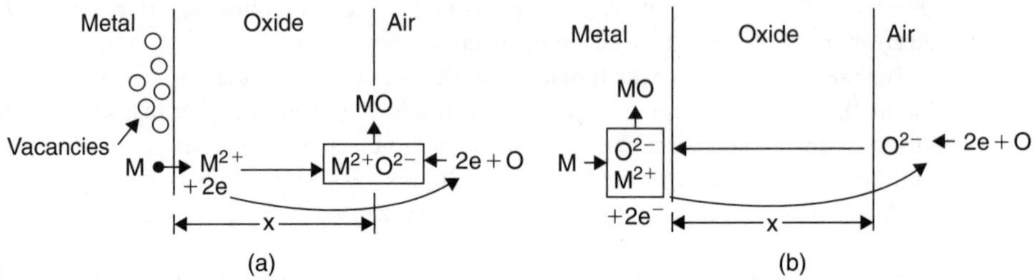

Figure 9.3 *Diffusion processes leading to oxide growth at: (a) oxide/air interface, e.g. Cu, Fe, and (b) metal/oxide interface, e.g. Ti, Zr (after Ashby and Jones, 2005).*

with the creation of vacancies in the zinc lattice. The migration of charged interstitial defects is accompanied by the migration of electrons and for thick oxide films it is reasonable to assume that the concentrations of the two migrating species are constant at the two surfaces of the oxide, i.e. oxide/gas and oxide/metal, governed by local thermodynamic equilibria. There is thus a constant concentration difference Δc across the oxide and the rate of transport through unit area will be $D\Delta c/x$, where D is a diffusion coefficient and x the film thickness. The rate of growth is then

$$\mathrm{d}x/\mathrm{d}t \propto D(\Delta c/x)$$

and the film thickens parabolically according to the relation

$$x^2 = \mathbf{k}t, \tag{9.6}$$

where **k** is a constant involving several structural parameters. Wagner has shown that the oxidation process can be equated to an ionic plus an electronic current, and obtained a rate equation for oxidation in chemical equivalents cm^{-2} s^{-1} involving the transport numbers of anions and electrons respectively, the conductivity of the oxide, the chemical potentials of the diffusing ions at the interfaces and the thickness of the oxide film. Many oxides thicken according to a parabolic law over some particular temperature range. It is a thermally activated process and the rate constant

$k = k_0 \times \exp[-Q/\mathbf{R}T]$, with Q equal to the activation energy for the rate-controlling diffusion process.

At low temperatures and for thin oxide films, a logarithmic rate law is observed. To account for this the Wagner mechanism was modified by Cabrera and Mott. The Wagner mechanism is only applicable when the concentrations of point defects and electrons are equal throughout the film; for thin oxide films this is not so, a charged layer is established at the oxide/oxygen interface. Here the oxygen atoms on the outer surface become negative ions by extracting electrons from the metal underneath the film and so exert an electrostatic attraction on the positive ions in the metal. When the oxide thickness is $\lesssim 10$ nm this layer results in an extremely large electric field being set up, which pulls the diffusing ions through the film and accelerates the oxidation process. As the film thickens the field strength decreases as the distance between positive and negative ions increases, and the oxidation rate approximates to that predicted by the Wagner theory.

As the scale thickens, according to a parabolic law, the resultant stress at the interface increases and eventually the oxide layer can fail either by fracture parallel to the interface or by a shear or tensile fracture through the layer. In these regions the oxidation rate is then increased until the build-up in stress is again relieved by local fracture of the oxide scale. Unless the scale fracture process occurs at the same time over the whole surface of the specimen, then the repeated parabolic nature of the oxidation rate will be smoothed out and an approximately linear law observed. This breakaway parabolic law is sometimes called paralinear, and is common in oxidation of titanium when the oxide reaches a critical thickness. In some metals, however, such as U, W and Ce, the linear rate process is associated with an interface reaction converting a thin protective inner oxide layer to a non-protective porous oxide.

Worked example

Zirconium oxidizes to ZrO_2. Given the density of Zr is 6.5 Mg m^{-3} and that of ZrO_2 5.9 Mg m^{-3}, calculate the Pilling–Bedworth ratio for the oxidation and indicate whether it would be protective.

Solution

The Pilling–Bedworth ratio $= \frac{\text{Volume of oxide}}{\text{Volume of metal}} = \frac{M_o \rho_M}{\rho_o M_M}$, where M_o is the molecular weight of ZrO_2, M_M is the weight of Zr, ρ_M is the density of Zr and ρ_o is the density of the oxide.

$$\text{P–B} = \frac{5.9 \times 123.22}{6.5 \times 91.22} = 1.5.$$

Since P–B > 1, the oxidation is protective.

Worked example

Corrosion of a galvanized steel sheet with a zinc coating 0.1 mm thick generates a corrosion current of 6×10^{-3} A m^{-2}. Determine if the coating is sufficient to give rust-free protection for 10 years (density of Zn = 7.13 Mg m^{-3}, atomic weight = 65.4).

Solution

$Zn \rightarrow Zn^{2+} + 2e$.

Number of electrons to produce $6 \times 10^{-3}\,A\,m^{-2}$ for 10 years

$$\begin{aligned} &= 6 \times 10^{-3} \times 10 \times 3.15 \times 10^{7}/1.6 \times 10^{-19} \\ &= 1.18 \times 10^{25} \\ &= 5.89 \times 10^{24} \text{ atoms of Zn} \\ \text{Mass of Zn} &= 5.89 \times 10^{24} \times 65.4/6.02 \times 10^{23} = 0.64\,\text{kg m}^{-2} \\ \text{Thickness} &= 0.64/7130 = 9 \times 10^{-5}\,\text{m} = 0.09\,\text{mm}. \end{aligned}$$

This is just smaller than the thickness of the coating (0.1 mm), so galvanizing is sufficient.

9.2.1.3 Parameters affecting oxidation rates

The Wagner theory of oxidation and its dependence on the nature of the defect structure has been successful in explaining the behavior of oxides under various conditions, notably the effects of small alloying additions and oxygen pressure variations. The observed effects can be explained by reference to typical *n*- and *p*-type semiconducting oxides. For oxidation of Zn to ZnO the zinc atoms enter the oxide interstitially at the oxide/metal interface ($Zn \rightarrow Zn_i^{2+} + 2e$), and diffuse to the oxide/oxygen interface. The oxide/oxygen interface reaction ($2Zn_i^{2+} + 4e + O_2 \rightarrow 2ZnO$) is assumed to be a rapid (equilibrium) process and, consequently, the concentration of defects at this interface is very small, and independent of oxygen pressure. This is found to be the case experimentally for oxide thicknesses in the Wagner region. By considering the oxide as a semiconductor with a relatively small defect concentration, the law of mass action can be applied to the defect species. For the oxide/oxygen interface reaction this means that

$$[Zn^{2+}]^2[e]^4 = \text{constant}.$$

The effect of small alloying additions can be explained (Wagner–Hauffe rule) by considering this equation. Suppose an alloying element is added to the metal that enters the oxide on the cation lattice. Since there are associated with each cation site, only two electron sites available in the valence band of the oxide, if the element is trivalent the excess electrons enter the conduction band, increasing the concentration of electrons. For a dilute solution the equilibrium constant remains unaffected and, hence, from the above equation, the net effect of adding the element will be to decrease the concentration of zinc interstitials and thus the rate of oxidation. Conversely, addition of a monovalent element will increase the oxidation rate. Experimentally it is found that Al decreases and Li increases the oxidation rate.

For Cu_2O, a *p*-type semiconductor, the oxide formation and cation vacancy ($Cu^+\square$) creation take place at the oxide/oxygen interface, according to

$$O_2 + 4Cu = 2Cu_2O + 4(Cu^+\square) + 4(e\square).$$

The defect diffuses across the oxide and is eliminated at the oxide/metal interface; the equilibrium concentration of defects is at the metal/oxide interface and the excess at the oxide/oxygen interface. It follows therefore that the reaction rate is pressure dependent. Applying the law of mass action to the oxidation reaction gives

$$[Cu^+\square]^4[e\square]^4 = \text{const } P_{O_2}$$

and, since electrical neutrality requires $[\text{Cu}^+\text{W}] = [\text{eW}]$, then

$$[\text{Cu}^+\square] = \text{const}\, P_{O_2}^{1/8} \tag{9.7}$$

and the reaction rate should be proportional to the 1/8th power of the oxygen pressure. In practice, it varies as $P_{O_2}^{1/7}$, and the discrepancy is thought to be due to the defect concentration not being sufficiently low to neglect any interaction effects. The addition of lower valency cations (e.g. transition metals) would contribute fewer electrons and thereby increase the concentration of holes, decreasing the vacancy concentration and hence the rate. Conversely, higher valency cations increase the rate of oxidation.

Worked example

Estimate the corrosion rate of iron in units of mm year^{-1} if the corrosion rate in an unstirred solution is controlled by the diffusion of oxygen. (F is 96 500 C mol^{-1}, the molar mass of iron is 55.8 g mol^{-1}, and its density is 7.9 g cm^{-3}.)

Solution

Fick's first law of diffusion: diffusion rate $\propto \frac{\Delta C}{x}$.
Diffusion-limited current density $i_L = \frac{nFD\Delta C}{\delta}$
D, diffusion coefficient: $\sim 2 \times 10^{-5}$ cm^2 s^{-1} for oxygen in room temperature in water.
ΔC, concentration difference: room temperature concentration of oxygen in water is ~0.25 mM (0.25×10^{-3} mol l^{-1}); it is assumed that the concentration is zero at the metal surface.
δ, diffusion distance: ~200 μm for natural convection in an unstirred solution.
This gives a value of $i_L \sim 100$ μA cm^{-2} for the diffusion-limited reduction of oxygen at a metal surface in an aqueous solution at room temperature.
Number of electrons to give a current of 100 μA cm^{-2} for one year

$$= \frac{10^{-4} \times 3.15 \times 10^{7}}{1.6 \times 10^{-19}} = 1.97 \times 10^{22} = 9.84 \times 10^{21} \text{ Fe atoms.}$$

$$\text{Mass of Fe} = \frac{9.84 \times 10^{21} \times 55.8}{6.02 \times 10^{23}} = 0.912 \text{ g cm}^{-2}$$

$$\text{Thickness} = \frac{0.912}{7.9} = 0.115 \text{ cm} = 1.15 \text{ mm.}$$

9.2.1.4 Oxidation resistance

The addition of alloying elements according to the Wagner–Hauffe rule just considered is one way in which the oxidation rate may be changed to give increased oxidation resistance. The alloying element may be added, however, because it is a strong oxide former and forms its own oxide on the metal surface in preference to that of the solvent metal. Chromium, for example, is an excellent additive, forming a protective Cr_2O_3 layer on a number of metals (e.g. Fe, Ni) but is detrimental to Ti, which forms an n-type anion-defective oxide. Aluminum additions to copper similarly improve the oxidation behavior by preferential oxidation to Al_2O_3. In some cases, the oxide formed is a compound oxide of both the solute and solvent metals. The best-known examples are the spinels with cubic structure

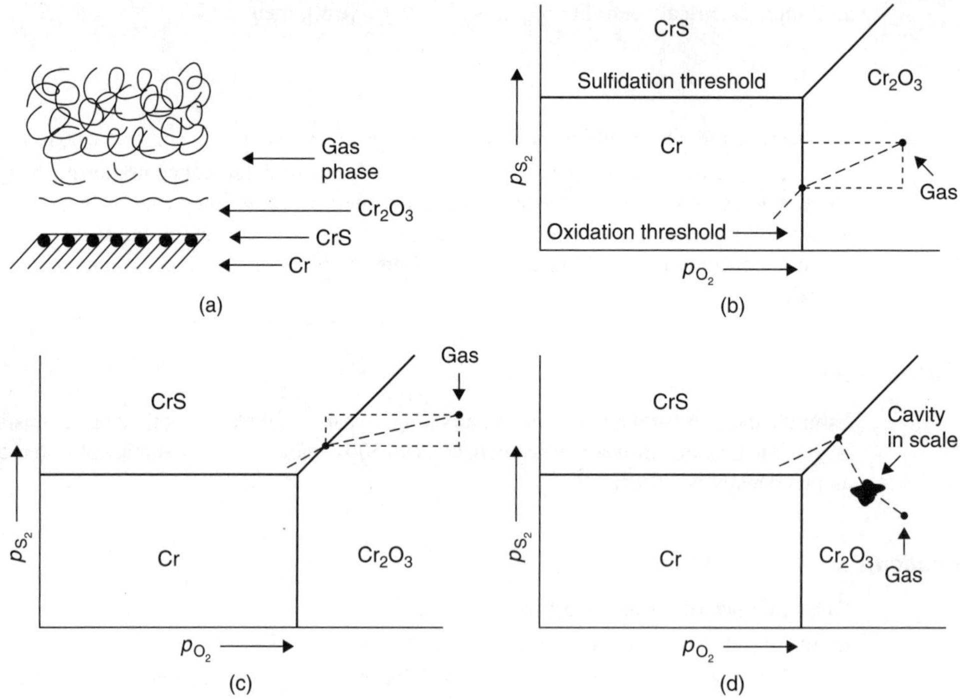

Figure 9.4 *Reaction paths for oxidation and sulfidation of chromium.*

(e.g. $NiO \cdot Cr_2O_3$ and $FeO \cdot Cr_2O_3$). It is probable that the spinel formation is temperature dependent, with Cr_2O_3 forming at low temperatures and the spinel at higher ones.

Stainless steels (ferritic, austenitic or martensitic) are among the best oxidation-resistant alloys and are based on Fe–Cr. When iron is heated above about 570°C the oxide scale which forms consists of wüstite, FeO (a *p*-type semiconductor) next to the metal, magnetite Fe_3O_4 (a *p*-type semiconductor) next and hematite Fe_2O_3 (an *n*-type semiconductor) on the outside. When Cr is added at low concentrations the Cr forms a spinel $FeO \cdot CrO_3$ with the wüstite and later with the other two oxides. However, a minimum Cr addition of 12% is required before the inner layer is replaced by Cr_2O_3 below a thin outer layer of Fe_2O_3. Heat-resistant steels for service at temperatures above 1000°C usually contain 18% Cr or more, and austenitic stainless steels 18% Cr, 8% Ni. The growth of Cr_2O_3 on austenitic stainless steels containing up to 20% Cr appears to be rate-controlled by chromium diffusion. Kinetic factors determine whether Cr_2O_3 or a duplex spinel oxide form. The nucleation of Cr_2O_3 is favored by higher Cr levels, higher temperatures and by surface treatments (e.g. deformation), which increase the diffusivity. Surface treatments which deplete the surface of Cr promote the formation of spinel oxide. Once Cr_2O_3 is formed, if this film is removed or disrupted, then spinel oxidation is favored because of the local lowering of Cr.

When chromium-bearing alloys, such as austenitic stainless steels, are exposed to the hot combustion products of fossil fuels, the outer layer of chromium oxide which forms is often associated with an underlying sulfide phase (Figure 9.4a). This duplex structure can be explained by using phase (stability) diagrams and the concept of 'reaction paths'. Accordingly, an isothermal section from the full phase diagram for the Cr–S–O system is shown in Figure 9.4b. The chemical activities of sulfur and oxygen in the gas phase are functions of their partial pressures (concentration). If the partial pressure of sulfur is relatively low, the composition of the gas phase will lie within the chromium oxide field and the alloy will oxidize (Figure 9.4b). Sulfur and oxygen diffuse through the growing

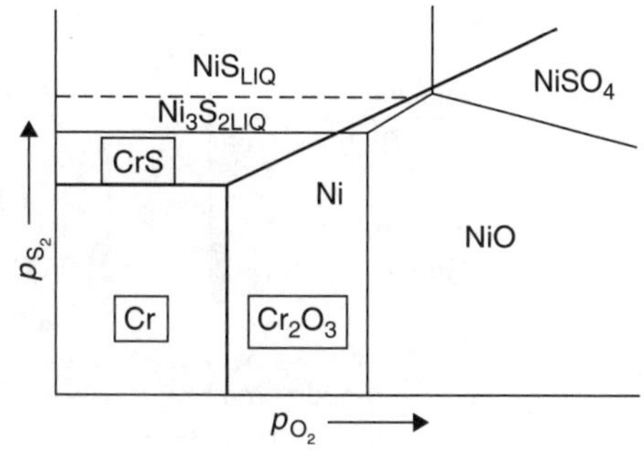

Figure 9.5 *Superimposition of isothermal sections from Cr–S–O and Ni–S–O systems.*

layer of oxide scale but S_2 diffuses faster than O_2; accordingly, the composition of the gas phase in contact with the alloy follows a 'reaction path', as depicted by the dashed line. Figure 9.4c shows the reaction path for gases with a higher initial partial pressure of sulfur. Its slope is such that first chromium oxide forms, and then chromium sulfide (i.e. $Cr + S = CrS$). Sometimes the oxide scale may crack or form voids. In such cases, the activity of S_2 may rise locally within the scale and far exceed that of the main gas phase. Sulfidation of the chromium then becomes likely, despite a low concentration of sulfur in the main gas stream (Figure 9.4d).

Relative tendencies of different metallic elements to oxidize and/or sulfidize at a given temperature may be gauged by superimposing their isothermal pS_2–pO_2 diagrams, as in Figure 9.5. For example, with the heat-resistant 80Ni–20Cr alloy (*Nichrome*), it can be reasoned that (1) Cr_2O_3 scale and CrS subscale are both stable in the presence of nickel, and (2) Cr_2O_3 forms in preference to NiO; that is, at much lower partial pressures of oxygen. The physical state of a condensed phase is extremely important because liquid phases favor rapid diffusion and thus promote corrosive reactions. Although nickel has a higher sulfidation threshold than chromium, the Ni–NiS eutectic reaction is of particular concern with Ni-containing alloys because it takes place at the relatively low temperature of 645°C.

9.2.2 Aqueous corrosion

9.2.2.1 *Electrochemistry of corrosion*

Metals corrode in aqueous environments by an electrochemical mechanism involving the dissolution of the metal as ions (e.g. $Fe \rightarrow Fe^{2+} + 2e$). The excess electrons generated in the electrolyte either reduce hydrogen ions (particularly in acid solutions) according to

$$2H^+ + 2e \rightarrow H_2$$

so that gas is evolved from the metal, or create hydroxyl ions by the reduction of dissolved oxygen according to

$$O_2 + 4e + 2H_2O \rightarrow 4OH^-.$$

The corrosion rate is therefore associated with the flow of electrons or an electrical current. The two reactions involving oxidation (in which the metal ionizes) and reduction occur at anodic and cathodic sites, respectively, on the metal surface. Generally, the metal surface consists of both anodic and cathodic sites, depending on segregation, microstructure, stress, etc., but if the metal is partially immersed there is often a distinct separation of the anodic and cathodic areas with the latter near the waterline where oxygen is readily dissolved (differential aeration). Figure 9.6 illustrates the formation

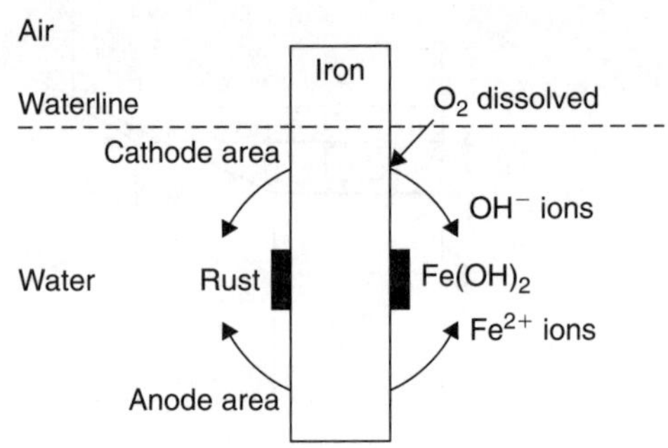

Figure 9.6 *Corrosion of iron by differential aeration.*

of such a differential aeration cell; Fe^{2+} ions pass into solution from the anode and OH^- ions from the cathode, and where they meet they form ferrous hydroxide, $Fe(OH)_2$. However, depending on the aeration, this may oxidize to $Fe(OH)_3$, red-rust $Fe_2O_3 \cdot H_2O$ or black magnetite, Fe_3O_4. Such a process is important when water, particularly sea water, collects in crevices formed by service, manufacture or design. In this form of corrosion the rate-controlling process is usually the supply of oxygen to the cathodic areas and, if the cathodic area is large, can often lead to intense local attack of small anode areas, such as pits, scratches, crevices, etc.

In the absence of differential aeration, the formation of anodic and cathodic areas depends on the ability to ionize. Some metals ionize easily, others with difficulty, and consequently anodic and cathodic areas may be produced, for example, by segregation or the joining of dissimilar metals. When any metal is immersed in an aqueous solution containing its own ions, positive ions go into solution until the resulting electromotive force (EMF) is sufficient to prevent any further solution; this EMF is the electrode potential or half-cell potential. To measure this EMF it is necessary to use a second reference electrode in the solution, usually a standard hydrogen electrode. With no current flowing, the applied potential cancels out the extra potential developed by the spontaneous ionization at the metal electrode over and above that at the standard hydrogen electrode. With different metal electrodes a table of potentials (E_0) can be produced for the half-cell reactions

$$M \rightarrow M^{n+} + ne, \tag{9.8}$$

where E_0 is positive. The usual convention is to write the half-cell reaction in the reverse direction so that the sign of E_0 is also reversed, i.e. E_0 is negative; E_0 is referred to as the standard electrode potential.

It is common practice to express the tendency of a metal to ionize in terms of this voltage, or potential, E_0, rather than free energy, where $\Delta G = -nFE_0$ for the half-cell reaction with nF coulombs of electrical charge transported per mole. The half-cell potentials are given in Table 9.2 for various metals, and refer to the potential developed in a standard ion concentration of one mole of ions per liter (i.e. unit activity), relative to a standard hydrogen electrode at 25°C, which is assigned a zero voltage. The voltage developed in any galvanic couple (i.e. two half-cells) is given by the difference of the electrode potentials. If the activity of the solution is increased, then the potential increases according to the Nernst equation, $E = E_0 + (\mathbf{R}T/nF) \ln a$.

The easily ionizable 'reactive' metals have large negative potentials and dissolve even in concentrated solutions of their own ions, whereas the noble metals have positive potentials and are deposited from solution. These differences show that the valency electrons are strongly bound to the positive core in the noble metals because of the short distance of interaction, i.e. $d_{\text{atomic}} \simeq d_{\text{ionic}}$. A metal will therefore displace from solution the ions of a metal more noble than itself in the series. When two dissimilar metals are connected in neutral solution to form a cell, the more metallic metal becomes

Table 9.2 *Electrochemical Series.*

Electrode reaction		*Standard electrode* electrode *potential E_0 (V)*
$Cs = Cs^+ + e$	Reactive metals ↑	−3.02
$Li = Li^+ + e$		−3.02
$K = K^+ + e$		−2.92
$Na = Na^+ + e$		−2.71
$Ca = Ca^{2+} + 2e$		−2.50
$Mg = Mg^{2+} + 2e$		−2.34
$Al = Al^{3+} + 3e$		−1.07
$Ti = Ti^{2+} + 2e$		−1.67
$Zn = Zn^{2+} + 2e$		−0.76
$Cr = Cr^{3+} + 3e$		−0.50
$Fe = Fe^{2+} + 2e$		−0.44
$Cd = Cd^{2+} + 2e$		−0.40
$Ni = Ni^{2+} + 2e$		−0.25
$Sn = Sn^{2+} + 2e$		−0.136
$Pb = Pb^{2+} + 2e$		−0.126
$H = 2H^+ + 2e$		0.00
$Cu = Cu^{2+} + 2e$	Noble metals ↑	+0.34
$Hg = Hg^{2+} + 2e$		+0.80
$Ag = Ag^+ + e$		+0.80
$Pt = Pt^{2+} + 2e$		+1.20
$Au = Au^+ + e$		+1.68

the anode and the metal with the lower tendency to ionize becomes the cathode. The Electrochemical Series indicates which metal will corrode in the cell, but gives no information on the rate of reactions. When an anode M corrodes, its ions enter into the solution initially low in M^+ ions, but as current flows the concentration of ions increases. This leads to a change in electrode potential known as polarization, as shown in Figure 9.7a, and corresponds to a reduced tendency to ionize. The current density in the cell is a maximum when the anode and cathode potential curves intersect. Such a condition would exist if the two metals were joined together or anode and cathode regions existed on the same metal, i.e. differential aeration. This potential is referred to as the corrosion potential and the current as the corrosion current.

In many reactions, particularly in acid solutions, hydrogen gas is given off at the cathode rather than the anode metal deposited. In practice, the evolution of hydrogen gas at the cathode requires a smaller additional overvoltage, the magnitude of which varies considerably from one cathode metal to another, and is high for Pb, Sn and Zn and low for Ag, Cu, Fe and Ni; this overvoltage is clearly of importance in electrodeposition of metals. In corrosion, the overvoltage arising from the activation energy opposing the electrode reaction decreases the potential of the cell, i.e. hydrogen atoms effectively shield or polarize the cathode. The degree of polarization is a function of current density and the potential E to drive the reaction decreases because of the increased rate of H_2 evolution, as shown in Figure 9.8 for the corrosion of zinc and iron in acid solutions. Corrosion can develop up to a rate given by the current when the potential difference required to drive the reaction is zero; for zinc this is i_{Zn} and for iron i_{Fe}. Because of its large overvoltage zinc is corroded more slowly than iron, even though there is

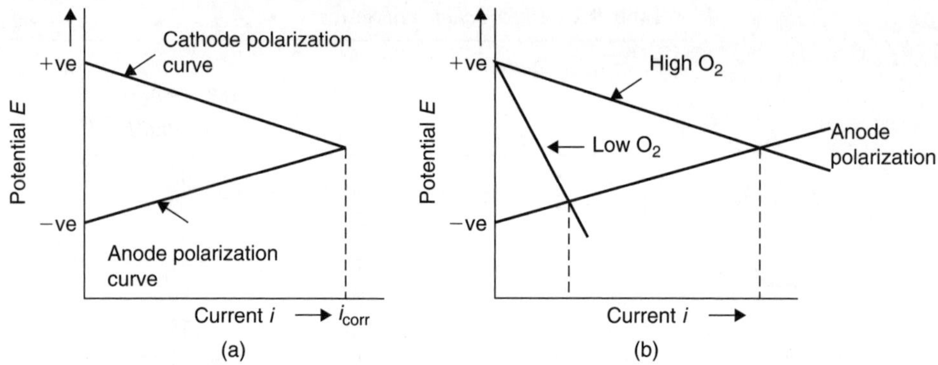

Figure 9.7 *Schematic representation of cathode and anode polarization curves (a) and influence of oxygen concentration on cathode polarization (b).*

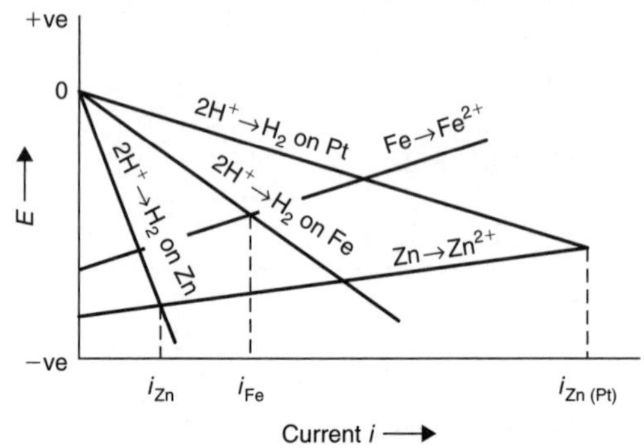

Figure 9.8 *Corrosion of zinc and iron and the effect of polarization.*

a larger difference between zinc and hydrogen than iron and hydrogen in the Electrochemical Series. The presence of Pt in the acid solution, because of its low overvoltage, increases the corrosion rate as it plates out on the cathode metal surface. In neutral or alkaline solutions, depolarization is brought about by supplying oxygen to the cathode area, which reacts with the hydrogen ions as shown in Figure 9.7b. In the absence of oxygen both anodic and cathodic reactions experience polarization and corrosion finally stops; it is well known that iron does not rust in oxygen-free water.

It is apparent that the cell potential depends on the electrode material, the ion concentration of the electrolyte, passivity and polarization effects. Thus, it is not always possible to predict the precise electrochemical behavior merely from the Electrochemical Series (i.e. which metal will be anode or cathode) and the magnitude of the cell voltage. Therefore, it is necessary to determine the specific behavior of different metals in solutions of different acidity. The results are usually displayed in Pourbaix diagrams, as shown in Figure 9.9. With stainless steel, for example, the anodic polarization curve is not straightforward as discussed previously, but takes the form shown in Figure 9.10, where the low-current region corresponds to the condition of passivity. The corrosion rate depends on the position at which the cathode polarization curve for hydrogen evolution crosses this anode curve, and can be quite high if it crosses outside the passive region. Pourbaix diagrams map out the regions of passivity for solutions of different acidity. Figure 9.9 shows that the passive region is restricted to certain conditions of pH; for Ti this is quite extensive, but Ni is passive only in very acid solutions and Al in neutral solutions. Interestingly, these diagrams indicate that for Ti and Ni in contact with each other in corrosive conditions then Ni would corrode, and that passivity has changed their order

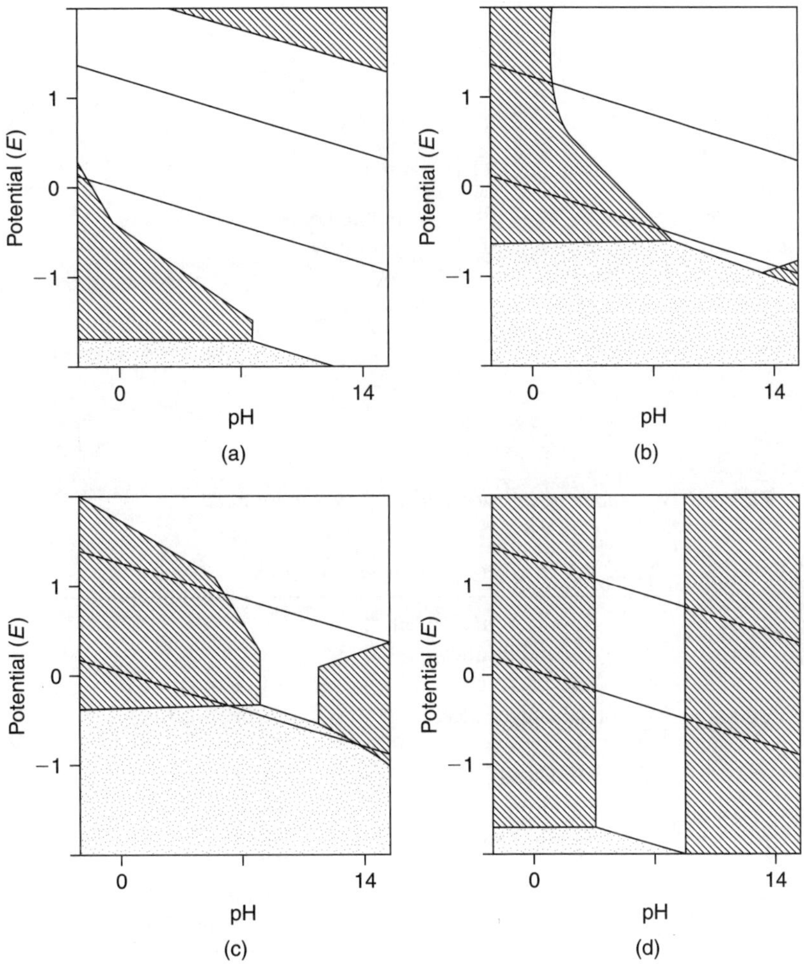

Figure 9.9 *Pourbaix diagrams for: (a) Ti, (b) Fe, (c) Ni, (d) Al. The clear regions are passive, the heavily shaded regions corroding and the lightly shaded regions immune. The sloping lines represent the upper and lower boundary conditions in service.*

in the Electrochemical Series. In general, passivity is maintained by conditions of high oxygen concentration but is destroyed by the presence of certain ions such as chlorides.

The corrosion behavior of metals and alloys can therefore be predicted with certainty only by obtaining experimental data under simulated service conditions. For practical purposes, the cell potentials of many materials have been obtained in a single environment, the most common being sea water. Such data in tabular form are called a Galvanic Series, as illustrated in Table 9.3. If a pair of metals from this series were connected together in sea water, the metal which is higher in the series would be the anode and corrode, and the further they are apart, the greater the corrosion tendency. Similar data exist for other environments.

9.2.2.2 Protection against corrosion

The principles of corrosion outlined above indicate several possible methods of controlling corrosion. Since current must pass for corrosion to proceed, any factor such as cathodic polarization which

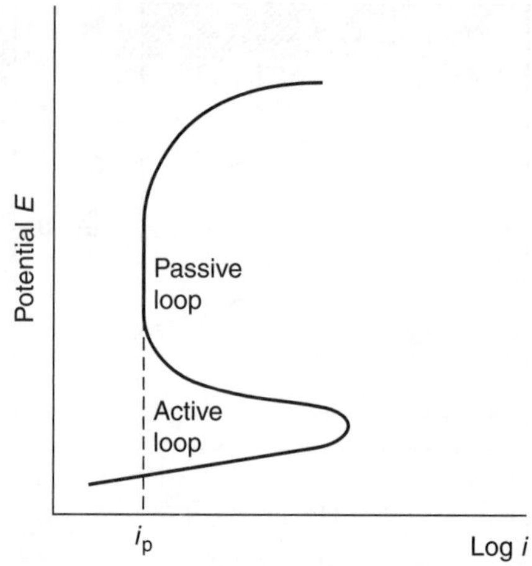

Figure 9.10 *Anode polarization curve for stainless steel.*

Table 9.3 *Galvanic Series in sea water.*

Increasing reactivity →	Anodic or most reactive		
	Mg and its alloys	*←Decreasing reactivity*	Cu
	Zn		Ni (active)
	Galvanized steel		*Inconel* (active)
	Al		Ag
	Mild steel		Ni (passive)
	Cast iron		*Inconel* (passive)
	Stainless steel (active)		*Monel*
	Pb		Ti
	Sn		Stainless steel (passive)
	Brass		Cathodic or most noble

reduces the current will reduce the corrosion rate. Metals having a high overvoltage should be utilized where possible. In neutral and alkaline solution, deaeration of the electrolyte to remove oxygen is beneficial in reducing corrosion (e.g. heating the solution or holding under a reduced pressure, preferably of an inert gas). It is sometimes possible to reduce both cathode and anode reactions by 'artificial' polarization (for example, by adding inhibitors which stifle the electrode reaction). Calcium bicarbonate, naturally present in hard water, deposits calcium carbonate on metal cathodes and stifles the reaction. Soluble salts of magnesium and zinc act similarly by precipitating hydroxide in neutral solutions.

Anodic inhibitors for ferrous materials include potassium chromate and sodium phosphate, which convert the Fe^{2+} ions to insoluble precipitates stifling the anodic reaction. This form of protection has no effect on the cathodic reaction and hence if the inhibitor fails to seal off the anode completely, intensive local attack occurs, leading to pitting. Moreover, the small current density at the cathode leads to a low rate of polarization and the attack is maintained. Sodium benzoate is often used as an anodic inhibitor in water radiators because of its good sealing qualities, with little tendency for pitting.

Some metals are naturally protected by their adherent oxide films; metal oxides are poor electrical conductors and so insulate the metal from solution. For the reaction to proceed, metal atoms have to diffuse through the oxide to the metal/liquid interface and electrons back through the high-resistance oxide. The corrosion current is very much reduced by the formation of such protective or passive oxide

Table 9.4 *Pre-coated automotive strip steel.*

Coated product	*Main characteristics*	*Typical applications*
Hot-dip zinc coated	Standard hot-dip product	Mainly non-visible parts
Galvannealed/iron–zinc alloy (heated to 500°C to allow interdiffusion of Fe and Zn)	Good weldability and paintability	Body panels and non-visible parts
Electro-zinc coated	Equivalent range of properties to CR; single-sided coatings are available	Body panels
Electro-zinc/nickel coated	Improved weldability	Body panels
Electro-zinc/nickel + organic	Improved corrosion resistance	Body panels

films. Al is cathodic to zinc in sea water even though the Electrochemical Series shows it to be more active. Materials which are passivated in this way are chromium, stainless steels, *Inconel* and nickel in oxidizing conditions. Reducing environments (e.g. stainless steels in HCl) destroy the passive film and render the materials active to corrosion attack. Certain materials may be artificially passivated by painting. The main pigments used are red lead, zinc oxide and chromate, usually suspended in linseed oil and thinned with white spirit. Slightly soluble chromates in the paint passivate the underlying metal when water is present. Red lead reacts with the linseed oil to form lead salts of various fatty acids, which are good anodic inhibitors.

Sacrificial or cathodic protection is widely used. A typical example is galvanized steel sheet when the steel is protected by sacrificial corrosion of the zinc coating. Any regions of steel exposed by small flaws in the coating polarize rapidly since they are cathodic and small in area; corrosion products also tend to plug the holes in the Zn layer. Zinc coatings are very important for corrosion protection in the car industry, as shown in Table 9.4. Zinc coating also acts as an important barrier layer in the use of steel panels for building cladding, one of the largest markets for steel. The zinc layer is followed by pretreatment layers of phosphate or chromate before a primer and decorative top coat (PVC plastisol paint up to 200 μm thick, polyester 25 μm or fluorocarbon 25 μm) is applied. Cathodic protection is also used for ships and steel pipelines buried underground. Auxiliary sacrificial anodes are placed at frequent intervals in the corrosive medium in contact with the ship's hull or pipe. Protection may also be achieved by impressing a d.c. voltage to make it a cathode, with the negative terminal of the d.c. source connected to a sacrificial anode.

Anodizing, the formation of a protective oxide, ~30 μm, may be achieved electrolytically with aluminum as the anode and electrolytes of either sulfuric acid or phosphoric acid. The anodic film produced may then be colored in a second-stage electrolytic process using metal salt solutions. The structure of the film produced by sulfuric acid consisting of long narrow pores is different to that produced by phosphoric acid, which is widened at the base of the pore (see Figure 9.11). The metal deposit in the wide pores gives rise to interference coloring as compared to conventional light scattering.

9.2.2.3 Corrosion failures

In service, there are many types of corrosive attack which lead to rapid failure of components. A familiar example is intergranular corrosion and is associated with the tendency for grain boundaries to undergo localized anodic attack. Some materials are, however, particularly sensitive. The common example of this sensitization occurs in 18Cr–8Ni stainless steel, which is normally protected by a passivating Cr_2O_3 film after heating to 500–800°C and slowly cooling. During cooling, chromium near the grain boundaries precipitates as chromium carbide. As a consequence, these regions are depleted

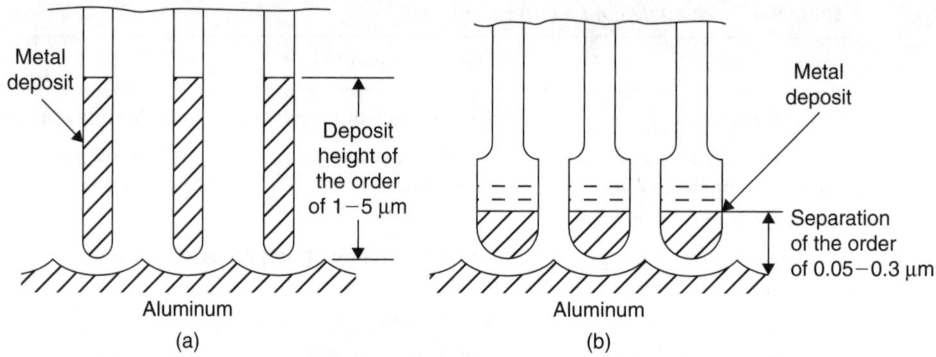

Figure 9.11 *Schematic microstructure of sulfuric acid (a) and phosphoric acid (b) anodizing treatment with metal deposits for coloration.*

in Cr to levels below 12% and are no longer protected by the passive oxide film. They become anodic relative to the interior of the grain and, being narrow, are strongly attacked by the corrosion current generated by the cathode reactions elsewhere. Sensitization may be avoided by rapid cooling, but in large structures that is not possible, particularly after welding, when the phenomenon (called weld decay) is common. The effect is then overcome by stabilizing the stainless steel by the addition of a small amount (0.5%) of a strong carbide former such as Nb or Ti, which associates with the carbon in preference to the Cr. Other forms of corrosion failure require the component to be stressed, either directly or by residual stress. Common examples include stress corrosion cracking (SCC) and corrosion fatigue. Hydrogen embrittlement is sometimes included in this category, but this type of failure has somewhat different characteristics and has been considered previously. These failures have certain features in common. SCC occurs in chemically active environments; susceptible alloys develop deep fissures along active slip planes, particularly alloys with low stacking-fault energy with wide dislocations and planar stacking faults, or along grain boundaries. For such selective chemical action the free energy of reaction can provide almost all the surface energy for fracture, which may then spread under extremely low stresses.

Stress corrosion cracking was first observed in α-brass cartridge cases stored in ammoniacal environments. The phenomenon, called season cracking since it occurred more frequently during the monsoon season in the tropics, was prevented by giving the cold-worked brass cases a mild annealing treatment to relieve the residual stresses of cold forming. The phenomenon has since extended to many alloys in different environments (e.g. Al–Cu, Al–Mg, Ti–Al), magnesium alloys, stainless steels in the presence of chloride ions, mild steels with hydroxyl ions (caustic embrittlement) and copper alloys with ammonia ions.

Stress corrosion cracking can be either transgranular or intergranular. There appears to be no unique mechanism of transgranular stress corrosion cracking, since no single factor is common to all susceptible alloys. In general, however, all susceptible alloys are unstable in the environment concerned but are largely protected by a surface film that is locally destroyed in some way. The variations on the basic mechanism arise from the different ways in which local activity is generated. Breakdown in passivity may occur as a result of the emergence of dislocation pile-ups, stacking faults, microcracks or precipitates (such as hydrides in Ti alloys) at the surface of the specimen, so that highly localized anodic attack then takes place. The gradual opening of the resultant crack occurs by plastic yielding at the tip and as the liquid is sucked in also prevents any tendency to polarize.

Many alloys exhibit coarse slip and have similar dislocation substructures (e.g. co-planar arrays of dislocations or wide planar stacking faults) but are not equally susceptible to stress corrosion. The observation has been attributed to the time necessary to repassivate an active area. Additions of Cr and Si to susceptible austenitic steels, for example, do not significantly alter the dislocation distribution but are found to decrease the susceptibility to cracking, probably by lowering the repassivation time.

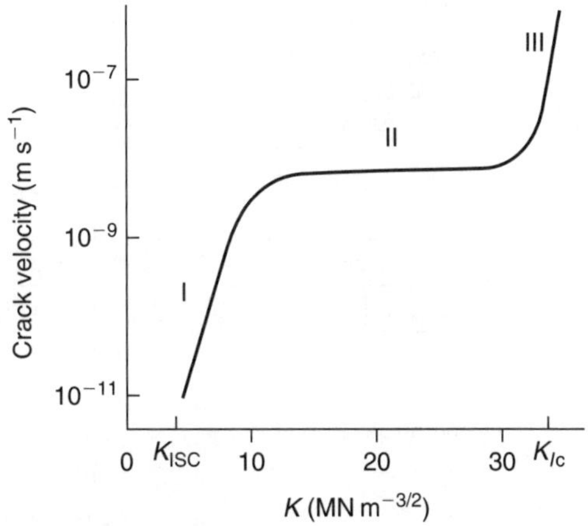

Figure 9.12 *Variation of crack growth rate with stress intensity during corrosion.*

The susceptibility to transgranular stress corrosion of austenitic steels, α-brasses, titanium alloys, etc., which exhibit co-planar arrays of dislocations and stacking faults may be reduced by raising the stacking-fault energy by altering the alloy composition. Cross-slip is then made easier and deformation gives rise to fine slip, so that the narrower, fresh surfaces created have a less severe effect. The addition of elements to promote passivation or, more importantly, the speed of repassivation should also prove beneficial.

Intergranular cracking appears to be associated with a narrow soft zone near the grain boundaries. In α-brass this zone may be produced by local dezincification. In high-strength Al alloys there is no doubt that it is associated with the grain boundary precipitate-free zones (i.e. PFZs). In such areas the strain rate may be so rapid, because the strain is localized, that repassivation cannot occur. Cracking then proceeds even though the slip steps developed are narrow, the crack dissolving anodically as discussed for sensitized stainless steel. In practice there are many examples of intergranular cracking, including cases (1) that depend strongly on stress (e.g. Al alloys), (2) where stress has a comparatively minor role (e.g. steel cracking in nitrate solutions) and (3) which occur in the absence of stress (e.g. sensitized 18Cr–8Ni steels); the last case is the extreme example of failure to repassivate for purely electrochemical reasons. In some materials the crack propagates, as in ductile failure, by internal necking between inclusions, which occurs by a combination of stress and dissolution processes. The stress sensitivity depends on the particle distribution and is quite high for fine-scale and low for coarse-scale distributions. The change in precipitate distribution in grain boundaries produced, for example, by duplex ageing can thus change the stress dependence of intergranular failure.

In conditions where the environment plays a role, the crack growth rate varies with stress intensity K in the manner shown in Figure 9.12. In region I the crack velocity shows a marked dependence on stress, in region II the velocity is independent of the stress intensity and in region III the rate becomes very fast as K_{IC} is approached. K_{ISC} is extensively quoted as the threshold stress intensity below which the crack growth rate is negligible (e.g. $\precsim 10^{-10}$ m s^{-1}) but, like the endurance limit in fatigue, does not exist for all materials. In region I the rate of crack growth is controlled by the rate at which the metal dissolves and the time for which the metal surface is exposed. While anodic dissolution takes place on the exposed metal at the crack tip, cathodic reactions occur at the oxide film on the crack sides, leading to the evolution of hydrogen, which diffuses to the region of triaxial tensile stress and hydrogen-induced cracking. At higher stress intensities (region II) the strain rate is higher, and then other processes become rate controlling, such as diffusion of new reactants into the crack tip region. In hydrogen embrittlement this is probably the rate of hydrogen diffusion.

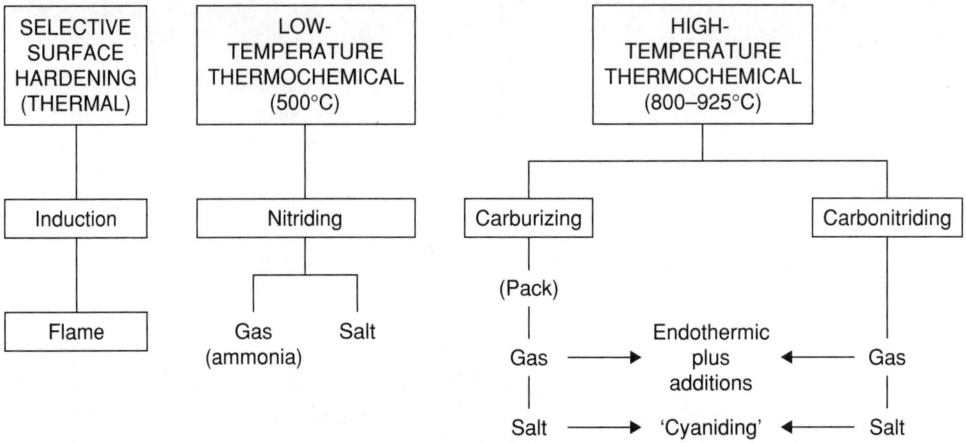

Figure 9.13 *Established methods of surface heat treatment.*

The influence of a corrosive environment, even mildly oxidizing, in reducing the fatigue life has been briefly mentioned in Chapter 6. The *S–N* curve shows no tendency to level out, but falls to low *S*-values. The damage ratio (i.e. corrosion fatigue strength divided by the normal fatigue strength) in salt water environments is only about 0.5 for stainless steels and 0.2 for mild steel. The formation of intrustions and extrusions gives rise to fresh surface steps which form very active anodic sites in aqueous environments, analogous to the situation at the tip of a stress corrosion crack. This form of fatigue is influenced by those factors affecting normal fatigue but, in addition, involves electrochemical factors. It is normally reduced by plating, cladding and painting, but difficulties may arise in localizing the attack to a small number of sites, since the surface is continually being deformed. Anodic inhibitors may also reduce the corrosion fatigue but their use is more limited than in the absence of fatigue because of the probability of incomplete inhibition, leading to increased corrosion.

Fretting corrosion, caused by two surfaces rubbing together, is associated with fatigue failure. The oxidation and corrosion product is continually removed, so that the problem must be tackled by improving the mechanical linkage of moving parts and by the effective use of lubricants.

With corrosion fatigue, the fracture mechanics threshold ΔK_{th} is reduced and the rate of crack propagation is usually increased by a factor of 2 or so. Much larger increases in crack growth rate are produced, however, in low-frequency cycling when stress corrosion fatigue effects become important.

9.3 Surface engineering

9.3.1 The coating and modification of surfaces

Surface heat treatments such as carburizing and nitriding are long-established industrial processes. They are still used in many industries, and rely on temperature and diffusion. These processes are summarized in Figure 9.13. The action of the new methods for coating or modifying material surfaces, such as vapor deposition and beam bombardment, can be highly specific and energy efficient. They allow great flexibility in controlling the chemical composition and physical structure of surfaces, and many materials which resisted conventional treatments can now be processed. Grain size and the degree of crystalline perfection can be varied over a wide range and beneficial changes in properties produced. The new techniques often eliminate the need for the random diffusion of atoms so that temperatures can be relatively low and processing times short. Scientifically, they are intriguing because their nature makes it possible to bypass thermodynamic restrictions on alloying and to form unorthodox solid solutions and new types of metastable phase.

Table 9.5 *Methods of coating and modifying surfaces (after R. F. Bunshah, 1984; by permission of Marcel Dekker).*

Atomistic deposition	*Particulate deposition*	*Bulk coatings*	*Surface modification*
Electrolytic environment	Thermal spraying	Wetting processes	Chemical conversion
Electroplating	Plasma spraying	Painting	Electrolytic
Electroless plating	Detonation gun	Dip coating	Anodizing (oxide)
Fused salt electrolysis	Flame spraying	Electrostatic spraying	Fused salts
Chemical displacement	Fusion coatings	Printing	Chemical – liquid
Vacuum environment	Thick film ink	Spin coating	Chemical – vapor
Vacuum evaporation	Enameling	Cladding	Thermal
Ion beam deposition	Electrophoretic	Explosive	Plasma
Molecular beam epitaxy	Impact plating	Roll bonding	Leaching
Plasma environment		Overlaying	Mechanical
Sputter deposition		Weld coating	Shot-peening
Activated reactive evaporation		Liquid-phase epitaxy	Thermal
Plasma polymerization			Surface enrichment
Ion plating			Diffusion from bulk
Chemical vapor environment			Sputtering
Chemical vapor deposition			Ion implantation
Reduction			Laser processing
Decomposition			
Plasma enhanced			
Spray pyrolysis			

The number and diversity of methods for coating or modifying surfaces makes general classification difficult. For instance, the energies required by the various processes extend over some five orders of magnitude. Illustrating this point, sputtered atoms have a low thermal energy (<1 eV), whereas the energy of an ion beam can be >100 keV. A useful introductory classification of methods for coating and modifying material surfaces appears in Table 9.5, which takes some account of the different forms of mass transfer. The first column refers to coatings formed from atoms and ions (e.g. vapor deposition). The second column refers to coatings formed from liquid droplets or small particles. A third category refers to the direct application of coating material in quantity (e.g. paint). Finally, there are methods for the near-surface modification of materials by chemical, mechanical and thermal means and by bombardment (e.g. ion implantation, laser processing).

Some of the methods that utilize deposition from a vapor phase or direct bombardment with particles, ions or radiation will be outlined: it will be apparent that each of the processes discussed has three stages: (1) a source provides the coating or modifying species, (2) this species is transported from source to substrate, and (3) the species penetrates and modifies the substrate or forms an overlay. Each stage is, to a great extent, independent of the other two stages, tending to give each process an individual versatility.

9.3.2 Surface coating by vapor deposition

9.3.2.1 Chemical vapor deposition

In the chemical vapor deposition (CVD) process a coating of metal, alloy or refractory compound is produced by chemical reaction between vapor and a carrier gas at or near the heated surface of a

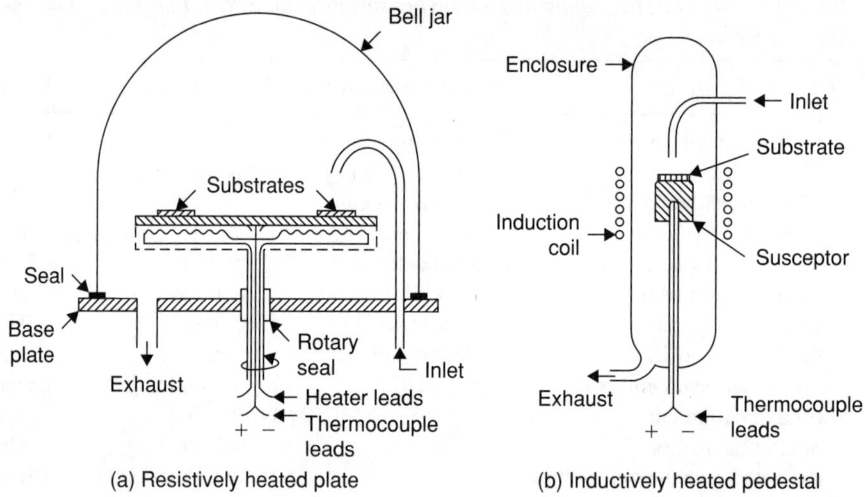

Figure 9.14 *Experimental CVD reactors (from Bunshah, 1984; by permission of Marcel Dekker).*

substrate (Figures 9.14a and b). CVD is not a 'line-of-sight' process and can coat complex shapes uniformly, having good 'throwing power'.[1] Typical CVD reactions for depositing boron nitride and titanium carbide respectively are:

$$BCl_{3(g)} + NH_{3(g)} \xrightarrow{500-1500^\circ C} BN_{(s)} + 3HCl_{(g)}$$

$$TiCl_{4(g)} + CH_{4(g)} \xrightarrow{800-1000^\circ C} TiC_{(s)} + 4HCl_{(g)}.$$

It will be noted that the substrate temperatures, which control the rate of deposition, are relatively high. Accordingly, although CVD is suitable for coating a refractory compound, like cobalt-bonded tungsten carbide, it will soften a hardened and tempered high-speed tool steel, making it necessary to repeat the high-temperature heat treatment. In one variant of the process (PACVD) deposition is plasma assisted by a plate located above the substrate which is charged with a radio-frequency bias voltage. The resulting plasma zone influences the structure of the coating. PACVD is used to produce ceramic coatings (SiC, Si_3N_4) but the substrate temperature of 650°C (minimum) is still too high for heat-treated alloy steels. The maximum coating thickness produced economically by CVD and PACVD is about 100 μm.

9.3.2.2 Physical vapor deposition

Although there are numerous versions of the physical vapor deposition (PVD) process, their basic design is either evaporation or sputter dependent. In the former case, the source material is heated by a high-energy beam (electron, ion, laser), resistance, induction, etc. in a vacuum chamber (Figure 9.15a). The rate of evaporation depends upon the vapor pressure of the source and the chamber pressure. Metals vaporize at a reasonable rate if their vapor pressure exceeds $1\,N\,m^{-2}$ and the chamber pressure is below $10^{-3}\,N\,m^{-2}$. The evaporant atoms travel towards the substrate (component), essentially following lines of sight.

[1] The term 'throwing power' conventionally refers to the ability of an electroplating solution to deposit metal uniformly on a cathode of irregular shape.

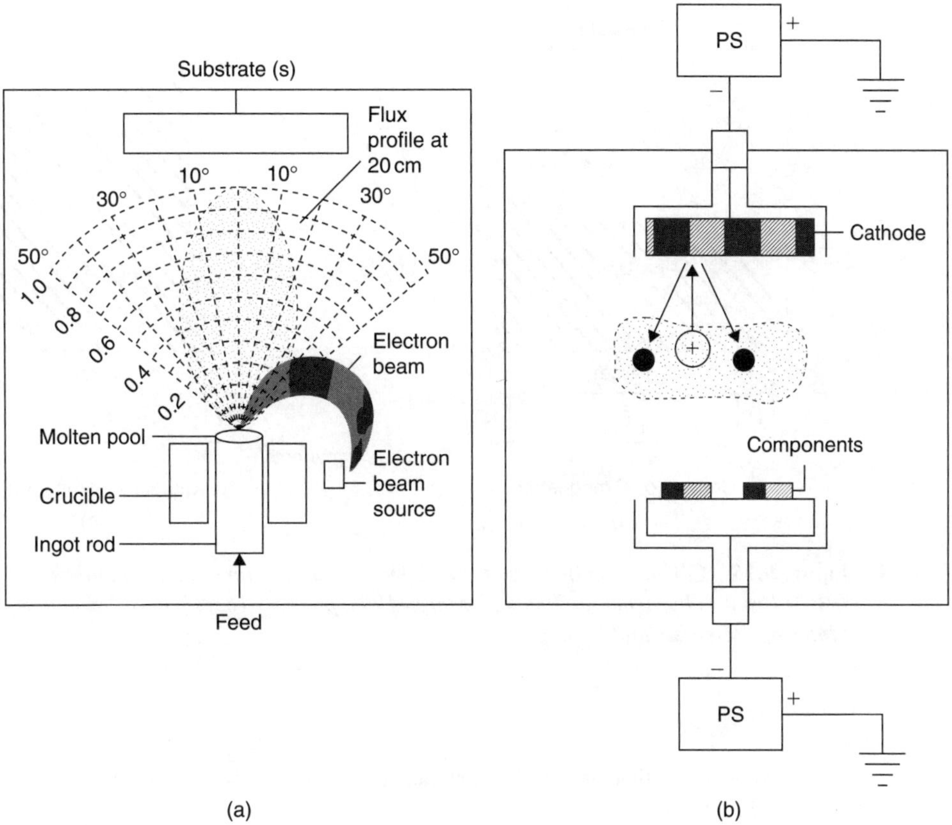

Figure 9.15 *Evaporation-dependent (a) and sputter-dependent (b) PVD (from Barrell and Rickerby, Aug 1989, pp. 468–473; by permission of the Institute of Materials, Minerals and Mining).*

When sputtering is used in PVD (Figure 9.15b), a cathode source operates under an applied potential of up to 5 kV (direct current or radio-frequency) in an atmosphere of inert gas (Ar). The vacuum is 'softer', with a chamber pressure of $1\text{–}10^{-2}\,\mathrm{N\,m^{-2}}$. As positive argon ions bombard the target, momentum is transferred and the ejected target atoms form a coating on the substrate. The 'throwing power' of sputter-dependent PVD is good and coating thicknesses are uniform. The process benefits from the fact that the sputtering yield (Y) values for metals are fairly similar. (Y is the average number of target atoms ejected from the surface per incident ion, as determined experimentally.) In contrast, with an evaporation source, for a given temperature, the rates of vaporization can differ by several orders of magnitude.

As in CVD, the temperature of the substrate is of special significance. In PVD, this temperature can be as low as 200–400°C, making it possible to apply the method to cutting and metal-forming tools of hardened steel. A titanium nitride (TiN) coating, $<5\,\mu\mathrm{m}$ thick, can enhance tool life considerably (e.g. twist drills). TiN is extremely hard (2400 HV), has a low coefficient of friction and a very smooth surface texture. TiN coatings can also be applied to non-ferrous alloys and cobalt-bonded tungsten carbide. Experience with the design of a TiN-coated steel has demonstrated that the coating/substrate system must be considered as a working whole. A sound overlay of wear-resistant material on a tough material may fail prematurely if working stresses cause plastic deformation of the supporting substrate. For this reason, and in accordance with the newly emerging principles of surface engineering, it has

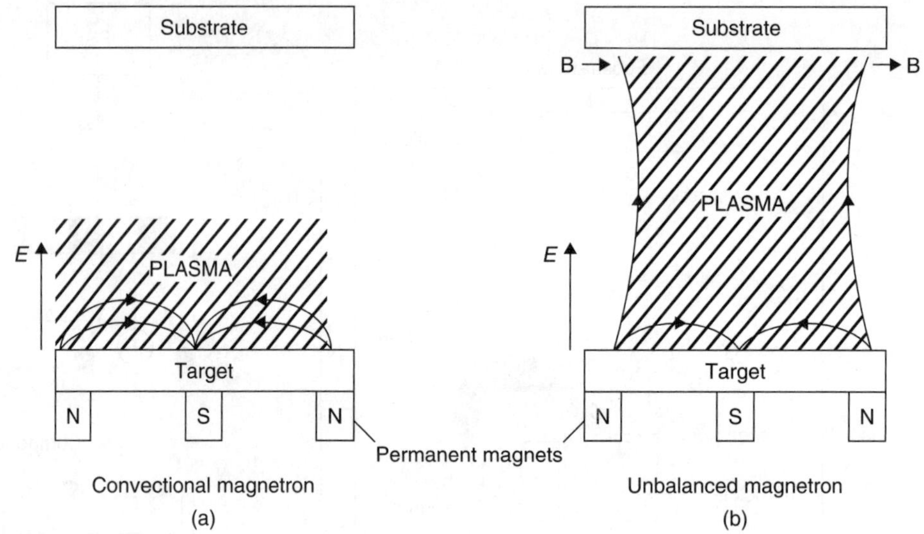

Figure 9.16 *Comparison of plasma confinement in conventional and unbalanced magnetrons (PVD) (from Kelly, Arnell and Ahmed, March 1993, pp. 161–165; by permission of the Institute of Materials, Minerals and Mining).*

been recommended that steel surfaces should be strengthened by nitriding before a TiN coating is applied by PVD.

Two important modifications of the PVD process are plasma-assisted physical vapor deposition (PAPVD) and magnetron sputtering. In PAPVD, also known as 'ion plating', deposition in a 'soft' vacuum is assisted by bombardment with ions. This effect is produced by applying a negative potential of 2–5 kV to the substrate. PAPVD is a hybrid of the evaporation-and sputter-dependent forms of PVD. Strong bonding of the PAPVD coating to the substrate requires the latter to be free from contamination. Accordingly, in a critical preliminary stage, the substrate is cleansed by bombardment with positive ions. The source is then energized and metal vapor is allowed into the chamber.

In the basic magnetron-assisted version of sputter-dependent PVD, a magnetic field is used to form a dense plasma close to the target. The magnetron, an array of permanent magnets or electromagnets, is attached to the rear of the target (water cooled) with its north and south poles arranged to produce a magnetic field at right angles to the electric field between the target and substrate (Figure 9.16a). This magnetic field confines electrons close to the target surface, increases the rate of ionization and produces a much denser plasma. The improved ionization efficiency allows a lower chamber pressure to be used; sputtered target atoms then become less likely to be scattered by gas molecules. The net effect is to improve the rate of deposition at the substrate. Normally the region of dense plasma only extends up to about 6 cm from the target surface. Development of unbalanced magnetron systems (Figure 9.16b) has enabled the depth of the dense plasma zone to be extended so that the substrate itself is subjected to ion bombardment. These energetic ions modify the chemical and physical properties of the deposit. (In one of the various unbalanced magnetron configurations, a ring of strong rare-earth magnetic poles surrounds a weak central magnetic pole.) This larger plasma zone can accommodate large, complex workpieces and rapidly forms dense, non-columnar coatings of metals or alloys. Target/substrate separation distances up to 20 cm have been achieved with unbalanced magnetron systems.

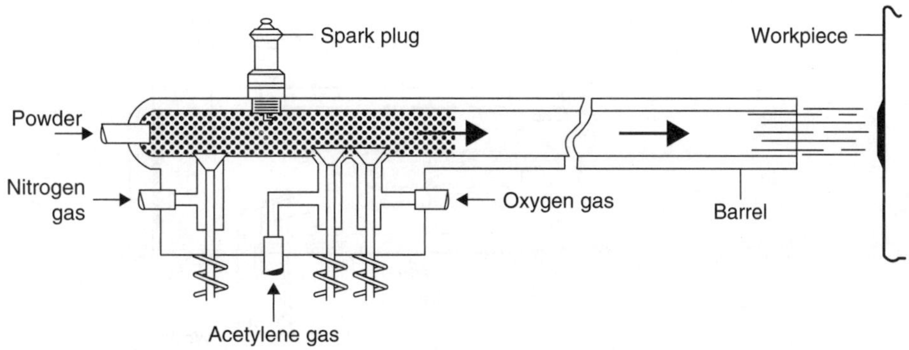

Figure 9.17 *Coating by detonation gun (from Weatherill and Gill, 1988; by permission of the Institute of Materials, Minerals and Mining).*

9.3.3 Surface coating by particle bombardment

Since the first practical realization of gas-turbine engines in the 1940s, the pace of engineering development has largely been prescribed by the availability of suitable high-temperature materials. Components in the most critical sections of the engine are exposed to hot products of combustion moving at high velocity. In addition, there are destructive agents passing through the engine, such as sea salt and sand. In this hostile environment, it is extremely difficult, if not impossible, to develop an alloy that combines the necessary high-temperature strength with corrosion resistance. Much effort has therefore been devoted to the search for alloy systems that will develop a thin, self-healing, 'protective' oxide scale. In practice, this outer layer does not prevent diffusing atoms from reaching and reacting with the alloy substrate, and it may also be subject to thinning by erosion. The difference in thermal expansion between the oxide (ceramic) scale and the metallic substrate can lead to rupture and spalling of the scale if the scale lacks plasticity or is weakly bonded to the alloy. Refractory coatings which resist wear and corrosion provide one possible answer to these problems.

The two established thermal spray methods[2] of coating selected here for brief description are used for gas-turbine components. In thermal spraying, powders are injected into very hot gases and projected at very high velocities onto the component (substrate) surface. On impact, the particles plastically deform and adhere strongly to the substrate and each other. The structure, which often has a characteristic lenticular appearance in cross-section, typically comprises refractory constituents, such as carbide, oxide and/or aluminide, and a binding alloy phase. Many types of thermally sprayed coatings can operate at temperatures $>1000°C$. They range in thickness from microns to millimeters, as required.

In the detonation-gun method (Figure 9.17) a mixture of metered quantities of oxygen and acetylene (C_2H_2) is spark-ignited and detonated. Powder of average diameter 45 mm is injected, heated by the hot gases and projected from the 1-m-long barrel of the gun onto the cooled workpiece at a velocity of roughly 750 m s^{-1}. Between detonations, which occur four to eight times per second, the barrel is purged with nitrogen. Typical applications, and coating compositions, for wear-resistant *D-Gun* coatings are bearing sealing surfaces (WC–9Co), compressor blades (WC–13Co) and turbine blade shroud interlocks ($Cr_3C_2/80Ni$–20Cr).

In the plasma-spray technique, powder is heated by an argon-fed d.c. arc (Figure 9.18) and then projected on to the workpiece at velocities of 125–600 m s^{-1}. A shielding envelope of inert gas (Ar)

[2] The Union Carbide Corporation has been granted patent rights for the *D-Gun* and plasma-spraying methods.

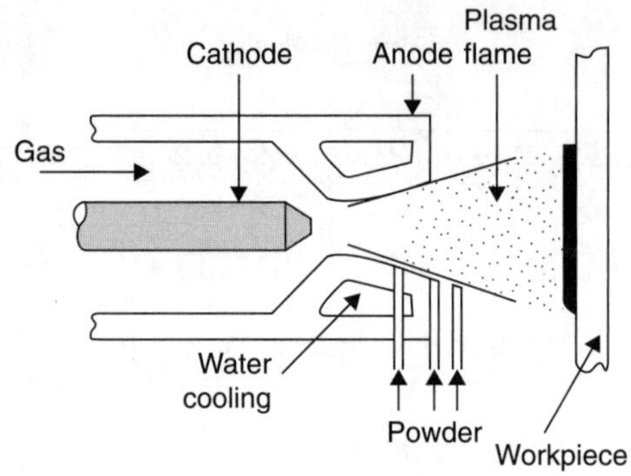

Figure 9.18 *Coating by plasma-spray torch (from Weatherill and Gill, 1988; by permission of the Institute of Materials, Minerals and Mining).*

is used to prevent oxidation of the depositing material. The process is used to apply MCrAlY-type coatings to turbine components requiring corrosion resistance at high temperatures (e.g. blades, vanes), where M signifies the high-m.p. metals Fe, Ni and/or Co. These coatings can accommodate much more of the scale-forming elements chromium and aluminum than superalloys (e.g. 39Co–32Ni–21Cr–7.5Al–0.5Y). They provide a reservoir of oxidizable elements and allow the 'protective' scale layer to regenerate itself. The small amount of yttrium improves scale adhesion. This particular composition of coating is used for hot gas path seals in locations where a small clearance between the rotating blades and the interior walls of the engine gives greater fuel efficiency. These coatings will withstand occasional rubbing contact.

9.3.4 Surface modification with high-energy beams

9.3.4.1 Ion implantation

The chemical composition and physical structure at the surface of a material can be changed by bombarding it, *in vacuo*, with a high-velocity stream of ions. The beam energy is typically about 100 keV; efforts are being made to increase the beam current above 5 mA so that process times can be shortened. Currently, implantation requires several hours. The ions may be derived from any element in the Periodic Table: they may be light (most frequently nitrogen) or heavy, even radioactive. Ion implantation[3] is a line-of-sight process: typically, a bombardment dose for each square centimeter of target surface is of the order of 10^{17}–10^{19} ions. These ions penetrate to a depth of 100–200 nm and their concentration profile in a plane normal to the surface is Gaussian. Beyond this modified region, the properties of the substrate are unaffected.

The beam usually has a sputtering effect which ejects atoms from the surface and skews the concentration profile. This effect is most marked when heavy ions or heavy doses are used. It is possible for a steady state to be achieved, with the rate of sputter erosion equal to the rate of implantation. Thus, depending upon the target, the type and energy of ion and the substrate material, sputter erosion is capable of limiting the amount of implantation possible. As a general guide, the maximum concentration of implanted ion is given, roughly, by the reciprocal of the sputtering yield (Y). As one would expect, Y increases in value with increases in ion energy. However, Y-values for pure metals are broadly similar, being about 1 or 2 for typical argon ion energies and not differing from each

[3] Pioneered by the UKAEA, Harwell, in the 1960s.

other by more than an order of magnitude. Thus, because of sputter, the maximum concentration of implanted ions possible is of the order of 40–50 at.%. In cases where it is difficult to attain this concentration, a thin layer of the material to be implanted is first deposited and then driven into the substrate by bombardment with inert gas ions (argon, krypton, xenon). This indirect method is called 'ion beam mixing'.

During ion bombardment each atom in the near-surface region is displaced many times. Various forms of structural damage are produced by the cascades of collisions (e.g. displacement spikes, vacancy/interstitial (Frenkel) pairs, dislocation tangles and loops, etc.). Damage cascades are most concentrated when heavy ions bombard target atoms of high atomic number (Z). The injection of atoms and the formation of vacancies tend to increase the volume of the target material so that the restraint imposed by the substrate produces a state of residual compressive stress. Fatigue resistance is therefore likely to be enhanced.

As indicated previously, the ions penetrate to a depth of about 300–500 atoms. Penetration is greater in crystalline materials than in glasses, particularly when the ions 'channel' between low-index planes. The collision 'cross-section' of target atoms for light ions is relatively small and ions penetrate deeply. Ion implantation can be closely controlled, the main process variables being beam energy, ion species, ion dose, temperature and substrate material.

Ion implantation is used in the doping of semiconductors, as discussed in Chapter 5, and to improve engineering properties such as resistance to wear, fatigue and corrosion. The process temperature is less than 150°C; accordingly, heat-treated alloy steels can be implanted without risk of tempering effects. Nitrogen implantation is applied to steel and tungsten carbide tools, and, in the plastics industry, has greatly improved the wear resistance of feed screws, extrusion dies, nozzles, etc. The process has also been used to simulate neutron damage effects in low-swelling alloys being screened for use in atomic fission and fusion reactors. A few hours' test exposure to an ion beam can represent a year in a reactor because the ions have a larger 'cross-section' of interaction with the atoms in the target material than neutrons. However, ions cannot simulate neutron behavior completely; unlike neutrons, ions are electrically charged and travel smaller distances (see Chapter 5).

9.3.4.2 Laser processing

Like ion implantation, the laser[4] process is under active development. A laser beam heats the target material locally to a very high temperature; its effects extend to a depth of 10–100 μm, which is about 1000 times greater than that for an ion beam. Depending on its energy, it can heat, melt, vaporize or form a plasma. The duration of the energy pulse can be 1 ns or less. Subsequent cooling may allow a metallic target zone to recrystallize, possibly with a refined substructure, or undergo an austenite/martensite transformation (e.g. automotive components). There is usually an epitaxial relation between the altered near-surface region and the substrate. Cooling may even be rapid enough to form a glassy structure (laser glazing). Surface alloying can be achieved by pre-depositing an alloy on the substrate, heating this deposit with a laser beam to form a miscible melt and allowing to cool. In this way, an integral layer of austenitic corrosion-resistant steel can be built on a ferritic steel substrate. In addition to its use in alloying and heat treatment, laser processing is used to enhance etching and electroplating (e.g. semiconductors).

The principal variables in laser processing are the energy input and the pulse duration (see Figure 9.19). For established techniques like cutting, drilling and welding metals, the rate of energy

[4] Light Amplification by Stimulated Emission of Radiation (LASER) devices provide photons of electromagnetic radiation that are in-phase (coherent) and monochromatic (see Chapter 5).

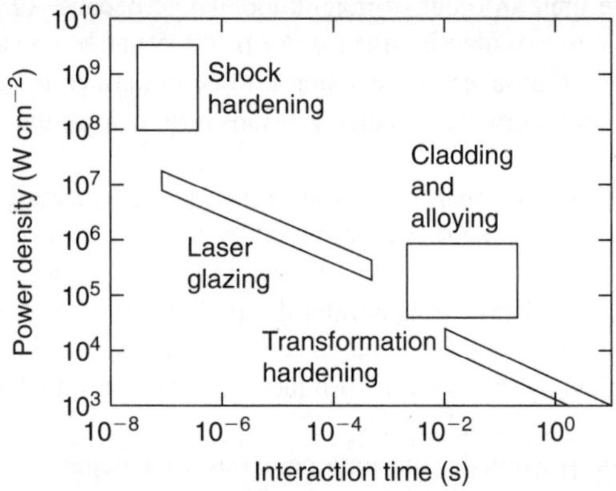

Figure 9.19 *Laser processing variables and applications.*

transfer per unit area ('power density') is of the order of 1 MW cm^{-2} and pulses are of relatively long duration (say, 1 ms). For more specialized functions, such as metal hardening by shock wave generation, the corresponding values are approximately 100 MW cm^{-2} and 1 ns. Short pulses can produce rapid quenching effects and metastable phases.

9.4 Thermal barrier coatings

Thermal barrier coatings used in gas turbine engines are 8% yttria/zirconia ceramics. This composition has a high melting point, low thermal conductivity and a reasonable coefficient of thermal expansion suitable for coating on nickel-based superalloys. The coating is applied either by electron beam physical vapor deposition (EBPVD) or by air plasma spraying (APS). EBPVD is superior, producing columnar grains from the component surface which are strongly bonded at their base but weakly bonded to each other. This microstructure allows thermal stresses to be easily accommodated. Generally a metallic bond coat layer between the ceramic and the superalloy is applied to provide adherence, together with a layer of alumina. Chemical bonding occurs between the bond coat, the alumina and the yttria/zirconia ceramic layer.

9.5 Diamond-like carbon

PVD coatings of TiN, TiCN and the like have provided good wear properties and improved machine tool life. However, the temperature of 450–500°C reached in the coating process has limited potential applications. Alternatively, carbon-based coatings have been developed, of which the diamond-like carbon coatings (DLC) have been successful. These coatings contain a mixture of graphitic and diamond bonds and have a process temperature ~150–200°C, a hardness ~2000 HV and a low coefficient of friction. The DLC coating may be deposited either by sputtering with argon ions or by dissociation with hydrocarbon gas.

9.6 Duplex surface engineering

It is well established that thin coatings of PVD TiN can provide a surface with improved tribological properties, e.g. low friction and resistance to wear, but fail under a high applied load. By contrast, deep hardened layers produced by energy beam surface alloying can sustain high contact stresses but still exhibit poor friction properties and high wear rates. It is the combination of both surface engineering technologies that is known as duplex surface engineering. Generally, the sequential application of two, or more, primary surface engineering technologies to produce a surface, or composite surface, with combined properties unobtainable with just one is duplex, or second generation, surface engineering.

Duplex surface engineering has been applied with considerable success to titanium alloys. This employs an oxygen diffusion (OD) deep case hardening via oxygen interstitial solid solution followed by a low-friction, high-wear-resistance diamond-like coating (DLC). To overcome the problem of poor coating adhesion to the substrate a graded TiN/TiC layer is developed between the hardened substrate and the DLC layer.

Problems

9.1 The weight gain for an oxidizing metal measured 47, 117 and 410 g m^{-2} after 20, 50 and 175 minutes respectively. Determine the weight gain after 500 minutes.

9.2 During service a steel sheet 1 mm thick is protected by tinning. If the tin plate is damaged over 0.5% of its area, generating a corrosion current of $2 \times 10^{-3}\ \text{A m}^{-2}$, determine whether the steel will rust through in 5 years. (Atomic weight of Fe = 55.9, density = $7.89\ \text{Mg m}^{-3}$.)

9.3 During the season cracking of brass, the crack growth rate is proportional to crack size at constant stress and to the square of stress at constant crack size. Determine the relationship between the crack growth rate and the stress intensity factor K. If the crack growth rate is $0.30\ \text{mm year}^{-1}$ at a stress of 4 MPa for a crack depth of 0.25 mm, calculate the constant in the relationship.

9.4 Assuming that oxidation of Fe to FeO obeys parabolic kinetics, calculate the weight of metal lost at 600°C after 1 year if the oxidation constant is $2 \times 10^{-7}\ \text{kg}^2\ \text{m}^{-4}\ \text{s}^{-1}$. What thickness does this correspond to?

9.5 During the electroplating of copper, 1 coulomb of electricity is passed through the cell, which has a cathode area of $10\ \text{mm}^2$. What thickness of copper is deposited?

9.6 A thin film of radioactive copper was electroplated on the end of a copper cylinder. After a high temperature anneal of 20 hours, the specimen was sectioned and the activity of each section counted. The following results were obtained:

Counts/minute	Distance from plated end (10^{-4} m)
5012	1
3981	2
2512	3
1413	4
525	5

Plot the data and determine the self-diffusion coefficient of Cu at the temperature of the experiment.

9.7 Explain the term second generation surface engineering. How are the properties of Ti–6Al–4V alloy improved by a diamond-like coating (DLC)?

9.8 Outline the important features of thermal barrier coatings for nickel-based superalloys.

9.9 Ion implantation followed by diffusion is a modern method of surface enrichment. If nitrogen ions are implanted to a depth of 0.1 μm producing a surface layer of 10 wt% N and a maximum hardness requires a nitrogen content of 0.25 wt%, calculate the time required to produce a

1 μm hardened layer by diffusion at 1000°C when $D = 9.6249 \times 10^{-12}$ m^2 s^{-1}. (Hint: the 'thin-surface-film' solution of Fick's second law is: $C_{x,t} = \frac{\alpha}{\sqrt{\pi D t}} \exp\left(\frac{-x^2}{4Dt}\right)$, where α is the amount of impurity per unit area present in the initial surface layer, D is the diffusivity, x is distance and t is time.)

9.10 Thermal barrier coating (TBC) systems consist of an outer zirconia-based layer of low thermal conductivity bonded to a superalloy substrate by an oxidation-resistant metallic bond coat. This bond coat is usually based on (Ni,Co)CrAlY or PtAl alloys, forms a protective alumina layer during service but in so doing prejudices the integrity of the TBC system. In a series of experiments, spallation of the outer ceramic layer occurred after the following exposure periods:

Temperature (°C)	Time to spallation (hours)
1300	4.71
1250	9.39
1200	19.6
1150	43.2
1100	101
1050	250
1000	668

Given that the thickness, h, of the alumina layer on the bond-coat surface grows parabolically with time, t, at a temperature T (in K) over the above range with a rate constant given by

$$k_p = \frac{h^2}{t} = 2.0 \times 10^{-6} \exp\left(-\frac{275\,000}{8.314T}\right) \text{ m}^2\,\text{s}^{-1},$$

what would you expect the spallation time to be at 975°C and why should you treat this prediction with caution?

9.11 The strain energy within a thin, flat oxide layer which remains attached to a thick metal substrate during cooling from the oxidation temperature, T_{ox}, is

$$W^* = E_{\text{ox}}(1 - \nu_{\text{ox}})(\Delta T)^2(\Delta\alpha)^2,$$

where W^* is the strain energy density (J m^{-3}) within the oxide, E_{ox} is the Young's modulus of the oxide (Pa), ν_{ox} is its Poisson's ratio, ΔT ($=[T_{\text{ox}} - T]$), where T is the current temperature, ΔT is the temperature drop (K) and $\Delta\alpha$ ($=[\alpha_{\text{metal}} - \alpha_{\text{ox}}]$), is the difference (in K^{-1}) between the linear thermal expansion coefficient of metal and oxide.

(i) Derive this equation for W^* assuming that the oxide is stress free at the oxidation temperature and that the oxide layer experiences equal biaxial, in-plane strains during cooling.

(ii) Show how the expression for W^* can be used to predict the critical temperature drop, ΔT_c, to initiate oxide spallation. [Hint: assume that there is no change in strain energy within the alloy substrate when spallation is initiated. Let the effective fracture energy of the oxide/metal interface be γ_f, in J m^{-2}, and the oxide thickness be h, in m.]

(iii) Calculate the value of γ_f using the following values: $E_{\text{ox}} = 380 \times 10^9$ Pa, $\nu_{\text{ox}} = 0.27$, $\Delta\alpha = 8.0 \times 10^{-6}$ K^{-1}, $h = 5.0 \times 10^{-6}$ m, $\Delta T_c = 671$ K. Why does this value of γ_f differ from the intrinsic value, of 2 J m^{-2}, for fracture of the oxide/metal interface?

9.12 Although chromia (Cr_2O_3) is thought to be thermodynamically more stable than the spinel $FeCr_2O_4$, demonstrate that it may be possible for iron to reduce chromia according to the reaction

$$Fe + 4/3Cr_2O_3 \rightarrow 2/3Cr + FeCr_2O_4. \qquad (A)$$

The standard free energy change ΔG° for this reaction is given in J by

$$\Delta G^\circ = -98\,115 + 36.0\,T,$$

where temperature, T, is in K.
What is the relevance of this calculation?
[Hint: assume the activity of a bulk phase is unity.]

9.13 The chromia layer formed by the selective oxidation of an austenitic steel grows parabolically with time, t, and reaches a thickness of 2 μm after the following exposure periods at different temperatures:

Temperature, °C (K)	Time to reach 2 μm thickness	
	Seconds	Hours
750 (1023)	2.85×10^7	7917
800 (1073)	6.94×10^6	1927
850 (1123)	1.92×10^6	533.3
900 (1173)	5.90×10^5	163.9
950 (1223)	2.00×10^5	55.6

Use this information to obtain an activation energy for the oxidation process and suggest a possible rate-controlling process.

Further reading

Bell, T. (1992). Surface engineering: its current and future impact on tribology. *J. Phys D.: Appl. Phys*. **25**, A297–306.

Bunshah, R. F. (1984). Overview of deposition technologies with emphasis on vapour deposition techniques. *In Industrial Materials Science and Engineering* Chap. 12. (edited by L. E. Murr), Marcel Dekker, New York.

Picraux, S. T. (1984). Surface modification of materials – ions, lasers and electron beams. *In Industrial Materials Science and Engineering* (edited by L. E. Murr), Chap. 11. Marcel Dekker, New York.

Shreir, L. L. (1976). *Corrosion*, Vols 1 and 2, 2nd edn. Newnes-Butterworth, London.

Trethewey, K. R. and Chamberlain, J. (1988). *Corrosion for Students of Science and Engineering*. Longman, Harlow.

Chapter 10

Non-metallics I – Ceramics, glass, glass-ceramics

10.1 Introduction

Ceramics are a class of important materials useful in both traditional and advanced applications. Ceramics represent a broad range of inorganic materials, typically oxides, borides, carbides and nitrides. Many ceramics are crystalline or have significant amounts of crystalline phases in their microstructure. Non-crystalline ceramics are generally termed glasses. Another important type of ceramic is carbon, which exists in a number of forms.

Many types of ceramics can be synthesized using unrefined raw materials (e.g. cement, porcelain, glass, refractories, etc.) and these are called 'coarse' ceramics. However, to meet the more demanding property requirements in modern engineering applications, certain special ceramics are made from highly refined raw materials using rigorously controlled composition and processing conditions. These advanced ceramics are called 'fine' or 'engineering' ceramics. Table 10.1 shows some examples of applications of engineering ceramics which cover a wide range of fields. One major category of usage of ceramics concerns their electric and electronic properties, with examples ranging from dielectricity, piezoelectricity, magnetism, semiconductivity and superconductivity (Section 5.7.5). The conductivity and internal reflection property for light of glass (amorphous SiO_2) also make it the choice of material for optical fibers (Section 5.10.2) which constitute an important element in today's telecommunication technologies (e.g. the internet). Being inorganic 'salts', many ceramics are chemically very inert, strong and light, making them suitable materials for biomedical implants (see Chapter 12). Ceramics are generally brittle and weak in tension in bulk, porous forms (see Section 10.6), but in particulate, coating or fully densed bulk forms, they can be very hard and wear resistant, and are excellent cutting and grinding materials for metals. The porosity of ceramics (see Section 10.2) plays an essential role not just in their mechanical strength but also in their thermal conductivity. Fully dense ceramics can be very efficient thermal conductors and porous ceramics very efficient heat insulators (Section 10.7).

10.2 Sintering of ceramic powders

Most engineering ceramics, being usually oxides, nitrides and carbides, have ionic and covalent bond character and are extremely high melting (typically $>2000^\circ C$), and so the traditional pouring-melt-into-mold fabrication techniques are not applicable. Ceramics are also generally very strong and brittle, and so they are difficult to be cut or machined. Instead, they are usually produced by 'powder metallurgy' methods based on sintering. Sintering is a process by which particles are bonded together by diffusion-related processes, usually carried out at temperatures well below the melting point of the particles, but are still high enough for diffusion to take place effectively. Densification is a natural consequence of sintering as the particles are bonded together, and in more complicated shapes non-uniform densification within the sample may lead to shape distortion and high residual stresses, which may cause internal cracking. Residual porosity, due to the free gaps between particles after sintering, reduces the strength σ quite seriously, often exponentially, as for alumina, when $\sigma = \sigma_0 \exp(-nP)$, where σ_0 is the strength for zero porosity, P is the volume fraction of porosity and n is an experimental constant. Porosity also reduces the modulus of elasticity and for some ceramics

Table 10.1 *Functions and applications of ceramics (after Ichinose, 1987).*

Functional group	*Oxide ceramics*			*Non-oxide ceramics*		
	Function	*Materials*	*Application*	*Function*	*Materials*	*Application*
Electric, electronic functions	Insulating	Al_2O_3, BeO	Substrates	Insulation	C, SiC	Substrates
	Dielectrical	$BaTiO_3$	Capacitor	Electrical conductivity	SiC, $MoSi_2$	Heat generator
	Piezo-electronics	$Pb(Zr_2, Ti_{1-2})O_3$, ZnO, SiO_2	Oscillator, ignition junction, surface elastic wave delaying junction	Semi-conductivity	SiC	Varistor, lightning shunt
	Magnetism	$Zn_{1-2}Mn_2$	Memory, operation junction	Electron emission	LaB_n	Electron-gun thermal anodes
	Semi-conductivity	Fe_3O_4	Magnetic core			
		SnO_2	Gas sensor			
		$ZnO–Bi_2O_3$	Varistor			
		$BaTiO_3$	Resistance junction			
	Ionic conductivity	βAl_2O_3,	NaS battery			
		Stable ZrO_2	Oxide sensor			
	Super conductivity	Y–Ba–Cu–O				
Mechanical functions	Wear resistance	Al_2O_3, ZrO_2	Polishing materials Grindstones	Wear resistance	B_3C, diamond	Wear-resistant materials, grindstones
	Machinability		Cutting tools	Machinability	C–BN, TiC, WC, TiN	Cutting tools
				Strength functions	Si_3N_4, SiC	Engines, heat resistors
				Lubricating functions	Sialon C, MoS_2, h-BN	Anticorrosives, tools Solid lubricants, mold-releasing agents
Optical functions	Fluorescence	Y_2O_2S: Eu	Fluorescent materials	Transparency	AION, nitrogen glass	Windows

	Transparency	Al_2O_3	Sodium lamp Mantle tube	Reflectiveness	TiN	Light collectors
	Optical polarization	PLZT	Optical polarization junction			
	Light conductivity	SiO_2 Multiple-constituent type glass	Optical communication fibers			
Thermal functions	Heat resistance	Al_2O_3	Structural refractories	Heat resistance	SiC, Si_3N_4, h-BN, C	Various refractories
	Heat insulation	K_2O-nTiO_2, CaO-nSiO_2, ZrO_2	Heat-insulating materials	Heat insulation	C, SiC	Various insulators
	Heat conductivity	BeO	Substrates	Heat conductivity	C, SiC	Substrates
Nuclear power-related functions	Nuclear reactor materials	UO_2 BeO	Nuclear fuel Moderator	Nuclear reactor materials	UC C, SiC C B_4C	Nuclear fuel, Coated nuclear fuel Moderator Control rod material
Biochemical functions	Teeth and bone materials	Al_2O_3, Ca_3(F, Cl) P_3O_{12}	Artificial teeth and bones	Corrosion resistance	h-BN, TiB_2 Si_3N_4, Sialon	Evaporation chamber Pump materials, others
	Carrier ability	SiO_2, Al_2O_3	Catalytic carriers		C, SiC	Anticorrosives

Table 10.2 *Powder synthesis methods (after Ichinose, 1987).*

Solid-state routes	Chemical reaction	Solid-state reaction Thermal decomposition of a solid Oxidization/reduction
	Crushing	Mechanical crushing (ball milling) Chemical pulverizing
Liquid-state routes	Liquefaction	Plasma jet
	De-solvent	Spray drying Freeze drying Emulsion Sol–gel
	Precipitation	Co-precipitation Electrolysis
Gas-state routes	Evaporation–condensation	PVD
	Gas-phase reaction	Thermal decomposition of gas Gas-phase oxidation Gas-phase synthesis

has been shown to obey an empirical relation of the form $E = E_0(1 - 1.9P + 0.9P^2)$, where E_0 is the modulus of a non-porous material.

The usual powder metallurgy route consists of three basic sequential stages: (i) powder processing, (ii) forming (shaping) and (iii) sintering (firing). After the powder particles are processed, they are mixed with a temporary binder, which is an organic wax-like material, to form a semi-solid slurry state. The slurry is then shaped into the desired geometry by a variety of methods, before finally sintered in a furnace. During the initial sintering stage, the binder material is removed by slowly heating to a temperature well below the final sintering temperature; too rapid heating may cause the binder to evaporate too rapidly, resulting in voids and cracks.

10.2.1 Powdering and shaping

A wide range of methods are available to produce ceramic powders, depending on the quality of the product desired. Cheap, coarse ceramic products such as porcelain and refractories are produced from powders made by crushing of natural minerals. Fine or engineering ceramics are sintered from high purity and extremely fine powders, which can be produced from a wide range of methods, as summarized in Table 10.2. In the solid route, mechanical crushing by means of ball milling is a widely used method for particle size reduction. In ball milling, the particles to be ground are placed inside a rotating container together with a grinding medium, usually another ceramic in the shape of balls or rods. As the mill turns, the grinding medium tumbles inside the container, thus grinding the particles into smaller sizes. Contamination from the mill lining as well as from the grinding medium is a common problem in ball milling. In the liquid route in Table 10.2, the material is turned into a liquid phase by either liquefaction or solution methods. A common method to liquefy ceramics is by plasma jet, and the resultant liquid trickle is then atomized to produce a fine powder upon solidification. Ceramic powders can also be synthesized from solution by dissolving the appropriate soluble salts into a solvent. The solvent can then be physically removed from the solution, by methods such as heating, spray drying, freeze drying, etc., to increase the concentration so as to precipitate out the solute. Alternatively, the solute can be separated from the solution by a suitable precipitation reaction. Another common method is 'sol–gel', in which a stable dispersion of submicron-sized particles (the 'sol') is first formed in a liquid. The sol is then allowed to form a 'gel', which is a softly bonded, polymer-like network throughout the gel, by methods such as evaporating part of the solvent, ageing or addition of a suitable electrolyte. Finally, the remaining solvent is removed or the solute precipitated out. Synthesis can also be done in the gaseous state. In the physical vapor deposition (PVD) method, the raw material is vaporized at a high temperature, followed by condensation to a fine powder form. In the chemical vapor deposition (CVD) method, the ceramic is produced by a chemical reaction of the vapor of metallic compounds.

Ceramic powder particles, after mixing with additives, including binders and sintering catalysts, can be compacted and shaped using methods such as die pressing, rubber mold pressing, extrusion and injection molding, and slip casting. In die pressing, the powder mixture is filled into a die and pressed into a strong solid product by uniaxial pressures between 200 and 2000 kg cm^{-2}. Die pressing is used for making refractories, tiles, electronic ceramics, nuclear fuel pallets and other products with relatively simple shapes and requirements for large quantities. Die pressing works well with mold shapes with a small aspect ratio, i.e. shapes that are short compared with their height. Long specimens are difficult to be compacted uniformly as the powder slurry has high viscosity, and internal friction between the particles and the die wall dissipates the pressure along the length of a long specimen, resulting in less compaction at the bottom part of the specimen, and thus non-uniform densification (shrinkage) and mechanical strength after sintering.

Long shapes such as rods or pipes can be produced by extrusion and more complicated shapes by injection molding. Slip casting is a technique for making shell-shaped objects with a small thickness. In slip casting, the ceramic slurry is poured into a hollow mold made of porous plaster. The mold absorbs water content from the slurry, causing the layer in contact with the mold surface to harden. The excessive slurry in the mold interior is then poured out, leaving only the hardened layer attached to the mold surface. After further drying, the casting is moved from the mold. Some applications such as electrical substrates (e.g. for capacitors), plate-fin heat exchangers and exhaust emission control devices require large area-to-volume ratio. Such sheet or tape shapes can be produced by a variant technique of slip casting, in which the slurry is spread on a flat surface. Drying of the slurry by, for example, blowing warm air on it causes the slurry sheet to harden.

10.2.2 Sintering

In the sintering process, the ceramic particles are bonded together under the action of heat and sometimes externally applied pressure. Figure 10.1 shows the contact region between two particles. During sintering, atoms on the convex surfaces of the two particles outside the contact region are transported to the gap region between the two particles, resulting in interparticle bonding there. Such material transport is due to the difference in surface curvature – compared to a flat surface, the convex surface curvature outside the gap region causes a higher internal pressure underneath the surface, as well as a higher vapor pressure above the surface, while the concave surface curvature in the gap region causes a lower internal pressure and vapor pressure. The difference in internal pressure and/or vapor pressure causes diffusion and viscous flow to occur at the sintering temperature. Thus, in the gaseous route, atoms after evaporating from the convex surface region diffuse towards the space above the gap region where the vapor pressure is lower and condense there. Inside the particles, the internal pressure gradient causes solid-state diffusion and/or liquid-state viscous flow, which transport atoms from the convex region to the concave, gap region. The rate of sintering depends critically on the sintering temperature but reduces gradually with time. For example, in the solid-state diffusion process, the linear shrinkage rate of the interparticle distance L is given by

$$\frac{\Delta L}{L} = \left(\frac{20\gamma\, a^3 D}{\sqrt{2}\mathbf{k}T}\right)^{2/5} r^{-6/5} t^{2/5},$$

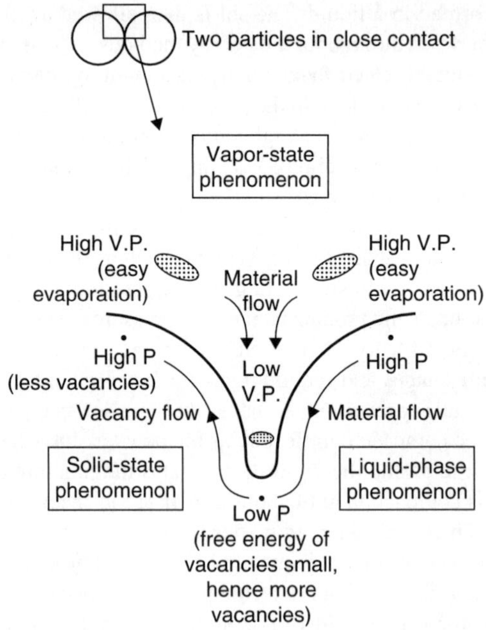

Figure 10.1 *Schematic illustration of sintering mechanisms.*

where γ is the surface energy, a the atomic size, D the self-diffusion coefficient, r the particle radius and t the time. The shrinkage rate rises with time but at a reducing rate, and hence the residual porosity can be rather easily controlled by the sintering time and temperature.

Common sintering processes include standard pressure sintering, hot pressing, hot isostatic pressing (HIP) and reaction sintering. Standard pressure sintering involves firing the materials at standard atmospheric pressure. It is a cheap but rather inefficient way for making traditional coarse ceramics such as porcelain, chinaware and refractories. In hot pressing, pressure is applied uniaxially on the green compact during sintering, which significantly increases the driving force for densification. Thus, compared to standard pressure sintering, hot pressing can reduce the densification time at the same temperature or, if the same time is used, it can allow the use of a lower sintering temperature, which would result in less grain growth.

While hot pressing is advantageous over standard pressure sintering, it is applicable only to simple product shapes. For complicated shapes, one has to resort to HIP. In HIP, gas pressure is applied uniformly over the workpiece while it is being sintered in a furnace. To achieve densification, the workpiece must first be evacuated and then sealed in a gas-impermeable envelope, often a glass capsule. If this is not done, any high-pressure gas will leak into the interior of the workpiece, resulting in pressure equalization and loss of the 'pressing' action. To apply the glass coating, the workpiece is first put inside a glass capsule, which is then evacuated and sealed. The whole piece is then heated until the glass softens and collapses under pressure onto the ceramic workpiece.

In reaction sintering, the ceramic is formed via a chemical reaction while it is being sintered at the same time. For example, in making Si_3N_4, a green compact of Si powder can be reaction sintered in a nitrogen flow to form the ceramic directly via the reaction $3Si + 2N_2 \rightarrow Si_3N_4$. Shrinkage and dimensional changes are typically small in reaction sintered products, although the resultant high porosity often leads to poor mechanical strength.

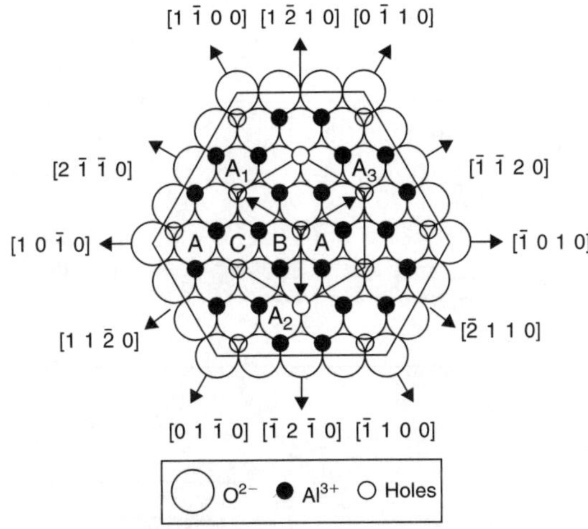

Figure 10.2 *Structure of α-alumina (corundum) viewed perpendicular to the (0 0 0 1) basal plane (from Hume-Rothery, Smallman and Haworth, 1988).*

10.3 Some engineering and commercial ceramics

10.3.1 Alumina

Alumina (Al_2O_3) is probably the most commonly used ceramic. It exists in two crystal forms, α-Al_2O_3 and γ-Al_2O_3, but α-alumina, or corundum, is of engineering importance and is shown in Figure 10.2. The structure is basically cph with layers of close-packed O^{2-} anions in **ABABAB**... sequence, with two-thirds of the octahedral sites filled with Al^{3+} cations to give the stoichiometric ratio and a coordination 6:4. γ-Alumina has the O^{2-} anions in an fcc arrangement giving a 'defect' spinel structure. It nevertheless has useful engineering properties in catalytic applications.

The mechanical strength, hardness and wear resistance, coupled with the low density of alumina, has been utilized in a wide variety of applications – for example, rotating seals in water pumps and washing machines, cutting tools, agricultural spray nozzles, shaft bearings, ball valves in general engineering and hip-joint prosthetic devices, dental implants in biomaterial applications (Chapter 12). The exceptional insulating properties of the ceramic are widely used in the electrical and electronic industries as substrates for circuitry, sealed packaging for semiconductor microcircuits, radio-frequency-transmitting valve envelopes, waveguide windows and transistor housings. The spark plug insulator exploits a wide range of alumina properties to fulfill its application in the very demanding conditions of engines, i.e. pressure, explosive temperatures, high voltage, corrosive gas environments, cyclic fatigue.

In all the above applications the alumina is processed to have a fine-grained structure (0.5–20 μm). Alumina may also be used in large tonnages in a coarse-grained form. Its high melting point (2050°C) and heat resistance make it appropriate for furnace refractories. In high-purity form it is used for thermocouple sheaths and for liquid metal filters.

10.3.2 Silica

Silica is a refractory ceramic widely used in steel and glass furnaces. The bricks are made by kiln-firing quartz around 1450°C to convert the quartz into a more open, less dense, form of mainly cristobalite, with tridymite. These structures are shown in Table 10.3. The high-temperature modification

Table 10.3 *Principal crystalline forms of silica.*

Form	*Range of stability (°C)*	*Modifications*	*Density (kg m^{-3})*
Cristobalite	1470–1723 (m.p.)	β (cubic)	2210
		α (tetragonal)	2330
Tridymite	870–1470	γ (?)	–
		β (hexagonal)	2300
		α (orthorhombic)	2270
Quartz	<870	β (hexagonal)	2600
		α (trigonal)	2650

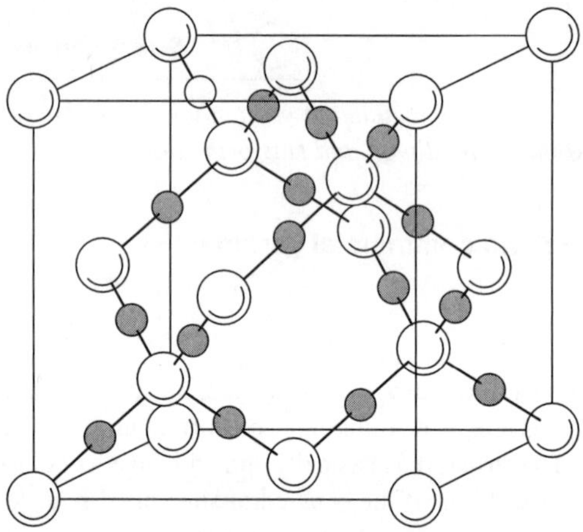

Figure 10.3 *Structure of β-cristobalite (from Kingery, Bowen and Uhlmann, 1976; by permission of Wiley-Interscience).*

β-cristobalite undergoes an $\alpha \rightarrow \beta$ transformation at 270°C accompanied by a significant 3% volume expansion. To avoid thermal stress cracking, the structure has to be cooled very slowly below 700°C and in furnace use is usually kept above this temperature throughout the life of the furnace lining.

The structure of β-cristobalite is shown in Figure 10.3, with small Si^{4+} cations located in a cubic arrangement similar to that of diamond. The larger O^{2-} anions form SiO_4^{4-} tetrahedra around each of the four occupied tetrahedral sites, with each Si^{4-} equidistant between two anions. The regular network structure of corner sharing tetrahedra has a fourfold coordination of anions around a cation, with CN = 4:2 the overall coordination of the structure.

As shown in Figure 10.3, the oxygen anions are bigger than the cations and their grouping in space determines the form of the structure. As the anion and cation sizes become more nearly equal, the paired coordination numbers change to 6:3 and then 8:4, typical of the structure groups represented by rutile (TiO_2) and fluorite (CaF_2) respectively.

10.3.3 Silicates

The silicates make up a variety of structures with a wide diversity of properties. A classification is shown in Table 10.4. Their structure is based on the SiO_4 tetrahedron.

Table 10.4 *Classification of Silicate structures.*

Type of silicate		$(Si^{4+} + Al^{3+}) : O^{2-}$*	*Arrangement of tetrahedra*†	*Examples*
Mineralogical name	*Chemical name*			
Nesosilicate	'Orthosilicate'	1:4	Isolated △ △	Zircon, olivines, garnets
Sorosilicate	'Pyrosilicate'	2:7	Pairing △△	Thortveitite
		1:3, 4:11	Linear chains 	Amphiboles, pyroxenes
Inosilicate	'Metasilicate'	3:9, 6:18, etc.	Rings 	Beryl
Phyllosilicate		2:5	Flat sheets 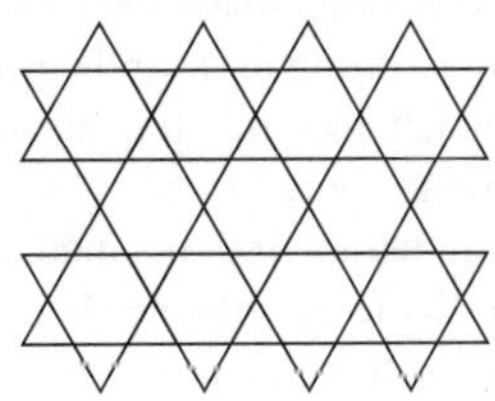	Micas, kaolin, talc
Tectosilicate		1:2	Framework	Feldspars, zeolites, ultramarines

*Only includes Al cations within tetrahedra.
†Δ represents a tetrahedron.

Zircon ($ZrSiO_4$) has the characteristic Si:O ratio of 1:4 for nesosilicates. Its structure comprises isolated SiO_4^{4-} tetrahedra. Its main industrial use is in the pottery industry, supporting ceramic ware during kiln firing. The olivines $(Mg^{2+},Fe^{2+})_2SiO_4^{4-}$ have compositions ranging from fosterite (Mg_2SiO_4, m.p. 1840°C) to Fe_2SiO_4 (fayalite, m.p. 1200°C) with useful refractory properties. The garnets $M_3^{2+}M_2^{3+}(SiO_4)_3$ have divalent cations Ca^{2+}, Mg^{2+}, Mn^{2+} or Fe^{2+} and trivalent cations Al^{3+}, Cr^{3+}, Fe^{3+} or Ti^{3+} and are extremely hard and colorful.

Representatives of the double-strand linear chains of SiO_4 tetrahedra include the fiber-forming asbestiform minerals, amosite (brown asbestos), crocidolite (blue asbestos) and chrysolite (white asbestos). The backbone of the chain of Si_4O_{11} units runs parallel to the fiber direction and allows cleavage parallel to that direction. Amosite fibers have been used for high-temperature insulation and are acid resistant, but are brittle. Chrysolite fibers are strong and flexible, and can be woven for brakes, conveyor belts and asbestos/cement composites. Nowadays, much more is known about the health hazards of these minerals and their industrial manufacture, and use is carefully regulated.

Kaolinite is an example of a silicate with a layered structure (Si:O = 2:5). It has the chemical formula $Al_2Si_2O_3(OH)_4$ and is the commonest clay mineral and a major constituent of china-clay. A schematic representation of the layers is shown in Figure 10.4. In each layer a sheet of SiO_4^{4-}

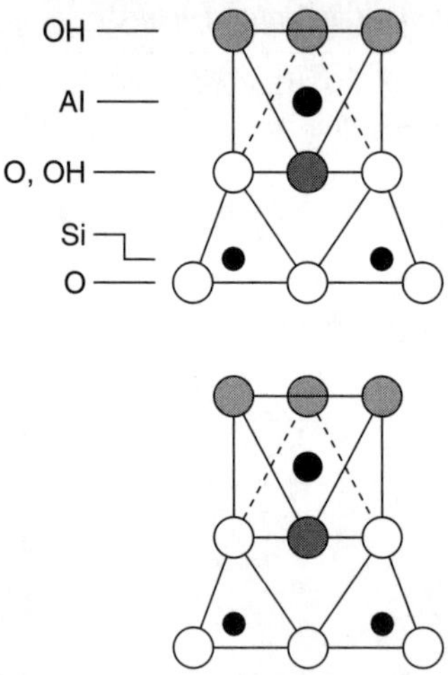

Figure 10.4 *Schematic representation of two layers of kaolinite structure (from Evans, 1966, by permission of Cambridge University Press).*

tetrahedra lies parallel to a sheet of $AlO_2(OH)_4$ octahedra, with the two sheets sharing common O^{2-} anions. There is strong ionic and covalent bonding within the layers but weak Van der Waals between the layers, which accounts for the softness, easy cleavage and moldability of the mineral. Another layered structure includes talc ($Mg_3Si_4O_{10}(OH)_4$), which has tetrahedral–tetrahedral layering, in contrast to the tetrahedral–octahedral layering of kaolinite. Another is mica, which allows perfect cleavage parallel to the silicate sheets.

Framework structures in which the SiO_4^{4-} tetrahedra share all four corners to form a regular and extended three-dimensional network (see Table 10.4) include the feldspars and zeolites. The zeolites have an unusual open structure with tunnels and/or polyhedral cavities. They are used as molecular sieves when the tunnel dimensions are controlled to separate molecules of different sizes in a flowing gaseous mixture. Overall the $(Al^{3+} + Si^{4+})$:O^{2-} ratio is 1:2 for zeolites to which water molecules (H_2O) are loosely bound. Dehydrated zeolites are easily produced by heating the mineral without altering the structure. In this condition it is used for absorbing gases (e.g. CO_2, NH_3). Dehydrated zeolites have a large surface area/mass ratio and are increasingly used as catalysts in the petrochemical industry.

10.3.4 Perovskites, titanates and spinels

There are many other complex oxides, of which the perovskites, titanates and spinels are industrially important. Although $CaTiO_3$ is a prototype perovskite, the development of the ceramic superconductors with a critical temperature $T_c \sim 35$ K, well above liquid hydrogen temperature, has brought the perovskite $YBa_2Cu_3O_{7-x}$ into prominence. The field has developed considerably in the last 10 years, with a wide range of new superconducting oxides (see Section 5.7.5) researched.

Barium titanate has also been widely studied for its electrical properties. It is cubic below 120°C and tetragonal above. The new structure exhibits marked ferroelectric characteristics. The basic spinel $MgAl_2O_4$ has also been widely developed to produce other valency II–III spinels, II–IV spinels and I–VI spinels (e.g. $Mg^{2+}Al_2^{3+}O_4^{2-}$, $Mg_2^{2+}Ge^{4+}O_4^{2-}$ and $Ag_2^{1+}Mo^{6+}O_4^{2-}$). Most spinels are of the II–III type. The II–III $MgAl_2O_4$ has a cubic unit cell comprising eight fcc sub-cells and overall contains

32 O^{2-} anions in fcc arrangement. The cations are distributed among the tetrahedral (CN = 4) and octahedral (CN = 6) interstices of these anions.

Ferrites ($NiFe_2O_4$ and $CoFe_2O_4$) form a spinel structure in which the allocation of cations to tetrahedral and octahedral sites is reversed. This reversal produces new and significant magnetic and electrical characteristics. Thus, $NiFe_2O_4$ may be written as $Fe^{3+}(Fe^{3+}Ni^{2+})O_4^{2-}$, which indicates that half the Fe^{3+} cations are in tetrahedral sites and the remainder of Fe^{3+} enter the octahedral sites with Ni^{2+}. The degree of reversal λ is influenced by heat treatment and cooling, varying between $\lambda = 0$ and 0.5. Magnetite is an 'inverse' spinel with a formula $Fe^{3+}(Fe^{3+}Fe^{2+})O_4^{2-}$ and $\lambda = 0.5$.

10.3.5 Silicon carbide

Silicon carbide is extremely hard ($H_v = 3300\,\mathrm{kg\,mm^{-2}}$), has outstanding high-temperature strength, abrasive resistance, chemical inertness and high thermal conductivity. It can be manufactured in a wide variety of shapes and forms from powders and polycrystalline products to fibers and single-crystal whiskers. It exists in two structural forms, non-cubic α-SiC (hexagonal) and cubic β-SiC up to 2100°C. The main source of the material is the production of powder via the Acheson carbothermic process, which reacts silica sand with coke at a temperature of 2600°C with a heavy current passed through a graphite conductor ($SiO_2 + 3C \rightarrow \alpha\text{-}SiC + 2CO$). There are several methods of forming the powder, including drypressing, HIPing, slip casting, extrusion and injection molding. The main methods of firing are hot pressing (HP SiC), pressureless sintering (S SiC) and reaction sintering (Si SiC).

Because of its chemical inertness and high temperature strength, SiC has been particularly applied to smelting processes, e.g. zinc and iron. In zinc smelting these are SiC components such as distillation retorts, trays and the rotating condensation impellers which have to withstand the action of molten zinc and zinc vapor. In iron making, silicon carbide has been used to line the water-cooled bosh and stack zones of iron-smelting blast furnaces, where its high thermal conductivity and abrasion resistance are very important. At ambient temperatures, SiC is used in machine components subjected to abrasive wear (e.g. mechanical seals, bearings, slurry pump impellers, wire dies, fiber spinnerets). In high-temperature engineering, silicon carbide is now regarded, together with silicon nitride and the sialons, as a leading candidate material for service in heat engine designs which involve operation at temperatures in excess of 1000°C (e.g. glow plugs, turbocharger rotors, turbine blades and vanes, rocket nozzles). Glow plugs minimize the hazards of 'flame-out' in the gas turbine engines of aircraft. Their function is to reignite the fuel/air mixture. They must withstand considerable thermal shock (Section 10.6.1), since temperature rises rapidly on start-up. An old established application makes use of the electrical conductivity in resistor elements for furnaces (e.g. Globars) which can operate in air up to 1650°C.

10.3.6 Silicon nitride

Silicon nitride (Si_3N_4) is usually prepared by the reaction of silicon powder with nitrogen at 1200°C according to $3Si + 2N_2 \rightarrow Si_3N_4$, an exothermic reaction which needs careful temperature control. It exists in two crystalline hexagonal forms, α and β, which are predominantly covalent. The α-form consists of interleaved corrugated sheets of eight- and 12-membered rings of silicon and nitrogen atoms. Each nitrogen atom is bonded to three silicon atoms in a trigonal configuration and each silicon tetrahedrally bonded to four nitrogen atoms. In β the sheets are stacked in a regular manner to give void channels parallel to the c-axis. The final product has a high modulus ($320\,\mathrm{GN\,m^{-2}}$), low thermal expansion coefficient ($3 \times 10^{-6}\,\mathrm{K^{-1}}$) and bulk density of $\sim 2700\,\mathrm{kg\,m^{-3}}$. It is strong, hard, and resistant to wear and thermal shock. Its applications include cutting tools, bearings, foundry equipment, furnace parts, metal molding dies and heat engine components.

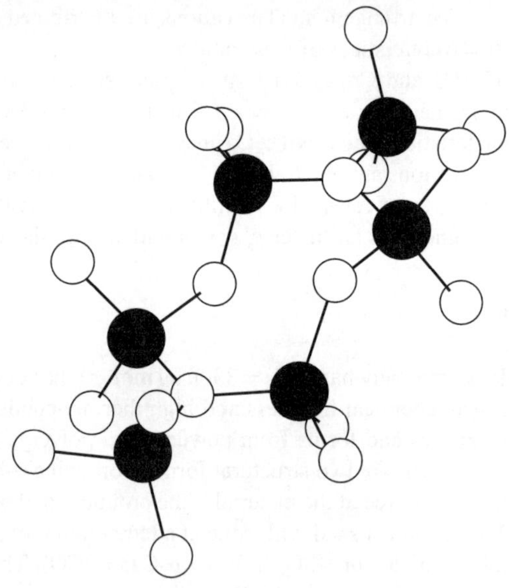

Figure 10.5 *The crystal structure of β-Si_3N_4 and β′-(Si, Al)$_3$(O, N)$_4$:(●) metal atom, (○) non-metal atom (from Jack, 1987, reprinted by permission of the American Ceramic Society).*

Industrially there are a number of production routes for silicon nitride, reaction bonded (RBSN), hot-pressed silicon nitride (HPSN) and hot isostatic pressed silicon nitride (HIPSN). For HPSN the silicon nitride powder is mixed with a fluxing agent (MgO, Y_2O_3 or Al_2O_3) and induction heated in graphite dies at 1850°C. Hot pressing increases the bulk density and improves strength and corrosion resistance. HIPing at pressures up to 300 MN m^{-2} improves the density even further and closes pores, but with a substantial amount of shrinkage. Producing a fully dense material to precise dimensional tolerances unfortunately leads to some engineering limitations, which the sialons have overcome.

10.3.7 Sialons

The sialons are a very important group of engineering ceramics consisting of phases in Si–Al–O–N and related systems with additions of Li, Be, Mg, Ca, Sc, Y or La to produce M–Si–Al–O–N. Structurally they are built up of (Si,Al)(O,N)$_4$ tetrahedra. β′-Sialon, a silicon–aluminum oxynitride, is isostructural with β-Si_3N_4 (Figure 10.5).

Four- and five-component systems are quite complex and are described in a behavior diagram represented in Figure 10.6, with all the concentrations expressed in equivalents.[1] The standard sialon square is derived from the tetrahedron made up of the four elements, Si, Al, O, N with the edges calibrated in equivalent percentages. The four components sit midway along a tetrahedron edge and form a square (Figure 10.7). The β′-sialon phase extends along a wide range of homogeneity and varying hexagonal unit cell dimensions. The general formula is $Si_{6-z}^{(24-4z)+}Al_z^{3z+}O_z^{2z-}N_{8-z}^{(24-3z)-}$, where z is the number of nitrogen atoms replaced by oxygen atoms. The value of z ranges from 0 to 4.

---

[1] Equivalent % oxygen = 100 (at.% O × 2)/(at.% O ×2 + at.% N × 3).
Equivalent % nitrogen = 100% − equivalent % oxygen.
Equivalent % aluminum = 100 (at.% Al × 3)/(at.% Al × 3 + at.% Si × 4).
Equivalent % silicon = 100% − equivalent % aluminum.

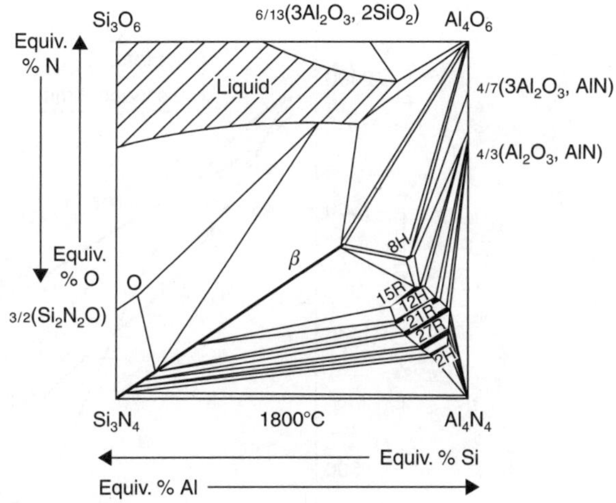

Figure 10.6 *Si–Al–O–N behavior diagram at 1800°C (from Jack, 1987, reprinted by permission of the American Ceramic Society).*

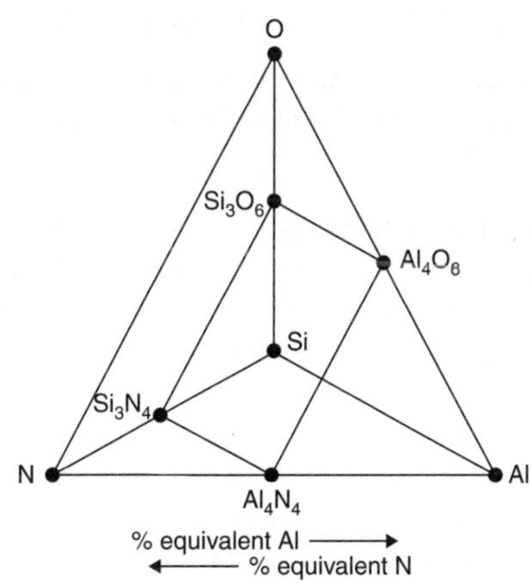

Figure 10.7 *Relation between Si–Al–O–N tetrahedron and square Si_3O_6–Al_4O_6–Al_4N_4–Si_3N_4 plane.*

Compositions in the Si–Al–O–N system between β'-sialon and the Al_3N_4 corner show six phases with structures based on wurtzite-type AlN. They are a series of polytypoids which exhibit changes in structural stacking in one dimension. These have the symbols 8H, 15H, 12H, 21R, 27R and 2H. The metals Li, Be, Mg, Sc and others can replace Si and Al in polytypoids, extending the range of homogeneity.

β'-Sialons may be prepared in several ways with different powder mixes: (i) Si_3N_4–AlN–Al_2O_3, (ii) Si_3N_4–AlN–SiO_2 and (iii) Si_3N_4–SiO_2–21R (a polytype). The nitride always has a thin surface oxide film so the problem is avoided with the polytypoid and produces a finer and more homogeneous product. Relatively simple fabrication techniques are available, similar to those used for oxide ceramics, and pressureless sintering enables dense complex shapes to be made.

Yttrium β'-sialons are very successful cutting tool materials and used for wire-drawing dies, bearings and other hard-wearing applications. After densification with Y_2O_3 low z-values of β'-sialon

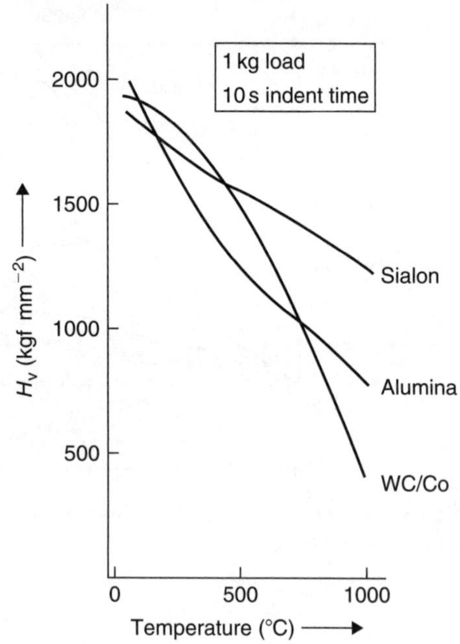

Figure 10.8 *Hot hardness of sialon, alumina and WC/Co cutting tool tips (from Jack, 1987, reprinted by permission of the American Ceramic Society).*

mixture when cooled from 1800°C give a β'-sialon with a grain boundary Y–Si–Al–O–N glass. This structure of (β'-sialon + glass) is strong and resists thermal shock at temperatures up to 1000°C. Controlled cooling, or heat treatment at about 1400°C, brings about a reaction of the matrix with the glass to give another β'-sialon and an intergranular Y–Al–garnet (YAG). This product has excellent oxidation and creep resistance.

The relative ease with which sialons can be manufactured into shapes and their ability to densify fully during sintering at temperatures ~1800°C without the need for pressure makes them very attractive engineering materials. Their strength, stability at high temperature, shock resistance, mechanical wear characteristics and ability to withstand molten metal are all special attributes. Figure 10.8 shows the superiority of sialon over cemented carbides for cutting tool tips, at higher cutting speeds and depths.

10.3.8 Zirconia

Zirconia (ZrO_2) exists in three crystalline forms, cubic, tetragonal and monoclinic (see Figure 10.9), and is able to harness the transitions to produce a material with reduced notch sensitivity and increased fracture toughness (~20 MN m$^{-3/2}$). The transformation-toughening property depends on retaining the tetragonal (t) phase to room temperature in a metastable condition. This is achieved by adding Y_2O_3, MgO or CaO. In the presence of a crack propagating into or near a metastable region, the stress field at the crack tip causes the t-crystals to transform to the stable but less dense monoclinic phase m-ZrO_2. The transformation is martensitic in character, with a 3–5% volume expansion, which helps to close the crack and relieve stresses at the tip.

The phase diagram for zirconia-rich solid solutions is shown in Figure 10.9 for ZrO_2–Y_2O_3. Three zirconia-based ceramics CSZ, TZP and PSZ are superimposed on the diagram. The CSZ ceramic is

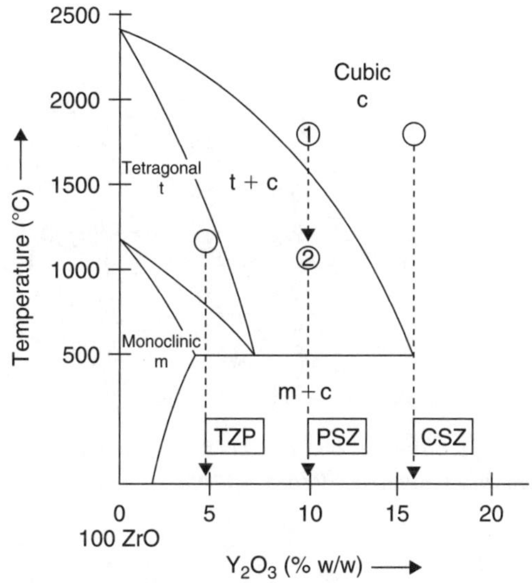

Figure 10.9 *Schematic phase diagram for $ZrO_2 - Y_2O_3$ system: all phases depicted are solid solutions. TZP = tetragonal zirconia polycrystal, PSZ = partially stabilized zirconia, CSZ = cubic-stabilized zirconia.*

Figure 10.10 *Electron micrograph of tetragonal zirconia polycrystal stabilized with 3 mol% yttria (with acknowledgement to M. G. Cain, Centre for Advanced Materials Technology, University of Warwick, UK).*

fully stabilized cubic and does not undergo transformation toughening but has other high temperature refractory uses. The TZP ceramic contains 2–4% Y_2O_3 and is produced in fine-grained form by sintering in the tetragonal phase field, which is retained at room temperature (Figure 10.10). The PSZ ceramic, partially stabilized, has 8–10% Y_2O_3. Sintering in the temperature range 1650–1850°C produces a cubic solid solution which is then modified by heating at 1100–1450°C to precipitate *t*-crystals within the cubic *c*-grains. The morphology of the precipitated crystals could be changed from platey to cuboid or lenticular by modifying with CaO or MgO. At temperatures around 900–1000°C, close to the *t–m* transition temperature, the toughening mechanism becomes less effective. Thermal cycling has the same effect of gradually reducing the fracture toughness. Fortunately, other

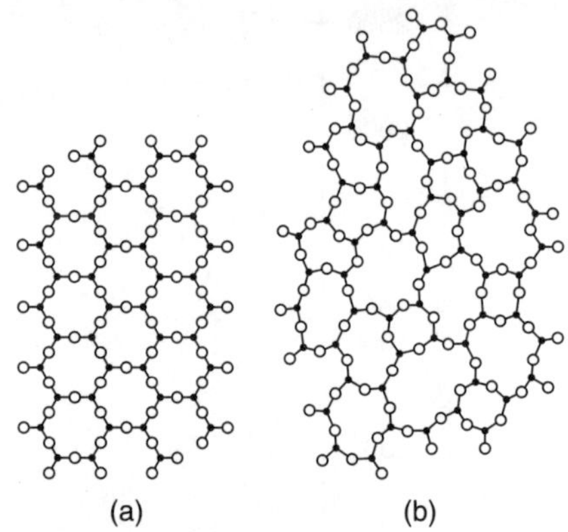

Figure 10.11 *Crystalline (a) and non-crystalline (b) forms of the same composition (from Kingery, Bowen and Uhlmann, 1976; by permission of Wiley-Interscience).*

forms of stabilization, e.g. HfO_2, have been developed which have better high-temperature fracture resistance properties.

Apart from the fracture-toughening properties, zirconia has other good high-temperature ceramic properties, e.g. high melting point, chemical durability, high strength and hardness, and has many uses based on them. Also above 1200°C, ZrO_2 becomes electrically conductive and is used as heating elements under oxidizing conditions. Its thermal expansion characteristics, which are similar to metallic alloys, allow for its use as components in heat engines.

10.4 Glasses

10.4.1 Structure and characteristics

Some materials can exist in both the crystalline and non-crystalline form, e.g. SiO_2, B_2O_3, as shown in Figure 10.11 for boric oxide. In the non-crystalline form the material is termed glassy, vitreous or amorphous and is characterized by the absence of any long-range order. In the figure, each triangular group (CN = 3) represents three oxygen anions around a small B^{3+} cation and, overall, forms a random network in three dimensions. For silica, it is SiO_4^{4-} tetrahedra that form a 3-D network. Oxides other than SiO_2 and B_2O_3 have the ability to form networks, e.g. P_2O_5, V_2O_5, while others such as Na_2O, K_2O, CaO have no tendency to form networks but are important network modifiers. Apart from fused silica, the two glass formers are added together, often with network modifiers, to produce commercial glasses (Vycor 96% SiO_2, 4% B_2O_3; Borosilicate 81% SiO_2, 13% B_2O_3, 4% Na_2O, 2% Al_2O_3). In general, the glass network becomes unstable and tends to crystallize if the additives increase the numerical ratio of oxygen to silicon ions above 2.5.

The rate of cooling from the molten or fused state is also important in controlling the glass formation in oxides. Rapid cooling restricts the time for ordering and favors glass formation.[2] Glassy materials do, however, cool in a different way to crystalline materials. The glass becomes more and more viscous with decreasing temperature in a continuous manner and there is no definite temperature at which the liquid transforms to a solid, as shown in Figure 10.12. The specific volume (volume per

[2] Cooling rates of 10^6 K s^{-1} by melt spinning or splat cooling have been used to produce metallic glasses with transition metal (Fe, Ni, Co, Pd, Mo) and metalloids (B, C, P, Si).

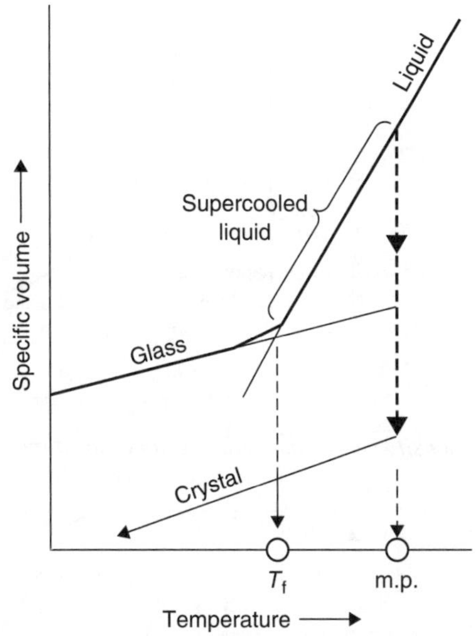

Figure 10.12 *Comparison of the formation of glass and crystals from a melt.*

unit mass, $m^3\ kg^{-1}$) versus temperature curve shows a decrease in slope at a 'fictive' temperature T_f, which may be considered a glass transition temperature, but it does depend on cooling rate. Viscous flow is given by the Newtonian equation

$$d\gamma/dt = \tau/\eta,$$

where $d\gamma/dt$ is the shear strain rate, τ is the applied shear stress and η the coefficient of viscosity. In its melting range, a typical SiO_2–Na_2O–CaO glass has a viscosity[3] of 5–50 N s m^{-2}, which may be compared with the viscosity of liquid metals of roughly 1 mN s m^{-2}. As glass is hot-drawn, its cross-sectional area decreases at a rate which depends solely upon the drawing force and viscosity, not upon area. For this reason, the glass extends uniformly and does not 'neck'. An increase in temperature and/or addition of network-modifying cations will reduce the viscosity. The temperature dependence of viscosity is given by an Arrhenius relation

$$\eta = A\exp(Q/\mathbf{k}T),$$

from which the activation energy Q for viscous flow may be obtained from a plot of log η against $1/T$. Addition of a nucleating agent, usually TiO_2, can reduce the activation energy and induce the formation of crystalline regions within the glassy matrix. This process of devitrification can result in the product having some desirable properties, i.e. high thermal and mechanical shock resistance. Controlled devitrification is now an accepted processing method for producing such glass-ceramic materials (see Section 10.4.3). Figure 10.13 shows a plot of log viscosity versus temperature for a typical SiO_2–Na_2O–CaO glass. Two practical operating temperatures are known as (1) the softening point and

[3] Absolute viscosity is the force required to move 1 m^2 of plane surface at a velocity of 1 m s^{-1} to a second plane surface which is parallel to the first and separated 1 m from it by a layer of the fluid phase. Kinematic viscosity = absolute viscosity/density. Unit of viscosity = 1 N s m^{-2} = 10 P (poise). Fluidity is the reciprocal of viscosity.

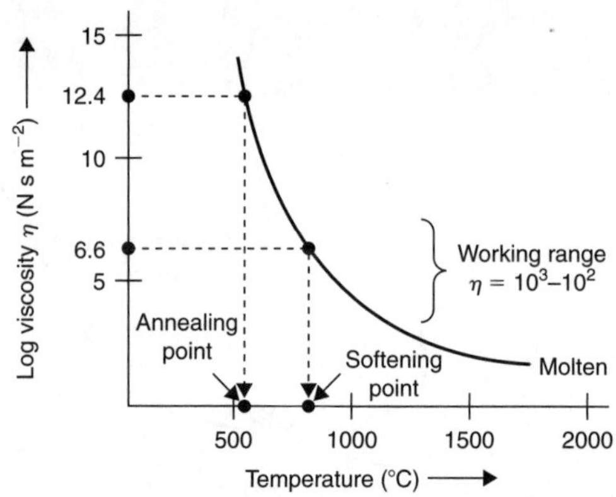

Figure 10.13 *Viscosity versus temperature curve for a typical SiO_2–Na_2O–CaO glass.*

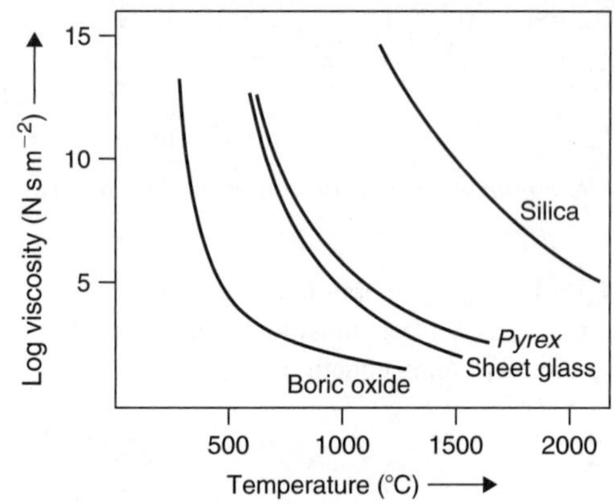

Figure 10.14 *Viscosity curves for typical glasses.*

(2) the annealing point. The softening point gives the maximum temperature at which the glass can be handled and for ordinary silica glass it is about 1000 K. At the annealing temperature, the ions are sufficiently mobile to allow residual stresses to be relieved in about 15 min. The point in the curve at which the slope is a maximum corresponds to the fictive or glass transition temperature. The working range for commercial silica glass corresponds to a viscosity range of 10^3–10^7 N s m^{-2}. The curve for this glass is quite steep, indicating that close temperature control is necessary during working (i.e. drawing, blowing, rolling, etc.). Figure 10.14 provides a comparison of the viscosity curves for different types of glass. Even at a temperature of 1300°C silica has a viscosity of about 10^{12} N s m^{-2}, which is still too high for working. Glasses with special chemical and physical properties often have a steep viscosity curve and tend to devitrify, presenting difficulties during traditional working temperatures. One solution is to hot extrude through a graphite die. High pressure (up to 10 GN m^{-2}) can overcome a high viscosity and reduce the extrusion temperature to prevent devitrification.

10.4.2 Processing and properties

The characteristic network structure imparts isotropic properties to the glass which change gradually with change in temperature. They can change shape without fracture, which affords them a number

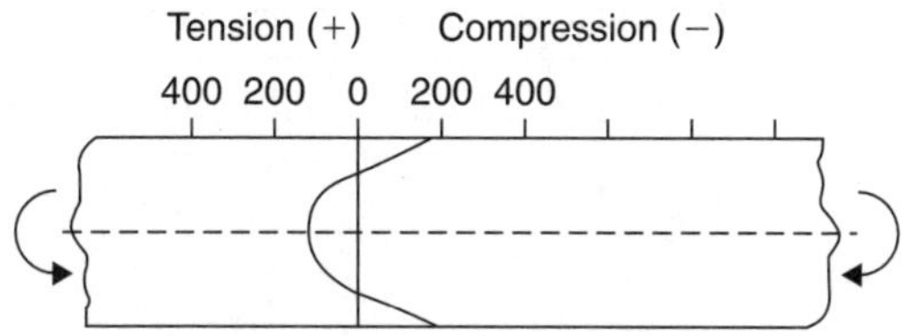

(a) Thermally toughened glass plate

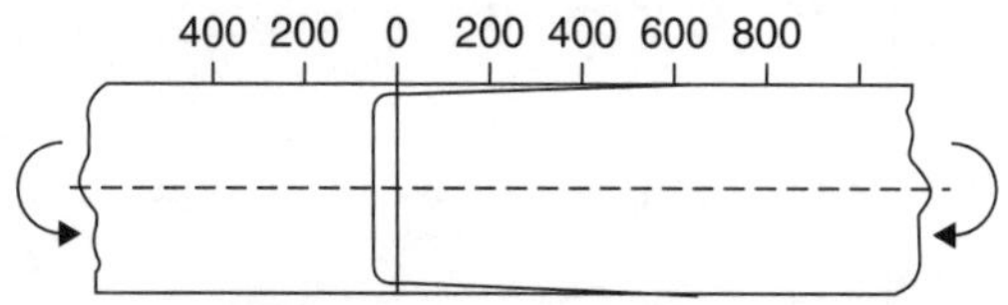

(b) Chemically toughened glass plate

Figure 10.15 *Distribution of residual stress in toughened glass plate (stress values in MN m^{-2}).*

of different forming methods, i.e. blowing, drawing, pressing, fiber forming and rolling, to produce a variety of products, e.g. tubes, sheets, filaments, coatings, etc. Glass blowing by 'individuals' is still fascinating to watch but most blowing is done by machine to produce bottles, jars and light bulbs, usually by pressing a hot glass 'blob' into a mold before blowing with compressed air to the finished shape. Continuous glass fibers are formed by drawing the molten glass through orifices in the base of a heating chamber under closely controlled temperature–viscosity conditions. After spray cooling and chemical sizing to improve the abrasion resistance, the filaments can be made into strands and yarns.

Glass sheets can be made by a novel float glass process. In this process, the molten glass is poured onto a bath of molten tin when the glass spreads uniformly across the surface to a controlled thickness as it cools from 1100 to 700°C. Subsequently the glass is reheated to 850°C to increase the viscosity and remove imperfections. The glass produced is perfectly parallel-sided and clear for plate glass without additional polishing, etc. Final treatment involves annealing to remove thermal stresses, followed by slow cooling to room temperature.

While glass is weak in tension it is quite strong in compression. This property is used in the manufacture of toughened glass by producing compressive stresses in the surface layers. This is achieved by thermal tempering when the glass is heated above T_f but below the softening temperature and cooled in a jet of air. The surface cools more rapidly than the interior, becoming rigid while the inside is still 'plastic'. On continued cooling the contraction of the inner region is restrained by the rigid outer layers. The stress distribution over a cross-section of a toughened glass plate is shown in Figure 10.15a. Surface stresses of 200 MN m^{-2} may be produced by this method, which has to be overcome to cause fracture of a toughened glass sheet with an additional tensile stress to initiate a surface crack. Toughening can also be produced by altering the surface composition by high-temperature diffusion to reduce its coefficient of thermal expansion. Lower temperature treatment involves the replacement of Na by large ions, such as K, by immersing the glass in a molten salt. On cooling from a relatively low temperature the core is rigid and the presence of large ions in the surface layers leads to compressive stresses, as shown in Figure 10.15b. The glass bottle manufacturing industry relies on this method of toughening. Sheet glass in a sandwich composite with polyvinyl butyrate (PVB) plastic interlayer is also used as a toughened material. On impact the PVB which is bonded to the glass by autoclaving at pressure at about 140°C deforms to absorb energy and tends to prevent the shattered glass from causing lacerations. Laminated glass can also withstand extreme changes of conditions as experienced in aircraft windows.

Table 10.5 *Composition and applications of some commercial glasses.*

Glass type	*SiO_2*	*Na_2O*	*CaO*	*Al_2O_3*	*B_2O_3*	*MgO*	*PbO*	
Vitreous silica	100	–	–	–	–	–	–	Low coefficient of expansion; optical components, fiber-optics, lamps
Soda lime	72	14	10	1	–	2	–	Convenient, easy working range 700–1000°C
Borosilicate (Pyrex)	81	4	–	2	13	–	–	Chemically durable, low coefficient of expansion, tolerant to thermal shock; used for laboratory glassware/chemical glassware
Lead glass	34	1					65	X-ray protection glass
Lead glass							23	Color TV protection glass
Lead/crystal glass							20–30	Relatively constant viscosity over large working temperature range; used for decorative glassware
Lead/optical glass	54		1				37	8% K_2O added as modifier; Pb increases refractive index and dispersion of SiO_2-based glass
Aluminosilicate glass	62	1	8	17	5	7	–	High softening temperature (>900°C) used for combustion tubes, cookware

Glass products, particularly sheets, must be free from flaws not only to withstand servicing stresses, e.g. toughening, but also quality control. The removal of heterogeneities during glass manufacture is termed fining. Bubbles are removed by flotation, which depends on Stokes's law, $v = \text{constant} \times g\rho r^2/\eta$, where ρ is the glass density, g the gravitational constant and r the radius of the bubble. The larger bubbles float out of the glass quicker than small ones. A number of bubble enlargement methods are available to aid the bubble removal. The small bubbles often disappear from molten glass by dissolution when the internal pressure increases according to $p = 2\gamma/r$. Controlling the composition is important for fining. Alumina, for example, is frequently added (1–2%) to silica-based glasses to improve the resistance to water and moisture. Unfortunately, it increases the melt viscosity and makes bubble removal more difficult.

Table 10.5 gives the approximate composition, characteristics and application of some common glasses.

10.4.3 Glass-ceramics

Glass-ceramics, a new type of engineering material developed in the past few decades, have found a wide variety of applications from domestic to high-tech scientific. Basically, they are glasses in which fine-grained crystalline regions are produced by controlled composition and heat treatment, without distortion and little dimensional change. Applications vary from hob plates for electric cookers to telescope mirrors and radar domes.

Controlled devitrification to make the glass-ceramic involves two stages of heat treatment following the cooling from the normal forming glass operation. The special glass to which a nucleating agent, or catalyst, has been added is held at a particular temperature to nucleate a large number of fine crystals throughout the glassy matrix. A second stage heat treatment raises the temperature of the glass to allow growth of the crystals and control the amount of residual glassy matrix. During heat treatment the rate of nucleation N (which should be high) and the rate of growth G (which should not be excessive) are both carefully controlled (in a similar manner to the austenite–pearlite transformation, see Section 7.3.2).

10.4.3.1 *β-Spodumene and β-quartz*

The SiO_2–Al_2O_3–LiO_2 system produces very versatile glass-ceramics, typically β-spodumene and β-quartz. β-Spodumene includes TiO_2 as a nucleating agent, and additions of MgO and ZnO as a cheaper substitute for Li_2O improve the working properties. First stage nucleation heat treatment is carried out at 780°C, when hexagonal β-quartz (β-eucryptite) precipitates titanite nuclei. In the second stage at 950°C, viscous flow relaxes the local stresses between the quartz and the matrix, as well as promotes the conversion of the metastable quartz to β-spodumene crystals ($\sim$1 μm). A final short heat treatment at 1125°C produces almost complete crystallization of a uniform fine grain size. This is helped by the small minor phases, e.g. TiO_2 and $Mg(Zn)Al_2O_4$ concentrated in the grain boundaries.

Glass-ceramic hobs for cookers are strong, smooth, easy to clean, stable and resistant to thermal shock. Careful balance of crystal products can control the expansion coefficient close to zero. Surface treatments (as discussed) can toughen the surface. One method employs the volatilization of elements (e.g. F) in the surface to produce compressive stresses. Another involves the ion exchange of Na^+ or K^+ for Li^+.

β-Quartz glass-ceramics are made with a mixture of nucleating agents, TiO_2 and ZrO_2, which enhance the β-quartz precipitation at lower processing temperatures (750°C for nucleation and 900°C for growth). This produces a finer-grained product with low thermal expansion used for telescope mirrors where dimensional stability is essential.

10.4.3.2 *Cordierite, mica and stagsitall glass-ceramics*

These glass-ceramics are produced from the MgO–Al_2O_3–SiO_2 ceramic system and combine high electrical resistivity with high strength and low thermal expansion. The nucleating agents TiO_2 and ZrO_2 produce a predominant cordierite phase after heat treatment at 1200°C, together with minor rutile and magnesium aluminum titanate.

In mica glass-ceramics based on fluorine-containing mica crystals, the microstructure shows a fine-grained structure of interlocking crystals. This produces high mechanical strength and high modulus values comparable to dense alumina. Mica crystals grow epitaxially on the early nuclei formed at 825°C and on further heating at 950°C results in highly anisotropic mica plates of $\sim$15 μm. Since the mica readily delaminates, the random crystals cause crack deflection, branching and blunting. This limitation of the fracture behavior accounts for the machinability, strength and shock resistance these

Table 10.6 *Selected properties of glass-ceramics.*

Property	*β-Spodumene*	*β-Quartz*	*Cordierite*	*Mica*	*Stagsitall*
RT–800°C Thermal expansion coefficient α ($10^{-7}\,K^{-1}$)	10–12.5	5–8	64–43	88–128	68–93
RT-600°C Thermal conductivity ($W\,m^{-1}\,K^{-1}$)	2.1–2.3	–	4–3	2.1–3.3	–
Modulus of elasticity (GPa)	80	93	119	65	95
Density ($g\,cm^{-3}$)	2.7	2.6	2.6	2.5	2.7
Fracture energy ($J\,m^{-2}$)	3.7	2.4	17.2	8.2	2.3
Critical stress intensity factor (Mpa $m^{1/2}$)	0.8	0.7	2.1	1.1	0.7

materials are noted for. Machining can be carried out to close tolerances without any precautions on special tooling.

Stagsitall glass-ceramics derived from steelmaking slags have found application in rolled form for building materials for flooring and cladding. The crystallinity is lower than the other glass-ceramics (~70%), and have excellent wear resistance and good durability, useful in industrial applications. They are mainly SiO_2 (58%), Al_2O_3 (7%) and CaO (23%), with small quantities of MnO_2, Fe_2O_3, etc. First stage heat treatment is 720°C for 2 h followed by crystallization at 850–900°C for 3 h. Nucleation occurs by sulfides and crystallization of calcium silicates and calcium, aluminum silicates. A successful commercial product is sheet material up to 3 m wide and 5–25 mm thick. The chemical complexity of slags makes crystallization heat treatment difficult to control but the use of cheap waste products is eco-friendly and reasonably economic.

Some glass-ceramic properties are shown in Table 10.6. Properties and applications are being developed all the time, because of their attractive properties. In heat engines and gas turbines a low thermal expansion coefficient is required for the regenerative heat exchanger, which is alternately heated and cooled by exhaust gases and combustion air. In glass-to-glass and metal-to-glass seals, precise matching of the expansion coefficients is possible by chemistry and heat treatment. In the mass color TV market, glass-ceramic solders based on the $PbO–ZnO–B_2O_3$ system have been used to seal the glass cone to the glass face plate at a relatively low temperature to avoid distortion before being transformed into a glass-ceramic by subsequent heating. Other applications are being found in chemical plant, aerospace equipment electrical circuits, nuclear engineering and general engineering.

10.5 Carbon

Pure carbon exists in three forms, including diamond, graphite and the newly discovered nanomolecular fullerene structures. Diamond and graphite in general are classified as ceramic materials because they are brittle and graphite materials are traditionally used as refractory materials. The discovery of fullerene-based structures is closely related to the birth of nanotechnology, which is described in more detail in Chapter 12.

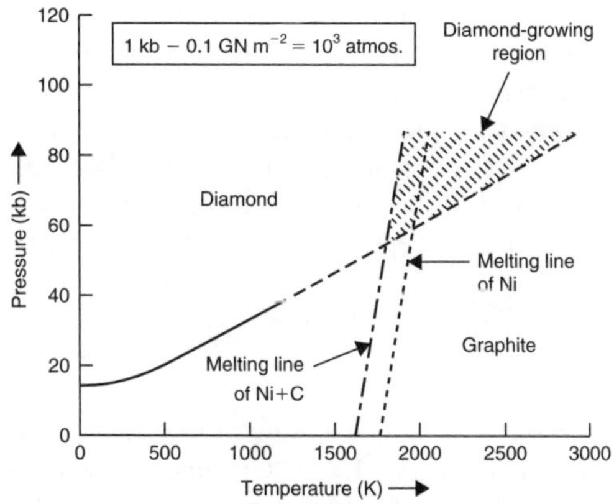

Figure 10.16 *Pressure–temperature diagram for carbon (after Bovenkerk* et al.*, 1959).*

10.5.1 Diamond

Well known for its hardness, inertness and beauty, diamond is a crystalline form of carbon exhibiting a cubic lattice structure (Figure 1.17a). The strong sp^3-hybridized covalent C–C bonds (0.1555 nm long) form a three-dimensional tetrahedral network, resulting in extraordinarily high rigidity ($E = 1035$ GPa–1 TPa), strength ($H_V = 9000\,\text{kg mm}^{-2}$) and melting point (3300°C). Natural, gemstone quality diamonds are precious and expensive, and are characterized by the so-called 'four Cs': carat (unit for mass; 1 carat = 0.2 g), clarity, color and cut. Their beauty and hence price are enhanced by meticulous cutting and polishing procedures. Natural diamonds of near-gem or lower 'boart' quality are flawed and imperfectly crystallized, and are therefore considerably cheaper. They are used for industrial applications including machining, polishing and as diamond indenters for hardness testing (Sections 4.2.2.4 and 4.8.4).

Diamonds can also be synthesized in the laboratory by applying a high pressure at a high temperature on graphite. The equilibrium P–T diagram of carbon shown in Figure 10.16 indicates that diamond is stable at high pressures and graphite is stable at low pressures. At ordinary pressures, diamond exists only as a metastable form relative to graphite. Transformation from graphite to diamond will therefore take place if a high pressure (>60 kb) is applied on graphite at a high temperature (e.g. 1800 K). Addition of a metal such as Ni, Cr, Mn, Fe or Co eases the process, since the metal acts as a molten solvent for carbon and a catalyst for diamond crystallization and growth. The growth region for diamond is therefore the shaded area in Figure 10.16, i.e. the region beyond the liquidus line for the C + Ni eutectic mixture and above the phase boundary between graphite and diamond. Bulk pieces of polycrystalline diamond are not available naturally, but can also be synthesized by sintering diamond crystallites at high temperature and pressure, again in the presence of a metal, which acts as a solvent and a catalyst.

High-purity diamonds, with impurity contents in parts per million, are rare and, with a band gap of 5.5 eV, are electrically non-conducting and highly transparent in the visible light range (which has photon energies in the range 1.68–3.32 eV; see Section 5.10.1). They are also extremely good conductors of heat (with thermal conductivities 1000–2000 W m^{-1} K^{-1}) as a result of the crystalline structure, which conducts heat efficiently by phonon propagation. These physical properties of diamond, however, are greatly affected by the presence of internal defects and impurities. Cracks or flaws mainly weaken the mechanical properties of diamonds. Inclusions in synthetic diamonds are mainly due to the metallic catalysts added, while inclusions in natural diamonds are mainly mineral particles.

Nitrogen is also a usual impurity, up to 0.1–0.2%, in both natural and synthetic diamonds. Natural diamonds often contain clusters or platelets of nitrogen atoms which formed under the geological periods of high temperatures and pressures, while synthetic diamonds often contain N atoms as singly dispersed, substitutional solutes. Introduction of nitrogen ($Z = 7$) into the carbon structure ($Z = 6$) brings in additional electrons which can now cause optical absorption and hence the transparency of the diamond is changed. Diamonds can also be made to be semiconducting by doping with boron ($Z = 5$), which introduces acceptor sites or 'holes' in the valence band.

10.5.2 Graphite

Graphite has a crystal structure consisting of hexagonal 'graphene' layers stacked in the sequence **ABABAB**, as shown in Figure 1.17b. Graphite is the most stable form of carbon at ordinary pressure and temperature, and exists in many forms which are either available naturally or synthesized for specific purposes. Natural graphite is traditionally used to make crucibles for metal melting, and also as lubricants, brake linings, bearings, pencil lead, etc. Synthetic carbons are usually produced by pyrolysis, i.e. heating at a high temperature in the absence of oxygen, of organic materials, which causes the carbon content to graphitize. Thus, charcoal is produced by pyrolysis of wood, when water and other volatile constituents are removed and the carbon content made to graphitize. Activated carbon differs from ordinary charcoal in that it contains a much larger area-to-volume ratio, usually achieved by a highly porous structure. Activated carbon is therefore used as filter materials for deodorization or purification purposes. Other types of synthetic carbons are used in a wide range of applications, including brushes for electric rotors, refractory linings for furnaces, conducting electrodes and nose cones for space vehicles (see Section 10.7), etc.

An important category of synthetic carbons are baked carbons, which are produced by baking a green compact of a mixture of carbonaceous filler, usually coke, and a binder such as coal-tar pitch or a thermosetting resin. The green compact is first formed by extrusion or molding, and the product is usually anisotropic as the elongated coke particles tend to align along the extrusion axis or perpendicular to the compression molding axis. Baking the green compact in a reducing atmosphere at ~1000–1200°C results in decomposition of the organic binder, producing carbon linkages between the filler particles and also evolution of volatiles, leaving behind a significant amount (20–30%) of porosity. The porosity, which is detrimental to both mechanical strength and electrical conductivity, can be reduced by impregnating with more pitch and rebaking. Further heating at ~2500–3000°C causes the carbon to graphitize, resulting in better electrical and thermal conductivities, as well as mechanical strength improvement through strain relief and defect annihilation. Both baked and graphitized carbons can be machined to accurate dimensions.

Another important type of synthetic carbon is pyrolytic carbon. Pyrolytic carbons are usually produced as thin coatings by pyrolysis of hydrocarbon gases on heated substrates. A hydrocarbon gas such as methane or propane is brought into contact with a heated substrate (e.g. at 2000°C) by an inert carrier gas. The hydrocarbon gas molecules are broken down thermally (e.g. $CH_4 \rightarrow C + 2H_2$) within the more stagnant boundary layer of the flow in contact with the substrate, resulting in deposition of carbon on the latter. The carbon deposit formed under such highly non-equilibrium conditions consists of graphite crystallites with structures deviating from the most stable form shown in Figure 1.17b. The stable graphite structure adopts the stacking sequence **ABABAB** and is anisotropic with an interlayer distance of 0.335 nm, much larger than the interatomic C–C bond lengths of 0.1415 nm in the graphene planes. The crystallites in pyrolytic carbons have their graphene planes randomly rotated about the *c*-axis, as shown in Figure 10.17. Such a random architecture of graphene planes, termed 'turbostratic' structure, is no longer describable by the stacking sequence **ABABAB**, and has an interplanar distance larger than the 0.335 nm of ordinary graphite, resulting in even higher anisotropy. Pyrolytic carbons

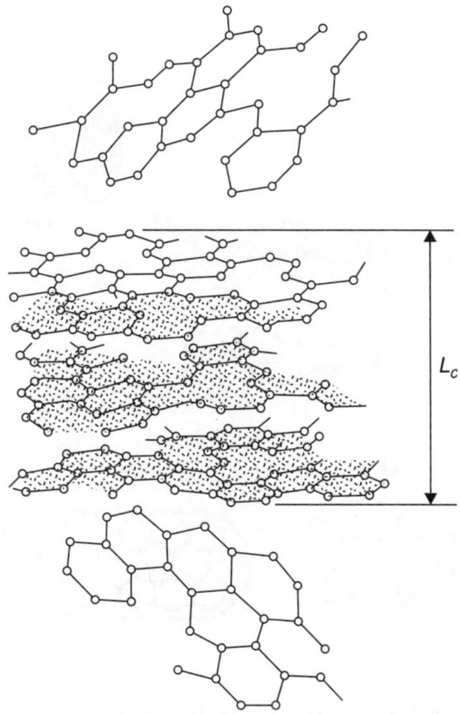

Figure 10.17 *Structure of turbostratic carbon (from Cahn and Harris, 1969).*

are usually polycrystalline, composing of crystallites characterized by their dimension L_c along the c-axis, as shown in Figure 10.17. They are mechanically stronger than ordinary graphite due to their usually small crystallite sizes ($L_c \sim 3$–10 nm), and are also less chemically active. One important use of pyrolytic carbons is to make medical prosthetic implants such as heart valves (see Section 12.2.4).

A third type of synthetic carbon is 'vitreous or glassy' carbon, which is obtained from the pyrolysis of a thermosetting polymer such as phenol formaldehyde resin. The polymer precursor can be shaped into the desired form, and is then carbonized by firing in an inert atmosphere or under vacuum at a high temperature. Removal of non-carbon contents through evolution of gases often results in significant shrinkage. The complex structure of the polymeric precursor inhibits graphitization and ordering, and so the resultant turbostratic crystallites ($L_c \sim 5$ nm) are much smaller than pyrolytic carbons. The interplanar distance of the graphene planes is large at ~0.35 nm, and cross-linking C–C bonds appear to exist between the planes, resulting in strengthening effects. Vitreous carbons are therefore hard and brittle, and because their crystallites are randomly oriented, the overall structure is isotropic, like glass. They are also more chemically inert than ordinary graphite and pyrolytic carbons, and are used in making inert wares, crucibles, furnace boats, crystal-growing and zone-refining components, etc.

10.5.3 Fullerenes and related nanostructures

In the mid-1980s, research on vaporized carbon led to the discovery of the C_{60} molecule, the third and a novel form of pure carbon. The C_{60} molecule has 60 carbon atoms forming a truncated icosahedron shell structure known as buckminsterfullerene (Figure 10.18). Fullerenes larger than C_{60}, such as C_{70} and molecules containing more than 200 C atoms, have also been found to exist. A related class of

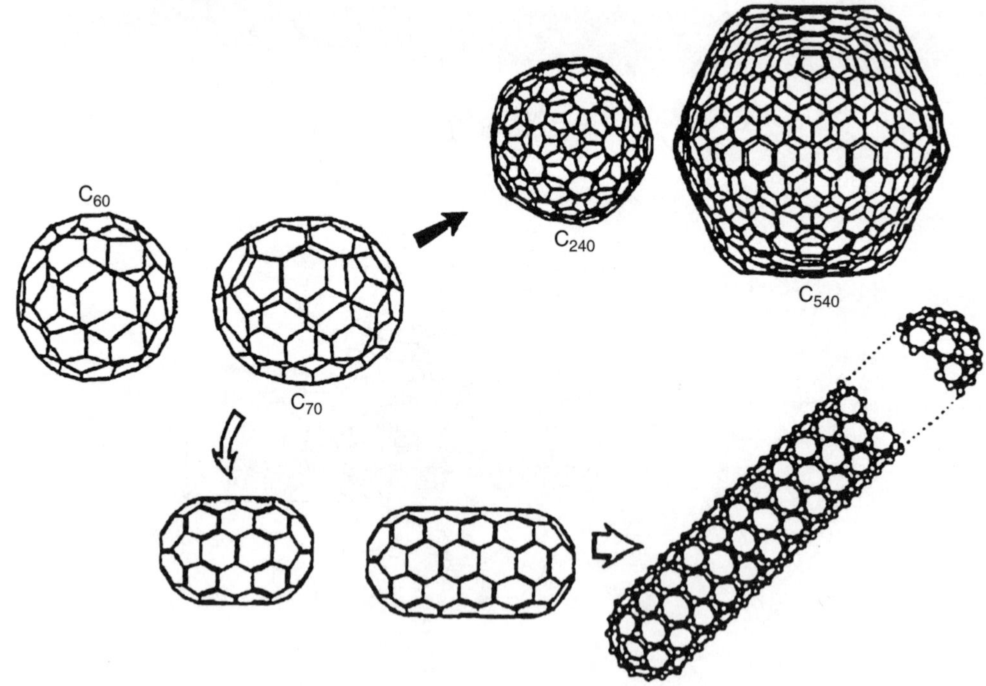

Figure 10.18 *Various fullerene-based structures (from Inagaki, 2000).*

important nanomaterials, known as carbon nanotubes (CNTs), is illustrated in Figure 10.19. CNTs can be imagined to be formed by rolling up a graphene sheet of carbon atoms arranged in a 2-D hexagonal honeycomb lattice. To form a CNT, a hoop vector, called the chiral vector, is first chosen to connect (0,0) and (m,n) in Figure 10.19d, and the tube is formed by rolling up the graphene sheet with this chiral vector forming a closed circle on the circumference of the tube. The chiral vector (m,n) thus defines the diameter as well as the structure of the tube; for example, an armchair configuration (Figure 10.19a) has a chiral vector (m,m) and a zigzag configuration (Figure 10.19b) $(m,0)$ or $(0,n)$. Another commonly used measure for the chirality is the chiral angle θ, which is the angle the chiral vector makes with the zigzag orientation on the graphene sheet. Thus, a zigzag tube has $\theta = 0°$ and an armchair tube has $\theta = 30°$, and a general chiral structure (Figure 10.19c) has a θ between 0° and 30°. Apart from the single-wall architecture in Figure 10.19, multi-walled CNTs also exist in which multiple graphene layers are rolled up to form an assembly of concentric tubes.

CNTs can be fabricated using a range of methods. In laser ablation, an intense pulsed laser beam is incident on a target of graphite doped with cobalt or nickel, which act as a catalyst. The target is heated to ~1200°C and the laser evaporates carbon from the target to form CNTs, which are carried away by a stream of argon gas and collected by a cooled substrate outside the furnace. In the electric-arc method, a voltage is applied between two graphite electrodes situated in a helium atmosphere. An electric arc forms and carbon atoms are ejected from the anode to the cathode, forming CNTs on the latter. If pure graphite electrodes are used, multi-walled CNTs will be formed, and to produce single-walled CNT, a metallic catalyst needs to be introduced onto the anode. Finally, in the chemical vapor deposition method, a hydrocarbon gas such as methane is decomposed at an elevated temperature. The resultant carbon atoms then form CNTs on a cooled surface containing a metallic catalyst. In all the processes mentioned above, a metallic catalyst in the form of nanopowder is involved. Figure 10.20 shows an *in situ* experiment in which a CNT was observed to grow by the catalytic action of a trapped

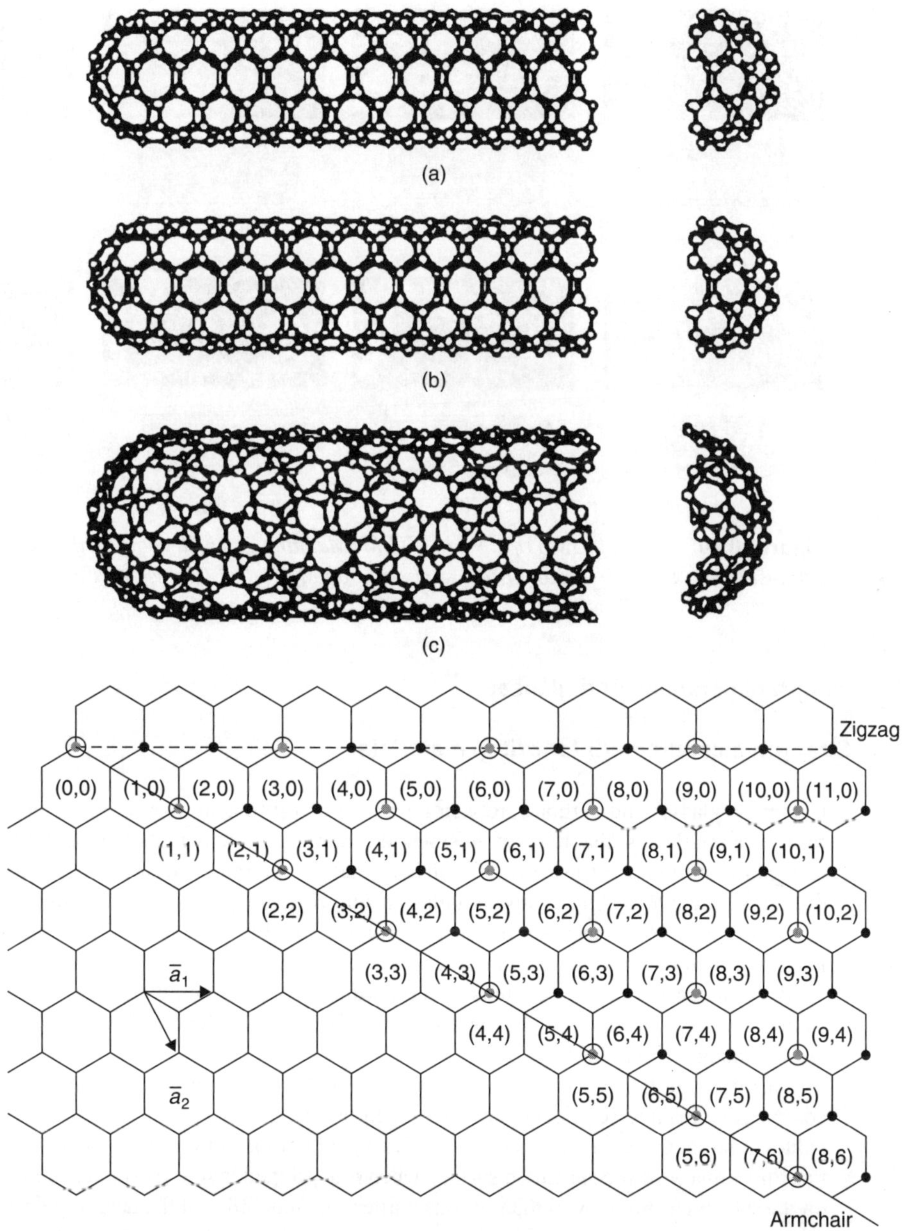

Figure 10.19 *Different forms of single-walled carbon nanotubes: (a) armchair structure, (b) zigzag structure, (c) chiral structure. (d) graphene sheet of carbons.*

Ni nanoparticle at its head, during decomposition of methane inside the TEM. The propulsion of the Ni nanoparticle drew out the CNT, as shown in Figure 10.20h. Fullerene-based structures and in particular CNTs are an important class of building block materials for nanotechnology, and more details are given in Chapter 12.

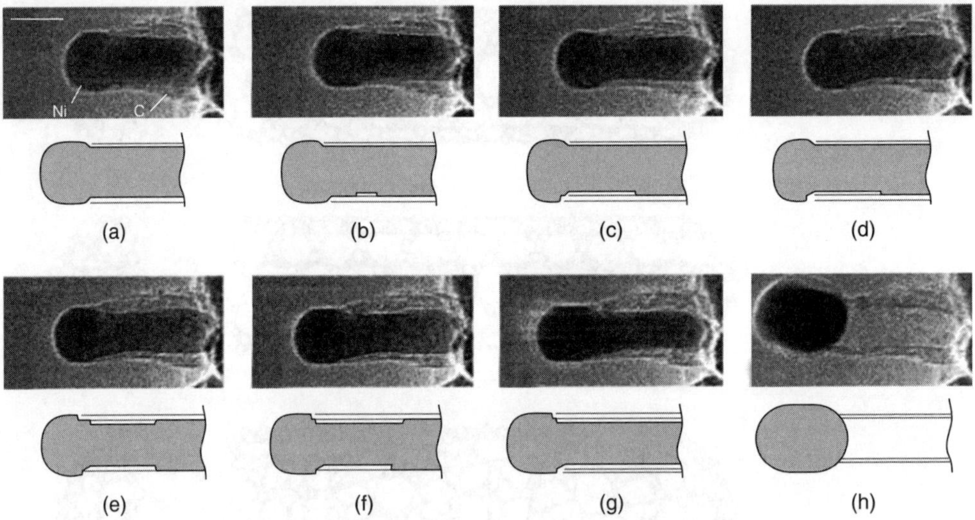

Figure 10.20 In situ *growth of CNT in a TEM. Methane gas is decomposed inside the TEM over a catalyst consisting of Ni nanoclusters supported on* $MgAl_2O_4$*. (Helveg and co-workers, 2004).*

10.6 Strength of ceramics and glasses

10.6.1 Strength measurement for brittle materials

Ceramics, glasses and carbons are well known to be brittle and there are two fundamental reasons for this. First, the residual porosity in sintered ceramics acts as potential nuclei for cracks. Secondly, dislocation motion is intrinsically difficult in crystalline ceramics and carbons (e.g. diamonds) because of their strong interatomic bonds, and glasses simply do not have dislocation plasticity as a possible mode of deformation because of their amorphous structures. Crack-tip blunting is therefore very limited in ceramics and glasses, and hence the fracture toughness of these materials is intrinsically low.

As discussed in Section 6.1, tensile and hardness tests are common methods for measuring the static strength of a material. For brittle materials, other methods include compression and bend tests. Amongst these, hardness and compression tests suppress fracture, which is the easier mode of damage in brittle ceramics and glasses, and unless the component is designed to carry only compressive loading (e.g. concrete columns in buildings), these two tests give unrealistically high values of strength. Tensile testing of ceramics and glasses is not generally favored since it is difficult to avoid damage of the specimen surface when it is gripped by the tensile machine. Flaws produced on the specimen surface will then seriously affect the reliability of the data. Bend or flexure tests are relatively easy to carry out, since very small pieces and simple shapes of specimens are required. Bend tests can be carried out in either the three-point support mode or the four-point support mode shown in Figure 10.21. The stress state in a bend test is non-uniform within the sample (Figure 10.22) and the strength indicator, known as the flexural strength or modulus of rupture (MoR), is the highest elastic stress in the sample when fracture just begins. The formulae to calculate the MoR are given in Figure 10.21.

A strength-related concern for using ceramics, especially in thermal applications, is their ability to withstand changes in temperature without fracture, called 'thermal shock resistance'. Thermal stresses in a component arise from the constraints on the tendency for the material to expand when heated or contract when cooled, due to any external stoppage or the geometry of the component itself (e.g.

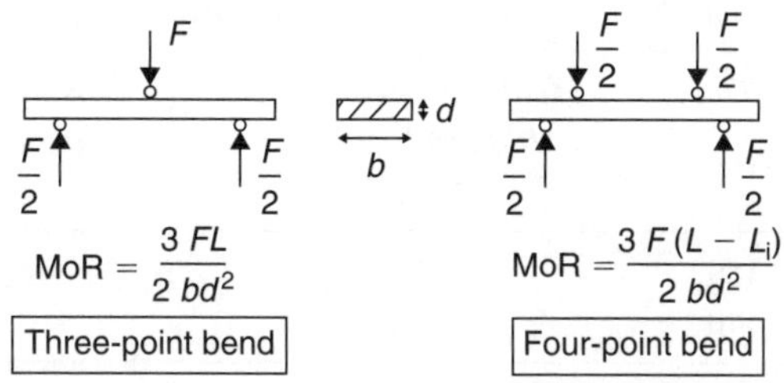

Figure 10.21 *Three-point and four-point bend test configurations. F = applied force, L = outer span, L_i = inner span, b = breadth of specimen, d = depth of specimen.*

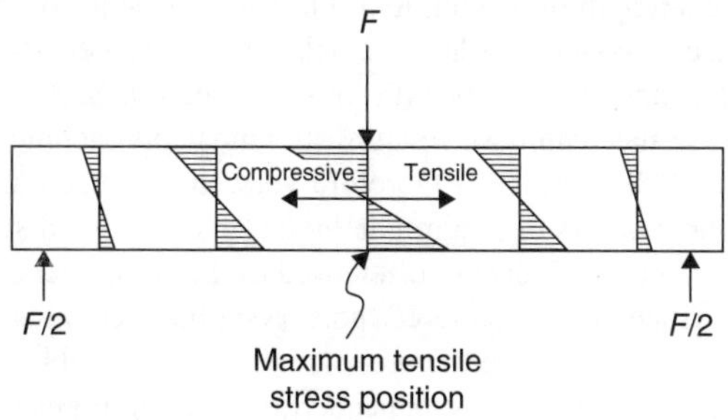

Figure 10.22 *Stress distribution in a three-point bend test.*

Table 10.7 *Thermal shock resistance of some ceramics (from Richerson, 1992).*

Al_2O_3	SiC	Si_3N_4 (reaction sintered)	Si_3N_4 (hot pressed)	β-Spodumene
96°C	230°C	570°C	650°C	4860°C

breakage of a cold teapot as hot water is poured into it). For shapes such as a bar constrained at both ends, the tensile stress developed due to a reduction in temperature ΔT is given by $\sigma = E\alpha\Delta T/(1-\nu)$, where E is the Young's modulus, ν the Poisson's ratio and α the linear expansion coefficient. The thermal shock resistance R is defined as the ΔT when σ reaches the fracture strength of the ceramic. Clearly, materials with lower values of α are more thermal-shock resistant. Table 10.7 gives the thermal-shock resistance of some ceramics. β-Spodumene has a very high thermal-shock resistance because of its extremely low thermal expansion coefficient ($-0.3 \times 10^{-6}\,K^{-1}$). Hot-pressed Si_3N_4 has higher thermal-shock resistance than reaction-sintered Si_3N_4 because of the higher fracture strength resulting from the hot-pressing condition, which reduces porosity.

10.6.2 Statistical nature and size dependence of strength

The fracture strength of brittle materials generally exhibits two unusual characteristics, namely statistical variations and size dependence. Since the fracture stress is inversely proportional to the square root of the crack size (see Chapter 7), a sample containing larger cracks will have a lower strength, and vice versa. Thus, macroscopically similar samples do not necessarily exhibit the same fracture strength, since their flaw distributions may not be the same. The fracture strengths of ceramics and

Table 10.8 *Strengths of ceramics in fiber and polycrystalline forms (from Richerson, 1992).*

	E (Gpa)	*Ideal fracture strength E/10 (GPa)*	*Measured strengths of fibers (GPa)*	*Measured strength of polycrystalline specimen (GPa)*
Al_2O_3	380	38	16	0.4
SiC	440	44	21	0.7

glasses are therefore not material constants, and on repeated measurements, they usually exhibit rather significant statistical scatter.

The fracture strength of a ceramic or glass also depends on the sample size. A larger sample has a higher chance of containing larger cracks, and so in general is weaker in tension than a smaller sample. Ceramic and glass fibers with sub-millimeter diameters have very low chances of containing big flaws, and so they can have fracture strengths approaching the ideal value of $E/10$ (Chapter 7; see also Table 10.8). Fibers are therefore very strong materials, able to withstand tensile stresses up to two to three orders of magnitude higher than the yield strength of steel, for example. This is the reason why ceramic fibers are often used as the strengtheners in making load-bearing composite materials (see Chapter 11). Sintered, polycrystalline ceramics in bulk forms, however, have much lower fracture strengths due to the presence of porosity (Table 10.8).

The statistical scatter and size dependence of strength pose a problem in the mechanical design of ceramic components. In the laboratory, one can only test small pieces of samples, but since these would have strengths higher than the real component, which is usually bigger, the strength data from the small samples cannot be used directly for the design. The strength data would also depend on the test method employed. In a three-point bend test (Figure 10.22), the stress is compressive (and hence will not contribute to fracture) within the upper half-thickness of the test bar, and is tensile within the lower half-thickness and increasingly so towards the bottom of the sample. In a longitudinal direction, the stress increases linearly from zero at the two lower supports to a peak value at the mid-span of the sample. Thus, if the largest crack in the test bar happens to occur near the mid-span at the bottom of the specimen, then the MoR value from the bend test would be close to the fracture strength recorded if the same bar is subjected to a tensile test. However, the chance for the largest crack to situate exactly at the mid-span and at the bottom is low and, hence, for the same material, the MoR values obtained from bend tests in general are higher, and will scatter more, than the fracture strengths obtained from tensile tests.

Weibull (1951) proposed a statistical approach to deal with the above problems concerning the design of ceramic components. Instead of treating the fracture strength as a deterministic quantity, Weibull proposed the use of a survival probability to describe the chance a given piece of ceramic will survive a given applied load. The survival probability was suggested to be

$$P_s(V) = \exp\left[-V(\sigma/\sigma_o)^m\right], \tag{10.1}$$

where V is the volume of the ceramic, σ is the applied tensile stress, and σ_o and m are material constants. The choice of the exponential function in equation (10.1) is to ensure the obvious condition $P_s(V_1 + V_2) = P_s(V_1) \cdot P_s(V_2)$ to be satisfied. To determine σ_o and m for a given material, small ceramic pieces of identical size are tested and their strength data, which should scatter for the reason explained above, are ranked in an ascending order. The survival probability of each tested sample is then assigned according to $P_s = 1 - i/(N+1)$, where i is the rank and N the total number of samples. A table between P_s and the fracture strength σ is thus obtained for the N samples, and from equation (10.1), a plot of $\ln(1/P_s)$ versus $\ln \sigma$ should yield a straight line with slope and y-intercept equal to m and $(\ln V - m \ln \sigma_o)$ respectively. The constants σ_o and m can therefore be determined, since the sample volume V is known. With σ_o and m determined this way, the fracture strength σ of any sample size V can be determined from equation (10.1) to correspond to a prescribed survival probability, say 0.95.

Equation (10.1) is for a uniform stress state σ, and in a varying stress situation, such as a three-point bend test, the Weibull formula can be generalized into

$$P_s(V) = \exp\left[-\int_V [\sigma(r)/\sigma_o]^m d^3r \right], \tag{10.2}$$

where $\sigma(r)$ is the tensile stress at point r in the material volume V. Substituting in the known stress function for a three-point bend test, for example, the ratio of the MoR from the bend test to the fracture stress σ_{tensile} from a tensile test at the same survival threshold can be shown to be $(MoR) : \sigma_{\text{tensile}} = [2(m+1)^2]^{1/m} : 1$. At a typical value of 10 for the Weibull modulus m, $(MoR) : \sigma_{\text{tensile}} = 1.73 : 1$, i.e. the bend test will overestimate the fracture strength by 73%, as a result of the small chance that the largest crack will occur near the mid-span and at the bottom of the test bar.

Worked example

Three-point bend tests in sintered Al_2O_3 give MoR values in the range of 350–580 MPa. Tensile tests on similar samples give tensile strengths from 200 to 310 MPa. Estimate the Weibull modulus of this material.

Solution

The mean MoR value is $(350+580)/2 = 465$ MPa. The mean tensile strength is $(200+310)/2 = 255$ MPa. Thus, $[2(m+1)^2]^{1/m} = 465/255 = 1.824$. Numerically or graphically solving this gives $m = 8.72$.

10.6.3 Stress corrosion cracking of ceramics and glasses

The strength of ceramics and glasses often has limited lifetime, due to the slow propagation of subcritical cracks to the critical crack size by stress corrosion at the crack tips (see Section 9.2.2.3). Thus, a specimen supporting a load σ may fracture in time τ, and increasing the load, the test temperature, the pH or even the moisture content of the test environment, which are all factors favoring crack-tip corrosion, may shorten the lifespan τ. The lifetime is governed by the crack growth velocity v which, at a constant test environment, varies with the instantaneous stress intensity factor according to Figure 9.12. Thus, in region I of Figure 9.12, $v \propto K^n$, where K is the mode I stress intensity factor and the exponent n is a material constant. Figure 10.23 shows the region I of a number of glasses and ceramics tested in water at 25°C. The lifetime τ of the sample is the time for a crack with an initial subcritical size c_i to propagate to the critical size c_c, i.e.

$$\tau = \int_{c_i}^{c_c} \frac{dc}{v}, \quad c = \frac{Y^2K^2}{\pi\sigma^2}, \quad c_c = \frac{Y^2K_c^2}{\pi\sigma^2},$$

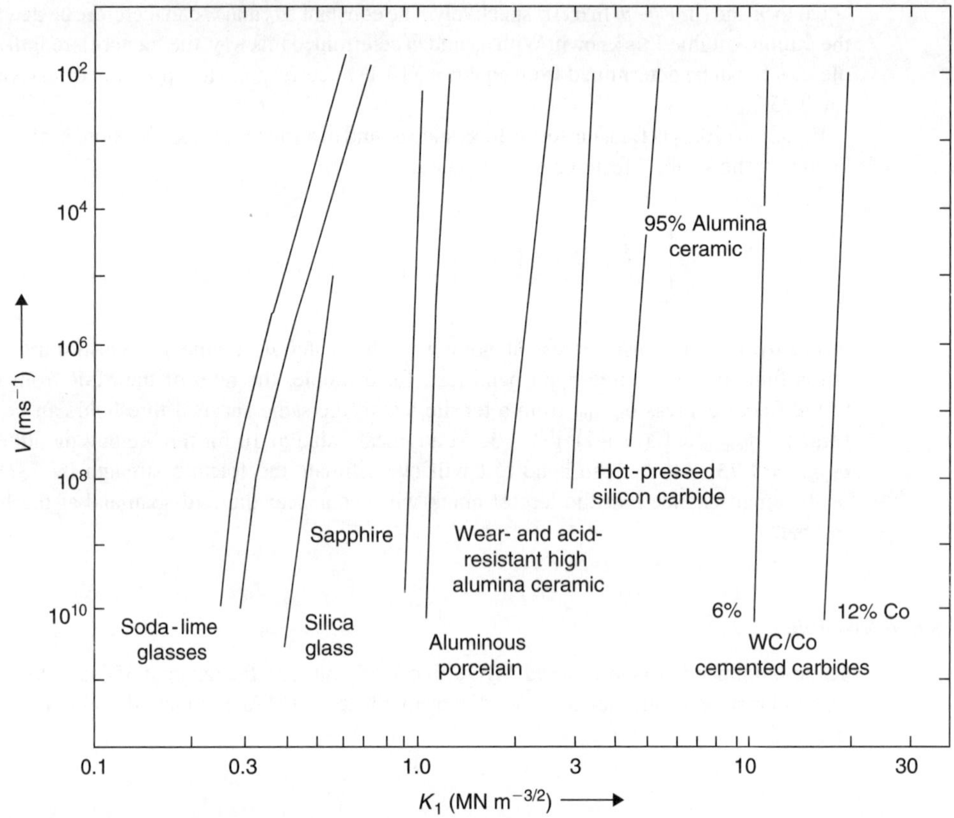

Figure 10.23 *Relation between crack velocity and stress intensity factor for various ceramics in neutral water at 25°C (after Creyke* et al., *1982).*

where Y is a geometrical constant and K_c the critical stress intensity factor. Substituting in the velocity law,

$$\tau \propto \frac{2Y^2}{\pi\sigma^2}\int_{K_i}^{K_c} K^{(1-n)}\mathrm{d}K = \frac{2Y^2}{\pi\sigma^2(n-2)}(K_i^{2-n} - K_c^{2-n}),$$

and since n is usually large (>10), K_c^{2-n} is negligible compared to K_i^{2-n}, so that

$$\tau \propto \frac{2Y^2K_i^{2-n}}{\pi\sigma^2(n-2)} = \frac{2Y^n}{(n-2)\pi^{n/2}c_i^{n/2-1}}\cdot\frac{1}{\sigma^n}.$$

The lifetime is therefore predicted to vary with the applied stress via an inverse power law. Figure 10.24 shows the strength–probability–time diagram for Al_2O_3 derived from bend tests at room temperature. At a given survival probability (say 80%), the lifetime indeed varies with an inverse power law of stress.

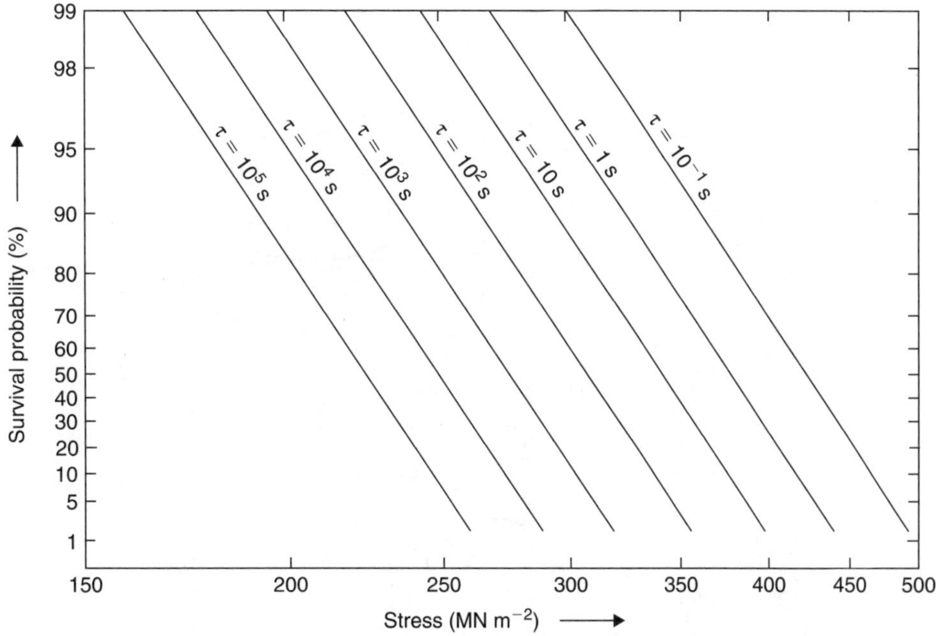

Figure 10.24 *Strength–probability–time diagram for alumina derived from bend tests at 293 K (after Davidge, 1986, by courtesy of Cambridge University Press).*

10.7 A case study: thermal protection system in space shuttle orbiter

Most ceramics are chemically very inert materials and their supreme integrity over metals and polymers at temperatures above 1000°C often makes them the only choice of candidate materials in applications at such high temperatures. By controlling the porosity during sintering, a wide range of ceramics can also be turned into very efficient heat conductors or insulators. Figure 10.25 shows the thermal conductivities of different ceramic materials. Crystalline and well-bonded ceramics such as graphite and bonded SiC are good conductors of heat because of effective phonon conduction in these materials. Porous, coarse ceramics such as firebrick or powdered ceramics are good insulators of heat because the heat transfer mechanism in these materials is mainly by radiation across the voids, which is an inefficient process.

The most well-known application of ceramic materials as high-temperature heat insulating materials is perhaps the thermal protection system in a space shuttle orbiter. A shuttle orbiter re-enters the earth's atmosphere at a certain yaw angle, so that it travels as a bluff aerodynamic body with respect to the surrounding air in order to achieve the appropriate drag for speed control. The frictional effects are so intense that the air surrounding the orbiter is turned into a plasma state. During this time, the temperature on the orbiter surface is generally above 400°C and can rise up to 1500°C at certain protruded parts, such as the nose and the wing and tail tips. To protect the Al- and Ti-based substructure from softening or even melting at such high temperatures, a thermal protection system (TPS), made principally of a range of ceramic materials, is constructed as an outermost skin of the orbiter. Localized damage of the TPS, however, can lead to the loss of the entire shuttle orbiter during re-entry, as witnessed by the *Columbia* tragedy on 1 February 2003.

In the NASA Space Shuttle program, the TPS is designed to last for 100 missions without major refurbishment. The TPS is an excellent example of ceramic engineering in that it involves the use of

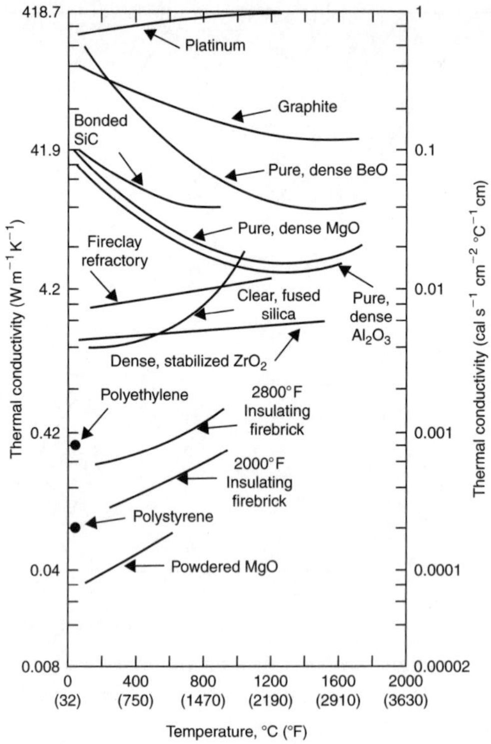

Figure 10.25 *Thermal conductivities of different materials (from Richerson, Modern Ceramic Engineering: Properties, Processing, and Use in Design, Marcel Dekker, 1992).*

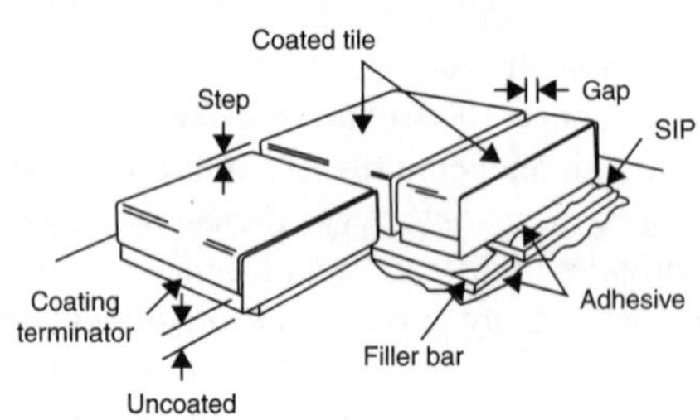

Tile – 352 kg m^{-3} (22 lb ft^{-3}) / 144 kg m^{-3} (9 lb ft^{-3}) Pure silica fiber – fired at ~1370 C (~2500 F)
Coating – Borosilicate (Glass) for waterproof and thermal properties – fired at 1150–1205 C (2100–2200 F)
SIP – Nomex felt
Filler – Coated Nomex felt
Adhesive – RTV silicone

Figure 10.26 *Silica tile system for thermal protection of a space shuttle orbiter (from Korb, Morant, Calland and Thatcher, 1981, by permission of the American Ceramic Society).*

different ceramic and glass materials, depending on the location to be protected. In areas exposed to the highest temperature during re-entry, i.e. the nose cap and the leading edges of the wings, reinforced carbon–carbon (RCC) is used, which can sustain temperatures up to 1650°C. RCC is a ceramic–ceramic composite consisting of a carbon matrix reinforced with graphite fibers, coated with silicon carbide and then impregnated with silica. RCC has sufficient strength and stiffness at temperatures up to 1650°C to withstand the airloads and thermal stresses as a result of the large thermal gradients across the upper to lower vehicle surfaces during re-entry. In the major part of the orbiter surface where the temperature is between 400 and 1260°C, the TPS is in the form of silica (SiO_2) tiles bonded to the substructure by adhesive. Figure 10.26 shows the typical construction of the silica tile system. The tiles are made from high-purity amorphous silica fibers, which are felted from a slurry and sintered at 1370°C into blocks. The tile material is highly porous (93% void) and, because it is amorphous, the thermal conductivity is very low (0.017–0.052 $W\,m^{-1}\,K^{-1}$, cf. Figure 10.25). It also has very low thermal expansivity and so thermal shock resistance is high. The silica has to have high purity (>99.62%) to avoid devitrification, which will increase the thermal expansivity. The exposed surfaces of each tile are coated with borosilicate glass to achieve waterproof and the proper emittance properties (Figure 10.26).

Problems

10.1 Glass is extremely susceptible to small cracks. If the glass is stressed to 50 MPa, determine the maximum permissible surface flaw to avoid fracture ($E = 70$ GPa and the surface energy $\gamma_s = 0.3\,J\,m^{-2}$).

10.2 A relationship for the Young's modulus E to the melting temperature T_m is given by the approximation

$$E = A\frac{\mathbf{k}T_m}{\Omega},$$

where A is a constant (~90). If diamond melts at 4200 K and has an atomic volume (Ω) $5.68 \times 10^{-30}\,m^3$, estimate the value for the modulus.

10.3 An expression relating porosity (p) to modulus is $E = E_0(1 - 1.9p + 0.91p^2)$. The modulus of elasticity for alumina (Al_2O_3) with 25% porosity is 220 GPa. Estimate the modulus for a non-porous ceramic.

10.4 The activation energy for viscous flow of a glass is 250 $kJ\,mol^{-1}$. If the strain rate at 1000°C (T_1) is $10^{-2}\,s^{-1}$, calculate the strain rate at a higher temperature of 1250°C (T_2).

10.5 Rubber strained 100% requires a stress of 20 MPa at room temperature. To maintain the strain over a period requires a drop in stress, i.e. relaxation. What is the stress after 10 days if the stress drops to exp(−1) of the original value after 30 days?

10.6 Diamond, silicon carbide and silicon have an energy gap of 5.6, 3.1 and 1.1 eV respectively. How is this variation in energy gap reflected in the appearance of each material?

10.7 How is the polymorphic form of zirconia exploited in toughening?

10.8 The flexural strength of porous alumina is $\sigma_{fs} = \sigma_o \exp(-np)$, where σ_o is the flexural strength for non-porous alumina, p is percentage porosity and n is a material constant. The room temperature flexural strength of alumina is 175 MPa for a specimen with 5% porosity and 75 MPa for one with 25% porosity. Determine the parameters σ_o and n in the relationship for the influence of porosity on flexural strength.

10.9 From the temperature dependence of the viscosity curve for SiO_2–Na_2O–CaO glass (shown in Figure 10.13), calculate the activation energy for the process over the intermediate temperature range 600–900°C.

10.10 From Figure 10.13 determine the maximum temperature to which an SiO_2–Na_2O–CaO glass may be heated if the deformation of a cylindrical specimen 100 mm long subjected to a tensile stress 50 kPa is to be less than 1% over a period of a year.

10.11 The following table shows the tensile strengths measured from a series of 15 dense Al_2O_3 (0–2% porosity) of identical sizes with a gauge section measuring 5 mm × 5 mm × 30 mm:

Rank, *i*	1	2	3	4	5	6	7	8	9	10	11	12	13	14	15
Strength (MPa)	192	203	212	224	230	235	243	250	253	260	276	282	286	295	301

(i) By plotting the data in an appropriate format, show that the data roughly obey the Weibull statistics.

(ii) Calculate the Weibull modulus m and the stress normalization constant σ_o for this batch of Al_2O_3.

(iii) If a survival probability of 95% is to be maintained, calculate the strength of a piece of Al_2O_3 specimen made using a similar method with dimensions measuring 50 mm × 50 mm × 300 mm.

Further reading

Binner, J. (1992). Processing of advanced ceramic powders. *Metals and Materials*, October, pp. 534–537. Institute of Materials.

Burkin, A. R. (ed.) (1987). *Production of Aluminium and Alumina*. Wiley, Chichester.

Cahn, R. W. and Harris, B. (1969). Newer forms of carbon and their uses. *Nature*, 11 January, **221**, 132–141.

Creyke, W. E. C., Sainsbury, I. E. J. and Morrell, R. (1982). *Design with Non-Ductile Materials*. Elsevier/Chapman & Hall, London.

Davidge, R. W. (1986). *Mechanical Behaviour of Ceramics*. Cambridge University Press, Cambridge.

Ichinose, N. (1987). *Introduction to Fine Ceramics – Applications in Engineering*. Wiley, New York.

Inagaki, M. (2000). *New Carbons – Control of Structure and Functions*. Elsevier, New York.

Jack, K. H. (1987). Silicon nitride, sialons and related ceramics. In *Ceramics and Civilisation*, Vol. 3. American Ceramic Society, New York.

Kingery, W. D., Bowen, H. K. and Uhlmann, D. R. (1976). *Introduction to Ceramics*, 2nd edn. Wiley-Interscience, New York.

Laminated Glass Information Centre (1993). *Laminated Glass*. LGIC, London.

McColm, I. J. (1983). *Ceramic Science for Materials Technologists*. Leonard Hill, Glasgow.

Mantell, C. L. (1968). *Carbon and Graphite Handbook*. Wiley-Interscience, New York.

Parke, S. (1989). Glass – a versatile liquid. *Metals and Materials*, January, pp. 26–32. Institute of Materials.

Rawson, H. (1980). *Properties and Applications of Glass*. Elsevier Science, Oxford.

Richerson, D. W. (1992). *Modern Ceramic Engineering: Properties, Processing, and Use in Design*. Marcel Dekker, New York.

Riley, F. L. (ed.) (1977). *Nitrogen Ceramics*. Noordhoff, Leiden.

Saito, S. (ed.) (1985). *Fine Ceramics*. Elsevier, New York.

Ubbelohde, A. R. J. P. and Lewis, F. A. (1960). *Graphite and its Crystal Compounds*. Clarendon Press, Oxford.

Wachtman, J. B. (ed.) (1989). *Structural Ceramics*, Vol. 29. Academic Press, New York.

Chapter 11

Non-metallics II – Polymers, plastics, composites

11.1 Polymer molecules

Naturally occurring polymers, e.g. wood and rubber, are in common engineering use, and proteins, enzymes and starches are of biological importance. Synthetic polymers are increasingly used in manufacturing for household and engineering applications. These plastics are polymers with additives and are based on long-chain molecules developed from alkenes, with the general formula C_nH_{2n} and a double carbon bond, which is opened up by temperature, pressure and/or catalysts, i.e. polymerization, or addition polymerization, since the molecules join together without losing any atoms. For the simplest alkane, ethane, the basic molecule or repeat unit of structure C_2H_4, known as a mer,[1] links up endwise to form a long-chain molecule or polymer $(C_2H_4)_n$ according to

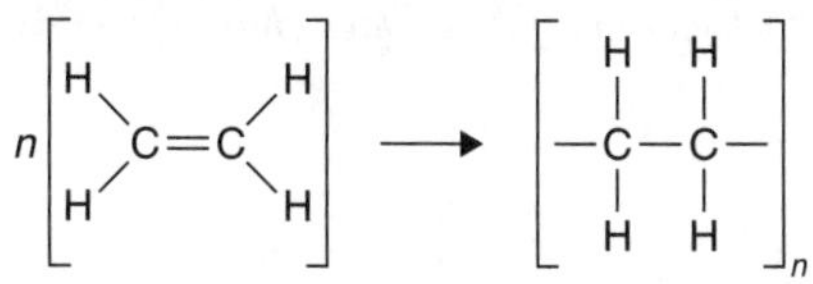

when the ethene becomes poly(ethene) or polyethylene (PE). The long-chain molecule has a zigzag backbone structure because of the tetrahedral bonding and contains thousands of mers which flex and twist (Figure 11.1). In bulk the chains make up a tangled mass.

In other common polymers (Table 11.1), one or more of the hydrogens is replaced by a different atom. This may be a Cl atom to make polyvinyl chloride (PVC), a methyl radical CH_3 to make polypropylene (PP), a benzene ring to make polystyrene (PS), or an acetate radical to make polyvinyl acetate (PVAc). Many of the other polymers are made up of more complicated mer structures, each containing several of these replacements, e.g. polycarbonate.

Condensation polymerization is another form of polymerization reaction involving more than one monomer species and the elimination of small molecules such as water. The process is used to produce polyesters, nylons and polycarbonates, where the reaction time is longer than for addition polymerization.

11.2 Molecular weight

As the length of molecules increases, so do some of the properties of the polymer, e.g. melting point, strength, viscosity, etc. If M_{mon} is the mass of the monomer, e.g. 28 for C_2H_4, the relative mass of a chain molecule is $M = n\,M_{mon}$. However, within the polymer there is a statistical distribution of chain lengths (Figure 11.2). In this case, the average molecular mass is $\overline{M}$, which may be defined as number average molecular mass $\overline{M}_N$ or weight average molecular mass $\overline{M}_W$. By dividing the chains into

[1] The term 'mer' is derived from the Greek word *meros*, meaning part; a single mer is a monomer with two free bonds and so is unsaturated and bifunctional. The addition of many mers make up the polymer.

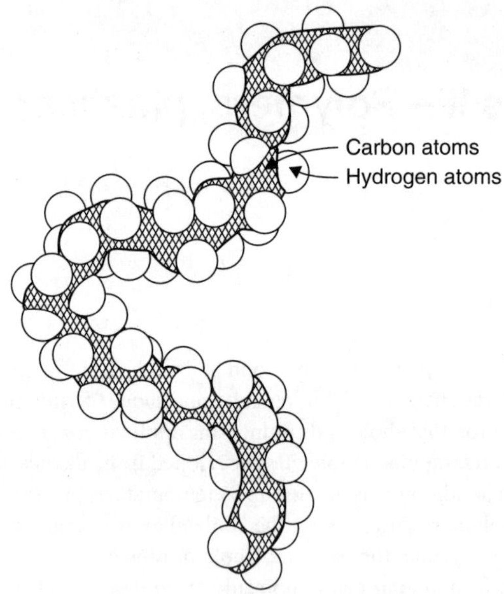

Figure 11.1 *A short section of a polymer chain (from Ashby and Jones, 2005).*

Table 11.1 *Some common polymers.*

Polymer	*Mer structure*	T_g*(°C)*	T_m*(°C)*
Polyethylene (PE)	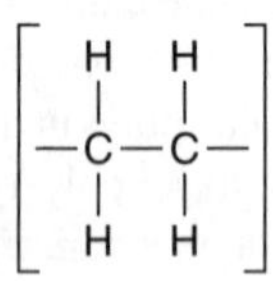	HD – 90 LD – 110	137 115
Polyvinyl chloride (PVC)	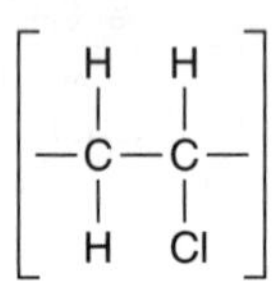	87	212
Polypropylene (PP)	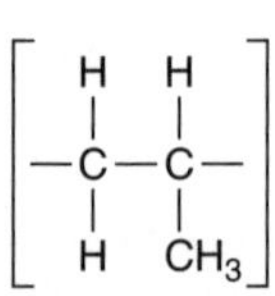	−10 (Isotactic) −18 (Atactic)	175
Polystyrene (PS)	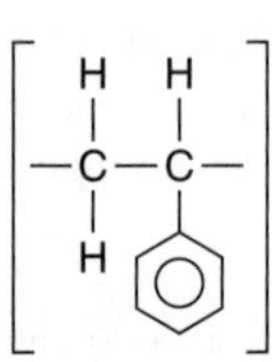	100	240

Table 11.1 *Continued*

Polymer	*Mer structure*	T_g(°C)	T_m(°C)
Polyvinyl acetate (PVAc)	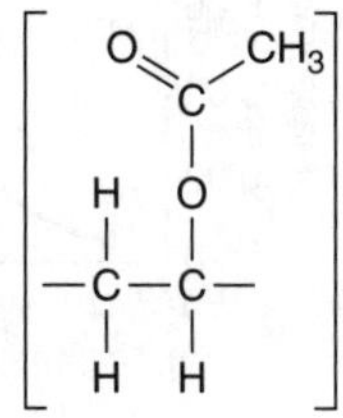		
Polycarbonate (PC)	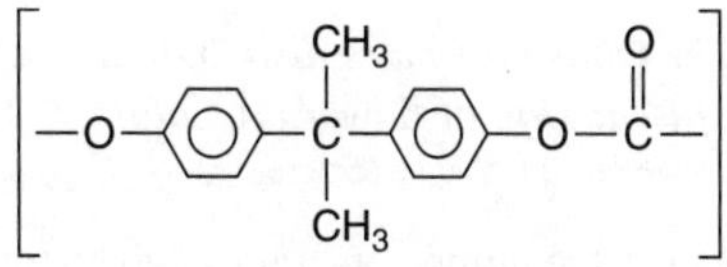	150	265
Polytetrafluoroethylene (PTFE)	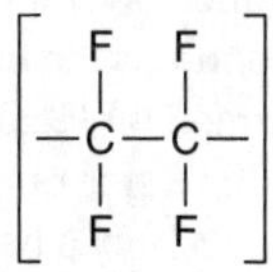	−97	327
Polyisoprene (natural rubber)	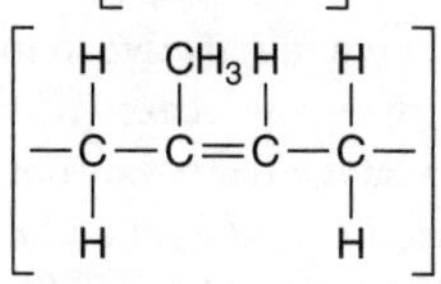	−73	28
Polychloroprene	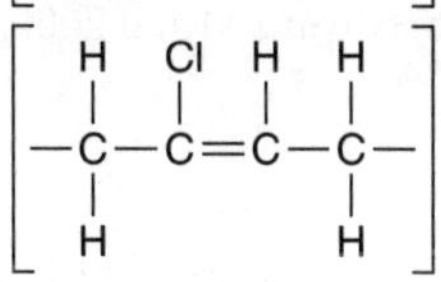	−50	80
Polybutadiene	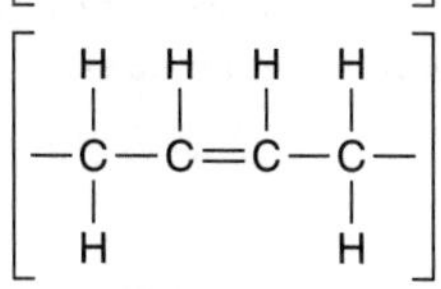	−90	154 120
Polydimethyl siloxane (silicone rubber)	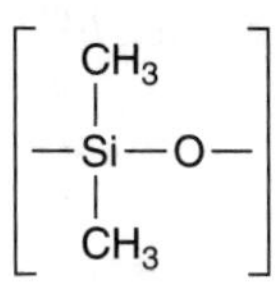	−123	−54
Phenol-formaldehyde (Bakelite)	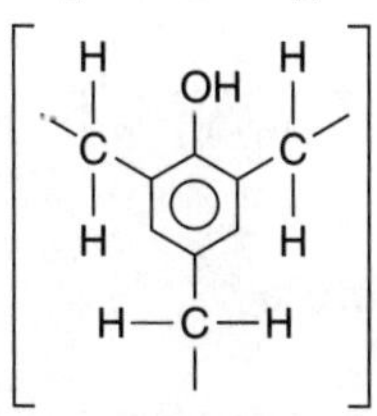		
Polymethyl methacrylate (PMMA)	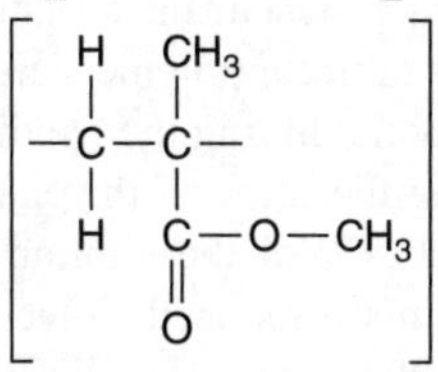	3	45

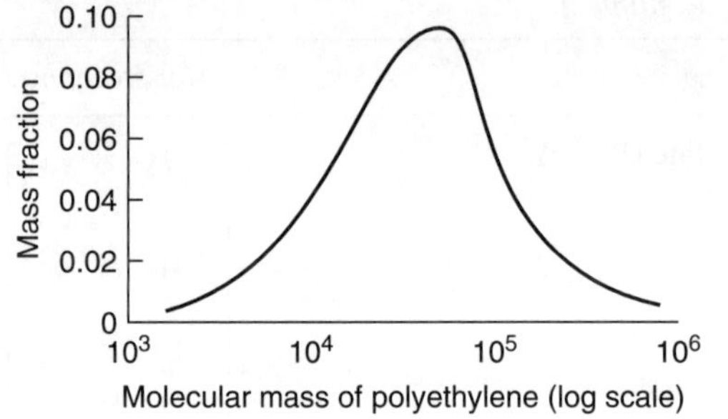

Figure 11.2 *The molecular mass distribution of a polyethylene, determined using GPC (from Mills, 1986; by permission of Edward Arnold).*

different size ranges, the number average is given by $\overline{M}_N = \sum x_i M_i$, where M_i is the mean molecular weight of the ith size range and x_i is the fraction of the total number of chains within the corresponding size range. The weight average molecular mass $\overline{M}_W$ is given by $\overline{M}_W = \sum w_i M_i$, where M_i is the mean molecular weight within a size range and w_i is the weight fraction of molecules within that size range. The values of $\overline{M}_W$ and $\overline{M}_N$ can be determined by gel permeation chromatography (GPC). $\overline{M}_W$ is always greater than $\overline{M}_N$ and sensitive to long-chain molecules. The ratio $\overline{M}_W/\overline{M}_N$ is known as the polydispersivity index, which increases with the spread of the molecular mass distribution (MMD). Generally there is an average degree of polymerization $\overline{n}$ for developing optimum molecular mass and resultant properties; $\overline{n}_N = \overline{M}_N/\overline{m}$ and $\overline{n}_W = \overline{M}_W/\overline{m}$, where $\overline{m}$ is the mer molecular weight. For most polymers, $\overline{n}$ lies between 200 and 2000, corresponding to molecular masses of 20 000–200 000.

Worked example

Calculate $\overline{M}_W$ and $\overline{M}_N$ if equal weight of polymers with molecular weights of respectively 10 000, 30 000 and 100 000 are mixed together.

Solution

Let the weight of each polymer be W.

$$\overline{M}_W = \frac{\sum w_i M_i}{\sum w_i} = \frac{W \times 10\,000 + W \times 30\,000 + W \times 100\,000}{3W} = 46\,666$$

$$\overline{M}_N = \frac{\sum N_i M_i}{\sum N_i} = \frac{\sum w_i}{\sum w_i/M_i} = \frac{3W}{W/10\,000 + W/30\,000 + W/100\,000} = 20\,930.$$

11.3 Polymer shape and structure

The properties of polymers depend on the nature of the polymer molecule, the molecular weight and its shape. The long chains of linear polymers are flexible and spaghetti-like, with van der Waals bonding between the chains. By modifying the conditions of polymerization it is possible to synthesize side reactions of sites along the spine of the chain molecule. Thus, chain molecules may acquire long, short or multiple branches. With the addition of branches the packing of the chains is less efficient and the density of the polymer is reduced. Polyethylene can be produced in both the low-density (LDPE) and high-density (HDPE) forms. The HDPE contains just a few short branches and is stronger and

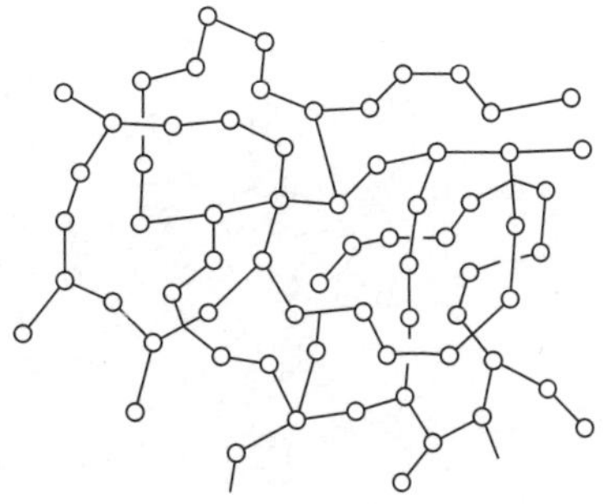

Figure 11.3 *Cross-linked polymer structure (from Ashby and Jones, 2005).*

more rigid than LDPE, with a melting point 22°C higher. The weak forces between adjacent chain molecules may be overcome by heating, and under pressure the chains straighten and slide past each other. Thus, molecular viscosity is common to thermoplastic polymers like polyethylene, which are melt extruded and injection molded. Branching reduces the molecular mobility. Cross-linking of adjacent chains at intervals along the chain by covalent bonds may also be introduced during synthesis. This is a feature of elastomers such as polyisoprene and polychloroprene (Neoprene), which can be stretched considerably. Elastomers behave viscoelastically at temperatures above glass transition (T_g) when stressed with both viscous, time-dependent, and elastic, instantaneous, strain characteristics, as molecular segments move and unravel. When the degree of cross-linking is highly developed, as for trifunctional mer units with three active covalent bonds, the structure becomes strong and rigid. The structure forms a network polymer and is the basis of the thermoset materials; the epoxies (Araldite) and phenol-formaldehyde (Bakelite) belong to this group (Figure 11.3). Other important chain structures are isotactic and atactic configurations. As the term implies, an isotactic configuration occurs when all the chlorine atoms in the PVC molecule, for example, lie on the same side of the chain. A fully randomized arrangement leads to an atactic form. Again, the configuration markedly affects the molecular mobility.

Addition polymerization is not confined to one type of mer. Copolymers are made by the polymerization of two monomers. Adding them randomly produces a random copolymer and in an ordered sequence a block copolymer. Styrene-butadiene rubber (SBR) is a common random copolymer formed from styrene and butadiene according to

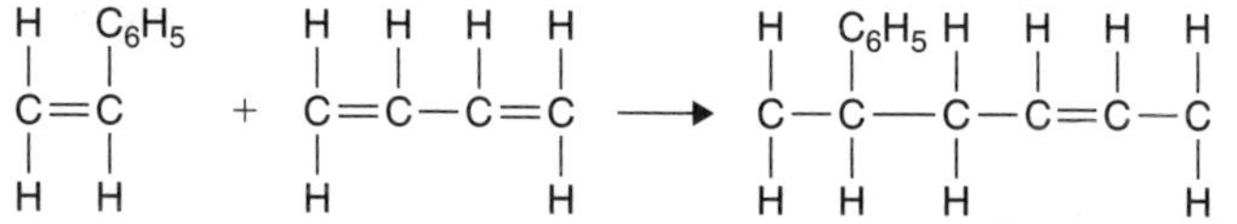

11.4 Polymer crystallinity

In most polymers the spaghetti-like chains of various lengths pack together in a random way, forming a tangled mass in the bulk materials. These polymers are therefore non-crystalline or amorphous. However, in some polymers the chain molecules form regions in which the repeat units are aligned in ordered arrays, in some cases folded backwards and forwards to resemble the 'jumping jack' firework. Such crystalline regions are usually quite small (~20 nm) and often associated with amorphous regions, as shown in Figure 11.4 for polyethylene. In this case, crystalline regions nucleate and grow during polymerization, approaching 50% for LDPE and 80% for HDPE. Clearly, the density of a crystalline polymer will be greater than that of an amorphous polymer of the same composition and

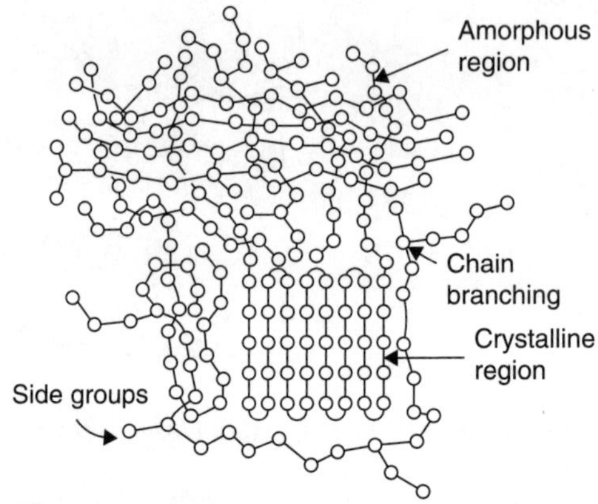

Figure 11.4 *Semi-amorphous structure of polyethylene (from Ashby and Jones, 2005).*

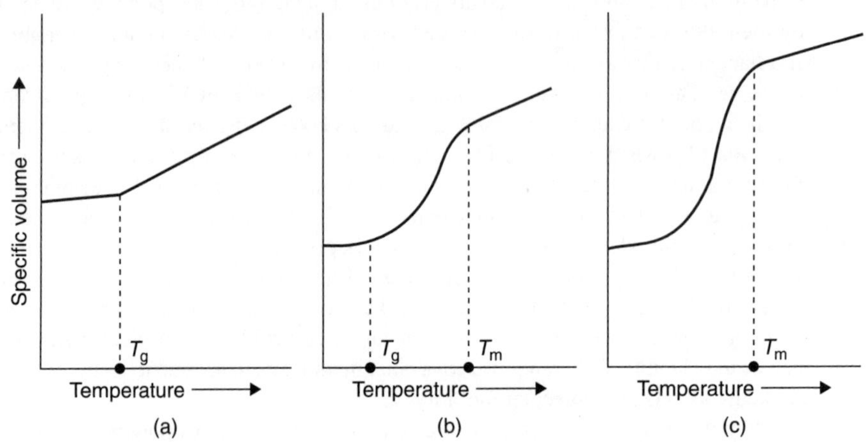

Figure 11.5 *Specific volume versus temperature plots for: (a) 100% amorphous polymer, (b) partially crystalline polymer, (c) 100% crystalline polymer.*

molecular weight. The degree of crystallinity by weight may be determined from accurate density measurements according to

$$\% \text{ crystallinity } \phi = \frac{\rho_c(\rho - \rho_a)}{\rho(\rho_c - \rho_a)} \times 100,$$

where ρ is the density of the polymer, ρ_c is the density of the fully crystalline polymer and ρ_a is the density of an amorphous polymer of the same material. Crystallization is favored by slow cooling rate, since this allows time for the packing and ordering of the chains. Application of stress also aids crystallization. In contrast, crystallization is hindered as molecules become longer and more complex, thereby reducing molecular mobility. Isotactic configurations favor crystallinity compared to atactic configurations. Matching configurations also favor crystallinity and are more likely in copolymers with regular block patterns of constituents than in random copolymers. Thermal degradation increases with increase in temperature and, above T_g, the glass transition temperature, the entangled molecules begin to lose their rigidity until, at T_m, the melting point, crystallinity breaks down completely. T_m increases with the degree of crystallinity and is 110°C for LDPE and 135°C for HDPE. Figure 11.5

shows the influence of temperature on the specific volume of a polymer with varying degree of crystallinity. Parameters such as specific volume, heat of fusion and density are influenced by the degree of crystallinity ϕ according to

$$p = \phi p_c + (1 - \phi) p_a,$$

where p is the overall experimental value of, for example, specific volume, p_c is the value for a crystalline polymer and p_a for the amorphous polymer. Differential scanning calorimetry (DSC) can provide the heat of fusion (see Figure 4.61c) and ϕ is given by

$$\phi = \text{Heat of fusion of sample/Heat of fusion of 100\% crystalline polymer.}$$

Worked example

Derive the formulae for % crystallinity by weight and by volume, for evaluation by density measurement of a polymer. High-density polyethylene (HDPE) has a density of 0.95 Mg m^{-3}. Calculate the degree of crystallinity by weight and by volume if the amorphous PE and fully crystalline PE have densities of 0.84 and 1.01 Mg m^{-3} respectively.

Solution

Let m, m_c and m_a be the total mass, mass of the crystalline phase and mass of the amorphous phase respectively of the polymer. Similarly, let V, V_c and V_a be the total volume, and volumes of the crystalline and amorphous phases. Furthermore, let $f = V_c/V$, i.e. crystallinity by volume, and $\phi = m_c/m$, i.e. crystallinity by weight.

$$\text{Since } m = m_c + m_a, \rho V = \rho_c V_c + \rho_a V_a, \text{ or } \rho = \rho_c f + \rho_a(1 - f).$$

$$\text{Hence, } f = \frac{\rho - \rho_a}{\rho_c - \rho_a}.$$

$$\text{Degree of crystallinity by weight } \phi = \frac{m_c}{m} = \frac{V_c \rho_c}{V \rho} = f\frac{\rho_c}{\rho} = \left(\frac{\rho - \rho_a}{\rho_c - \rho_a}\right)\frac{\rho_c}{\rho}.$$

Alternatively, since $V = V_c + V_a, m/\rho = m_c/\rho_c + m_a/\rho_a$, giving

$$1/\rho = \phi/\rho_c + (1 - \phi)/\rho_a. \text{ Thus, } \phi = \frac{1/\rho - 1/\rho_a}{1/\rho_c - 1/\rho_a}, \text{ which is the same as the above equation.}$$

$$\text{For HDPE, } f = \frac{\rho - \rho_a}{\rho_c - \rho_a} = \frac{0.95 - 0.84}{1.01 - 0.84} = 0.647 = 64.7\%;$$

$$\phi = \left(\frac{\rho - \rho_a}{\rho_c - \rho_a}\right)\frac{\rho_c}{\rho} = \frac{(0.95 - 0.84) \times 1.01}{(1.01 - 0.84) \times 0.95} = 0.688 = 68.8\%.$$

11.5 Polymer crystals

By analogy with fundamental studies carried out on metals, single crystals of polymers have been produced and have considerably advanced our knowledge of polymer structures. Crystals of polyethylene

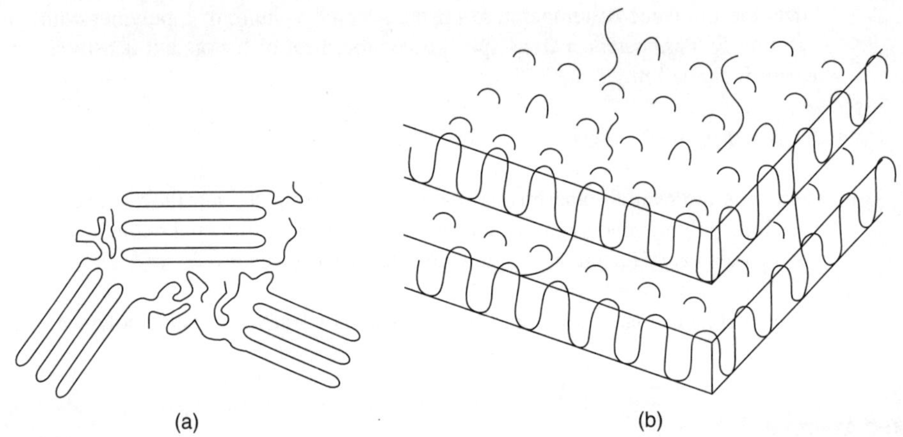

Figure 11.6 *Folded chain model for crystallinity in polymers shown in two dimensions (a) and three dimensions (b) (after Askeland, 1990, p. 534; by permission of Chapman & Hall).*

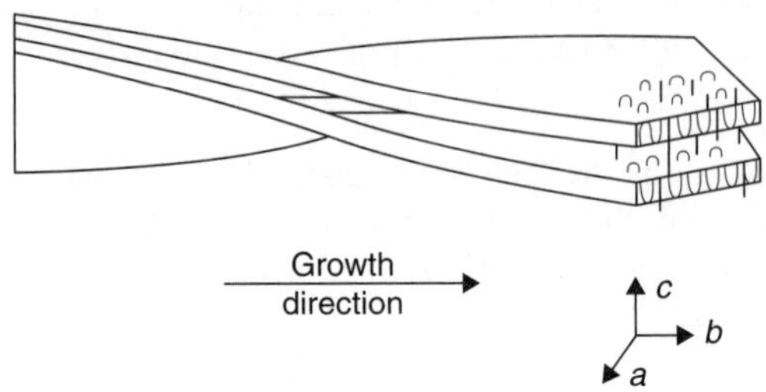

Figure 11.7 *Schematic representation of a possible model for twisted lamellae in spherulitic polyethylene, showing chain folds and intercrystalline links (courtesy R. Young).*

are prepared by slowly cooling a dilute solution of PE in xylene from 135°C. The small PE crystals that crystallize out are regularly shaped thin platelets (or lamellae) 10–20 nm thick and 10 μm or so across. Multiple chain folding occurs with the axes of the chain molecules approximately perpendicular to the large lamellae faces, as shown in Figure 11.6. The measured density of these crystals is usually less than the theoretical value, indicating the presence of defects and irregular arrangements.

Polymers crystallized from the melt under normal cooling conditions often form a spherulitic morphology.[2] Heterogeneous nucleation occurs throughout the melt and radial growth occurs until they impinge on one another (somewhat like grains in metals). Each spherulite is composed of many ribbon-like chain-folded lamellae making up 70–80% of the spherulite, between which are regions of amorphous material, as shown in Figure 11.7. The spherulite boundaries are clearly visible in the micrograph (Figure 11.8) when the lamellae twist as they grow radially. The refractive index gradually changes for each lamella, causing incident plane polarized light to become elliptically polarized in four quadrants of each spherulite to display the characteristic 'Maltese cross' pattern. Chain folding is a common feature of the polymer structure and is thought to be aided by the process of reptation, when a long-chain molecule is able to 'creep' through the free space within the tangled mass (Figure 11.9).

[2] Often, the degree of crystallization can be represented by the Avrami equation (6.44) in the primary stage when spherulites are growing at the expense of the melt.

Figure 11.8 *Polarized light micrograph of two-dimensional spherulites grown in a thin film of poly-l-lactide acid (courtesy of W.Y. Zhou and W.L. Cheung).*

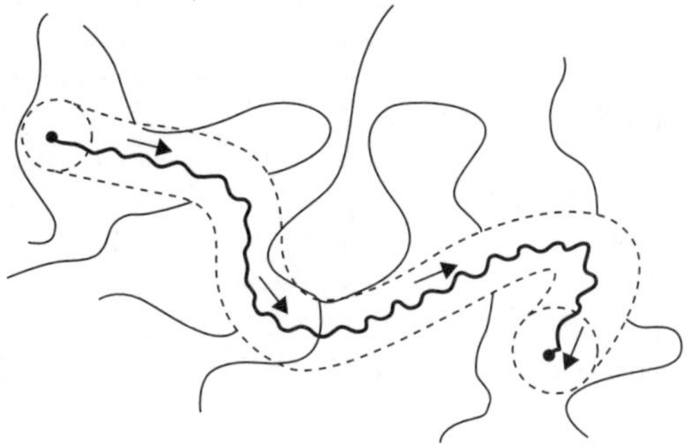

Figure 11.9 *Movement of a polymer molecule by reptation.*

11.6 Mechanical behavior

11.6.1 Deformation

At a basic level, the mechanical behavior of polymers may be compared directly with that of metals, in that stress–strain properties, modulus, tensile strength, elongation, etc. are measured. The broad classification of polymers into thermoplastics, thermosets and elastomers gives rise to distinctly different stress–strain curves, as shown in Figure 11.10. While these curves are somewhat similar to brittle, ductile and superplastic metals, polymer materials are neither as strong nor as stiff as metals. The glass transition temperature T_g is an important parameter in defining the mechanical properties, which are extremely sensitive to changes of temperature within the vicinity of room temperature. Table 11.2 gives the mechanical characteristics of some of the common polymers.

The slope of the stress–strain curve increases with increasing strain rate, and decreasing temperature has a similar effect. Modulus is particularly dependent on temperature and usually divided into three regions, namely rigid, leathery and rubbery, as shown in Figure 11.11. Increasing the molecular weight tends to increase the rubbery region, while an increase of crystallinity raises the melting point and thus the modulus of the rubbery region.

With reference to Figure 11.12, during the elastic deformation of a semicrystalline polymer, it is mostly the chain molecules which elongate. This is followed by an upper yield point, when a small

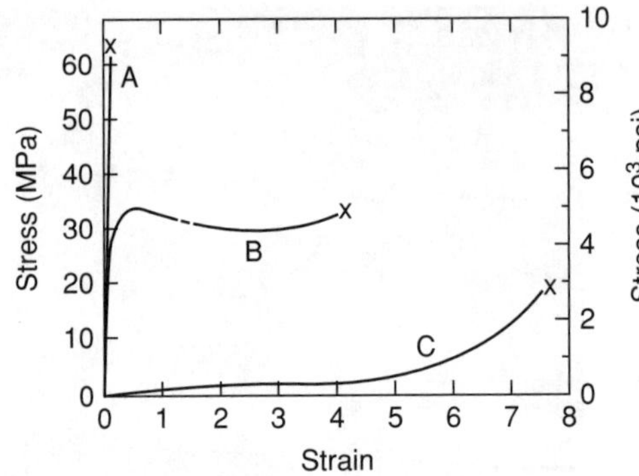

Figure 11.10 *Stress–strain behavior of brittle (A), plastic (B) and elastomeric (C) polymers.*

Table 11.2 *Mechanical properties of some polymers.*

Polymer	*Tensile modulus GPa*	*Yield MPa*	*Tensile Strength MPa*	*Elong % at break*	*Density Mgm^{-3}*
Polyethylene (LDPE)	0.25	12	18	305	0.91
(HDPE)	1.1	26	30	600	0.95
Polyvinyl chloride (PVC)	3.2	42	46	60	1.3
Polypropylene (PP)	1.4	34	36	350	0.9
Polystyrene (PS)	2.8	50	50	2	1.0
Polycarbonate (PC)	2.5	62	67	130	1.2
Polytetrafluoroethylene (PTFE)	0.47	–	27	300	2.3
Polyisoprene	–	–	8	1000	0.83
Phenol-formaldehyde (Bakelite)	3.8	–	48	10	1.35
Polymethacrylate (PMMA)	2.8	65	65	4	1.2

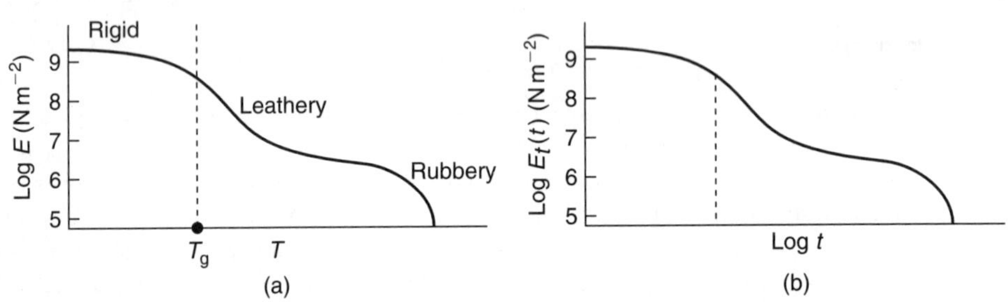

Figure 11.11 *Temperature dependence of tensile modulus (a) and time dependence of relaxation modulus (b) in thermoplastic polymeric solid (from Hertzberg, 1989; by permission of John Wiley and Sons).*

neck forms and locally strengthens as the chains align (curve B in Figure 11.10). A lower yield point follows when the deformed region spreads along the tensile gauge length. During plastic deformation the amorphous regions between the lamellae stretch as they attempt to align with the tensile axis, following which the lamella regions attempt to reorient themselves. With increasing deformation,

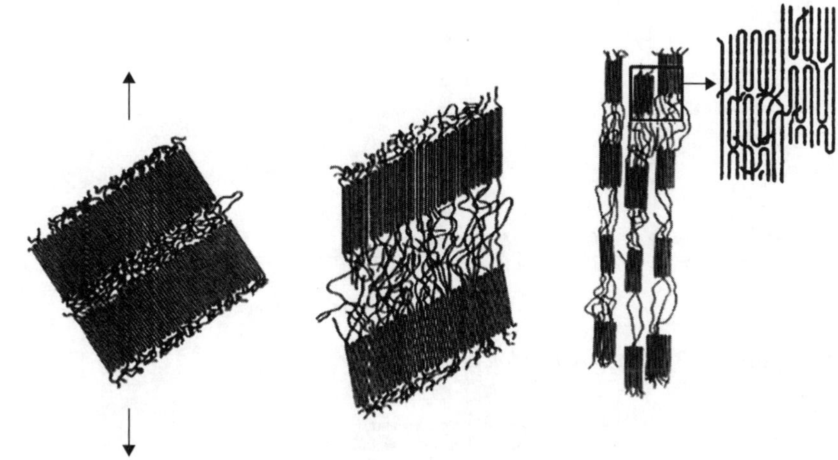

Figure 11.12 *Structural change in a semicrystalline polymer during tensile elongation (after Schultz, 1974).*

more and more lamella regions become aligned with the stress axis until after a significant deformation the polymer is highly oriented. The lamellae are broken down into smaller 'blocks', making up microfibrils, as shown in Figure 11.12. The individual blocks retain their chain-folding structure and are linked together by the molecules from the unfolding of the original lamellae. With this process the spherulite structure undergoes a considerable shape change and is very much broken up by extensive deformation. Nevertheless, annealing of the polymer close to the melting point tends to recover some of the original structure.

11.6.2 Viscoelasticity

Under certain conditions most polymers will behave in a viscoelastic manner when stressed and to exhibit both viscous (i.e. time-dependent) and elastic (instantaneous) strain characteristics. Viscoelastic behavior is dependent on time and temperature, and may be studied using a stress relaxation technique when a specimen is strained in tension to a predetermined, relatively low, level, as shown in Figure 11.13. The stress necessary to maintain the strain ε_o is found to decrease with time, due to molecular relaxation processes, and the relaxation modulus E_t at a time t is given by

$$E_t = \sigma_t/\varepsilon_o.$$

The variation of log E_t with time is shown in Figure 11.11b and is similar to the tensile elastic modulus versus temperature curve (Figure 11.11a).

When the stress level is maintained constant, viscoelastic creep takes place. The results are displayed, as for metals, in a strain versus time plot. The creep rate is temperature dependent and increases sharply as the temperature approaches T_g. The effects of viscoelasticity can also be easily demonstrated in a nanoindentation test during unloading (Section 4.8.4). During slow unloading from the peak load (Figure 11.14), even though the load is reducing the indenter may still keep sinking further into the polymer, resulting in a 'nose'-shaped unloading curve. The viscous strain component always increases as long as a positive load is acting, and so if it dominates over the elastic component, the overall strain will still be increasing even though the load is reducing. However, at a high unloading rate, the viscous component may become a lot stiffer than the elastic component, so that the overall response may become very elastic.

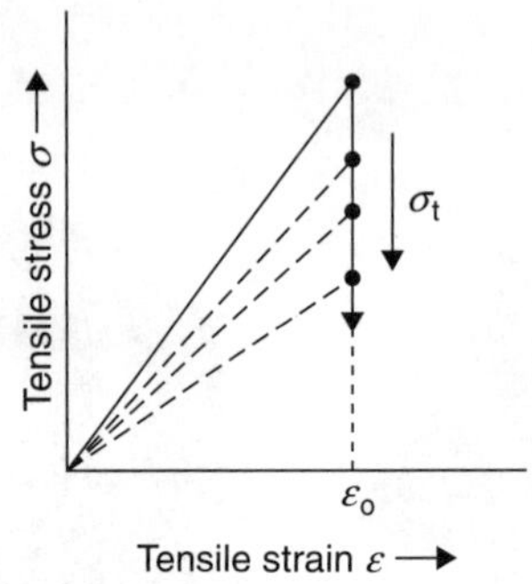

Figure 11.13 *Stress relaxation at constant strain.*

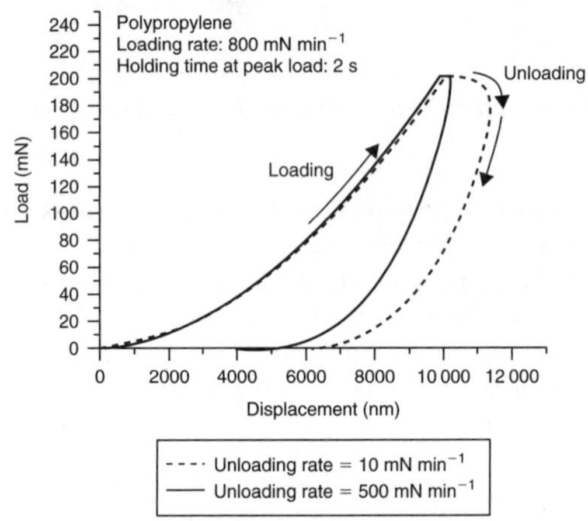

Figure 11.14 *Nanoindentation load–displacement curves of polypropylene at different unloading rates at room temperature. At a small unloading rate, the unloading curve becomes nose shaped, with a negative apparent unloading stiffness (Tang and Ngan, 2003).*

11.6.3 Fracture

Overall, the fracture toughness of polymers is quite low ($K_c \sim 1\,\text{MN}\,\text{m}^{-3/2}$), because of their low modulus. However, thermoplastic polymers exhibit both brittle and ductile modes of fracture and can display a brittle–ductile transition. Lowering temperature, increase of strain rate, sharp notches, etc. all favor brittle behavior. Brittle amorphous polymers are characterized by crazing, which can form at a stress much lower than yield stress. Crazes are narrow zones of highly deformed polymer containing microvoids ~10–20 nm separated by bridges 10–40 nm in diameter of oriented molecular chains. They are usually formed in glassy polymers (PMMA, PS) but may occur in semicrystalline polymers such as PP. With increasing stress, microvoids grow and coalesce to resemble a crack. The craze differs from the final crack in that it can still support some load across the craze (Figure 11.15). Crazes are easily visible even to the naked eye and may be induced by chemical agents and stress. The stress for crazing is temperature dependent and as mobility increases so the stress for crazing decreases. Chemical agents increase the susceptibility and the resultant fracture is termed environment stress cracking (ESC), with amorphous polymers being more susceptible than semicrystalline polymers. Although not a proper crack, crazing can cause leakage if the specific polymer is used for pipes.

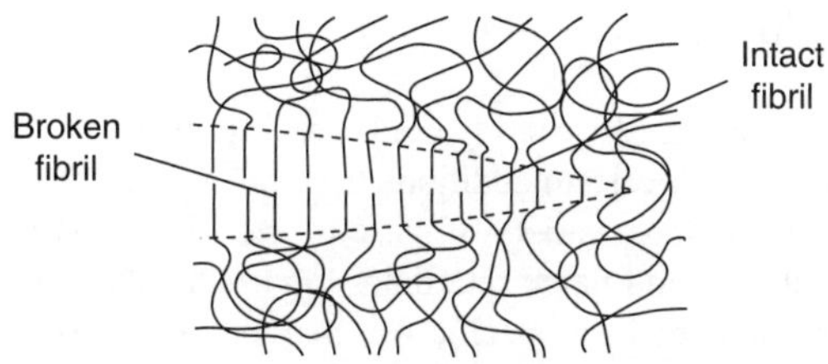

Figure 11.15 *Schematic structure of a craze.*

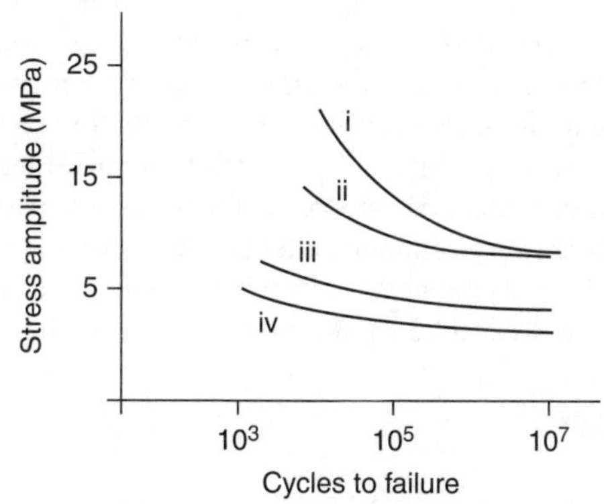

Figure 11.16 *Schematic fatigue curves for: (i) PS, (ii) PMMA, (iii) PP and (iv) PE.*

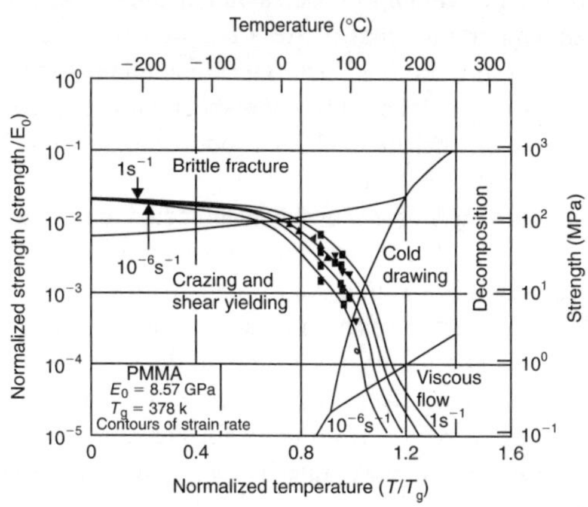

Figure 11.17 *Deformation map for PMMA showing deformation regions as a function of normalized stress versus normalized temperature (from Ashby and Jones, 1986).*

The craze can become a proper crack and propagate as a result of the stress concentration at the tip. Polymers like the epoxy resins have very low toughness values because the cross-linking prevents the flow of material at the tip to blunt the crack.

The fatigue strength of most polymers shows the same features as metals when the stress amplitude is plotted against cycles to failure (Figure 11.16). Fatigue occurs at low stress levels relative to the yield strength. Polyethylene has relatively poor fatigue properties compared to polypropylene and polyester (PET). Using the concept of deformation maps, Ashby has shown it is possible to portray the strength–temperature characteristics of polymers (Figure 11.17).

11.7 Plastics and additives

The polymers as described are rarely used in real applications. The plastics with which we are familiar are polymers to which certain additions have been made. The additive ranges from 0.1% to a few percent and is made for a variety of reasons, including environment protection, ease of processing toughening, coloring and flame retarding. Environmentally, sunlight can alter and degrade plastics and stabilizers are added for specific wavelengths. Plasticizers are added after polymerization and before processing to increase flexibility, ductility and toughness. They are generally liquids of low molecular weight which increase the separation of the chain molecules by occupying sites between the chains and effectively lowering T_g. These are needed when processing for sheets and tubing. Filler materials are added to improve strength, abrasion resistance and toughness, but they also lower the cost of production. For thermosets, mica, glass fiber and even fine sawdust are added. Fillers, e.g. clay or barium sulfate, are particularly important for vulcanizing rubber by facilitating handling. Carbon black is used as a filler for car tyres to improve tear strength, hardness and resistance to road abrasion. Flame retardants are extremely important, particularly in textiles and children's toys. Finally, colorants are very common either as dyes, when the molecules in the dye become part of the polymer molecule, or as pigments, which have a refractive index similar to the polymer and remain as a separate phase. Table 11.3 shows some common applications of different polymers.

11.8 Polymer processing

Processing is extremely important in polymer technology in order to develop the optimum properties for the product. Starting with polymer powder/granules, it is the control of the various processes which develops the individual molecules and arranges them to maximum advantage. High rates of shear and close temperature control are a feature of all the methods shown in Figure 11.18 which are commonly used in thermoplastics. These include (a) injection molding of thermoplastics (e.g. PE, PS), (b) extrusion into bars, tubes and sheet, and (c) thermoforming into sheet and thin-walled hollow shapes, such as buckets, baths, car body components, boat parts and aeroplane accessories, with ABS, PS, PVC and PMMA.

Thermoplastics are highly viscous when subjected to the shear stress conditions during processing as they flow through the die. Typical rates of shear are $10–10^3 s^{-1}$ for extrusion and $10^{-5}–10^{-3} s^{-1}$ for injection molding. Flow is non-Newtonian and the shear stress/shear strain ratio (apparent shear viscosity) falls, giving rise to a pseudo-plastic behavior. Fluidity as measured by a melt flow index test (MFI)[3] indicates that at the high stress a broad distribution of molecular mass leads to less pseudo-plastic behavior. Quantitatively $\tau = c\gamma^n$, where c and n are constants, and since $\tau = \eta\gamma$, the viscosity $\eta = c\gamma^{n-1}$. The index n ranges from unity for Newtonian flow to <0.2 depending on the polymer. The index decreases as the shear rate increases and the thermoplastic melt behaves increasingly pseudo-plastic.

In blow molding and filament drawing it is tensile stresses rather than shear stresses that predominate. In this case it is tensile viscosity, i.e. tensile stress/strain rate, which governs the dimensional stability of the product. During blow molding of LDPE the tensile viscosity rises with increase of stress and counteracts any tendency for wall thinning, whereas for HDPE and PP the reverse is true and wall thinning and rupture are a possibility. Dimensional stability can be a problem for extruded products, as strain energy is released on exiting from the die and elastic recovery takes place. This has to be allowed for in product design. For polymers with a wide molecular mass distribution, the elastic shear modulus is low and elastic recovery is appreciable but slow. For a narrow molecular mass distribution recovery is less but quicker.

[3] The MFI is the weight (g) collected in 10 minutes when the heated polymer is extruded through a standard die.

Table 11.3 *Common applications of polymers.*

	Application	*Useful properties*
Polyethylene		
LDPE	Packaging	
HDPE	Blow molded containers, Injection molded crates, Extruded pipes	Chemical resistant, electrically insulating, tough, low coefficient of friction
Linear LDPE	Agriculture, horticulture and construction (silage sheets, tunnel housing, cloches, damp-proof membranes)	
Ultra-high MPE	Artificial joints, surgical prostheses	Purity, wear resistant, tough
PVC	Building and construction (window frames, cladding, underground pipes)	Stiff, hard, low thermal conductivity, weather resistance
PB	Piping for underfloor heating	Heat resistant, degraded by Cl^-
PCarbonate	Transparent roofing	Dimensionally stable, water resistant, good impact properties
PP	Automotive components (bumpers, panels, fans, ducts, etc.)	Fatigue strength, stiff, tough, resistant to heat, creep and chemicals
PS	Rigid heat-insulating foam	Dimensional stability, cheap, resists crazing
SSR	Synthetic rubbers, car tyres	Resistant to chemical attack
Polychloroprene	Automotive seal, water circuit pipes	Resistant to oil-heat
Silicone rubber	Medical implants, gaskets seals, coatings	Resilience, chemical stability, insulating

Defects can occur in polymer products as a result of manufacture. Heating to the processing temperature significantly reduces the polymer density and on cooling shrinkage defects may occur, particularly in crystalline polymers, which pack more closely than amorphous polymers. Thick sections are particularly prone to shrinkage cavities. Voids may also form when the surface cools faster and becomes more rigid. Polymers have a long thermal conductivity and this accentuates relative rates of cooling to produce both surface and internal defects.

Thermosets are more difficult to process because of the need to incorporate a curing process. Compression molding is similar in principle to mounting a metallographic specimen. In transfer molding, which can produce more complex shapes, resin is melted in a primary vessel and then transferred to a final, vented mold for final curing. In reaction injection molding (RIM) with polyurethyane, or nylon, polymerization takes place during forming in about 30 seconds after two or more streams of chemical reactants are injected into a mixing chamber. Moldings manufactured for 'higher' temperature service are stabilized or post-cured by heating at 100°C for a short time. The RIM process has now been developed to such an extent that cycle times are comparable to those for thermoplastics.

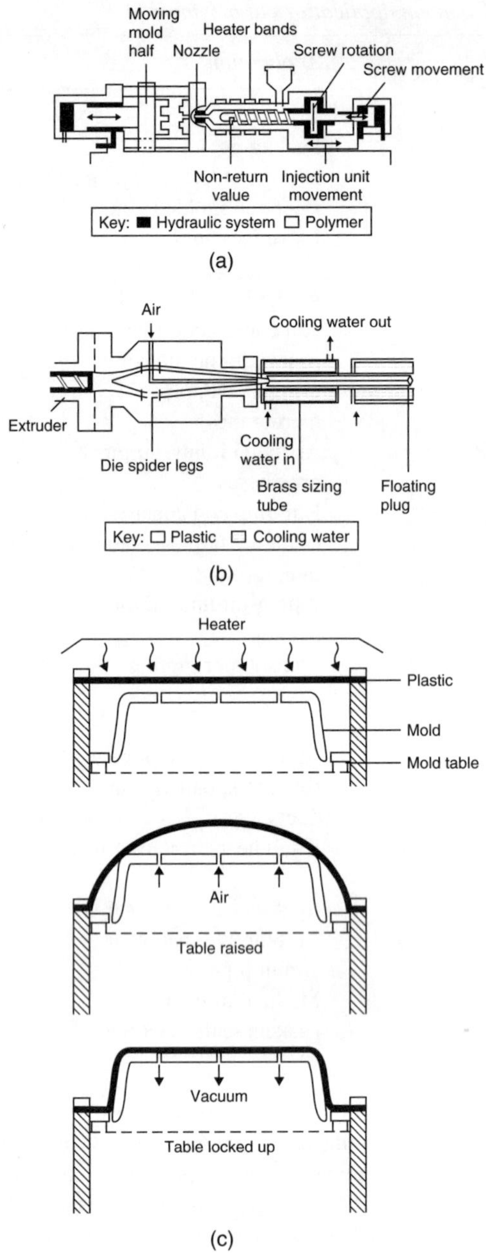

Figure 11.18 *(a) Injection-molding machine. (b) Production of plastic pipe by extrusion. (c) Thermoforming of plastic sheet (from Mills, 1986; by permission of Edward Arnold).*

11.9 Electrical properties

Polymers with their strong covalent and weak van der Waals bonds are insulators. Some basic properties are given in Table 11.4.

Table 11.4 *Electrical properties of some common polymers.*

Polymer	*Electrical resistance (Ωm)*	*Dielectric strength ($V\,m^{-1}$)*	*Dielectric constant*
PE	10^{15}	1.5×10^8	2.3
PVC	10^{14}	0.5×10^8	4.0
PP	10^{15}	1×10^8	2.3
PS	10^{14}	1.3×10^8	2.5

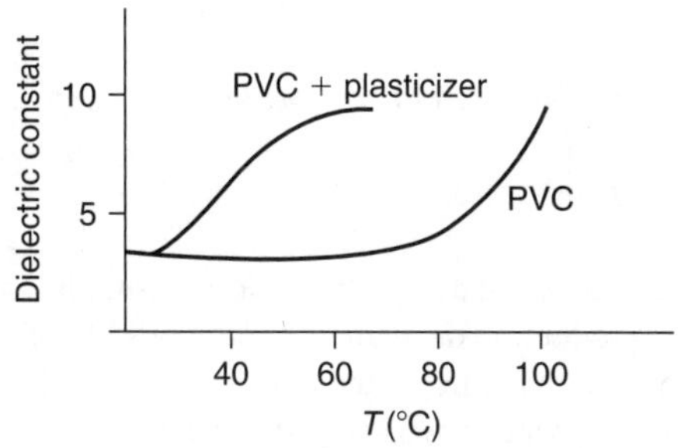

Figure 11.19 *Schematic showing variation of dielectric constant with temperature in PVC.*

The small amount of conduction is by movement of electrons or ions from molecule to molecule and is temperature dependent. The breakdown electric field is $10^7\,V\,m^{-1}$ but this deteriorates with degradation of the polymer. Polymers can be made conducting either by using a conducting material as additive, e.g. carbon black, or by incorporating some ionic bonding into the molecular structure.

With AC properties, molecular rotation becomes important and to contribute to polarization and thus the dielectric constant, the relevant molecule must be charged. This is a condition favored in PVC with its strong electronegative chlorine atom (see Table 11.1). Figure 11.19 shows how temperature influences the dielectric constant for PVC and mechanically strengthened PVC.

Polymers are, of course, universally used for making capacitors with capacitance ranging from 0.1 pF to 100 μF. A thin film of PS or PP is coated with aluminum before being rolled up and sealed in an epoxy resin.

11.10 Composites

11.10.1 Particulate composites

A composite, as the name implies, is made up from two (or more) materials to produce a combination of properties unobtainable by either of the constituent materials by themselves. The principle has been practiced for years with the use of concrete, which uses an aggregate (of particles) bonded together with cement. Reinforcing the concrete with pre-stressed steel rods further improves the mechanical strength in the longitudinal direction and the fracture behavior in the transverse directions. Dispersion-strengthened metal alloys have been discussed in Chapters 7 and 8 with nickel-based alloys, dual-phase steels and mechanically alloyed steels, nickel and aluminum. These may be considered as particle-strengthened composites. The moduli of these particulate composites lie between the upper and lower curves shown in Figure 11.20, which are for fiber composites. Such particulate composites are much easier to produce, e.g. fillers in polymers, than fiber composites and are therefore cheaper. In addition, they are largely isotropic, which can be an advantage, and are commonly used for many everyday applications (see Section 11.10.7).

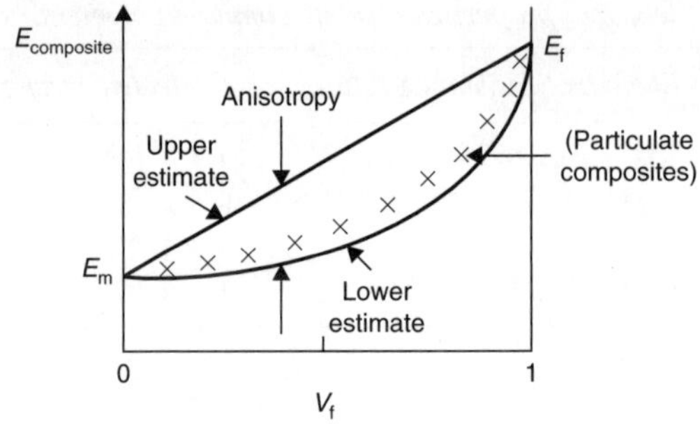

Figure 11.20 *Young's modulus of composite (from Ashby and Jones, 2005).*

11.10.2 Fiber-reinforced composites

Fiber-reinforced composites have been developed with the various types of matrix, polymer, ceramic and metal. During production the matrix, continuous phase, envelops the fibers and forms a coherent bond, usually chemical in nature, at the fiber/matrix interface. During deformation, the matrix transfers stress to the fibers, which are usually stronger and have a higher elastic modulus. The response of the composite then depends on the relative properties of the fiber and the matrix, the volume fraction of each phase, the fiber length and orientation. Failure of individual fibers is local and takes place without destroying the integrity of the whole composite. If debonding of the matrix/fiber interface takes place as a result of a propagating crack, a significant amount of strain energy is required.

For a composite with a volume fraction V_f of fibers parallel to the stress axis, the strain in the fibers and the matrix is the same and the stress carried by the composite is

$$\sigma_c = V_f\sigma_f + (1 - V_f)\sigma_m,$$

and since Hooke's law applies $\sigma = E\varepsilon$,

$$\sigma_c = V_f E_f \varepsilon + (1 - V_f)E_m\varepsilon$$

and the modulus of the composite is

$$E_c = V_f E_f + (1 - V_f)E_m. \tag{11.1}$$

For a composite with fibers perpendicular to the stress axis then the stress in the fiber and matrix are equal, and

$$\varepsilon = V_f\varepsilon_f + (1 - V_f)\varepsilon_m = V_f\frac{\sigma}{E_f} + (1 - V_f)\frac{\sigma}{E_m}$$

and the modulus of the composite is then

$$E_c = \frac{\sigma}{\varepsilon} = \frac{1}{\dfrac{V_f}{E_f} + \dfrac{(1 - V_f)}{E_m}} = \frac{E_f E_m}{V_f E_m + (1 - V_f)E_f}. \tag{11.2}$$

These two equations are plotted in Figure 11.20 as the upper and lower limits of the composite modulus. This anisotropy in the behavior is sometimes exploited in applications, e.g. golf shafts, vaulting poles. In some applications it would be detrimental and hence fibers have to be introduced in a criss-cross configuration to avoid the problem.

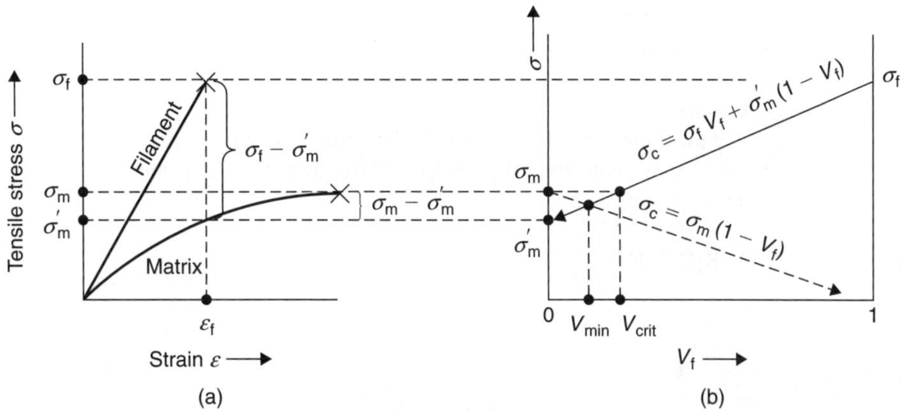

Figure 11.21 *(a) Stress–strain curves for filament and matrix. (b) Dependence of composite strength on volume fraction of continuous filaments.*

The composite loses its effectiveness at a critical strain corresponding to the strain at fracture of the fibers ε_f. At this point, the stress in the matrix σ'_m is the lower limit of the composite strength which lies between σ'_m and σ_f depending on the volume fraction of fibers: the strength σ_c falls as V_f falls, as shown in Figure 11.21b, and the volume fraction of fibers must exceed V_{crit} if the composite is to benefit from the matrix/fiber combination. At the critical point $\sigma_m = \sigma_c$ and $V_{crit} = V_f$, so that

$$V_{crit} = (\sigma_m - \sigma'_m)/(\sigma_f - \sigma'_m). \tag{11.3}$$

In practice, a low V_{crit} and high σ_f are manufacturing aims. A strong work-hardening matrix, which increases the numerator $(\sigma_m - \sigma'_m)$, requires a large volume fraction of fibers V_f to be effective.

Worked example

A fiber composite material consists of long parallel tungsten wires embedded in a copper matrix. Given the data below, calculate the critical volume fraction of fibers (V_{crit}) above which the tensile strength of the composite exceeds that of the matrix alone.

Fracture strength of tungsten wires = 3900 MPa
Tensile strength of copper = 600 MPa
Flow stress of copper at the strain at which the tungsten wires break = 450 MPa.

Solution

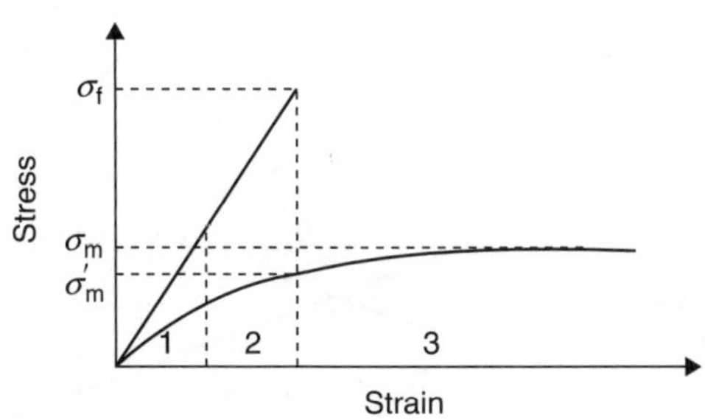

The above diagram shows the stress–strain behavior of the matrix and fibers. In region 1, both the matrix and fiber are elastic and at the end of region 1 the matrix yields. In region 2 the fiber is still elastic and the matrix work hardens. At the onset of region 3 the fiber breaks and there are two possibilities. Either the matrix can work harden sufficiently to withstand the applied load or it cannot and the composite fails. As a result of these two possibilities the composite will initially weaken when fibers are added.

Let σ_c be the composite strength and V_f the volume fraction of fibers.

If the matrix can withstand the load when the fiber fails then the strength of the composite is decided by the matrix:

$$\sigma_c = \sigma_m(1 - V_f).$$

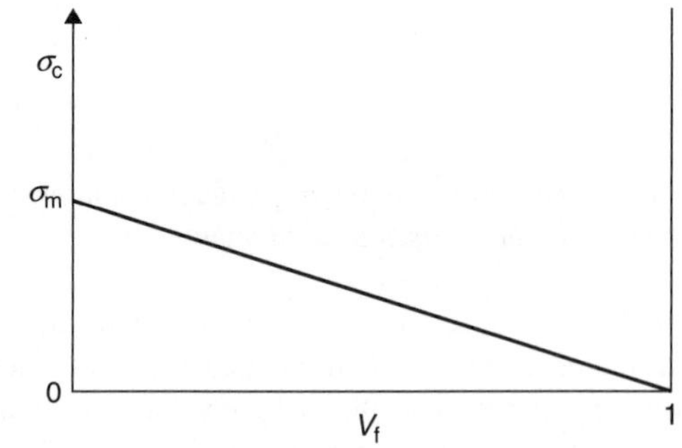

In the second situation when the composite fails, the strength of the composite is given by:

$$\sigma_c = \sigma_f V_f + \sigma_m'(1 - V_f).$$

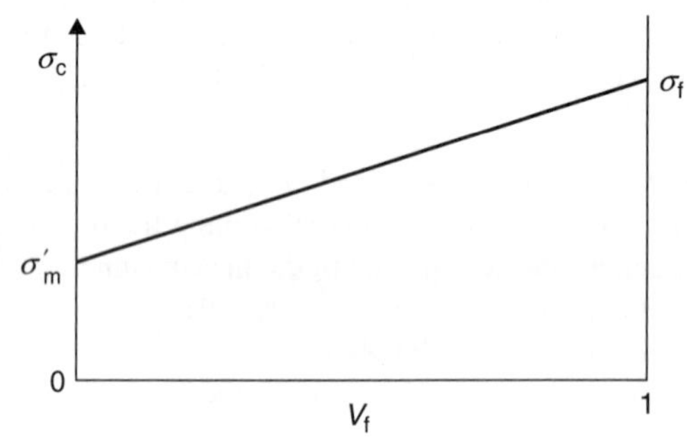

Combining the two diagrams we can see that there is a critical volume of fibers above which strengthening occurs:

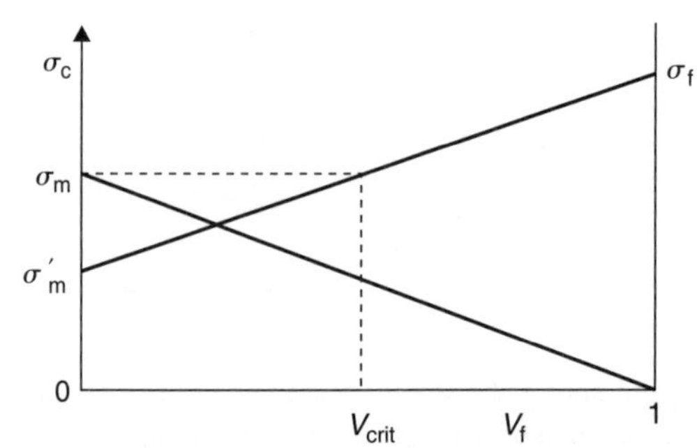

It is given by:

$$\sigma_c = \sigma_m = \sigma_f V_{crit} + \sigma'_m (1 - V_{crit})$$

$$\therefore V_{crit} = \frac{\sigma_m - \sigma'_m}{\sigma_f - \sigma'_m}.$$

$$\text{Hence, } V_{crit} = \frac{600 - 450}{3900 - 450} = 0.043.$$

11.10.3 Fiber orientations

Composites with continuous fibers oriented parallel to the stress axis are anisotropic in mechanical properties with a strength as shown in Figure 11.21b. For fibers oriented at an angle ϕ to the stress axis the composite strength is reduced and given by

$$\sigma_c = \eta V_f \sigma_f + (1 - V_f)\sigma_m, \tag{11.4}$$

where η is an orientation factor when $\eta < 1$ for $\phi > 0$. Figure 11.22 shows how the composite strength in the direction of the fibers σ_{cl} varies with fiber orientation ϕ. The strength parallel to the fibers is controlled by tensile fracture of the fibers and is given by $\sigma_c = \sigma_f \sec^2 \phi$. For shear failure on a plane parallel to the fiber $\sigma_c = 2\tau_m$ cosec 2ϕ and for debonding of the fiber/matrix interface $\sigma_c = \sigma_f$ cosec$^2\phi$. It can be seen that the composite strength falls quite sharply with increase in ϕ up to 10°.

Worked example

Laminates are formed from a stack of laminae (plies) not only at 0° and 90° but at other angles. A criterion for the failure of an angle-ply lamina is

$$\left(\frac{\sigma_{||}}{\sigma^*_{||}}\right)^2 + \left(\frac{\sigma_{\perp}}{\sigma^*_{\perp}}\right)^2 + \left(\frac{\tau_{\#}}{\tau^*_{\#}}\right)^2 = 1,$$

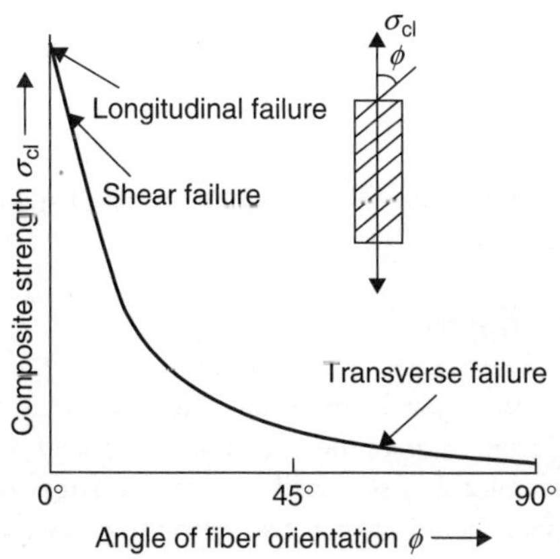

Figure 11.22 *Relation between failure mode, strength and fiber orientation (schematic diagram for unidirectional continuous-fiber composite).*

where the asterisks refer to critical values of the tensile stresses parallel ($\sigma_{||}$) and normal ($\sigma_{\perp}$) to fibers or to the fiber/matrix shear stress ($\tau_{\#}$). If $\sigma_{||}^*$ is 300 MPa, $\sigma_{\perp}^*$ is 30 MPa and $\tau_{\#}^*$ is 60 MPa, what value of applied tensile stress is required to produce failure in a lamina whose fibers are inclined at 60° to the tensile axis?

Solution

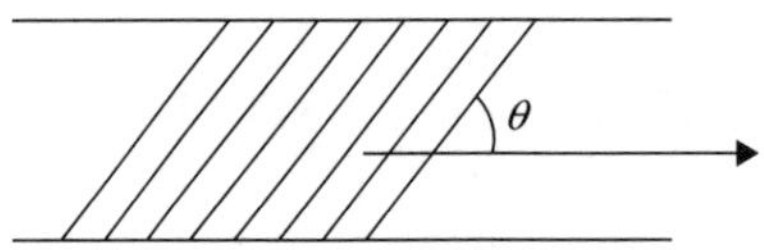

$$\sigma_{||} = \sigma \cos^2\theta = \sigma/4$$

$$\sigma_{\perp} = \sigma \sin^2\theta = 3\sigma/4$$

$$\tau_{\#} = \sigma \sin\theta \cos\theta = \sqrt{3}\sigma/4$$

$$\left(\frac{\sigma}{1200}\right)^2 + \left(\frac{3\sigma}{120}\right)^2 + \left(\frac{\sqrt{3}\sigma}{240}\right)^2 = 1,$$

so $\sigma = 38.4$ MPa.

If $\sigma_{\perp}$ and $\tau_{\#}$ are ignored, $\sqrt{900}\sigma = 1200$, so $\sigma = 40$ MPa.

11.10.4 Influence of fiber length

The mechanical behavior of a composite depends upon the respective properties of the matrix and reinforcing fibers and the volume fraction, length and orientation of fibers. For a given diameter of fiber an increase in length will increase the bonding at the fiber/matrix interface and favor the transfer of stress. For effective strengthening and stiffening the fibers have to be longer than a critical length l_c which depends on the diameter of the fiber, its tensile strength σ_f and the fiber–matrix bond strength or shear yield strength of the matrix, whichever is the lower value, τ, according to

$$l_c = \sigma_f d/2\tau.$$

If an increment of tensile stress $\delta\sigma$ is applied to an element of fiber δl, the balance between tensile force and shear force is

$$\delta\sigma(\pi d^2/4) = \tau(\pi d\, \delta l),$$

hence the gradient $\delta\sigma/\delta l$ for the build up of tensile stress is $4\tau/d = \sigma_f/(l_c/2)$. The critical length l_c is therefore $\sigma_f d/2\tau$. In terms of the critical aspect ratio $l_c/d \geq \sigma_f/2\tau$. When a composite with fibers shorter than l_c is subjected to a tensile stress, the stress profile is as shown in Figure 11.23a; the difference in displacement between the matrix and fiber causes a shear stress τ at the cylindrical interface towards each fiber end. A corresponding tensile stress σ builds up within the fiber. At each end of the fiber these stresses change over a distance known as the transfer length $l/2$, i.e. the tensile stress increases as the interfacial shear stress decreases. When the fiber length is increased to l_c the

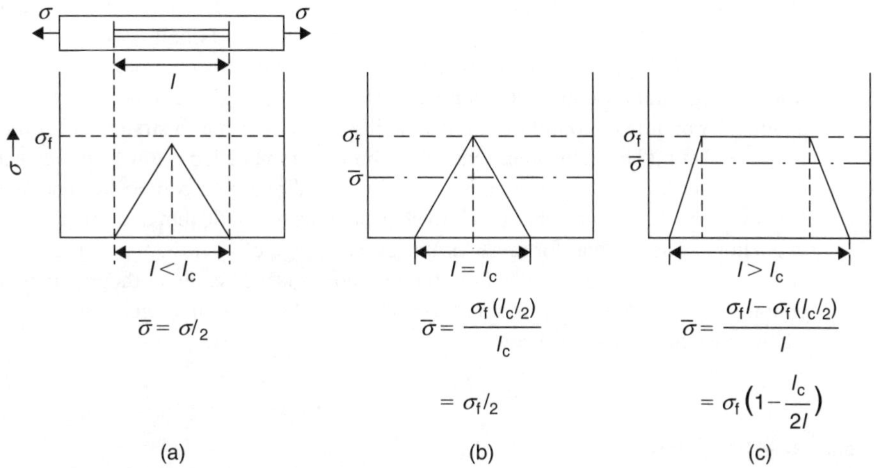

Figure 11.23 *Distribution of tensile stress in a short fiber. σ_f = fracture stress of fiber in tension, $\overline{\sigma}$ = mean tensile stress.*

transfer length becomes $l_c/2$ and the peak tensile strength coincides with the fracture stress for the fiber (Figure 11.23b). If the fiber is below the critical length, fiber failure cannot occur. At the critical condition the peak stress is σ_f but the average tensile stress is $\overline{\sigma} = \sigma_f(l_c/2)/l_c = \sigma_f/2$. For fibers above the critical length a plateau develops in the stress profile (Figure 11.23c). The average tensile stress on the fiber is given by

$$\overline{\sigma} = [\sigma_f l - \sigma_f(l_c/2)]/l = \sigma_f(1 - l_c/2l)$$

and approaches the fracture stress σ_f as the fiber length increases above l_c. Then, provided the shear stresses do not cause 'pull-out' of the fiber, fracture will eventually occur in the mid region of the fiber.

11.10.5 Composite fibers

11.10.5.1 Types of fibers

Fibers may take the form of whiskers, actual fibers or wires. Whiskers are extremely small in diameter, dislocation free and consequently extremely strong. Fibers which are more commonly used are either polycrystalline or amorphous and have a modestly small diameter. They commonly exist as polymers or ceramics, e.g. polymer amides, glass, aluminum oxide, carbon, boron or silicon carbides. Wires have a relatively large diameter and are commonly steel, molybdenum or tungsten. Interest in whisker reinforcement, once great, has tended to decline because of the carcinogenic risks associated with handling them during preparation and composite manufacture.

11.10.5.2 Glass fibers

Glass fibers are made from molten glass by passing it through the nozzles of an electrically heated bushing made from Pt–10Rh alloy. (The number of nozzle holes in the base of the bushing is 204, or a multiple of 204.) The filaments emerge at a velocity of 50–100 m s^{-1} and are rapidly cooled to prevent crystallization, hauled over a 'size' applicator and finally collected by a rotating cylinder (collet). 'Sizing' applies a coating which loosely bonds the fibers, protects their fragile glass surfaces from damage and introduces a surface-modifying 'coupling' agent to promote eventual fiber/matrix bonding. A bundle of continuous, untwisted fibers is known as a strand, and contains 204, 408, 816 or more fibers. (The equivalent unit for carbon fibers is called a tow.) Strands can be combined to form a larger bundle (roving), and are used for unidirectional composites and as yarn for weaving. Alternatively, they can be chopped into short lengths of 25–50 mm and oriented randomly in a plane (e.g. chopped strand mat for press molding). The range of fiber diameters is 5–20 μm. E-glass is the principal material for the production of continuous glass fibers but other compositions are available, giving a higher modulus of elasticity (H-modulus glass), greater tensile strength (S-glass) or better alkali resistance (AR-glass), etc.

11.10.5.3 Carbon fibers

Carbon fibers, 5–10 μm diameter, are available in untwisted tows containing 1000, 3000, 6000, 12 000 or 120 000 filaments and as pre-pregs with resin. In the UK and the USA they are mostly produced from the textile polyacrylonitrile (PAN) and its copolymers. Many types of PAN-derived carbon fiber are produced (e.g. commodity, high modulus, high extension, etc.).

The three-stage process for manufacturing carbon fiber is based on the controlled degradation or pyrolysis of spun fibers of PAN. Hot stretching is a central feature of processing: it counteracts the tendency of the fibers to shrink and induces a high degree of molecular orientation. The tow is first oxidized under tension at a temperature of 200°C and forms a stable, cross-linked 'ladder' structure. In the second stage, heating in as inert atmosphere at temperatures between 800 and 1600°C carbonizes the structure, releasing hydrogen and nitrogen, reducing the original mass by 40–50%. The oriented carbon fibers are heat treated at temperatures in the range 1300–2000°C; higher temperatures encourage graphitization and improves the elastic modulus but lowers the strain to failure. Finally, the fibers are surface treated (e.g. electrolytic oxidation) to improve subsequent bonding to the matrix and 'sized' to facilitate handling.

Each carbon fiber produced is very pure and consists essentially of interwoven 'ribbons' of turbostratic graphite (Figure 11.24a) and some amorphous carbon. The ribbons are aligned parallel to the fiber axis. Being an imperfect structure, the amount of porosity is appreciable. In general terms, *a*-axes of the planar crystallites are parallel to the fiber axis, the other *a*-axis is radial or circumferential, and the *c*-axes are perpendicular to the fiber axis. As the structure becomes more truly graphitic, the ribbon orientation approaches that of the fiber axis and the axial modulus increases. The fiber structure is highly anisotropic: the moduli of elasticity along the fibers and perpendicular to the fibers are 200–800 and 10–20 GN m^{-2} respectively. (The modulus for E-glass is about 73 GN m^{-2}.)

11.10.5.4 Aramid fibers

Polymer-reinforcing aramid fibers (*Kevlar 29* and *49, Twaron*) are based on aromatic polyamides. Their linear molecular structure is produced from spun fiber by a process of drawing and heating under tension at a temperature of ~550°C. This structure (Figure 11.24b) contrasts with the more planar structure of carbon fibers, giving aramid fibers a fibrillar character that can absorb considerable amounts of impact energy. Impacting aramid fibers splits them into numerous microfibrils, giving exceptional 'stopping power'. Aramid fibers and aramid/resin laminates are therefore used for ballistic applications (e.g. armor). They have a stable elastic modulus of 50–130 GN m^{-2} at temperatures

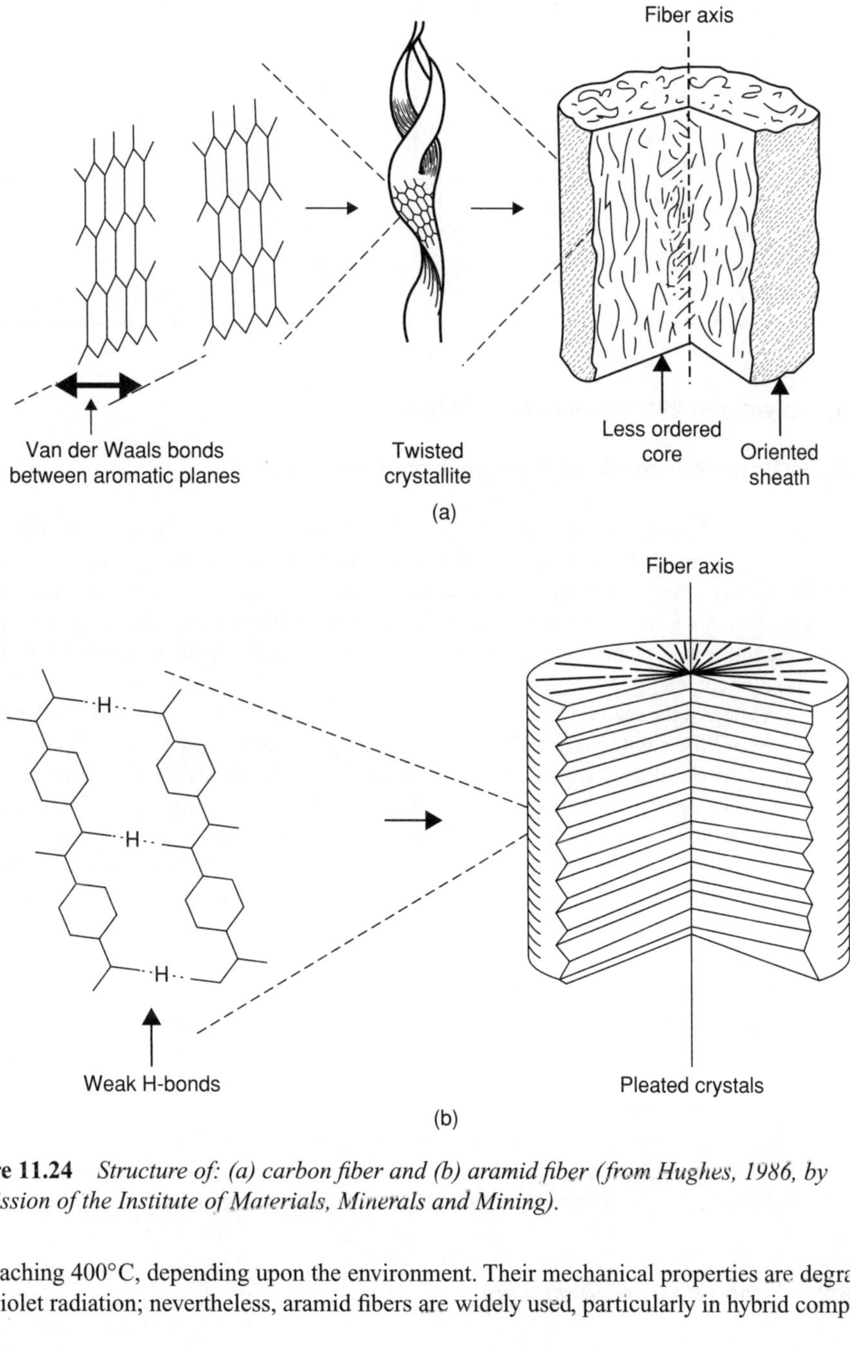

Figure 11.24 *Structure of: (a) carbon fiber and (b) aramid fiber (from Hughes, 1986, by permission of the Institute of Materials, Minerals and Mining).*

approaching 400°C, depending upon the environment. Their mechanical properties are degraded by ultraviolet radiation; nevertheless, aramid fibers are widely used, particularly in hybrid composites.

11.10.5.5 Boron fibers

These have been developed for aerospace applications and sport equipment, and have high strength and stiffness. They are produced by vapor deposition when the chemical reaction with the reactant gases is $2BCl_3 + 3H_2 \rightarrow 2B + 6HCl$, resulting in crystal size of about 2 nm.

Table 11.5 *Properties of some reinforcing fibers.*

Fiber	*Density ($g\,cm^{-3}$)*	*Ultimate tensile strength (GPa)*	*Young's modulus (GPa)*	*Specific strength ($GPa\,g^{-1}\,cm^3$)*	*Specific modulus ($GPa\,g^{-1}\,cm^3$)*	*Maximum usable temperature (°C)*
E-glass	2.54	3.4	72	1.34	28	550
Carbon	1.9	2.3	277	1.21	177	2500
Kevlar	1.45	3.6	130	2.48	90	150
Boron	2.6	3.5	420	1.35	160	700

11.10.6 Polymer–matrix composites (PMCs)

11.10.6.1 Glass fiber-reinforced polymer composite (GFRP)

Polyester-based matrices remain the principal choice for polymer–matrix composites. Dough molding compounds (DMC) and sheet molding compounds (SMC) contain roughly equal volume fractions of polyester resin, inert filler particles and chopped glass fibers. When heated, these compounds rapidly become fluid, reproduce the contours and details of the molding dies accurately, and then cure. This technique is used for domestic articles, panels and doors of vehicles, cabinets for office and electronic equipment, etc.

As part of the search for cheaper fabrication methods, much effort has been devoted to the development of PMCs with thermoplastic matrices (e.g. nylon 66, PP, PTFE, PET, polyether sulfone (PES), etc.). For example, water boxes of car radiators and shell housings for street lamps have been made from a composite of 33% glass fibers in a nylon 66 matrix (*Maranyl*). The general advantages of a thermoplastic matrix are its toughness, indefinite shelf life and, in the absence of curing, a shorter time cycle for fabrication. However, during the necessary heating, the viscosity of the matrix is higher than that of a thermoset resin, making infiltration between fibers more difficult. Furthermore, the pre-pregs are stiff and lack the drapability of thermoset pre-pregs, which enables them to bend easily into shape.

Randomly oriented short fibers of glass are commonly used in sheets and in three-dimensional moldings. In fact, fiber misorientation is quite common in composites, frequently being an unavoidable result of fabrication. For instance, when resins loaded with short fibers are injection molded, the mixture follows complex flow paths. If the finished molding is sectioned, fibers will clearly indicate the patterns of flow: these patterns are determined by melt viscosity, mold contours and processing conditions. Fortunately, flow patterns are repeatable from molding to molding. Close to the surface of the molding, the short fibers tend to follow streamline paths; in the central core, where flow is more turbulent, fibers are likely to assume transverse orientations.

11.10.6.2 Carbon fiber reinforcement

For the exacting requirements of aerospace and high-performance aircraft, the principal PMCs have epoxide matrices reinforced with continuous fibers of carbon or aramids (*Kevlar*). The basic advantage of epoxides is that they can be used at higher service temperatures than polyester matrices. Although the T_g value of a polymer provides an indication of its temperature ceiling, it is substantially higher than the maximum temperature for safe service under load. For instance, the maximum temperature for a load-bearing epoxide matrix is about 160°C, whereas the corresponding T_g values lie in the range 200–240°C (depending upon the method of determination). The search for matrices with a higher temperature capability has led to the development of bismaleimides (BMI) and polyimides (PI). These and other new polymers raised the ceiling temperature closer to 200°C, but have sometimes introduced a brittleness problem and they can be difficult to process.

Carbon fiber-reinforced polymers (CFRPs) are favored as construction materials for specialized and demanding applications (e.g. helicopter rotors, monocoque chassis of racing cars, aircraft floor panels, spacecraft components, sports goods, high-speed loom components). Laminates of continuous carbon fibers (*Grafil*) are widely used. Carbon fibers are also used in metal–matrix and ceramic–matrix composites. Frequently, they are combined with other types of fiber to form hybrid composites (e.g. glass and carbon, aramid (*Kevlar*) and carbon). For the rapid deceleration of racing cars and landing aircraft, which develop very high frictional forces and temperatures in braking systems, composites with a carbonaceous matrix reinforced with carbon fibers are used. These carbon–carbon composites combine the refractory potential of carbon with the high specific strength/stiffness of carbon fibers.

11.10.6.3 Fabrication of PMCs

There are quite a number of different methods of fabricating polymer–matrix composites:

(i) Hand lay-up or spray up – chopped-strand mats, woven roving and other fabrics made from the fibers are placed on the mold and impregnated with resin by painting and rolling. Layers are built up until design thickness is achieved. The molding cures without heat or pressure. In spray-up, chopped rovings and resin are sprayed simultaneously into a prepared mold and rolled before the resin cures.

(ii) Vacuum molding – layers of fibers, usually unidirectional sheets, are pre-impregnated with resin and partially cured (β-staged) to form a pre-preg. The pre-preg sheets are stacked on the mold surfaces in predetermined orientations, covered with a flexible bag and consolidated using a vacuum or pressure bag in an autoclave at the required curing temperature.

(iii) Filament winding – continuous rovings or strands of fibers are fed over rollers and guides through a bath of resin and then wound, using a program-controlled machine, onto a mandrel at predetermined angles. The resin is partially or completely cured before removing the component, usually a tube, from the mandrel.

(iv) Hot-press molding – heated matched dies or tools are loaded with raw material (sheet molding compound, SMC, dough molding compound, DMC, cloth or unidirectional pre-preg) pressed to the shape of the cavity and cured.

(v) Centrifugal casting – mixtures of the fibers and resin are introduced into a rotating mold and allowed to cure *in situ*.

(vi) Polymer-reinforced reaction injection molding (PRIM) – a rapid curing resin system involving two components which are mixed immediately before injection is used. Fibers are either placed in the closed mold before resin is injected or added as short chopped fibers to one of the resin components to form a slurry before injection.

(vii) Pultrusion – a continuous feed of fibers, in pre-selected orientations, is impregnated with resin and pulled through a heated die to give the shape of the final section (e.g. tubes or I-beams). Partial or complete cure occurs during passage through the die.

(viii) Cold press molding – a low-pressure, low-temperature process in which fibers are impregnated with resin and then pressed between matched dies. Heat is generated during the cure.

(ix) Injection molding, transfer molding – molten or plasticized polymer mixed with short fibers is injected, usually at high pressure, into the cavity of a split mold and allowed to solidify or cure.

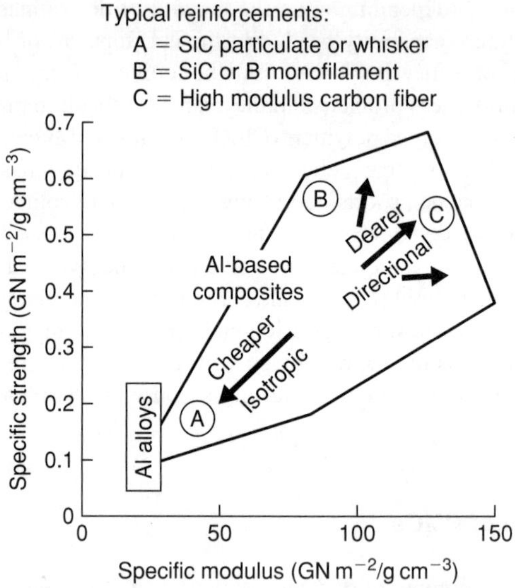

Figure 11.25 *Range of longitudinal tensile specific strengths and moduli achieved in aluminum-based composites (from Feest, 1988, by permission of the Institute of Materials, Minerals and Mining).*

In each of the above methods there is need to prevent air or vapors becoming trapped in the composite and forming weakening voids. Voids form preferentially at fiber/matrix interfaces and between the plies of layered composites. Low-viscosity resins, outgassing and high pressures are some of the means used to minimize this porosity.

11.10.7 Metal–matrix composites (MMCs)

The largest industrial application of metal–matrix composites is with particle reinforcement, being essentially isotropic and easier to process. Matrices based upon low-density elements are predominant, principally aluminum and its alloys, titanium and its alloys, and magnesium. Cheap particles and fibers of silicon carbide are the commonest choice of reinforcement material for aluminum and its alloys. Alumina (*Saffil*) reinforcement is also used in many MMCs. Figure 11.25 illustrates the considerable improvements in specific tensile strength (longitudinal) and specific modulus that result when aluminum alloys are reinforced. These property changes cannot be matched by orthodox alloy development, although the improvement is at a cost. Anisotropy is also introduced; the ratio of longitudinal strength to transverse strength can be 15:1 or more for MMCs. This is an important consideration when service conditions frequently involve multi-axial stresses. Table 11.6 shows some typical examples of MMC applications.

11.10.7.1 Fabrication of MMCs

11.10.7.1.1 Particle reinforcement

A wide variety of methods is available for producing MMCs, taking into consideration the health hazards associated with fine particles and fibers. They usually involve either melting of the matrix metal, powder blending or vapor/electrodeposition.

Table 11.6 *MMC applications.*

MMC	*Applications*
Ti–6Al–V, TiAl or Ti_3Al matrix, reinforced with coated SiC fibers	Aerospace: compressor disks, blades, engine pylons, stabilizers
Al alloy matrix, SiC fiber	Components for high-speed carpet-making equipment, textile production, food packaging
Al alloy matrix, 12% v/v alumina particles, 9% v/v carbon fibers	Overall weight saving for car engine cylinder blocks; particles give strength, fibers lubrication
Al alloy matrix, 20% v/v alumina fibers 2–4 μm diameter, 200–500 μm long	Squeeze casting diesel engine piston body with fiber-reinforced bowl for higher temperature (300–350°C), higher pressure combustion chamber

(i) Powder blending techniques are favored, rather than melting, when the metallic matrix has a high melting point, thus minimizing the fiber/matrix interaction problem. After the critical blending operation, the MMC powder is canned, vacuum degassed and consolidated by hot pressing or HIPing. Finally, the MMC is worked and shaped (e.g. extrusion, forging). Particle volume fractions of 0.25–0.50 are typical.

(ii) Mechanical alloying (see Section 8.4.3) is broadly similar to powder blending; the essential difference is that alloying of pure metal powders is achieved in high-speed ball mills.

(iii) In the melt-stirring technique, the presence of particles raises the melt viscosity. Possible difficulties include non-wetting of particles, agglomeration, and/or gravity settling of particles and unwanted particle/metal interactions.

(iv) Spray co-deposition is based upon a spray process originally developed for building deposits of steels that are difficult to cast and nickel-based superalloys. A stream of induction-melted metal or alloy is broken into fine droplets by relatively cold inert gas (nitrogen). Droplets begin to freeze before striking a movable collector placed in the line of flight. In the MMC variant, reinforcement particles are injected into this stream. Fine-grained MMC deposits of SiC/aluminum alloy and Al_2O_3/steel can be built as plate, tube, billets for hot working, cladding, etc.

11.10.7.1.2 Fiber reinforcement

The fabrication of continuous fiber-reinforced MMCs is technically demanding. A high volume fraction of fibers is difficult to 'wet' while low V_f preforms are weak to handle. A high melt temperature lowers viscosity and assists infiltration but increases fiber/metal reaction. (SiC, Al_2O_3 and carbon fibers react with aluminum alloy matrices at a temperature of 500°C.) There is also a risk that secondary working operations will damage the fibers.

(i) Diffusion bonding: used to consolidate metal with continuous fibers of SiC, Al_2O_3, boron and carbon. Fibers are pre-coated with matrix metal or carbon, and processing is a balance between promotion of solid-state diffusion and limitation of fiber/matrix reactions.

(ii) Squeeze casting: a single-shot process combining gravity die casting with closed-die forging. It is mainly used for discontinuous-fiber MMCs. Melt containing short fibers is fed into a lower die and then compressed at high pressure (35–70 $MN\,m^{-2}$) by an upper die. Applied pressure has an undercooling effect which, together with the loss of heat through the dies, favors rapid solidification. The high pressure also discourages the nucleation of gas bubbles. The final matrix structure produced is fine-grained and dense.

(iii) Liquid pressure forming: a preheated fiber preform is placed within split dies, and air pressure in the cavity reduced below 1 mb. Molten aluminum alloy under pressure is then infiltrated upwards through the preform and allowed to solidify. LPF uses a lower pressure (<8 MN m^{-2}) than squeeze casting and there is less limitation on size and shape of product. Particle-reinforced and continuous, discontinuous MMCs can be produced.

11.10.7.2 Composite failure

The introduction of short stress-concentrating fibers into a metal or alloy tends to reduce the ductility and toughness below that of the unreinforced matrix. When a short-fiber MMC is deformed by applied stress, the amount of strain in fiber and matrix may differ substantially, leading to rupture of the interfacial bonding at the end regions of fibers. This debonding results in the nucleation and rapid growth of voids. Void formation is the dominant mode of tensile failure in SiC/aluminum composites. A difference in the coefficients of thermal expansion of fiber and matrix can produce the same effect, possibly during the cooling stages of composite fabrication or during thermal excursions and cycling in service. Thus, because $\alpha_f < \alpha_m$, fibers and particles develop residual compressive stress and the matrix is left in a state of tension. This disparity in expansion characteristics between the reinforcement phase and the matrix also helps to generate a high density of dislocations in the matrix and a high rate of strain hardening.

Hard reinforcement particles or fibers can benefit wear resistance in service but give difficulties during the finish-machining operations that are usually needed for MMC products. SiC particles act as chip breakers during machining of reinforced aluminum alloys and reduce cutting forces but, unfortunately, shorten tool life significantly. Machining also tends to generate subsurface damage.

11.10.7.3 Melt-grown composites

The microstructures of alloy systems such as eutectics and eutectoids have already been discussed in Chapter 2 when fine lamellae or fibers are produced. These may be considered as metal–matrix composites formed by controlled solidification. To form an intimate combination of different phases at the micron level of structure with unique physical and mechanical properties, such composites are usually produced from the molten alloy by controlled slow solidification. Most industrially important eutectic alloys grow with an essentially planar interface. One disadvantage of the binary alloys is the low volume fraction of the filament phase that can be produced. In the Al–Ni system, for example, only 0.1 v/v of Al_3Ni can be produced in the aluminum matrix. More complex alloy systems can give higher volume fractions, such as in Co–Cr–C with 30% v/v carbide fibers in a matrix of cobalt and chromium. Aligned two-phase structures can be grown from the melt in nickel- and cobalt-based alloys and are useful for high-temperature applications. NITAL is such a nickel-based alloy with a ductile nickel matrix and a fine precipitate of γ' (Ni_3Al) reinforced with a carbide phase of TaC, Cr_3C_2 or NbC.

11.10.8 Ceramic–matrix composites (CMCs)

11.10.8.1 Composite fibers

The commonest CMCs are probably those based on concrete or cement with steel reinforcement used for the construction industries. More modern versions include glass and carbon reinforcement, but any improved properties are outweighed by the high cost of manufacture. Development of the steel-reinforced concrete matrix includes the inclusion of low-carbon steel or austenitic steel wires (0.25–0.5 μm diameter) with an aspect ratio of 100:1. This 'Wirand' concrete has improved toughness, greater flexural properties and increased compressive strength, with a good resistance to salt water and abrasion. It is therefore used in sea defence constructions.

Silicon carbide, alumina and carbon fibers have been used in various reinforced materials, particularly inorganic glasses and glass ceramics, but also engineering ceramics such as alumina and silicon nitride.

11.10.8.2 Composite manufacture

As with most composites the fabrication process is labor intensive, demanding and expensive. During fabrication of a continuous-fiber CMC, it is necessary to infiltrate the fibers, which may be in the form of woven cloth, with a fluid form of the matrix-building material. The most common method uses a slurry of ceramic particles in water or alcohol, an organic binder and often a surface-wetting agent. After drying, the shape is hot-pressed. Alternative production routes now include sol–gel reaction, melt infiltration, chemical vapor deposition, reaction sintering and transfer molding. Fiber/matrix interface reactions have to be avoided as well as fiber oxidation, which can occur for carbon and SiC fibers around 400°C. Coatings and an inert environment help, but with increased cost. For that reason glassy-ceramic matrices are favored but these have a relatively low useful working temperature (500–800°C).

11.10.8.3 Fiber reinforcement

Large tonnages of cement and concrete reinforced composites are used in construction. However, in general engineering CMCs should in principle be strong candidates for high-temperature applications requiring good strength and stability, e.g. cutting tools, advanced engine components. Ceramics tend to be notch sensitive and brittle, and the reinforcements are designed to improve the toughness and reliability of the composite. Inorganic glasses and glass-ceramics are acceptable matrices and 'engineering' ceramics such as alumina and silicon nitride are finding favor. Applications include SiC whiskers/polycrystalline alumina for cutting-tool tips, graphite fibers/borosilicate glass for substrate of laser mirrors, and graphite/glass-ceramic for bearings, seals and brakes. The SiC fiber/lithium alumina silicate (LAS) glass composite is a likely material for heat engines.

11.10.8.4 Composite failure

Reinforcement of ceramics aims to improve toughness and prevent catastrophic failure. The main mechanisms of failure include cracking of the matrix, followed by matrix/fiber interfacial failure and fiber pull-out, and finally fiber failure. Hopefully, these are sequential events leading to a more gradual failure of the ceramic composite. Figure 11.26 shows such a failure sequence during a tensile stress–strain curve for a CMC. The improvement in fracture strain compared to the matrix is known as the overload capability. At the end of the initial failure stage, the matrix shows microcracking and the load is increasingly transferred to the fibers. A high modulus ratio E_f/E_m and high volume fraction of fibers (~0.5) is desirable. Strong bonding between the fiber and the matrix also assists the transfer of load and increases the amount of energy absorbed during fiber pull-out. Modifying and controlling the surface character of the fibers is particularly important, since its behavior under stress determines the path of any propagating crack. For very strong interfaces the crack may propagate through matrix

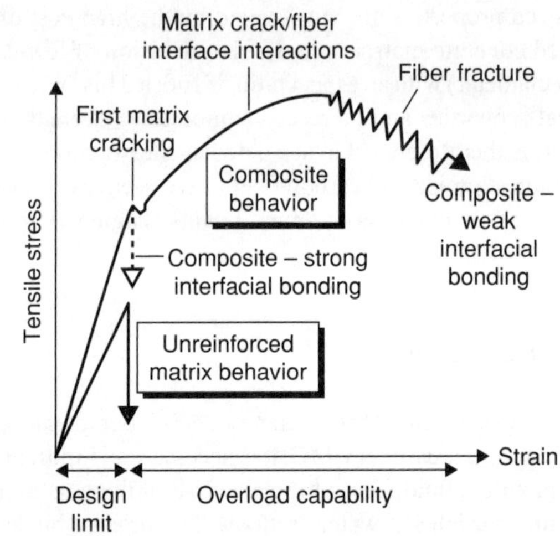

Figure 11.26 *Stages of failure in CMC reinforced with continuous fibers (from King, 1989, by permission of the Institute of Materials, Minerals and Mining).*

and fiber, whereas for much weaker interfaces the crack could travel along the interfacial region. Matching the thermal expansion coefficient (α) of fiber and matrix is also important. A mismatch favors crack deflection and can introduce stresses that reduce crack growth; compressive stresses in the matrix introduced on cooling when $\alpha_f > \alpha_m$ has a toughening effect on propagating microcracks.

Problems

11.1 Calculate the number average and weight average molecular mass of a polymer blend formed by mixing together five types of polystyrene with different molecular mass: 1 g of 15 000, 1.5 g of 23 000, 0.8 g of 34 000, 2 g of 45 000 and 1.2 g of 52 000.

11.2 If 1 wt% of styrene monomer is added to the polymer blend in Problem 11.1, calculate the new number average and weight average molecular mass. Comment on any significant difference with the original values.

11.3 (a) Crystalline polyethylene (PE) has an orthorhombic unit cell ($\alpha = \beta = \gamma = 90°$) with lattice constants 0.736 nm, 0.492 nm and 0.254 nm, containing a [–CH_2–CH_2–CH_2–CH_2] unit. Calculate the density of crystalline PE.

(b) A certain semicrystalline PE has a density of 0.91 Mg m^{-3}. Given that amorphous PE has a density of 0.84 Mg m^{-3}, determine the degree of crystallinity by weight.

11.4 (a) If the creep of polymers is defined by the viscous flow equation $\dot{\varepsilon} = C\sigma \exp(-Q/\mathbf{R}T)$, show that the viscosity during tensile deformation is given by $\eta = [\exp(Q/\mathbf{R}T)]/(3C)$.

(b) How can creep resistance of polymers be improved?

11.5 A rubber is stretched around a parcel with a stress of 7 MPa. This stress relaxes and is only 6.8 MPa after one month. What will be the stress after one year?

11.6 The parcel in Problem 11.5, normally kept at RT (~20°C), is moved to the greenhouse and exposed to a temperature of 35°C when, after a further month, the stress drops from 6.8 to 5.5 MPa. Calculate the activation energy for viscous flow in the rubber.

11.7 Write down an expression for the Young's modulus of a composite material E_c in terms of the matrix Young's modulus E_m and modulus of the fiber E_f and the volume fraction of parallel fibers V_f. Hence calculate E_c for a carbon fiber epoxy resin with $V_f = 0.5$, $E_f = 390$ GPa and $E_m = 3$ GPa.

11.8 Write down an expression for the density of a composite ρ_c in terms of the matrix density and density of the fiber material. Hence calculate the density for $V_f = 0.5$, $\rho_m = 1.15$, $\rho_f = 2.55\ \text{Mg m}^{-3}$ for a glass fiber–polyester resin.

11.9 A composite bar of nylon reinforced by discontinuous carbon fibers (15% volume fraction, 500 μm long, 6 μm diameter) aligned along the axis of the bar is pulled to failure. The tensile strength of the fibers is 3 GPa and the fiber–matrix bond strength is 30 MPa. Calculate the mean stress in the fibers when the bar fails.

Further reading

Ashbee, K. H. G. (1993). *Fundamental Principles of Fiber Reinforced Composites*, 2nd edn. Technomic, Lancaster, USA.

Ashby, M. F. and Jones, D. R. H. (1986). *Engineering Materials*, Vol. 2. Elsevier Science, Oxford.

Brook, R. J. and MacKenzie, R. A. D. (1993). Nanocomposite materials. *Materials World*, January, pp. 27–30. Institute of Materials.

Feest, E. A. (1988). Exploitation of the metal–matrix composites concept. *Metals and Materials*, May, pp. 273–278. Institute of Materials.

Harris, B. (1986). *Engineering Composite Materials.* Institute of Metals, London.

Hertzberg, R. W. (1989). *Deformation and Fracture Mechanics of Engineering Materials*, 3rd edn. Wiley, Chichester.

Hughes, J. D. H. (1986). *Metals and Materials*, June, pp. 365–368. Institute of Materials.

Hull, D. (1981). *An Introduction to Composite Materials.* Cambridge University Press, Cambridge.

Imperial Chemical Industries, Plastics Division (1974). *Thermoplastics: Properties and Design* (edited by R. M. Ogorkiewicz), Chap. 11 by P. C. Powell. Wiley, Chichester.

Kelly, A. (1986). *Strong Solids.* Clarendon Press, Oxford.

King, J. E. (1989). *Metals and Materials*, December, pp. 720–726. Institute of Materials.

Lemkey, F. D. (1984). Advanced *in situ* composites. In *Industrial Materials Science and Engineering* (edited by L. E. Murr), Chap. 14. Marcel Dekker.

Mascia, L. (1989). *Thermoplastics: Materials Engineering*, 2nd edn, Elsevier Applied Science, London.

McCrum, N. G., Buckley, C. P. and Bucknall, C. B. (1991). *Principles of Polymer Engineering.* Oxford University Press, Oxford.

McLean, M. (1983). *Directionally-Solidified Materials for High-Temperature Service.* Metals Society, London.

Metals and Materials (1986). Set of articles on composites, Institute of Materials:

Hancox, N. L. Pt I. Principles of fibre reinforcement (May).

Hughes, J. D. H. Pt II. Fibres (June).

Hancox, N. L. Pt III. Matrices (July).

Bowen, D. H. Pt IV. Manufacturing methods (September).

Davidson, R. Pt V. Performance (October).

Bowen, D. H. Pt VI. Applications of polymer–matrix composites (December).

Mills, N. J. (1986). *Plastics: Microstructure, Properties and Applications.* Edward Arnold, London.

Morley, J. G. (1987). *High-Performance Fibre Composites.* Academic Press, New York.

Morton-Jones, D. H. (1989). *Polymer Processing.* Chapman & Hall, London.

Powell, P. C. (1974). *Thermoplastics: Properties and Design* (edited by R. M. Ogorkiewicz), Chap. 11. Wiley-Interscience.

Shah, V. (1984). *Handbook of Plastics Testing Technology.* Wiley, Chichester.

Wood, A. K. (1989). Advances in engineering plastics manufacture. *Metals and Materials*, May, pp. 281–284. Institute of Materials.

Chapter 12

Case examination of biomaterials, sports materials and nanomaterials

12.1 Introduction

In recent years a number of areas have emerged of special technological importance which depend to a significant extent on the already well-established science of metals and materials. Such areas include the successful applications of metals, polymers, ceramics and composites for biomaterials used in medicine and dentistry, and the increasing use of new materials for sports equipment. An even newer field is that of nanotechnology, where developments are continually being sought for the exciting novel properties of materials with structures below the micron level. These areas are all high-tech, requiring special attention to quality and manufacture. As a consequence, they are generally high-cost applications.

In this chapter a number of practical cases are examined which include (i) biomaterials, (ii) sports materials and (iii) nanomaterials.

12.2 Biomaterials

12.2.1 Introduction and bio-requirements

Biomaterials are materials used in medicine and dentistry that are intended to come in contact with living tissue. The familiar tooth filling is where most humans first encounter biomaterials, but increasingly many people now rely on more critical implants such as joint replacements, particularly hips, and cardiovascular repairs. Undoubtedly, these biomaterial implants improve the quality of life for an increasing number of people each year, not just for an ageing population with greater life expectancy, but for younger people with heart problems, injuries or inherited diseases.

The requirements for a biomaterial are extremely demanding. Replacement or repair of a body feature, tissue, organ or function often necessitates the material used to have specialized mechanical, physical and chemical properties. However, the very first requirement is biocompatibility with the human body, i.e. the ability of the material to perform with an appropriate host response. Unfortunately, no material is universally biocompatible, since a material may be biocompatible in one application but not with another. Biocompatibility is therefore application specific.

For the successful use of the biomaterial, consideration has to be given to the appropriate material selection, engineering design and manufacturing process. While proper design and manufacture is essential, it is particularly important to select the correct material to provide the appropriate properties as well as being biocompatible, recognizing that the combined influence of mechanical and chemical factors can be quite serious, e.g. causing fatigue, corrosion fatigue, stress corrosion, wear, fracture. It is also important to recognize that the biological environment is not constant, and that oxygen levels, availability of free radicals and cellular activity will vary. Corrosion and degradation can lead to loss of integrity of the implant and, of course, release ions into the body, often setting up an allergic reaction.

Table 12.1 *Composition of orthopedic implant alloys (wt%); from Bonfield, 1997.*

Element	*Cobalt-based alloys*			*Stainless steel*		*Titanium alloys*	
	ASTM F75 cast	*ASTM F90 wrought*	*ASTM F563 isostatically pressed*	*ASTM F138/A*	*ASTM F138/9B*	*Commercial purity titanium*	*Ti–6Al–4V*
Co	Balance	Balance	Balance	–	–	–	–
Cr	27–30	12–19	18–22	17–20	17–20	–	–
Fe	0.75 max	3.0 max	4–6	Balance	Balance	0.3–0.5	0.25 max
Mo	5–7	–	3–4	2–4	2–4	–	–
Ni	2.5 max	9–11	15–25	10–14	10–14	–	–
Ti	–	–	0.5–3.5	–	–	Balance	Balance
Al	–	–	–	–	–	–	5.5–6.5
V	–	–	–	–	–	–	3.5–4.5
C	0.35 max	0.05–0.15	0.05 max	0.03 max	0.08 max	0.01 max	0.08 max
Mn	1.0 max	2.0 max	1.0 max	2.0 max	2.0 max	–	–
P	–	–	–	0.03 max	0.025 max	–	–
S	–	–	0.01 max	0.03 max	0.01 max	–	–
Si	1.0 max	1.0 max	0.5 max	0.75 max	0.75 max	–	–
O	–	–	–	–	–	0.18–0.40	0.13 max
H	–	–	–	–	–	0.01–0.015	0.012 max
N	–	–	–	–	–	0.03–0.05	0.05 max

Biomaterial applications make use of all classes of material, metals, ceramics, polymers and composites, divided roughly into three user types. These are (i) inert or relatively inert with minimal host response, (ii) bioactive, which actually stimulates bonding to the surrounding tissue, and (iii) biodegradable, which resorb in the body over a period of time. Metals are generally chosen for their inert qualities, whereas ceramics and polymers may offer bioactivity or resorption.

The most common metallic materials used are austenitic stainless steels, cobalt–chromium alloys or titanium; typical compositions are shown in Table 12.1. Recently, titanium alloys, particularly Ti–6Al–4V, have been introduced because of their corrosion resistance, strength and elastic modulus (see Table 12.2), but poor tribology can still be a problem. It is also favored for its superior biocompatibility and, unlike Co–Cr or stainless steel, does not cause hypersensitivity. Of the ceramics, aluminum oxide, calcium phosphate, apatite, carbon/graphite and bioglass are in use mainly for their inertness, good wear characteristics, high compressive strength and in some cases bioactivity. Their poor tensile properties and fracture toughness are design limitations. Polymers are widely used, both alone and in combination with ceramics or metals. These include: polymethyl methacrylate (PMMA) for cement and lenses; polyethylene for orthopedics; polyurethane as blood contact material, e.g. vascular tubing, cardiovascular devices, catheters; polysiloxanes in plastic surgery, maxillofacial and cardiovascular surgery; polyesters and polyamides in wound closure management. Composites such as ultra-high-molecular-weight polyethylene reinforced with either carbon fibers or the ceramic hydroxyapatite are increasingly being considered for applications involving high contact stress and wear resistance.

12.2.2 Introduction to bone and tissue

An understanding of the structure of bone is important when considering the repair of fractures and the replacement of joints. In repairing bone fractures it is necessary to hold the bone together while

Table 12.2 *The mechanical properties of some natural and biomaterials.*

Material	*Elastic modulus ($GN\,m^{-2}$)*	*Tensile strength ($MN\,m^{-2}$)*	*Elongation (%)*	*Fracture toughness ($MN\,m^{-3/2}$)*	*Fatigue strength ($MN\,m^{-2}$)*
Austenitic stainless steel	200	200–1100	40	100	200–250
Cobalt–Chromium	230	450–1000	10–30	100	600
Ti–6Al–4V	105–110	750–1050	12	80	350–650
Alumina	365	–	<1	–	400
Hydroxyapatite	85	40–100	–	–	
Glass fiber	70	2000	2	1–4	
PMMA	2.8	55	8	–	20–30
Bone cement	2.3–3	1.5	1–2	400	
Polyethylene	1	20–30		1–4	16
Nylon 66	4.4	700	25		
Silicone rubber	6×10^{-3}	1.4			
Polycarbonate	2	60			
Bone (cortical)	7–25	50–150	–	2–12	
Bone (cancellous)	0.1–1.0	50–150		2–12	
Tooth enamel	13	240	–	–	
Tooth dentine	–	135		–	
Collagen, tendon, wet	2	100	10	–	

the natural healing process takes place, usually for a few months. Metallic alloys have been universally used for plates, pins, screws, etc. They are considered to have the required strength and rigidity together with sufficient corrosion resistance. Generally, these fixtures are considered temporary until the bone has joined, after which they are removed. The main reason for this is that the very rigidity with which the bone is held to allow healing will eventually lead to progressive weakening of the bone structure.

The biological structure of bone is reproduced in Figure 12.1. In material science terms, the apparent complexity of bone can be described as a composite made up of a matrix of collagen (polymer) reinforced with approximately 50% volume fraction of hydroxyapatite (ceramic) nanometer-scale crystals. Most bones are made of a porous cellular structure (cancellous bone) covered with a denser compact shell. The porosity and density of the cancellous bone varies with location in the body, depending on the stress state in that region. Low-density regions have a relatively open cell structure and high-density regions more closed. The density determines the strength and stiffness of the bone which grows and develops to support the stress imposed on it. This may be uniaxial, when the cell walls will be oriented and thicker along this direction, or more uniform when the cells are roughly equiaxed.

Bone is not a static entity but dynamic in nature, continually undergoing remodeling, where 'old' bone is resorbed and replaced by new bone. Various factors control the process but extremely important in stimulating the bone-producing cells (osteoblasts) is the application of stress. In bone, the mineral phosphate material is slowly dissolved and resorbed by the body and when subjected to normal stress is replenished by new bone material synthesized by the osteoblasts. This recycling process ensures a healthy, strong bone structure with ageing. Bones for which the major stress is carried by metal implants do not show the same tendency for replenishment, so that the bone surrounding the implant is resorbed without replenishment, leading to a loosening of the implant. A rigid metal plate attached to a bone that has healed will nevertheless carry the majority of the load. To avoid this 'stress shielding'

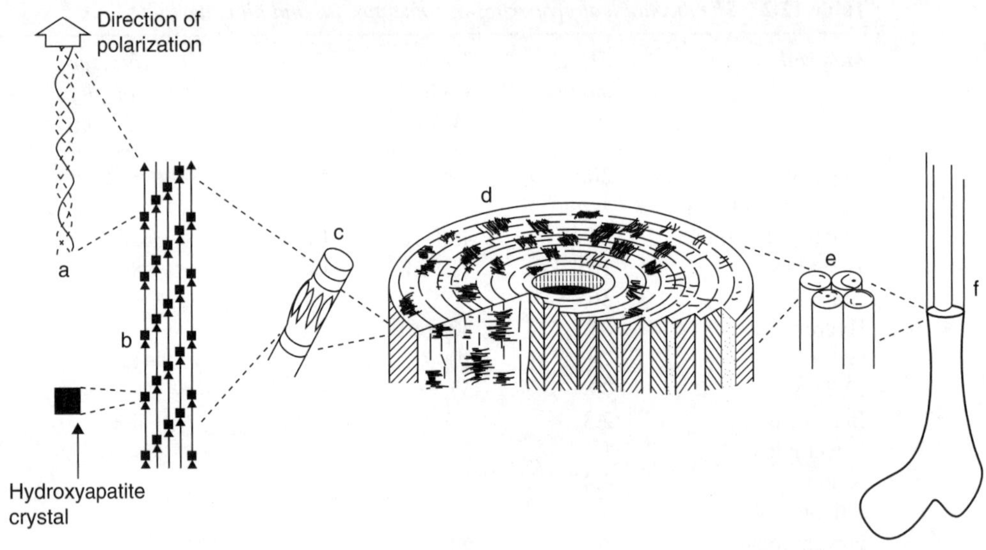

Figure 12.1 *The hierarchy of structure in bone. At the molecular level (a), polarized triple helix of tropocollagen molecules assembles into a microfibril (b), with small gaps between the ends of the molecules into which small (5 by 30 nm) crystals of hydroxyapatite later form. These microfibrils assemble into larger fibrils (c), which then form the layers in the osteon (d – part cut away to show the alternating orientation of fibers in the annular layers). Osteons form in association with each other (e), forming bone (f). The cells which are responsible for most of this process, the osteocytes, are shown sitting between the layers of the osteon (d) (after Vincent, 1990; courtesy Institute of Materials, Minerals and Mining).*

leading to bone weakness, it is recommended that the surgeon removes the holding plates soon after the fracture has healed.

With the aim of trying to avoid the removal operation, alternative approaches are being considered for bone plates, screws, etc., including the use of biodegradable, or resorbable, materials such as polymers or composites, which could be resorbed into the surrounding tissue or dissolve completely over a period of time after the fracture has healed. However, the combination of strength, ductility, toughness, rigidity and corrosion resistance of metals is hard to match with the non-metallics.

The large difference in elastic modulus between competing biomaterials and bone (see Table 12.2) is evident. Of the various metals, titanium and its alloys are clearly the most suitable and are being increasingly used. Titanium does, however, have a tribological weakness but the application of coatings and surface engineering is being increasingly adopted to overcome this problem.

Tissues include skin, tendons, ligaments and cartilages. Skin has the dual property of keeping the body fluids in while allowing the outward movement of moisture through a porous membrane, which is important in cooling and maintaining the body temperature. Skin also protects against infections, such as bacteria, but is not, of course, particularly strong. It is made up of layers including an outer epidermis and an inner dermis, a dense network of nerve and blood vessels. It is therefore virtually impossible to make an artificial skin from biomaterials to match this complexity. Nevertheless, skin replacements have been made from polymers with an open structure which provides a basic framework onto which real skin is able to grow. Moreover, with a biodegradable polymer the framework degrades as the new tissue regrows. The porous film can be coated with silicone rubber to provide infection protection and retain fluids while the skin grows. When sutured in place, tissue-forming cells (fibroblasts) migrate into the porous polymer framework to generate new skin layers. For severe burns, artificial skin can

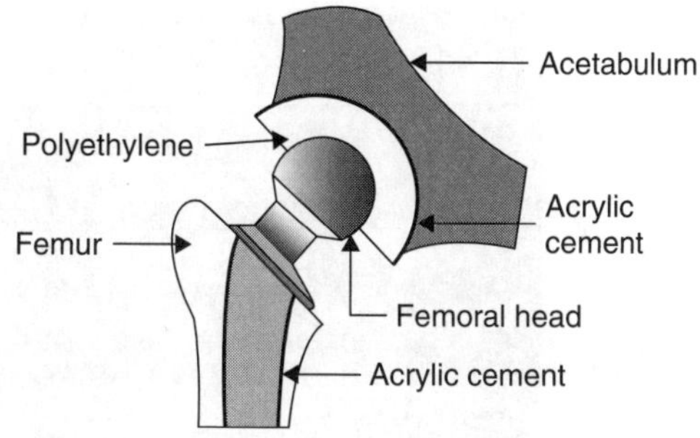

Figure 12.2 *Schematic diagram of a replacement hip joint.*

be made by growing epidermal skin cells within a biodegradable collagen mesh in a culture medium. The synthetic skin can then be grafted onto the patient. Other biodegradable products include the copolymers lactic acid–glycolic acid and lysine–lactic acid. The adhesion of the polymer framework can be improved by incorporating an adhesive protein fibronectin.

Other tissues such as ligaments and cartilages are largely elastic filaments of fibrous proteins. Synthetic substitutes have included *Dacron* polyesters, PTFE fibers and pyrolyzed carbon fibers, with mixed success. The fibers may be coated with polylactic acid polymer, which breaks down in the body to be replaced by collagen. At this stage such techniques are relatively new, but it does suggest that in future the growth of cells in a culture vessel may possibly supply complex biomaterials for various implanted functions. This approach is termed tissue engineering and, together with biometrics, i.e. the mimicking of the working of biological systems, offers a way of producing materials which totally synergize with the human body.

12.2.3 Case consideration of replacement joints

12.2.3.1 Hip joints

The basic design of an artificial hip joint includes an alloy femoral stem, with a metallic or ceramic femoral head moving in an ultra-high-molecular-weight polyethylene (UHMWPE) acetabular cup, as shown in Figures 12.2 and 12.3. The average life of the joint is better than 10 years, but implants tend to loosen as a result of bone resorption due to modulus mismatch. Friction and wear also cause wear debris between the cancellous, i.e. more porous, bone and the cup, and also between the femoral head and the softer UHMWPE. Overall the failure rate is now about 1–2% per year, as shown in Figure 12.4.

Friction and wear problems are being improved by using ultra-hard materials as the bearing material. A good resistance to frictional wear together with biocompatibility makes alumina well suited for the femoral head. Unfortunately, Al_2O_3 has a low impact strength and is liable to failure when subjected to stresses introduced by extra activity, e.g. jumping. Research has also shown that ion implantation of the UHMWPE in a nitrogen atmosphere at 10^{-5} mbar at 80 keV to a dose of 1×10^{17} ions cm^{-2} reduces the wear behavior to virtually zero. The enchanced surface properties, together with higher hardness and elastic moduli, would very much improve the interaction problems. A compromise solution is to have a titanium stem with stainless steel head coated with a thin layer of ceramic.

The problem of loosening of the joint due to bone resorption is being tackled by second generation biomaterials and implants which mimic the body's tissues. Thus, while continuing with improved bioinert materials, development has focused on bioactive materials which influence the biological

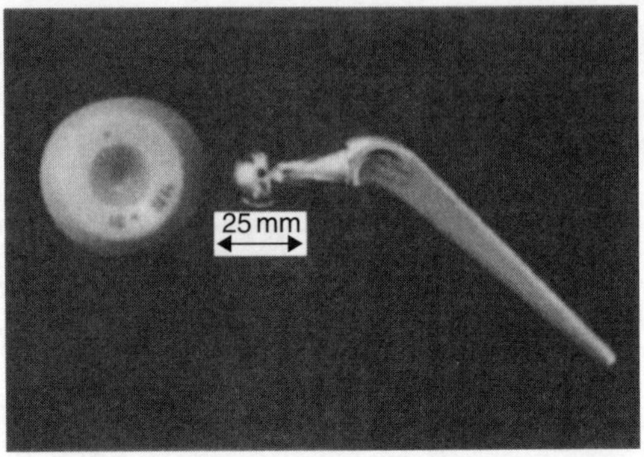

(a)

(b)

Figure 12.3 *(a) Conventional hip joint made of stainless steel and UHMWPE acetabular cup (courtesy R. Grimer, Royal Orthopaedic Hospital, Birmingham). (b) Modular artificial hip made of titanium stem, alumina head and UHMWPE acetabular cup (courtesy P. Marquis, Dental School, Birmingham).*

response in a positive way, e.g. encourage bonding to surrounding tissue with stimulation of new bone growth. With this bioactive approach the interface between the body cells and the implant is critical and the materials science of the biomaterials surface extremely important. One such material is a composite of hydroxyapatite (HA) in high-density polyethylene. The HA provides the strengthening and stiffness reinforcement to the tough polyethylene matrix. *HAPEX*, as it is called, is biocompatible and bioactive, encouraging bone growth onto the implant material which, of course, contains HA. The biological response has been demonstrated by cell culture studies with human oesteoblasts in which the cells grew and spread over the composite initiated at the HA particles. *In vivo* testing has shown a stable interface between the implant and bone.

12.2.3.2 Shoulder joints

These joints, like hip joints, are increasingly being replaced by metal prostheses (Figure 12.5). Initially, similar principles were applied to shoulder joints as to hip joints, using a stainless steel ball and socket with polyethylene cup. However, because it is not a load-bearing joint, the trend in shoulder replacement is simply to resurface the worn part on the humerus side with a stainless steel cap rather than replace both sides of the joint.

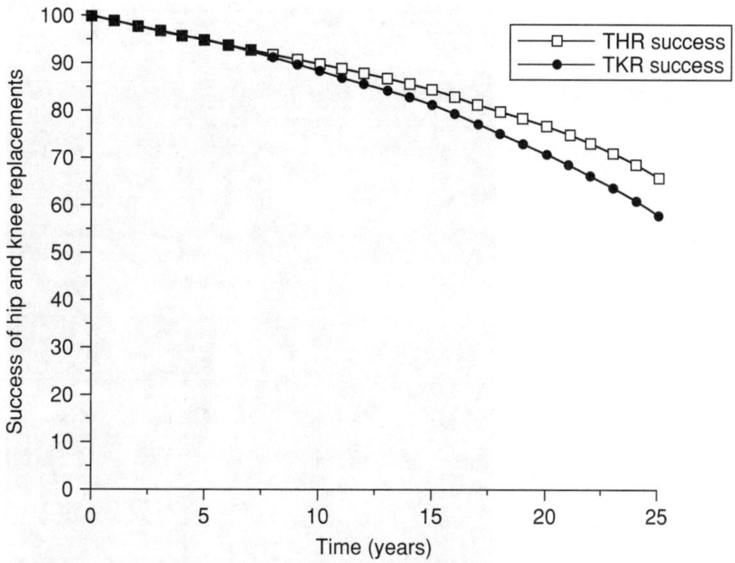

Figure 12.4 *The failure rate for a total hip replacement (THR) and total knee replacement (TKR) (courtesy R. Grimer, Royal Orthopaedic Hospital, Birmingham).*

12.2.3.3 Knee joints

Total knee replacements (TKR) are increasingly being made to give pain-free improved leg function to many thousands of patients suffering from osteoarthritis or rheumatoid arthritis. As with hip replacements, TKR involves removing the articulating surfaces of the affected knee joint and replacing them with artificial components made from biomaterials. Such operations have a successful history and a failure rate of less than 2% per year (Figure 12.4). A typical replacement joint consists of a tibial base plate or tray, usually made of stainless steel, Co–Cr or titanium alloy, with a tibial insert (UHMWPE) that acts as the bearing surface (see Figure 12.6). The femoral component largely takes the shape of a natural femoral condyle made from the above alloys, and articulates with the bearing surface, together with the kneecap or patella. The patella may be all polyethylene or metal backed. The components are either fixed by cement or uncemented. Cemented components use acrylic cement (polymethyl methacrylate). Uncemented components rely on bone ingrowth into the implant. Titanium used on its own shows evidence of bone ingrowth, but more recently this has been improved by use of hydroxyapatite (HA) coating. Problems include wear debris from the PTFE and possible fracture of the component.

One of the continuing challenges is improved design against failure, currently about 2%, as a result of overloading or increased physical activity, and possibly bone resorption.

12.2.3.4 Finger joints

Replacement finger and hand joints are far more complex than other joints because of the degree of flexibility required through large angles while maintaining overall stability. Early designs were based on the movement of a metal 'hinge', but had a problem of fatigue and/or corrosion. Nowadays silicone rubber is more commonly used, showing little tendency to fatigue failure while being biocompatible with an ability to absorb lipids and fatty acids.

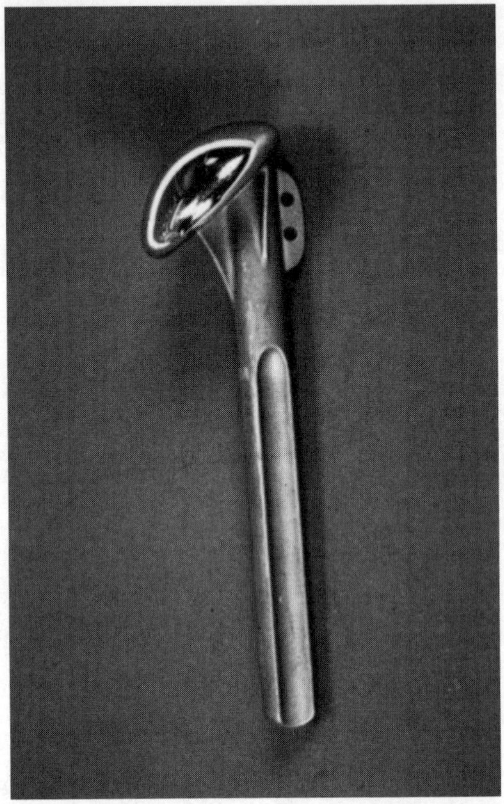

Figure 12.5 *Shoulder joint prosthesis (courtesy P. Marquis, Dental School, Birmingham).*

Polysiloxane is used in hand surgery to form a 'tunnel' to allow transplanted tendons to slide back and forward. It is also used to replace carpal bones. It is, however, not suitable for longer joints or bones unless reinforced.

12.2.4 Biomaterials for heart repair

12.2.4.1 *Heart valves*

Heart disease is one of the major killers in the developed world. Many of the serious conditions arise from the strain imposed on the heart by obstructions to the flow of blood either in the main passages of the circulatory system or as a result of valvular disease. The heart acts as a blood pump with four valves which open in response to a unidirectional flow of blood. Two of the valves allow the blood into the heart and the other two control the blood leaving the heart. Problems can arise with the heart valves if they become structurally damaged by disease affecting the opening and closing mechanisms.

The early prostheses developed in the 1960s for valve replacement were based on a stainless steel ball-in-cage or polysiloxane ball in a Co–Cr alloy cage (see Figure 12.7). These restrict the blood flow, even in the open position, and were superseded by a tilting disk device which opens and closes to the beat of the heart to allow the required blood flow. The major problem with valves arose from the tendency to initiate a blood clot and thus it was found necessary to give valve recipients anticoagulant

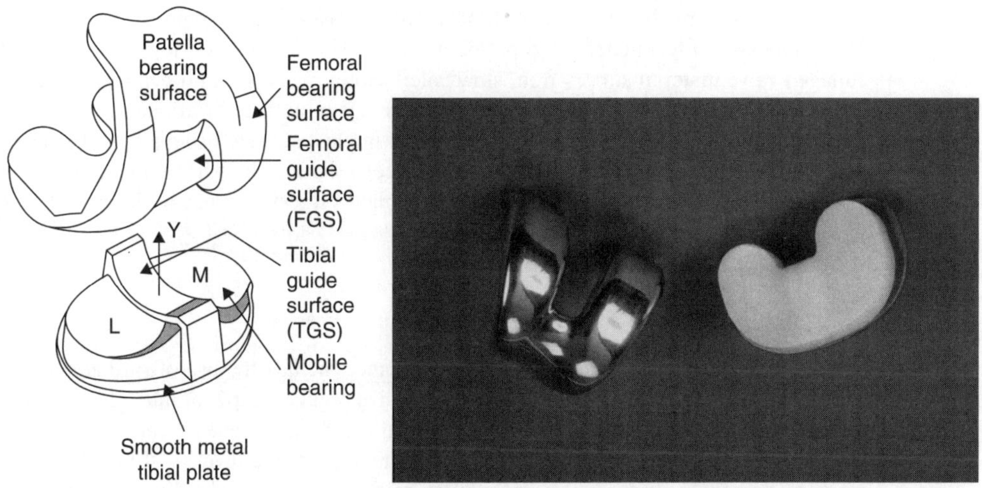

(a) (b)

Figure 12.6 *A total knee replacement joint. (a) Schematic diagram (after Walker and Sathasiwan, 1999). (b) Photograph of a stainless steel prosthesis (courtesy R. Grimer, Royal Orthopaedic Hospital, Birmingham).*

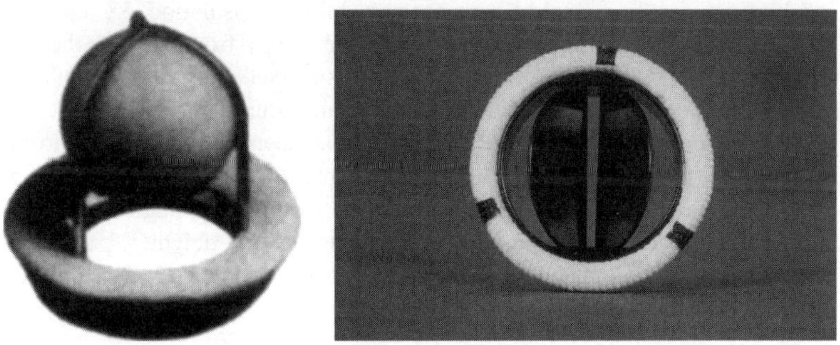

(a) (b)

Figure 12.7 *Photographs of replacement heart valves. (a) Ball-in-a-cage type (courtesy Institute of Materials, Minerals and Mining). (b) Pyrolitic carbon disk type (courtesy P. Marquis, Dental School, Birmingham).*

drugs. Nowadays, advanced valves are made with Co–Cr or titanium bodies with metal or graphite disks, or occluders, coated with pyrolytic carbon. These coatings are made by heating a hydrocarbon, such as methane, to about 1500°C, depositing the carbon vapor on the graphite surface to a thickness of about 1 mm. Small amounts of silane mixed with the CH_4 add Si to the deposit, increasing its strength. Pyrolytic carbon is strong and wear resistant but, more importantly, resists the formation of blood clots on its surface. The disks are attached by a coated Ti metal arm to a fabric ring made of polymer (PTFE), which is sewn to the tissue of the heart valve opening. *Dacron* cloth has also been used and encourages tissue growth with better anchorage and thromboembolic resistance.

The prosthesis has moving parts and thus catastrophic failure by fracture is a finite possibility. As an alternative, the construction of artificial valves from biological tissue has been developed. Collageneous tissue from the heart wall of cows and heart valves from pigs have been used to make these 'bioprosthetic' valves. These valves are naturally biocompatible with a reduced risk of thrombosis. Unfortunately, two problems have hindered this development. The first is that the collagen valve material suffers from slow calcification, whereby hard deposits form on the surface of the valves, causing them to stick and tear. The use of anticalcification drugs is a possibility, but the second problem has thrown the whole area of implant surgery using animal tissue in doubt. This problem is the emergence of BSE (mad cow disease) in cattle, which has led to restrictions and worries in the use of animal tissue for reconstructive implant surgery because of the fear of transmission of viral illness from animal tissue to humans.

12.2.4.2 Pacemakers

Other developments in improving heart performance include the use of cardiac pacemakers which produce a 5 V electrical impulse for 1/500 second at regular heartbeat rate. These devices have been available for some time but have improved significantly over recent years. The basic requirement is to provide electrical signals to the heart at the appropriate level to stimulate the patient's own electrical activity to produce the proper physiological change, normally linking the pacemaker into the cardiac system so that it works when needed. The biomaterial aspects of the pacemaker are, however, also important, not only in overcoming the problems introduced by any device/body environmental interaction, but also in designing the proper electrical supply and insulation. Power supplies have advanced considerably in the last few years and lithium cells are now exclusively used. Titanium is again the most common biomaterial to encase the device, manufactured and electron beam welded to seal it hermetically. Polymers have been used for encapsulation, e.g. epoxy resin or silicone rubber, but these materials do not completely prevent moisture from entering the pacemaker, shortening the lifetime of the device. These have now been superseded by titanium alloys because of their better strength and environmental properties. These are sutured into the aorta with a *Dacron* sleeve.

Another problem area is that provided by the electrodes which have to flex with every heartbeat and hence are liable to fatigue failure (see Chapter 6). Good design and choice of electrode materials can minimize this problem. The electrical supply passes through the titanium casing via a ceramic insulator and the leads to the heart are insulated with a polymer (polyetherurethane). Degradation with time is still a possibility and has to be considered in an effective design.

12.2.4.3 Artificial arteries

In branches of surgery, particularly cardiovascular surgery, there is often a need to replace arteries blocked by atherosclerosis. Sometimes this can be achieved by using tissue grafts from the patients, thereby avoiding any immune response. In other cases it is necessary to use artificial arteries made from polymers; such arteries must be tough and flexible enough to avoid kinking, with the added requirement of avoiding the formation of blood clots. In modern surgery, the blood clotting tendency can be removed by anticoagulants, such as heparin, but the ultimate goal is to provide artificial arteries with natural clot resistance. Several different polymers have been used to make blood vessels, but none is entirely satisfactory.

The polyester *Dacron* can make small tubes but these have porous walls which have to be sealed. This is achieved by treating with the protein albumin and heating it to form a coagulated coating. In the body, the albumin degrades and is replaced by the natural protein collagen, forming a smooth lining (pseudointima). This process leads to some initial inflammation, which is one disadvantage of this biomaterial. Woven *Dacron* is quite rigid and unsuitable for small arteries and is difficult to suture; it is mostly used for resected aortic aneurysms. Knitted *Dacron* is easier to suture. It may be coated with polyurethane, tetrafluoroethylene or heparin to reduce the thrombogenic tendency. The use of poly(hydroxyethyl methacrylate) coating establishes an endothelial-like cell layer in a few weeks.

Other arterial polymers include PTFE and polyurethanes. PTFE is used as an expanded foam to form the porous tubes. These rapidly develop a smooth neointima layer and thus acquire blood compatibility. Polyurethanes have a natural compatibility with blood, and are tough and flexible in tube form, but unfortunately slowly degrade in the body, producing toxic products. PTFE coatings on polyester and polyurethane vessels have also been tried. Silicone-lined tubing has been used for extracorporeal circulation during open-heart surgery.

To produce artificial arteries with built-in clot resistance, heparin molecules have been attached to their surface either directly by chemical bonds or by cross-linking to form a polymerized heparin film. To mimic total thrombosis resistance, however, requires not only anticoagulation, but also avoidance of platelet deposition normally achieved by the endothelial cells lining the blood vessel releasing the protein prostacyclin. Ideal artificial arteries should have both of these anticlotting agents attached at their surface.

12.2.5 Reconstructive surgery

12.2.5.1 Plastic surgery

The many different biomaterials used in plastic surgery include collagen, silicone (polydimethylsiloxane), *Teflon*, polyethylene, *Dacron* and polyglocolic acid. Polysiloxanes are widely used in reconstructive plastic surgery because of their lack of any tissue reactivity, mechanical properties and structure. Polysiloxane may be either heat vulcanized or vulcanized at room temperature for more delicate structures and is ideal for soft tissue replacement where repeated flexure occurs. In applications where strength is also required, steel-wire reinforcement is necessary. The material is used not only for implants but also for explants, where it is molded to fit a specific shape, colored to give a skin match and fixed with tissue adhesive. Common examples are ear prosthesis or facial reconstruction after loss due to injury or cancer surgery. Silicone mammary implants are now widely used for cosmetic reasons or after mastectomy for breast cancer. These have an outer shell with the appropriate shape and resilience covering an inner volume of silicone gel. In these reconstructions *Dacron* may be used as a lining for the ear and as a backing in a breast implant to provide a better fastening of the implant to the surrounding tissue.

12.2.5.2 Maxillofacial surgery

Disease or injury to the oral and facial area may be repaired by implants from a range of biomaterials. Often the effective prostheses can improve the confidence and well-being of the patient beyond the immediate functional repair. The jaw, and jaw bone area, can utilize many of the metal implants already discussed. Commercially pure titanium in perforated sheet form has been used because of its biocompatibility and ease of manipulation and fixing.

Polymers, particularly silicones and polyurethanes, may be used to replace flexible tissues of the nose, cheek and ear regions of the face. Polysiloxanes have been used for onlays in the area of the molar bone in the lateral side of the mandible or over the forehead to smooth it. Reinforced with stainless steel wires, they can also replace the mandible. A porous composite of polytetrafluoroethylene (PTFE) strengthened with carbon fibers may be used to replace damaged bone structures. Both PTFE and carbon are biocompatible, and fibrous tissue growth into the pores ensures bonding of the artificial and natural bone. Composites of hydroxyapatite in a polyethylene matrix *HAPEX* have been used for patients who had either fractured the orbital floor supporting the eye or had lost an eye. A great advantage of such material implants is that they can be shaped during the operation and inserted on the base of the eye socket and bonded firmly to the supporting bone.

12.2.5.3 Ear implants

HAPEX has been also used successfully for other clinical replacements, mainly middle-ear implants. These transmit sound from the outer to inner ear, where the vibrations are translated into electrical signals to be processed by the brain. Middle-ear malfunction can lead to deafness (conduction deafness) which may be cured by implant surgery. Otosclerosis (middle-ear deafness) may result from fibrosis of the middle ear caused by repeated infections or from an hereditary disease. A stapedectomy removes a small amount of bone and immobilized tissue and replaces it with biomaterial. The implant has a hydroxyapatite head on a *HAPEX* shaft which can be trimmed and shaped in the operating theatre using an ordinary scalpel to fit the individual patient.

12.2.5.4 Other surgical applications

Polymers are also used in a wide variety of general surgical applications. Polyurethane has good tissue and blood contact properties and is used in both short-term applications, e.g. catheters, endotracheal tubes, vascular tubing, hemodialysis parts, and long-term applications, e.g. heart-assist devices. *Dacron* is used in composite form with a poly(2-hyroxyethyl methacrylate) matrix for orthopedic tendon reconstruction. Varying the composite mix can alter the properties to match the requirements. Reinforced *Dacron* fabric is used for reconstructing the trachea and, in woven *Dacron* form, for small bowel repair or replacement. It is also used in the genitourinary systems, in mesh form, in repairing hernias and abdominal wall defects. Polysiloxanes are used in neurological surgery, e.g. in valves to drain fluids produced intercranially and also as tubes to drain other canals such as the middle ear. Not having a porous structure, these tubes resist bacterial contamination.

Shape–memory–effect (SME) alloys (see Chapter 8, Section 8.6.4), particularly Ni–Ti, have been used in several biomedical applications because of their unique behavior and for their biocompatibility. In orthopedics, for example, pre-stressed fracture bone plates can be made to shrink on heating to provide a rigid, compressive load fixing. By contrast, Ni–Ti rods can be programmed to provide traction on local heating. In other applications Ni–Ti has been used in artificial heart muscles, teeth-straightening devices, intrauterine contraceptive devices and as a filter in the vena cava (inserted cold the filter opens its mesh at the temperature of the deoxygenated blood flowing back to the heart).

In spinal surgery, commercially pure titanium cables and screws have been used for the correction of scoliosis by gradual tightening of the cable to straighten the spine. A big advantage of Ti for these devices is its resistance to crevice corrosion.

12.2.6 Ophthalmics

The main usage of biomaterials in ophthalmics is for hard contact lenses, soft contact lenses, disposable contact lenses and artificial interocular lenses. Apart from being easier to wear, soft contact lenses have several advantages over hard lenses. Soft lenses ride on a thinner tear film, fitting closer to the shape of the cornea, generally giving better corneal health, less irritation from dust under the lens and less likelihood of being dislodged.

For satisfactory lens performance the biomaterial must be inert, dimensionally stable to retain its optical properties, have a low density ($\sim 1\,g\,cm^{-3}$), a refractive index close to that of the cornea (1.37) and good oxygen permeability to maintain a healthy cornea. Hard contact lenses first became available using PMMA because it was light (density $1.19\,g\,cm^{-3}$), easy to shape, has a refractive index of 1.49 and is reasonably biocompatible. They are, however, difficult to wear for long periods because PMMA is (i) hydrophobic and not easily wetted by eye fluids, and (ii) has a very low oxygen permeability. These have now been superseded by soft lenses.

A wide variety of materials have been used for soft lenses, including silicones, hydroxyethyl methacrylate (HEMA), and copolymers with HEMA as the major component and vinyl pyrrolidine to increase the water content. Most soft lenses are made from hydrogels; poly(2-hydroxyethyl methacrylate) is still the most popular, containing a small amount of ethylene glycol dimethacrylate to act as a cross-linking agent. These hydrogel materials have excellent compatibility and other properties, but even with good oxygen permeability (100 times better than PMMA) additional oxygen transplant via tear exchange is necessary. To increase the permeability the water content of the lens is raised, but too high a water content can result in reduced strength and poor handleability.

After cataract surgery and removal, polysiloxane coated with wetting agents such as polyvinyl pyrrolidine (PVP) has been used. For intraocular problems, silicone injectable has been used.

12.2.7 Dental materials

12.2.7.1 Cavity fillers

Dentistry has always been very dependent on the use of biomaterials, and particularly receptive to the application of new developments in metals, ceramics, polymers and composites.

Dental amalgams have been used in cavity fillings for more than 100 years, initially with silver–mercury amalgams, later modified by tin additions to control the amount of expansion. These amalgams produced the weak, corrodible intermetallic γ_2 phase, Sn_7Hg, and hence the modern dental amalgam now also contains copper (>12%) in order to suppress this phase. The amalgam is made by mixing silver, tin, copper alloy powder with mercury and this mixture is packed into the cavity, where it hardens to produce a strong, corrosion-resistant, biocompatible filling. There is some evidence that even this filling may be susceptible to corrosion as a result of the Cu_6Sn_5 (η') phase and the addition of Pd has been advocated. Attempts to replace the Hg amalgam by gallium, indium, silver, tin and copper pastes have not yet been completely successful.

Alternative resin-based composite filling materials have been continuously developed since they were first introduced in the 1960s. These composite fillings have a strength similar to amalgams but poorer wear properties. The paste is created by mixing a dimethacrylate monomer with resin and adding a filler of micron-sized silane-coated ceramic particles. The paste is activated by strong light when the resin polymerizes. Bonding of the composite resin to the tooth structure employs a phosphoric acid etch of the tooth enamel. This produces mini-chasms into which the resin material flows and locks to form a strong mechanical bond. This technique is not successful, however, for bonding to the dentine in the tooth cavity (Figure 12.8) and so, in the absence of enamel, dentine bonding agents have to be used. These are primers containing bifunctional compounds with (i) hydrophilic molecules which form links with the wet dentine in the tooth cavity and (ii) hydrophobic molecules which form links with the resin in the composite.

Cavities in front teeth are usually filled with glass cements to match the color and translucency of the enamel. Silicate cements are formed when phosphoric acid displaces metal ions from an alumina–silica glass, containing metal oxides and fluorides. The cement sets when aluminum phosphate is precipitated between the glass particles. Developments based on this basic chemistry employ polymeric acids with carboxylate groups. In this case, the metal ions displaced from the glass cross-link with the polymeric acid chain, causing the cement to set. In addition, the acids undergo an ion-exchange reaction with the calcium phosphate in the apatite of the dental material. These glass ionomer cements therefore form direct chemical bonds to the tooth material. Resin-modified versions are also available which have improved durability; these contain carboxylate groups to give a good bond to the tooth and dimethacrylate, as in the composite resin.

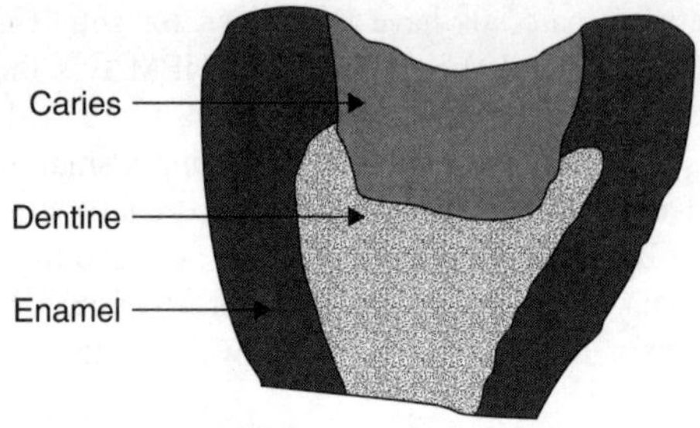

Figure 12.8 *Schematic diagram of a tooth.*

12.2.7.2 Bridges, crowns and dentures

Missing teeth may be replaced by artificial teeth in a number of different ways. For a group of missing teeth, removal partial dentures (RPDs) may be the answer; they consist of a cast metal framework of Co–Cr or Ni–Cr alloy carrying the artificial teeth and having end clasps to retain it to good natural teeth nearby. Fixed partial dentures (FPDs or bridges) may be used for a few missing teeth. Sometimes the supporting teeth are cut down to accommodate a close-fitting artificial tooth casting, which is cemented into place. In other cases, the alloy framework carrying the artificial teeth is bonded to acid-etched teeth to avoid cutting down good teeth. The teeth are acid etched and the metal framework electrolytically etched to produce structural grooves and chasms which allow strong mechanical bonds to be formed with resin-based composite cements. In some situations, etching can be avoided when the oxides on the metal framework can be treated with bifunctional primers to form chemical links to the cement.

Over the last 20 years or so the quality of bonded restorations, i.e. porcelain-covered metal castings, has been refined to combine the impact strength of a metal substructure with the appearance of dental porcelain. These porcelains have good mechanical properties and their thermal expansion characteristics are matched to the metal in order to avoid interfacial stresses and cracking.

Dental porcelains are basically vitrified feldspar with metallic oxide pigments to simulate natural tooth enamel. They are usually supplied to the dental laboratories in the reacted and ground forms for final fabrication by the technician; this involves mixing the powder with distilled water to form a paste which is used to make the crown, then drying and firing in order to sinter and densify the crown material. Generally, firing is carried out in stages, starting with the innermost structure of the crown, followed by the body, and finally the outer glaze and surface staining. Developments have included the strengthening of the inner core material with alumina to prevent cracking and the addition of magnesia to form a magnesia–alumina spinel, which has a low shrinkage on firing. Glass-ceramics have also been used either to fabricate the crown by casting followed by heat treatment to produce crystallization, or by machining from a pre-fired block of glass-ceramic under CAM/CAD conditions.

For complete replacement dentures, the basic material, which has existed for many decades, is methyl methacrylate. Substitute materials have been limited, and most improvements and developments have occurred in the processing technology. While the mechanical properties of denture base resins are not particularly good (modulus of elasticity 3×10^9 N m^{-2}, tensile strength $\sim$100 kN m^{-2}, elongation $\sim$3%), they do have suitable surface and abrasion properties, and are chemically inert, non-toxic and cheap. Improvements have been forthcoming in elastomers used for taking impressions; these now include vinyl addition silicone and polyether elastomers.

12.2.7.3 Dental implants

Dental implants have been far less developed than those associated with body implants (see hip joints, etc.). Probably the simplest forms are posts, Co–Cr, stainless steel, titanium alloy or gold

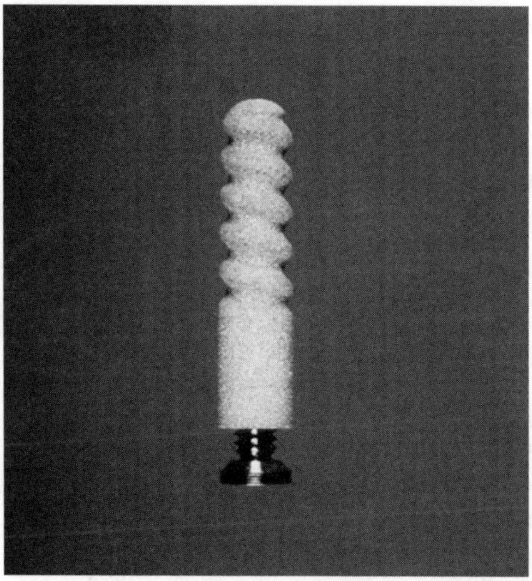

Figure 12.9 *Hydroxyapatite-coated titanium root implant (courtesy P. Marquis, Dental School, Birmingham).*

alloys cemented, or even screwed, into the tooth canal after the tooth has been root treated to remove the nerve. Dental porcelain caps may then be cemented onto the root post. Ti implants have been screwed into the bone beneath extracted teeth. After some time the passive surface layer of the titanium implants becomes osseo-integrated with the bone and can be used as a strong base onto which a titanium mini structure can be fitted, complete with tooth assembly. Osseo-integration is improved by using a coating on the titanium implant such as hydroxyapatite or bioglass (Figure 12.9). Ceramic and carbon implants set into the bone have been used with sapphire single crystals and pyrolitic graphite as favored materials.

12.2.8 Drug delivery systems

Oral administration of tablets is a familiar method for taking medicine. Less familiar is regular injection. Both methods have many disadvantages, not least of which is that the drug is being used as a general body medicine when, ideally, it is required only at some specific site in the body. A second disadvantage is the variation in drug level with time, as it is metabolized from a high to an insignificant rate rather than a steady, more moderate rate. Nowadays, these disadvantages are being addressed by developing controlled drug delivery; in some cases to specified tissues or organs.

One form of controlled drug delivery and targeting system uses polymeric biomaterial to contain the drug, which then escapes by diffusion. The diffusion pathway through the polymer is provided by the gaps in the chain-like structure, which may be varied to control the release rate. One such release system uses a copolymer of lactic and glycolic acids to target peptide-based drugs to reduce prostate cancer. Another approach uses a biodegradable material chemically bonded to the drug. In the body the drug is gradually released as the material degrades, giving a continued release over a known degradation period at a specific site within the body.

A more ambitious system delivers a drug in response to blood chemistry levels, such as the release of insulin in response to glucose level, and has been used with some success. The implanted micro-infusion system consists of a titanium reservoir of insulin together with a micro-pump which delivers the insulin via a fine catheter. In principle, a glucose sensor could provide a feedback control to the reservoir to complete the creation of an artificial pancreas.

12.3 Sports materials

12.3.1 Introduction

Formula 1 motor racing and ocean-going yacht racing have always attracted large sums of money for the latest materials development in the search for lighter, faster and more energy-efficient progress. However, in the last two decades there has been an enormous growth in the application of advanced materials to sports equipment for individual use, i.e. tennis rackets, golf clubs, bicycles, etc. The application of these high-performance materials is again high cost, but the competitive nature of both the professional sportsman and amateur has driven a growth in the sports equipment market which is quite remarkable. Golf clubs and tennis rackets which used to cost a few tens of pounds now cost several hundreds, but the incurred benefits in terms of performance and achievement justifies the extra financial outlay.

In this section, a case examination is made of the materials specifically developed for golf, tennis, cycling, skiing, archery and fencing, together with a consideration of the materials developed for protection when participating in these sports.

12.3.2 Golf equipment

A golf club consists essentially of two parts, the shaft and the head. Design considerations for the shaft include its weight, bending stiffness and torsional stiffness. During the drive the shaft tends to bend and twist a few degrees. Torsional stiffness keeps the club head face 'square' to the direction of flight on impact and is therefore desirable. However, off-center impact is often related more to the non-planar nature of the swing than torsional stiffness, off-center impact causing the ball either to hook or to slice. Bending of the shaft in a typical swing is shown in Figure 12.10. Stiff shafts tend to reduce the bending moment. The bend point of the shaft (or shaft region with the minimum radius of curvature during bending) is also important. A bend point higher up the shaft is favored by the big hitters.

A variety of materials have been used to make the shaft, which is usually hollow to minimize the weight. These include alloy steel, (7XXX) aluminum alloys, Ti–6Al–4V and more recently composites. Epoxy matrices with carbon, Kevlar, glass or boron are used for reinforcement in composite shafts. The so-called 'graphite' shafts are CFRP hollow shafts and are extremely light with a high damping capacity for the impact vibrations which travel along the shaft to the grip. They are made by wrapping pre-preg sheets of carbon fiber around a tapered mandrel, by filament winding around a mandrel or resin transfer molding (see Chapter 11). Composite manufacture allows some manipulation of the torsional stiffness and bend point. The balance point of the shaft can also be manipulated to meet the demands of the variety of club heads being designed.

Early driver heads were traditionally made of persimmon with a plastic face insert, sole plate and incorporating a head plug to adjust the center of gravity and weight to around 200 g. Such a small head required great skill and accuracy to strike the 'sweet spot' on the club face. A revolution in driver head design has been made with the introduction of a larger hollow metal volume, which enables the larger sweet spot to be located easily. Metal processing can produce these large-headed 'woods' in stainless steel, maraging steels, aluminum alloys, Ti–6Al–4V and other titanium alloys (Ti–15Cr–6Zr). A zirconium-based alloy, Zr–13.2 Cu–11 Ti–9.8 Ni–3.4 Be, in the amorphous state

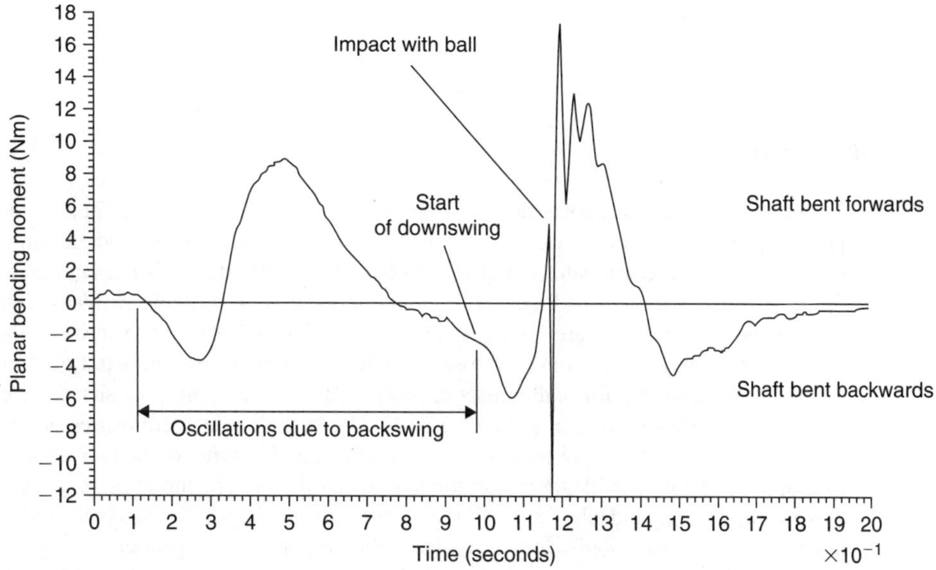

Figure 12.10 *Shaft bending in typical swing of golf club (after Horwood, 1994).*

has also been developed – a glassy metallic club surface can increase the coefficient of restitution at the impact with the golf ball. Most 'oversized' driver heads are now made with titanium alloys, metastable β-Ti alloys or $(\alpha + \beta)$ Ti alloys, incorporating a crown, face and sole plate. An integrated design can be investment cast and the precision casting may be welded by tungsten inert gas (TIG) or electron beam welding (EBW) to form a thin-walled hollow shell, which may be filled with low-density foam. With hollow cavity heads the moment of inertia is greater and the center of gravity is located closer to the strike face. Some designs still use separate inserts, e.g. TriMetal clubs, which have a maraging steel face and Cu/W sole plate to lower the center of gravity. Such innovative features confer a more forgiving club head with regard to inaccurate striking. A greater moment of inertia reduces head twisting when the impact is off the sweet spot, thereby reducing ball spin.

Generally, the components should be as thin as possible, have a high yield stress and low elastic modulus. Such a σ_Y/E index minimizes any plastic deformation by impact and maximizes elastic deformation. It therefore follows that components fabricated by forging or rolling will be superior to cast because of their finer microstructure coupled with finer precipitates, resulting in high yield strength. Welding of the high-strength alloys may require post-weld heat treatment.

The club heads of 'irons' are also undergoing development. Commonly made of stainless steels, 17Cr–14Ni, AISI Types 431 and 304 by either forging or investment casting, the traditional blade is hollowed out at the back to produce 'cavity-back' iron heads. This produces peripheral weighting, which is more 'forgiving' for off-center shots. Putters have less dynamic demands than woods and irons, and most design features are for weight distribution, balance and sweet spot. The traditional flat-faced, hollow-backed 'ping putter' made with cast manganese bronze (Cu–Zn–Mn) is still fashionable but more putters are now large flat-bottomed made of stainless steel.

With modern materials golf balls have developed into a three-layer composite. The inner core is generally made of solid or liquid filled rubber which is covered with a polybutediene layer. The outer coating is made of an ionomer (surlyn) or polyurethane and is dimpled. These dimples (~400) are either round or hexagonal in shape and cover more than three-quarters of the ball surface. The dimples alter the aerodynamic drag on the ball by producing a boundary layer on the ball which is turbulent at a lower velocity than on a smooth ball. This leads to a reduced pressure drag which remains fairly constant with increased speed.

12.3.3 Tennis equipment

Tennis racket frames have, over the years, developed from wood through aluminum to graphite composite materials. Frames of wood and aluminum are now out of favor and are almost exclusively graphite. Graphite composite frames are stronger and stiffer than other frames and can be manufactured with a larger head. Impact and vibration decrease with frame stiffness, favoring graphite frames and, because they are lighter, may be swung somewhat faster. Almost all the top graphite rackets are manufactured from thermoset carbon fiber composites with extra glass fiber reinforcement and the racket tip for high impact resistance. Basic carbon pre-preg sheets are cut at different angles and widths to produce layers (Figure 12.11). In the lay-up, zero degree pre-preg is used for bending stiffness and 45° pre-preg for torsional stiffness. The tube for the racket frame is formed by folding around a polyamide thermoplastic nylon 66 foil tube. The throat piece is made by wrapping pre-preg around expandable foam. All the pieces, frame, pre-preg layer handle and throat, are then assembled inside the mold, closed and given the appropriate temperature-curing cycle. Finishing includes removing excess resin, sanding, drilling the grommet holes for the strings, lacquering and hardening.

Natural gut from the intestines of cows is probably the top choice of professionals for strings, but is less readily available and extremely expensive. Strings made from nylon are by far the most used, but Kevlar and polyester are also used. The strings must be resistant to UV radiation, abrasion, moisture, creep and any chemical agents such as sweat, oils, etc. However, the major requirement is to reduce the impact force on the hand and arm, and provide ball control and impact power. Strings are typically 1.3 mm in diameter, made from a braid composed of plaited and intertwined bundles of strong fibers. After immersion in an elastomeric solution, the string is given a polymeric coating to protect it from abrasion and wear. They are strung at a tension typically of 250–300 N, when high tension leads to improved ball control and lower tension to increased power. Being viscoelastic, the strings lose tension with time from stringing, which has to be allowed for in manufacture.

Tennis balls intended for fast courts such as grass consist of a hollow rubber compound core covered by a felt fabric. The core is made by gluing segments together with an adhesive, which is subsequently cured. The ball is slightly pressurized with a gas which gradually leaks after removal from the pressurized container. Thus, the balls lose bounce and have to be replaced after a number of games. Pressureless balls having a thicker core are used by amateurs and weekend players. All balls must conform to standards of size and bounce. Tennis balls have a relatively high drag coefficient, as a result of the 'fuzz' on the felt covering, which decreases as the fuzz wears with games played.

12.3.4 Bicycles

Almost everybody has grown up having owned a bicycle. They are probably the most energy-efficient means of travel and have rarely failed us. Nevertheless, the stress distribution in the frame is both in-plane and out-of-plane and, being quite complex, can lead to failure. For most bikes the failure is likely to be a fatigue type, originating at the usual places of high stress. Low-carbon steel is the most common material for bike frames, being cheap, tough, weldable and easy to paint, but the relative merits of other materials are shown in Table 12.3 based on simple deflecting beam

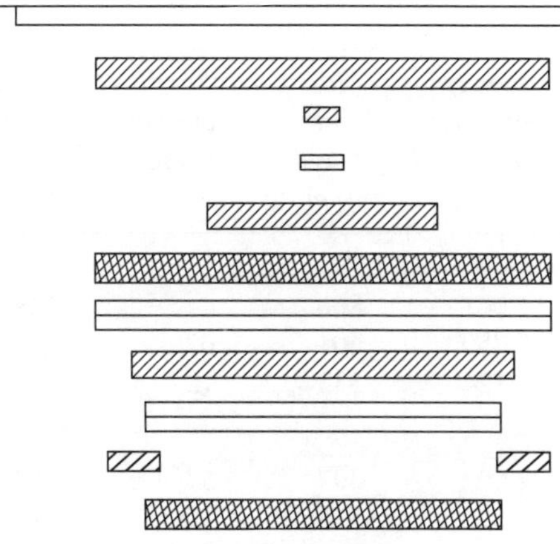

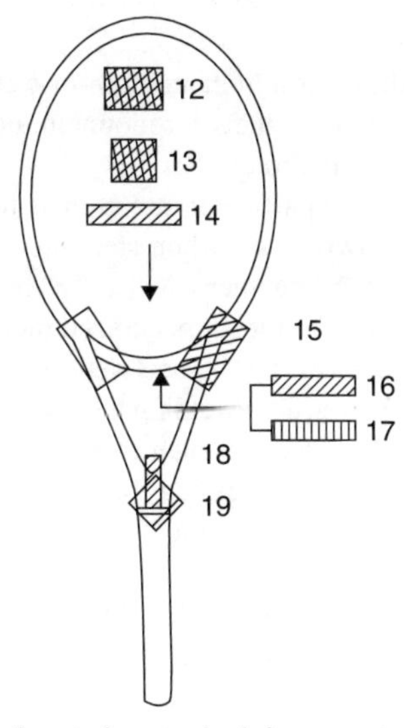

Pre-preg layers	Parts	Material	Fiber angle
The main tube	1	PA foil	Direction
	2	Glass	±20°
	3	Carbon	±30°
	4	Carbon	0
	5	Glass	0
	6	Carbon	±30°
	7	Carbon	0
	8	Carbon	±30°
	9	Carbon	0
	10	Glass	±30°
	11	Carbon	±30°
The throat piece and reinforcements	12	Carbon	±30°
	13	Carbon	±30°
	14	Carbon	±30°
	15	Glass	±20°
	16	Carbon	±30°
	17	Carbon	90°
	18	Glass	±30°
	19	Glass	±30°

Figure 12.11 *A typical lay-up for a composite racket (Jenkins, 2003, courtesy of CRC Press).*

equations.[1] It is desirable to lower the tensile stress σ_{max} in the convex surface of the tube and maximize the radius of bending r. Reducing the tube wall thickness $(D-d)$ increases the surface stress σ_{max} but cold drawing to reduce thickness increases the strength and reduces weight. In many bikes the front wheel forks, which experience severe bending stresses, are elliptical in shape, giving a higher moment of inertia but with the major ellipse diameter able to reduce the stress in the appropriate

[1] $M/I=\sigma/y=E/r$, where M = bending moment, I = moment of inertia of beam section, σ = stress at a point, y = distance of point from neutral axis of beam, E = modulus of elasticity and r = the radius of curvature of the beam shape.

Table 12.3 *Comparison of weight and bending characteristics of four metallic frame materials: r and σ_{max} calculated for tubes subjected to a bending moment of 100 N m (from McMahon and Graham, 1992).*

Material	D *(mm)*	d *(mm)*	*Moment of inertia I (cm^4)*	*Mass per unit length (g m^{-1})*	*Radius of curvature r(mm)*	*Maximum stress σ_{max} (MN m^{-2})*	σ_{max}/σ_y
Racing cycles							
C steel	28.70	26.16	1.041	860		138	0.58
Cr–Mo steel	28.78	26.93	0.750	601	67	190	0.39
Al alloy (6061-T6)	28.80	25.91	1.124	332	127	128	0.50
Mountain cycles (top tube)							
Al alloy (6061-T6)	38.10	35.56	2.50	397	57	76	0.30
Ditto	34.93	30.81	2.87	575	49	61	0.24
Ti–3Al–2.5V	31.75	29.47	1.41	549	64	45	0.18

plane. Of the alloys listed, the titanium alloy has a high strength-to-weight ratio and with excellent corrosion resistance is now competing with alloy steels for mountain bikes, which have to be tough enough to withstand sudden impact shock on the rough terrain.

Joining the tubes to form the frame requires particular attention as the joints often coincide with the highest bending moments. The cold-drawn, low-carbon steel tubes of mass-produced standard bicycles are joined by brazing. Reinforcing lugs together with ($\alpha + \beta$) brass brazing alloy are inserted in the tube ends and heated to around 875°C. For more specialized racing frames butted alloy tubes are fillet brazed with a silver brazing alloy with an oxyacetylene torch at a lower temperature to reduce any heat-affected zones (HAZ). However, TIG welding has tended to replace brazing for alloy steel frames. The hardenability of the Cr–Mo steels is such that a strong mixture of alloy carbides form in the weld fillet on cooling after welding, increasing the hardness in the HAZ. The resultant structure has a better fatigue resistance than other steels and its favorable strength/weight ratio makes it competitive with the titanium alloy. When joining the titanium alloy tubes the weld pool has to be shrouded with flowing argon to prevent the pick-up of embrittling gases, e.g. oxygen, nitrogen, hydrogen. Similarly, aluminum alloy tubes have to be guarded against overageing and softening. Carbon fiber-reinforced polymers may also be used for high-performance bicycles. The monocoque (single shell) frame, shown in Figure 12.12, is one such construction and features a cantilever seat, a disk rear wheel and three-spoke front wheel, all made from CFRP.[2] Joining problems may be overcome by the use of epoxy adhesives, which are well established in the aircraft industry. These structural adhesives produce bonds which are strong and durable, damp vibrations, save weight and reduce assembly costs. Normally, thermoplastic and/or elastomeric constituents are added to the thermosetting component to improve the toughness. The CFRP wheels clearly have a much lower aerodynamic drag than multi-spoke wheels. They are basically a contoured sandwich structure consisting of two carbon/epoxy facings, possibly with a polystyrene foam core, bonded onto a light aluminum alloy hub and to a carbon/epoxy rim. With the more conventional wheel the spokes are pre-tensioned (~400 MN m^{-2}) plain carbon steel, sometimes coated with sacrificial zinc for corrosion protection. The fatigue endurance of the carbon steel is acceptable for the cyclic stress conditions. The

[2] Chris Boardman won the 400 m individual pursuit at the 1992 Olympic Games in Barcelona with a Lotus prototype.

Figure 12.12 *High-performance Zipp bicycle with monocoque frame (courtesy of Julian Ormandy, School of Metallurgy and Materials, University of Birmingham, UK).*

rim is usually made of extruded precipitation-hardened aluminum alloy, bent to shape and joined. With the spokes, the aluminum (6001-T6) is sufficiently strong and light with good corrosion resistance.

12.3.5 Skiing materials

Skis that were originally made from a single piece of hickory wood have now developed into a rather complex multi-component, multi-material piece of equipment. The main requirements of the ski are good strength and flexibility along the length with torsional stiffness, so that the skier's weight is properly distributed while traversing the irregular snow contours. In addition, it is necessary to dampen the ski structure to absorb dynamic impact loading. To meet these demands a multi-layer structure has evolved consisting of a base, usually polyurethane, a shock-absorbing core, usually natural ash or hickory wood, fiberglass and an elastomeric secondary core, steel or high-strength aluminum alloy edges, side walls of glass materials, top layer and reinforcing damping layers. An early design is shown in Figure 12.13, but this has evolved with the exact design and processing commercially guarded in a very competitive industry. In parallel with material development, modern skis have been reduced in length from around 200 cm in 1990 to around 160 cm.

The ski is only part of the equipment necessary for skiing. Boots, bindings and ski poles are also needed. Ski boots must provide a firm grip on the skier's ankles to allow proper control of the ski. External Hytrel–Kevlar components have been used to stiffen the ski boot for this purpose (Hytrel is a thermoplastic polyester elastomer). Binding, i.e. the attachment between boot and ski, has also developed significantly, with sophisticated release mechanisms. These release instantly under too severe twisting, while providing foot stability and control. An acetal (highly crystalline) polymer, which is tough, has a low glass transition temperature T_g and good fatigue resistance, is used for the locking bar, heel release lever, heel block and front swivel plate, while glass-reinforced nylon is used for the front block and two base plates. These polymers are tough at low temperatures and both UV and moisture resistant. Other release bindings employ titanium and plastic components. Figure 12.14 shows the variety of different materials used in the binding for snowboards. Finally, ski poles are tubular with high specific stiffness and toughness. They are usually made of CFRP or CFRP–GRP hybrids.

12.3.6 Archery

Every schoolboy having read, or watched, Robin Hood is familiar with the longbow, probably knows it was made from yew and that considerable force (35–70 kgf) was needed to draw it. When drawn,

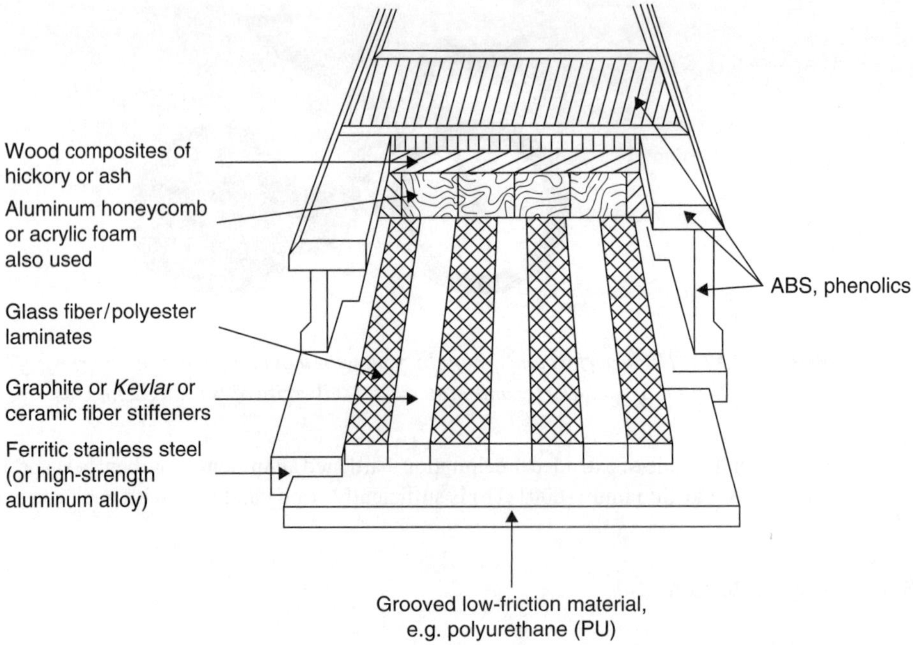

Figure 12.13 *Transverse section showing multi-component structure of a downhill ski (from Easterling, 1990, by permission of the Institute of Materials, Minerals and Mining).*

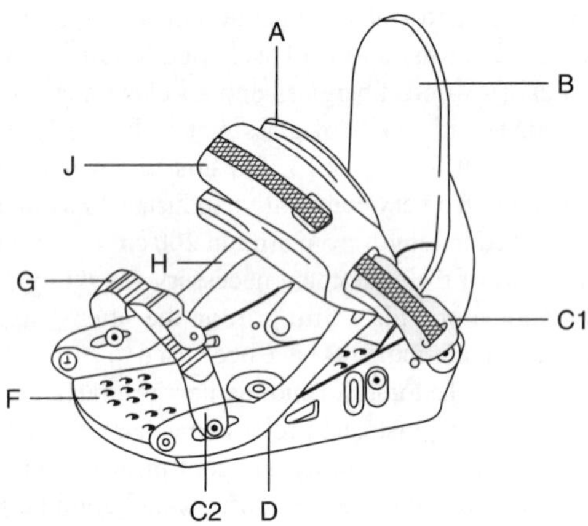

Figure 12.14 *Snowboard binding utilizing: thermoplastic elastomer (Hytrel) – ankle strap A, spoiler B, ratchet strap G; nylon (Zytrel) – side frames D and H, base and disk F, top frame J; acetal homopolar (Delrin) – strap buckles C1 and C2 (courtesy of Fritschi Swiss Bindings AG and Du Pont UK Ltd).*

the strain energy stored in the 'string' and the two (upper and lower) limbs of the bow is transferred to the arrow, which accelerates up to a velocity of $50\,m\,s^{-1}$. Yew was favored for longbows, from the various woods around, because of its excellent bending strength and capacity to store energy.

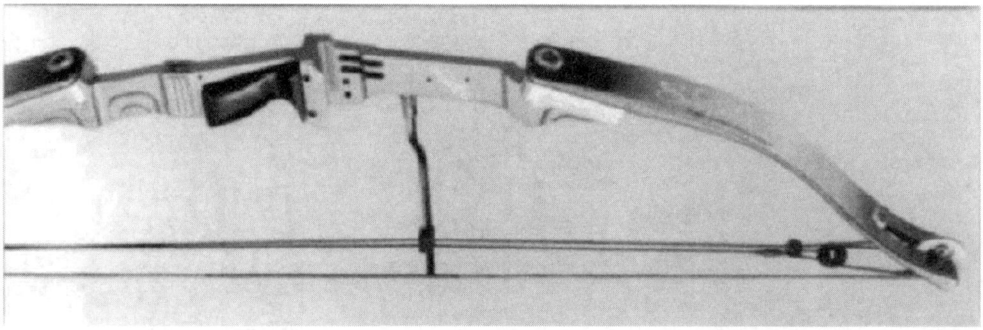

Figure 12.15 *Modern competition bow, compound type: laminated upper limb (wood, GFRP, CFRP) and CNC-machined central riser (Al–Mg–Si alloy 6082) (courtesy of Merlin Bows, Loughborough, UK).*

Modern bows are either the standard recurve (Olympic) bow or the multi-material compound bow. The recurve bow has limb sections which are wide, flat and thin, and has a particular resistance to any twisting during use. The compound bow is shown in Figure 12.15. The mid-section (riser or grip) is usually made of a light, strong forged aluminum alloy, while laminated wood, GFRP or CFRP is used for the upper and lower limbs. The release of an arrow from the bow involves a combination of three arrow characteristics, namely length, mass and stiffness. Correct matching of these features is necessary for the arrow to clear the bow properly. Arrow shafts are usually made of tubular, drawn and anodized aluminum (7079-T9, 7178-T9) or similar alloys bonded to a smooth outer wrap of CFRP. Barreling from the middle to the end of the shaft removes any undesirable flexing introduced by the variable bending moment of the constant-diameter shafts. Finally, the feathers or fletching are nowadays made from polyethylene terephthalate (Mylar) and provide the arrows with stability during flight. In smooth polymeric form they are durable and weather resistant, with low aerodynamic drag. The same material in stranded form (Dacron) is used for the bow string.

12.3.7 Fencing foils

It is evident that sports materials are very much influenced by the developments of the newer composites, based on polymers. However, one area where old-fashioned metallurgy comes into its own is the sport of fencing, at which the UK has often excelled. Traditionally, the foils are made from medium-carbon alloy spring steels, hot worked, then oil quenched and tempered to develop a martensitic structure with a yield strength around 1600 MN m^{-2}. The foil is just less than a meter long, tapering to a rectangular section of 4 mm × 3 mm. With this type of spring steel the foil is able to bend quite easily when parried during a match. Unfortunately, continual striking by an opponent's foil leads to surface notches, which can result in fatigue failure, producing a rather dangerous broken blade. Nowadays, highly alloyed maraging steels (see Chapter 8) are used for competition fencing. These very-low-carbon steels were developed for their exceptional high strength and toughness, and are based on Fe–18% Ni with additions of 5% Mo and 8% Co. They are solution treated at 815°C, air cooled to produce low-carbon martensite and aged at 485°C to produce precipitates of intermetallic compounds. These steels are expensive but well established and reliable, so there is little incentive for development in metal–matrix composites.

The fencing mask is one area where polymer materials outperform metals. Mask design has been revolutionized by the application of the polycarbonate thermoplastic (Lexan). It is transparent and has high impact strength, resisting local deformation under conditions where metal masks deform.

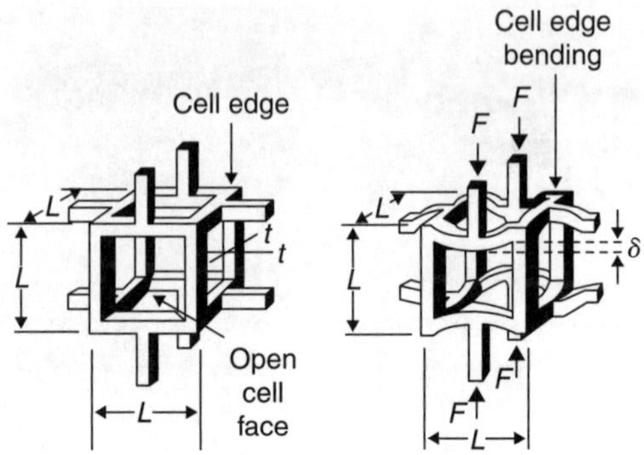

Figure 12.16 *Deformation of an open-cell foam structure (from M. F. Ashby, 2005).*

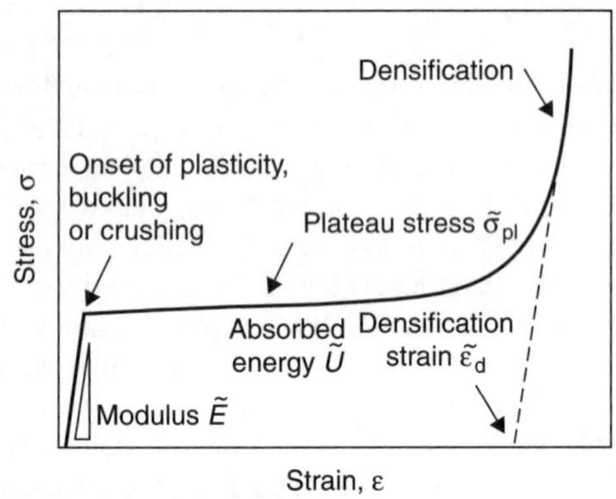

Figure 12.17 *Compressive stress–strain behavior of foam materials (from M. F. Ashby, 2005).*

12.3.8 Sports protection

The development of the science and manufacture of polymeric foams has revolutionized the field of protection equipment for sport. Cycle and motor cycle crash helmets, body protection, football shin pads, etc. have all quickly improved. Most thermoplastic and thermosetting resins can be formed as foams with physical or chemical blowing agents. The mechanical behavior of the foam depends on the basic polymer properties, the relative density of the foam (e.g. 0.1 means the cell walls occupy 10% of the total volume) and the shape of the cell. Foams are, however, very much bending-dominated structures, as shown in Figure 12.16. Under a compressive stress the foam is linear elastic, as shown in Figure 12.17, up to the elastic limit, after which the cell walls start to buckle and collapse at almost constant stress until the cell walls impinge on one another, when the stress rises rapidly. The 'constant' stress collapse stage is important for energy absorption and is given by $U \approx \overline{\sigma}\,\overline{\varepsilon}_d$, where $\overline{\sigma}$ is the plateau stress (yield, buckling or fracture stress of the foam) and $\overline{\varepsilon}_d$ is the densification strain. Some foams collapse by elastic buckling, while brittle foams collapse by fracture of the cell walls. The densification strain is important since the energy absorption behavior has effectively 'bottomed out'. This situation is undesirable for protective gear and indicates an inadequate thickness of foam protection.

For football shin and ankle guards the foam is covered with a plastic shell and energy is absorbed by compression of the foam between the shell and the leg. The currently used foam is ethylene vinyl

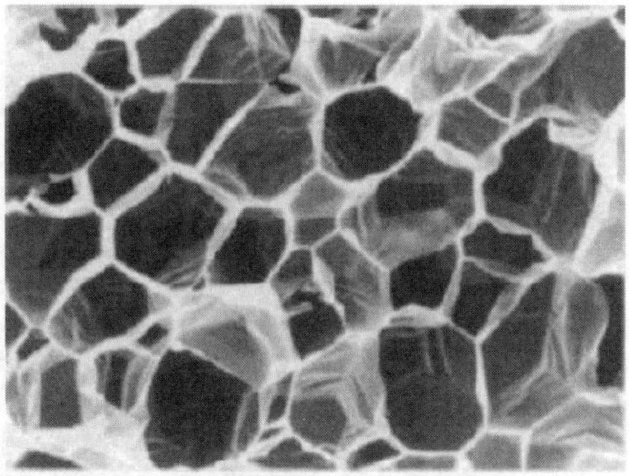

Figure 12.18 *Cell structure of polystyrene foam, as used for shock-absorbent packaging: average cell diameter 100 μm (courtesy of Chris Hardy, Department of Metallurgy and Materials, University of Birmingham, UK).*

acetate (EVA) or ethylene–styrene interpolymers (ESI). The shell is usually polycarbonate (PC), since LDPE tends to buckle, causing the foam to 'bottom out'. For crash helmets the outer shell is made of a strong, durable polymer, e.g. ABS or GRP, which has a reasonably high fracture toughness to resist shattering on impact. Some newer helmet designs have an additional inner microshell of PVC shaped to the inner foam. Glassy polymers having a high bulk compressive yield stress are used for the foam liners with glassy polypropylene (EPP) or glassy polystyrene (EPS) liners most favored. Figure 12.18 shows a micrograph of a polystyrene foam. Some liners can recover after not too violent impact, while others undergo permanent structural damage. EPP, for example is less brittle than EPS and recovers better. Whether the helmet should be discarded after an impact depends on the nature of the impact. In some sports the player is more liable to experience a number of minor impacts, e.g. cricket helmets, while for cycling the impact is likely to be more serious, when the helmet should be replaced.

12.4 Materials for nanotechnology

12.4.1 Introduction

Based on the success of microtechnology (integrated circuits, etc.) since the 1970s, scientists began in the 1980s to enquire about the possibility of developing technologies in the next level down, at the nanometer length scale. The outcome was nanotechnology, involving materials and structures of dimensions in the range of about 1–100 nm. Apart from a desire to further miniaturize from microtechnology, a strong motivation for nanotechnology is to exploit the various new physical phenomena and novel properties of materials when their sizes shrink to below ~1 micron. The development of nanotechnology has been greatly facilitated by continuous breakthrough along two major fronts beginning from the 1980s, namely (i) the invention of novel microscopy techniques such as STM and AFM, that enable the observation, characterization, as well as manipulation of nanometer-sized materials and structures, and (ii) the development of new synthetic techniques that allow nanometer-sized materials and structures to be made. The microscopy aspects have already been dealt with in Chapter 4, and in this section the focus is on the materials aspects.

Nanotechnology is based on building blocks including individual atoms, groups of atoms or molecules, ultra-small solid shapes or bulk solids that are nanostructured. Examples include

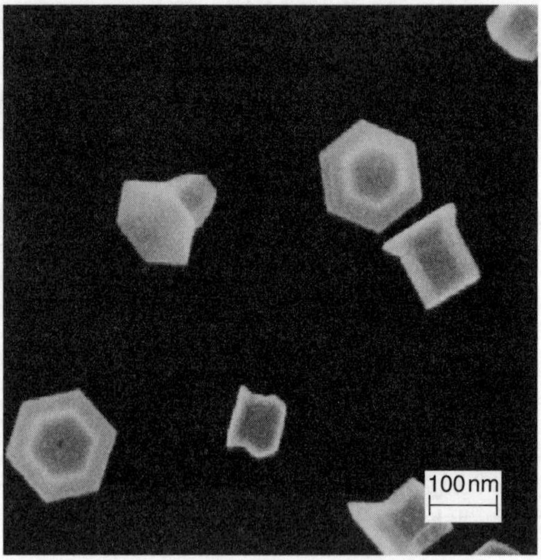

Figure 12.19 *Metallic zinc nanocrystals produced by thermal evaporation. The crystals adopt overall hexagonal shapes based on the hcp crystal structure of zinc, an indication of the strong anisotropy of surface energy (Ng, Muley, Ngan and co-workers, 2006).*

fullerenes, nanotubes and nanowires, colloids, confined atomic clusters such as nanometer-sized islands on supporting substrates, nanocrystalline solids, etc. To discover new nanomaterials and to develop methods to make them is an ongoing, intensive research field, and so a complete account is not possible at this stage. However, many common nano-sized materials are fabricated using either the 'top-down' or 'bottom-up' approach. A top-down method is one that evolves from a microfabrication technique such as lithography, nanoimprinting, etc. Microlithography methods (Figure 12.19) make use of radiation and so the manufactured feature size is limited by the wavelength of the radiation. However, by exploiting the interference of light (laser), for example, sub-wavelength periodic features can be produced. The bottom-up approach relies primarily on a material's ability to self-assemble to form nano-sized shapes. A common self-assembly technique is known as heteroepitaxy, in which a semiconductor thin film deposited on a slightly mismatching substrate relaxes to form a periodic array of islands upon annealing. While this technique is material dependent, a certain degree of control is possible through the choice of the substrate, controlling the temperature and/or straining the substrate during the self-assembly process.

12.4.2 Nanoparticles

Nanoparticles from a few to a few dozen nanometers are an important form of materials for nanotechnology. Sufficiently small particles may exhibit the 'quantum confinement' effect in which the electronic states and the associated properties of the particle are altered relative to the bulk situation (see Section 12.4.4). In addition to this, nanometer-sized particles often behave differently from their bulk counterparts as a result of their much higher fraction of surface atoms present, since many physical and chemical properties of matter are due to the surface atoms. A currently much exploited nanotechnology is based on the photocatalytic action of TiO_2, which produces disinfection and deodorization effects. Under the action of sunlight, TiO_2 acts as a catalyst for the decomposition of water molecules in air into hydroxyl radicals and superoxide anions. The latter two are

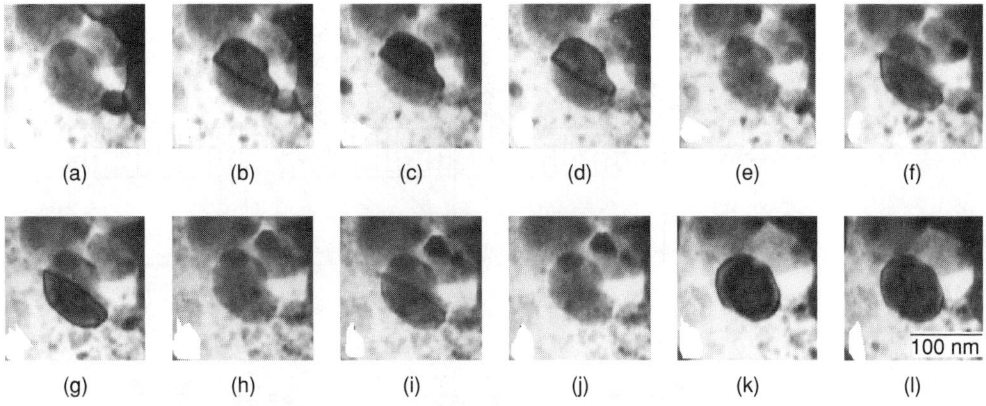

Figure 12.20 *Sequence of grain coalescence recorded from an Ni–25 at.% Al sputter-deposited film with low density of nanocrystals, during* in situ *TEM annealing at 800°C (Ng and Ngan, 2002, courtesy of Materials Research Society).*

strong oxidation reagents and can in turn decompose toxic organic substances through oxidation, thus achieving disinfection effects. Although such a photocatalytic property of TiO_2 is well known in the past, using the TiO_2 in nanoparticle form greatly improves the catalytic efficiency, since catalysis is a surface reaction effect. In therapeutic applications, silver nanoparticles (of the order of 10 nm in size) have also been found to be effective in wound healing, exhibit anti-HIV activities and are effective against proliferation of different types of bacteria, including the deadly *E. coli*.

There are numerous methods, both chemical and physical, for producing nanoparticles in different shapes and sizes. Metallic nanoparticles can be routinely prepared by physical vapor deposition methods, such as thermal evaporation, magnetron sputtering, etc. Figure 12.19 shows metallic Zn nanocrystals prepared by thermal evaporation. The shapes of the crystals in this case are heavily affected by the crystallographic anisotropy of the surface energy, again an indication of the dominating effects of surface atoms. Because of the energetic surface atoms, nanocrystals are less stable and exhibit fluctuations even at relatively low temperatures. Figure 12.20 shows a series of snapshots of the coalescence process between two nano-sized Ni–25 at.% Al grains recorded during *in situ* heating in the TEM at 800°C. The change in the bright-field contrast of the two participating grains from (b)–(d) to (f)–(j) indicates thermal fluctuations of the granular orientations before final coalescence in (k)–(l). Nano-silver particles can be prepared by a chemical route such as dissolving $AgNO_3$ in HEPES buffer. Oxides (e.g. different forms of ZnO) can be produced rather easily by oxidation of the metal in controlled atmospheres, or through a chemical route such as a precipitation reaction (see Chapter 10).

12.4.3 Fullerenes and nanotubes

The various forms of fullerene-based molecules and carbon nanotubes (CNTs) have been introduced in Section 10.5.3 as a recently discovered form of carbon. Fullerenes and in particular the CNTs have attracted a lot of attention as an important class of building blocks for nanotechnology, because their properties can be engineered via chemical or physical manipulation. The C_{60} molecules can be made to bond to one another to form an fcc crystal, with each lattice site occupied by a C_{60}. Such a crystalline state is an insulator, but upon doping with alkaki atoms such as potassium it becomes an electrical conductor. The electronic properties of CNTs depend on their diameter as well

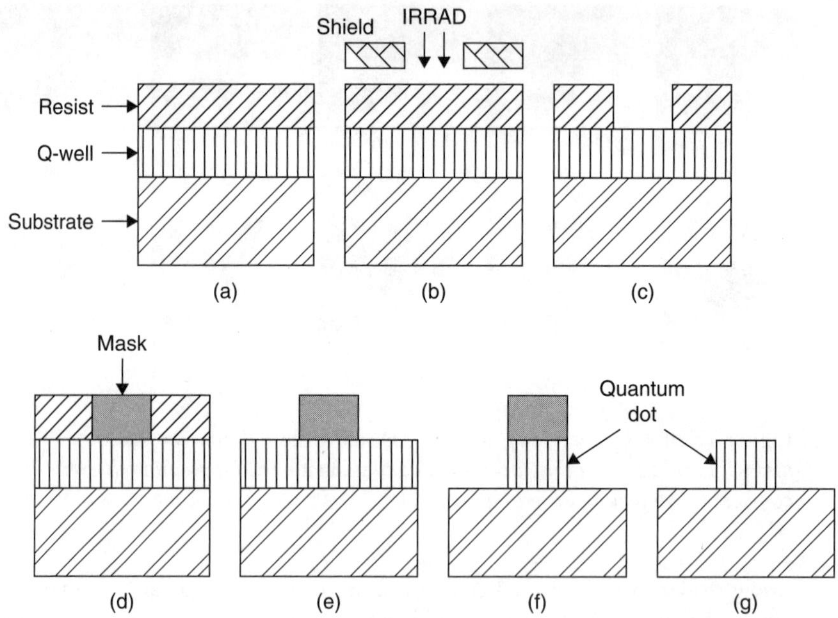

Figure 12.21 *Steps for making a quantum dot using lithography (Poole and Owens, 2003, courtesy of Wiley-Interscience).*

as their chirality. Armchair CNTs tend to be metallic, and for a given chirality, smaller tubes tend to be semiconducting, while larger tubes tend to be metallic. CNTs are also known to exhibit field-emission effects, when a small electric field applied along the tube axis causes an intensive emission of electrons from the end of the tube. This enables CNTs to be considered for applications such as electron sources for flat-panel displays. Bridging two metallic electrodes by a CNT also forms a nano-scale field-effect transistor, in which a small change in voltage applied across the electrodes can cause a huge change in the current through the CNT by several orders of magnitude. Apart from finding potential applications in future high-speed computers, the I–V characteristics of a CNT field-effect transistor are also strongly affected by the gaseous environment in which the transistor is situated, allowing potential applications as gas sensors. CNTs are also potential lithium- and hydrogen-storage materials for fuel cell applications. CNTs are also highly flexible – when bent, they are able to recover their original shape without permanent changes. They are ultra strong and stiff, with breakage stress exceeding tens of GPa and Young's modulus of the order of TPa. CNTs are therefore being explored as strengthening fibers in composites. The French Babolat tennis rackets, made of CNT-reinforced composites, are said to be several times stiffer than current carbon–composite rackets. The mechanical properties of small material volumes are further discussed in Section 12.4.6.

12.4.4 Quantum wells, wires and dots

These are usually semiconductors (e.g. GaAs) deposited on substrates, in the form of thin layers (wells), lines (wires) or small islands (dots). These structures are typically produced by lithography or by self-assembly on a substrate. In the lithography method, shown in Figure 12.21, a photo-resist is first deposited on a quantum well structure. Irradiation (optical light, UV, electrons or even neutrons) is then shone onto the region of the resist where the nanostructure (e.g. a quantum dot) is to be made, either through a mask or shield, as shown in Figure 12.21a, or, in the case of electron-beam

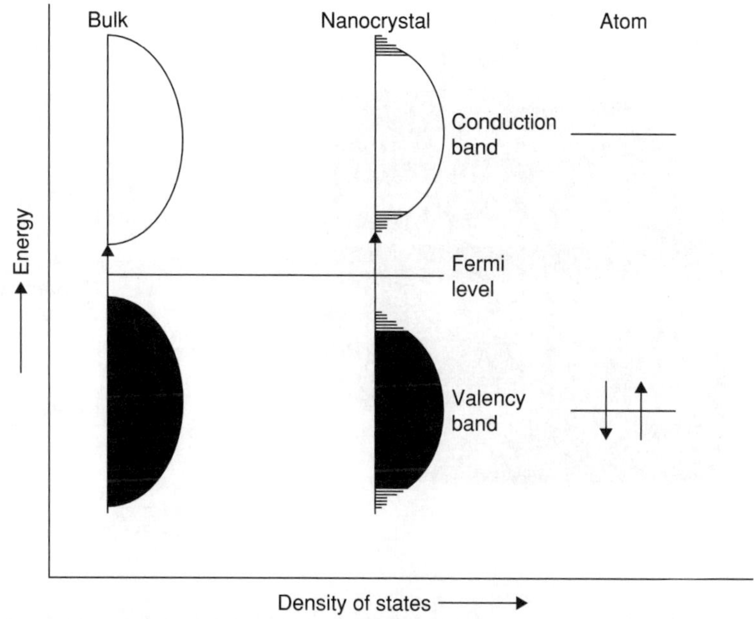

Figure 12.22 *Schematic showing the dependence of band gap on size of a semiconductor.*

lithography, by directing the electron beam accordingly. After washing away the exposed region of the resist with a developer, an etching mask is deposited back on the feature region. The remaining photo-resist not covered by the etching mask is then removed to form the desired pattern.

Unlike free electrons in a metal, which are unconfined, the electrons in quantum wells, wires and dots are spatially confined. As an example, the discrete energy states of an infinite potential well of width L are given by $E = n^2h^2/(8\,mL^2)$ (see worked example in Section 1.5) and, likewise, the states in other confined structures such as finite wells, wires and dots are confined and are different from those in the bulk condition. Physical properties that depend on the density of states of electrons, such as the specific heat capacity, magnetic susceptibility, etc. are also expected to be different from bulk behavior. In semiconductors a measure of the confinement effect is the radius a_e of an exciton. An exciton is a bound state of an electron–hole pair, and its radius a_e measures the effective range of Coulomb-type interaction between the electron and the hole. For typical semiconductors such as GaAs, a_e is of the order of 10 nm, and so a particle with size comparable to a_e will exhibit strong quantum confinement effects. In strongly confined situations, the band gap (Section 5.7.2) of the semiconductor will increase relative to the bulk condition (Figure 12.22). The band gap of a bulk semiconductor must be smaller than the difference between the occupied and unoccupied energy levels in the free atomic state (see also Figure 1.4), and since a nanocrystal can be thought of as an intermediate case between the bulk form and the single atom state, its band gap is higher than that of the bulk state.

Semiconductors are important light-emitting materials and one important development focus is to fine-tune their band gaps to cover the entire visible wavelength spectrum, i.e. the so-called 'band-gap engineering'. Visible light has a wavelength range of 375–740 nm, corresponding to a photon energy range of 1.68–3.32 eV. Pure GaN, for example, has a direct band gap of about 3.4 eV, which falls in the ultraviolet range, but alloying with InN and AlN allows the band gap to be tunable between 1.9 and 6.2 eV, which now overlaps with the optical range. An alternative method of band-gap engineering is to make use of the quantum confinement effect shown in Figure 12.22, in which the increase of the band gap as the material size decreases causes 'blue shift' of the emitting light. In practice, a mixture of the alloying and quantum confinement techniques is used in band-gap engineering.

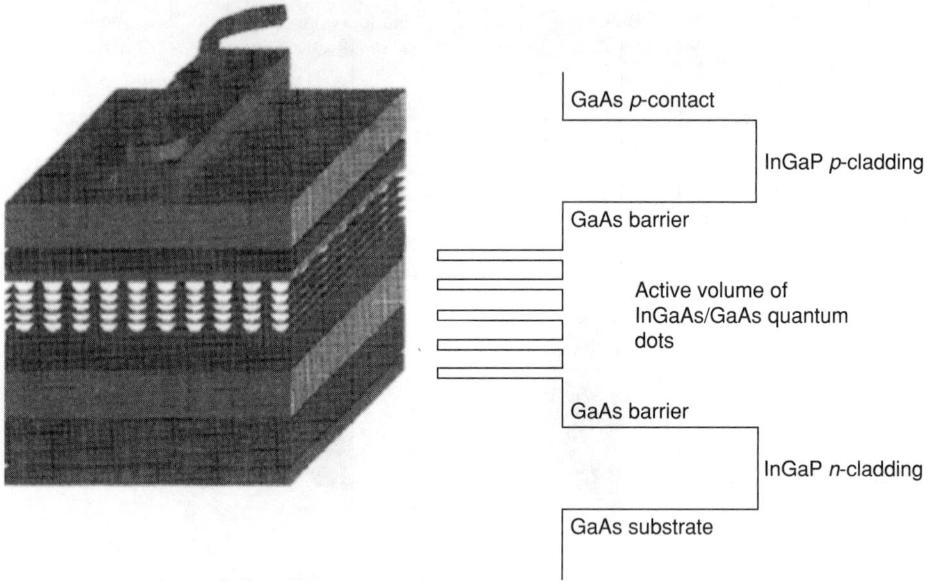

Figure 12.23 *Schematic of a quantum-dot laser (from Center for Quantum Devices, Northwestern University).*

An example being intensively investigated is the Zn_xCd_{1-x} Se alloy system. Nanoparticles with a core–shell structure, in which a core of higher band-gap ZnSe (2.7 eV) is surrounded by a shell of lower band-gap CdSe (1.75 eV), is able to emit in the range between red and blue for x between 0.1 and 0.5.

The discrete nature of the density of states and the widened band gap in a size-confined semiconductor have other applications, including infrared detectors and lasers. As an infrared detector, the incoming radiation excites electrons in the lower bound states to higher energy conduction states. Once in the conduction states, the electrons can conduct electricity and so the power of the radiation is measurable in terms of the resultant current. As discussed in Section 5.10.3, a conventional laser works by virtue of the existence of (i) discrete energy levels, the difference of which corresponds to the laser emission, and (ii) a 'population inversion' mechanism by which more electrons can reside at the upper energy level than the lower one so that, upon triggering, they fall down simultaneously to give a coherent emission of radiation. In a quantum-dot laser source, the discrete energy levels in the quantum dots provide these two requirements. Figure 12.23 shows the layout of a quantum-dot laser, in which the active volume consists of multiple layers of InGaAs/GaAs quantum dots.

12.4.5 Bulk nanostructured solids

Another important class of nanomaterials includes solids that have nano-scale microstructures. Examples include nanocrystalline solids, and tailor-made structures such as photonic crystals. Nanocrystalline solids are simply polycrystalline materials with nanometer-sized grains. They can be made by powder metallurgy techniques and thin films can also be made using physical deposition techniques such as magnetron sputtering, but low-porosity nanocrystalline metals with grain sizes approaching the nm limit are more commonly made using electrodeposition. Nanocrystalline solids have attracted considerable interest in the past because they exhibit unusual mechanical properties when compared to ordinary polycrystalline materials (see Section 12.4.6 for details). In the emerging field of photonics, light, as opposed to electrons in electronics, is used to transmit signals through

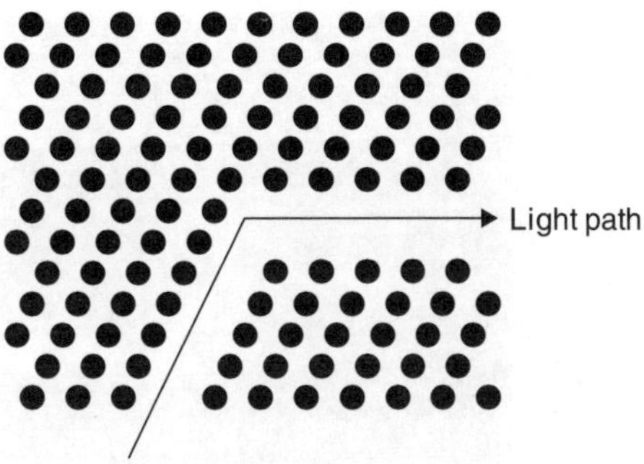

(a)

(b)

Figure 12.24 *(a) Schematic illustrating a defected photonic crystal being used as a waveguide bend. (b) A proposed 3-D photonic crystal (Povinelli and co-workers, 2001, by permission of American Physical Society).*

waveguides. Just as electrons are scattered by crystals and exhibit phenomena such as Bragg reflection governed by dispersion surfaces, similar properties can be exploited for monochromatic light, but then the 'crystal' concerned must have a periodicity comparable to the wavelength of light. A photonic crystal is therefore a tailor-made 2-D or 3-D periodic structure of dielectric material, with a lattice periodicity comparable to that of the optical wavelength. A simple application is a waveguide bend, shown in Figure 12.24a, which is a 2-D lattice of particles of a dielectric material (silica, alumina, etc.), but with missing particles along a bent path. The spacing of the particles is such that strong Bragg reflection is experienced by the incident light, so that the latter is prevented from entering the lattice. The light is thus forced to travel along the curved path of missing particles. More complicated 3-D structures have also been proposed (Figure 12.24b) which exhibit interesting and useful dispersion properties.

12.4.6 Mechanical properties of small material volumes

The strength of submicron-sized metals and ceramics is considerably higher than their bulk forms. In Chapter 10, we have seen that the fracture strength of brittle glasses and ceramics is strongly size

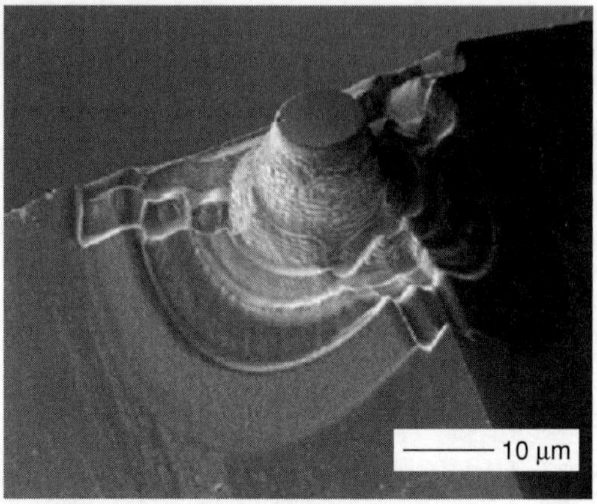

Figure 12.25 *Flat-end diamond punch used for microcompression, manufactured by focused-ion milling.*

dependent, due to the decreasing chance for the material to contain large flaws as its size reduces. Thus, glass and carbon fibers are orders of magnitude stronger than their bulk counterparts, and they are highly flexible so they are effective strengtheners and stiffeners in plastic-based composites (Section 11.10). Compared to fibers, nanotubes and nanowires are even stronger and more flexible, although accurate evaluation of fracture stresses in these materials is difficult. Carbon nanotubes are known to have high fracture strains of 10–30% and exceptionally high Young's moduli of the order of TPa.

For metals, it was discussed in Section 6.4.5 that sub-millimeter-sized whiskers may exhibit sharp yield points even when bulk-sized samples of the same material do not (e.g. Cu). The occurrence of the sharp yield point in this case is due to the necessity to generate dislocations from the perfect lattice, and the upper yield point (Figure 6.24) may approach the theoretical yield strength of $\sim E/10$. Similar observations were also made in compression experiments on micron- to submicron-sized metallic crystals using a flat-end diamond punch (Figure 12.25) in a nanoindenter (Section 4.6.4). Figure 12.26 shows that as the column size falls into the submicron regime in the case of Ni_3Al, the elastic-to-plastic transition becomes very abrupt. The GPa level yield stress here is an order of magnitude higher than the yield stress of the bulk sample, which does not exhibit a sharp yield phenomenon.

While the yield stress of a submicron-sized crystal may be as high as several GPa, its thermal stability may not be great. Under a GPa-level elastic stress, although the perfect crystal environment is momentarily maintained, the atomic planes sandwiching a potential slip plane may be sheared very close to the instability point (Figure 3.10). Such a small margin to instability can easily be overcome by thermal agitation. Figure 12.27a shows results from a nanoindentation experiment on a well-annealed aluminum crystal at room temperature. The stressed region underneath the indenter was of the order of 100 nm across and the average initial dislocation spacing in the annealed state ($>>10\,\mu m$) was such that no dislocation could be expected to be present inside the stressed volume. In the first loading cycle, no plasticity was created during the $\sim$600 s holding at 120 μN, and the stressed volume recovered elastically upon load removal. However, during the next cycle at 200 μN, the stressed volume survived elastically only for about 450 s, after which plasticity occurred followed by creep. The delay of $\sim$450 s in this case is thought to be due to the occurrence of a big enough lattice wave to trigger instability at some point within the stressed volume. Such a waiting time in general increases as the holding load decreases, as shown in Figure 12.27b in the case of Ni_3Al.

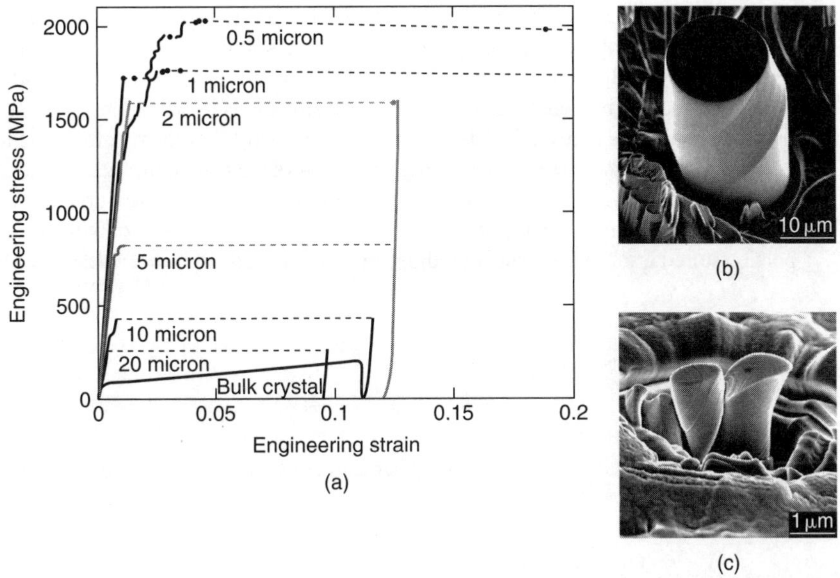

Figure 12.26 *(a) Stress–strain curves of Ni_3Al microcolumns. (b) SEM image of a microcolumn produced by focused-ion milling. (c) Post-compression image of a column (Uchic* et al.*, 2005, courtesy of the American Association for the Advancement of Science).*

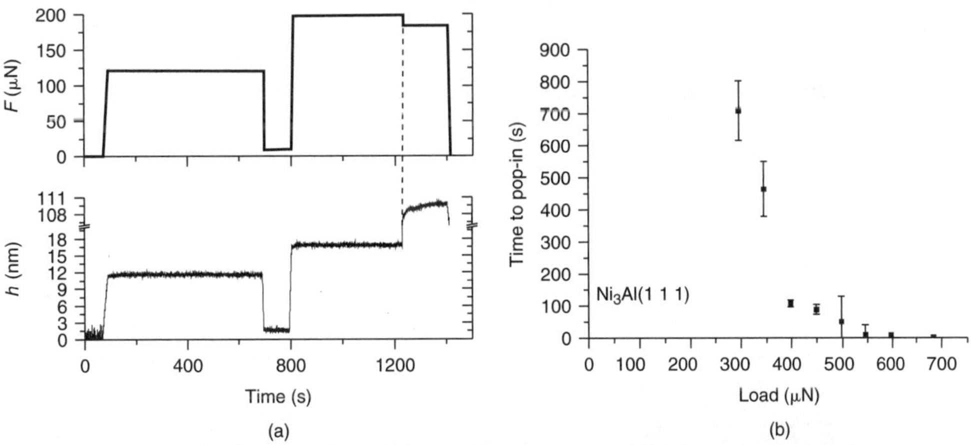

Figure 12.27 *(a) Load (*F*) and tip displacement (*h*) vs time data in a nanoindentation experiment at room temperature on an annealed aluminum crystal using a sharp Berkovich tip (Feng and Ngan, 2001). (b) Relationship between waiting time for plasticity and holding load in Ni_3Al (Chiu and Ngan, 2001).*

The size and time dependence of the elastic-to-plastic transition is unique to small crystals, and is a phenomenon that must be sufficiently understood before the high-strength advantage of small crystals can be utilized in applications.

In addition to the onset of plasticity, the flow resistance of submicron crystals in situations with strong strain gradients (e.g. torsion, indentation) also exhibits a remarkable strengthening effect as the material size decreases. The torsional strength of metallic whiskers, as well as the nanoindentation hardness on crystalline specimens, often exhibit a significant rise as the material size decreases (Figure 12.28a). Both deformation situations feature strong plastic strain gradients which have to be accommodated by geometrically necessary dislocations (Section 6.6.2.4). In the Nix–Gao model, the rise in hardness in a small indent is modeled as a work-hardening effect due to the geometrically necessary dislocations (GNDs) in the plastic zone. As shown in Figure 12.28b, the GNDs produce steps of average spacing $s = ba/h$ on the specimen surface, where b is the Burgers vector, and a and h are the radius and depth of the indent (assumed to be conical). If λ is the total line length of the GNDs, the portion stored in an annulus ring of radius r and thickness dr is $d\lambda = 2\pi r(dr/s) = 2\pi r(h/ba)dr$, and so

$$\lambda = \int_0^a \frac{h}{ba} 2\pi r dr = \frac{\pi h a}{b}.$$

Assuming that all GNDs reside inside a hemisphere with the same radius a as the indent, the density of the GNDs is given by

$$\rho_G = \frac{\lambda}{(2/3)\pi a^3} = \frac{3h}{2ba^2} = \frac{3}{2bh}\tan^2\theta. \tag{12.1}$$

The amount of work hardening is given by Taylor's model (Section 6.6.2.1) to be

$$\tau = \alpha\mu b\sqrt{\rho_G + \rho_s}, \tag{12.2}$$

where ρ_s is the density of the 'statistically stored' dislocation population. Assuming that the direct stress $\sigma = \sqrt{3}\tau$ and that the hardness $H = 3\sigma$ (Section 6.1.3), with equations (12.1) and (12.2), the hardness is given by

$$\frac{H}{H_0} = \sqrt{1 + \frac{h^*}{h}}, \tag{12.3}$$

where H_0 and h^* are constants. Equation (12.3) predicts that the hardness increases as the indent size h decreases. This equation was found to rather accurately describe the hardness data such as those in Figure 12.28a within a certain indent depth range.

Despite the impressive strength of small crystal volumes, their plastic deformation following the first yield point is usually very jumpy, exhibiting sporadic, discrete strain bursts. Figure 12.29a shows the stress-strain curve of an aluminum micro-column compressed according to the load schedule shown in the inset diagram. The discrete strain bursts are seen as large discontinuities in the stress-strain graph. The post-deformation micrograph in Figure 12.29b reveals many slip steps on the column's free surface, and each of these is likely to be associated with a strain burst in Figure 12.29a. Small crystalline materials may thus be difficult to be formed into smooth shapes by plastic deformation.

The creep deformation of small crystal volumes, even under low temperatures relative to melting, also needs special attention. The lower panel of Figure 12.27a indicates that immediately after the onset of plasticity at ~1205 s, the indenter displacement h increased gradually with time, implying that the specimen underwent creep deformation under fixed indentation force at room temperature.

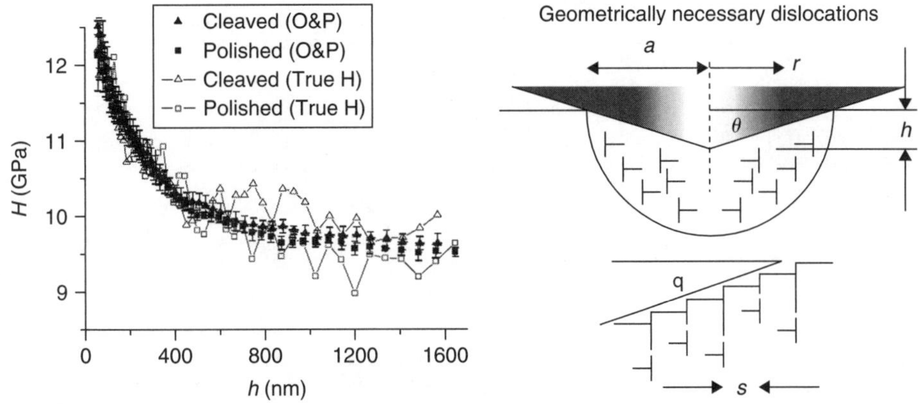

(a) (b)

Figure 12.28 *(a) Nanoindentation hardness at varying indent depths for MgO (Feng and Nix, 2004). (b) Geometrically necessary dislocations within the plastic zone of an indent (Nix and Gao, 1998).*

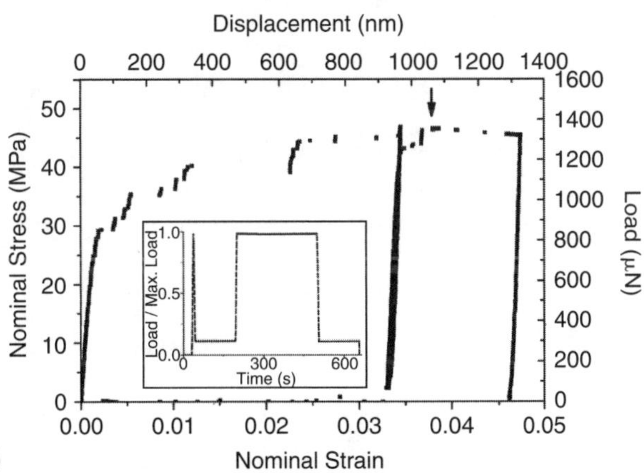

(a)

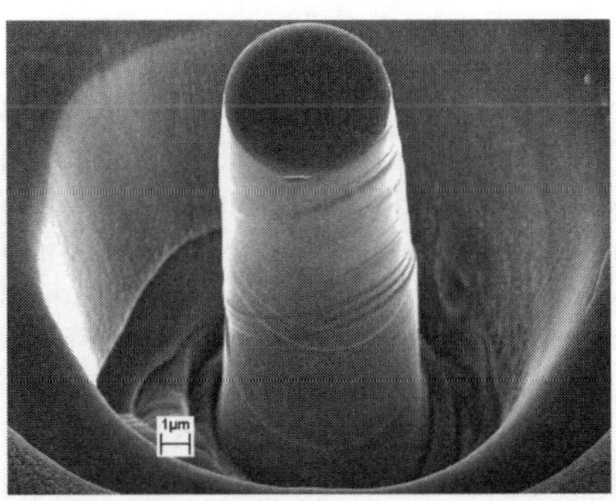

(b)

Figure 12.29 *(a) Stress-strain curve of an aluminum micro-column during compression experiment by a flat-ended nanoindenter. The load schedule used is shown in the inset. (b) SEM micrograph of the deformed column (K.S. Ng and A.H.W. Ngan, unpublished).*

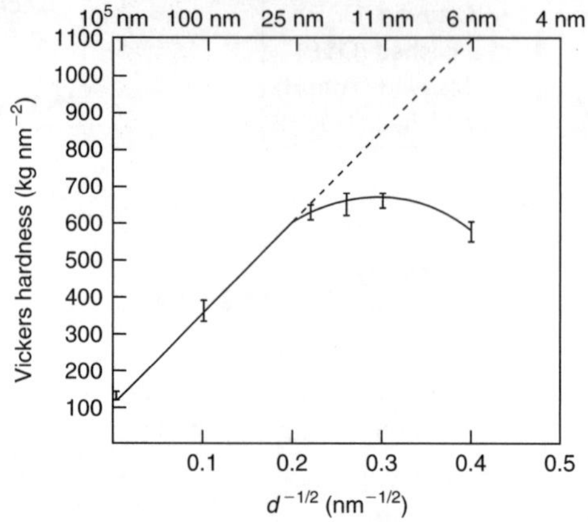

Figure 12.30 *Inverse Hall–Petch behavior in electrodeposited nanocrystalline Ni (Erb, 1995).*

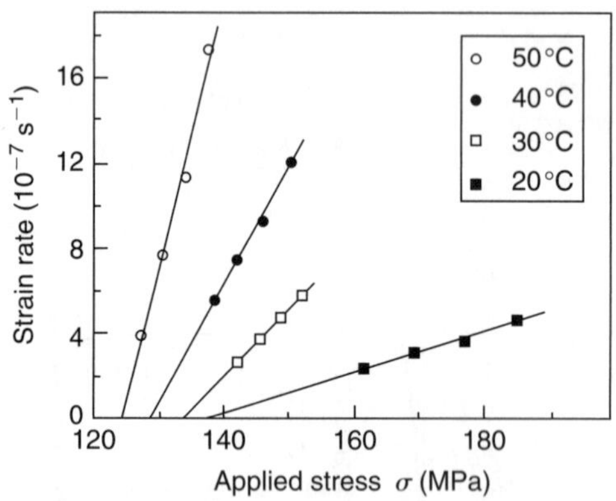

Figure 12.31 *Flow stress versus strain rate in electrodeposited nanocrystalline Cu (Cai, Lu and co-workers, 1999).*

That diffusion is important in the low-temperature deformation of nanocrystalline materials can be seen from the few unusual mechanical characteristics of these materials as compared with their ordinary polycrystalline counterparts. First, their grain-boundary fraction is a lot higher than in ordinary polycrystalline solids, causing their density and elastic modulus to be significantly lower. Dislocation activities are heavily confined as the grain size approaches 1 nm, but the yield strength does not increase according to the Hall–Petch relation (equation (6.18)). Instead, at grain sizes smaller than about 10 nm, the 'inverse Hall–Petch' behavior is observed where the strength drops instead of rises with decreasing grain size (Figure 12.30). A mechanism dominated by grain boundary activities, possibly Coble creep type diffusion, is thought to be responsible. A related phenomenon is that nanocrystalline metals typically deform with high strain-rate sensitivity (approaching Newtonian flow behavior) once a certain threshold stress is surpassed. Figure 12.31 shows the creep results in electrodeposited nanocrystalline copper at low deformation temperatures, where it can be seen that the creep flow stress is given by $\sigma = \sigma_o + K\dot{\varepsilon}$, where σ_o is a threshold stress, K is a temperature-dependent constant and $\dot{\varepsilon}$ is the strain rate. The linear dependence term $K\dot{\varepsilon}$ again suggests a diffusional creep mechanism at low temperatures. Nanocrystalline metals can also exhibit

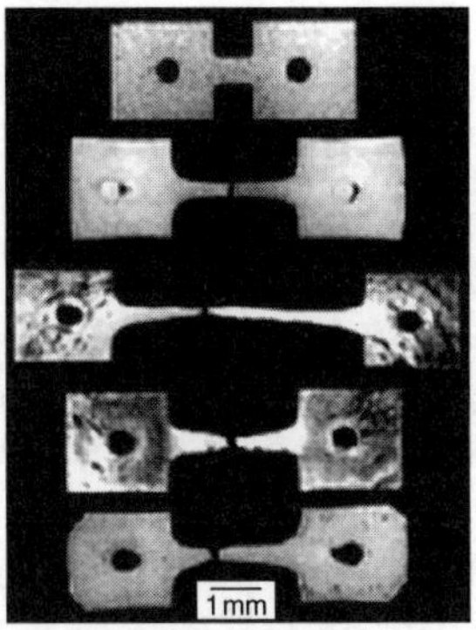

Figure 12.32 *Superplasticity in electrodeposited nanocrystalline Ni (Mukherjee, 2002).*

superplasticity (Figure 12.32), which is a phenomenon characteristic of a high strain-rate sensitivity (Section 6.4.12).

12.4.7 Bio-nanotechnology

Nanotechnology is increasingly being used in medicine, via biomaterials. These include carbon nanotubes in an ultra-sensitive DNA detector, release of drug molecules controlled by nanoporous membranes with pores marginally larger than the drug molecules, and surface nanoreceptors to provide drug release specifically to damaged tissue. Diabetes is one condition becoming more prevalent worldwide and where it is aimed to implant drugs beneath the skin to deliver as needed to maintain a steady blood glucose level. Infection is also receiving particular attention. Biodegradable nanobeads coated with specific molecules are able to mimic the ability of white blood cells to reduce inflamed blood cell walls by traveling through the bloodstream and target the inflammation site. Transport wound dressings have been developed from nano-structured membrane material which protects the skin from bacterial infection. Cells may be grafted onto the dressing to promote tissue regeneration. Another innovation is to incorporate drug delivery in the membrane to provide controlled medication. Bacteria infection itself may be treated with a new class of material called 'peptide nanotubes'. These are about 3 nm in diameter and 6 nm long and are made to perforate bacterial membranes without harming healthy cells. In oncology nanoparticles can be directed at a tumor with accuracy to have negligible effect on the other tissues, and allow higher doses to be administered. Breast cancer is one such treatment where anti-cancer drugs are attached to the protein, albumin

Problems

12.1 Outline three applications of shape memory alloys (SMAs) in biomedical surgery.

12.2 Suggest ways in which the performance of hip joints could possibly be improved.

12.3 What is 'bioglass'? What are the main advantages and disadvantages of 'bioglass' as an implant material?

12.4 Assuming the tubes of a bicycle, radius r, of wall thickness t, act like cantilever beams and the deflection is given by the equation

$$\text{deflection } \Delta = Fl^3/3E\pi r^3 t,$$

where F is the force and E Young's modulus. Determine which material, (i) steel, (ii) aluminum or (iii) CFRP, provides the best stiffness/weight ratio (E is 200, 69 and 270 GN m^{-2} respectively).

12.5 Assume that an archery bow behaves like a thin beam loaded in three-point bending. Compare the suitability of spring steel, forged aluminum and a composite. (The elastic stored energy of the bow per unit volume is $\sim\frac{1}{2}\sigma^2/E$.)

12.6 As an idealized model, consider that a nanocrystalline structure consists of identical cubic grains of size 5 nm separated by a grain boundary phase of width 0.2 nm. Calculate the volume fraction of the grain boundary phase. Furthermore, by treating the nanocrystalline structure as a composite of the bulk phase and the grain-boundary phase, estimate the overall Young's modulus of the structure, assuming that the intrinsic Young's modulus of the bulk phase is 200 GPa.

12.7 Derive an expression for the diameter D of a single-walled carbon nanotube (CNT) with chiral vector (m, n), in terms of the C–C bond distance a (=0.143 nm). A CNT has a diameter of 0.956 nm. Find its chiral vector. What is the configuration of this CNT called, and what is its electrical conductivity likely to be?

12.8 A device to detect virus is made by a silicon cantilever produced by e-beam lithography. The cantilever is rectangular in shape and has dimensions ($L \times t \times w$) 500 nm × 50 nm ×50 nm. The natural frequency of the cantilever is given by

$$f = \frac{1.875^2}{2\pi}\frac{t}{L^2}\sqrt{\frac{E}{12\rho}},$$

where the Young modulus E for silicon is 190 GPa and density ρ is 2330 kg m^{-3}. The device detects the virus by noting the change in the natural frequency of the cantilever, after the virus molecules have attached on the surface of the latter. If each virus to be detected weighs 10^{-23} kg, estimate the change in the cantilever frequency for every 1000 copies of virus attached to the cantilever surface.

Further reading

Biomaterials

Ball, P. (1998). Spare parts – biomedical materials. In *Measure for Measure*, Chap. 5. Princeton University Press, Princeton, NJ, USA.

Bhumbra, R. S. *et al.* (1998). Enhanced bone regeneration and formation around implants. *J. Biomed. Mater. Res. (Appl. Biomater.)*, **43**, 162–167.

Bonfield, W. (1992). Can materials stimulate advances in orthopaedics? *Science of New Materials*, p. 168. Blackwell, Oxford, UK.

Bonfield, W. (1997). Biomaterials – a new generation. *Materials World*, January, p. 18. Institute of Materials.

Brown, D. (1994). Polymers in dentistry. *Progress in Rubber and Plastics Technology*, **10**, 185.

Brown, D. (1996). Filling the gap. *Materials World*, May, p. 259. Institute of Materials.

Helson, J. E. F. A. and Jürgan Brime, H. (eds) (1998). *Metals as Biomaterials.* John Wiley.

Ratner, B. (ed.) (1996). *Biomaterials Science.* Academic Press, New York.

Vincent, J. (1990). Materials technology from nature. *Metals and Materials*, June, p. 395. Institute of Materials.

Williams, D. F. (1991). Materials for surgical implants. *Metals and Materials*, January, p. 24. Institute of Materials.

Sports materials

Ashby, M. F. (2005). *Materials Selection in Mechanical Design*. Elsevier, Butterworth-Heinemann.

Baker, T. J. (1989). Fencing blades – a materials challenge. *Metals and Materials*, Dec., pp. 715–718. Institute of Materials.

Blyth, P. H. and Pratt, P. L. (1992). The design and materials of the bow/the arrow. Appendices to *Longbow: A Social and Military History*, 3rd edn by Robert Hardy. Patrick Stephens, Cambridge.

Cochran, A. (ed.) (1994). *Golf: The Scientific Way*. Aston, Hemel Hempstead, Herts, UK.

Easterling, K. E. (1993). *Advanced Materials for Sports Equipment*. Chapman & Hall, London.

Gibson, L. J. and Ashby, M. J. (1988). *Cellular Solids – Structure and Properties*. Pergamon Press.

Jenkins, M. (ed.) (2003). *Materials for Sports Equipment*. CRC Press.

Knowles, S., Fletcher, G., Brooks, R. and Mather, J. S. B. (1998). Development of a superior performance cricket helmet. In *The Engineering of Sport* (edited by S. J. Haake). Blackwell Science, Oxford.

Lees, A. W. (ed.) (1989). *Adhesives and the Engineer*. Mechanical Engineering Publications, London.

McMahon, C. J. and Graham, C. D. (1992). *Introduction to Materials: The Bicycle and the Walkman*. Merion Books, Philadelphia.

Pearson, R. G. (1990). In *Engineering Polymers* (edited by R. W. Dyson), Chap. 4 on foams, pp. 76–100. Blackie, Glasgow.

Shields, J. (1984). *Adhesives Handbook*, 3rd edn. Butterworths, Oxford.

Nanomaterials

Poole, C. P. Jr and Owens, F. J. (2003). *Introduction to Nanotechnology*. Wiley Interscience, New Jersey.

Stokes, R. J. and Evans, D. F. (1997). *Fundamentals of Interfacial Engineering*. Wiley-VCH, New York.

Wilson, M., Kannangara, K., Smith, G., Simmons, M. and Raguse, B. (2002). *Nanotechnology: Basic Science and Emerging Technologies*. CRC Press, Florida.

Numerical answers to problems

Chapter 1

1.1 (i) K, Ca, Sc, Ti, V, Cr, Mn, Fe, Co, Ni
(ii) K^+: $1s^2\ 2s^2\ 2p^6\ 3s^2\ 3p^6$, Br^-: $1s^2\ 2s^2\ 2p^6\ 3s^2\ 3p^6\ 3d^{10}\ 4s^2\ 4p^6$
1.2 0.236 nm, 5.2×10^{11} Pa, −5.3 eV
1.3 2.49×10^{-10} m, 2.7 eV
1.4 5.27×10^{-10} m
1.5 0.34, 0.118 nm
1.6 $[\frac{-1}{4}, \frac{-1}{4}, \frac{-1}{4}]; [\frac{-1}{4}, \frac{1}{4}, \frac{1}{4}]; [\frac{1}{4}, \frac{-1}{4}, \frac{1}{4}]; [\frac{1}{4}, \frac{1}{4}, \frac{-1}{4}]$
1.7 0.3524 nm, 8.91 Mg m^{-3}
1.8 4 Mg^{2+} and 4 O^{2-}

Chapter 2

2.1 41.23%, 20.24%, 12.15%; 1.4, 1.4, 1.24
2.2 29%
2.5 38% liquid containing 32% Ni; 62% solid containing 45% Ni.
2.6 43.9% α containing 19% Sn; 56.1% β containing 2.5% Pb.
2.9 (a) Cementite 11.7%, ferrite 88.3%; (b) cementite 2.6%, pearlite 22.6%

Chapter 3

3.5 1/2[1 1 0]; 1/2[1 $\bar{1}$ 0]; 1/2[1 0 1]; 1/2[1 0 $\bar{1}$]; 1/2[0 1 1]; 1/2[0 1 $\bar{1}$].
3.6 Screw [1 $\bar{1}$ 0], edge [1 1 2]
3.7 ($\bar{1}$ 1 1), (1 1 $\bar{1}$)
3.8 [1 2 $\bar{1}$]
3.9 (1 $\bar{1}$ 0), (1 0 $\bar{1}$), (0 1 $\bar{1}$)
3.10 (a) a[1 0 0]; (b) $a/2$[1 1 0]; (c) [0 0 0 1]
3.11 (a) $a\sqrt{3}/2$; (b) $a/\sqrt{6}$
3.12 61.9°, [$\bar{1}$ 1 0]
3.13 11.5 GPa
3.14 (iii) ~0.5 J m^{-2}
3.15 (a) 64 mJ m^{-2}; (c) $\frac{a}{2}[1\,0\,\bar{1}] \rightarrow \frac{a}{6}[2\,\bar{1}\,\bar{1}] + \frac{a}{6}[1\,1\,\bar{2}]$; (d) BC → B$\delta$ + δC
3.16 17 atom spacings

Chapter 4

4.1 27.7°
4.2 0.154 nm

4.3 {1 1 1}, {2 0 0}, {2 2 0}
4.4 13.5 mm
4.5 80.7 mm
4.6 0.361 nm
4.7 CuK_β at 8.904 keV; CuL_α at 0.94 keV
4.8 $Mg_{0.435}O_{0.565}$
4.9 74.1 wt% Fe, 16.9 wt% Cr, 8.8 wt% Ni
4.10 (i) $\pm\frac{a}{2}[\bar{1}\,1\,0]$; (ii) (1 1 1) or $(1\,1\,\bar{1})$
4.11 $(\bar{1}\,1\,1)$ with $g = 0\,\bar{2}\,2$, $(1\,1\,\bar{1})$ with $g = 2\,\bar{2}\,0$
4.12 $[\bar{1}\,\bar{0}\,1\,3]$

Chapter 5

5.1 (i) 10^{-16}, 10^{-8}, 4.6×10^{-6}; (ii) $10^{-20}\,m^2\,s^{-1}$, $10^{-12}\,m^2\,s^{-1}$, $4.6 \times 10^{-10}\,m^2\,s^{-1}$; (iii) $10^{-36}\,m^2\,s^{-1}$, $10^{-20}\,m^2\,s^{-1}$, $2.1 \times 10^{-15}\,m^2\,s^{-1}$; (iv) 60 μm; (v) 6 nm; (vi) 1.1 eV
5.2 62.5 kJ mol^{-1} (or 0.65 eV/atom), $2.36 \times 10^{-7} m^2\,s^{-1}$
5.3 0.253 K, 4.2×10^{-3}
5.4 (a) 0.67 eV; (b) 1861 nm
5.5 1.083×10^{-3}
5.6 $1.4 \times 10^{-15}\,\Omega$

Chapter 6

6.1 (b) $[\bar{1}\,1\,0]$, 0.49σ
6.2 $(1\,1\,1)[1\,0\,\bar{1}]$, $(1\,1\,1)[0\,1\,\bar{1}]$, $(1\,1\,\bar{1})[1\,0\,1]$, $(1\,1\,\bar{1})[0\,1\,1]$
6.3 UYS: $2.8 \times 10^4\,N\,cm^{-2}$, yield drop: $0.58 \times 10^4\,N\,cm^{-2}$
6.4 0.25 μm
6.5 85 nm
6.6 (ii) 44.3 MPa; (iii) 667 K for H–N/Coble creep boundary, $\sigma = 0.0508\ \exp(4.5 \times 10^3/T)$ MPa for the dislocation creep/Coble creep boundary
6.7 76.9 kJ mol^{-1}
6.8 1282 K
6.9 5.2×10^8 cycles

Chapter 7

7.1 90.3 h, 22 MPa
7.2 66.06 kJ mol^{-1}, 78.93 kJ mol^{-1}
7.3 23.2 days, 33.3 years
7.4 603 s
7.6 319 MPa
7.7 4.8×10^6 cycles
7.8 (b) 2.55 mm

Chapter 8

8.2 0.06% C, 1.5% Mn, 0.85% Si
8.3 (b) 1 mm
8.4 0.027 s, 0.082 s
8.5 ~50 MPa

Chapter 9

9.1 1175 $g m^{-2}$
9.3 8.77 $kg m^{-2}$, 1.11 mm
9.5 3.686 μm
9.6 $4.045 \times 10^{-13} m^2 s^{-1}$
9.9 0.4743 s
9.10 1125 h
9.11 (iii) 40 $J m^{-2}$
9.13 257.9 $kJ mol^{-1}$

Chapter 10

10.1 5.3 μm
10.2 1000 GPa
10.3 472.1 GPa
10.4 0.48 s^{-1}
10.5 14.3 MPa
10.8 216.3 MPa, 0.0424
10.9 541.4 $kJ mol^{-1}$
10.10 ~500°C
10.11 (ii) 7.5733, 634.2 MPa if V in mm^3 or 41.4 MPa if V in m^3; (iii) 71.8 MPa

Chapter 11

11.1 29 156, 35 246
11.2 7742, 34 898
11.3 (a) 1.011 $Mg m^{-3}$; (b) 45.5%
11.5 5.0 MPa
11.6 $1.0 \times 10^5 J mol^{-1}$
11.7 197 GPa
11.8 1.85 $Mg m^{-3}$
11.9 2.1 GPa

Chapter 12

12.6 11.1%, 178 GPa
12.7 $D = (a/\pi)\sqrt{3(m^2 + mn + n^2)}$; (7,7)
12.8 −1.33 MHz

Appendix 1
SI units

The Système Internationale d'Unités (SI) was introduced in the UK in the late 1960s. Historically, the SI can be traced from the metric enthusiasms of Napoleonic times, through a centimeter–gram (c.g.) system, a centimeter–gram–second (c.g.s.) system, a meter–kilogram–second (MKS) system in 1900 and a meter–kilogram–second–ampere (MKSA Giorgi) system in 1950. Table A1 lists the seven basic units and Table A2 lists the prefixes.

The SI is 'rational, comprehensive and coherent'. Coherency means that the product or quotient of *basic* units gives an appropriate *derived* unit of the resultant quantity. A coherent system facilitates

Table A1

Quantity	*Unit*	*Symbol*
Length	meter	m
Mass	kilogram	kg
Time	second	s
Electric current	ampere	A
Temperature	degree Kelvin	K
Luminous intensity	candela	cd
Amount of substance	mole	mol

Table A2

Factor	*Prefix*	*Symbol*
10^{12}	tera	T
10^{9}	giga	G
10^{6}	mega	M
10^{3}	kilo	k
10^{2}*	hecto*	h*
10^{1}*	deca*	da*
10^{-1}*	deci*	d*
10^{-2}*	centi*	c*
10^{-3}	milli	m
10^{-6}	micro	μ
10^{-9}	nano	n
10^{-12}	pico	p
10^{-15}	femto	f
10^{-18}	atto	a

*Discouraged.

Table A3

Physical quantity	*SI unit*	*Definition of unit*
Volume	cubic meter	m^3
Force	newton (N)	kg m s
Pressure, stress	newton per square meter	$N\ m^{-2}$
Energy	joule (J)	N m
Power	watt (W)	$J\ s^{-1}$
Electric charge	coulomb (C)	A s
Electric potential	volt (V)	$W\ A^{-1}$
Electric resistance	ohm (Ω)	$V\ A^{-1}$
Electric capacitance	farad (F)	$A\ s\ V^{-1}$
Frequency	hertz (Hz)	s^{-1}

manipulation of units, checking the dimensions of equations and, most importantly, the correlation of different disciplines. Some of the more frequently used derived units are given in Table A3.

The force unit, the newton, is the cornerstone of the SI. Appropriately, the gravitational attraction for an apple is roughly 1 newton. The SI unit of stress is $N\ m^{-2}$: the pascal (Pa) is an orphan, being non-SI and non-coherent. Energy is defined in mechanical terms, being the work done when the point of application of a force of 1 N is displaced through a distance of 1 m in the direction of the force.

The surprising frequency with which SI units are misused in textbooks, learned papers, reports, theses and even examination papers justifies a reminder of some rules:

1. Try to locate basic units after the solidus: $1\ MN/m^2$ preferred to $1\ N/mm^2$.
2. A space is significant: ms is not the same as m s.
3. Prior to calculations, convert to basic SI units: 1 mm becomes 1×10^{-3} m.
4. Where possible, work in steps of $10^{\pm 3}$.
5. If possible, group digits in threes and avoid commas: 37 532 rather than 37,532 because a comma means decimal point in some countries.
6. When selecting a prefix, arrange for the preceding number to lie between 0.1 and 1000: use 10 mm rather than 0.01 m. In a comparison, one may break this rule, e.g. 'increase from 900 kN to 12 000 kN'.
7. Do not use double prefixes: pF, not μμF.
8. Avoid multiples of the solidus: acceleration written as m/s/s self-destructs.

Appendix 2

Conversion factors, constants and physical data

Quantity		*Symbol*	*Traditional units*	*SI units*
1 atmosphere (pressure)		atm		$101.325\ \text{kN m}^{-2}$
Avogadro constant		N_A		$0.602 \times 10^{24}\ \text{mol}$
1 angstrom		Å	$10^{-8}\ \text{cm}$	$10^{-10}\ \text{m}$
1 barn		b	$10^{-24}\ \text{cm}^2$	$10^{-28}\ \text{m}^2$
1 bar		bar or b		$10^5\ \text{N m}^{-2}$
Boltzmann constant		**k**		$1.380 \times 10^{-23}\ \text{J 1}$
1 calorie		cal		4.1868 J
1 dyne		dyn	$0.224\,809 \times 10^{-5}\ \text{lbf}$	$10^{-5}\ \text{N}$
1 day			86 400 s	86.4 ks
1 degree (plane angle)			0.017 45 rad	17 mrad
Electron rest mass		m_e	$9.109\,56 \times 10^{-28}\ \text{g}$	$9.109\,56 \times 10^{-31}$
1 erg (dyn cm)			$6.242 \times 10^{11}\ \text{eV}$	$10^{-7}\ \text{J}$
			$2.39 \times 10^{-8}\ \text{cal}$	
1 erg/cm^2			$6.242 \times 10^{11}\ \text{eV cm}^{-2}$	$10^{-3}\ \text{J m}^{-2}$
Gas constant		**R**	$8.314\,3 \times 10^7\ \text{erg K}^{-1}\ \text{mol}^{-1}$	$8.3143\ \text{J K}^{-1}\ \text{mol}$
			$1.987\ \text{cal K}^{-1}\ \text{mol}^{-1}$	
Density	Al		$2.71\ \text{g cm}^{-3}$	$2710\ \text{kg cm}^{-3}$
	Fe	ρ	$7.87\ \text{g cm}^{-3}$	$7870\ \text{kg cm}^{-3}$
	Cu		$8.93\ \text{g cm}^{-3}$	$8930\ \text{kg cm}^{-3}$
	Ni		$8.90\ \text{g cm}^{-3}$	$8900\ \text{kg cm}^{-3}$
Electronic charge		e	$1.602 \times 10^{-20}\ \text{emu}$	0.1602 aC
1 electron-volt		eV	$3.83 \times 10^{-20}\ \text{cal}$	0.1602 aJ
			$1.602 \times 10^{-12}\ \text{erg}$	
Faraday		$F = N_A e$		$9.6487 \times 10^4\ \text{C mol}^{-1}$
1 inch		in	2.54 cm	25.4 mm
1 kilogram		kg	2.204 62 lb	1 kg
1 kilogram-force/cm^2		kgf/cm^2	14.22 lbf/in^2	
1 litre		l	0.220 gal	1 dm^3
1 micron		μm	10^4 angstrom	$10^{-6}\ \text{m}$
			$10^{-4}\ \text{cm}$	
Minute (angle)			2.908×10^{-4} radian	$2.908 \times 10^{-4}\ \text{rad}$
Modulus of elasticity	Al			$70\ \text{GN m}^{-2}$
(average)	Fe			$210\ \text{GN m}^{-2}$
	Ni	E		$209\ \text{GN m}^{-2}$
	Cu			$127\ \text{GN m}^{-2}$
	Au			$79\ \text{GN m}^{-2}$
Planck's constant		**h**	$6.6262 \times 10^{-27}\ \text{erg s}$	$6.6262 \times 10^{-34}\ \text{J s}$

Continued

Quantity		*Symbol*	*Traditional units*	*SI units*
Poisson's ratio	Al			0.34
	Au			0.44
	Cu			0.35
	Mg	ν		0.29
	Pb			0.44
	Ti			0.36
	Zn			0.25
pound		lb	453.59 g	0.453 kg
pound (force)		lbf		4.448 22 N
psi		lbf/in^2	7.03×10^{-2} kgf/cm^2	6 894.76 $N\,m^{-2}$
radian		rad	57.296 degrees	1 rad
Shear modulus	Al		2.7×10^{11} $dyn\,cm^{-2}$	27 $GN\,m^{-2}$
(average)	Fe		8.3×10^{11} $dyn\,cm^{-2}$	83 $GN\,m^{-2}$
	Ni	μ	7.4×10^{11} $dyn\,cm^{-2}$	74 $GN\,m^{-2}$
	Cu		4.5×10^{11} $dyn\,cm^{-2}$	45 $GN\,m^{-2}$
	Au		3.0×10^{11} $dyn\,cm^{-2}$	30 $GN\,m^{-2}$
ton (force)		1 tonf		9.964 02 kN
tsi		1 $tonf/in^2$	1.574 9 kgf/mm^2	15.444 3 $MN\,m^{-2}$
tonne		t	1000 kg	10^3 kg
torr		torr	1 mm Hg	133.322 $N\,m^{-2}$
Velocity of light (*in vacuo*)		c	$2.997\,925 \times 10^{10}$ cm/s	$2.997\,925 \times 10^8$ $m\,s^{-1}$

Index